MOLECULAR NUTRITION AND DIABETES

MOLECULAR NUTRITION AND DIABETES

A VOLUME IN THE MOLECULAR NUTRITION SERIES

Edited by

Didac Mauricio
Department of Endocrinology and Nutrition, CIBER of Diabetes and Associated Metabolic Diseases (CIBERDEM), University Hospital & Health Sciences Research Institute Germans Trias i Pujol, Badalona, Spain

AMSTERDAM • BOSTON • HEIDELBERG • LONDON
NEW YORK • OXFORD • PARIS • SAN DIEGO
SAN FRANCISCO • SINGAPORE • SYDNEY • TOKYO
Academic Press is an imprint of Elsevier

Academic Press is an imprint of Elsevier
125 London Wall, London EC2Y 5AS, UK
525 B Street, Suite 1800, San Diego, CA 92101-4495, USA
225 Wyman Street, Waltham, MA 02451, USA
The Boulevard, Langford Lane, Kidlington, Oxford OX5 1GB, UK

ISBN: 978-0-12-801585-8

British Library Cataloguing in Publication Data
A catalogue record for this book is available from the British Library

Library of Congress Cataloging-in-Publication Data
A catalog record for this book is available from the Library of Congress

For Information on all Academic Press publications
visit our website at http://store.elsevier.com/

Publisher: Mica Haley
Acquisition Editor: Tari K. Broderick
Editorial Project Manager: Jeff Rossetti
Production Project Manager: Chris Wortley
Designer: Victoria Pearson

Typeset by TNQ Books and Journals
www.tnq.co.in

Printed and bound in the United States of America

Dedication

To Aurora and Carla for taking care of me so lovingly when disease suddenly came into my life.

Contents

II
MOLECULAR BIOLOGY OF THE CELL

21. MicroRNA and Diabetes Mellitus

SOFIA SALÖ, JULIAN GEIGER, ANJA E. SØRENSEN, AND LOUISE T. DALGAARD

22. Diabetes Mellitus and Intestinal Niemann-Pick C1—Like 1 Gene Expression

POOJA MALHOTRA, RAVINDER K. GILL, PRADEEP K. DUDEJA, AND WADDAH A. ALREFAI

23. Dietary Long Chain Omega-3 Polyunsaturated Fatty Acids and Inflammatory Gene Expression in Type 2 Diabetes

JENCY THOMAS AND MANOHAR L. GARG

24. Polymorphism, Carbohydrates, Fat, and Type 2 Diabetes

JOSE LOPEZ-MIRANDA AND CARMEN MARIN

25. Genetic Basis Linking Variants for Diabetes and Obesity with Breast Cancer

SHAIK MOHAMMAD NAUSHAD, M. JANAKI RAMAIAH, AND VIJAY KUMAR KUTALA

26. Vitamin D Status, Genetics, and Diabetes Risk

DHARAMBIR K. SANGHERA AND PIERS R. BLACKETT

27. NRF2-Mediated Gene Regulation and Glucose Homeostasis

YOKO YAGISHITA, AKIRA URUNO, AND MASAYUKI YAMAMOTO

28. Hepatic Mitochondrial Fatty Acid Oxidation and Type 2 Diabetes

ABDELHAK MANSOURI, WOLFGANG LANGHANS, JEAN GIRARD, AND CARINA PRIP-BUUS

29. Current Knowledge on the Role of Wnt Signaling Pathway in Glucose Homeostasis

TIANRU JIN

Contributors

Waddah A. Alrefai Division of Gastroenterology and Hepatology, Department of Medicine, University of Illinois at Chicago, Chicago, IL, USA; The Jesse Brown VA Medical Center, Chicago, IL, USA

Jaime Amaya-Farfan Food and Nutrition Department, Protein Resources Laboratory, Faculty of Food Engineering, University of Campinas (UNICAMP), Campinas, São Paulo, Brazil

Giovanni Annuzzi Department of Clinical Medicine and Surgery, Federico II University, Naples, Italy

Julie C. Antvorskov The Bartholin Institute, Rigshospitalet, Copenhagen, Denmark

Anna Ardévol MoBioFood Research Group, Departament de Bioquímica i Biotecnologia, Universitat Rovira i Virgili, Tarragona, Spain

Knud Erik Bach Knudsen Department of Animal Science, Aarhus University, Tjele, Denmark

Silvia Berciano Nutritional Genomics of Cardiovascular Disease and Obesity, IMDEA-Food Institute, CEI UAM+CSIC, Madrid, Spain

Piers R. Blackett Department of Pediatrics, Section of Endocrinology, College of Medicine, University of Oklahoma Health Sciences Center, Oklahoma City, OK, USA

Mayte Blay MoBioFood Research Group, Departament de Bioquímica i Biotecnologia, Universitat Rovira i Virgili, Tarragona, Spain

Lutgarda Bozzetto Department of Clinical Medicine and Surgery, Federico II University, Naples, Italy

Karsten Buschard The Bartholin Institute, Rigshospitalet, Copenhagen, Denmark

Lu Cai Department of Pediatrics, University of Louisville, Louisville, KY, USA; Wendy L. Novak Diabetes Care Center, University of Louisville, Louisville, KY, USA

Younan Chen School of Biomedical Sciences, CHIRI Biosciences Research Precinct, Faculty of Health Sciences, Curtin University, Perth, WA, Australia

Fausto Chiazza Department of Drug Science and Technology, University of Turin, Turin, Italy

Carla Beatriz Collares-Buzato Department of Biochemistry and Tissue Biology, Institute of Biology, University of Campinas, Campinas, SP, Brazil

Massimo Collino Department of Drug Science and Technology, University of Turin, Turin, Italy

Giuseppina Costabile Department of Clinical Medicine and Surgery, Federico II University, Naples, Italy

Vinicius F. Cruzat School of Biomedical Sciences, CHIRI Biosciences Research Precinct, Faculty of Health Sciences, Curtin University, Perth, WA, Australia

Leticia Cuéllar (Literature search in Chapter 1) Servicio de Evaluación del Servicio Canario de la Salud (SESCS), Santa Cruz de Tenerife, Spain

Lidia Daimiel-Ruiz Nutritional Genomics of Cardiovascular Disease and Obesity, IMDEA-Food Institute, CEI UAM+CSIC, Madrid, Spain

Louise T. Dalgaard Department of Science, Systems and Models, Roskilde University, Roskilde, Denmark

Suzanne M. de la Monte Department of Medicine, Rhode Island Hospital, Providence, RI, USA; Department of Pathology, Rhode Island Hospital, Providence, RI, USA; Department of Neurology, Rhode Island Hospital, Providence, RI, USA; Department of Neurosurgery, Rhode Island Hospital, Providence, RI, USA; The Liver Research Center, Rhode Island Hospital, Providence, RI, USA; Rhode Island Hospital, Providence, RI, USA; The Warren Alpert Medical School of Brown University, Providence, RI, USA

Nathalia Romanelli Vicente Dragano Laboratory of Cell Signaling, Obesity and Comorbidities Research Center (OCRC), Faculty of Medical Sciences (FCM-Unicamp), University of Campinas, Campinas, SP, Brazil

Pradeep K. Dudeja Division of Gastroenterology and Hepatology, Department of Medicine, University of Illinois at Chicago, Chicago, IL, USA; The Jesse Brown VA Medical Center, Chicago, IL, USA

Eduardo Esteve Unit of Diabetes, Endocrinology and Nutrition, Biomedical Research Institute (IDIBGi), Hospital 'Dr. Josep Trueta' of Girona, Girona, Spain; CIBERobn Fisiopatología de la Obesidad y Nutrición, Girona, Spain

José Manuel Fernández-Real Unit of Diabetes, Endocrinology and Nutrition, Biomedical Research Institute (IDIBGi), Hospital 'Dr. Josep Trueta' of Girona, Girona, Spain; CIBERobn Fisiopatología de la Obesidad y Nutrición, Girona, Spain

Lidia García Servicio de Evaluación del Servicio Canario de la Salud (SESCS), Santa Cruz de Tenerife, Spain; Red de Investigación en Servicios de Salud en Enfermedades Crónicas (REDISSEC), Madrid, Spain

Manohar L. Garg Nutraceuticals Research Group, School of Biomedical & Sciences and Pharmacy, University of Newcastle, Newcastle, NSW, Australia

Rosa Gasa Diabetes and Obesity Research Laboratory, Institut d'Investigations Biomediques August Pi i Sunyer, Barcelona, Spain; Centro de Investigación Biomédica en Red de Diabetes y Enfermedades Metabólicas Asociadas (CIBERDEM), Barcelona, Spain

Julian Geiger Department of Science, Systems and Models, Roskilde University, Roskilde, Denmark

Ravinder K. Gill Division of Gastroenterology and Hepatology, Department of Medicine, University of Illinois at Chicago, Chicago, IL, USA

Jean Girard INSERM, Institut Cochin, Paris, France; CNRS, Paris, France; Université Paris Descartes, Paris, France

Ramon Gomis Diabetes and Obesity Research Laboratory, Institut d'Investigations Biomediques August Pi i Sunyer, Barcelona, Spain; Centro de Investigación Biomédica en Red de Diabetes y Enfermedades Metabólicas Asociadas (CIBERDEM), Barcelona, Spain; University of Barcelona, Hospital Clínic, Barcelona, Spain

Noemí González-Abuín MoBioFood Research Group, Departament de Bioquímica i Biotecnologia, Universitat Rovira i Virgili, Tarragona, Spain

Luis Goya Department of Metabolism and Nutrition, Institute of Food Science and Technology and Nutrition (ICTAN-CSIC), Madrid, Spain

Ettore Griffo Department of Clinical Medicine and Surgery, Federico II University, Naples, Italy

Merete Lindberg Hartvigsen Department of Endocrinology and Internal Medicine, Aarhus University Hospital, Aarhus, Denmark

Mette Skou Hedemann Department of Animal Science, Aarhus University, Tjele, Denmark

Kjeld Hermansen Department of Endocrinology and Internal Medicine, Aarhus University Hospital, Aarhus, Denmark

Susan Huse Department of Pathology, Rhode Island Hospital, Providence, RI, USA; The Warren Alpert Medical School of Brown University, Providence, RI, USA

Tianru Jin Division of Advanced Diagnostics, Toronto General Research Institute, University Health Network, Toronto, ON, Canada

Knud Josefsen The Bartholin Institute, Rigshospitalet, Copenhagen, Denmark

Miran Kim Department of Medicine, Rhode Island Hospital, Providence, RI, USA; The Liver Research Center, Rhode Island Hospital, Providence, RI, USA; Rhode Island Hospital, Providence, RI, USA; The Warren Alpert Medical School of Brown University, Providence, RI, USA

Vijay Kumar Kutala Department of Clinical Pharmacology and Therapeutics, Nizam's Institute of Medical Sciences, Hyderabad, India

Wolfgang Langhans Physiology and Behavior Laboratory, Institute of Food, Nutrition and Health, ETH Zurich, Schwerzenbach, Switzerland

Gilbert C. Liu Department of Pediatrics, University of Louisville, Louisville, KY, USA

Pablo C.B. Lollo Food and Nutrition Department, Protein Resources Laboratory, Faculty of Food Engineering, University of Campinas (UNICAMP), Campinas, São Paulo, Brazil; Faculty of Health Sciences, Federal University of Grande Dourados (UFGD), Dourados, Mato Grosso do Sul, Brazil

Jose Lopez-Miranda Lipids and Atherosclerosis Unit, IMIBIC/Reina Sofıa University Hospital, University of Cordoba and CIBER Fisiopatologia Obesidad y Nutricion (CIBERobn), Instituto de Salud Carlos III, Cordoba, Spain

Laura López Ríos (Figure in Chapter 1) Endocrinology and Nutrition Department, Complejo Hospitalario Universitario Insular Materno-Infantil de Gran Canaria, Las Palmas de Gran Canaria, Spain; Instituto Universitario de Investigaciones Biomédicas y Sanitarias, Universidad de Las Palmas de Gran Canaria, Las Palmas de Gran Canaria, Spain

Valeriya Lyssenko Steno Diabetes Center A/S, Gentofte, Denmark; Department of Clinical Sciences, Diabetes and Endocrinology Unit, Lund University Diabetes Center, Lund University, Sweden

Pooja Malhotra Division of Gastroenterology and Hepatology, Department of Medicine, University of Illinois at Chicago, Chicago, IL, USA

Abdelhak Mansouri Physiology and Behavior Laboratory, Institute of Food, Nutrition and Health, ETH Zurich, Schwerzenbach, Switzerland

Carmen Marin Lipids and Atherosclerosis Unit, IMIBIC/Reina Sofıa University Hospital, University of Cordoba and CIBER Fisiopatologia Obesidad y Nutricion (CIBERobn), Instituto de Salud Carlos III, Cordoba, Spain

Anne y Castro Marques Universidade Federal do Pampa-Campus Itaqui, Itaqui, RS, Brazil

Maria Ángeles Martin Department of Metabolism and Nutrition, Institute of Food Science and Technology and Nutrition (ICTAN-CSIC), Madrid, Spain

Xiao Miao Department of Ophthalmology, The Second Hospital of Jilin University, Changchun, China

Victor Mico Nutritional Genomics of Cardiovascular Disease and Obesity, IMDEA-Food Institute, CEI UAM+CSIC, Madrid, Spain

Marciane Milanski Laboratory of Metabolic Disorders, Faculty of Applied Sciences, University of Campinas – UNICAMP, Limeira, São Paulo, Brazil

Priscila N. Morato Food and Nutrition Department, Protein Resources Laboratory, Faculty of Food Engineering, University of Campinas (UNICAMP), Campinas, São Paulo, Brazil; Faculty of Health Sciences, Federal University of Grande Dourados (UFGD), Dourados, Mato Grosso do Sul, Brazil

Carolina S. Moura Food and Nutrition Department, Protein Resources Laboratory, Faculty of Food Engineering, University of Campinas (UNICAMP), Campinas, São Paulo, Brazil

Shaik Mohammad Naushad School of Chemical and Biotechnology, SASTRA University, Thanjavur, India

Philip Newsholme School of Biomedical Sciences, CHIRI Biosciences Research Precinct, Faculty of Health Sciences, Curtin University, Perth, WA, Australia

Anna Novials Diabetes and Obesity Research Laboratory, Institut d'Investigations Biomediques August Pi i Sunyer, Barcelona, Spain; Centro de Investigación Biomédica en Red de Diabetes y Enfermedades Metabólicas Asociadas (CIBERDEM), Barcelona, Spain

Jose M. Ordovas Nutrition and Genomics Laboratory, JM-USDA-Human Nutrition Research Center on Aging, Tufts University, Boston, MA, USA; Nutritional Genomics of Cardiovascular Disease and Obesity, IMDEA-Food Institute, CEI UAM+CSIC, Madrid, Spain

Montserrat Pinent MoBioFood Research Group, Departament de Bioquímica i Biotecnologia, Universitat Rovira i Virgili, Tarragona, Spain

Carina Prip-Buus INSERM, Institut Cochin, Paris, France; CNRS, Paris, France; Université Paris Descartes, Paris, France

M. Janaki Ramaiah School of Chemical and Biotechnology, SASTRA University, Thanjavur, India

Sonia Ramos Department of Metabolism and Nutrition, Institute of Food Science and Technology and Nutrition (ICTAN-CSIC), Madrid, Spain

Wifredo Ricart Unit of Diabetes, Endocrinology and Nutrition, Biomedical Research Institute (IDIBGi), Hospital 'Dr. Josep Trueta' of Girona, Girona, Spain; CIBERobn Fisiopatología de la Obesidad y Nutrición, Girona, Spain

David Sala Development, Aging and Regeneration Program (DARe), Sanford-Burnham Medical Research Institute, La Jolla, CA, USA

Sofia Salö Department of Science, Systems and Models, Roskilde University, Roskilde, Denmark; Danish Diabetes Academy, Odense, Denmark

Rosa M. Sánchez Hernández Endocrinology and Nutrition Department, Complejo Hospitalario Universitario Insular Materno-Infantil de Gran Canaria, Las Palmas de Gran Canaria, Spain; Instituto Universitario de Investigaciones Biomédicas y Sanitarias, Universidad de Las Palmas de Gran Canaria, Las Palmas de Gran Canaria, Spain

Dharambir K. Sanghera Department of Pediatrics, Section of Genetics, College of Medicine, University of Oklahoma Health Sciences Center, Oklahoma City, OK, USA; Department of Pharmaceutical Sciences, College of Pharmacy, University of Oklahoma Health Sciences Center, Oklahoma City, OK, USA

Joan-Marc Servitja Diabetes and Obesity Research Laboratory, Institut d'Investigations Biomediques August Pi i Sunyer, Barcelona, Spain; Centro de Investigación Biomédica en Red de Diabetes y Enfermedades Metabólicas Asociadas (CIBERDEM), Barcelona, Spain

Anja E. Sørensen Department of Science, Systems and Models, Roskilde University, Roskilde, Denmark; Danish Diabetes Academy, Odense, Denmark

Jian Sun Cardiovascular Center, The First Hospital of Jilin University, Changchun, China

Jency Thomas Department of Human Biosciences, LaTrobe University, Victoria, Australia

Adriana Souza Torsoni Laboratory of Metabolic Disorders, Faculty of Applied Sciences, University of Campinas – UNICAMP, Limeira, São Paulo, Brazil

Marcio Alberto Torsoni Laboratory of Metabolic Disorders, Faculty of Applied Sciences, University of Campinas – UNICAMP, Limeira, São Paulo, Brazil

Akira Uruno Department of Medical Biochemistry, Tohoku University Graduate School of Medicine, Sendai, Japan

Ana M. Wägner Endocrinology and Nutrition Department, Complejo Hospitalario Universitario Insular Materno-Infantil de Gran Canaria, Las Palmas de Gran Canaria, Spain; Instituto Universitario de Investigaciones Biomédicas y Sanitarias, Universidad de Las Palmas de Gran Canaria, Las Palmas de Gran Canaria, Spain

Shudong Wang Cardiovascular Center, The First Hospital of Jilin University, Changchun, China; Department of Pediatrics, University of Louisville, Louisville, KY, USA

Yonggang Wang Cardiovascular Center, The First Hospital of Jilin University, Changchun, China

Julia C. Wiebe Endocrinology and Nutrition Department, Complejo Hospitalario Universitario Insular Materno-Infantil de Gran Canaria, Las Palmas de Gran Canaria, Spain; Instituto Universitario de Investigaciones Biomédicas y Sanitarias, Universidad de Las Palmas de Gran Canaria, Las Palmas de Gran Canaria, Spain

Kupper A. Wintergerst Department of Pediatrics, University of Louisville, Louisville, KY, USA; Wendy L. Novak Diabetes Care Center, University of Louisville, Louisville, KY, USA

Gemma Xifra Unit of Diabetes, Endocrinology and Nutrition, Biomedical Research Institute (IDIBGi), Hospital 'Dr. Josep Trueta' of Girona, Girona, Spain; CIBERobn Fisiopatología de la Obesidad y Nutrición, Girona, Spain

Yoko Yagishita Department of Medical Biochemistry, Tohoku University Graduate School of Medicine, Sendai, Japan

Masayuki Yamamoto Department of Medical Biochemistry, Tohoku University Graduate School of Medicine, Sendai, Japan

Antonio Zorzano Institute for Research in Biomedicine (IRB Barcelona), Barcelona, Spain; Departament de Bioquímica i Biologia Molecular, Facultat de Biologia, Universitat de Barcelona, Barcelona, Spain; CIBER de Diabetes y Enfermedades Metabólicas Asociadas (CIBERDEM), Instituto de Salud Carlos III, Madrid, Spain

Preface

The life of mankind has changed dramatically, from the industrial society of the twentieth century to the age of technology invading our daily lives during the past few years. This major change has occurred only in a few decades' time. Although malnutrition is still a major nutritional challenge in some countries, the epidemic of obesity and type 2 diabetes mellitus has emerged as a major global issue related to dietary intake changes. Furthermore, although not affecting as many people, type 1 diabetes also creates a major burden because of its impact on younger patients and its increasing incidence in different regions around the world. Nutritional factors are clearly involved from the beginning in different sequences of the pathogenetic process that leads to diabetes mellitus. Therefore, we are in great need of gaining much more insight into the molecular pathways and the molecular mediators involved in the pathogenesis of diabetes and associated conditions. On the other hand, medical nutrition therapy is one of the mainstays of the treatment of diabetes mellitus. This has important implications in terms of introducing the best evidence-based approaches to the treatment of diabetes that should be based on high-quality research findings. But, again, although the body of evidence on which the nutritional management of the disease is based has increased significantly especially in recent years, there is still a long way to go.

The content of the book is organized into three sections. The introductory section includes chapters that aim to address the general concepts of nutrition related to diabetes. The initial introductory chapter addresses general aspects of the clinical nutritional approach to the management of diabetes. The remaining chapters in this section focus on the molecular mechanisms related to nutrition involved in the pathogenesis of hyperglycemia (i.e., insulin secretion and insulin action) including diet–gene interactions. The second section deals with the molecular biology of diabetes and focuses on areas such as oxidative stress, mitochondrial function, insulin resistance, high-fat diets, nutraceuticals, and lipid accumulation. The final section explores the genetic machinery behind diabetes and its metabolic-associated disturbances, including signaling pathways, gene expression, genome-wide association studies, and specific gene expression. It is not possible to include in one book all the potential areas of interest to those professionals working in the field of nutrition and diabetes. Nevertheless, in this first edition, the aim is to keep the focus of this series on molecular nutrition and cover important areas of interest in the field.

In conclusion, I truly hope that the content of the book will attract the interest of readers that need to gain insight on the different issues and challenges of their daily work. The book represents a timely and useful contribution to the rapidly expanding field of molecular nutrition in diabetes. I hope that its content not only helps the reader to answer questions, but also helps induce the passion to generate and solve important research questions that should ultimately contribute to the prevention and treatment of the large population of people with diabetes mellitus. I also hope that the content attracts the interest of a wide array of professional profiles. It has been a privilege to serve as the editor of this book that features contributions of researchers from all around the globe providing state-of-the-art reviews.

Didac Mauricio, MD, PhD
Editor

Acknowledgments

I thank Professor Victor R. Preedy for trusting me as the lead editor of this book. I am also grateful to all authors for accepting to participate in this endeavor and for contributing their expertise. I also wish to express my gratitude to the editorial team at Elsevier, Inc. in San Diego for their support, especially Jeff Rossetti for being always on the other side of my mailbox and for his excellent work as Editorial Project Manager.

SECTION 1

GENERAL AND INTRODUCTORY ASPECTS

CHAPTER

1

Nutrition and Diabetes: General Aspects

Julia C. Wiebe[1,2], *Rosa M. Sánchez Hernández*[1,2], *Lidia García*[3,4], *Ana M. Wägner*[1,2],
Figure by Laura López Ríos[1,2], *Literature search by Leticia Cuéllar*[3]

[1]Endocrinology and Nutrition Department, Complejo Hospitalario Universitario Insular Materno-Infantil de Gran Canaria, Las Palmas de Gran Canaria, Spain; [2]Instituto Universitario de Investigaciones Biomédicas y Sanitarias, Universidad de Las Palmas de Gran Canaria, Las Palmas de Gran Canaria, Spain; [3]Servicio de Evaluación del Servicio Canario de la Salud (SESCS), Santa Cruz de Tenerife, Spain; [4]Red de Investigación en Servicios de Salud en Enfermedades Crónicas (REDISSEC), Madrid, Spain

1. INTRODUCTION

Overweight and obesity are currently associated with more deaths worldwide than underweight, according to a report by the World Health Organization in 2014.[1] Obesity, a problem previously prevalent only in rich regions, is now a public health challenge also in low- and middle-income countries. Worldwide, 44% of diabetes can be attributed to overweight and obesity.[1] Indeed, diabetes incidence has also increased in a parallel manner to obesity, reaching epidemic proportions. In 2014, the estimated global prevalence of diabetes was 9% among people aged 18 years and older.[2]

Nutrition is a key component of diabetes management, where it fulfills general (adequate growth and development, weight maintenance) and specific (cardiovascular (CV) protection, glycemic control) purposes. At the same time, eating has strong cultural implications and changes, and limitations to food intake have a great impact on quality of life.[3,4] Thus, when giving dietary advice, it is crucial to focus on those recommendations whose benefits are based on strong, clinical evidence.

When the terms "nutrition and diabetes" are used to start a search in PubMed, they result in 18,920 hits (February 9, 2015). If "nutrition OR diet" are combined with diabetes, the number increases to 55,575 hits. Nevertheless, when the search is limited to randomized controlled trials, only 1049 and 3686 hits are found, respectively. Thus, although the interest and research on nutrition and diabetes are extensive, the highest level of clinical evidence represents only a minor fraction of the published studies.

The aim of this chapter is to review the relevant, clinical evidence available about the effects of nutritional interventions on the control of type 1 diabetes (T1D) and type 2 diabetes (T2D) and their complications. For this purpose, the most recently published international guidelines on the subject have been considered (Section 3) and a systematic review of randomized controlled trials has been performed (Section 4). In addition, to put present evidence into context, a historical description of nutritional recommendations and their changes in the past century is provided (Section 2).

2. HISTORICAL PERSPECTIVE

Nutritional recommendations for diabetes have changed dramatically in the past century. Before the discovery of insulin, patients with T1D were advised to fast in order to obtain sugar-free urine. Once this was achieved, dietary carbohydrate content was increased by 10 g/day until persistent glycosuria appeared.[5] This was done using green, bulky vegetables and, depending on the patient's "tolerance," also some garden vegetables and sometimes even potatoes and cereal. Fruit generally remained a minor fraction of carbohydrate intake in these patients, preferably used as dessert. The carbohydrate tolerance of each patient was used to design his or her maintenance diet, where the maximum carbohydrate allowance was the highest amount at

Molecular Nutrition and Diabetes
http://dx.doi.org/10.1016/B978-0-12-801585-8.00001-4

which the urine remained sugar-free. Protein intake was calculated to consist of 1.0–1.5 g/kg; the rest of the calories were accounted for by fat. The diet was adapted to the severity of diabetes and to the presence/absence of acidosis.[5] The progression from a "vegetable day," containing only 5 g of carbohydrate, is described by Hill as follows: "...carbohydrate 15 g, protein 25 g, fat 150 g. From this, the diet is slowly raised, increasing first the fat, then the protein and lastly the carbohydrate. The fat is never raised above 200 g and the calories seldom above 2200. On this, the patients hold their weight, feel well, and usually remain sugar-free."[6] After the discovery of insulin, the carbohydrate allowance increased progressively, as did the recommended total caloric intake[7] (Table 1). In 1933, Elliott Joslin recommended the following: "At present the diet I give my patients is approximately carbohydrate 140 g, protein 70 g, fat 90 g. Children need much more protein and if they require more calories I am inclined to give these calories equally divided between carbohydrate and fat." Indeed, advocates for a normal diet for the affected children started their campaigns, based on better nutritional results, fewer acute complications, and last, but not least, better acceptance.[8] The 1940s and 1950s witnessed a debate on whether patients with diabetes should be on a controlled or a free diet.[9,10] Several studies showed similar results on weight and hypoglycemia, although larger glucose fluctuations were observed with the free diet.[11] In the discussion, Forsyth et al. stated "From our experience of a group of 50 diabetics given liberal diets and insulin over a period of five years we are satisfied that, if adolescents and obese diabetics are excluded, clinical control, as defined earlier in this paper, can be attained in most patients. However, the degree of hyperglycemia and glycosuria and the daily fluctuation of blood-sugar levels are undoubtedly greater in such patients than in those on controlled diets." Nevertheless, soon, evidence appeared to support that hyperglycemia was associated with a higher risk of retinopathy and vessel calcification.[12,13] Thus, emphasis was put on the degree of glycemic control, while still supporting a relatively "free" diet. According to Forsyth herself: "By the term 'free diet' we imply liberty rather than license. Simple instructions are given to ensure the quality of the diet, and regular timing of meals is considered essential. Concentrated carbohydrates, such as table sugar, jam, chocolate, and sweets, are restricted." The American Diabetes Association (ADA) released its first exchange lists to facilitate constant dietary composition.[14] The link between dietary fat and atherosclerosis was recognized and, thus, fat intake was progressively reduced, especially at the expense of saturated fat.[15,16]

The first oral agents (sulfonylureas and phenformin) were available for the treatment of diabetes from the

TABLE 1 Historical Changes in Macronutrient Intake (% of Total Daily Caloric Intake) Recommendations for People with Diabetes

Year	CH	P	F	Cal/kg	References
1915	10	22	68	18	7
1927	23	15	62	28	7
1933	34	17	49		7
1941	30–33	NS	NS	1200–2500	17
1955	58–62	27–30	11–12	1600–2200, to achieve target weight	139
1963 (ADA diet)	41	17	42	Individualized	16
1963 (low-fat diet)	64	16	20	Individualized	16
1968 (free diet)	NS. Regular intake, 6 meals, avoid simple sugars	NS	NS	Hypocaloric	21
1969 ("controlled diet")	40; exchangeable 10-g portions	NS	NS	Weight loss in the overweight	a
1992	50–55	10–15	30–35 (≤10% saturated)	Calorie restriction if overweight	23
2002	NS	15–20	<7–10% saturated	Calorie restriction, focusing on fat reduction (<30%)	28
2008	NS	NS	<7% saturated	Weight loss if overweight	27

ADA, American Diabetes Association; CH, carbohydrate (% of total daily caloric intake); F, fat; P, protein; NS, not specified.
[a] *Chance GW. Outpatient management of diabetic children. BMJ 1969;2:493–5.*

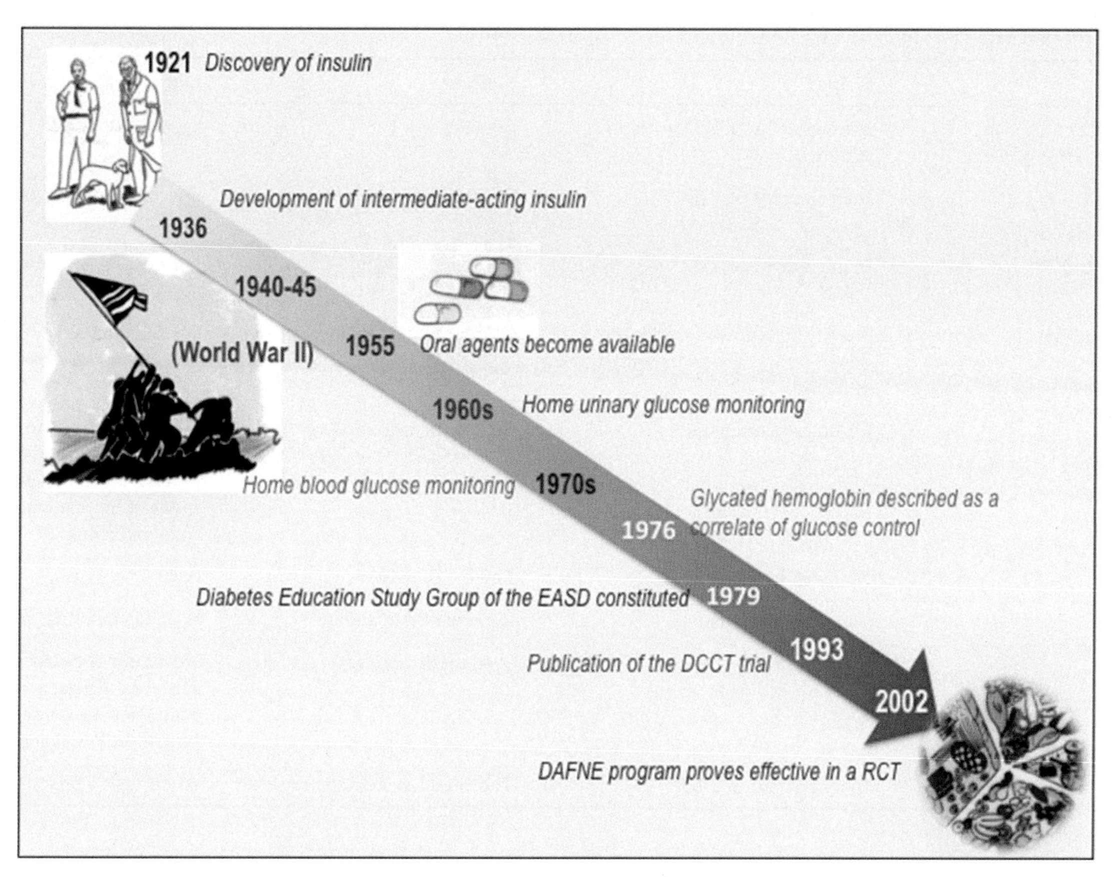

FIGURE 1 Selected historical milestones conditioning the (nutritional) treatment of diabetes.

late 1950s,[17] adding tools to the treatment of "mild," adult-onset diabetes (Figure 1). Home monitoring of glucose in the urine in the 1960s led to home blood glucose monitoring in the 1970s and, thus, gave patients better chances of improving their glycemic control.[18] The discovery in 1976 and later implementation of glycated hemoglobin (HbA1c) as a marker of glycemic control and predictor of chronic diabetes complications was another important breakthrough that paved the way for what was to be achieved less than two decades later.[19] The importance of diabetes education was acknowledged by the constitution of the Diabetes Education Study Group within the European Association of the Study of Diabetes in 1979, under the leadership of Jean-Philippe Assal.

Evidence published in the 1980s[20] questioned the need to refrain from simple sugars such as sucrose,[21,22] a recommendation that had settled like a dogma and is sometimes still heard today.

In the 1990s, calorie restriction for overweight patients was emphasized, fat was limited to 30–35% of total caloric intake, and saturated fat to 10%, especially in people with T2D.[23] Children with diabetes are advised to eat a normal diet, taking carbohydrate content into consideration and distributing food regularly throughout the day to avoid hyper- and hypoglycemia. Special "diabetes foods" were not considered necessary.[24] Indeed, Ingrid Mühlhauser and Michael Berger's group in Germany proved that, with adequate training, a free diet was possible, without deterioration of glycemic control.[25] The publication of the results of the Diabetes Control and Complications Trial in 1993 set the standards of treatment for T1D that are valid still today.[26]

At the turn of the twenty-first century, dietary recommendations for diabetes focus on fat restriction (<30% of total caloric intake), especially at the expense of saturated fat (<7–10%), and on achieving a healthy weight. Patients on intensive insulin treatment are advised to quantify carbohydrate content in their meals and adjust their premeal insulin dose thereafter.[27,28]

3. GUIDELINES

To assess present recommendations, available diabetes guidelines published between 2009 and 2014 have been reviewed. Table 2 summarizes their main features and Table 3 displays the recommendations based on review of the evidence and the level of evidence attributed to each statement, according to the cited document.

TABLE 2 Main Features of the Diabetes Guidelines (2009–2014) Examined

	Scope	Method	Nutritional advice
ALAD 2010[145]	Management of T2D in Latin America	Consensus	Specific, detailed
Chile 2010[a,32]	Management of T2D	Review of evidence	Specific, detailed, translation into food consumption
SNS-Spain 2012[a,34]	Management of T1D	Systematic review	For general health and diabetes-specific
Diabetes UK 2011[a,35]	Nutrition in adults at risk for developing diabetes or with T1D or T2D	Review of evidence	Specific, detailed, comprehensive
IDF 2012[146]	Diabetes management	Review of evidence	Individualize, reduce calories, control foods rich in added fats and sugar, adjust timing to medication, management of hypoglycemia
AACE[a,37]	Diabetes management	Review of evidence	Specific, detailed
ADA 2014[a,29]	Nutrition in diabetes	Review of evidence	Specific, detailed, comprehensive
AAP 2013[31]	Management at onset of T2D in children 10–18 years	Systematic review	Academy of Nutrition and Dietetics' Pediatric Weight Management Evidence-Based Nutrition Practice Guidelines
Diabetes Australia 2014–2015[30]	Management of T2D	Thorough, systematic review	Australian Dietary Guidelines

T1D, type 1 diabetes; T2D, type 2 diabetes.
[a]*Those based on review of the evidence and providing diabetes-specific recommendations, as well as level of evidence, are described in more detail in Table 3.*

Most diabetes management guidelines include some sort of nutritional recommendation, which ranges from very general to more specific advice. The ADA updated its recommendations recently,[29] and its 2013 position statement is probably the most extensive and detailed guideline currently available. A few months after its publication, the American Heart Association (AHA)/American College of Cardiology (ACC) guidelines were released after thorough, systematic review of the available evidence. The AHA/ACC recommendations are not specifically aimed at people with diabetes, but it is reasonable to use them as a complement to diabetes guidelines where specific studies in diabetes do not provide enough evidence.

Most current guidelines emphasize the importance of a healthy diet, similar to that recommended for the nondiabetic population, aimed at improving CV protection. Indeed, several guidelines refer to general nutritional recommendations as part of diabetes management.[30,31] For example, the Australian guidelines for the management of T2D, read: "The diet recommended for a person with diabetes [...] is qualitatively little different from the Mediterranean (MED) diet, or that recommended for all people (irrespective of whether they have diabetes, hypertension or dyslipidemia)."[30] Nevertheless, specific recommendations for diabetes are also available from several guidelines (Table 3).

3.1 Caloric Intake

Present guidelines agree that one of the aims of nutrition therapy is to adapt energy intake to patients' needs,[29,32] which will mainly depend on age, baseline body weight, and physical activity. The ADA recommends: "For overweight or obese adults with type 2 diabetes, reducing energy intake while maintaining a healthful eating pattern is recommended to promote weight loss."[29]

3.2 Carbohydrate

Unlike older recommendations, most present guidelines do not define an optimal amount of carbohydrate intake. Furthermore, food quality, rather than type of carbohydrate is emphasized: fresh, fiber-rich foods are advised to replace processed foods, containing high amounts of added sugar and fat (and, thus, calories). Glycemic index (GI) and glycemic load are more controversial concepts, which is reflected by the lack of agreement in the different guidelines.

TABLE 3 Recommendations and Evidence Rating (When Available and Understandable) According to Different Guidelines

Guideline	Energy	Carbohydrate	Fat	Protein	Sweeteners	Supplements
Chile 2010	Reduce intake if overweight (A)	50–60% TCI Limit sucrose Choose low-GI foods	<35% TCI <7% TCI saturated	–	Distinguishes caloric from noncaloric	–
AACE 2011[a]	Attain and maintain healthy weight	Advise about healthy sources (NE)	Limit saturated and *trans* fat (NE)	Choose sources low in saturated fat; avoid/limit processed meat[2]	–	Not advised (NE)
Diabetes UK 2011	Weight reduction for the overweight or obese person with T2D is effective in improving glycemic control and cardiovascular risk factors. (A)	Adjust insulin dose to intake in T1D (A). Low-GI diets may reduce HbA1c up to 0.5% (A)	Saturated fats should be limited and replaced by unsaturated, especially MUFA (A). Limit *trans* fat (C) Omega-3 can safely improve triglyceride (B). Mediterranean diet improves, blood pressure high-density lipoprotein cholesterol and triglyceride (B)	No clear evidence of benefit of protein restriction in diabetic nephropathy	Sucrose similar to other carbohydrate. Nonnutritive sweeteners are safe	Not routinely advised
Spain 2012	–	Adjust insulin dose to intake (A) Foods rich in sucrose can be replaced by other sources of carbohydrate (A) 25–30 g (7–13 g soluble) fiber (A)	–	Reduce to less than 0.8 g/kg in case of nephropathy (caution in case of renal failure and hypoalbuminemia) (A)	Choose those without an impact on glycemia (B) Limit drinks and foods sweetened with fructose (B)	–
ADA 2013	Reduce intake if overweight (A)	Adjust amount and insulin dose (B) Choose healthy sources (fruit, vegetables, whole grains) (B) Fiber and whole grains as general population (C)	A Mediterranean-style eating pattern, rich in monounsaturated fat, may benefit glycemic control and cardiovascular risk (B) Omega-3 (B) and saturated fats (C) recommended as for general population	Reducing amount below usual intake does not alter the course of nephropathy (A)	Avoid sugar-sweetened beverages (B) Nonnutritive sweeteners have the potential to reduce total carbohydrate and calorie intake (B)	Routine supplementation with vitamin E, C, carotene or omega-3 is not advised (A) For other supplements, there is insufficient evidence (C)

AACE, American Association of Clinical Endocrinologists; GI, glycemic index; MUFA, monounsaturated fatty acids; NE, no evidence; T1D, type 1 diabetes; T2D, type 2 diabetes; TCI, total caloric intake. A, B and C show decreasing levels of evidence, as self-defined by the individual guidelines. See Table 1 of *Diabetes Care* 2008;**31**(Suppl. 1):S1–S2.
[a]*The AACE uses a four-category, numerical quality of evidence rating.*

Carbohydrate counting for insulin adjustment is central in the management of T1D, and this recommendation is present in most documents counseling about this disease.[29,33–35] The ADA guidelines give specific advice regarding carbohydrate quantification and distribution according to glucose-lowering treatment to prevent hypoglycemia, as well as to treat the latter.[29]

3.3 Fat

Modifications in fat intake are aimed at improving CV health. Thus, advice in diabetes guidelines is comparable to that found for the general population[36] and is based more on the type than the total quantity of fat consumed. Patients are advised to reduce saturated and *trans* fat intake and to consume sources rich in polyunsaturated (PUFA), omega-3 fatty acids, and monounsaturated fatty acids (MUFAs).[29,35]

3.4 Protein

Recommendations on protein intake have remained rather stable over the years (Table 1) and across guidelines (Table 3). The only exception is the management of diabetic nephropathy, in which protein restriction was the rule until recently. Presently, there is disparity in this recommendation.

3.5 Sweeteners

Sugar-sweetened beverages are discouraged because of their high caloric content and their association with obesity. Noncaloric sweeteners (e.g., aspartame, cyclamate) are described as an option to reduce calorie and carbohydrate intake.[29,32,34]

3.6 Supplements

Only three of the reviewed guidelines specifically mention micronutrient and/or herbal supplements and none of them recommends them.[29,35,37]

3.7 Sodium Intake

Salt intake recommendations do not differ from those made to the general population. Unspecific[32,35] or specific (to less than 2.3 g/day) sodium intake reduction[29] is recommended, which is consistent with the 2013 AHA/ACC guideline to reduce CV risk.[36]

3.8 Translation into Food Intake

Most previous recommendations could be summarized as the following advice to patients: control portion size if weight is an issue; base your daily diet on vegetables, fruit, whole grains, legumes, and low-fat dairy products; and eat fish (preferably oily) at least twice per week. Choose fresh products whenever possible to prepare you own meals and avoid processed foods, especially meats, sweets, and sugar-containing beverages.

4. EVIDENCE FROM CLINICAL TRIALS

In this section, we summarize evidence published in systematic reviews and meta-analyses of clinical studies, updated with more recent randomized controlled trials (RCTs), on the effect of different types of diets or nutritional supplements on glycemic control and other CV risk factors. Our main sources of information have been: (1) A systematic review and meta-analysis[38] including 16 RCTs (3037 participants) of at least 6 months' duration assessing the effect of various diets on glycemic control, lipids, and weight loss, following Cochrane guidelines. (2) A systematic review[39] also including nonrandomized and smaller randomized controlled trials as well as observational studies and case–control studies, published between 2001 and 2010, with at least 10 participants per study group. (3) These two systematic reviews were updated by a literature search (MEDLINE, MEDLINE in process, EMBASE, Cochrane Central Register of Controlled Trials, Cochrane Database of Systematic Reviews, and Database of Abstracts of Reviews of Effects), in October 2014, by two of the authors (J.C.W., R.M.S.), focused on RCTs (at least 6 months' duration and 50 participants per group), which resulted, after screening 2801 references, in five additional studies of interest not mentioned in the systematic reviews on macronutrients: one RCT assessing GI,[40] two on low-carbohydrate diets,[41,42] and two on high-protein diets.[43,44] To analyze supplements, micronutrients, and specific diets, our systematic literature search was complemented by other studies, identified by a hand search. For the review of the different eating patterns and carbohydrate counting in T1D, a systematic review[38] and the literature search were complemented by studies identified by a hand search.

The interventions were classified into five groups, in which four included almost exclusively studies in T2D and the last one reported on nutritional education in patients with T1D.

4.1 Macronutrient Composition

According to the ADA's "Recommendations for the Management of Adults With Diabetes" position statement, there is little evidence for an ideal caloric combination of carbohydrate, protein, and fat.[29] Thus, the ADA advises that distribution of calories among macronutrients should be based on individual eating patterns, preferences, and metabolic goals.[29]

However, numerous studies have attempted to identify the optimal mix of macronutrients for the meal plans of people with diabetes.

4.1.1 Protein Content

Diets in which 20–30% or more of the total caloric intake come from proteins are considered "high protein." The two studies analyzed in a recent meta-analysis[38] compared a high-protein (26.5–30%) diet with a control diet (15–19% of calories from proteins).[45,46] Even though no significant differences were found in glycemic control, weight, or lipids in the single studies, combined analysis of pooled data showed significantly lower HbA1c concentrations (−0.28%; $P < 0.00001$) in the high-protein diet group and no change in urinary albumin excretion.[45] A systematic review[39] including three smaller, short (4–16 weeks) RCTs[47–49] showed contradictory results on glycemic control and CV risk factors. Finally, an additional, much larger RCT (418 participants randomized),[44] comparing two diets with 15% versus 30% protein, with the difference accounted for by carbohydrate, showed no differences between groups for weight, HbA1c, lipids, blood pressure, or renal function at one year.

4.1.1.1 Protein Intake in Diabetic Kidney Disease

Moderate protein restriction is recommended for patients with diabetic nephropathy in several recent guidelines,[27,34] although the last position statement by the ADA[29] does not give this advice. The reason for this is the publication of new evidence on this issue.

In patients with early renal disease (urinary albumin/creatinine ratio 3–30 mg/mmol), no differences were found in renal function, albuminuria, or blood pressure 1 year after following a high-protein (30%, 1.2 g/kg) diet, when compared with a standard-protein (20%, 0.9 g/kg) diet.[43] Three earlier systematic reviews and a meta-analysis studied the effect of modified or low-protein diets on glycemic control, CV risk factors, and renal function in people with T1D or T2D.[39,50,51] Four long (1–4 years), albeit relatively small, RCTs included patients (23–47 per group) with microalbuminuria,[52] macroalbuminuria,[53,54] or both.[55,39] In two of the studies, daily protein intake was reduced compared with the control group (0.86–0.89 g vs 1.02–1.24 g),[53,54] whereas in the other two, protein intake turned out to be higher in the planned lower-protein group than in the control groups.[52,55] No significant differences for glycemic control, CV risk factors, or renal function (glomerular filtration rate, proteinuria, serum albumin) were found between the groups. The studies analyzed in the mentioned systematic review[39] were also included in another systematic review and in a meta-analysis, which reached similar conclusions.[50,51]

4.1.2 Carbohydrate Content and Quality

Macronutrient distribution of a standard diet ranges from 55 to 65% carbohydrate, <30% fat, and 10–20% protein,[32] but no official definition of "low- (or high-) carbohydrate diets" exists. For the purpose of this review, a low-carbohydrate diet was defined as 20–40% of total caloric intake and a high-carbohydrate diet as >65%.[38]

A recent systematic review[39] of seven RCTs and two meta-analyses examined the effects of moderate- or high-carbohydrate diets on glycemic control in patients with T2D[45,47,49,56–60] or T1D.[61] Comparison of the studies was difficult, given the differences in fat and protein contents of the control diets, duration of follow-up (5–74 weeks), and the number of participants.[10–99] Four studies found no significant differences in glycemic control,[45,49,56,57] one RCT found significantly lower HbA1c in the lower-carbohydrate diet, whereas another study found improved LDL cholesterol with a high-carbohydrate diet and two studies found improved triglyceride with a low-carbohydrate diet.[39] A systematic review,[38] which assessed the outcome of two additional RCTs comparing a high-carbohydrate diet with a diet high in MUFAs and a high-fiber diet to a low-fat diet, found no significant differences in weight, glycemic control, or lipids.[62,63]

A systematic review[39] on low carbohydrate diets included 11 studies, seven of which with a very low-carbohydrate content[64–70] and four with a moderately low content.[71–74] The analyzed clinical trials[3] and RCTs[8] included adults with T2D with varying duration of follow-up (14 days–1 year) and sample sizes (10–55 participants). HbA1c decreased in four of the very low and two of the moderate carbohydrate groups, but studies were small, of short duration, not all of them were RCTs, and dropout rates were high. High-density lipoprotein (HDL) cholesterol seemed to improve with reduction of total carbohydrate intake, but an impact from differences in caloric intake and weight loss on the results is possible.[39] A meta-analysis including several of the mentioned RCTs[65,67,68,74–78] showed that, overall, low-carbohydrate diets had no significant effects on weight, reduced HbA1c modestly (0.12% points), and increased HDL cholesterol by 0.11 mmol/L (4.2 mg/dL).[38]

Results of two independent studies comparing very low-carbohydrate to low-fat diets showed no differences between groups in HbA1c or associated CV risk factors.[41,42]

TABLE 4 Description of Dietary Interventions Mentioned in the Text and Their Effect on Glycemic Control and CV Risk Factors

Diet	Description	Effect
High-protein content	>20–30% of total caloric intake from protein	Little evidence. Possible, modest reduction in HbA1c
Protein restriction	0.9 g/kg versus 1.0–1.2 g/kg in patients with diabetic nephropathy	No effect on albuminuria, glomerular filtration rate, or blood pressure
Carbohydrate content	Low carbohydrate content: 20–40% of total caloric intake from carbohydrates	Modest reduction in HbA1c (0.12%). Increase in HDL cholesterol (4.2 mg/dL)
	High-carbohydrate content: >65% of total caloric intake from carbohydrates	No significant differences in weight, glycemic control, or lipids
Low GI	GI < 55: Whole wheat or pumpernickel bread, oatmeal (rolled or steel-cut), oat bran, muesli, pasta (al dente), converted rice, barley, bulgur, sweet potato, corn, yam, lima/butter beans, peas, legumes and lentils, most fruits, nonstarchy vegetables, and carrots[147]	Modest reduction in HbA1c (0.14%). Increase in HDL cholesterol
Low total-fat diets	<30% of total caloric intake from fat	No evidence of benefit
Supplements	Herbal products, micronutrients, specific foods	No solid evidence of benefit, except for triglyceride reduction with n-3 PUFA
Mediterranean	High proportion: monounsaturated fat (mainly from olive oil), plant foods like fruit, vegetables, breads, cereals, and nuts. Moderate consumption: fish and poultry. Low intake: dairy products (mostly cheese and yogurt), red and processed meats and sweets Low content in saturated fat (7–8% of energy); total fat intake: <25% to >35% of daily intake	Reduction in HbA1c (0.47%), weight (1.84 kg) and triglycerides (0.21 mmol/L; 18 mg/dL). Increase in HDL cholesterol (0.04 mmol/L; 1.5 mg/dL), no changes in LDL cholesterol.
Vegetarian and vegan	Vegan diets: no flesh foods and animal-derived products Vegetarian diets: similar to vegan diets but include eggs (ovo) and dairy (lacto) products	Vegan diets: no consistent improvement of glycemic control or CV risk factors Vegetarian diets: limited research, conclusion difficult
DASH	High intake: fruits, vegetables, low-fat dairy products, nuts Reduced intake: sodium, saturated fat, sweets, red meat, and sugar-containing beverages	Healthy people: Improved blood pressure People with diabetes: limited evidence
DAFNE	Intensive flexible insulin therapy education program in T1D. Objective: adjusted insulin doses based on carbohydrate intake	Reduction in clinical admissions and HbA1c; improved quality of life

CV, cardiovascular; GI, glycemic index; HbA1c, glycated hemoglobin; HDL, high-density lipoprotein; LDL, low-density lipoprotein; PUFA, polyunsaturated fatty acids; T1D, type 1 diabetes.

4.1.2.1 Low GI

GI is defined by the postprandial glucose excursion occurring after the ingestion of a carbohydrate-containing food compared with white bread or glucose. Many fresh, unprocessed, and fiber-rich foods have a low GI (see Table 4). A meta-analysis included three RCTs comparing low GI (defined as GI between 39 and 77) with other diets.[74,79,80] A modest (0.14% points) but significant decrease in HbA1c was seen, as was an increase in HDL in the low-GI diet, but no significant reductions in LDL cholesterol, triglyceride, or weight.[38] However, when compared with a low-carbohydrate diet, a low-GI diet seems to be inferior in reducing HbA1c.[68] Smaller, shorter studies do not add significant knowledge to the mentioned meta-analysis.[39]

4.1.3 Fat Content

Low-fat (<30% of total calories) diets are often assigned to the comparison group in studies assessing low-carbohydrate diets.[38] A systematic review assessed seven RCTs on low-fat eating patterns in people with diabetes,[39] with only one of them in T1D.[61] HbA1c was reduced in one study,[81] whereas no significant effects were seen on lipids.[49,56,57,61,81–83] Vegan and vegetarian diets are also relatively low in fat and are described later (see Section 4.2.2).

Although there is limited research on this issue in diabetes, reductions in saturated fat below 7% have proved beneficial for the reduction of CV risk in earlier studies, especially at the expense of the reduction in low-density lipoprotein (LDL) cholesterol.[36] Mediterranean-style eating patterns, which are rich in MUFA, are discussed later (see Section 4.4.1).

4.1.4 Summary

A meta-analysis of RCTs showed small but significant effects of low-carbohydrate and low-GI diets on HbA1c. Assessment of high-protein diets was based on two studies only, which limits the conclusions that can be drawn. Currently, there is no evidence to restrict protein intake below what is recommended for the general population in patients with diabetic nephropathy.

4.2 Supplements

Even though a well-balanced diet is essential for the treatment of T2D, there are data showing insufficient intake of some micronutrients,[84,85] as well as supplementation with others in more than half of the patients.[84] The most frequently consumed supplements in a Polish study were magnesium, herbs, antioxidant vitamins, B-group vitamins, and omega-3 fatty acids.[84] In this section, we try to shed light on the effects of supplements of specific interest in diabetic nutrition.

4.2.1 Herbal Products

There is currently no solid evidence that herbal products improve glycemic control. A systematic review included 27 studies assessing individual herbs, of which 15 were RCTs, mostly short term and small, with limited quality. The authors concluded that evidence was insufficient to draw conclusions.[86] A high-quality, very well-performed RCT showed promising effects of cinnamon, with 0.5% point reductions in HbA1c and 5/3 mmHg in systolic/diastolic blood pressure,[87] but a more recent, systematic review including 10 RCTs did not confirm their effects.[88]

4.2.2 Whole Grains

Results from two large prospective cohort studies including almost 120,000 patients indicated that higher whole grain consumption (defined as foods containing the entire grain seed) was associated with lower total and CV mortality, independent of other dietary and lifestyle factors.[89] To assess the effect of whole grain consumption on glycemic control and CV risk markers in patients with diabetes, two small (15–20 participants), short (5–12 weeks) RCTs comparing whole grains with fiber in people with T2D, were performed.[90,91] Consumption of whole grains was not associated with any changes in HbA1c and neither study found any significant differences in CV risk markers.[39]

4.2.3 Omega-3 Fatty Acids

High triglyceride concentrations are considered a CV risk factor, and n-3 PUFA consumption reduces them. However, results on safety of n-3 PUFAs are contradictory. Although some older studies using high doses ($\geq$10 g/day fish oil) have reported unfavorable effects on glucose metabolism, recent trials using lower doses (2–4 g/day) have shown that n-3 PUFA supplementation improves triglyceride levels without impairing glucose metabolism in hypertriglyceridemic patients with T2D.[92]

In a systematic review assessing the effects of n-3 PUFA,[39] one of six RCTs[93] found a significant decrease in HbA1c and three[94–96] showed increases in HDL cholesterol, when compared with corn or olive oil.

4.2.4 Nuts

Most tree nuts and peanuts are rich in PUFA, whereas walnuts and pine nuts are rich in MUFA. However, effects of nut-enriched diets in people with diabetes are contradictory.[97–104] They do not seem to alter glycemia and results regarding lipid changes are mixed.[39]

4.3 Micronutrients

Evidence of the effect of micronutrients on glycemic control and CV risk factors is limited by variation in micronutrient dosing, baseline and achieved micronutrient levels and heterogeneity in study quality and methodology. A systematic review of RCTs found limited evidence for chromium supplementation on glucose metabolism and lipids.[105] Results of clinical studies evaluating the effect of magnesium on glycemic control are contradictory. Although one RCT restored serum magnesium levels and improved metabolic control in T2D patients with decreased serum magnesium concentrations,[106] another one performed in patients with normal concentrations did not have an effect on glycemic control or plasma lipid concentrations.[107] Studies assessing the effect of vitamin D on glycemic control also showed conflicting results,[108–112] and an RCT assessing the influence of dairy calcium intake did not affect HbA1c or CV risk factors.[113]

4.4 Eating Patterns

Eating patterns or dietary patterns are food groups consumed in characteristic combinations.[114] Several factors influence these eating patterns, such as the availability of different kinds of foods, cultural patterns, traditions, health beliefs, and economy.[115] The most relevant eating patterns in the context of diabetes and CV health, reviewed in this section, are the MED style, vegetarian, and Dietary Approaches to Stop Hypertension (DASH) dietary patterns.[29]

4.4.1 *Mediterranean (MED) Style*

The MED dietary pattern is characterized by a high proportion of MUFA, mainly from olive oil; abundant amounts of plant foods such as fruit, vegetables, breads, cereals, and nuts; moderate consumption of fish and poultry; and low intake of dairy products (mostly cheese and yogurt), red and processed meats, and sweets. MED typically has a low content of saturated fat (7–8% of energy) and a total fat intake varying between <25% and >35% of total daily intake, depending on the region. It has mainly been studied in Mediterranean countries.[116]

The CV benefits of MED have been identified in observational cohort studies,[117,118] in secondary prevention in a 4-year trial after a first myocardial infarction,[119] and, recently, in primary CV prevention.[120] In the latter study, adherence to MED was associated with fewer CV events in participants with diabetes. In a recent meta-analysis,[38] a comparison of MED and three other diets[78,121,122] showed a reduction in HbA1c (0.47%), weight (1.84 kg), and triglyceride (0.21 mmol/L; 18 mg/dL) and an increase in HDL cholesterol (0.04 mmol/L; 1.5 mg/dL), but no changes in LDL cholesterol. Another study[123] compared two variations of MED (supplemented with nuts or olive oil) with a lower fat diet in 1224 participants with high CV risk (819 with diabetes). The most relevant finding was a decrease in triglyceride levels in the MED group with nut supplementation.

4.4.2 *Vegetarian and Vegan*

Vegan diets can be defined as diets avoiding all flesh foods and animal-derived products, whereas vegetarian diets are similar to vegan diets but include eggs (ovo) and dairy (lacto) products.

Research in vegetarian and vegan diets is limited. Six vegetarian and low-fat vegan studies of diverse follow-up duration (12–74 weeks)[56,124–127] in patients with T2D were reviewed.[29,38,39] No consistent improvement in glycemic control or CV risk factors was found, except when energy intake was restricted and patients lost weight. Indeed, in all but one study,[124] vegetarian diets were associated with weight loss. Research regarding vegetarian eating patterns is limited and methodological variation among studies makes conclusive remarks difficult.[39] However, the two longest RCTs suggest that this type of dietary pattern may improve glycemic control and lipid concentrations.[56,127]

4.4.3 *Dietary Approaches to Stop Hypertension (DASH)*

The DASH style is focused on high intake of fruits and vegetables, whereas sodium is reduced. Consumption of low-fat dairy products and nuts is high and intake of saturated fat, sweets, red meat, and sugar-containing beverages is low.[128]

The DASH eating plan has been demonstrated to improve blood pressure and could be associated with lower CV risk in people without diabetes.[128,129] However, in people with diabetes, the evidence is limited. An 8-week, randomized, crossover, controlled trial[130] in 31 patients with T2D, comparing a DASH diet with a control diet, showed improvements in glycemic control, weight loss, lipids, and blood pressure at the end of the DASH diet period.

4.5 Carbohydrate Counting in T1D

The long-term benefits of reducing HbA1c concentrations by intensive insulin therapy in T1D were demonstrated more than 20 years ago.[26] However, the reduction in the onset and progression of microvascular complications was counterbalanced by an increase in the incidence of severe hypoglycemia and weight gain, and required continuous, expensive support from healthcare staff.[26] Through a 5-day structured training program, which promoted dietary flexibility and insulin dose adjustment, Mühlhauser and Berger's group in Düsseldorf improved patients' glycemic control without increasing their weight or their risk for hypoglycemia. Furthermore, the need for long-term support from the health care team was minimal.[131] In 2002, this program was translated into English, adapted, implemented, and evaluated in an RCT, under the name of Dose Adjustment For Normal Eating (DAFNE). It improved glycemic control, reduced hospital admissions, and increased quality of life and dietary freedom in patients with T1D.[132]

Many other studies have demonstrated the benefits of adjusting insulin to carbohydrate intake on HbA1c.[132–137] No changes in weight or lipids have been observed, except in one study performed in patients with continuous subcutaneous insulin infusion, which showed a reduction in body mass index and waist circumference.[133] A study in children with T1D comparing strict carbohydrate prescription and counting with a more flexible option based on individual serving sizes including training in the choice of low-GI foods showed a significant reduction in HbA1c and an increase in quality of life in the more flexible group.[138]

A long-term evaluation of the impact of the DAFNE course was performed by Speight et al.[139] They showed that, after 4 years, the improvement in quality of life was still maintained and HbA1c remained better than at baseline, even though it had started to deteriorate a year after the program ended. Retrospective data also support durable HbA1c reductions in these kinds of educational programs.[140,141]

Based on the high level of evidence, for T1D patients, the ADA guidelines recommend the participation in an intensive flexible insulin therapy education program using the carbohydrate-counting meal-planning approach.[29] The benefits of these educational programs in T1D have been demonstrated, whether they are performed individually or in groups.[29]

5. FURTHER RESEARCH

Although there is increasing knowledge on dietary interventions in diabetes, much of the existing evidence is conflicting. Studies, when randomized, are often limited by their size and duration as well as by the confounding effects of calorie restriction, weight loss, and the presence of several, simultaneous changes in diet composition. Additional evidence is needed to confirm the conclusions that can be drawn from the presently published studies. It is important to try to isolate the effect of a concrete intervention from the rest of the diet in sufficiently large, long-term RCTs. This could be applied to the most promising, albeit insufficiently proven, interventions in diabetes, such as the MED and DASH eating patterns and their variations.

An excellent example of this is the randomized, crossover, controlled trial designed to assess the effect of GI and carbohydrate content within a healthful DASH-style diet in overweight patients. Four 5-week interventions were compared: a high-GI (65%), high-carbohydrate (58% of caloric intake) diet; a low-GI (40%), high-carbohydrate diet; a high-GI, low-carbohydrate (40%) diet; and a low-GI, low-carbohydrate diet. No benefits of the low-GI diets were seen on insulin sensitivity, lipids, or blood pressure.[142]

6. CONCLUSIONS

Given that a variety of dietary modifications may have a positive effect on glycemic control and CV risk, it is reasonable to adapt advice to the patient's preference and focus our advice on improving CV health. Overall, results seem to be most positive with dietary patterns such as the MED or DASH diets, or even vegan/vegetarian. Effects of herbal products and micronutrients supplements are contradictory, but patients should report the use of such products to their practitioners because they might interact with other medications and an excessive intake may have negative health effects.[29,143,144]

Recommendations should focus (as guidelines do) on encouraging the intake of fresh fruits and vegetables, low-fat dairy products, nuts, and nontropical vegetable oils and fish. Highly processed foods, such as meats, sugar-sweetened beverages, and other sweets should be limited, although there is no evidence to specifically avoid sucrose in the diet.

References

1. World Health Organization. Available from: http://www.who.int/features/factfiles/obesity/en; 2014 [cited January 26, 2015].
2. World Health Organization. Available from: http://www.who.int/mediacentre/factsheets/fs312/en/ [cited January 26, 2015].
3. Bradley C, Speight J. Patient perceptions of diabetes and diabetes therapy: assessing quality of life. *Diabetes Metab Res Rev* September–October 2002;**18**(Suppl. 3):S64–9.
4. Speight J, Sinclair AJ, Browne JL, Woodcock A, Bradley C. Assessing the impact of diabetes on the quality of life of older adults living in a care home: validation of the ADDQoL Senior. *Diabet Med* January 2012;**30**(1):74–80.
5. Allen FM, Stillman E, Fitz R. *Total dietary regulation in the treatment of diabetes*. Monograph no. 11. New York: Rockefeller Institute for Medical Research; 1919.
6. Hill LW, Sherrick JL. Report on the allen treatment of diabetes. *Bosten Med Surg J* 1915;**172**:696–700.
7. Joslin E. Fat and the diabetic. *N Engl J Med* 1933;**209**:519–28.
8. Medovy H. the treatment of diabetes in children by means of a normal type of diet. *Can Med Assoc J* December 1933;**29**(6):605–9.
9. Hunt B. Why diet the diabetic? *Med J Aust* January 20, 1951;**1**(3): 114–7.
10. Tolstoi E. The free diet for diabetic patients. *Am J Nurs* October 1950;**50**(10):652–4.
11. Forsyth CC, Kinnear TW, Dunlop DM. Diet in diabetes. *Br Med J* May 19, 1951;**1**(4715):1095–101.
12. Forsyth CC, Payne WW. Free diets in the treatment of diabetic children. *Arch Dis Child* August 1956;**31**(158):245–53.
13. Root HF, Keiding N. Results of treatment with free diet and controlled diet in diabetes mellitus. *Trans Am Clin Climatol Assoc* 1952;**63**:26–36.
14. Caso EK. Acceptance of the ADA diets and exchange lists. *Diabetes* July–August 1953;**2**(4):275–6.
15. Singh I. Low-fat diet and therapeutic doses of insulin in diabetes mellitus. *Lancet* February 26, 1955;**268**(6861):422–5.
16. Stone DB, Connor WE. The prolonged effects of a low cholesterol, high carbohydrate diet upon the serum lipids in diabetic patients. *Diabetes* March–April 1963;**12**:127–32.
17. Walker JB. Carbohydrate consumption and diabetes. *Proc Nutr Soc* 1964;**23**:143–9.
18. Kühl C. From Hagedorn-Norman Jensen's blood glucose method to insulin analogues. In: Binder C, Deckert T, Nerup J, editors. *Diabetes and Denmark*. Copenhagen: Gad Publishers; 2007. p. 14–26.
19. Christiansen JS. The relation between blood glucose and HbA1c-A Danish pioneering effort. In: Binder C, Deckert T, Nerup J, editors. *Diabetes and Denmark*. Copenhagen: Gad Publishers; 2007. p. 52–60.
20. Bantle JP, Laine DC, Castle GW, Thomas JW, Hoogwerf BJ, Goetz FC. Postprandial glucose and insulin responses to meals containing different carbohydrates in normal and diabetic subjects. *N Engl J Med* July 7, 1983;**309**(1):7–12.

21. Crapo PA, Olefsky JM. Food fallacies and blood sugar. *N Engl J Med* July 7, 1983;**309**(1):44–5.
22. Wood Jr FC, Bierman EL. Is diet the cornerstone in management of diabetes? *N Engl J Med* November 6, 1986;**315**(19):1224–7.
23. Nutrition Subcommittee of the British Diabetic Association's Professional Advisory Committee. Dietary recommendations for people with diabetes: an update for the 1990s. *Diabet Med* March 1992;**9**(2):189–202.
24. Magrath G, Hartland BV. Dietary recommendations for children and adolescents with diabetes: an implementation paper. British Diabetic Association's Professional Advisory Committee. *Diabet Med* November 1993;**10**(9):874–85.
25. Mühlhauser I, Bott U, Overmann H, Wagener W, Bender R, Jörgens V, et al. Liberalized diet in patients with type 1 diabetes. *J Intern Med* June 1995;**237**(6):591–7.
26. The DCCT study. The effect of intensive treatment of diabetes on the development and progression of long-term complications in insulin-dependent diabetes mellitus. The Diabetes Control and Complications Trial Research Group. *N Engl J Med* September 30, 1993;**329**(14):977–86.
27. American Diabetes Association. Standards of medical care in diabetes. *Diabetes Care* 2008;**31**(Suppl. 1):12–54.
28. American Diabetes Association. Evidence-based nutrition principles and recommendations for the treatment and prevention of diabetes and related complications. *Diabetes Care* January 2002; **25**(1):202–12.
29. Evert AB, Boucher JL, Cypress M, Dunbar SA, Franz MJ, Mayer-Davis EJ, et al. Nutrition therapy recommendations for the management of adults with diabetes. *Diabetes Care* November 2013; **36**(11):3821–42.
30. *The Royal Australian College of General Practitioners and Diabetes Australia. General practice management of type 2 diabetes – 2014–15.* Melbourne. 2014.
31. Copeland KC, Silverstein J, Moore KR, Prazar GE, Raymer T, Shiffman RN, et al. Management of newly diagnosed type 2 Diabetes Mellitus (T2DM) in children and adolescents. *Pediatrics* February 2013;**131**(2):364–82.
32. *Ministerio de Salud. Guía clínica diabetes mellitus tipo 2.* Santiago: Minsal; 2010.
33. Chiang JL, Kirkman MS, Laffel LM, Peters AL, Type 1 Diabetes Sourcebook Authors. Type 1 diabetes through the life span: a position statement of the American Diabetes Association. *Diabetes Care* 2014;**37**:2034–54.
34. *Grupo de trabajo de la Guía de Práctica Clínica sobre Diabetes mellitus tipo 1. Guía de Práctica Clínica sobre Diabetes mellitus tipo 1. Plan de Calidad para el Sistema Nacional de Salud del Ministerio de Sanidad y Política Social. Agencia de Evaluación de Tecnologías Sanitarias del País Vasco-Osteba.* Guías de Práctica Clínica en el SNS: OSTEBA no. 2009/10; 2012.
35. Dyson PA, Kelly T, Deakin T, Duncan A, Frost G, Harrison Z, et al. Diabetes UK evidence-based nutrition guidelines for the prevention and management of diabetes. *Diabet Med* November 2011; **28**(11):1282–8.
36. Eckel RH, Jakicic JM, Ard JD, de Jesus JM, Houston Miller N, Hubbard VS, et al. 2013 AHA/ACC guideline on lifestyle management to reduce cardiovascular risk: a report of the American College of Cardiology/American Heart Association Task Force on Practice Guidelines. *Circulation* June 24, 2013;**129**(25 Suppl. 2):S76–99.
37. Handelsman Y, Mechanick JI, Blonde L, Grunberger G, Bloomgarden ZT, Bray GA, et al. American Association of Clinical Endocrinologists Medical Guidelines for clinical practice for developing a diabetes mellitus comprehensive care plan. *Endocr Pract* March–April 2011;**17**(Suppl. 2):1–53.
38. Ajala O, English P, Pinkney J. Systematic review and meta-analysis of different dietary approaches to the management of type 2 diabetes. *Am J Clin Nutr* March 2013;**97**(3):505–16.
39. Wheeler ML, Dunbar SA, Jaacks LM, Karmally W, Mayer-Davis EJ, Wylie-Rosett J, et al. Macronutrients, food groups, and eating patterns in the management of diabetes: a systematic review of the literature, 2010. *Diabetes Care* February 2012; **35**(2):434–45.
40. Fabricatore AN, Wadden TA, Ebbeling CB, Thomas JG, Stallings VA, Schwartz S, et al. Targeting dietary fat or glycemic load in the treatment of obesity and type 2 diabetes: a randomized controlled trial. *Diabetes Res Clin Pract* April 2011;**92**(1):37–45.
41. Goldstein T, Kark JD, Berry EM, Adler B, Ziv E, Raz I. The effect of a low carbohydrate energy-unrestricted diet on weight loss in obese type 2 diabetes patients e A randomized controlled trial. *Eur E J Clin Nutr Metab* 2011;**6**:178–86.
42. Guldbrand H, Dizdar B, Bunjaku B, Lindstrom T, Bachrach-Lindstrom M, Fredrikson M, et al. In type 2 diabetes, randomisation to advice to follow a low-carbohydrate diet transiently improves glycaemic control compared with advice to follow a low-fat diet producing a similar weight loss. *Diabetologia* August 2012;**55**(8): 2118–27.
43. Jesudason DR, Pedersen E, Clifton PM. Weight-loss diets in people with type 2 diabetes and renal disease: a randomized controlled trial of the effect of different dietary protein amounts. *Am J Clin Nutr* August 2013;**98**(2):494–501.
44. Krebs JD, Elley CR, Parry-Strong A, Lunt H, Drury PL, Bell DA, et al. The Diabetes Excess Weight Loss (DEWL) Trial: a randomised controlled trial of high-protein versus high-carbohydrate diets over 2 years in type 2 diabetes. *Diabetologia* April 2012; **55**(4):905–14.
45. Brinkworth GD, Noakes M, Parker B, Foster P, Clifton PM. Long-term effects of advice to consume a high-protein, low-fat diet, rather than a conventional weight-loss diet, in obese adults with type 2 diabetes: one-year follow-up of a randomised trial. *Diabetologia* October 2004;**47**(10):1677–86.
46. Larsen RN, Mann NJ, Maclean E, Shaw JE. The effect of high-protein, low-carbohydrate diets in the treatment of type 2 diabetes: a 12 month randomised controlled trial. *Diabetologia* April 2011;**54**(4):731–40.
47. Gannon MC, Nuttall FQ, Saeed A, Jordan K, Hoover H. An increase in dietary protein improves the blood glucose response in persons with type 2 diabetes. *Am J Clin Nutr* October 2003; **78**(4):734–41.
48. Parker B, Noakes M, Luscombe N, Clifton P. Effect of a high-protein, high-monounsaturated fat weight loss diet on glycemic control and lipid levels in type 2 diabetes. *Diabetes Care* March 2002;**25**(3):425–30.
49. Wycherley TP, Noakes M, Clifton PM, Cleanthous X, Keogh JB, Brinkworth GD. A high-protein diet with resistance exercise training improves weight loss and body composition in overweight and obese patients with type 2 diabetes. *Diabetes Care* May 2010;**33**(5):969–76.
50. Pan Y, Guo LL, Jin HM. Low-protein diet for diabetic nephropathy: a meta-analysis of randomized controlled trials. *Am J Clin Nutr* September 2008;**88**(3):660–6.
51. Robertson L, Waugh N, Robertson A. Protein restriction for diabetic renal disease. *Cochrane Database Syst Rev* 2007; (4):CD002181.
52. Pijls LT, de Vries H, van Eijk JT, Donker AJ. Protein restriction, glomerular filtration rate and albuminuria in patients with type 2 diabetes mellitus: a randomized trial. *Eur J Clin Nutr* December 2002;**56**(12):1200–7.
53. Meloni C, Tatangelo P, Cipriani S, Rossi V, Suraci C, Tozzo C, et al. Adequate protein dietary restriction in diabetic and nondiabetic patients with chronic renal failure. *J Ren Nutr* October 2004; **14**(4):208–13.
54. Hansen HP, Tauber-Lassen E, Jensen BR, Parving HH. Effect of dietary protein restriction on prognosis in patients with diabetic nephropathy. *Kidney Int* July 2002;**62**(1):220–8.

55. Dussol B, Iovanna C, Raccah D, Darmon P, Morange S, Vague P, et al. A randomized trial of low-protein diet in type 1 and in type 2 diabetes mellitus patients with incipient and overt nephropathy. *J Ren Nutr* October 2005;**15**(4):398–406.
56. Barnard ND, Cohen J, Jenkins DJ, Turner-McGrievy G, Gloede L, Green A, et al. A low-fat vegan diet and a conventional diabetes diet in the treatment of type 2 diabetes: a randomized, controlled, 74-wk clinical trial. *Am J Clin Nutr* May 2009;**89**(5):1588S–96S.
57. Gerhard GT, Ahmann A, Meeuws K, McMurry MP, Duell PB, Connor WE. Effects of a low-fat diet compared with those of a high-monounsaturated fat diet on body weight, plasma lipids and lipoproteins, and glycemic control in type 2 diabetes. *Am J Clin Nutr* September 2004;**80**(3):668–73.
58. Kirk JK, Graves DE, Craven TE, Lipkin EW, Austin M, Margolis KL. Restricted-carbohydrate diets in patients with type 2 diabetes: a meta-analysis. *J Am Diet Assoc* January 2008;**108**(1):91–100.
59. Kodama S, Saito K, Tanaka S, Maki M, Yachi Y, Sato M, et al. Influence of fat and carbohydrate proportions on the metabolic profile in patients with type 2 diabetes: a meta-analysis. *Diabetes Care* May 2009;**32**(5):959–65.
60. Rodriguez-Villar C, Perez-Heras A, Mercade I, Casals E, Ros E. Comparison of a high-carbohydrate and a high-monounsaturated fat, olive oil-rich diet on the susceptibility of LDL to oxidative modification in subjects with Type 2 diabetes mellitus. *Diabet Med* February 2004;**21**(2):142–9.
61. Rosenfalck AM, Almdal T, Viggers L, Madsbad S, Hilsted J. A low-fat diet improves peripheral insulin sensitivity in patients with Type 1 diabetes. *Diabet Med* April 2006;**23**(4):384–92.
62. Brehm BJ, Lattin BL, Summer SS, Boback JA, Gilchrist GM, Jandacek RJ, et al. One-year comparison of a high-monounsaturated fat diet with a high-carbohydrate diet in type 2 diabetes. *Diabetes Care* February 2009;**32**(2):215–20.
63. Milne RM, Mann JI, Chisholm AW, Williams SM. Long-term comparison of three dietary prescriptions in the treatment of NIDDM. *Diabetes Care* January 1994;**17**(1):74–80.
64. Daly ME, Paisey R, Millward BA, Eccles C, Williams K, Hammersley S, et al. Short-term effects of severe dietary carbohydrate-restriction advice in Type 2 diabetes—a randomized controlled trial. *Diabet Med* January 2006;**23**(1):15–20.
65. Davis NJ, Tomuta N, Schechter C, Isasi CR, Segal-Isaacson CJ, Stein D, et al. Comparative study of the effects of a 1-year dietary intervention of a low-carbohydrate diet versus a low-fat diet on weight and glycemic control in type 2 diabetes. *Diabetes Care* July 2009;**32**(7):1147–52.
66. Dyson PA, Beatty S, Matthews DR. A low-carbohydrate diet is more effective in reducing body weight than healthy eating in both diabetic and non-diabetic subjects. *Diabet Med* December 2007;**24**(12):1430–5.
67. Stern L, Iqbal N, Seshadri P, Chicano KL, Daily DA, McGrory J, et al. The effects of low-carbohydrate versus conventional weight loss diets in severely obese adults: one-year follow-up of a randomized trial. *Ann Intern Med* May 18, 2004;**140**(10):778–85.
68. Westman EC, Yancy Jr WS, Mavropoulos JC, Marquart M, McDuffie JR. The effect of a low-carbohydrate, ketogenic diet versus a low-glycemic index diet on glycemic control in type 2 diabetes mellitus. *Nutr Metab (Lond)* 2008;**5**:36.
69. Yancy Jr WS, Foy M, Chalecki AM, Vernon MC, Westman EC. A low-carbohydrate, ketogenic diet to treat type 2 diabetes. *Nutr Metab (Lond)* 2005;**2**:34.
70. Boden G, Sargrad K, Homko C, Mozzoli M, Stein TP. Effect of a low-carbohydrate diet on appetite, blood glucose levels, and insulin resistance in obese patients with type 2 diabetes. *Ann Intern Med* March 15, 2005;**142**(6):403–11.
71. Haimoto H, Sasakabe T, Wakai K, Umegaki H. Effects of a low-carbohydrate diet on glycemic control in outpatients with severe type 2 diabetes. *Nutr Metab (Lond)* 2009;**6**:21.
72. Jonsson T, Granfeldt Y, Ahren B, Branell UC, Palsson G, Hansson A, et al. Beneficial effects of a Paleolithic diet on cardiovascular risk factors in type 2 diabetes: a randomized cross-over pilot study. *Cardiovasc Diabetol* 2009;**8**:35.
73. Miyashita Y, Koide N, Ohtsuka M, Ozaki H, Itoh Y, Oyama T, et al. Beneficial effect of low carbohydrate in low calorie diets on visceral fat reduction in type 2 diabetic patients with obesity. *Diabetes Res Clin Pract* September 2004;**65**(3):235–41.
74. Wolever TM, Gibbs AL, Mehling C, Chiasson JL, Connelly PW, Josse RG, et al. The Canadian Trial of Carbohydrates in Diabetes (CCD), a 1-y controlled trial of low-glycemic-index dietary carbohydrate in type 2 diabetes: no effect on glycated hemoglobin but reduction in C-reactive protein. *Am J Clin Nutr* January 2008;**87**(1): 114–25.
75. Haimoto H, Iwata M, Wakai K, Umegaki H. Long-term effects of a diet loosely restricting carbohydrates on HbA1c levels, BMI and tapering of sulfonylureas in type 2 diabetes: a 2-year follow-up study. *Diabetes Res Clin Pract* February 2008;**79**(2):350–6.
76. Iqbal N, Vetter ML, Moore RH, Chittams JL, Dalton-Bakes CV, Dowd M, et al. Effects of a low-intensity intervention that prescribed a low-carbohydrate vs. a low-fat diet in obese, diabetic participants. *Obesity (Silver Spring)* September 2009;**18**(9):1733–8.
77. Samaha FF, Iqbal N, Seshadri P, Chicano KL, Daily DA, McGrory J, et al. A low-carbohydrate as compared with a low-fat diet in severe obesity. *N Engl J Med* May 22, 2003;**348**(21): 2074–81.
78. Elhayany A, Lustman A, Abel R, Attal-Singer J, Vinker S. A low carbohydrate Mediterranean diet improves cardiovascular risk factors and diabetes control among overweight patients with type 2 diabetes mellitus: a 1-year prospective randomized intervention study. *Diabetes Obes Metab* March 2010;**12**(3):204–9.
79. Ma Y, Olendzki BC, Merriam PA, Chiriboga DE, Culver AL, Li W, et al. A randomized clinical trial comparing low-glycemic index versus ADA dietary education among individuals with type 2 diabetes. *Nutrition* January 2008;**24**(1):45–56.
80. Jenkins DJ, Kendall CW, McKeown-Eyssen G, Josse RG, Silverberg J, Booth GL, et al. Effect of a low-glycemic index or a high-cereal fiber diet on type 2 diabetes: a randomized trial. *JAMA* December 17, 2008;**300**(23):2742–53.
81. Coppell KJ, Kataoka M, Williams SM, Chisholm AW, Vorgers SM, Mann JI. Nutritional intervention in patients with type 2 diabetes who are hyperglycaemic despite optimised drug treatment—Lifestyle Over and Above Drugs in Diabetes (LOADD) study: randomised controlled trial. *BMJ* 2010;**341**:c3337.
82. Li Z, Hong K, Saltsman P, DeShields S, Bellman M, Thames G, et al. Long-term efficacy of soy-based meal replacements vs an individualized diet plan in obese type II DM patients: relative effects on weight loss, metabolic parameters, and C-reactive protein. *Eur J Clin Nutr* March 2005;**59**(3):411–8.
83. Yip I, Go VL, DeShields S, Saltsman P, Bellman M, Thames G, et al. Liquid meal replacements and glycemic control in obese type 2 diabetes patients. *Obes Res* November 2001;**9**(Suppl. 4):341S–7S.
84. Zablocka-Slowinska K, Dzielska E, Gryszkin I, Grajeta H. Dietary supplementation during diabetes therapy and the potential risk of interactions. *Adv Clin Exp Med* November–December 2015;**23**(6): 939–46.
85. Walker AF. Potential micronutrient deficiency lacks recognition in diabetes. *Br J Gen Pract* January 2007;**57**(534):3–4.
86. Yeh GY, Eisenberg DM, Kaptchuk TJ, Phillips RS. Systematic review of herbs and dietary supplements for glycemic control in diabetes. *Diabetes Care* April 2003;**26**(4):1277–94.
87. Akilen R, Tsiami A, Devendra D, Robinson N. Glycated haemoglobin and blood pressure-lowering effect of cinnamon in multiethnic Type 2 diabetic patients in the UK: a randomized, placebo-controlled, double-blind clinical trial. *Diabet Med* October 2010;**27**(10):1159–67.

88. Leach MJ, Kumar S. Cinnamon for diabetes mellitus. *Cochrane Database Syst Rev* 2012;**9**:CD007170.
89. Wu H, Flint AJ, Qi Q, van Dam RM, Sampson LA, Rimm EB, et al. Association between dietary whole grain intake and risk of mortality: two large prospective studies in US men and women. *JAMA Intern Med* January 5, 2015;**175**(3):373–84.
90. Lu ZX, W KZ, Muir JG, OD K. Arabinoxylan fibre improves metabolic control in people with Type II diabetes. *Eur J Clin Nutr* 2004; **58**:621–8.
91. Vuksan V, Whitham D, Sievenpiper JL, Jenkins AL, Rogovik AL, Bazinet RP, et al. Supplementation of conventional therapy with the novel grain Salba (*Salvia hispanica* L.) improves major and emerging cardiovascular risk factors in type 2 diabetes: results of a randomized controlled trial. *Diabetes Care* November 2007; **30**(11):2804–10.
92. Goncalves Reis CE, Camara Landim K, Santos Nunes AC, Dullius J. Safety in the hypertriglyceridemia treatment with N-3 polyunsaturated fatty acids on glucose metabolism in subjects with type 2 diabetes mellitus. *Nutr Hosp* 2015;**31**(2):570–6.
93. Pooya S, Jalali MD, Jazayery AD, Saedisomeolia A, Eshraghian MR, Toorang F. The efficacy of omega-3 fatty acid supplementation on plasma homocysteine and malondialdehyde levels of type 2 diabetic patients. *Nutr Metab Cardiovasc Dis* June 2009;**20**(5):326–31.
94. Petersen M, Pedersen H, Major-Pedersen A, Jensen T, Marckmann P. Effect of fish oil versus corn oil supplementation on LDL and HDL subclasses in type 2 diabetic patients. *Diabetes Care* October 2002;**25**(10):1704–8.
95. Pedersen H, Petersen M, Major-Pedersen A, Jensen T, Nielsen NS, Lauridsen ST, et al. Influence of fish oil supplementation on in vivo and in vitro oxidation resistance of low-density lipoprotein in type 2 diabetes. *Eur J Clin Nutr* May 2003;**57**(5):713–20.
96. Woodman RJ, Mori TA, Burke V, Puddey IB, Watts GF, Beilin LJ. Effects of purified eicosapentaenoic and docosahexaenoic acids on glycemic control, blood pressure, and serum lipids in type 2 diabetic patients with treated hypertension. *Am J Clin Nutr* November 2002;**76**(5):1007–15.
97. Gillen LJ, Tapsell LC, Patch CS, Owen A, Batterham M. Structured dietary advice incorporating walnuts achieves optimal fat and energy balance in patients with type 2 diabetes mellitus. *J Am Diet Assoc* July 2005;**105**(7):1087–96.
98. Li TY, Brennan AM, Wedick NM, Mantzoros C, Rifai N, Hu FB. Regular consumption of nuts is associated with a lower risk of cardiovascular disease in women with type 2 diabetes. *J Nutr* July 2009;**139**(7):1333–8.
99. Lovejoy JC, Most MM, Lefevre M, Greenway FL, Rood JC. Effect of diets enriched in almonds on insulin action and serum lipids in adults with normal glucose tolerance or type 2 diabetes. *Am J Clin Nutr* November 2002;**76**(5):1000–6.
100. Ma Y, Njike VY, Millet J, Dutta S, Doughty K, Treu JA, et al. Effects of walnut consumption on endothelial function in type 2 diabetic subjects: a randomized controlled crossover trial. *Diabetes Care* February 2009;**33**(2):227–32.
101. Tapsell LC, Gillen LJ, Patch CS, Batterham M, Owen A, Bare M, et al. Including walnuts in a low-fat/modified-fat diet improves HDL cholesterol-to-total cholesterol ratios in patients with type 2 diabetes. *Diabetes Care* December 2004;**27**(12):2777–83.
102. Tapsell LC, Batterham MJ, Teuss G, Tan SY, Dalton S, Quick CJ, et al. Long-term effects of increased dietary polyunsaturated fat from walnuts on metabolic parameters in type II diabetes. *Eur J Clin Nutr* August 2009;**63**(8):1008–15.
103. Wien M, Oda K, Sabate J. A randomized controlled trial to evaluate the effect of incorporating peanuts into an American Diabetes Association meal plan on the nutrient profile of the total diet and cardiometabolic parameters of adults with type 2 diabetes. *Nutr J* 2014;**13**:10.
104. Parham M, Heidari S, Khorramirad A, Hozoori M, Hosseinzadeh F, Bakhtyari L, et al. Effects of pistachio nut supplementation on blood glucose in patients with type 2 diabetes: a randomized crossover trial. *Rev Diabet Stud* Summer 2014;**11**(2):190–6.
105. Balk EM, Tatsioni A, Lichtenstein AH, Lau J, Pittas AG. Effect of chromium supplementation on glucose metabolism and lipids: a systematic review of randomized controlled trials. *Diabetes Care* August 2007;**30**(8):2154–63.
106. Rodriguez-Moran M, Guerrero-Romero F. Oral magnesium supplementation improves insulin sensitivity and metabolic control in type 2 diabetic subjects: a randomized double-blind controlled trial. *Diabetes Care* April 2003;**26**(4):1147–52.
107. de Valk HW, Verkaaik R, van Rijn HJ, Geerdink RA, Struyvenberg A. Oral magnesium supplementation in insulin-requiring Type 2 diabetic patients. *Diabet Med* June 1998;**15**(6): 503–7.
108. Jorde R, Figenschau Y. Supplementation with cholecalciferol does not improve glycaemic control in diabetic subjects with normal serum 25-hydroxyvitamin D levels. *Eur J Nutr* September 2009; **48**(6):349–54.
109. Patel P, Poretsky L, Liao E. Lack of effect of subtherapeutic vitamin D treatment on glycemic and lipid parameters in Type 2 diabetes: a pilot prospective randomized trial. *J Diabetes* March 2010;**2**(1):36–40.
110. Parekh D, Sarathi V, Shivane VK, Bandgar TR, Menon PS, Shah NS. Pilot study to evaluate the effect of short-term improvement in vitamin D status on glucose tolerance in patients with type 2 diabetes mellitus. *Endocr Pract* July–August 2010;**16**(4): 600–8.
111. Nikooyeh B, Neyestani TR, Farvid M, Alavi-Majd H, Houshiarrad A, Kalayi A, et al. Daily consumption of vitamin D- or vitamin D + calcium-fortified yogurt drink improved glycemic control in patients with type 2 diabetes: a randomized clinical trial. *Am J Clin Nutr* April 2011;**93**(4):764–71.
112. Soric MM, Renner ET, Smith SR. Effect of daily vitamin D supplementation on HbA1c in patients with uncontrolled type 2 diabetes mellitus: a pilot study. *J Diabetes* March 2011;**4**(1):104–5.
113. Shahar DR, Abel R, Elhayany A, Vardi H, Fraser D. Does dairy calcium intake enhance weight loss among overweight diabetic patients? *Diabetes Care* March 2007;**30**(3):485–9.
114. Schwerin HS, Stanton JL, Smith JL, Riley Jr AM, Brett BE. Food, eating habits, and health: a further examination of the relationship between food eating patterns and nutritional health. *Am J Clin Nutr* May 1982;**35**(Suppl. 5):1319–25.
115. Jones-McLean EM, Shatenstein B, Whiting SJ. Dietary patterns research and its applications to nutrition policy for the prevention of chronic disease among diverse North American populations. *Appl Physiol Nutr Metab* April 2010;**35**(2):195–8.
116. Willett WC, Sacks F, Trichopoulou A, Drescher G, Ferro-Luzzi A, Helsing E, et al. Mediterranean diet pyramid: a cultural model for healthy eating. *Am J Clin Nutr* June 1995;**61**(Suppl. 6):1402S–6S.
117. Sofi F, Abbate R, Gensini GF, Casini A. Accruing evidence on benefits of adherence to the Mediterranean diet on health: an updated systematic review and meta-analysis. *Am J Clin Nutr* November 2010;**92**(5):1189–96.
118. Serra-Majem L, Roman B, Estruch R. Scientific evidence of interventions using the Mediterranean diet: a systematic review. *Nutr Rev* February 2006;**64**(2 Pt 2):S27–47.
119. de Lorgeril M, Salen P, Martin JL, Monjaud I, Delaye J, Mamelle N. Mediterranean diet, traditional risk factors, and the rate of cardiovascular complications after myocardial infarction: final report of the Lyon Diet Heart Study. *Circulation* February 16, 1999;**99**(6): 779–85.
120. Estruch R, Ros E, Martinez-Gonzalez MA. Mediterranean diet for primary prevention of cardiovascular disease. *N Engl J Med* August 15, 2013;**369**(7):676–7.

121. Toobert DJ, Glasgow RE, Strycker LA, Barrera Jr M, Radcliffe JL, Wander RC, et al. Biologic and quality-of-life outcomes from the Mediterranean Lifestyle Program: a randomized clinical trial. *Diabetes Care* August 2003;**26**(8):2288–93.
122. Esposito K, Maiorino MI, Ciotola M, Di Palo C, Scognamiglio P, Gicchino M, et al. Effects of a Mediterranean-style diet on the need for antihyperglycemic drug therapy in patients with newly diagnosed type 2 diabetes: a randomized trial. *Ann Intern Med* September 1, 2009;**151**(5):306–14.
123. Salas-Salvado J, Fernandez-Ballart J, Ros E, Martinez-Gonzalez MA, Fito M, Estruch R, et al. Effect of a Mediterranean diet supplemented with nuts on metabolic syndrome status: one-year results of the PREDIMED randomized trial. *Arch Intern Med* December 8, 2008;**168**(22):2449–58.
124. Turner-McGrievy GM, Barnard ND, Cohen J, Jenkins DJ, Gloede L, Green AA. Changes in nutrient intake and dietary quality among participants with type 2 diabetes following a low-fat vegan diet or a conventional diabetes diet for 22 weeks. *J Am Diet Assoc* October 2008;**108**(10):1636–45.
125. Nicholson AS, Sklar M, Barnard ND, Gore S, Sullivan R, Browning S. Toward improved management of NIDDM: a randomized, controlled, pilot intervention using a lowfat, vegetarian diet. *Prev Med* August 1999;**29**(2):87–91.
126. Tonstad S, Butler T, Yan R, Fraser GE. Type of vegetarian diet, body weight, and prevalence of type 2 diabetes. *Diabetes Care* May 2009;**32**(5):791–6.
127. Kahleova H, Matoulek M, Malinska H, Oliyarnik O, Kazdova L, Neskudla T, et al. Vegetarian diet improves insulin resistance and oxidative stress markers more than conventional diet in subjects with Type 2 diabetes. *Diabet Med* May 2011;**28**(5):549–59.
128. Sacks FM, Svetkey LP, Vollmer WM, Appel LJ, Bray GA, Harsha D, et al. Effects on blood pressure of reduced dietary sodium and the Dietary Approaches to Stop Hypertension (DASH) diet. DASH-Sodium Collaborative Research Group. *N Engl J Med* January 4, 2001;**344**(1):3–10.
129. Appel LJ, Moore TJ, Obarzanek E, Vollmer WM, Svetkey LP, Sacks FM, et al. A clinical trial of the effects of dietary patterns on blood pressure. DASH Collaborative Research Group. *N Engl J Med* April 17, 1997;**336**(16):1117–24.
130. Azadbakht L, Fard NR, Karimi M, Baghaei MH, Surkan PJ, Rahimi M, et al. Effects of the Dietary Approaches to Stop Hypertension (DASH) eating plan on cardiovascular risks among type 2 diabetic patients: a randomized crossover clinical trial. *Diabetes Care* January 2010;**34**(1):55–7.
131. Mühlhauser I, Jörgens V, Berger M, Graninger W, Gurtler W, Hornke L, et al. Bicentric evaluation of a teaching and treatment programme for type 1 (insulin-dependent) diabetic patients: improvement of metabolic control and other measures of diabetes care for up to 22 months. *Diabetologia* December 1983;**25**(6):470–6.
132. DAFNE Study Group. Training in flexible, intensive insulin management to enable dietary freedom in people with type 1 diabetes: dose adjustment for normal eating (DAFNE) randomised controlled trial. *BMJ* October 5, 2002;**325**(7367):746.
133. Laurenzi A, Bolla AM, Panigoni G, Doria V, Uccellatore A, Peretti E, et al. Effects of carbohydrate counting on glucose control and quality of life over 24 weeks in adult patients with type 1 diabetes on continuous subcutaneous insulin infusion: a randomized, prospective clinical trial (GIOCAR). *Diabetes Care* April 2011;**34**(4):823–7.
134. Rossi MC, Nicolucci A, Di Bartolo P, Bruttomesso D, Girelli A, Ampudia FJ, et al. Diabetes Interactive Diary: a new telemedicine system enabling flexible diet and insulin therapy while improving quality of life: an open-label, international, multicenter, randomized study. *Diabetes Care* January 2009;**33**(1): 109–15.
135. Kulkarni K, Castle G, Gregory R, Holmes A, Leontos C, Powers M, et al. Nutrition Practice Guidelines for Type 1 Diabetes Mellitus positively affect dietitian practices and patient outcomes. The Diabetes Care and Education Dietetic Practice Group. *J Am Diet Assoc* January 1998;**98**(1):62–70. quiz 1–2.
136. Graber AL, Elasy TA, Quinn D, Wolff K, Brown A. Improving glycemic control in adults with diabetes mellitus: shared responsibility in primary care practices. *South Med J* July 2002;**95**(7):684–90.
137. Scavone G, Manto A, Pitocco D, Gagliardi L, Caputo S, Mancini L, et al. Effect of carbohydrate counting and medical nutritional therapy on glycaemic control in Type 1 diabetic subjects: a pilot study. *Diabet Med* April 2010;**27**(4):477–9.
138. Gilbertson HR, Brand-Miller JC, Thorburn AW, Evans S, Chondros P, Werther GA. The effect of flexible low glycemic index dietary advice versus measured carbohydrate exchange diets on glycemic control in children with type 1 diabetes. *Diabetes Care* July 2001;**24**(7):1137–43.
139. Speight J, Amiel SA, Bradley C, Heller S, Oliver L, Roberts S, et al. Long-term biomedical and psychosocial outcomes following DAFNE (Dose Adjustment For Normal Eating) structured education to promote intensive insulin therapy in adults with sub-optimally controlled Type 1 diabetes. *Diabetes Res Clin Pract* July 2010;**89**(1):22–9.
140. McIntyre HD, Knight BA, Harvey DM, Noud MN, Hagger VL, Gilshenan KS. Dose adjustment for normal eating (DAFNE) - an audit of outcomes in Australia. *Med J Aust* June 7, 2010;**192**(11): 637–40.
141. Sämann A, Muhlhauser I, Bender R, Kloos C, Muller UA. Glycaemic control and severe hypoglycaemia following training in flexible, intensive insulin therapy to enable dietary freedom in people with type 1 diabetes: a prospective implementation study. *Diabetologia* October 2005;**48**(10):1965–70.
142. Sacks FM, Carey VJ, Anderson CA, Miller 3rd ER, Copeland T, Charleston J, et al. Effects of high vs low glycemic index of dietary carbohydrate on cardiovascular disease risk factors and insulin sensitivity: the OmniCarb randomized clinical trial. *JAMA* December 17, 2014;**312**(23):2531–41.
143. Tariq SH. Herbal therapies. *Clin Geriatr Med* May 2004;**20**(2): 237–57.
144. Verkaik-Kloosterman J, McCann MT, Hoekstra J, Verhagen H. Vitamins and minerals: issues associated with too low and too high population intakes. *Food Nutr Res* 2012:56.
145. Guzman JR, Lyra R, Aguilar-Salinas CA, Cavalcanti S, Escano F, Tambasia M, et al. Treatment of type 2 diabetes in Latin America: a consensus statement by the medical associations of 17 Latin American countries. Latin American Diabetes Association. *Rev Panam Salud Publica* December 2010;**28**(6):463–71.
146. International Diabetes Federation (IDF) CGTF. Global guideline for type 2 diabetes. Available from: www.idf.org [cited December 29, 2014].
147. www.diabetes.org/food-and-fitness/food/what-can-i-eat/understanding-carbohydrates/glycemic-index-and-diabetes.html [cited January 31, 2015].

CHAPTER

2

Dietary Patterns and Insulin Resistance

Adriana Souza Torsoni, Marciane Milanski, Marcio Alberto Torsoni

Laboratory of Metabolic Disorders, Faculty of Applied Sciences, University of Campinas – UNICAMP, Limeira, São Paulo, Brazil

1. INTRODUCTION

The World Health Organization has highlighted the persistent increase in the prevalence of diabetes and obesity among the population. Type 2 diabetes (T2D) accounts for approximately 90% of all diabetes cases with obesity closely linked to this condition.

For individuals with diabetes, lifestyle changes and nutritional-combined therapy are important for preventing the development of insulin resistance, the main component in the pathophysiology of T2D. During the initial phase of the pathogenesis of T2D, insulin resistance leads to increased blood glucose levels. At present, it is widely accepted by researchers in this field that the increased prevalence of obesity in the world significantly contributes to this framework. Excessive weight gain promotes the induction of inflammation in different tissues and stimulates the migration of macrophages and monocytes, mainly in the adipose tissue. These cells can secrete a variety of inflammatory factors such as interleukin-6, interleukin-1β, and tumor necrosis factor that impair insulin signaling in target tissues.[1,2] As a result of this change in peripheral insulin signaling, diets rich in carbohydrates and with a high rate of glucose can further contribute to increased glucose levels that stimulate insulin production as well as enhance insulin secretion from β cells in the pancreas, leading to hyperinsulinemia. Moreover, this excessive activity promotes the damage of the structure and function of β cells; in the long-term, it can lead to cell failure and T2D development. Therefore, reduction in insulin signaling, such as that in insulin resistance, prevents the efficient control of postprandial glycemia. Thus, controlling the intake of different macronutrients via diet is undoubtedly an important factor for preventing T2D and complications arising from high blood glucose levels.

In addition to carbohydrates, many other nutritional constituents are part of various diets prescribed by nutritionists to individuals with diabetes or to those at a risk of developing this disease, such as obese individuals. In this chapter, we discuss the role of carbohydrates, lipids, and proteins in the prevention and pathogenesis of insulin resistance development. At present, there is no consensus for the most efficient type of diet for both controlling blood glucose levels and promoting weight loss. However, when choosing a particular type of diet, one must always consider the potential of the dietary components in reducing intestinal absorption, particularly glucose, low glycemic index, and the presence of components with antioxidant and anti-inflammatory properties. The properties of these dietary components might facilitate the control of blood glucose levels, plasma lipids levels, and blood pressure. In this context, nutritional therapy may act at different levels; it prevents the development of insulin resistance and, consequently, prevents that of diabetes; it also prevents and controls metabolic disorders and complications associated with diabetes. In all prescribed diets for individuals with diabetes or for those at a risk of developing this condition, using a diet that also promotes weight loss is an important criterion. Diets with few simple carbohydrates, Mediterranean diet, diets that are low lipid/energy-restricted, and diets with a high-protein and low-carbohydrate content/caloric restriction contribute to the reduction of body weight in varying degrees as well as to the improvement of lipid profiles and insulin resistance.

Although it seems quite logical that diets with high carbohydrate content and high glycemic index are harmful for individuals with diabetes, the results obtained from studies involving different diets are still controversial. In the study by Black et al., no changes in carbohydrate metabolism were observed in nonobese

http://dx.doi.org/10.1016/B978-0-12-801585-8.00002-6

individuals on isocaloric diets with varying quantities of sucrose.[3] Similarly, the use of other weight loss strategies, such as diets with a low glycemic index and those that are low calorie, resulted in no difference in insulin sensitivity after 3 months.[4] However, in a study involving 17 men consuming four different types of diet for 24 days, the authors found that a diet with a high glycemic index increased resistance to insulin when evaluated by homeostatic model assessment-insulin resistance (a method used to quantify insulin resistance and β-cell function) as compared with other diets, such as a low glycemic index diet, a high-fat diet, and a high glycemic index diet rich in sucrose.[5] As pointed out by Deer et al., the interpretation of this study was impaired because of a reduction in body weight presented by individuals on low-glycemic diet, as compared with the body weight of those on other three evaluated diets.[6] In another study, the authors also showed that a 7-day diet based on a low glycemic index diet improved the glucose profile of individuals with diabetes.[7] In these studies, the weight reduction caused by the diet may have significantly contributed to this result. The association of obesity with inflammatory pathway activation is largely known, and the low-grade inflammation observed in an obese individual is closely associated with the development of insulin resistance by mechanisms, including increased levels of ceramide and diacylglycerol,[8] activation of proinflammatory pathways via Toll-like receptor 4 (TLR4), and nuclear factor-kB activation of serine kinases.[9] All these factors can contribute to a reduction in the phosphorylation and activation of proteins, such as insulin receptor substrate (IRS)-1, phosphoinositide 3-kinase, and AKT[10,11] as well as to an increase in the phosphorylation of serine/threonine IRS-1 protein.[2] Thus, the consumption of diet that promotes weight loss can reduce the activation of inflammatory pathways and decrease insulin resistance.

2. CARBOHYDRATES

Among the different macronutrients in a diet, carbohydrates represent the nutrient with a greater potential to quickly and directly modulate blood glucose levels. Some carbohydrates have received more attention than others, including glucose and fructose, which are found in small quantities in fruits and honey; sucrose, a disaccharide comprising glucose and fructose joined by a glycosidic linkage and mainly found in cane sugar; and starch, a glucose polymer found in large quantities in potatoes, maize, and rice. The role of foods containing carbohydrates in increasing blood glucose level is affected by different factors, such as the type of carbohydrate, the method of preparation, and the presence of components that can alter digestion and intestinal absorption.

The use of these sugars and the mechanisms of digestion and intestinal absorption are different. Sucrose and starch, for example, require the action of enzymes to allow the basic units glucose and fructose to be released and absorbed. Until the 1960s, the use of sugars in food was restricted to sucrose, but via starch hydrolysis and fructose synthesis by an isomerization reaction carried out by the industry, corn syrup was created. Although use of this component has not increased significantly in the past 30 years, a significant reduction in sucrose consumption in the United States was observed when it was initially introduced.[12]

At present, different studies have pointed out the adverse health effects associated with high carbohydrate consumption. Among the investigated carbohydrates, fructose and its association with metabolic damages, such as the metabolic syndrome, have received much attention.[13–16] However, we must carefully analyze the data in the literature describing the damage caused by fructose consumption. Many studies have evaluated the effects of pure fructose in the diet and compared them with those of glucose alone in the diet.[17–19] This type of analysis does not reflect human nutrition because the consumption of glucose or fructose alone in the diet is low in humans.[15,16]

Fructose and glucose follow different paths in the body. These simple carbohydrates have different intestinal absorption rates; therefore, from its intake they influence differently to blood glucose. Fructose is captured by the isoform of glucose transporter 5 (GLUT5), which is expressed on the apical region of the enterocytes. This mechanism is independent of adenosine triphosphate (ATP) consumption and sodium absorption. Upon entry into the enterocytes, the transport of fructose into the bloodstream is mediated by GLUT2 located at the basal area of the enterocyte.[20] Moreover, part of the fructose in the enterocyte can convert into lactate, which is released into the portal system as well as into glucose through gluconeogenesis. The function of intestinal fructose metabolism is still unknown, but it has been suggested that it participates in the control of peripheral metabolism and ingestion through neural reflexes by glucose sensors.[12,21]

In the circulation, fructose is almost completely extracted by the liver via the GLUT2 transporter. Furthermore, glucose is also captured by other tissues in large quantities, such as adipose and muscle tissues, by the GLUT4 transporter. In the hepatocyte, the initial steps of fructose metabolism differ from those of glucose. Fructose is a substrate of fructokinase, and glucose is subjected to the action of hexokinase IV. These enzymes have different affinities for their substrates. Hexokinase IV has a high Km (low affinity) for glucose, and the pathway to produce pyruvate from glucose (the end-product of the metabolic pathway) is controlled by the availability of insulin and ATP in the cell. On the other hand, fructokinase has a high affinity (low Michaelis constant or Km) for fructose, and the metabolic route to pyruvate is shorter and independent of the availability of both ATP and insulin. The pyruvate produced from fructose can produce fatty acids in the hepatocytes. This was observed in rats undergoing the administration of C14 fructose. The authors observed an increase in the incorporation of the isotope C14 in lipids. In humans, the acute administration of fructose was also shown to stimulate the incorporation of acetate labeled with C13 in very low-density lipoprotein.[22] In addition to promoting lipogenesis, fructose can also inhibit the oxidation of lipids.

Fructose is metabolized independently of insulin and has a low glycemic count. Because of these characteristics, replacing part of the glucose with fructose in the diet of an individual with T2D could be interesting. However, most studies investigating this type of diet indicate an increase in plasma triglyceride levels, reduced high-density lipoprotein levels, insulin resistance, weight gain, hypertension, and dyslipidemia.[23–25] Fructose consumption's effect on the induction of ectopic deposition of lipids in various tissues, including those of the liver, muscle, and endocrine cells, should also be considered. Such deposits are closely linked to the development of insulin resistance. The deposition of lipids in these tissues are associated with the stimulation of lipogenesis and a reduction in the oxidation of fatty acids by a mechanism dependent on PGC-1β activation, a coactivator of SREBP-1c. In this regard, rodents using a sucrose-rich diet for a few weeks demonstrated lipid accumulation in muscle fibers and insulin resistance in muscle tissues.[26,27] In humans, although less well-studied, the consumption of a diet rich in fructose (approximately 3 g/kg/body weight) for a short period (1 week) resulted in deleterious effects, such as increased fat deposition in the liver and muscle tissues,[28] associated with insulin resistance. As previously mentioned, the serine phosphorylation of insulin signaling pathway components is an important mechanism for hormone resistance development. The c-Jun N-terminal kinase (JNK) is a serine kinase that can be activated by fructose by changing the redox state of the hepatocyte[29,30] and by increasing the serine phosphorylation of the IRS-1 protein, thereby reducing insulin signaling.

Furthermore, similar to the diets rich in fatty acids, intestinal fructose may lead to increased blood endotoxin levels.[31,32] Endotoxemia is the change in the permeability of the intestinal flora, which allows the passage of lipopolysaccharide derived from intestinal bacteria into the bloodstream. This contributes to a state of low-grade endotoxemia, the activation of TLR4, the induction of pro-inflammatory pathways, and the development of insulin resistance.[33–35]

Although observed in rodents, it is clear that the use of diets high in fructose impair insulin signaling, and the effects in humans appear to be the side effects of mainly fructose, including the ectopic deposition of fat lipotoxicity and oxidative stress.

The use of carbohydrate in the diet of individuals with T2D seems to be particularly important regarding the use of resistant starch and fiber. These types of dietary components may pass intact through the stomach and reach the colon, where they serve as a substrate for fermentation by the bacteria comprising the intestinal flora. Fermentation in the colon produces short chain fatty acids that are capable of activating G-coupled receptor 34 proteins, which reduces the release of free fatty acids from the adipose tissue, possibly through the inhibition of the enzyme hormone-sensitive lipase.[36–40] In this context, the consumption of approximately 15 g/day of resistant starch increased insulin sensitivity in patients with T2D. In another study, oat fiber intake resulted in a similar effect. Additionally, some types of fibers inside the intestine produce a more viscous solution hindering the digestion of starch and the absorption of glucose. This has an important effect in reducing the potential of food to rapidly increase glycemia.[36–40]

T2D is a disease characterized by high blood glucose levels caused by changes in insulin sensitivity, high-hepatic glucose production, and damage to pancreatic β cells. Excessive body weight gain is one of the main contributors to the development of this pathology; behavioral interventions for body weight reduction are important to avoid the appearance of new cases among the population. However, along with this type of behavior, the use of diets rich in fiber and cereal starch will slow down the digestion; a reduced amount of refined sugar, fatty acids of vegetable origin, and protein from lean sources will surely help reduce the risk of T2D development.

3. LIPIDS

Dietary lipids comprise different types of fatty acids, the molecules of which are made up of hydrocarbon chains that are differentiated by the number of carbons (short, medium, or long chain) and by the absence (saturated) or presence of unsaturation (monounsaturated and polyunsaturated). These are some of the features that determine the function of these nutrients in metabolism. In recent years, the concern for an adequate lipid intake is increasing among the population because of the numerous studies that have shown the effects of different sources of this substance in the body. Although T2D is mainly influenced by lifestyle, diet composition can affect the development of disease as well as that of any associated complications. The molecular mechanisms of nutrients involved in the induction of insulin resistance, particularly the different types of fatty acids in a diet are still poorly understood. Despite the view that a higher intake of total fat in the diet may contribute to the development of diabetes both directly by inducing insulin resistance and indirectly by promoting body weight gain, the results of studies involving humans using metabolic markers do not support the idea that high-fat diets themselves have an adverse effect on insulin sensitivity.[41–44] It is clear that excess energy arising from a diet high in total calories will be used for the de novo synthesis of fatty acids that will be mobilized from the liver into the bloodstream in the form of very low-density lipoprotein and transported to the extrahepatic tissues where they will be taken up mainly into the adipose tissue and stored as triglycerides. In addition to low-density lipoprotein, dietary fat contributes to an increase in the circulating lipid levels in the form of triglyceride-rich chylomicrons, which are also captured by peripheral tissues. In pathological conditions, such as obesity and T2D, the accumulation of ectopic fatty acids increases because of both reduction in fatty acid oxidation and enhanced fatty acid uptake.[45–48] The increase in lipids droplets in cells correlates inversely with insulin sensitivity in the muscle, liver, and adipocytes.[49–53]

Chronic exposure to excess lipids over time in metabolically active tissues and organs such as muscle, liver, and pancreas can lead to significant dysfunctions often involving the activation of inflammatory pathways and resistance to systemic insulin action. Functional impairments associated with increased circulating levels of lipids and their induced metabolic alterations in fatty acids utilization and intracellular signaling have been broadly termed lipotoxicity.[54,55] However, it seems that ectopic lipid accumulation in tissues not prepared to store fat can only be used as a marker for the onset of insulin resistance and cannot be considered a direct cause, because athletes display high insulin sensitivity but also present increased levels of intramuscular fatty acids.[56] Even if they do not seem to be directly involved, fatty acids contribute to insulin resistance because they lead to the synthesis of many intermediates derived from lipids, such as diacylglycerol and ceramide. Over the years, several studies have provided compelling evidence that ceramide plays a key role in the progression of insulin resistance in tissues sensitive to this hormone, mainly through the inhibition of the insulin signaling pathway.[53]

In vitro studies confirm that ceramide synthesis from palmitate is decreased in the presence of the monounsaturated fatty acid oleate, which prevents the accumulation of ceramide and subsequent insulin resistance.[57,58] Because the quality of dietary fat and the plasma lipid profile is correlated, there is no doubt that dietary intervention can influence insulin resistance.

On considering the ideal dietary composition of lipids for preventing insulin resistance, the evidence shows that fat quality is more important than total fat intake. Thus, to achieve and maintain a healthy metabolic profile, diets that provide fatty acids of vegetable origin seem to be more advantageous as compared with those provide fats of animal origin.[43,59,60] Moreover, depending on the degree of saturation, free fatty acids may have different effects on insulin signaling. Studies have shown that saturated fatty acids, such as palmitate (16:0) and stearic (18:0), impair insulin signaling in muscle,[61,62] whereas monounsaturated or polyunsaturated fatty acids have no effect or they may even improve insulin action.[63–65] Although the reasons for these differences are not yet clear, it has been suggested that unsaturated fatty acids can be preferably used for the synthesis and storage of triglycerides, whereas saturated fatty acids can be used for the synthesis of lipid intermediates deleterious to the cells, such as ceramide and diacylglycerol.[53]

In addition to supplying energy functions to the organism, fatty acids exert signaling functions as eicosanoid precursors, and they regulate cellular functions via the activation of G-protein coupled receptors, including the secretion of hormones and the regulation of immune response. Saturated fatty acids are known to activate some members of the toll-like receptor family (mainly TLR4). Activation of these receptors triggers the innate immune response through the activation of nuclear factor-kB and JNK and the subsequent release of inflammatory mediators, such as tumor necrosis factor. Proteins belonging to inflammatory pathways cause insulin resistance by disturbing the transduction of intracellular insulin signal through a physical blockage or serine phosphorylation of the insulin receptor substrate, as mentioned earlier in this chapter. Moreover, there is evidence that unsaturated fatty acids could minimize the effects of pro-inflammatory from saturated fats, by the activation of G-coupled receptor 120 as well as by decreasing the biosynthesis of ceramides.[66]

Clinical studies that prospectively analyzed the diet, as in the case of the Nurses' Health Study, involving 5672 women with T2D, showed that increased consumption of saturated fat and cholesterol and a lower polyunsaturated:saturated ratio were associated with an increased risk of cardiovascular diseases. Replacing saturated with monounsaturated fat was more effective in reducing such risks as compared with replacement with carbohydrates.[67] Moreover, increased consumption of foods rich in omega-3 fatty acid has been associated with lower incidence of coronary diseases.[68] On the other hand, a study involving 12,536 individuals with hyperglycemia who received omega-3 capsules daily or a placebo demonstrated no decreases in cardiovascular event rates.[69] Based on the results of these studies, it is recommended that omega-3 intake is in the form of foods rich in this type of essential fat, such as fish and vegetable oils. No evidence of benefit in the supplementation of omega-3 polyunsaturated fatty acid for individuals with diabetes or those with glucose intolerance has yet been observed.[59,70] For patients with diabetes or those at a risk of developing diabetes, replacing part of the carbohydrates or saturated fat-rich foods with monounsaturated fatty acids (MUFA) in a Mediterranean-style diet would be beneficial from the standpoint of glycemic control. In the Dietary Intervention Randomized Controlled Trial, 322 adults with an average body mass index of 31 kg/m^2 and T2D or patients with coronary heart disease were randomly assigned one of the three types of diets, including a low-fat diet, a low-carbohydrate diet, or a Mediterranean diet. After 2 years, the results showed that for weight loss, both the Mediterranean diet and the low-carbohydrate diet were more effective than the diet low in lipids. However, the Mediterranean diet resulted in a significant decrease in C-reactive protein concentration in plasma, glucose levels, and insulin resistance.[71] Evidence supporting the beneficial effects of the Mediterranean diet was also shown by Esposito and colleagues in a review that showed better glycemic control and lower insulin resistance in patients who followed this type of diet as compared with controls.[72]

The Mediterranean diet is rich in whole grains, vegetables, fruits, and MUFA (like those found in olive oil), with limited amounts of poultry, fish, dairy products, red wine, and red meat.[73] The positive effects on cardiovascular disease have been often attributed to its high MUFA content rather than the saturated fatty acid content.[74] Olive oil and other ingredients of the Mediterranean diet, such as red wine, fruits, and vegetables, are known for their anti-inflammatory properties owing to the presence of phenolic compounds.[75] Additionally, both oil (rich in MUFA) and fatty fish, such as sardines (rich in n-3 fatty acids) found in the Mediterranean diet, are known to increase blood adiponectin levels that correlate with insulin sensitivity.[76–78]

In addition to the Mediterranean-style diet, other diets have been involved in the improvement of metabolic conditions, including a low-carbohydrate/high-protein diet, a vegan diet, and a vegetarian diet; however, most studies show that there are individual variations in the degree of improvement in response to different diets tested.

The most recent recommendation from the American Diabetes Association on the dietary management of patients with diabetes in 2014 was that the individualization of macronutrient distribution should be based on current eating patterns, habits, preferences, and metabolic goals.[59] In 2013, the Canadian Diabetes Association not only provided macronutrient distribution ranges ideal for nutritional care in diabetes, but also emphasized the importance of the individualization of objectives and on the quality of the prescribed macronutrients.[70]

Therefore, we suggest that the adoption of a Mediterranean diet can help prevent T2D, improve glycemic control, and reduce cardiovascular risk in individuals with established diabetes. However, it seems to be obvious that there is no standard dietary treatment that is more efficient for this population. These patients should then receive individualized nutritional guidance to achieve the goals proposed for their treatment.

4. PROTEINS

The widespread acceptance of low-carbohydrate, high-protein diets, such as the Atkins diet, Zone diet, and South Beach diet, derives from the fact that they are apparently safe in the short term and demonstrate efficiency in weight loss.[79,80]

In patients who are overweight, obese, and with T2D, consumption of a diet with a high proportion of calories from protein increases weight loss and facilitates an improvement in numerous risk factors for cardiovascular and metabolic diseases, such as glucose homeostasis and lipid metabolism.[80]

However, the weight loss caused by any dietary formulation is dependent on a reduction in the caloric intake. Low percentage dietary protein (10–15%) increases energy intake, whereas the increased protein intake may help reduce energy consumption.[81]

From a metabolic point of view, dietary proteins are effective in inducing satiety and are important modulators of insulin secretion by the β cells of the pancreas. Thus, the consumption of large amounts of protein from meals during the day can induce a prolonged satiety effect as compared with the consumption of meals with a standard amount of protein or the consumption of those rich in carbohydrates and lipids.[82] The satiety signal seems to be involved in several gut hormone modulations associated with a reduction in food intake. Studies suggest that after the consumption of high quantities of protein, there is an increase in the plasma levels of anorectic hormones, such as cholecystokinin and incretins, including glucagon-like peptide 1 and glucose-dependent insulinotropic polypeptide, which are not only associated with the regulation of food intake but also potentiate insulin secretion from β cells.[82–84]

Dietary proteins exhibit an insulinotropic effect that promotes increased plasma glucose clearance.[85] Studies conducted in healthy individuals showed an increment in plasma insulin level after meals rich in casein, soy protein, or dietary milk serum protein (whey protein) supplement as compared with plasma insulin level after a high-glucose meal, with an effect on secretion remaining high for more than 5 h.[86] In contrast, some amino acids can decrease signal pathways and the amplification of insulin secretion[85]; thus, they negatively modulate the function of pancreatic β cells.

However, the effect of high-protein diets on insulin sensitivity is still controversial. Furthermore, when the dietary protein content is increased by decreasing carbohydrate content, it is difficult to determine the change responsible for the effects observed on insulin sensitivity. In addition, the regular consumption of a low-carbohydrate, high-protein diet without a suitable selection of protein source and carbohydrate type may correlate with an increased risk of cardiovascular disease and T2D.[87]

In this context, the Diet, Obesity, and Genes European multicenter trial revealed that a high-protein intake may increase meta-inflammation and contribute to insulin resistance development.[88]

The consumption of a high-protein diet, which induces moderate weight loss (5–10%) with decreased intramyocellular or omental fat, is enough to induce an increase in insulin sensitivity.[79,86]

Short-term studies (less than 6 months of intervention) conducted in patients who were overweight, obese, or with T2D, using high-protein diets (>20%) and energy restriction showed an improvement in insulin sensitivity that was partially dependent on weight loss. A recent meta-analysis that evaluated high-protein diets using energy restriction concluded that there is no effect of protein intake on glucose homeostasis.[86] Similarly, short-term studies (less than 6 months of intervention) conducted in healthy subjects and patients who were overweight, or with T2D, using high-protein (>20%) and energy balanced diets without weight loss showed that the effects of high-protein diets on insulin sensitivity are inconclusive.

In overweight individuals, after an initial period of weight loss, a low-fat diet (24%) supplemented with whey protein or casein (protein, 35%) was not able to alter insulin sensitivity compared with a high-carbohydrate diet (protein, 16%; carbohydrate, 63%). Only in acute high-protein diets (35%, 1 week) associated with very low carbohydrates (5%), a decrease in daily postprandial plasma insulin levels was observed. Weickert et al. demonstrated a reduction in insulin sensitivity after 6 weeks of a high-protein (25–30%) and low-carbohydrate diet (40–45%).[86,89] However, the diet included large amounts of vegetables and dairy products and was compared with a diet rich in fiber (protein, 15%; carbohydrate, 55%).

Observational studies suggest that the consumption of high-protein diets in the long-term have been associated with an increased risk of developing T2D and metabolic syndrome. In long-term studies (greater than 6 months of intervention) conducted in healthy subjects using high protein, the development of insulin resistance and an impaired glucose tolerance was observed in energy-balanced diets (24%) compared with a normal protein diet (10%).[86]

However, a recent comparison between the effect of diets rich in protein (>25% of protein) and low in protein content (10–15%) conducted for more than 1 year in healthy and insulin-resistant individuals showed no one effect on glycemic control.[90]

The variability factors in an experimental design involving humans, including sample size, study duration, study population, and macronutrient composition used in the study may contribute to a huge difference observed in the results.[90]

Similarly, in a high-protein diet, there are numerous factors that may be important for metabolic effects, such as the absolute amount of protein per meal, the number of protein-containing meals during the day, and the type of protein intake. Special attention should be given to the amino acid composition because the amino acids glutamine, alanine, and arginine have an insulinotropic effect, inducing an increase in insulin secretion and modulating the signaling of this hormone.[82]

Glutamine and its metabolites aspartate and glutamate, particularly when administered in combination with leucine, increase the exocytosis of insulin via the activation of glutamate dehydrogenase. Alanine acts by numerous multifactorial mechanisms that include converting pyruvate, glutamate, aspartate, and lactate. Arginine-mediated insulin secretion is dependent on the changes in plasma membrane potential and calcium influx. However, high concentrations of arginine have shown an opposite effect on insulin secretion by stimulating inducible nitric oxide synthase, which overwhelm the antioxidant defenses of pancreatic β cells.[85]

Recent research indicates that branched chain amino acids (BCAA), such as leucine, isoleucine, and valine, are a group of amino acids that play an important role in insulin exocytosis. In particular, leucine has the potential to improve postprandial glycemic control because of its ability to stimulate the endogenous release of insulin.[91] A study conducted by Lu et al. has shown that dietary supplementation with BCAAs have a beneficial effect on the recovery of islet function in rats with diabetes induced by streptozotocin. In this model, researchers evaluated the expression of genes involved in insulin secretion (PKD1, Pdx1, and JNK) and observed that dietary supplementation with BCAA animals led to an increase in insulin secretion.[92]

BCAAs compose a significant proportion of dietary amino acids. Their content in mixed sources of protein is approximately 20%, and almost 80% of the total ingested content reaches the bloodstream. High concentrations of BCAA can be found in dairy products and supplements such as whey protein. Consumption of dairy products has been associated with improvements in weight loss and T2D management[85]; consumption of whey protein hydrolysates have been suggested for improving fasting insulin levels and release, and glycemic control in obese and T2D individuals.[84,93]

Supplementation of BCAAs or a BCAA-rich diet is believed to improve metabolic health. Accordingly, an increase in the recommended daily protein intake has been proposed, which increase dietary and serum levels of BCAAs. However, increased serum levels of BCCA and their catabolites are positively correlated with whole-body insulin resistance. In clinical studies, aromatic amino acids and BCAAs are regarded as independent predictors of insulin resistance, diabetes, and cardiovascular events.[86] Newgard et al. found a positive correlation between increased plasma levels of BCAAs and decreased insulin sensitivity in obese adults, as determined by an increase in the homeostatic model assessment index.[81,86,94] However, in young individuals, the plasma concentration of amino acids was positively correlated with the function of pancreatic β cells, indicating an effect opposite to that described previously.[81]

The BCAA catabolism pathway leads to the production of branched alpha-keto acids that are oxidized and decarboxylated in an irreversible step of the pathway catalyzed by the mitochondrial multienzyme branched chain α-ketoacid dehydrogenase complex (BCKDC). BCKDC is an enzyme inhibited by insulin and high concentrations of free fatty acids as well as by the phosphorylation of the residue Ser293. Dephosphorylation of the serine residue in the protein catalyzed by mitochondrial protein phosphatase 2C leads to its activation. The products of BCKDC activity are acyl-coenzyme-chain branched A species that are further metabolized by multiple enzymatic steps in the mitochondrial matrix.[91,95]

Recent studies have demonstrated that reduced BCKDC activity in the liver and adipose tissues of obese and diabetic animals as well as of humans results in higher plasma levels of BCAAs and their metabolites. The visceral adipose tissue, in particular, may have an important role in this regard.[95]

Two potential mechanisms have been considered to explain how BCAAs may contribute to the development of insulin resistance. The first mechanism proposes that an excess of dietary BCAAs, particularly leucine, activates the mammalian target of rapamycin complex 1 signaling pathway. Persistent activation of mammalian target of rapamycin complex 1 and serine kinase S6K1 promote insulin resistance through the serine phosphorylation of IRS-1 and IRS-2, inhibiting the association of the insulin receptor and finally leading to a decrease in insulin signal and T2D.[81,91] The second mechanism proposes that increased levels of BCAAs operate only as biomarker of impaired insulin signaling, but are not responsible for the development of insulin resistance or T2D in individuals with BCAA dysmetabolism (impaired BCAA metabolism). Impairment in BCAA metabolism could result in its accumulation as well as mitotoxic metabolites that induce β-cell mitochondrial dysfunction, leading to stress signaling and eventually to the apoptosis of β cells.[91] In addition, several studies have shown that patients who were obese and with T2D have reduced plasma levels of adiponectin, an important regulator of lipid and glucose metabolism, and a reduced level of AMPK activation in many tissues. AMPK is considered the main energy sensor involved in the regulation of glucose metabolism and lipids. Adiponectin is able to reverse the reduction of catabolism of BCAAs by stimulating the expression of mitochondrial protein phosphatase 2C dependent on AMPK. This stimulation leads to an increase in BCKDC activity and consequently leads to a decrease in the circulating levels of BCAAs.[95]

Otherwise, insulin resistance may promote aminoacidemia by increasing protein degradation that is normally suppressed by insulin or by inducing dysfunction in BCAA metabolism.[91] It has been suggested that hyperaminoacidemia promotes hyperinsulinemia. This scenario, when prolonged in healthy subjects, may be deleterious, leading to a decrease in insulin sensitivity or pancreatic β-cell exhaustion.[86]

5. CONCLUDING REMARKS

Different types of diets have been shown to improve metabolic disorders associated with T2D. Different dietary nutrients can collaborate to lose body weight and to reduce the inflammatory pathways activation in tissue that play a fundamental role in glucose homeostasis. Currently, research in different fields of nutrition, medicine, and basic science have investigated the role of epigenetic modulation in the development of diabetes and obesity. However, there is a need to improve the quality of randomized studies in short- and long-term studies, including personalized epigenetics aspects. This could contribute to better comprehension of the mechanism that can increase the possibility of the patient to reach his or her objective and improve the outcome of nutritional therapy.

References

1. Coppack SW. Pro-inflammatory cytokines and adipose tissue. *Proc Nutr Soc* 2001;**60**(3):349–56.
2. Shulman GI. Cellular mechanisms of insulin resistance. *J Clin Invest* 2000;**106**(2):171–6.
3. Black RN, Spence M, McMahon RO, Cuskelly GJ, Ennis CN, McCance DR, et al. Effect of eucaloric high- and low-sucrose diets with identical macronutrient profile on insulin resistance and vascular risk: a randomized controlled trial. *Diabetes* 2006;**55**(12):3566–72.
4. Melanson KJ, Summers A, Nguyen V, Brosnahan J, Lowndes J, Angelopoulos TJ, et al. Body composition, dietary composition, and components of metabolic syndrome in overweight and obese adults after a 12-week trial on dietary treatments focused on portion control, energy density, or glycemic index. *Nutr J* 2012;**11**:57.
5. Brynes AE, Mark Edwards C, Ghatei MA, Dornhorst A, Morgan LM, Bloom SR, et al. A randomised four-intervention crossover study investigating the effect of carbohydrates on daytime profiles of insulin, glucose, non-esterified fatty acids and triacylglycerols in middle-aged men. *Br J Nutr* 2003;**89**(2):207–18.
6. Deer J, Koska J, Ozias M, Reaven P. Dietary models of insulin resistance. *Metab Clin Exp* 2015;**64**(2):163–71.
7. Brynes AE, Lee JL, Brighton RE, Leeds AR, Dornhorst A, Frost GS. A low glycemic diet significantly improves the 24-h blood glucose profile in people with type 2 diabetes, as assessed using the continuous glucose MiniMed monitor. *Diabetes Care* 2003;**26**(2):548–9.
8. Galadari S, Rahman A, Pallichankandy S, Galadari A, Thayyullathil F. Role of ceramide in diabetes mellitus: evidence and mechanisms. *Lipids Health Dis* 2013;**12**:98.
9. Wang N, Wang H, Yao H, Wei Q, Mao XM, Jiang T, et al. Expression and activity of the TLR4/NF-kappaB signaling pathway in mouse intestine following administration of a short-term high-fat diet. *Exp Ther Med* 2013;**6**(3):635–40.
10. Areias MF, Prada PO. Mechanisms of insulin resistance in the amygdala: influences on food intake. *Behav Brain Res* 2015;**282C**:209–17.
11. Guo S. Insulin signaling, resistance, and the metabolic syndrome: insights from mouse models into disease mechanisms. *J Endocrinol* 2014;**220**(2):T1–23.
12. Tappy L, Le KA. Metabolic effects of fructose and the worldwide increase in obesity. *Physiol Rev* 2010;**90**(1):23–46.
13. Lowndes J, Kawiecki D, Pardo S, Nguyen V, Melanson KJ, Yu Z, et al. The effects of four hypocaloric diets containing different levels of sucrose or high fructose corn syrup on weight loss and related parameters. *Nutr J* 2012;**11**:55.
14. Malik VS, Popkin BM, Bray GA, Despres JP, Willett WC, Hu FB. Sugar-sweetened beverages and risk of metabolic syndrome and type 2 diabetes: a meta-analysis. *Diabetes Care* 2010;**33**(11):2477–83.
15. Rippe JM, Angelopoulos TJ. Sucrose, high-fructose corn syrup, and fructose, their metabolism and potential health effects: what do we really know? *Adv Nutr* 2013;**4**(2):236–45.
16. Rippe JM, Kris Etherton PM. Fructose, sucrose, and high fructose corn syrup: modern scientific findings and health implications. *Adv Nutr* 2012;**3**(5):739–40.
17. Stanhope KL, Schwarz JM, Keim NL, Griffen SC, Bremer AA, Graham JL, et al. Consuming fructose-sweetened, not glucose-sweetened, beverages increases visceral adiposity and lipids and decreases insulin sensitivity in overweight/obese humans. *J Clin Invest* 2009;**119**(5):1322–34.
18. Teff KL, Elliott SS, Tschop M, Kieffer TJ, Rader D, Heiman M, et al. Dietary fructose reduces circulating insulin and leptin, attenuates postprandial suppression of ghrelin, and increases triglycerides in women. *J Clin Endocrinol Metab* 2004;**89**(6):2963–72.
19. Teff KL, Grudziak J, Townsend RR, Dunn TN, Grant RW, Adams SH, et al. Endocrine and metabolic effects of consuming fructose- and glucose-sweetened beverages with meals in obese men and women: influence of insulin resistance on plasma triglyceride responses. *J Clin Endocrinol Metab* 2009;**94**(5):1562–9.
20. Douard V, Ferraris RP. Regulation of the fructose transporter GLUT5 in health and disease. *Am J Physiol Endocrinol Metab* 2008;**295**(2):E227–37.
21. Mithieux G, Misery P, Magnan C, Pillot B, Gautier-Stein A, Bernard C, et al. Portal sensing of intestinal gluconeogenesis is a mechanistic link in the diminution of food intake induced by diet protein. *Cell Metab* 2005;**2**(5):321–9.
22. Parks EJ, Skokan LE, Timlin MT, Dingfelder CS. Dietary sugars stimulate fatty acid synthesis in adults. *J Nutr* 2008;**138**(6):1039–46.
23. Barcelo-Fimbres M, Seidel Jr GE. Effects of either glucose or fructose and metabolic regulators on bovine embryo development and lipid accumulation in vitro. *Mol Reprod Dev* 2007;**74**(11):1406–18.
24. Bizeau ME, Pagliassotti MJ. Hepatic adaptations to sucrose and fructose. *Metab Clin Exp* 2005;**54**(9):1189–201.
25. Havel PJ. Dietary fructose: implications for dysregulation of energy homeostasis and lipid/carbohydrate metabolism. *Nutr Rev* 2005;**63**(5):133–57.
26. Pagliassotti MJ, Kang J, Thresher JS, Sung CK, Bizeau ME. Elevated basal PI 3-kinase activity and reduced insulin signaling in sucrose-induced hepatic insulin resistance. *Am J Physiol Endocrinol Metab* 2002;**282**(1):E170–6.
27. Pagliassotti MJ, Prach PA, Koppenhafer TA, Pan DA. Changes in insulin action, triglycerides, and lipid composition during sucrose feeding in rats. *Am J Physiol* 1996;**271**(5 Pt 2):R1319–26.
28. Le KA, Ith M, Kreis R, Faeh D, Bortolotti M, Tran C, et al. Fructose overconsumption causes dyslipidemia and ectopic lipid deposition in healthy subjects with and without a family history of type 2 diabetes. *Am J Clin Nutr* 2009;**89**(6):1760–5.
29. Wei Y, Pagliassotti MJ. Hepatospecific effects of fructose on c-jun NH2-terminal kinase: implications for hepatic insulin resistance. *Am J Physiol Endocrinol Metab* 2004;**287**(5):E926–33.
30. Wei Y, Wang D, Topczewski F, Pagliassotti MJ. Fructose-mediated stress signaling in the liver: implications for hepatic insulin resistance. *J Nutr Biochem* 2007;**18**(1):1–9.
31. Bergheim I, Weber S, Vos M, Kramer S, Volynets V, Kaserouni S, et al. Antibiotics protect against fructose-induced hepatic lipid accumulation in mice: role of endotoxin. *J Hepatol* 2008;**48**(6):983–92.
32. Spruss A, Kanuri G, Wagnerberger S, Haub S, Bischoff SC, Bergheim I. Toll-like receptor 4 is involved in the development of fructose-induced hepatic steatosis in mice. *Hepatology* 2009;**50**(4):1094–104.
33. Cani PD, Amar J, Iglesias MA, Poggi M, Knauf C, Bastelica D, et al. Metabolic endotoxemia initiates obesity and insulin resistance. *Diabetes* 2007;**56**(7):1761–72.
34. Saito T, Hayashida H, Furugen R. Comment on: Cani et al. (2007) Metabolic endotoxemia initiates obesity and insulin resistance: Diabetes 56:1761–1772. *Diabetes* 2007;**56**(12):e20. author reply e1.
35. Thuy S, Ladurner R, Volynets V, Wagner S, Strahl S, Konigsrainer A, et al. Nonalcoholic fatty liver disease in humans is associated with increased plasma endotoxin and plasminogen activator inhibitor 1 concentrations and with fructose intake. *J Nutr* 2008;**138**(8):1452–5.
36. Maki KC, Davidson MH, Witchger MS, Dicklin MR, Subbaiah PV. Effects of high-fiber oat and wheat cereals on postprandial glucose and lipid responses in healthy men. *Int J Vitam Nutr Res* 2007;**77**(5):347–56.

37. Maki KC, Galant R, Samuel P, Tesser J, Witchger MS, Ribaya-Mercado JD, et al. Effects of consuming foods containing oat beta-glucan on blood pressure, carbohydrate metabolism and biomarkers of oxidative stress in men and women with elevated blood pressure. *Eur J Clin Nutr* 2007;**61**(6):786–95.
38. Maki KC, Pelkman CL, Finocchiaro ET, Kelley KM, Lawless AL, Schild AL, et al. Resistant starch from high-amylose maize increases insulin sensitivity in overweight and obese men. *J Nutr* 2012;**142**(4):717–23.
39. Robertson MD, Bickerton AS, Dennis AL, Vidal H, Frayn KN. Insulin-sensitizing effects of dietary resistant starch and effects on skeletal muscle and adipose tissue metabolism. *Am J Clin Nutr* 2005; **82**(3):559–67.
40. Tarini J, Wolever TM. The fermentable fibre inulin increases postprandial serum short-chain fatty acids and reduces free-fatty acids and ghrelin in healthy subjects. *Appl Physiol Nutr Metab* 2010;**35**(1):9–16.
41. Ley SH, Hamdy O, Mohan V, Hu FB. Prevention and management of type 2 diabetes: dietary components and nutritional strategies. *Lancet* 2014;**383**(9933):1999–2007.
42. Riserus U, Willett WC, Hu FB. Dietary fats and prevention of type 2 diabetes. *Prog Lipid Res* 2009;**48**(1):44–51.
43. Hu FB, van Dam RM, Liu S. Diet and risk of Type II diabetes: the role of types of fat and carbohydrate. *Diabetologia* 2001;**44**(7): 805–17.
44. Halton TL, Liu S, Manson JE, Hu FB. Low-carbohydrate-diet score and risk of type 2 diabetes in women. *Am J Clin Nutr* 2008;**87**(2): 339–46.
45. Kim JY, Hickner RC, Cortright RL, Dohm GL, Houmard JA. Lipid oxidation is reduced in obese human skeletal muscle. *Am J Physiol Endocrinol Metab* 2000;**279**(5):E1039–44.
46. Kelley DE, He J, Menshikova EV, Ritov VB. Dysfunction of mitochondria in human skeletal muscle in type 2 diabetes. *Diabetes* 2002;**51**(10):2944–50.
47. Ritov VB, Menshikova EV, He J, Ferrell RE, Goodpaster BH, Kelley DE. Deficiency of subsarcolemmal mitochondria in obesity and type 2 diabetes. *Diabetes* 2005;**54**(1):8–14.
48. Heilbronn LK, Gan SK, Turner N, Campbell LV, Chisholm DJ. Markers of mitochondrial biogenesis and metabolism are lower in overweight and obese insulin-resistant subjects. *J Clin Endocrinol Metab* 2007;**92**(4):1467–73.
49. Koyama K, Chen G, Lee Y, Unger RH. Tissue triglycerides, insulin resistance, and insulin production: implications for hyperinsulinemia of obesity. *Am J Physiol* 1997;**273**(4 Pt 1):E708–13.
50. Phillips DI, Caddy S, Ilic V, Fielding BA, Frayn KN, Borthwick AC, et al. Intramuscular triglyceride and muscle insulin sensitivity: evidence for a relationship in nondiabetic subjects. *Metab Clin Exp* 1996;**45**(8):947–50.
51. Szendroedi J, Roden M. Ectopic lipids and organ function. *Curr Opin Lipidol* 2009;**20**(1):50–6.
52. Thomas EL, Fitzpatrick JA, Malik SJ, Taylor-Robinson SD, Bell JD. Whole body fat: content and distribution. *Prog Nucl Magn Reson Spectrosc* 2013;**73**:56–80.
53. Hage Hassan R, Bourron O, Hajduch E. Defect of insulin signal in peripheral tissues: important role of ceramide. *World J Diabetes* 2014;**5**(3):244–57.
54. Wende AR, Symons JD, Abel ED. Mechanisms of lipotoxicity in the cardiovascular system. *Curr Hypertens Rep* 2012;**14**(6):517–31.
55. Aon MA, Bhatt N, Cortassa SC. Mitochondrial and cellular mechanisms for managing lipid excess. *Front Physiol* 2014;**5**:282.
56. Goodpaster BH, He J, Watkins S, Kelley DE. Skeletal muscle lipid content and insulin resistance: evidence for a paradox in endurance-trained athletes. *J Clin Endocrinol Metab* 2001;**86**(12): 5755–61.
57. Gao D, Griffiths HR, Bailey CJ. Oleate protects against palmitate-induced insulin resistance in L6 myotubes. *Br J Nutr* 2009; **102**(11):1557–63.
58. Gao D, Pararasa C, Dunston CR, Bailey CJ, Griffiths HR. Palmitate promotes monocyte atherogenicity via de novo ceramide synthesis. *Free Radic Biol Med* 2012;**53**(4):796–806.
59. Evert AB, Boucher JL, Cypress M, Dunbar SA, Franz MJ, Mayer-Davis EJ, et al. Nutrition therapy recommendations for the management of adults with diabetes. *Diabetes Care* 2014;**37**(Suppl. 1): S120–43.
60. Estruch R, Ros E, Martinez-Gonzalez MA. Mediterranean diet for primary prevention of cardiovascular disease. *N Engl J Med* 2013; **369**(7):676–7.
61. Chavez JA, Summers SA. Characterizing the effects of saturated fatty acids on insulin signaling and ceramide and diacylglycerol accumulation in 3T3-L1 adipocytes and C2C12 myotubes. *Arch Biochem Biophys* 2003;**419**(2):101–9.
62. Powell DJ, Turban S, Gray A, Hajduch E, Hundal HS. Intracellular ceramide synthesis and protein kinase Czeta activation play an essential role in palmitate-induced insulin resistance in rat L6 skeletal muscle cells. *Biochem J* 2004;**382**(Pt 2):619–29.
63. Dimopoulos N, Watson M, Sakamoto K, Hundal HS. Differential effects of palmitate and palmitoleate on insulin action and glucose utilization in rat L6 skeletal muscle cells. *Biochem J* 2006;**399**(3): 473–81.
64. Cao H, Gerhold K, Mayers JR, Wiest MM, Watkins SM, Hotamisligil GS. Identification of a lipokine, a lipid hormone linking adipose tissue to systemic metabolism. *Cell* 2008;**134**(6):933–44.
65. Storlien LH, Kraegen EW, Chisholm DJ, Ford GL, Bruce DG, Pascoe WS. Fish oil prevents insulin resistance induced by high-fat feeding in rats. *Science* 1987;**237**(4817):885–8.
66. Glass CK, Olefsky JM. Inflammation and lipid signaling in the etiology of insulin resistance. *Cell Metab* 2012;**15**(5):635–45.
67. Tanasescu M, Cho E, Manson JE, Hu FB. Dietary fat and cholesterol and the risk of cardiovascular disease among women with type 2 diabetes. *Am J Clin Nutr* 2004;**79**(6):999–1005.
68. Hu FB, Cho E, Rexrode KM, Albert CM, Manson JE. Fish and long-chain omega-3 fatty acid intake and risk of coronary heart disease and total mortality in diabetic women. *Circulation* 2003;**107**(14): 1852–7.
69. Bosch J, Gerstein HC, Dagenais GR, Diaz R, Dyal L, Jung H, et al. n-3 fatty acids and cardiovascular outcomes in patients with dysglycemia. *N Engl J Med* 2012;**367**(4):309–18.
70. Sievenpiper JL, Dworatzek PD. Food and dietary pattern-based recommendations: an emerging approach to clinical practice guidelines for nutrition therapy in diabetes. *Can J Diabetes* 2013; **37**(1):51–7.
71. Shai I, Schwarzfuchs D, Henkin Y, Shahar DR, Witkow S, Greenberg I, et al. Weight loss with a low-carbohydrate, Mediterranean, or low-fat diet. *N Engl J Med* 2008;**359**(3):229–41.
72. Esposito K, Maiorino MI, Ceriello A, Giugliano D. Prevention and control of type 2 diabetes by Mediterranean diet: a systematic review. *Diabetes Res Clin Pract* 2010;**89**(2):97–102.
73. Kastorini CM, Panagiotakos DB. Mediterranean diet and diabetes prevention: myth or fact? *World J Diabetes* 2010;**1**(3):65–7.
74. Martinez-Gonzalez MA, de la Fuente-Arrillaga C, Nunez-Cordoba JM, Basterra-Gortari FJ, Beunza JJ, Vazquez Z, et al. Adherence to Mediterranean diet and risk of developing diabetes: prospective cohort study. *BMJ* 2008;**336**(7657):1348–51.
75. Cardeno A, Sanchez-Hidalgo M, Alarcon-de-la-Lastra C. An update of olive oil phenols in inflammation and cancer: molecular mechanisms and clinical implications. *Curr Med Chem* 2013; **20**(37):4758–76.
76. Ortega R. Importance of functional foods in the Mediterranean diet. *Public Health Nutr* 2006;**9**(8A):1136–40.
77. Mantzoros CS, Williams CJ, Manson JE, Meigs JB, Hu FB. Adherence to the Mediterranean dietary pattern is positively associated with plasma adiponectin concentrations in diabetic women. *Am J Clin Nutr* 2006;**84**(2):328–35.

78. Lihn AS, Pedersen SB, Richelsen B. Adiponectin: action, regulation and association to insulin sensitivity. *Obes Rev* 2005;**6**(1):13–21.
79. Lara-Castro C, Garvey WT. Diet, insulin resistance, and obesity: zoning in on data for Atkins dieters living in South Beach. *J Clin Endocrinol Metab* 2004;**89**(9):4197–205.
80. Khazrai YM, Defeudis G, Pozzilli P. Effect of diet on type 2 diabetes mellitus: a review. *Diabetes Metab Res Rev* 2014;**30**(Suppl. 1):24–33.
81. Russell WR, Baka A, Bjorck I, Delzenne N, Gao D, Griffiths HR, et al. Impact of diet composition on blood glucose regulation. *Crit Rev Food Sci Nutr* 2013.
82. Bendtsen LQ, Lorenzen JK, Bendsen NT, Rasmussen C, Astrup A. Effect of dairy proteins on appetite, energy expenditure, body weight, and composition: a review of the evidence from controlled clinical trials. *Adv Nutr* 2013;**4**(4):418–38.
83. Sousa GT, Lira FS, Rosa JC, de Oliveira EP, Oyama LM, Santos RV, et al. Dietary whey protein lessens several risk factors for metabolic diseases: a review. *Lipids Health Dis* 2012;**11**:67.
84. Jakubowicz D, Froy O. Biochemical and metabolic mechanisms by which dietary whey protein may combat obesity and Type 2 diabetes. *J Nutr Biochem* 2013;**24**(1):1–5.
85. Newsholme P, Cruzat V, Arfuso F, Keane K. Nutrient regulation of insulin secretion and action. *J Endocrinol* 2014;**221**(3):R105–20.
86. Rietman A, Schwarz J, Tome D, Kok FJ, Mensink M. High dietary protein intake, reducing or eliciting insulin resistance? *Eur J Clin Nutr* 2014;**68**(9):973–9.
87. Lagiou P, Sandin S, Lof M, Trichopoulos D, Adami HO, Weiderpass E. Low carbohydrate-high protein diet and incidence of cardiovascular diseases in Swedish women: prospective cohort study. *BMJ* 2012;**344**:e4026.
88. Astrup A, Raben A, Geiker N. The role of higher protein diets in weight control and obesity-related comorbidities. *Int J Obes (Lond)* 2014.
89. Weickert MO, Roden M, Isken F, Hoffmann D, Nowotny P, Osterhoff M, et al. Effects of supplemented isoenergetic diets differing in cereal fiber and protein content on insulin sensitivity in overweight humans. *Am J Clin Nutr* 2011;**94**(2):459–71.
90. Wycherley TP, Moran LJ, Clifton PM, Noakes M, Brinkworth GD. Effects of energy-restricted high-protein, low-fat compared with standard-protein, low-fat diets: a meta-analysis of randomized controlled trials. *Am J Clin Nutr* 2012;**96**(6):1281–98.
91. Lynch CJ, Adams SH. Branched-chain amino acids in metabolic signalling and insulin resistance. *Nat Rev Endocrinol* 2014;**10**(12): 723–36.
92. Lu M, Zhang X, Zheng D, Jiang X, Chen Q. Branched-chain amino acids supplementation protects streptozotocin-induced insulin secretion and the correlated mechanism. *Biofactors* 2014;**41**(2): 127–33.
93. Graf S, Egert S, Heer M. Effects of whey protein supplements on metabolism: evidence from human intervention studies. *Curr Opin Clin Nutr Metab Care* 2011;**14**(6):569–80.
94. Newgard CB, An J, Bain JR, Muehlbauer MJ, Stevens RD, Lien LF, et al. A branched-chain amino acid-related metabolic signature that differentiates obese and lean humans and contributes to insulin resistance. *Cell Metab* 2009;**9**(4):311–26.
95. Lian K, Du C, Liu Y, Zhu D, Yan W, Zhang H, et al. Impaired adiponectin signaling contributes to disturbed catabolism of branched-chain amino acids in diabetic mice. *Diabetes* 2015;**64**(1): 49–59.

CHAPTER

3

β-Cell Metabolism, Insulin Production and Secretion: Metabolic Failure Resulting in Diabetes

Younan Chen, Vinicius F. Cruzat, Philip Newsholme

School of Biomedical Sciences, CHIRI Biosciences Research Precinct, Faculty of Health Sciences, Curtin University, Perth, WA, Australia

1. INTRODUCTION TO PANCREATIC β-CELL METABOLISM AND METABOLIC LINKS TO INSULIN SECRETION

Pancreatic β cells are often referred to as "fuel sensors" because they continually monitor and respond to circulating nutrient levels, plus additional neurohormonal signals, to secrete insulin to best meet the needs of the organism. β-cell nutrient sensing involves activation of glycolytic and oxidative metabolism, resulting in production of coupling signals that promote insulin biosynthesis and secretion. Islet β cells are particularly responsive to elevations in circulating glucose, coupling nutrient, and other stimuli with the insulin-secretory machinery. In writing this chapter, we are fully aware that most studies cited have used rat-, mouse-, or hamster-derived insulinoma β-cell lines for in vitro studies. It is difficult to maintain primary rodent islet β-cell mass and function for more than a few days in vitro and there is scarcity of human islets for research purposes. The generation of functional and stable human β-cell lines has also proved problematic. Nevertheless, the major rodent β-cell lines have provided substantial data and insights into cell function in normal or pathogenic situations.[1,2] The most widely used cell lines include INS 1, MIN 6, RINm5F, and BRIN-BD11.[3] It is important to state that intact islet structures (comprising α, β, and δ cells, which secrete glucagon, insulin, and somatostatin, respectively) are required to maintain appropriate and pulsatile hormone secretion in response to nutrient stimuli in vivo.

Increased β-cell glycolytic flux results in a rapid increase in production of adenosine triphosphate (ATP) and reducing equivalents, an increased activity of shuttle mechanisms (responsible for transferring electrons to the mitochondrial matrix), and tricarboxylic acid (TCA) cycle activity. The outcome is an enhanced cytoplasmic ATP to adenine diphosphate (ADP) ratio, which prompts closure of ATP-sensitive K^+ (K_{ATP}) channels in the plasma membrane evoking membrane depolarization, and subsequent opening of voltage-gated Ca^{2+} channels (VDCC). This results in an increase in cellular Ca^{2+} influx—a primary driver of the insulin secretory mechanism (Figure 1).[4]

2. THE ROLE OF GLUCOSE METABOLISM, FATTY ACID METABOLISM, AND AMINO ACID METABOLISM IN THE GENERATION OF METABOLIC STIMULUS–SECRETION COUPLING FACTORS

The metabolism of fatty acids and amino acids, in addition to glucose, can result in optimal and coordinated generation of metabolic stimulus-secretion coupling factors (MCFs), such as nicotinamide adenine dinucleotide phosphate (reduced form, NADPH) and acyl coenzyme A (CoA) in addition to ATP, achieving appropriately coordinated levels of insulin release.[5]

Molecular Nutrition and Diabetes
http://dx.doi.org/10.1016/B978-0-12-801585-8.00003-8

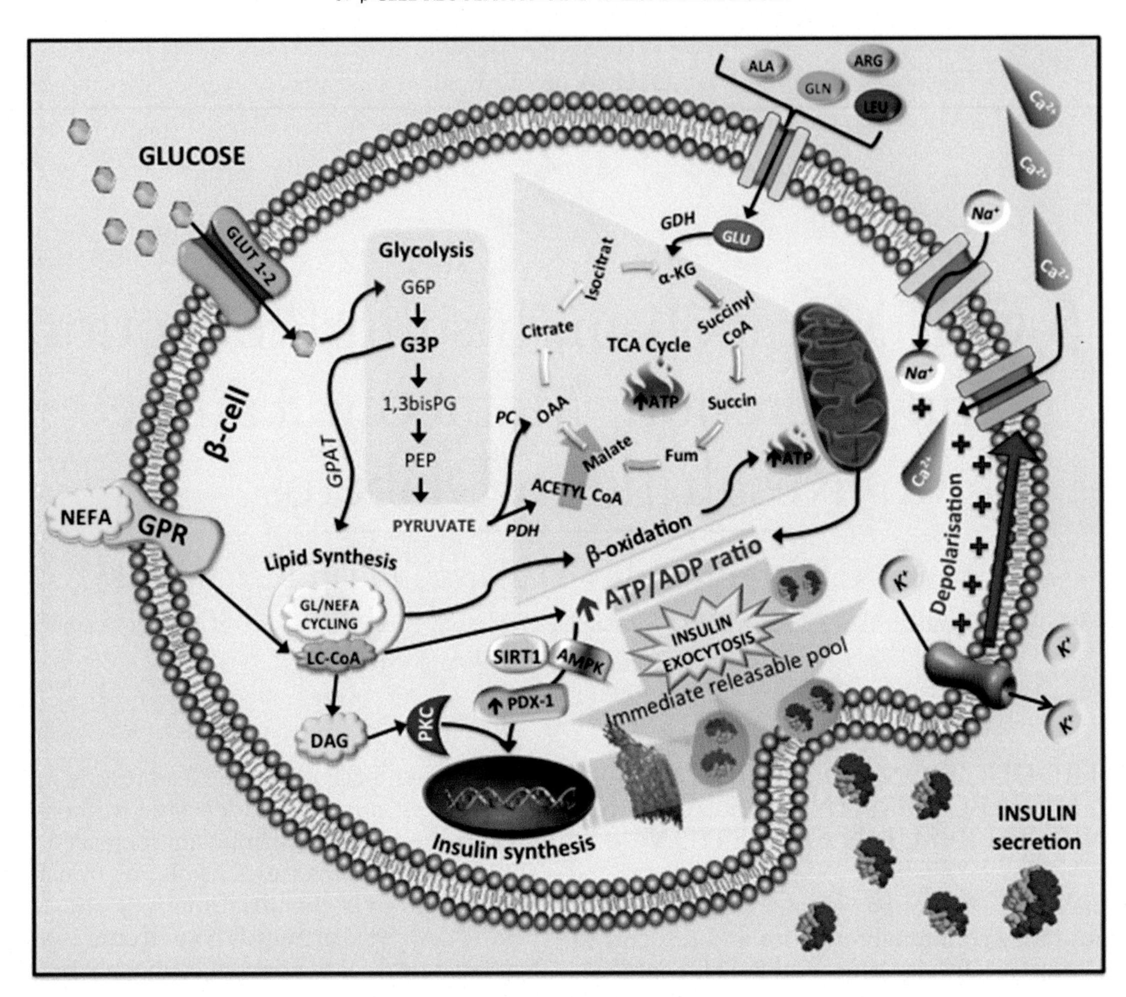

FIGURE 1 β-cell insulin secretion induced by nutrients from the diet. Under appropriate conditions, β cells respond to glucose, amino acids, and nonesterified fatty acids (NEFA). α-KG, α-ketoglutarate; ALA, alanine; ARG, arginine; 1,3bisPG, 1,3- bisphosphoglycerate; DAG, diacylglycerol; G3P, glyceraldehyde 3-phosphate; G6P, glucose 6-phosphate; GDH, glutamate dehydrogenase; GLN, glutamine; GLU, glutamate; GLUT 1-2, glucose transporter 1-2; GPAT, glycerol-3-phosphate acyltransferase; LC-CoA, long-chain CoA; LEU, leucine; OAA, oxaloacetate; PC, pyruvate carboxylase; PEP, phosphoenolpyruvate; PDH, pyruvate dehydrogenase.

2.1 Glucose Metabolism and Insulin Secretion

2.1.1 Glucose Transporters and Glucokinase

As a metabolic, survival, and differentiation factor of the β cell, glucose provides the main stimulus to drive insulin exocytosis via glycolysis and mitochondrial oxidation as well as [illegible] insulin gene expression and protein production. The β cell is equipped with abundant glucose transporters (GLUTs), GLUT2 in rodent, and GLUT1, 2, and 3 in human cells that mediate the rapid equilibrium of intra- and extracellular glucose concentration in an insulin-independent manner.[6,7] The internalization of glucose quickly initiates glycolysis and glucose is phosphorylated at the first step to produce glucose-6-phosphate. This reaction is rapidly catalyzed by glucokinase (GCK). In light of its low affinity (high Michaelis constant or Km, approximately 6 mM) and high kinetic capacity, GCK serves as the rate-limiting factor for β-cell glycolytic flux in high glucose conditions.[4,8,9]

2.1.2 The First Phase of Insulin Secretion Is K_{ATP}^{+} Channel–Dependent

The classic theory of insulin exocytosis is based on an increase in the intracellular ATP/ADP ratio in β cells, following elevated glucose metabolism via glycolysis and mitochondrial glucose oxidation.[4,10] Pyruvate derived from the penultimate step of glycolysis is normally converted to lactate by pyruvate dehydrogenase in many other cells. However, β cells are associated with very low expression levels of lactic acid dehydrogenase, so most of the pyruvate is oxidized in the mitochondria to generate ATP and additional metabolic coupling factors required for insulin secretion.[11]

The main metabolic steps for pyruvate oxidation are[8,10]: (1) oxidative decarboxylation by pyruvate dehydrogenase to produce acetyl-CoA, which can condense with oxaloacetate to enter the TCA cycle as citrate; and (2) anaplerotic metabolism by pyruvate carboxylase to replenish the carbon pool of the TCA cycle, and subsequent efflux of specific TCA cycle intermediates including citrate.[12] Some molecules derived in these processes are important MCFs, including NADPH, malonyl-CoA, and glutamate. Pancreatic β cells express high levels of mitochondrial nicotinamide adenine dinucleotide (NADH)/NADPH (redox) shuttles such as pyruvate/malate, pyruvate/citrate, malate/aspartate, and Gly-3-P shuttles to transport these molecules between cytosol and mitochondria.[9,11] Efficient oxidation of pyruvate in the mitochondria quickly generates NADH and flavin adenine dinucleotide ($FADH_2$), that indirectly stimulate ATP generation through the mitochondrial electron transport chain (ETC) and activity of ATP synthase. The enhanced ATP/ADP ratio induces plasma membrane depolarization by closure of β-cell K_{ATP} channels, and subsequently the opening of L-type VDCC. The influx of extracellular Ca^{2+} leads to insulin exocytosis from a readily releasable pool of insulin-containing vesicles.[4,8,9,11,13,14]

2.1.3 *The Second Phase of Chronic Insulin Secretion Is $K_{ATP}{}^{+}$ Independent*

This triggering mechanism of K_{ATP}-dependent glucose-stimulated insulin secretion (GSIS) is responsible for the first phase of the insulin secretion, which occurs over the first 5–10 min, but the second, more sustained phase of insulin release over a subsequent period of 30–60 min is absolutely dependent on metabolic stimulus–secretion coupling to produce amplifying signals that contribute to pulsatile $[Ca^{2+}]c$ changes that potentiate insulin granule translocation from a reserve pool to the readily releasable pool followed by docking and priming.[14] Metabolic oscillations are critical for the oscillations of $[Ca^{2+}]c$, which induce pulses of insulin secretion by individual islets.[15,16]

Glucagon-like peptide-1 (GLP-1) robustly enhances nutrient-induced insulin secretion in a glucose-dependent manner. GLP-1 binds to specific G protein–coupled receptors (GPRs) on β-cell membrane resulting in an increase in intercellular cyclic adenosine monophosphate (cAMP) that augments the action of increased intracellular $[Ca^{2+}]$ through protein kinase A and the Epac2/Rap1 signaling cascade, resulting in increased size of the readily releasable pool of insulin granules.[17,18] The amplification action of GLP-1 is resistant to diazoxide (which results in opening of $K_{ATP}{}^{+}$ channels), and therefore is independent of K_{ATP} channel closure. In the past decade, it has become clear that small G proteins and their regulatory proteins are required for the regulatory events responsible for insulin release and have provided additional information related to our understanding of the amplifying signals in the second phase of insulin secretion.[19,20] There are distinct temporal activation processes linked to the small Rho guanosine triphosphatases (GTPases), Cdc42, and Rac1in β cells, and both activated Cdc42 and Rac1 stimulate the translocation of insulin secretory vesicles to the plasma membrane through their effects on cytoskeletal remodeling. Accumulated evidence now supports the concept that Rho GTPases and their regulation by guanine nucleotide dissociation inhibitors have a fundamental importance in regulating the second phase of insulin secretion.[21]

2.2 Fatty Acid Metabolism and Insulin Secretion

2.2.1 *Free Fatty Acids and β Cells*

Fatty acids are believed to be one of the major endogenous energy sources for β cells. Free fatty acids (FFAs) gain entry to the β cell by freely diffusing through the plasma membrane because of their hydrophobic nature.[9] It is noteworthy that the exposure time of β cells to, and the chemical structure of, FFAs result in different effects on β cells, and in some cases their roles are shifted from protection to toxicity or from secretagogue to secretion inhibitor.[22,23]

2.2.2 *β oxidation of FFAs in β-Cell Mitochondria*

In a fasting/starvation state, FFAs are metabolized by β oxidation in the mitochondria to provide energy critical for cell function and basal insulin release. First, cytosolic FFAs are metabolized to long-chain acyl-CoA by acyl-CoA synthase and then transported to the mitochondrial matrix by carnitine palmitoyltransferase-1 (CPT-1).[24] Here, long-chain CoA (LC-CoA) is oxidized to produce CO_2, NADH, $FADH_2$, and ultimately ATP. The ATP so generated may contribute to sustained basal release of insulin in a traditional K_{ATP}-dependent manner.[9,11]

2.2.3 *A Critical Shift in FFA Metabolism Occurs after Glucose Influx: The "Trident Model" Hypothesis*

After ingestion of carbohydrate, followed by digestion and glucose influx into β cells, a shift occurs in β-cell lipid metabolism from oxidative "fuel" to indirect amplification of GSIS. The three major arms of FFA metabolism include TCA/malonyl-CoA metabolic signaling, glycerolipid/nonesterified fatty acid (NEFA) cycling, and activation of GPRs.[25]

2.2.3.1 TCA/Malonyl-CoA Metabolic Signaling

Citrate, produced by glycolysis and subsequent TCA metabolism, is converted to malonyl-CoA by acetyl-CoA carboxylase (ACC), and malonyl-CoA allosterically inhibits CPT-1, thus preventing further FFA transport into the mitochondria for β oxidation. Accumulation of cytosolic FFA can amplify insulin secretion by regulating ion channel proteins, increasing Ca^{2+} influx, and generating insulinotropic lipids such as LC-CoA and diacylglycerol (DAG).[11,25] DAG activates protein kinase C (PKC) and some key insulin vesicle docking and priming proteins such as MUnc-13, synaptosomal-associated protein 25, and synaptotagmin, thus insulin exocytosis is enhanced.[3,11,25,26]

AMP-activated protein kinase (AMPK) is a key molecular player in energy homeostasis. Generally, activated AMPK stimulates catabolic pathways (glycolysis, fatty acid oxidation, and mitochondrial biogenesis) and inhibits anabolic pathways (gluconeogenesis, glycogen, fatty acid, and protein synthesis).[27] AMPK is sensitive to the cellular energy status. When the AMP/ATP ratio is high, AMPK can increase FFA β oxidation by inhibiting ACC so, therefore reducing malonyl-CoA levels and additionally enhancing malonyl-CoA decarboxylase activity.[9,25]

2.2.3.2 Glycerolipid/Fatty Acid Cycling

The cycling of glycerolipid/fatty acids includes the formation of Gly-3-P from inbound glucose (approximately 25% of glucose carbon) via glycolysis, triacylglycerol formation via esterification reactions involving FFAs, and, on the other hand, generation of glycerol and FFAs from triglycerides by lipolysis.[11] This cycle is a convergence point of glucose and lipid metabolism in β cells. It can either provide LC-CoA and DAG from triacylglycerol hydrolysis to potentiate insulin exocytosis or generate MCFs such as NADH in β oxidation and TCA cycle–dependent process.[9,25]

2.2.3.3 G Protein–Coupled Receptor

Over the past several decades, cell-surface receptors have been demonstrated to play key role in FFA action on cellular function. This signaling pathway is separate from the cytosolic and mitochondrial metabolism of FFAs and contributes to the acute FFA-induced insulin secretion response. Recently, GPRs have been successfully identified as multiple cell surface receptors for FFAs, also named as FFA receptors (FFARs).[28] Among these, GPR40 (FFAR1), GPR 41 (FFAR3), GPR 119, and GPR 120 (FFAR4) are important in β-cell physiology.[22] GPR 40 is expressed in pancreatic β cells and mediates insulin secretion upon long-chain FFA stimulation (e.g., palmitate, palmitoleate, oleate). Overexpression of GPR40 or application of GPR40 agonists augments GSIS in normal or diabetic mice, whereas GPR40 knockout impairs glucose- and arginine-stimulated insulin secretion without affecting intercellular fuel metabolism.[28–30] It is believed that GRP40 can couple to the Ca^{2+}-mobilizing G protein, $G\alpha_{(q/11)}$, resulting in PKC and protein kinase D activation, both important for regulating cytoplasmic Ca^{2+} levels and insulin secretion.[31,32]

GPR120 displays 10% homology with GPR40 and has a high affinity for long-chain saturated and unsaturated FFAs. Recent studies have found messenger RNA (mRNA) expression of GPR120 in many β-cell lines and mouse pancreatic islets, suggesting its role may be related to sensing dietary fat and regulating insulin secretion.[33] Activation of GPR120 results in secretion of GLP-1, GIP, and CCK from enteroendocrine intestinal L and K cells through the phosphatidylinositide 3-kinase (PI3K)/Akt pathway, which promotes insulin secretion and enhances β-cell mass. However, recent reports suggest that further work is required to fully understand these regulatory pathways.[34–36]

2.3 Amino Acid Metabolism and Insulin Secretion

In addition to glucose and lipids, amino acids are the third class of nutrients that are essential for GSIS. However, individual and physiological concentrations of individual amino acids do not enhance GSIS, but when the concentration is elevated or in specific combinations, they can positively or negatively regulate both the triggering and amplification phases of GSIS. The insulinotropic mechanisms of amino acids mostly affect the increase in ATP, the generation of mitochondrial-derived stimulus secretion coupling factors, and the direct or indirect depolarization of plasma membrane.[4,13]

L-glutamine and L-alanine are quantitatively the most abundant amino acids in blood and extracellular fluids, closely followed by the branched chain amino acids.[37] Numerous in vitro and in vivo studies have demonstrated that L-alanine is a powerful insulin secretagogue acting by electrogenic Na^+ cotransport induced membrane depolarization and generation of ATP, glutamate, aspartate, and lactate.[38,39]

In rat islets, the metabolism of glutamine can generate aspartate (through TCA cycle metabolism and generation of oxaloacetate) and glutamate (via glutaminase).[13,40] In combination with leucine (an allosteric activator of glutamate dehydrogenase), insulin exocytosis is enhanced by increased TCA cycle activity, ATP production, and MCFs.[11] It is noteworthy that glutamine also contributes to β-cell antioxidant defense through providing glutamate for the γ-glutamyl cycle, thus potentiating glutathione synthesis.[41,42] Another important cellular antioxidant is cysteine. It is a direct substrate for glutathione synthesis. However, higher concentration of cysteine may generate excessive hydrogen sulfide (H_2S), which is reported to inhibit GSIS via multiple actions on the insulin secretory process, such as decreased ATP levels, disappearance of $[Ca^{2+}]_i$ oscillations, and attenuated glucose-induced hyperpolarization of the mitochondrial membrane potential.[13,43]

Arginine is a positively charged amino acid and can enter the β cell via the electrogenic transporter mCAT2A, resulting in direct membrane depolarization and subsequent Ca^{2+} influx. However, arginine-derived nitric oxide (NO) generated by activation of inducible nitric oxide synthase (iNOS) is a potential stimulant, but at higher concentrations is detrimental to insulin secretion.[13,44]

3. NUTRIENT REGULATION OF β-CELL GENE EXPRESSION

Glucose controls pancreatic β-cell function by regulating gene expression, enabling mammals to adapt metabolic activity to changes in nutrient supply. In pancreatic β cells, in addition to a fundamental role in the regulation of insulin secretion as described previously, glucose serves as a principal physiological regulator of insulin gene expression.[4,9] Glucose is known to control transcription factor recruitment, level of transcription, alternative splicing, and stability of insulin mRNA.[12] The following gives an overview of some notable aspects of this complex area of study.

In β cells, three transcriptional factors bind to the insulin promoter to regulate insulin gene expression: pancreatic and duodenal homeobox 1 (Pdx-1), neurogenic differentiation 1, and MafA (v-maf avain musculo aponeurotic fibrosarcoma oncogene homolog A), acting in synergy so stimulating insulin gene expression in response to increasing plasma glucose.[12,45,46,47] However, consistent with the detrimental β-cell response to prolonged exposure to high glucose concentrations, impairments of Pdx-1 and MafA binding to the insulin promoter have been noted, leading to decreased insulin biosynthesis, content, and secretion capacity. Similarly, prolonged exposure to high fatty acid levels can impair insulin gene expression, this time accompanied by an accumulation of triacylglycerol in β cells—particularly palmitate rich triacylglycerol—where a negative effect may be attributable to ceramide formation.[3,48] Moreover, palmitate is known to induce a decrease in binding activity of transcriptional factors with respect to the insulin promoter.[49]

An important role of amino acids with respect to gene expression has recently been highlighted.[50] In an Affymetrix microarray study using BRIN-BD11 β cells, prolonged (24 h) exposure to alanine or glutamine upregulated β-cell gene expression, particularly genes involved in metabolism, signal transduction, and oxidative stress.[50] Interestingly, 24-h exposure of BRIN-BD11 cells to glutamine strongly increased calcineurin (catalytic and regulatory subunits) mRNA expression, and this Ca^{2+}-binding protein has been reported to play a role in regulation of insulin exocytosis in mouse pancreatic β cells. Glutamine also increased Pdx-1 and ACC mRNA expression, both essential for insulin secretion via elevation of insulin transcription and generation of lipid signaling factors, respectively.

4. METABOLIC FAILURE IN β-CELL DYSFUNCTION AND ONSET OF DIABETES

4.1 A Brief Introduction to Type 1 and Type 2 Diabetes

Pancreatic β-cell failure is a major contributing factor to the onset and progression of both type 1 and type 2 diabetes (T1D and T2D). β-cell dysfunction refers to a reduction of insulin secretion and/or a failure to respond to plasma glucose, possibly leading to progressive β-cell death.[9]

T1D is caused by specific T cell—mediated autoimmune attack against β cells.[51] It is clearly demonstrated that islet infiltrated T cells and macrophages respond to islet-associated antigens, such as glutamic acid decarboxylase, insulinoma-associated antigen-2, zinc transporter 8, triggering pancreatic β-cell damage and associated reductions in insulin secretion capacity. Circulating B lymphocyte—generated anti-islet autoantibodies targeting the latter antigens are commonly detected in T1D.[52] Development of T1D may be also influenced by dietary factors including early infant dietary protein, low vitamin D, omega 3 polyunsaturated fatty acid intake, and duration of exposure to gluten.[53] Human leukocyte antigen DR-DQ haplotypes are another key issue,[54] possibly contributing up to 40% of T1D risk.[8]

In contrast, the development of T2D is more complex and takes a longer time from the initiating stage of compensatory response to metabolic excesses to the end-stage of endocrine and physiologic decompensation, characterized by loss of β-cell mass and function, resulting in excessive plasma glucose concentrations.[9]

Though the pathogenesis of T1D and T2D is different, there is some convergence in cellular signaling pathways responsible for reactive oxygen species (ROS) and reactive nitrogen species (RNS) generation, cytokine and transcriptional factor expression, and cell death. In this section, we will briefly address the metabolic failure of β cells and the onset of diabetes with a focus on T2D.

4.2 Glucotoxicity

All diabetic patients will share the common problem of abnormally high glucose levels. At early stages, even though fasting glucose and hemoglobin A1c levels might be within the normal range, postprandial glucose levels are often in excess of the normal range. The continual exposure to modest increases in blood glucose over a long period has detrimental effects on β cells, a process defined as "glucotoxicity."[55,56]

4.3 β-Cell Exhaustion

Increased metabolic load stimulates excessive insulin secretion so maintaining normo-glycemia, which is considered as a compensation response in the first stage of T2D. Constant entry of glucose into the β cell leads to a state of insensitivity to glucose, a process of "metabolic desensitization."[56] β-cell exhaustion happens when the insulin stores are permanently depleted after prolonged, chronic stimulation with glucose or other secretagogues. The important distinction between β-cell exhaustion and glucotoxicity is that the exhausted islets still maintain insulin synthesis to facilitate cell function, leading to possible cell recovery.[56] Nevertheless, glucose toxicity implies the gradual, time-dependent establishment of irreversible damage to the insulin production and secretion capacity, involving increased small heterodimer partner nuclear receptor mRNA, downregulation of GCK mRNA, reduced PDX-1 mRNA and insulin promoter binding activity, compromised mitochondrial function, and induction of apoptosis.[56–58] It is probable that dysfunctional β cells rather than death of the cells contribute more to the high levels of plasma glucose associated with diabetes.

4.4 Reactive Oxygen Species

Extensive studies have established that oxidative stress results in hyperglycemia-induced β-cell dysfunction. Physiologically, ROS such as superoxide anion ($O_2^{\bullet -}$), hydrogen peroxide (H_2O_2), and hydroxyl radicals, plus the related peroxynitrite ($ONOO^-$), contribute to the attenuation of key metabolic and physiologic events, including mitochondrial function.[59,60] ROS and RNS also have potential to damage cell membranes, DNA and protein structures, modulation of transcriptional factors such as nuclear factor-κB (NF-κB) leading to cell apoptosis.[8,60,61]

Pancreatic β cells are considered to be particularly vulnerable to oxidative damage. It has been reported that β cells express relatively low levels of antioxidant enzymes, including catalase and glutathione (GSH) peroxidase (GPx).[56,62,63] Increased glucose uptake stimulates increased glycolytic flux in β cells and subsequently generates metabolic coupling factors (as NADPH) to stimulate insulin secretion.[11,60,61] These accumulated reducing equivalents could also lead an enhancement of ROS production in β cells after chronic glucose exposure. Experimentally, increased hyperglycemia-linked superoxide production results in increased poly[ADP-ribose] polymerase (PARP)-1 activity, GAPDH ribosylation, and decreased GAPDH activity.[64] The inhibition of GAPDH results in activation of alternate glucose metabolism such as polyol, hexosamine, PKC activation, and advanced glycation end-product pathways,[65] driving oxidative damage and disrupting superoxide scavenging by antioxidants (e.g., GSH).[8]

Mitochondria are another source of ROS production.[66–68] Mitochondria membrane potential ($\Delta\psi$) is elevated following enhanced glucose metabolism, reducing equivalent formation (e.g., NADH), mitochondrial inner membrane electron transport, proton translocation, and increase in ATP synthase activity.[55] When electron transport exceeds threshold, the electron transfer in complex III is blocked. These electrons then escape the ETC to reduce molecular oxygen to superoxide. Further reaction with nitric oxide can generate peroxynitrite.

However, an additional enzyme-based system that can generate superoxide is NADPH oxidase (NOX), and NOX2 is the predominant isoform expressed in pancreatic β cells.[59] The in vivo experiments found that a reduction of islet NOX2 expression constituted an adaptive response to high level of metabolic overload. However, if the stimulus is maintained for long period, the failed adaptive response may be replaced by positive feedback, enhancing NOX2 expression and causing oxidative stress.[59]

The oxidative and nitrosative stresses activate the cytokine and apoptosis signaling pathways, thus causing dysfunction of β cells. The ROS-activated c-Jun N-terminal kinase (JNK) can phosphorylate Ser^{307} of insulin receptor substrate (IRS)-1, leading to impaired insulin–insulin receptor–IRS–PI3K-Akt signaling cascade, which results in an increase of Foxo-1–dependent gene expression and ultimately reduced PDX-1 translocation to nucleus.[8,69,70]

4.5 Maturity Onset Diabetes of the Young

In addition to T1D and T2D, relatively rare monogenic forms of diabetes, named maturity onset diabetes of the young (MODY),[71] have been described with up to 80% cases characterized by heterozygous mutations in *GCK* or *HNF1A/4A* genes.[72] GCK-MODY manifests a mild and nonprogressive fasting hyperglycemia from birth, requiring little or no treatment outside of diet and exercise to avoid late complications. Heterozygous loss-of-function mutations in GCK, the glucose "sensor" and the rate-limiting enzyme in glycolysis, result in a decreased rate of glucose phosphorylation and subsequently the glycemic threshold for insulin release is not reached.[73,74] To date, more than 600 significant mutations have been identified throughout the 10 exons and the promoter of the pancreatic β-cell GCK gene. Nevertheless, the compensatory activity of the unaffected GCK allele can maintain the glucose level under homeostatic control.[75,76]

Another type of MODY is HNF1A-MODY, which refers to heterozygous mutations in the gene of transcription factor hepatocyte NF-1 (*HNF1A*),[77] resulting in abnormal gene expression for proteins involved in glucose transport and glucose metabolism in the cytosol and the mitochondria. Reduced β-cell proliferation and exacerbated apoptosis result in progressive decline in islet insulin secretion. These patients have a normal glucose level at birth, but undergo a progressive deficit of insulin secretion accompanied with time.[72]

4.6 Lipotoxicity

We have already discussed FFAs from the perspective of insulin secretagogue action in pancreatic β cells. However, when lipids are ectopically accumulated in nonadipose tissue such as islets, muscle, and liver, the resulting cytotoxic effects are termed "lipotoxicity."[3,9,78] The molecular findings associated with the process of lipotoxicity may help to explain the links between obesity and T2D. It is noteworthy that lipotoxicity encompasses toxicity not only from an excess of neutral lipids such as triacylglycerol, but also from endogenous lipid synthesis from glucose through the process of lipogenesis.

4.6.1 Lipotoxicity is Dependent on Lipid Structure and Exposure Time

Lipotoxicity is dependent on the fatty acid chain length. Long-chain saturated NEFAs, such as palmitic acid, exhibit a strong cytotoxic effect upon insulin-producing cells, whereas short-chain as well as unsaturated NEFAs are well-tolerated. Moreover, long-chain unsaturated NEFAs such as arachidonic or oleic acid counteract the toxicity of palmitic acid. This protective potency decreases with shortening of the chain length.[22,79,80,81]

Different types of FFAs are metabolized at different organelles. Short- and medium-chain (C4–C8) fatty acids are exclusively β-oxidized in the mitochondria, whereas long-chain (C10–C16) fatty acids are β-oxidized in the mitochondria and the peroxisomes (C14–C16). Very long-chain (C17–C24) fatty acids are preferentially processed by peroxisomes.[22,82,83] The effects of long-chain saturated NEFAs are also time-dependent, which means that in β cells the short-term (1 to several hours) effect is via an enhancement of GSIS via GPR40-mediated signaling pathways[28] (discussed in last section) and the long-term (1–3 days) effect is the induction of lipotoxicity via an enhancement of mitochondrial and peroxisomal metabolism, indirectly yielding high levels of H_2O_2, as well as inducing endoplasmic reticulum (ER) stress, ultimately leading to apoptosis. Interestingly, evidence from in vitro and animal studies demonstrated that GPR40 and GPR120 are not involved in the toxicity of NEFAs. Studies with GPR40 KO mice demonstrated that after 3 days of exposure to fatty acids, the level of lipotoxicity was not reduced.[84]

4.6.2 Reactive Oxygen Species

The most important intracellular sites of $O_2^{\bullet-}$ generation are complexes I and III of the mitochondrial respiratory chain (ETC). When NEFAs undergo β oxidation in mitochondria, they induce electron flux through the ETC. Chronic lipid exposure facilitates mitochondrial-associated one-electron reduction of O_2 to $O_2^{\bullet-}$. In addition, NEFAs cause detachment of cardiolipin-bound cytochrome *c* from the outer side of the inner membrane, thus further impeding electron transport along the ETC.[22,85]

As mentioned in the mechanism of glucotoxicity, plasma membrane NOX is stimulated by exposure to various metabolic intermediates and coupling factors.[86] It has been reported that high levels of palmitic acid also increase the activity of NOX via PKC-dependent activation. In a clonal rat pancreatic β-cell line, BRIN-BD11, one of the NOX components ($p47^{phox}$) was elevated after a 24-h treatment with palmitic acid, and ROS was remarkably increased in dose- and time-dependent manner.[3,60]

In addition to ECT and NOX-induced ROS, another important ROS production site (in response to long-chain NEFAs) is the peroxisomes. Long- and very long-chain NEFAs are more toxic than others. The peroxisomes contain the enzymes for lipid oxidation and biosynthesis of glycerolipids, and H_2O_2 is a by-product of the oxidation process.[22] The detoxification of H_2O_2 is mediated either through GPx in the cytosol and mitochondria, or catalase in the peroxisomes. However, both GPx and catalase are expressed at low levels in pancreatic β cells.[87] It is therefore possible that the increased H_2O_2 production via NEFA metabolism can react in an iron-catalyzed reaction with the superoxide radical ($O_2^{\bullet-}$) yielding the highly reactive hydroxyl radical ($OH^{\bullet}$).

4.7 ER Stress

The ER is an important organelle in terms of the biosynthesis of insulin and other secreted proteins and engagement in a wide range of cellular processes such as transcriptional activity, stress responses, and apoptosis.[88] Pancreatic β cells are extremely sensitive to imbalance of the ER system because the ER in β cells operates close to capacity because of insulin synthesis, folding, and secretion. When the chaperone system becomes overloaded, the accumulated misfolded proteins trigger the unfolded protein response (UPR), ultimately leading to cell apoptosis (Figure 2). Over the past decade, studies have reported that ER stress is a key factor in lipotoxicity-associated β-cell death and is the most important link between insulin resistance and β-cell loss, the two fundamental drivers of T2D.[89]

The classic pathway of ER stress and UPR is widely documented.[8,88] Three major signaling molecules are involved in UPR: protein kinase RNA-like ER kinase (PERK), activating transcription factor 6 (ATF6), and the serine/threonine-protein kinase/endoribonuclease IRE1.[89] They sense the accumulated misfolded proteins via their lumenal domain. ATF6 is activated by release of Bip from the protein chaperone Bip/Grp78, and then transfers from ER to Golgi for proteolytic cleavage, which allows it to release its basic leucine zipper domain and migrates into nucleus to

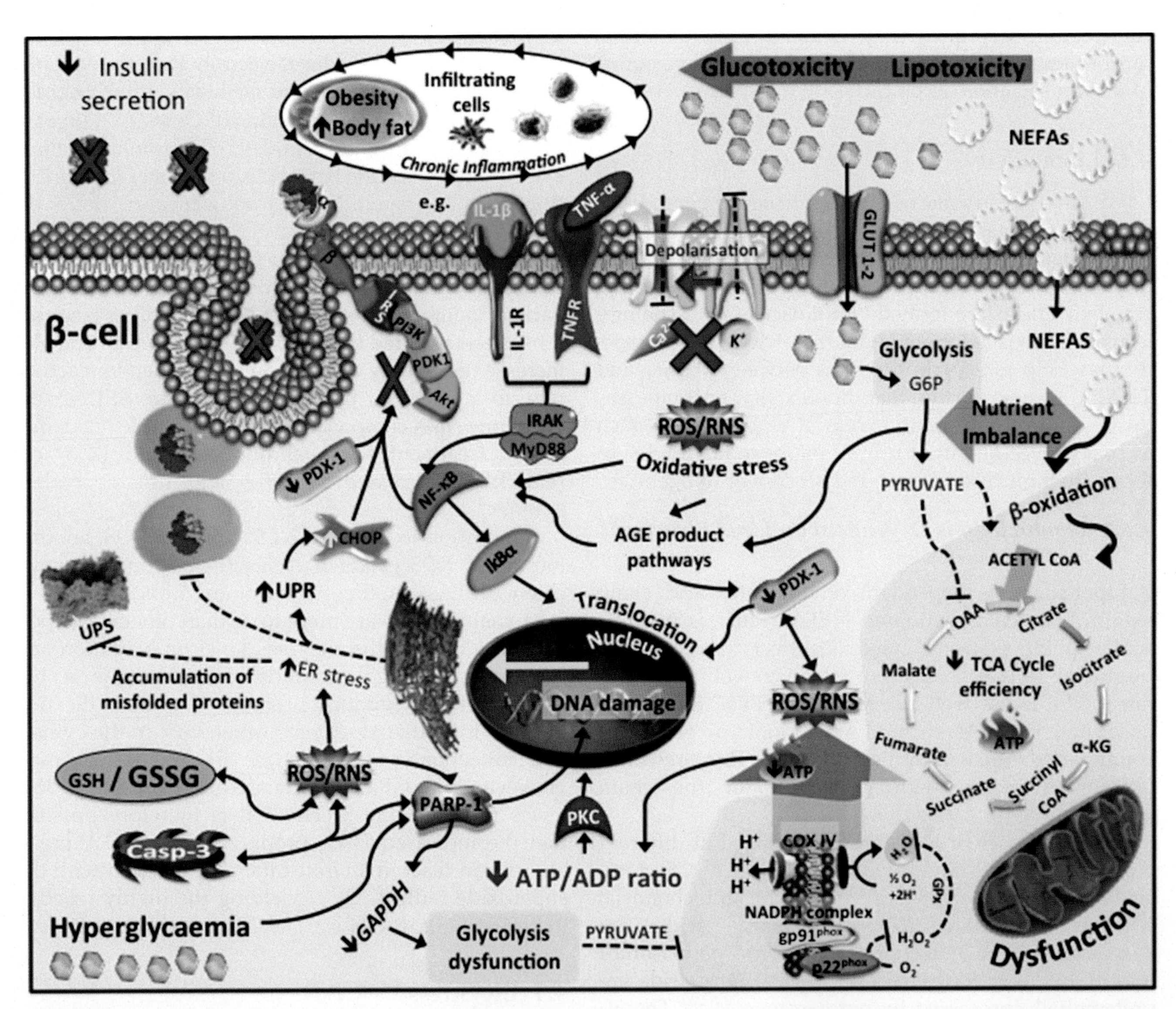

FIGURE 2 β-cell dysfunction mediated by glucotoxicity and lipotoxicity. Lipids and glucose in excess, as observed in obesity-induced diabetes, promote inflammation and directly affects β-cell function until its exhaustion for insulin secretion. Over time, β cells, through apoptotic pathways, lose the ability to secrete insulin (see the text for more details). AKT, protein kinase B; CASP-3, caspase 3; COX IV, cytochrome oxidase IV; ER, endoplasmic reticulum; G6P, glucose 6-phosphate; GAPDH, glyceraldehyde 3-phosphate dehydrogenase; GLUT 1-2, glucose transporter 1 or 2; GSH, glutathione; GSSG, oxidized glutathione; H^+, hydrogen ions; H_2O_2, hydrogen peroxide; IκBα, nuclear factor of kappa light polypeptide gene enhancer in B-cells inhibitor alpha; IL-1β and receptor, interleukin-1β; IRAK, IL-1 receptor–associated kinase 1; IRS, insulin receptor substrates; MYD88, myeloid differentiation primary response gene 88; NEFA, nonesterified fatty acid; NF-κB, nuclear factor-κB; O_2^-, anion superoxide; OAA, oxaloacetate; PARP-1, poly [ADP-ribose] polymerase 1; PDK1, 3-phosphoinositide dependent protein kinase-1; PDX-1, pancreatic and duodenal homeobox 1; PI3K, phosphatidylinositide 3-kinase; PKC, protein kinase C; TNF, tumor necrosis factor; TNFR, tumor necrosis factor receptor; UPR, unfolded protein response; UPS, ubiquitin-proteasome system.

stimulate gene transcription. IRE1 is directly activated and subsequently alternatively splices and nuclear translocates X-box binding protein 1, another transcription factor. After PERK is activated, it phosphorylates the translational factor eukaryotic initiation factor 2 (eIF2)-α and transiently blocks general translation. The initial stage of UPR is an adaptive enhancement of ER capacity through upregulation of protein-folding chaperones and ER-associated degradation. If ER stress is exacerbated, apoptosis is triggered as the terminal stage of UPR. The reported downstream molecular mechanism includes activation of transcription factor CCAAT/-enhancer-binding protein homologous protein (CHOP), protease, caspase 12, and the signaling pathway including tumor necrosis factor (TNF) receptor-associated factor 2 and JNK.[90]

Many ER stressors, including nitric oxide, cytokines, and glucotoxic stimuli may cause a decrease in levels of the calcium pump SERCA2 and depletion of ER Ca^{2+} content, thus compromising the folding capacity of ER. However, the exact mechanism of chronic lipid exposure impaired ER to Golgi protein trafficking is still controversial. Recent studies revealed that localized reductions in sphingomyelin rather than phospholipids play the key role in this process. Particularly, a loss of sphingomyelin disrupts ER lipid rafts that are essential for the correct packaging of secretory cargo into export vesicles. This effect contributes to defective protein trafficking, ER stress, and apoptosis in β cells.[91]

5. THE CROSS-TALK OF APOPTOSIS WITH ROS AND ER STRESS IN β-CELL DYSFUNCTION

Diverse stimuli, such as glucose, lipids, oxidative and nitrosative stress, and autoimmune-associated antibodies (in T1D) can induce cell apoptosis, contributing to the loss of pancreatic β cells in both T1D and T2D.[92,93] The final stages of glucotoxicity and lipotoxicity in T2D result in progressive cell death, but the mechanisms involved are different from autoimmune response induced apoptosis in T1D.

There are extrinsic and intrinsic pathways responsible for apoptosis.[94,95] In the extrinsic or mitochondrial-independent pathway, the immune factors Fas ligand, TNF-α, or TNF-related apoptosis-inducing ligand bind to the Fas and TNF-αR1 cell-surface death receptors, leading to the formation of a homotrimeric ligand—receptor complex that recruits further cytosolic factors, such as Fas-associated death domain protein and caspase-8 to form an oligomeric death-inducing signaling complex (DISC). DISC activates the initiator caspase, caspase 8, which then cleaves the effector caspase-3.

The intrinsic or mitochondrial-dependent pathway is triggered in response to a wide range of death stimuli that are generated from within cell. The BH3-only proteins (suchasBIDandBIM)transducethesignaltomitochondria after receiving apoptotic stimuli, then the pro-apoptotic members of the BCL2 family mediate the release of cytochrome*c* from the mitochondrial membrane; subsequently, caspase-9 is activated resulting in the activation of caspase-3. In T1D, β-cell apoptosis is caused by direct contact with islet-associated immune cells that secret soluble mediators including oxygen free radicals, NO, and cytokines such as interleukin (IL)-1β, interferon-γ, and TNF-α. IL-1β can activate NF-κB, contributing to upregulation of iNOS and chemokines.[8,93]

Hyperglycemia shares some common pathways with cytokine induced β-cell apoptosis, including C-Myc, A20, and hemooxygenase expression. However, iNOS and nuclear factor of kappa light polypeptide gene enhancer in B-cells inhibitor alpha, two NF-κB—dependent genes induced by IL-1β, are not induced (at least in rodent β cells) exposed to high glucose, suggesting that glucotoxicity-related β-cell apoptosis is NF-κB—independent.[93,96] Extensive evidence indicates that hyperglycemia-induced apoptosis is mainly due to the oxidative stress-induced cytokines causing activation of JNK that phosphorylates signal transducer and activator of transcription 1, affecting caspases and Bcl-2 family proteins, ultimately resulting in intrinsic β-cell apoptosis.[97] Recent studies on glucolipotoxicity proposed that NOX2 in the β cell (NADPH oxidase) is the "trigger" of cellular dysfunction of the β cell. Activation of Rac-1 and NOX2 under glucose and lipid overload conditions generate ROS, then activating stress kinases (p38, JNK1/2, and p53 kinase) that result in loss of membrane potential and release of cytochrome-*c* in mitochondria, activating intrinsic apoptosis.[69,92,98]

The mechanisms linking ER stress and β-cell apoptosis remain elusive.[89] CHOP represses the antioxidant genes and induces growth arrest and DNA damage-inducible protein (GADD34) expression, which promotes dephosphorylation of eIF2α.[99] The subsequent increased oxidative stress not only activates cell death, but also further compromises ER capacity for protein folding. Other downstream mechanisms of ER stress involve loss of myeloid cell leukemia sequence 1, an antiapoptotic member of the BH3 family, and upregulation of p53-upregulated modulator of apoptosis and DP5/Hrk, which are downstream of the PERK pathway.[100] The transcription profile of palmitate-regulated genes in β-cell dysfunction revealed that many of these genes were responsive to NF-κB transcription regulation,[49] in contrast to glucotoxicity.

Other studies also demonstrate that during stearic acid—induced apoptosis, JNK inhibition neither decreased the rate of apoptosis or rate of activation of caspases-8, -9, -7, or -2, or PARP cleavage. Furthermore, inhibition of JNK activity did not affect CHOP expression. These controversial results suggest that different mechanisms are involved in glucotoxicity and lipotoxicity induced β-cell dysfunction and cell death.

6. CONCLUDING REMARKS

Pancreatic β cells are sensitive to nutrient supply (and indeed oversupply) and work in a precise and coordinated manner to maintain glucose homeostasis in the body. Glucose, lipid, and amino acid metabolism play a key role in insulin secretion by production of key MCFs, regulation of insulin gene expression, protein synthesis, and the mechanism of insulin vesicle translocation and exocytosis. The K_{ATP}^{+} – dependent pathway of GSIS controls the first phase of the insulin secretion and is dependent on rapid glycolytic metabolism. In contrast, the amplifying signals that contribute to potentiation of insulin release in the second phase are still under investigation, involving G protein and protein kinase signaling and generation of MCFs in the mitochondria. Defects in insulin production, secretion, and defects in viability and mass of β cells lead to diabetes mellitus. Constant nutrient overload and diminished physical activity have led to an astonishing increased prevalence of T2D all over the world. Therefore, a better understanding of the complexities of β-cell dysfunction in T2D is required. In this chapter, we briefly summarized the basic knowledge and recent advances in term of the pathogenesis of β-cell dysfunction. Nutrient-dependent generation and elevations in ROS/RNS from mitochondrial and nonmitochondrial sources, ER stress, and elevated apoptosis are the most important events leading to glucotoxicity and lipotoxicity in T2D. Hence, therapeutic strategies targeting the molecules involved in excess nutrient dependent β-cell failure are promising for the development of novel antidiabetic treatment modalities.

References

1. McClenaghan NH, Flatt PR. Engineering cultured insulin-secreting pancreatic B-cell lines. *J Mol Med* 1999;**77**(1):235–43.
2. McClenaghan NH, Flatt PR. Physiological and pharmacological regulation of insulin release: insights offered through exploitation of insulin-secreting cell lines. *Diabetes Obes Metab* 1999;**1**(3):137–50.
3. Newsholme P, Keane D, Welters HJ, Morgan NG. Life and death decisions of the pancreatic beta-cell: the role of fatty acids. *Clin Sci* 2007;**112**(1):27–42.
4. Newsholme P, Krause M. Nutritional regulation of insulin secretion: implications for diabetes. *Clin Biochem Rev* 2012;**33**(2):35–47.
5. Salvucci M, Neufeld Z, Newsholme P. Mathematical model of metabolism and electrophysiology of amino acid and glucose stimulated insulin secretion: in vitro validation using a beta-cell line. *PloS One* 2013;**8**(3):e52611.
6. Rorsman P, Braun M. Regulation of insulin secretion in human pancreatic islets. *Annu Rev Physiol* 2013;**75**:155–79.
7. De Vos A, Heimberg H, Quartier E, et al. Human and rat beta cells differ in glucose transporter but not in glucokinase gene expression. *J Clin Invest* 1995;**96**(5):2489–95.
8. Fu Z, Gilbert ER, Liu D. Regulation of insulin synthesis and secretion and pancreatic Beta-cell dysfunction in diabetes. *Curr Diabetes Rev* 2013;**9**(1):25–53.
9. Keane K, Newsholme P. Metabolic regulation of insulin secretion. *Vitam Horm* 2014;**95**:1–33.
10. Jensen MV, Joseph JW, Ronnebaum SM, Burgess SC, Sherry AD, Newgard CB. Metabolic cycling in control of glucose-stimulated insulin secretion. *Am J Physiol Endocrinol Metab* 2008;**295**(6): E1287–97.
11. Newsholme P, Cruzat V, Arfuso F, Keane K. Nutrient regulation of insulin secretion and action. *J Endocrinol* 2014;**221**(3):R105–20.
12. Martens GA, Pipeleers D. Glucose, regulator of survival and phenotype of pancreatic beta cells. *Vitam Horm* 2009;**80**:507–39.
13. Newsholme P, Abdulkader F, Rebelato E, et al. Amino acids and diabetes: implications for endocrine, metabolic and immune function. *Front Biosci (Landmark Ed)* 2011;**16**:315–39.
14. Komatsu M, Takei M, Ishii H, Sato Y. Glucose-stimulated insulin secretion: a newer perspective. *J Diabetes Invest* 2013;**4**(6):511–6.
15. Henquin JC. Regulation of insulin secretion: a matter of phase control and amplitude modulation. *Diabetologia* 2009;**52**(5): 739–51.
16. Henquin JC. The dual control of insulin secretion by glucose involves triggering and amplifying pathways in beta-cells. *Diabetes Res Clin Pract* 2011;**93**(Suppl. 1):S27–31.
17. Meloni AR, DeYoung MB, Lowe C, Parkes DG. GLP-1 receptor activated insulin secretion from pancreatic beta-cells: mechanism and glucose dependence. *Diabetes Obes Metab* 2013;**15**(1):15–27.
18. McIntosh CH, Widenmaier S, Kim SJ. Glucose-dependent insulinotropic polypeptide signaling in pancreatic beta-cells and adipocytes. *J Diabetes Invest* 2012;**3**(2):96–106.
19. Tsushima RG. Second-phase insulin secretion gets cool. *Am J Physiol Endocrinol Metab* 2011;**301**(6):E1070–1.
20. Ullrich S. Glucose-induced insulin secretion: is the small G-protein Rab27A the mediator of the K(ATP) channel-independent effect? *J Physiol* 2008;**586**(Pt 22):5291.
21. Wang Z, Thurmond DC. Differential phosphorylation of RhoGDI mediates the distinct cycling of Cdc42 and Rac1 to regulate second-phase insulin secretion. *J Biol Chem* 2010;**285**(9):6186–97.
22. Gehrmann W, Elsner M, Lenzen S. Role of metabolically generated reactive oxygen species for lipotoxicity in pancreatic beta-cells. *Diabetes Obes Metab* 2010;**12**(Suppl. 2):149–58.
23. Dhayal S, Welters HJ, Morgan NG. Structural requirements for the cytoprotective actions of mono-unsaturated fatty acids in the pancreatic beta-cell line, BRIN-BD11. *Br J Pharmacol* 2008;**153**(8): 1718–27.
24. Berne C. The metabolism of lipids in mouse pancreatic islets. The biosynthesis of triacylglycerols and phospholipids. *Biochem J* 1975;**152**(3):667–73.
25. Nolan CJ, Madiraju MS, Delghingaro-Augusto V, Peyot ML, Prentki M. Fatty acid signaling in the beta-cell and insulin secretion. *Diabetes* 2006;**55**(Suppl. 2):S16–23.
26. Nolan CJ, Prentki M. The islet beta-cell: fuel responsive and vulnerable. *Trends Endocrinol Metab* 2008;**19**(8):285–91.
27. Lim CT, Kola B, Korbonits M. AMPK as a mediator of hormonal signalling. *J Mol Endocrinol* 2010;**44**(2):87–97.
28. Feng XT, Leng J, Xie Z, Li SL, Zhao W, Tang QL. GPR40: a therapeutic target for mediating insulin secretion (review). *Int J Mol Med* 2012;**30**(6):1261–6.

29. Del Guerra S, Bugliani M, D'Aleo V, et al. G-protein-coupled receptor 40 (GPR40) expression and its regulation in human pancreatic islets: the role of type 2 diabetes and fatty acids. *Nutr Metab Cardiovasc* 2010;**20**(1):22–5.
30. Latour MG, Alquier T, Oseid E, et al. GPR40 is necessary but not sufficient for fatty acid stimulation of insulin secretion in vivo. *Diabetes* 2007;**56**(4):1087–94.
31. Schnell S, Schaefer M, Schofl C. Free fatty acids increase cytosolic free calcium and stimulate insulin secretion from beta-cells through activation of GPR40. *Mol Cell Endocrinol* 2007;**263**(1–2): 173–80.
32. Fujiwara K, Maekawa F, Yada T. Oleic acid interacts with GPR40 to induce Ca2+ signaling in rat islet beta-cells: mediation by PLC and L-type Ca2+ channel and link to insulin release. *Am J Physiol Endocrinol Metab* 2005;**289**(4):E670–7.
33. Moran BM, Abdel-Wahab YH, Flatt PR, McKillop AM. Evaluation of the insulin-releasing and glucose-lowering effects of GPR120 activation in pancreatic beta-cells. *Diabetes Obes Metab* 2014; **16**(11):1128–39.
34. Greenhill C. Diabetes: Gpr120 agonist has anti-inflammatory and insulin-sensitizing effects. *Nat Rev Endocrinol* 2014;**10**(9):510.
35. Oh da Y, Walenta E, Akiyama TE, et al. A Gpr120-selective agonist improves insulin resistance and chronic inflammation in obese mice. *Nat Med* 2014;**20**(8):942–7.
36. Sanchez-Reyes OB, Romero-Avila MT, Castillo-Badillo JA, et al. Free fatty acids and protein kinase C activation induce GPR120 (free fatty acid receptor 4) phosphorylation. *Eur J Pharmacol* 2014;**723**:368–74.
37. Dixon G, Nolan J, McClenaghan N, Flatt PR, Newsholme P. A comparative study of amino acid consumption by rat islet cells and the clonal beta-cell line BRIN-BD11-the functional significance of L-alanine. *J Endocrinol* 2003;**179**(3):447–54.
38. Keane D, Newsholme P. Saturated and unsaturated (including arachidonic acid) non-esterified fatty acid modulation of insulin secretion from pancreatic beta-cells. *Biochem Soc Trans* 2008; **36**(Pt 5):955–8.
39. Brennan L, Shine A, Hewage C, et al. A nuclear magnetic resonance-based demonstration of substantial oxidative L-alanine metabolism and L-alanine-enhanced glucose metabolism in a clonal pancreatic beta-cell line: metabolism of L-alanine is important to the regulation of insulin secretion. *Diabetes* 2002;**51**(6):1714–21.
40. McClenaghan NH, Barnett CR, O'Harte FP, Flatt PR. Mechanisms of amino acid-induced insulin secretion from the glucose-responsive BRIN-BD11 pancreatic B-cell line. *J Endocrinol* 1996; **151**(3):349–57.
41. Brennan L, Corless M, Hewage C, et al. 13C NMR analysis reveals a link between L-glutamine metabolism, D-glucose metabolism and gamma-glutamyl cycle activity in a clonal pancreatic beta-cell line. *Diabetologia* 2003;**46**(11):1512–21.
42. Cruzat VF, Keane KN, Scheinpflug AL, Cordeiro R, Soares M, Newsholme P. Alanyl-Glutamine improves pancreatic beta-cell function following ex vivo inflammatory challenge. *J Endocrinol* 2014.
43. Kaneko Y, Kimura Y, Kimura H, Niki I. L-cysteine inhibits insulin release from the pancreatic beta-cell: possible involvement of metabolic production of hydrogen sulfide, a novel gasotransmitter. *Diabetes* 2006;**55**(5):1391–7.
44. Sener A, Best LC, Yates AP, et al. Stimulus-secretion coupling of arginine-induced insulin release: comparison between the cationic amino acid and its methyl ester. *Endocrine* 2000;**13**(3): 329–40.
45. Andrali SS, Sampley ML, Vanderford NL, Ozcan S. Glucose regulation of insulin gene expression in pancreatic beta-cells. *Biochem J* 2008;**415**(1):1–10.
46. Shao S, Fang Z, Yu X, Zhang M. Transcription factors involved in glucose-stimulated insulin secretion of pancreatic beta cells. *Biochem Biophys Res Commun* 2009;**384**(4):401–4.
47. Im SS, Kim SY, Kim HI, Ahn YH. Transcriptional regulation of glucose sensors in pancreatic beta cells and liver. *Curr Diabetes Rev* 2006;**2**(1):11–8.
48. Shao S, Liu Z, Yang Y, Zhang M, Yu X. SREBP-1c, Pdx-1, and GLP-1R involved in palmitate-EPA regulated glucose-stimulated insulin secretion in INS-1 cells. *J Cell Biochem* 2010;**111**(3):634–42.
49. Choi HJ, Hwang S, Lee SH, et al. Genome-wide identification of palmitate-regulated immediate early genes and target genes in pancreatic beta-cells reveals a central role of NF-kappaB. *Mol Biol Rep* 2012;**39**(6):6781–9.
50. Corless M, Kiely A, McClenaghan NH, Flatt PR, Newsholme P. Glutamine regulates expression of key transcription factor, signal transduction, metabolic gene, and protein expression in a clonal pancreatic beta-cell line. *J Endocrinol* 2006;**190**(3):719–27.
51. Kawasaki E. Type 1 diabetes and autoimmunity. *Clin Pediatr Endocrinol* 2014;**23**(4):99–105.
52. Wallberg M, Cooke A. Immune mechanisms in type 1 diabetes. *Trends Immunol* 2013;**34**(12):583–91.
53. Littorin B, Blom P, Scholin A, et al. Lower levels of plasma 25-hydroxyvitamin D among young adults at diagnosis of autoimmune type 1 diabetes compared with control subjects: results from the nationwide Diabetes Incidence Study in Sweden (DISS). *Diabetologia* 2006;**49**(12):2847–52.
54. Pugliese A. Genetics of type 1 diabetes. *Endocrinol Metab Clin North Am* 2004;**33**(1):1–16. vii.
55. Fridlyand LE, Philipson LH. Does the glucose-dependent insulin secretion mechanism itself cause oxidative stress in pancreatic beta-cells? *Diabetes* 2004;**53**(8):1942–8.
56. Robertson RP, Harmon J, Tran PO, Tanaka Y, Takahashi H. Glucose toxicity in beta-cells: type 2 diabetes, good radicals gone bad, and the glutathione connection. *Diabetes* 2003;**52**(3):581–7.
57. Robertson RP, Zhang HJ, Pyzdrowski KL, Walseth TF. Preservation of insulin mRNA levels and insulin secretion in HIT cells by avoidance of chronic exposure to high glucose concentrations. *J Clin Invest* 1992;**90**(2):320–5.
58. Olson LK, Redmon JB, Towle HC, Robertson RP. Chronic exposure of HIT cells to high glucose concentrations paradoxically decreases insulin gene transcription and alters binding of insulin gene regulatory protein. *J Clin Invest* 1993;**92**(1):514–9.
59. Newsholme P, Rebelato E, Abdulkader F, Krause M, Carpinelli A, Curi R. Reactive oxygen and nitrogen species generation, antioxidant defenses, and beta-cell function: a critical role for amino acids. *J Endocrinol* 2012;**214**(1):11–20.
60. Newsholme P, Morgan D, Rebelato E, et al. Insights into the critical role of NADPH oxidase(s) in the normal and dysregulated pancreatic beta cell. *Diabetologia* 2009;**52**(12):2489–98.
61. Newsholme P, Haber EP, Hirabara SM, et al. Diabetes associated cell stress and dysfunction: role of mitochondrial and non-mitochondrial ROS production and activity. *J Physiol* 2007; **583**(Pt 1):9–24.
62. Ivarsson R, Quintens R, Dejonghe S, et al. Redox control of exocytosis: regulatory role of NADPH, thioredoxin, and glutaredoxin. *Diabetes* 2005;**54**(7):2132–42.
63. Lenzen S. Oxidative stress: the vulnerable beta-cell. *Biochem Soc Trans* 2008;**36**(Pt 3):343–7.
64. Robertson RP. Chronic oxidative stress as a central mechanism for glucose toxicity in pancreatic islet beta cells in diabetes. *J Biol Chem* 2004;**279**(41):42351–4.
65. Tajiri Y, Moller C, Grill V. Long-term effects of aminoguanidine on insulin release and biosynthesis: evidence that the formation of advanced glycosylation end products inhibits B cell function. *Endocrinology* 1997;**138**(1):273–80.

66. Jezek P, Olejar T, Smolkova K, et al. Antioxidant and regulatory role of mitochondrial uncoupling protein UCP2 in pancreatic beta-cells. *Physiol Res* 2014;**63**(Suppl. 1):S73–91.
67. Blake R, Trounce IA. Mitochondrial dysfunction and complications associated with diabetes. *Biochim Biophys Acta* 2014; **1840**(4):1404–12.
68. Maechler P. Mitochondrial signal transduction in pancreatic beta-cells. *Best Pract Res Clin Endocrinol Metab* 2012;**26**(6):739–52.
69. Syed I, Kyathanahalli CN, Jayaram B, et al. Increased phagocyte-like NADPH oxidase and ROS generation in type 2 diabetic ZDF rat and human islets: role of Rac1-JNK1/2 signaling pathway in mitochondrial dysregulation in the diabetic islet. *Diabetes* 2011; **60**(11):2843–52.
70. Arora DK, Machhadieh B, Matti A, Wadzinski BE, Ramanadham S, Kowluru A. High glucose exposure promotes activation of protein phosphatase 2A in rodent islets and INS-1 832/13 beta-cells by increasing the posttranslational carboxylmethylation of its catalytic subunit. *Endocrinology* 2014;**155**(2):380–91.
71. Tattersall RB, Fajans SS. A difference between the inheritance of classical juvenile-onset and maturity-onset type diabetes of young people. *Diabetes* 1975;**24**(1):44–53.
72. McDonald TJ, Ellard S. Maturity onset diabetes of the young: identification and diagnosis. *Ann Clin Biochem* 2013;**50**(Pt 5):403–15.
73. Ellard S, Bellanne-Chantelot C, Hattersley AT. European Molecular Genetics Quality Network Mg. Best practice guidelines for the molecular genetic diagnosis of maturity-onset diabetes of the young. *Diabetologia* 2008;**51**(4):546–53.
74. Velho G, Petersen KF, Perseghin G, et al. Impaired hepatic glycogen synthesis in glucokinase-deficient (MODY-2) subjects. *J Clin Invest* 1996;**98**(8):1755–61.
75. Osbak KK, Colclough K, Saint-Martin C, et al. Update on mutations in glucokinase (GCK), which cause maturity-onset diabetes of the young, permanent neonatal diabetes, and hyperinsulinemic hypoglycemia. *Hum Mutat* 2009;**30**(11):1512–26.
76. Gloyn AL. Glucokinase (GCK) mutations in hyper- and hypoglycemia: maturity-onset diabetes of the young, permanent neonatal diabetes, and hyperinsulinemia of infancy. *Hum Mutat* 2003;**22**(5): 353–62.
77. Lopez-Garrido MP, Herranz-Antolin S, Alija-Merillas MJ, Giralt P, Escribano J. Co-inheritance of HNF1a and GCK mutations in a family with maturity-onset diabetes of the young (MODY): implications for genetic testing. *Clin Endocrinol* 2013;**79**(3):342–7.
78. Unger RH, Clark GO, Scherer PE, Orci L. Lipid homeostasis, lipotoxicity and the metabolic syndrome. *Biochim Biophys Acta* 2010; **1801**(3):209–14.
79. Ortsater H. Arachidonic acid fights palmitate: new insights into fatty acid toxicity in beta-cells. *Clin Sci* 2011;**120**(5):179–81.
80. Dixon G, Nolan J, McClenaghan NH, Flatt PR, Newsholme P. Arachidonic acid, palmitic acid and glucose are important for the modulation of clonal pancreatic beta-cell insulin secretion, growth and functional integrity. *Clin Sci* 2004;**106**(2):191–9.
81. Keane DC, Takahashi HK, Dhayal S, Morgan NG, Curi R, Newsholme P. Arachidonic acid actions on functional integrity and attenuation of the negative effects of palmitic acid in a clonal pancreatic beta-cell line. *Clin Sci* 2011;**120**(5):195–206.
82. Eaton S, Bartlett K, Pourfarzam M. Mammalian mitochondrial beta-oxidation. *Biochem J* 1996;**320**(Pt 2):345–57.
83. Wanders RJ, Waterham HR. Biochemistry of mammalian peroxisomes revisited. *Annu Rev Biochem* 2006;**75**:295–332.
84. Tan CP, Feng Y, Zhou YP, et al. Selective small-molecule agonists of G protein-coupled receptor 40 promote glucose-dependent insulin secretion and reduce blood glucose in mice. *Diabetes* 2008; **57**(8):2211–9.
85. Rial E, Rodriguez-Sanchez L, Gallardo-Vara E, Zaragoza P, Moyano E, Gonzalez-Barroso MM. Lipotoxicity, fatty acid uncoupling and mitochondrial carrier function. *Biochim Biophys Acta* 2010;**1797**(6–7):800–6.
86. Schrauwen P, Hesselink MK. Oxidative capacity, lipotoxicity, and mitochondrial damage in type 2 diabetes. *Diabetes* 2004;**53**(6): 1412–7.
87. Tiedge M, Lortz S, Drinkgern J, Lenzen S. Relation between antioxidant enzyme gene expression and antioxidative defense status of insulin-producing cells. *Diabetes* 1997;**46**(11):1733–42.
88. Hetz C, Chevet E, Harding HP. Targeting the unfolded protein response in disease. *Nat Rev Drug Discov* 2013;**12**(9):703–19.
89. Biden TJ, Boslem E, Chu KY, Sue N. Lipotoxic endoplasmic reticulum stress, beta cell failure, and type 2 diabetes mellitus. *Trends Endocrinol Metab* 2014;**25**(8):389–98.
90. Liu H, Cao MM, Wang Y, et al. Endoplasmic reticulum stress is involved in the connection between inflammation and autophagy in type 2 diabetes. *Gen Comp Endocrinol* 2015;**210C**:124–9.
91. Boslem E, Weir JM, MacIntosh G, et al. Alteration of endoplasmic reticulum lipid rafts contributes to lipotoxicity in pancreatic beta-cells. *J Biol Chem* 2013;**288**(37):26569–82.
92. Anuradha R, Saraswati M, Kumar KG, Rani SH. Apoptosis of beta cells in diabetes mellitus. *DNA Cell Biol* 2014;**33**(11):743–8.
93. Cnop M, Welsh N, Jonas JC, Jorns A, Lenzen S, Eizirik DL. Mechanisms of pancreatic beta-cell death in type 1 and type 2 diabetes: many differences, few similarities. *Diabetes* 2005;**54**(Suppl. 2): S97–107.
94. Riedl SJ, Shi Y. Molecular mechanisms of caspase regulation during apoptosis. *Nat Rev Mol Cell Biol* 2004;**5**(11):897–907.
95. Creagh EM, Conroy H, Martin SJ. Caspase-activation pathways in apoptosis and immunity. *Immunol Rev* 2003;**193**:10–21.
96. Elouil H, Cardozo AK, Eizirik DL, Henquin JC, Jonas JC. High glucose and hydrogen peroxide increase c-Myc and haeme-oxygenase 1 mRNA levels in rat pancreatic islets without activating NFkappaB. *Diabetologia* 2005;**48**(3):496–505.
97. Barthson J, Germano CM, Moore F, et al. Cytokines tumor necrosis factor-alpha and interferon-gamma induce pancreatic beta-cell apoptosis through STAT1-mediated Bim protein activation. *J Biol Chem* 2011;**286**(45):39632–43.
98. Carrizzo A, Forte M, Lembo M, Formisano L, Puca AA, Vecchione C. Rac-1 as a new therapeutic target in cerebro- and cardio-vascular diseases. *Curr Drug Targets* 2014;**15**(13):1231–46.
99. Song B, Scheuner D, Ron D, Pennathur S, Kaufman RJ. Chop deletion reduces oxidative stress, improves beta cell function, and promotes cell survival in multiple mouse models of diabetes. *J Clin Invest* 2008;**118**(10):3378–89.
100. Han J, Back SH, Hur J, et al. ER-stress-induced transcriptional regulation increases protein synthesis leading to cell death. *Nature Cell Biol* 2013;**15**(5):481–90.

CHAPTER

4

Diet–Gene Interactions in the Development of Diabetes

Jose M. Ordovas[1,2], *Silvia Berciano*[2], *Victor Mico*[2], *Lidia Daimiel-Ruiz*[2]

[1]Nutrition and Genomics Laboratory, JM-USDA-Human Nutrition Research Center on Aging, Tufts University, Boston, MA, USA; [2]Nutritional Genomics of Cardiovascular Disease and Obesity, IMDEA-Food Institute, CEI UAM+CSIC, Madrid, Spain

1. EARLY HISTORY OF THE DISEASE AND THE SEESAW OF THE DIETARY THERAPIES

The current epidemic of diabetes has its roots buried deep in the history of humankind. This disease has attracted the attention of our ancestors for thousands of years. This is reflected in the papyrus Ebers found in a tomb in Thebes and was bought in Luxor by Edwin Smith in 1862, who sold it in 1872 to the renowned German Egyptologist Georg Ebers. This papyrus has been dated c.1550 BC, but there is evidence suggesting that it was copied from much older documents dating as far back as 3400 BC. This document represents the most comprehensive medical papyrus found so far and contains descriptions and therapies about different medical conditions. Diabetes is described as "to eliminate urine which is too plentiful." The prescription described for this condition reads: "A measuring glass filled with water from the Bird pond, elderberry, fibres of the asit plant, fresh milk, beer-swill, flower of the cucumber, and green dates. Make it into one, strain and take for four days." (Other authors have translated this sentence differently: "Mix cakes, wheat grains, fresh grits, green lead, earth, and water. Let stand moist, then strain, then take for four days.")[1] The benefits of olive oil did not escape the physicians of the time, and a recipe to alleviate urinary troubles in adults included rectal injections of olive oil, honey, sweet beer, sea salt, and seeds of the wonderfruit. It is also in Egypt where the first skeletal remains associated with diabetes mellitus (DM) have been found dating c.2055–1650 BC. The skeleton belonging to a 40- to 45-year-old male shows numerous pathological conditions consistent with type 2 diabetes (T2D).

The next historical reference to diabetes comes from the Indian subcontinent in an Ayurveda essay attributed to Sushruta, who lived around the sixth century BC. (others have placed the works of this physician several centuries later).[2] Sushruta called diabetes the disease of the sweet urine. About 300 years later, another Ayurveda physician, Charaka, emphasized the excess urination that characterizes the disease. The latter also emphasizes the key role of knowledge and prevention to alleviate human suffering: "A physician who fails to enter the body of a patient with the lamp of knowledge and understanding can never treat diseases. He should first study all the factors, including environment, which influence a patient's disease, and then prescribe treatment. It is more important to prevent the occurrence of disease than to seek a cure." The contributions of these Indian physicians to the knowledge of diabetes are numerous, but sometimes masked by the more widespread literature surrounding Greek and Roman contributions. Indian physicians described the sweetness of the urine; they identified two types of diabetes, one having a more genetic contribution and the other being more environmental. Moreover, they characterized the environmental factors as being associated with a sluggish lifestyle (e.g., sitting, excessive sleep, lack of physical activity) and unhealthy diet (e.g., overindulging in sweet and fatty foods), most of them associated with those who were rich at the time. Therefore, they encouraged abstaining from sweet foods and reducing the intake of rice and other grains as part of the treatment of diabetes. Moreover, they recognized the benefits of physical exercise as part of the lifestyle changes.

As indicated earlier, the Western medical literature has echoed more intensely the contributions to diabetes of the Greek and Roman physicians. However, Hippocrates,

Molecular Nutrition and Diabetes
http://dx.doi.org/10.1016/B978-0-12-801585-8.00004-X

"the father of medicine," did not refer to diabetes in his writings, although this has been debated given his mention of excessive urinary flow with wasting of the body. Nevertheless, Hippocrates was also an advocate of the notion of preventive medicine and emphasized the impact of diet, exercise, and lifestyle on health. Conversely, some of his disciples, such as Galen and Aretaeus (or Areteus), took a much greater interest on diabetes, which was then considered as a rare disease. Galen described diabetes as a "diarrhea of urine" and "the thirsty disease" and considered diabetes a kidney disease. We probably owe to Aretaeus of Cappadocia the coining of the term "diabetes" derived from the ionic Greek "siphon."[3] However, other authors attribute the term to Apollonius of Memphis around 230 BC or to Demetrius of Apamea (first or early second century BC). Regardless of the origin of the term, Aretaeus provides a very precise description of the disease: "an affliction that is not very frequent ... being a melting down of the flesh and limbs into the urine ... life is short, disgusting and painful ... thirst unquenchable ... the kidneys and bladder never stop making water ... it may be something pernicious, derived from other diseases, which attack the bladder and kidneys." Despite his key contributions to the field of diabetes, little is known about his life and even in which century he lived, but is estimated between the first and the third century AD. His dietary prescription against diabetes included dates, raw quinces, and gruel.

Following the seminal contributions described here, the knowledge of diabetes experienced incremental steps during the ensuing centuries. Some of these milestones include the sweet taste of urine that has been attributed to Avicenna (~1000 AD)[4] and Thomas Willis (1674)[5]; the differentiation between DM and diabetes insipidus by Johann Peter Franks (1792, although this had been pointed out by Ayurvedic physicians two millennia earlier)[6]; and the identification of glucose as the agent giving the sweet taste to the urine of diabetics by Mathew Dabson (1776)[7] and Chevreul (1807).[8] It is interesting to highlight the dietary approach of John Rollo,[9] the first to use the term "mellitus" (1809) associated with diabetes. Rollo developed the first effective treatment for diabetes (1797): "Breakfast, 1.5 pints of milk and 0.5 pints of lime-water, mixed together; and bread and butter. For noon, plain blood puddings, made of blood and suet only. Dinner, game, or old meats, which have been long kept; and as far as the stomach may bear, fat and rancid old meats, as pork. To eat in moderation. Supper, the same as breakfast." In summary, a 1500-calorie diet high in animal food and low in vegetables (grains and breads), sometimes supplemented with antimony, opium, and digitalis. This dietary approach represented a departure from other remedies used until that century that included more witchcraft remedies such as "gelly of viper's flesh," broken red coral, sweet almonds, and "fresh flowers of blind nettles" as well as the diet recommended by Thomas Willis, about one century earlier, based on "delectable Milk and barley-water boiled with bread." The early twentieth century introduced low energy—low carbohydrate diets, comprising 70% fat, 10% carbohydrates, and 20% protein. However, later in the twentieth century, the recommendations completely reversed to a low-fat, high-carbohydrate diet, consistent with the crusade against dietary fat occurring during these more recent times.

2. NUTRITIONAL MANAGEMENT OF DIABETES IN THE TWENTY-FIRST CENTURY

The disease that was considered rare by the "fathers" of Western medicine has reached epidemic proportions today, and the current predictions do not anticipate any relief in the near future. Some current estimates reveal that 18.8 million people in the United States have been diagnosed with diabetes and other 7 million may be unknowingly living with the disease. Moreover, 79 million people in the United States may have blood glucose levels that place them at a very high risk of developing diabetes. Overall, 100 million people, the equivalent to one-third of the entire US population, are already suffering or at risk of developing the complications of diabetes.

As in the past, medical societies, and specifically the American Diabetes Association, acknowledge the key role of nutrition in diabetes prevention and management, but the recommendations have shifted from the one-size-fits-all numeric targets to more individualized eating plans aimed at reaching specific targets for biomarkers and body weight. It also places a lot of emphasis on education, both for the public and for the health care providers. It also acknowledges the many faces of diabetes that go well beyond type 1 and type 2 and that makes necessary a better understanding of the etiology and the molecular basis of the disease. This knowledge is becoming available, thanks to the progress being made on the genetics of diabetes.

3. DIABETES, A COMPLEX DISEASE WITH A SIGNIFICANT GENETIC COMPONENT

Indian physicians observed the clustering of diabetes in families as early as the sixth century BC. However, the first documented description of inheritance has been attributed to Richard Morton (1637–1689),[10] who described how, within a family of seven members, four became diabetic at the age of dentition. From the behavioral perspective, Morton also defined diabetes as "a continual flow of nutritive juices pouring out through the kidneys, which frequently befalls intellectual persons, and drinkers of brandy and diuretic liquors." Likewise, 100 years later, Isenflamm reported seven siblings from healthy parents who developed diabetes at age 8–9 years.[11] The evidence supporting the genetic component of diabetes then kept accruing in the literature at a faster pace. However, even during the early years of the twentieth century, some authors disputed such evidence. Studies in twins and extended families during the 1930s and 1940s finally provided unquestionable support to the inheritance of diabetes and the presence of different genetic entities. The next major step came during the mid-1970s when the genetics of diabetes was linked to the human leukocyte antigen (HLA) system. Shortly thereafter, with the newly acquired ability of cloning and sequencing, the gene for insulin was characterized in humans and mapped to the short arm of chromosome 11.[12] Using Southern blotting and restriction fragment length polymorphism analysis, a series of reports began to identify a series of restriction fragment length polymorphisms in insulin and other candidate genes that were associated with T2D[13]; however, these techniques were very rudimentary and did not allow analysis of the large sample sizes needed to obviate the type I and type II errors during the following two decades in genetic studies of such complex diseases as diabetes. This problem was finally resolved with the advent of genome-wide association studies that included thousands of individuals.

3.1. Current Knowledge about the Genetics of Diabetes

One of the biggest challenges that researchers have faced when attempting to elucidate the genetic contribution to diabetes risk is the heterogeneity of the disease. Although there are some common factors, such as hyperglycemia that results from impaired insulin secretion and/or action, DM embodies a wide group of metabolic disorders with different genetic and pathophysiological features. However, there is ample evidence that all types of DM have an important genetic component. Twin studies have estimated the heritability of type 1 diabetes (T1D) and T2D at 0.88 and 0.64, respectively.[14] Over the past two decades, numerous studies have aimed to identify genetic markers that could provide the information we need to better characterize disease subtypes, understand their cause and underlying mechanisms, and therefore improve both diagnosis and therapeutic strategies. Different classification criteria have been used over time, evolving from age of onset (juvenile/adult) to treatment (insulin-dependent/noninsulin-dependent) and finally to the current—less simplistic—cause-based approach that divides the different DM entities into four groups: other types of diabetes, T1D, T2D, and gestational diabetes (GDM).

3.2. Monogenic Forms of Diabetes

Most cases of DM are due to a combination of genetic and environmental factors, although monogenic forms account for 1–5% of all cases of diabetes in young people. Neonatal DM is the rarest monogenic form, occurring in only 1 in 100,000–500,000 live births, and often is caused by a single mutation in adenosine triphosphate–sensitive K^+ channel genes KCNJ11/ABCC8,[15] INS, GCK, PDX1, PTF1A, FOXP3, RFX6, ZFP57, GLIS3, NEUROD1, NEUROG3, PAX6, HNF1B, IER3IP1, SLC19A2, SLC2A2, and EIF2AK3 or methylation defects at the imprinted HYMAI/ZAC domain that occur in most transient neonatal DM cases.[16,17]

Approximately 70% of the cases of familial diabetes—also known as maturity-onset diabetes of the young (MODY), the most common form of monogenic diabetes—result from mutations in HNF1A (MODY3) or GCK (MODY2). More than 200 mutations that lead to those subtypes have been identified in these two genes. Many of them are unique, and a precise diagnosis of MODY may require sequencing.[18] Other known MODY mutations are harbored in genes such as HNF4A, HNF1B, PAX4, PDX1, KLF11 (involved in the PDX1-mediated inhibition of the insulin promoter), CEL, INS, WFS1, and NEUROD1.[16,19–21]

Additionally, around 1% of all diabetes cases are due to mutations in the mitochondrial genes MT-TK, MT-TE, or MT-TL1 (A3243G being the most common mutation), causing the condition known as maternally inherited diabetes and deafness.[22]

Despite the monogenic origin of these syndromes, their clinical and phenotypic expression varies dramatically across subjects. It has been suggested that this variability may depend upon the position of the mutation within the respective locus, as well as gene–gene and gene–environment interactions. Additional knowledge is required for more precise risk prediction and improved management through diet, lifestyle, or pharmacological agents.[16]

3.3. Type 1 Diabetes

T1D is characterized by an absolute insulin deficiency resulting from T1D autoimmune-mediated destruction of pancreatic β cells. Some patients show no evidence of autoimmunity: they are said to suffer from idiopathic

T1D. T1D represents 5–10% of the total cases, affecting more than 20 million people worldwide.

The etiology of T1D has been studied for decades now and it is not completely elucidated, although it is recognized that disease risk is defined by a combination of environmental and genetic factors. The HLA region on chromosome 6p21 harbored the first T1D autoimmune mediated–associated genetic variants identified through linkage studies, haplotypes of the class II genes HLA-DRB1, HLA-DQA1, and HLA-DQB1, accounting for about 50% of the familial clustering.[23] The combination of two susceptible alleles, DR3/DR4, was associated to the highest risk.[24]

After the demonstration in 2009 of the independent effects of HLA-A, HLA-B, and HLA-DPB1 in the modulation of T1D risk, later studies using candidate genes identified additional variants associated with T1D at INS, CTLA4, PTPN22, TCF7, and IL2RA.[25] With regard to the insulin gene, INS contains multiple polymorphisms, but T1D is most strongly associated not to one of its single-nucleotide polymorphisms but to a highly polymorphic variable number tandem repeat. About 80% of T1D cases are homozygous for the short class I allele (26–63 repeats), whereas the longer class III alleles are thought to confer protection against this disease. The intermediate class II allele is very rare in Caucasians. Recent studies[26,27] have explored the role of copy number variants and deletions in T1D susceptibility but they have not been able to identify novel associated loci. Genome-wide scans for linkage to T1D were not successful in identifying novel risk regions because of their limited sensitivity, but one locus on chromosome 21 (UBASH3A) was confirmed.[28]

A recent genome-wide association study with 4075 T1D cases and 2604 controls identified 452 genes associated with T1D including 4 non-HLA genes (RASIP1, STRN4, BCAR1, and MYL2) that were replicated in independent populations.[29] More than 50 non–HLA-related loci have been demonstrated to influence T1D susceptibility, many of them (including CTRB1/2, IFIH1, GLIS3, and PTPN2) are expressed in β cells, meaning that not only the immune system but also β-cell function could be affected. However, the effect of these variants on T1D susceptibility is very modest and, as we gain knowledge on how prenatal and early nutrition affect T1D risk, it is important that we investigate the role of gene–environment interactions that could account for an important part of the missing heritability of the disease.

3.4. Type 2 Diabetes

Approximately 90% of the cases fall into the T2D group, a broad category characterized by insulin resistance occasionally combined with abnormal insulin secretion. PPARG was the first candidate gene identified and reproducibly associated to T2D. Its most common variant in European populations, Pro12Ala, has been associated with increased insulin sensitivity and lower T2D risk.[30] From the late 1990s, important efforts were placed into linkage studies of T2D, but only two susceptibility genes—CAPN10 and TCF7L2—were identified.[31,32] With the advent of high-throughput genotyping techniques, genome-wide association studies have identified more than 70 susceptibility variants associated to T2D,[33] which will be thoroughly reviewed in Part 3 of this book. However, all together they seem to explain only about 10% of the heritability of this condition. Knowing the important contribution of lifestyle factors like nutrition and physical activity to T2D risk, it is critical that we continue to study the interplay between genetic and environmental factors to understand how this type of interaction influences T2D development.

3.5. Gestational Diabetes

GDM has been defined as glucose intolerance with onset during pregnancy. Approximately 7% (1–14% depending on ancestry: African-Americans, Native Americans, Hispanics, and South Asians show higher incidences) of all pregnancies are complicated by GDM. Risk factors for this disorder include high maternal age (>35 years old), high body mass index (BMI), short stature, family history of T2D, and previous birth of a macrosomic baby.[34] Transgenerational effects have been observed, and women who presented a low or high weight at birth have an increased risk of GDM and/or T2D, and so may their daughters. Genetic variants linked to this GDM have been found in more than 20 loci, most notably TCF7L2, HHEX, KCNJ11, PPARG, KCNQ1,CDKN2A/B, IGF2BP2, SLC30A8, and FTO.[35] These findings were not surprising, given the similarities of GDM and T2D.

In summary, a deeper knowledge of the genetic mechanisms involved in diabetic conditions is key to the development of efficient therapies. Despite the fact that environmental factors play an important role in the vast majority of the cases of diabetes, understanding the underlying genetic causes may help define which elements of this environment are more crucial to different subsets of diabetic patients.

4. THE ROLE OF GENE–DIET INTERACTIONS IN DIABETES RISK

Despite the knowledge gained toward a better understanding of the genetic predisposition to DM summarized in the previous section,[36,37] most of the potential inheritance remains "missing." This "missing heritability" has been attributed to unknown genetic variants, epigenetic mechanisms, gene–gene interactions, and to gene–environment interactions.[18] This suggests that environmental factors influence diabetes incidence and development. Diet is one of the environmental factors that most contributes to modulate risk of development of DM. Dietary factors that influence T1D incidence and development include the use of breast milk versus infant formula,[38] highly hydrolyzed infant formula versus conventional infant formula,[39] and early/late exposure to gluten[40] and vitamin D.[41] Gut microbiota also plays a role in the development of T1D through its influence in gut immunity,[42] and it has been shown that long-term dietary habits contribute to define the gut microbial composition, which affects gut immunity.[43] Thus, gut microbiota is one of the factors that may mediate the dietary modulation of T1D predisposition. The same scenario is true for T2D. Both association and genome-wide association studies have identified more than 100 single-nucleotide polymorphisms in more than 70 genes associated with T2D-related traits including fasting insulin and glucose levels, homeostatic model assessment (HOMA) index or β-cell function.[44] But only transcription factor-7–like 2 (TCF7L2) was shown to clearly contribute to T2D risk.[45–48] The contribution of those gene variants to T2D and related traits appears to be poor, and lifestyle is suggested to play a crucial role in T2D development. Nutrigenetic studies describing gene–diet interactions as modulators of T2D incidence are scarce, but suggest that genetic predisposition to T2D may be partially or nearly completely abolished by a healthy lifestyle or lifestyle modifications.[49] In 2011, Lee et al. developed a database of gene–environment interactions relevant to nutrition, blood lipids, cardiovascular disease, and T2D.[50] Such a database contains more than 550 such interactions along with other 1430 instances where a lack of statistical significance was found. A recent study has aimed to calculate how much gene–nutrient interactions contribute to T2D-related traits and has shown that 25.1% and 24.2% of fasting insulin and HOMA-IR heritability could be explained by the interaction between gene variants and carbohydrate intake. Similarly, 39% of β-cell function could be explained by interactions between gene variants and n-6 polyunsaturated fatty acid (PUFA) intake.[51]

4.1. Lifestyle as a Modulator of the Genetic Risk of T2D: Gene–Diet Interactions

An "unhealthy lifestyle" can be defined as the lack of physical activity (PA) and the maintenance of an unhealthy dietary pattern (frequently linked to overnutrition). Epidemiological studies have shown that food overconsumption and a sedentary lifestyle are associated with a higher risk of developing T2D. In this regard, it has been demonstrated that lack of PA and a Western dietary pattern increase the odds of developing T2D, whereas high levels of PA and prudent food consumption decrease it.[52–55] Furthermore, interventional studies have demonstrated that the adoption of a healthier lifestyle may prevent T2D.[49] The Chinese Da Qing Diabetes Prevention Study reported a 42% reduction in diabetes incidence after a 6-year lifestyle intervention in individuals with impaired glucose tolerance. The lifestyle intervention included diet, PA, or both and the incidence of T2D in the dietary intervention group was 23.9% lower than in the control group.[56] Consistently, the Finnish Diabetes Prevention Study and the US Diabetes Prevention Program reported a 58% reduction in diabetes incidence as a result of the lifestyle modification.[57,58] Similar results were found in Japanese and Indian populations (67.4% and 28.5% reduction in diabetes incidence, respectively).[59,60] Interestingly, Knowler et al.[57] and Ramachandran et al.[60] showed that the intervention on lifestyle was more efficient that metformin in the prevention of T2D in their respective interventional studies. In most of the studies,[61–63] except for the Chinese Da Qing Diabetes Prevention Study,[62] the reduction in T2D incidence was correlated with a reduction in body weight. Additionally, these studies have shown that, in a long-term follow-up after discontinuation of the intervention, diabetes risk still remained substantially reduced, even if body weight was partially or totally regained. These results highlight the role of obesity as a T2D risk factor, but also suggest that the benefit of a healthy lifestyle is mediated, not only by a reduction in body weight, but also through the improvement in metabolic traits, including insulin sensitivity, blood glucose control, and lipid profile.

Lifestyle could interact with genes predisposing to T2D to modulate risk of developing the disease. However, current studies show controversial results. Qi et al.[64] generated a Genetic Risk Score based on 10 T2D-related polymorphisms and observed that a Western dietary pattern was associated with T2D in subjects with a high Genetic Risk Score ($\geq$12 risk alleles), but not in those with a low Genetic Risk Score ($<$12 risk alleles).[34] These results support the notion that genetics contributes to T2D risk and that the diet is effective in the context of a high genetic predisposition. However, a recent meta-analysis of 15 Cohorts for Heart and Aging Research in Genomic Epidemiology consortium studies failed to find gene–diet interactions associated to fasting glucose and insulin levels and suggested that a favorable diet influence on T2D incidence regardless of genotype.[65] Therefore, the potential of the diet as modulator of genetic predisposition to T2D is controversial.

Studies focusing on the well-known T2D risk gene, TCF7L2, have also shown contradictory results. The US Diabetes Prevention Program study showed that the risk-conferring TT genotype at TCF7L2 rs7903146 is associated to T2D development, but this association is abolished in the lifestyle intervention group.[46] Interestingly, although metformin treatment reduced the effect of this polymorphism on T2D risk, lifestyle intervention resulted more effective. Despite these results, the authors did not find a significant interaction between the TCF7L genotype and the US Diabetes Prevention Program intervention, probably because these interventions succeeded primarily by improving insulin sensitivity[66] and these variants affect insulin secretion. Bo et al.[67] and Haupt et al.[68] also failed to demonstrate an interaction between lifestyle and this polymorphism in modulating T2D risk. However, the case-control study carried out by Fisher et al.[69] in the European Prospective Investigation into Cancer and Nutrition study showed that although whole grain intake was inversely correlated with T2D risk, the T-allele of this polymorphism abolished the protective effect of whole-grain intake. On the other hand, the Finnish Diabetes Prevention Study showed that the rs12255372TT genotype was significantly associated with an increased risk of incidence of T2D (2.85-fold risk) in the control group, but not in the intervention group,[70] which was based on dietary counseling from a nutritionist and increase in PA.[71] Moreover, T2D risk conferred by this polymorphism depended on the glycemic index of the diet.[72]

The major allele of PPARγ Pro12Ala gene variant has been associated with an increased risk for T2D.[73] Conversely, the minor allele has been associated with a higher BMI, total and low-density lipoprotein cholesterol, and obesity.[74,75] In this scenario, Ala12 seems to be associated to a lower risk of T2D in healthy subjects, but this may not be the case in obese subjects.[76] Thus, the effect of the Ala12 allele on T2D risk is complex and context-dependent. It has been shown that diet affects BMI, body composition, and metabolic parameters in Ala carriers more than in Pro/Pro homozygotes.[77] Lifestyle modification has been suggested to attenuate the negative effect of the Ala12 allele on metabolic profile, body weight, and diabetes risk in obese subjects. In the Finnish Diabetes Prevention Program, Lindi et al.[78] showed that in obese participants with impaired glucose tolerance, the Ala12 allele was associated to a higher risk of developing T2D. However, homozygous carriers of the Ala12 allele lost more weight during the lifestyle intervention and did not develop T2D, compared with those with the same genotype in the control group. In the Diabetes Prevention Program conducted in nondiabetic subjects with elevated fasting and postchallenge glucose levels, Pro12 carriers gained weight in the control group but lost it during the lifestyle intervention. Ala12 lost weight in both groups, but such loss was higher in the lifestyle intervention group.[79] Overall, the current evidence suggests that obese carriers of the Ala12 allele benefit more from a lifestyle-based intervention.

The FTO gene has been consistently associated with obesity risk. However, its association with T2D is controversial and whether it is associated with T2D development independent of obesity needs further clarification. A recent case-control study in 3430 T2D cases and 3622 controls with no differences in BMI found a gene–diet interaction between FTO rs9939609 polymorphism and adherence to the Mediterranean Diet with T2D.[80] The authors reported that adherence to the Mediterranean Diet abolished the T2D risk attributed to the risk allele. This polymorphism was also associated with increased fat and decreased fiber intake in diabetic patients.[81]

4.2. Gene–Nutrient Interactions in the Modulation of T2D Risk: Macronutrients and Micronutrients

Unlike dietary patterns, other studies have considered the interaction of specific nutrients (macro and micronutrients) with the genotypic predisposition to T2D. We have previously described the effect of the interactions between variants in TCF7L2 and PPARG genes and lifestyle in T2D incidence. Specific dietary factors also have shown to interact with these genes to modulate T2D risk. The interaction between TCF7L2 rs12573128 and dietary fat intake was shown to influence insulin sensitivity and glucose tolerance.[82] Using family-based association methods, Nelson et al. showed that the Pro12 allele in the PPARG gene was associated with T2D only when consumption of PUFA was high in a population of non-Hispanic white participants form the Gene Environment Interaction study in Colorado. However, this interaction was not significant when they conducted generalized estimation equation analyses.[83] Similarly, results from the Data from an Epidemiological Study on the Insulin Resistance Syndrome study conducted in a French population have shown that high-fat consumption was associated with an increased T2D risk in homozygous for the Pro allele, but not in Ala carriers, even although Ala homozygous had a higher BMI than Pro carriers among high-fat consumers.[84] The study of Luan et al.[85] also showed that Ala carriers benefit more from a high intake of PUFA, because the BMI is higher in Ala carriers compared with those homozygous for the Pro allele when PUFA: saturated fatty acids (SAT) ratio is low, but when PUFA:SAT ratio is high, Ala carriers show a lower BMI. However, another study failed to find any interaction between the PUFA:SAT ratio and the Pro12Ala polymorphism in modulating BMI in an Asian population.[86] Collectively, these studies suggest that Ala12 carriers may be more responsive to the beneficial effects of unsaturated fat and less responsive to the harmful effect of saturated fats than Pro carriers on cardiometabolic traits and T2D risk.

Variants in genes involved in dietary fat metabolism and mobilization also modulate diabetes-related traits by interacting with nutrients. The Thr allele of the Ala54Thr polymorphism in the intestinal fatty acid–binding protein (FABP2) gene was associated with a decrease in insulin sensitivity when saturated fatty acids (SFAs) were replaced by monounsaturated fatty acids and carbohydrates.[87] The rs2270188G > T polymorphism in the caveolin 2 (CAV2) gene interacts with dietary fat, because homozygous individuals of the rare T allele doubled the risk of T2D if daily fat intake increased from 30% to 40% energy, the interaction was even more pronounced with an increase in dietary SFA from 10% to 20% energy.[88] In addition, polymorphisms G11482A and A14995T in the perilipin1 gene (PLIN1) showed significant interactions with SFA and carbohydrate intake that modulate insulin resistance. Thus, homozygote carriers of the minor alleles had a higher HOMA-IR index if SFA intake was high, but HOMA-IR index decreased if carbohydrate intake increased. Interestingly, these associations were only found in women.[89]

Dysregulation of the circadian system has also been associated with diabetes. Thus, gene variants in circadian genes could increase T2D risk and interact with nutrients.[90] An interaction between the CLOCK rs1801260 polymorphism and dietary fat that modulates insulin sensitivity was observed in the Coronary Diet Intervention With Olive Oil and Cardiovascular Prevention study.[91] After 12 months of low-fat intervention, those homozygous for the major allele (TT) displayed higher insulin sensitivity compared with carriers of the minor allele C. On the other hand, a high carbohydrate intake was associated with higher insulin resistance only in homozygous carriers of the minor allele C of the CRY1 rs2287161 polymorphism.[92]

Other studies have established gene–nutrient interactions in relation with T2D risk in other susceptibility genes such as SLC30A8, IRS1, GCKR, or ADIPOQ.[7] In the case of the SLC30A8 gene, a meta-analysis has associated high total zinc intake with lower fasting glucose levels. However, the SLC30A8 rs11558471 polymorphism modulated this effect so the inverse association between total zinc intake and fasting glucose was stronger in individuals carrying the glucose-raising A allele compared with individuals who do not carry it.[93] Similar results were found by Shan et al. in relation to the SLC30A8 rs13266634 variant.[94]

Polymorphisms in the insulin receptor substrate 1 (IRS1) also interact with macro- and micronutrients to modulate insulin sensitivity. For instance, nonesterified fatty acid and steady-state plasma glucose levels were lower in G/R subjects of the G972R polymorphism at IRS1 (rs1801278) compared with G/G subjects only in a low-fat, high-carbohydrate diet, but not in an SFA or a monounsaturated fatty acid diet.[95] Another polymorphism in IRS1 (rs294364) showed significant interactions with 25-hydroxyvitamin D to modulate insulin resistance. The minor T allele was associated with lower insulin resistance with higher circulating 25-hydroxyvitamin D levels in the Multi-Ethnic Study of Atherosclerosis.[96] GCKR inhibits glucokinase in the liver and pancreatic islet cells and is considered a susceptibility gene candidate for a form of MODY. The variant rs781194 is associated with fasting glucose and insulin levels. Nettleton et al.[97] reported an interaction between this variant and dietary whole grain intake by which in carriers of the insulin-raising allele, a greater whole grain intake was associated with a smaller reduction of fasting insulin. Other reports have shown interactions between the adiponectin gene (ADIPOQ) and carbohydrate intake. The G276T variant has been associated with T2D and related metabolic traits in Asians,[98–101] but this association can be modulated by carbohydrate intake. Data from a prospective study including 673 patients with T2D in Korea showed that the G allele was associated with higher fasting blood glucose only in subjects consuming a low-carbohydrate diet (<55% of energy). When carbohydrate intake was intermediate (55–65%), carriers of the T allele had greater fasting blood glucose and hemoglobin A1C concentrations, and when carbohydrate intake was high (>65%), carriers of the T allele had greater high-density lipoprotein cholesterol concentrations.

5. CONCLUDING REMARKS

In summary, several studies have demonstrated significant gene–diet interactions modulating T2D. However, most findings lack consistency across populations and more solid experimental approaches will be needed to clarify the role of gene–diet interactions in the development of T2D. This knowledge will contribute to better risk prediction as well as the implementation of more precise and efficacious personalized dietary recommendations for its prevention and therapy.

References

1. King KM. Diabetes: classification and strategies for integrated care. *Br J Nurs* 2003;**12**:1204–10.
2. Frank LL. Diabetes mellitus in the texts of old Hindu medicine (Charaka, Susruta, Vagbhata). *Am J Gastroenterol* 1957;**1**:76–95.
3. Laios K, Karamanou M, Saridaki Z, Androutsos G. Aretaeus of Cappadocia and the first description of diabetes. *Hormones (Athens)* 2012;**1**:109–13.

4. Siahpoosh M, Ebadiani M, Shah Hosseini G, et al. Avicenna the first to describe diseases which may be prevented by exercise. *Iran J Public Health* 2012;**41**:98–101.
5. Molnar Z. *Nat Rev Neurosci* 2004;**5**:329–35.
6. von Engelhardt E. Outlines of historical development. In: *Diabetes: its medical and cultural history.* New York: Springer-Verlag; 1989. p. 3–10.
7. Lasker SP, McLachlan CS, Wang L, et al. Discovery, treatment and management of diabetes. *J Diabetol* 2010;**1**:1.
8. Barthold SW. Introduction: unsung heroes in the battle against diabetes. *ILAR J* 2004;**45**:227–30.
9. Marble A. John Rollo. *Diabetes* 1956;**5**:325–7.
10. Hitman GA, Niven MJ. Genes and diabetes mellitus. *Br Med Bull* 1989;**45**:191–205.
11. Kennedy S. Hereditary diabetes mellitus. *JAMA* 1931;**96**:241–5.
12. Owerbach D, Bell GI, Rutter WJ, Shows TB. The insulin gene is located on chromosome 11 in humans. *Nature* 1980;**286**:82–4.
13. Rotwein P, Chyn R, Chirgwin J, et al. Polymorphism in the 5′-flanking region of the human insulin gene and its possible relation to type 2 diabetes. *Science* 1981;**213**:1117–20.
14. van Dongen J, Slagboom PE, Draisma HH, et al. The continuing value of twin studies in the omics era. *Nat Rev Genet* 2012;**13**: 640–53.
15. Flanagan SE, Patch AM, Mackay DJ, et al. Mutations in ATP-sensitive K+ channel genes cause transient neonatal diabetes and permanent diabetes in childhood or adulthood. *Diabetes* 2007; **56**:1930–7.
16. Vaxillaire M, Bonnefond A, Froguel P. The lessons of early-onset monogenic diabetes for the understanding of diabetes pathogenesis. *Best Pract Res Clin Endocrinol Metab* 2012;**26**:171–87.
17. Hattersley A, Bruining J, Shield J, et al. The diagnosis and management of monogenic diabetes in children and adolescents. *Pediatr Diabetes* 2009;**10**(Suppl. 12):33–42.
18. Groop L, Pociot F. Genetics of diabetes–are we missing the genes or the disease? *Mol Cell Endocrinol* 2014;**382**:726–39.
19. Perakakis N, Danassi D, Alt M, et al. Human Krüppel-like factor 11 differentially regulates human insulin promoter activity in β-cells and non-β-cells via p300 and PDX1 through the regulatory sites A3 and CACCC box. *Mol Cell Endocrinol* 2012;**363**:20–6.
20. Murphy R, Ellard S, Hattersley AT. Clinical implications of a molecular genetic classification of monogenic beta-cell diabetes. *Nat Clin Pract Endocrinol Metab* 2008;**4**:200–13.
21. Neve B, Fernandez-Zapico ME, Ashkenazi-Katalan V, et al. Role of transcription factor KLF11 and its diabetes-associated gene variants in pancreatic beta cell function. *Proc Natl Acad Sci USA* 2005; **102**:4807–12.
22. Maassen JA, 'T Hart LM, Van Essen E, et al. Mitochondrial diabetes: molecular mechanisms and clinical presentation. *Diabetes* 2004;**53**(Suppl. 1):S103–9.
23. Ounissi-Benkalha H, Polychronakos C. The molecular genetics of type 1 diabetes: new genes and emerging mechanisms. *Trends Mol Med* 2008;**14**:268–75.
24. Thomson G, Robinson WP, Kuhner MK, et al. Genetic heterogeneity, modes of inheritance, and risk estimates for a joint study of Caucasians with insulin-dependent diabetes mellitus. *Am J Hum Genet* 1988;**43**:799–816.
25. Pociot F, Akolkar B, Concannon P, et al. Genetics of type 1 diabetes: what's next? *Diabetes* 2010;**59**:1561–71.
26. Zanda M, Onengut-Gumuscu S, Walker N, Shtir C, et al. A genome-wide assessment of the role of untagged copy number variants in type 1 diabetes. *PLoS Genet* 2014;**10**:e1004367.
27. Cooper NJ, Shtir CJ, Smyth DJ, et al. Detection and correction of artefacts in estimation of rare copy number variants and analysis of rare deletions in type 1 diabetes. *Hum Mol Genet* 2015;**24**: 1774–90.
28. Concannon P, Chen WM, Julier C, et al. Genome-wide scan for linkage to type 1 diabetes in 2,496 multiplex families from the Type 1 Diabetes Genetics Consortium. *Diabetes* 2009;**58**: 1018–22.
29. Qiu YH, Deng FY, Li MJ, Lei SF. Identification of novel risk genes associated with type 1 diabetes mellitus using a genome-wide gene-based association analysis. *J Diabetes Invest* 2014;**5**:649–56.
30. Deeb SS, Fajas L, Nemoto M, et al. A Pro12Ala substitution in PPARgamma2 associated with decreased receptor activity, lower body mass index and improved insulin sensitivity. *Nat Genet* 1998;**20**:284–7.
31. Duggirala R, Blangero J, Almasy L, et al. Linkage of type 2 diabetes mellitus and of age at onset to a genetic location on chromosome 10q in Mexican Americans. *Am J Hum Genet* 1999;**64**: 1127–40.
32. Reynisdottir I, Thorleifsson G, Benediktsson R, et al. Localization of a susceptibility gene for type 2 diabetes to chromosome 5q34-q35.2. *Am J Hum Genet* 2003;**73**:323–35.
33. Sun X, Yu W, Hu C. Genetics of type 2 diabetes: insights into the pathogenesis and its clinical application. *BioMed Res Int* 2014: 926713.
34. Ben-Haroush A, Yogev Y, Hod M. Epidemiology of gestational diabetes mellitus and its association with type 2 diabetes. *Diabet Med* 2004;**21**:103–13.
35. Lauenborg J, Grarup N, Damm P, et al. Common type 2 diabetes risk gene variants associate with gestational diabetes. *J Clin Endocrinol Metab* 2009;**94**:145–50.
36. Doria A, Patti ME, Kahn CR. The emerging genetic architecture of type 2 diabetes. *Cell Metab* 2008;**8**:186–200.
37. Ntzani EE, Kavvoura FK. Genetic risk factors for type 2 diabetes: insights from the emerging genomic evidence. *Curr Vasc Pharmacol* 2012;**10**:147–55.
38. Virtanen SM, Knip M. Nutritional risk predictors of beta cell autoimmunity and type 1 diabetes at a young age. *Am J Clin Nutr* 2003; **78**:1053–67.
39. Knip M, Virtanen SM, Becker D, et al. Early feeding and risk of type 1 diabetes: experiences from the Trial to Reduce Insulin-dependent diabetes mellitus in the Genetically at Risk (TRIGR). *Am J Clin Nutr* 2011;**94**(Suppl. 6):1814S–20S.
40. Norris JM, Barriga K, Klingensmith G, et al. Timing of initial cereal exposure in infancy and risk of islet autoimmunity. *JAMA* 2003;**290**:1713–20.
41. Hypponen E, Läärä E, Reunanen A, et al. Intake of vitamin D and risk of type 1 diabetes: a birth-cohort study. *Lancet* 2001;**358**: 1500–3.
42. McLean MH, Dieguez Jr D, Miller LM, Young HA. Does the microbiota play a role in the pathogenesis of autoimmune diseases? *Gut* 2015;**64**:332–41.
43. Wu GD, Chen J, Hoffmann C. Linking long-term dietary patterns with gut microbial enterotypes. *Science* 2011;**334**:105–8.
44. Berná G, Oliveras-López MJ, Jurado-Ruíz E, Tejedo J, Bedoya F, Soria B, et al. Nutrigenetics and nutrigenomics insights into diabetes etiopathogenesis. *Nutrients* 2014;**6**:5338–69.
45. Grant SF, Thorleifsson G, Reynisdottir I, et al. Variant of transcription factor 7-like 2 (TCF7L2) gene confers risk of type 2 diabetes. *Nat Genet* 2006;**38**:320–3.
46. Florez JC, Jablonski KA, Bayley N, et al. TCF7L2 polymorphisms and progression to diabetes in the Diabetes Prevention Program. *N Engl J Med* 2006;**355**:241–50.
47. Voight BF, Scott LJ, Steinthorsdottir V, et al. Twelve type 2 diabetes susceptibility loci identified through large-scale association analysis. *Nat Genet* 2010;**42**:579–89.
48. Peng S, Zhu Y, Lü B, et al. TCF7L2 gene polymorphisms and type 2 diabetes risk: a comprehensive and updated meta-analysis involving 121,174 subjects. *Mutagenesis* 2013;**28**:25–37.

49. Temelkova-Kurktschiev T, Stefanov T. Lifestyle and genetics in obesity and type 2 diabetes. *Exp Clin Endocrinol Diabetes* 2012; **120**:1–6.
50. Lee YC, Lai CQ, Ordovas JM, Parnell LD. A database of gene-environment interactions pertaining to blood lipid traits, cardiovascular disease and type 2 diabetes. *J Data Min Genomics Proteomics* 2011;**2**:106.
51. Zheng JS, Arnett DK, Lee YC, et al. Genome-wide contribution of genotype by environment interaction to variation of diabetes-related traits. *PLoS One* 2013;**8**:e77442.
52. van Dam RM, Rimm EB, Willett WC, et al. Dietary patterns and risk for type 2 diabetes mellitus in U.S. men. *Ann Intern Med* 2002;**136**:201–9.
53. Kriska AM, Saremi A, Hanson RL, et al. Physical activity, obesity, and the incidence of type 2 diabetes in a high-risk population. *Am J Epidemiol* 2003;**158**:669–75.
54. Meisinger C, Löwel H, Thorand B, et al. Leisure time physical activity and the risk of type 2 diabetes in men and women from the general population. The MONICA/KORA Augsburg Cohort Study. *Diabetologia* 2005;**48**:27–34.
55. Stefanov TS, Vekova AM, Kurktschiev DP, Temelkova-Kurktschiev TS. Relationship of physical activity and eating behaviour with obesity and type 2 diabetes mellitus: Sofia Lifestyle (SLS) study. *Folia Med (Plovdiv)* 2011;**53**:11–8.
56. Pan XR, Li GW, Hu YH, et al. Effects of diet and exercise in preventing NIDDM in people with impaired glucose tolerance. The Da Qing IGT and Diabetes Study. *Diabetes Care* 1997;**20**:537–44.
57. Knowler WC, Barrett-Connor E, Fowler SE, et al. Reduction in the incidence of type 2 diabetes with lifestyle intervention or metformin. *N Engl J Med* 2002;**346**:393–403.
58. Tuomilehto J, Lindström J, Eriksson JG, et al. Prevention of type 2 diabetes mellitus by changes in lifestyle among subjects with impaired glucose tolerance. *N Engl J Med* 2001;**344**:1343–50.
59. Kosaka K, Noda M, Kuzuya T. Prevention of type 2 diabetes by lifestyle intervention: a Japanese trial in IGT males. *Diabetes Res Clin Pract* 2005;**67**:152–62.
60. Ramachandran A, Snehalatha C, Mary S, et al. The Indian Diabetes Prevention Programme shows that lifestyle modification and metformin prevent type 2 diabetes in Asian Indian subjects with impaired glucose tolerance (IDPP-1). *Diabetologia* 2006;**49**: 289–97.
61. Lindstrom J, Ilanne-Parikka P, Peltonen M, et al. Sustained reduction in the incidence of type 2 diabetes by lifestyle intervention: follow-up of the Finnish Diabetes Prevention Study. *Lancet* 2006; **368**:1673–9.
62. Li G, Zhang P, Wang J, et al. The long-term effect of lifestyle interventions to prevent diabetes in the China Da Qing Diabetes Prevention Study: a 20-year follow-up study. *Lancet* 2008;**371**:1783–9.
63. Diabetes Prevention Program Research Group, Knowler WC, Fowler SE, et al. 10-year follow-up of diabetes incidence and weight loss in the Diabetes Prevention Program Outcomes Study. *Lancet* 2009;**374**:1677–86.
64. Qi L, Cornelis MC, Zhang C, et al. Genetic predisposition, Western dietary pattern, and the risk of type 2 diabetes in men. *Am J Clin Nutr* 2009;**89**:1453–8.
65. Nettleton JA, Hivert MF, Lemaitre RN, et al. Meta-analysis investigating associations between healthy diet and fasting glucose and insulin levels and modification by loci associated with glucose homeostasis in data from 15 cohorts. *Am J Epidemiol* 2013;**177**: 103–15.
66. Kitabchi AE, Temprosa M, Knowler WC, et al. Role of insulin secretion and sensitivity in the evolution of type 2 diabetes in the diabetes prevention program: effects of lifestyle intervention and metformin. *Diabetes* 2005;**54**:2404–14.
67. Bo S, Gambino R, Ciccone G, Rosato R, et al. Effects of TCF7L2 polymorphisms on glucose values after a lifestyle intervention. *Am J Clin Nutr* 2009;**90**:1502–8.
68. Haupt A, Thamer C, Heni M, et al. Gene variants of TCF7L2 influence weight loss and body composition during lifestyle intervention in a population at risk for type 2 diabetes. *Diabetes* 2010;**59**: 747–50.
69. Fisher E, Boeing H, Fritsche A, et al. Whole-grain consumption and transcription factor-7-like 2 (TCF7L2) rs7903146: gene-diet interaction in modulating type 2 diabetes risk. *Br J Nutr* 2009; **101**:478–81.
70. Wang J, Kuusisto J, Vänttinen M, et al. Variants of transcription factor 7-like 2 (TCF7L2) gene predict conversion to type 2 diabetes in the Finnish Diabetes Prevention Study and are associated with impaired glucose regulation and impaired insulin secretion. *Diabetologia* 2007;**50**:1192–200.
71. Lindström J, Louheranta A, Mannelin M, et al. The Finnish Diabetes Prevention Study (DPS): lifestyle intervention and 3-year results on diet and physical activity. *Diabetes Care* 2003; **26**:3230–6.
72. Cornelis MC, Qi L, Kraft P, Hu FB. TCF7L2, dietary carbohydrate, and risk of type 2 diabetes in US women. *Am J Clin Nutr* 2009;**89**: 1256–62.
73. Altshuler D, Hirschhorn JN, Klannemark M, et al. The common PPARgamma Pro12Ala polymorphism is associated with decreased risk of type 2 diabetes. *Nat Genet* 2000;**26**:76–80.
74. Yao YS, Li J, Jin YL, et al. Association between PPAR-gamma2 Pro12Ala polymorphism and obesity: a meta-analysis. *Mol Biol Rep* 2014;**13**:13.
75. Li Q, Chen R, Bie L, et al. Association of the variants in the PPARG gene and serum lipid levels: a meta-analysis of 74 studies. *J Cell Mol Med* 2015;**19**:198–209.
76. Gouda HN, Sagoo GS, Harding AH, et al. The association between the peroxisome proliferator-activated receptor-gamma2 (PPARG2) Pro12Ala gene variant and type 2 diabetes mellitus: a HuGE review and meta-analysis. *Am J Epidemiol* 2010;**171**:645–55.
77. Lapice E, Vaccaro O. Interaction between Pro12Ala polymorphism of PPARgamma2 and diet on adiposity phenotypes. *Curr Atheroscler Rep* 2014;**16**:462.
78. Lindi VI, Uusitupa MI, Lindström J, et al. Association of the Pro12Ala polymorphism in the PPAR-gamma2 gene with 3-year incidence of type 2 diabetes and body weight change in the Finnish Diabetes Prevention Study. *Diabetes* 2002;**51**:2581–6.
79. Franks PW, Jablonski KA, Delahanty L, et al. The Pro12Ala variant at the peroxisome proliferator-activated receptor gamma gene and change in obesity-related traits in the Diabetes Prevention Program. *Diabetologia* 2007;**50**:2451–60.
80. Ortega-Azorín C, Sorlí JV, Asensio EM, et al. Associations of the FTO rs9939609 and the MC4R rs17782313 polymorphisms with type 2 diabetes are modulated by diet, being higher when adherence to the Mediterranean diet pattern is low. *Cardiovasc Diabetol* 2012;**11**:137.
81. Steemburgo T, Azevedo MJ, Gross JL, et al. The rs9939609 polymorphism in the FTO gene is associated with fat and fiber intakes in patients with type 2 diabetes. *J Nutrigenet Nutrigenomics* 2013;**6**: 97–106.
82. Ruchat SM, Elks CE, Loos RJ, et al. Evidence of interaction between type 2 diabetes susceptibility genes and dietary fat intake for adiposity and glucose homeostasis-related phenotypes. *J Nutrigenet Nutrigenomics* 2009;**2**:225–34.
83. Nelson TL, Fingerlin TE, Moss LK, et al. Association of the peroxisome proliferator-activated receptor gamma gene with type 2 diabetes mellitus varies by physical activity among non-Hispanic whites from Colorado. *Metabolism* 2007;**56**:388–93.

84. Lamri A, Abi Khalil C, Jaziri R, et al. Dietary fat intake and polymorphisms at the PPARG locus modulate BMI and type 2 diabetes risk in the D.E.S.I.R. prospective study. *Int J Obes (Lond)* 2012;**36**:218–24.
85. Luan J, Browne PO, Harding AH, et al. Evidence for gene-nutrient interaction at the PPARgamma locus. *Diabetes* 2001;**50**:686–9.
86. Tai ES, Corella D, Deurenberg-Yap M, et al. Differential effects of the C1431T and Pro12Ala PPARgamma gene variants on plasma lipids and diabetes risk in an Asian population. *J Lipid Res* 2004;**45**:674–85.
87. Marín C, Pérez-Jiménez F, Gómez P, et al. The Ala54Thr polymorphism of the fatty acid-binding protein 2 gene is associated with a change in insulin sensitivity after a change in the type of dietary fat. *Am J Clin Nutr* 2005;**82**:196–200.
88. Fisher E, Schreiber S, Joost HG, et al. A two-step association study identifies CAV2 rs2270188 single nucleotide polymorphism interaction with fat intake in type 2 diabetes risk. *J Nutr* 2011;**141**:177–81.
89. Corella D, Qi L, Tai ES, et al. Perilipin gene variation determines higher susceptibility to insulin resistance in Asian women when consuming a high-saturated fat, low-carbohydrate diet. *Diabetes Care* 2006;**29**:1313–9.
90. Marcheva B, Ramsey KM, Buhr ED, et al. Disruption of the clock components CLOCK and BMAL1 leads to hypoinsulinaemia and diabetes. *Nature* 2010;**466**:627–31.
91. Garcia-Rios A, Gomez-Delgado FJ, Garaulet M, et al. Beneficial effect of CLOCK gene polymorphism rs1801260 in combination with low-fat diet on insulin metabolism in the patients with metabolic syndrome. *Chronobiol Int* 2014;**31**:401–8.
92. Dashti HS, Smith CE, Lee YC, et al. CRY1 circadian gene variant interacts with carbohydrate intake for insulin resistance in two independent populations: Mediterranean and North American. *Chronobiol Int* 2014;**31**:660–7.
93. Kanoni S, Nettleton JA, Hivert MF, et al. Total zinc intake may modify the glucose-raising effect of a zinc transporter (SLC30A8) variant: a 14-cohort meta-analysis. *Diabetes* 2011;**60**:2407–16.
94. Shan Z, Bao W, Zhang Y, et al. Interactions between zinc transporter-8 gene (SLC30A8) and plasma zinc concentrations for impaired glucose regulation and type 2 diabetes. *Diabetes* 2014;**63**:1796–803.
95. Marín C, Pérez-Martínez P, Delgado-Lista J, et al. The insulin sensitivity response is determined by the interaction between the G972R polymorphism of the insulin receptor substrate 1 gene and dietary fat. *Mol Nutr Food Res* 2011;**55**:328–35.
96. Zheng JS, Parnell LD, Smith CE, et al. Circulating 25-hydroxyvitamin D, IRS1 variant rs2943641, and insulin resistance: replication of a gene-nutrient interaction in 4 populations of different ancestries. *Clin Chem* 2014;**60**:186–96.
97. Nettleton JA, McKeown NM, Kanoni S, et al. Interactions of dietary whole-grain intake with fasting glucose- and insulin-related genetic loci in individuals of European descent: a meta-analysis of 14 cohort studies. *Diabetes Care* 2010;**33**:2684–91.
98. Hara K, Boutin P, Mori Y, et al. Genetic variation in the gene encoding adiponectin is associated with an increased risk of type 2 diabetes in the Japanese population. *Diabetes* 2002;**51**:536–40.
99. Jang Y, Lee JH, Kim OY, et al. The SNP276G>T polymorphism in the adiponectin (ACDC) gene is more strongly associated with insulin resistance and cardiovascular disease risk than SNP45T>G in nonobese/nondiabetic Korean men independent of abdominal adiposity and circulating plasma adiponectin. *Metabolism* 2006;**55**:59–66.
100. Yang WS, Yang YC, Chen CL, et al. Adiponectin SNP276 is associated with obesity, the metabolic syndrome, and diabetes in the elderly. *Am J Clin Nutr* 2007;**86**:509–13.
101. Hwang JY, Park JE, Choi YJ, et al. Carbohydrate intake interacts with SNP276G>T polymorphism in the adiponectin gene to affect fasting blood glucose, HbA1C, and HDL cholesterol in Korean patients with type 2 diabetes. *J Am Coll Nutr* 2013;**32**:143–50.

CHAPTER

5

Pathogenesis of Type 1 Diabetes: Role of Dietary Factors

Julie C. Antvorskov, Karsten Buschard, Knud Josefsen
The Bartholin Institute, Rigshospitalet, Copenhagen, Denmark

1. DIETARY FACTORS INVOLVED IN TYPE 1 DIABETES DEVELOPMENT

Many different dietary factors have been suspected to play a role in the pathogenesis of type 1 diabetes (T1D)[1]: nitrates and nitrites,[2] low vitamin D,[3] short-term breastfeeding,[4] and early exposure to different dietary proteins, as cow milk proteins and cereals, or to fruit- and berry juice,[5] and meat.[6] Exposure to cow milk proteins has been a main focus for many years. Studies have found that early intake of infant formula, and thus exposure to cow milk proteins, increases the risk of developing antibodies raised against the β cells and subsequent development of T1D.[7,8] Thus, it has been suggested that dietary intake of bovine insulin breaks neonatal tolerance to self-insulin.[9] However, in a recent study in infants at risk of T1D, the use of a hydrolyzed infant formula did not reduce the incidence of diabetes-associated autoantibodies after 7 years.[10] This is opposed to studies describing that hydrolyzed casein diet only both lowers the incidence of diabetes in animal models of T1D (the nonobese diabetic (NOD) mice)[11] and decreased by approximately 50% the cumulative incidence of one or more diabetes-associated autoantibodies, in infants with a family member affected by T1D or carrying a risk-conferring human leukocyte antigen (HLA) genotype. This finding was confirmed in a follow-up study at 10 years of age.[12] After weaning, high intake of cow milk has in some studies been found to increase the risk for T1D,[13,14] although in other studies conflicting data are found.[15,16] Overall, the results about cow milk intake and its connection to T1D pathogenesis could therefore be regarded as conflicting, and our major focus in this chapter will therefore be on the developing evidence of a connection between gluten intake and T1D.[17]

2. T1D, CELIAC DISEASE, AND GLUTEN INTAKE

During T1D development, the insulin-producing cells of the pancreas are destroyed by invading T lymphocytes reacting to self-antigens. Although far from all cells are destroyed, the remaining cells are not capable of regulating the blood glucose level. Because the cells are starved intracellularly of glucose, the metabolism is redirected toward fatty acid breakdown, which may cause potentially lethal ketoacidosis. Even if appropriately treated with exogenous insulin, T1D is a serious disease that may result in blindness, amputation, kidney failure, nephropathy, and other complications.

The incidence of T1D is highly variable across the world with high occurrence in Scandinavia, especially in Finland where 57.4 per 100,000 of at-risk population develop the disease per year. The disease is increasing in Europe by 3–4% per year, especially among young children, but also in other parts of the world as in Kuwait, for example, which now has an incidence of 22.3/100,000.[18]

The rise in T1D incidence is occurring more rapidly than can be accounted for by genetic change, elucidating the influence of environmental factors. Several observations supports the role of the environment in the pathogenesis of T1D: low concordance rate of T1D in monozygotic twins (30–40%), rapid increase in incidence, conspicuous increase in incidence in nonmigratory populations during a short course of time, studies in transmigratory populations, and difference in incidence between genetic similar populations.[19,20] Environmental factors that could play a role in disease susceptibility are, besides dietary factors: enteroviruses,[21] changes in the composition of gut microbiota,[22] stressful life events, and the "hygiene hypothesis"

Molecular Nutrition and Diabetes
http://dx.doi.org/10.1016/B978-0-12-801585-8.00005-1

suggesting that exposure to a large number of infections early in life prevents development of autoimmunity because of appropriately priming of the adaptive immune system[23] (Figure 1).

Celiac disease (CD) is an inflammatory intestinal disease with autoimmune features triggered by exposure to dietary gluten and related cereal proteins. The proteins induce an inflammatory immune response in the intestine, and withdrawal of them results in disease remission. The intestinal inflammation can lead to complete destruction of the intestinal epithelium, which results in crypt hyperplasia, loss of the villous structure, and infiltration of lymphocytes, mainly in the proximal part of the small intestine, with consequent malabsorption of nutrients, vitamins, and minerals.[24]

Screening studies indicate that CD has a high prevalence (approximately 1%) in many Western societies.[25] The classic disease picture of CD is characterized by children with severe symptoms including chronic diarrhea or constipation, abdominal distension and pain, and poor weight gain. Today, this is rarely observed, and CD is diagnosed because of unspecific abdominal discomfort or extraintestinal symptoms (e.g., anemia).[26] Moreover, studies suggest that a large group of people have undetected CD,[27,28] which has led to the concept of a "celiac iceberg" depicting different undiagnosed, silent forms of CD.[29]

Although patients with CD and T1D often have similar HLA haplotypes, other genes are also involved in the susceptibility to both CD and T1D; in fact, up to 15 risk alleles contribute to both diseases.[30] Thus, about half of the identified risk alleles are shared between the two diseases,[31] and concordantly, there is a high prevalence of patients with both diseases. The average prevalence of CD among children with T1D is 2–12%, and patients with CD have an earlier onset of diabetes than T1D patients without CD.[32,33]

3. DIETARY GLUTEN

Of the different taxonomic classes of cereals, wheat, rye, and barley are in the same tribe.[34] Wheat is currently the primary stable food for almost one-third of the world's population, including industrialized countries. This is probably due to the ability of wheat to give high yields under different cultivation conditions. But the most important factor is the unique properties of wheat that allow it to be processed into many different food products: bread, cakes, pasta, noodles, etc. These properties of wheat are described as viscoelasticity, which is attributed to the network formed when wheat flour is mixed with water to form dough. This network forms because of a protein called gluten (Figure 2), which accounts for up to 80% of the proteins in wheat. Gluten consists of gliadins (α-, γ-, and ω-gliadin, which are storage proteins classified as prolamins) and glutenins. Gliadins are monomers and confers elasticity to dough, whereas glutenins form large polymeric structures[34,35] that are responsible for extensibility.

There are storage proteins in wheat, rye, and barley other than prolamins, but the prolamins are special: The *amount* of prolamins is 40–50% in wheat, rye, and barley compared with 10% in oat, for example. Their *molecular mass* is higher (30,000–90,000) than in other

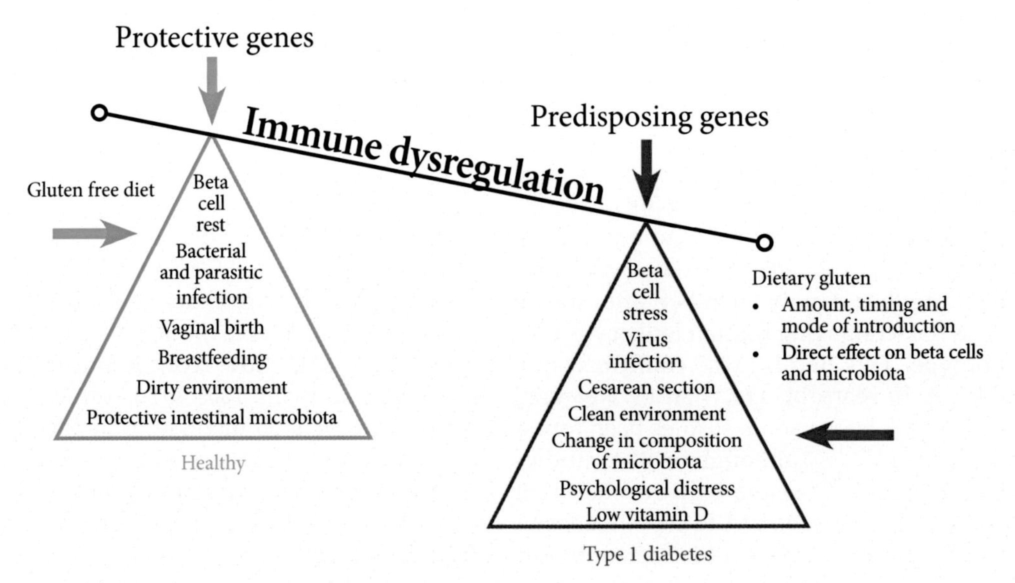

FIGURE 1 The intake of a diet containing gluten affects the pathogenesis of type 1 diabetes (T1D). T1D is a multifactorial disease dependent on a genetic predisposition and different environmental factors that affect the immune system toward dysregulation, resulting in susceptibility toward development of disease.

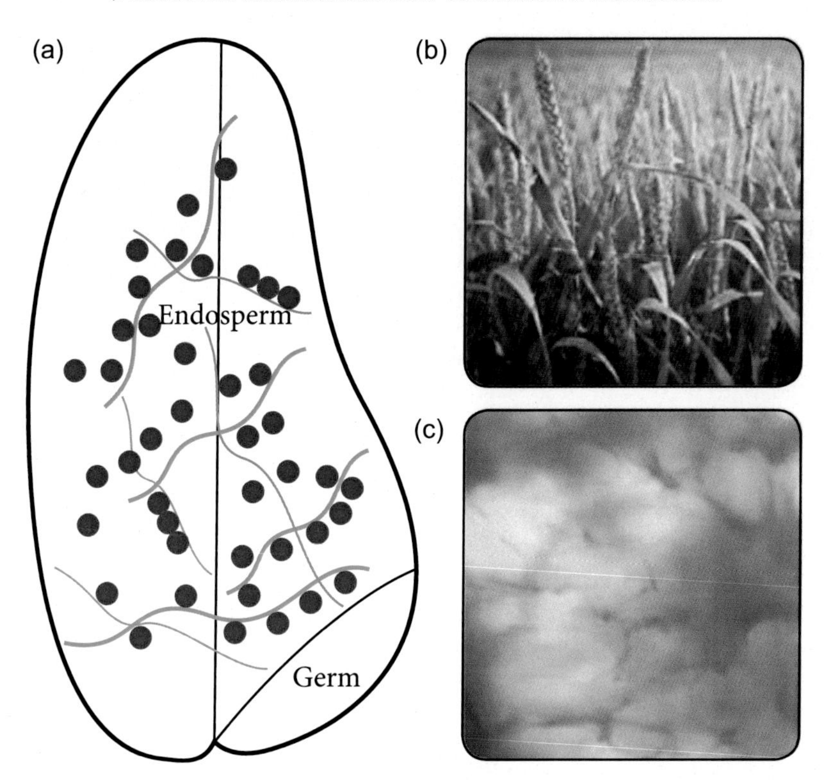

FIGURE 2 Gluten proteins differ from other cereal proteins because of a number of properties: A higher level of prolamins, higher molecular mass, high level of proline and glutamine residues, and multiple repeats (PQQPFPQQ). (a) The wheat grain, with the gluten network composed of gliadins (red) and glutenins (blue). (b) Wheat is currently the primary stable food for almost one third of the world's population, including industrialized countries. (c) The gluten network in a dough made from wheat flour.

cereal tribes (10,000−16,000 in rice), and the prolamins also differ in their *amino acid sequence*, with high proportions of proline and glutamine. The high level of these specific amino acids is the result of two characteristics: repeats based on short peptide motifs (e.g., PQQPFPQQ) and regions rich in one or more amino acids. These repeats vary in extent (e.g., 25% in maize, to more than 90% in wheat, rye and barley) and reflect the overall content of glutamine (40 residues/mol) and proline (30 residues/mol) in the prolamins of wheat, rye, and barley. This is in contrast to other cereals in which the repeats primarily consist of other amino acids, as in maize with a high content of leucine and alanine. Finally, the prolamins of wheat, rye, and barley are insoluble in water (highly hydrophobic) because of the presence of the repeated sequences, its propensity to form hydrogen bonds, and to the predominantly hydrophobic amino acid composition.[34,36,37]

In this section, we described wheat, rye, and barley storage proteins, which are essential for the ability of the crop to be processed into different, important food products. Most studies have been carried out on wheat gluten, which will therefore be our primary focus in the next section, but because of the similar structure of rye and barley, many of the mentioned results can also be applied to these cereals.

4. GLUTEN PEPTIDES ARE RESISTANT TO INTESTINAL DEGRADATION

Proteins are usually efficiently hydrolyzed into single amino acids or short peptides of 2 or 3 residues in the digestive tract. These products are efficiently absorbed, but are too short to be immunogenic.

The digestion of dietary proteins starts in the stomach, with denaturation by gastric acid and where proteolysis is initiated by gastric pepsins, which catalyze specific cleavage at phenylalanine or tyrosine. The next phase is intestinal where pancreatic enzymes (trypsin, chymotrypsin, elastase, and carboxypeptidase A and B) further reduce the majority of proteins into oligopeptides and free amino acids. Trypsin, chymotrypsin, and elastase are endopeptidases, which catalyze the hydrolysis of peptide bonds inside the protein chain. The final steps of digestion take place in the intestinal lumen and are associated with small intestinal, epithelial cells. The epithelial cells' brush-border membrane is facing the intestinal lumen and contains a number of peptidases. These peptidases are usually dimeric, integral membrane proteins with the catalytic site projecting into the lumen of the intestine. The brush-border membrane also contains peptidases, which rapidly hydrolyze peptides containing proline residues. The specificities of these enzymes are complementary to those of the pancreatic proteases, which have little or no ability to hydrolyze peptide bonds involving proline.[38,39] The result of the enzymatic processing is the release of free amino acids into the blood via the basolateral membrane.

As mentioned, gluten and related proteins from rye and barley are rich in proline, suggesting some resistance to gastric pepsins and pancreatic proteases. In a study of the digestion of α- and γ-gliadins, this was confirmed.[40] Different peptides derived from α- and γ-gliadin were incubated with a cocktail of gastric and pancreatic enzymes (pepsin, trypsin, chymotrypsin, elastase, carboxypeptidase A), which revealed that all peptides (except one) were highly resistant (more than 90% of starting material remained) to digestion.[40] Subsequently, the peptides were incubated with rat intestinal brush-border peptidases, which caused digestion of some of the peptides (to levels lower than 10% of starting material). Interestingly, this degradation required that the peptides were incubated 4 h with the brush-border membrane preparations. This prolonged exposure to the brush-border surface may not occur under normal physiological conditions. The results indicated two levels of resistance to enzymatic processing, one at the gastric/pancreatic level and one at the brush-border peptidase level.[40] The study showed that some gliadin peptides have a high resistance to degradation by the pancreatic enzymes while degradation by the brush-border enzymes was detected, but at a slow rate. The slow cleavage rate implies that there could be an intestinal, luminal accumulation of gliadin peptides following a gluten-containing meal.

Other studies have confirmed that gluten proteins are highly resistant to digestion. In vitro experiments showed that intact α2-gliadin after prolonged treatment with gastric and pancreatic enzymes and brush-border membrane fractions resulted in a stable proteolytic fragment, the 33-mer peptide (LQLQPFPQPQLPYPQPQL-PYPQPQLPYPQPQPF). Although control dietary peptides (myoglobin and other gluten peptides) were rapidly proteolyzed (half-life of digestion approximately 60 min), the 33-mer peptide remained intact (half-life of digestion >20 h).[41] To validate these studies, the 33-mer peptide and control peptides were incubated with human brush-border membrane. The control peptides were almost completely proteolyzed within 1–5 h, but the 33-mer peptide remained intact for at least 15 h. The results demonstrated that the 33-mer peptide is stable and remain intact throughout the digestive process. Homolog sequences to the 33-mer were found in rye and barley, but not in oat or other cereals.[41] The lower content of proline residues in other cereals (e.g., oat) results in higher susceptibility for degradation by proteases. This was shown in a study of the gastrointestinal enzyme degradation of storage proteins from barley, rye, and oat where only proteins from oat were clearly sensitive to digestion.[42]

Recently, studies have shown that microbial enzymes are capable of degrading gluten proteins.[43,44] Preliminary data suggest that CD patients have a higher level of gliadin-metabolizing enzymes with bacterial origin than healthy controls. The authors suggest that the induction of gliadin proteolysis by intestinal bacteria could initiate CD because of microbial formation of immunogenic gluten peptides. The possible involvement of the intestinal bacteria microflora in the pathogenesis of CD is supported by studies showing that CD patients have changes in the composition of the duodenal microbiota of the gut.[45,46] Involvement of the microflora in the pathogenesis of T1D has also been described[22] and will be discussed later.

Thus, proteins from wheat, rye, and barley can be expected to be comparably resistant to gastric and intestinal proteolysis, although some degradation might be performed by microbial enzymes. Ingestion of gluten is therefore likely to result in a sustained high intestinal concentration of nondegradable gluten peptides.

5. DIETARY GLUTEN INFLUENCES THE DEVELOPMENT OF T1D

Studies suggest that dietary gluten is involved in the pathogenesis of T1D in humans and animals.

The highest incidence of diabetes in experimental animal models of T1D is found in animals on a wheat-based diet,[47] and a cereal-based diet promotes T1D development in both the NOD mice and BioBreeding (BB) rat.[48,49]

Because of the disease-promoting effect of a gluten-containing diet, studies of the effect of a diet without gluten were conducted. It was demonstrated that a GF nonpurified diet largely prevented diabetes onset in NOD mice as the effect of a GF diet for 320 days caused an incidence reduction from 64% in control NOD mice to 15% in NOD mice on the GF diet.[50] Also, diabetes onset was delayed and the incidence reduced in female NOD mice that received wheat and barley protein-free diet throughout life (45% by age 32 weeks vs 88% in control mice).[51] Thus, dietary gluten and the risk of diabetes development are highly correlated.

In continuation of the animal studies, the effect of a GF diet on humane prediabetic patients was studied, but on a limited number of patients: 17 first-degree relatives (positive for at least two β-cell autoantibodies) received a GF diet for 6 months followed by 6 months of a gluten-containing diet. The autoantibody titers did not show significant changes according to the diet. However, the insulin response in a glucose tolerance test significantly increased ($P = 0.04$) in 12 of 14 patients on the GF diet. After return to a normal diet, the acute insulin response decreased ($P = 0.07$) in 10 of 13 patients. Furthermore, the patients' insulin sensitivity measured according to the homeostasis model of insulin resistance, improved after the GF diet and subsequently decreased ($P < 0.005$) after 6 months of normal diet.[52] The effect of a GF diet for 1 year was also tried in seven T1D-predisposed children positive for β-cell autoantibodies. The GF diet did not influence the autoantibody titers.[53] In these studies, the beneficial effect of a GF diet was investigated in subjects with immunity against the islets of Langerhans (i.e., β-cell autoantibodies), suggesting that the role of gluten (and the gluten-free (GF) diet) should be studied in the early events that initiate the T1D autoimmune process. In studies of patients with both CD and T1D, a GF diet mediates clinical improvements for the patients, such as an increase in weight, height, hemoglobin, corpuscular volume, and diabetic control,[33,54,55] although other studies find no effect.[56,57] Recently, the effect of a GF diet was studied in a case of a newly diagnosed child. The patient initiated a GF diet right after T1D diagnosis, which resulted in a reduction in hemoglobin A1c from 7.8% at diagnosis to 5.8–6.0% and a fasting blood glucose in the normal range without insulin therapy. Sixteen months after diagnosis, the fasting blood glucose was still normal, and the patient without daily insulin therapy. Glutamate decarboxylase autoantibodies did not change during the trial. The authors suggest that a GF diet prolongs remission.[58] Thus, a GF diet might preserve the function of the insulin-producing β cells in individuals at high risk of developing T1D, although the mechanism is not clear,[52] and it should be investigated in a human intervention trial if individuals at high risk of developing T1D might also benefit from a GF diet.

6. THE IMMUNE RESPONSE TO GLUTEN IN T1D PATIENTS

Rectal gluten exposure is used to investigate abnormal reactivity to gluten in both CD and T1D patients. In one-fifth of tTG antibody–negative T1D patients with CD-associated HLA haplotypes, but normal mucosal histology, gluten causes mucosal lymphocyte infiltration.[59] These results are supported by a study of 17 children with T1D and CD-associated HLA haplotypes, but with a normal mucosal architecture and no serologic markers of CD. Jejunal biopsies from the T1D patients showed signs of mucosal inflammation and lymphocyte infiltration and an increased immune response to gliadin. In the nondiabetic controls, no signs of mucosal inflammation were observed, even in those with CD-associated genetic susceptibility.[60] Also, 20 of 42 tTG antibody–negative T1D patients showed increased T-cell proliferation to wheat proteins, resulting in secretion of interferon (IFN)-γ, interleukin (IL)-17A and tumor necrosis factor (TNF), which are pro-inflammatory cytokines. This was not due to a higher frequency of CD-associated (HLA-DQB1*02) haplotypes among the T1D patients, indicating that the response to wheat proteins must be diabetes-specific.[61] A similar result was seen in an earlier study where 24% of newly diagnosed T1D patients reacted with proliferation when stimulated with wheat proteins.[62] Thus, at least in a subgroup of T1D patients, hypersensitivity is found, but because it is not linked to CD-associated HLA haplotype, it is most likely a diabetes-specific effect.

Diabetes patients show immune reactivity to other wheat antigens than gluten. Screening a wheat complementary DNA (cDNA) library with serum from diabetic rats identified the presence of antibodies against a wheat storage globulin (Glo-3A), and the study found a correlation between the reactivity of immunoglobulin G antibodies to Glo-3A and islet inflammation. Moreover, newly diagnosed diabetic patients had antibodies against Glo-3A, but such were absent from nondiabetic controls,[63,64] and antibody levels were elevated in patients suffering both CD and T1D.[65] Increased Glo-3A immune reactivity was not found in a different study investigating children with islet autoantibodies.[66]

Thus, the evidence is considerable that gluten and possibly other wheat proteins can participate in the pathogenesis of T1D.

7. THE EFFECT OF GLUTEN ON T1D DEPENDS ON DOSE, CONTEXT, AND TIMING

The effect of gluten in the pathogenesis of T1D depends on the *dose* of dietary gluten and on the *context* and *timing* of gluten introduction neonatally.

Regarding the *dose* of dietary gluten, both low and high amounts of gluten protect against diabetes development in NOD mice. A GF diet resulted in only 16% diabetes development, and a similar/comparable reduction was seen in a diet containing four times more than the standard diet.[67] These results were later confirmed.[68]

The diabetogenic effect of gluten is also dependent on the *context* in which it is introduced. The diabetes incidence depends on the diet the mothers are fed during and after pregnancy. If mothers are fed a gluten-containing diet during lactation, the diabetes incidence is reduced in the offsprings.[69] This result is thus similar to those obtained in the human Diabetes and Autoimmunity Study in the Young study[70] and in a study in NOD mice, in which late introduction of gluten (i.e., not immediately after weaning) delayed diabetes onset and resulted in lower levels of insulin autoantibodies and insulitis in the offsprings.[51]

Early and late neonatal exposure to gluten both increase the risk of developing islet autoantibodies, thus emphasizing the *timing* of gluten introduction. This was found in two prospective cohort studies. In the BABYDIAB study, gluten exposure before the age of 3 months resulted in an increased risk of development of islet antibodies (hazard ratio = 4.0) when compared with a control group that was breastfed only.[71] Similarly, the Diabetes and Autoimmunity Study in the Young study found that any introduction of cereals before the age of 3 months, whether or not they contained gluten, was associated with a risk of developing autoantibodies (hazard ratio = 4.32), although reduced if breastfeeding was continued during gluten introduction. This was also the case when introduced late, after 7 months of age (hazard ratio = 5.6).[70] The risk of late introduction was not found in the BABYDIAB study.

Together, these studies suggest that a "time window" exists during which the introduction of gluten poses the lowest risk of autoantibody development[72] and that the time of gluten introduction is critical for the immune tolerance to food antigens during adult life.[73]

That dose, context, and timing are important in oral tolerance induction to gluten is also seen in CD.

During 1985–1987, the incidence rate of CD in children less than 2 years old increased four-fold, then returned to normal. These fluctuations were correlated to specific changes in gluten introduction, as gluten exposure had been delayed to 6 months, that larger amounts of gluten were used and that breastfeeding was terminated when gluten was introduced.[74] Breastfeeding was previously shown to protect children from CD[75] and T1D[4] through modulation of the developing cellular immune system.[76]

Differences in gluten introduction can thus influence the effect of gluten on the development of CD and T1D.

8. GLUTEN INTAKE, T1D, AND THE INTESTINAL MICROFLORA

It has been discussed whether the effect of gluten on the development of T1D is a direct effect of gluten or an indirect effect caused by gluten-induced changes of the intestinal microflora. First of all, gluten intake changes the composition of the intestinal microbiota, and NOD mice on a gluten-containing diet have a higher level of Gram-positive and cecal bacteria compared with NOD mice fed a GF diet.[77] Another study found that specific species are increased by the intake of a GF diet such as *Akkermansia*, whereas *Bifidobacterium*, *Barnesiella*, and *Tannerella* was increased during intake of gluten.[78] That changes in the microflora could influence T1D pathogenesis is also illustrated by results from BB rats showing that specific antibiotic treatment (fusidic acid[79] and vancomycin[80]) directed against Gram-positive bacteria results in a reduction of T1D. Some studies also showed an increased level of diabetes development in mice raised in germ-free animal facilities, possibly caused by changes in innate immune parameters affecting disease outcome, while others did not.[22] In humans, the establishment of a natural microflora early in life seems to influence children risk for chronic immune disorders. Thus, children delivered by cesarean section have significantly increased risk of development of disease[81] and a 20% increase in the risk of childhood-onset type T1D after cesarean section delivery.[82] The mechanism behind this effect of cesarean section seems to be the long-term systemic effect on the regulatory immune system. In NOD mice, the first exposure to microorganisms after birth seems to be important for the early life intestinal colonization pattern and the subsequent priming of regulatory immune system in mice. Adult NOD mice born by cesarean section had lower proportions of Foxp3$^+$ regulatory T cells, tolerogenic CD103$^+$ dendritic cells (DCs), and lower anti-inflammatory IL-10 gene expression both in intestinal lymphoid compartments (mesenteric lymph nodes) as well as in systemic lymphoid organs (spleens) implying a long-term systemic effect on the regulatory immune system.[83] However, if gluten effects on T1D development is mediated through these diet-induced changes in intestinal microflora or a more direct effect of gluten intake is the course, is questioned by studies showing that it is a microflora-independent mechanism that is responsible for the protective effects of the GF diet. Patrick et al. showed that a cereal-based diet is a stronger promoter of T1D than gut microflora, and stronger disease protection was seen in animals fed a hydrolyzed casein-based diet than could be obtained by changing the microflora.[84]

9. INTESTINAL ALTERATIONS IN ANIMAL MODELS OF T1D AND HUMAN PATIENTS

The intestine is continuously exposed to a wide variety of substances, some of which are potentially harmful to the host. In order to counteract the entry of molecules with antigenic properties, the intestine possesses several mechanisms (e.g., digestive enzymes, tight junctions, microbiota) that provide an integrated and functionally effective barrier against the environment. If this tightly regulated trafficking of molecules is disturbed, molecules with antigenic properties cross the intestinal barrier and gain access to the intestinal submucosa, which could increase intestinal inflammatory activity.

In BB rats, the intestine is more leaky and the mucosal crypts flattened compared with control rats. Further, before insulitis and overt diabetes, extensive epithelial cell proliferation and pro-inflammatory activity are seen.[85–88]

Similar findings have been obtained in humans. In 81 patients with different levels of islet autoimmunity, all had increased intestinal permeability as measured by lactulose-mannitol testing before the onset of T1D.[89] In 22 of 46 strictly nonceliac T1D patients similar changes in intestinal barrier structure and function were found,[90] with higher gut-permeability and changes in microvilli and tight junctions. In pediatric diabetic patients with HLA-DQ2, higher intestinal permeability was also seen, which could facilitate contact of food antigens with the mucosal immune system.[91] These patients are therefore more likely to develop pathological immune responses toward food antigens, which is similar to findings in CD, where changes in the jejunal tight junction and therefore barrier function are seen early in the disease.[92] Therefore, in both T1D and CD patients, the barrier function of the intestine seems critical for disease development.

The mechanism of increased intestinal permeability probably involves the protein zonulin, which opens intestinal tight junctions. When compared with diabetes-resistant controls, BB rats have a 35-fold higher intestinal intraluminal zonulin levels, and blockade of the zonulin receptor reduces the diabetes incidence by 70%.[93] The importance of the barrier function for disease development is illustrated in a study in BB rats in which diabetes is prevented following restoration of the impaired intestinal barrier,[94] and by studies in T1D patients, where 70% of prediabetic T1D patients had increased serum zonulin compared with controls.[95] The results suggest that zonulin participates in the pathogenesis of T1D, as the protein is upregulated and the intestinal permeability increased before the onset of T1D. Interestingly, a direct effect of gliadin on intestinal permeability has been found. Gliadin is able to bind to the chemokine receptor CXCR3, which leads to MyD88-dependent zonulin release in enterocytes, thus directly increasing intestinal permeability.[96]

These studies show that changes in intestinal morphology and permeability are seen before the development of T1D.

10. THE NUMBER OF PANCREAS-INFILTRATING AUTOREACTIVE T CELLS IS INCREASED IN THE INTESTINAL TISSUE

Much evidence suggests that the intestinal immune system is involved in the development of T1D and that pancreas-infiltrating autoreactive T cells are likely to become activated in the lymphoid tissue in the gut.

Immune-activation has been demonstrated in structurally normal intestine of patients with T1D, both in patients with and without CD-associated HLA genotype. In children diagnosed with T1D, but not with CD, the mucosal expression of major histocompatibility complex class II antigens[97,98] and intercellular adhesion molecule-1 were higher.[98] Also, the biopsies from jejunum showed higher mucosal densities of IFN-γ–, TNF-α–, IL-1α–, and IL-4–positive cells than controls.[98] During intestinal inflammation, the pro-inflammatory cytokine IL-1α is secreted by monocytes and epithelial cells. In contrast, the Th2-cytokine IL-4 enhances epithelial permeability.[98] Interestingly, these findings were also seen in patients not carrying the CD-associated HLA-DQ2 genotype, suggesting that the intestinal inflammation is part of T1D pathogenesis. Thus, immune cells in the gut mucosa seem activated in T1D patients, a feature that cannot be explained by genotypes shared with CD patients.

In NOD mice, the lymphocytes in the insulitis process express α4β7 integrin, which is specific for the gut-associated lymphoid tissue. The α4β7 integrin is found to be predominantly expressed by lymphocytes in the prediabetic phase, and adoptive transfer experiments of diabetogenic splenocytes elucidated that the first lymphocytes that accumulate in the pancreas express α4β7 integrin. Diabetes development can be prevented by antibodies against α4β7 integrin,[99,100] and mesenteric lymphocytes can transfer diabetes in passive transfer experiments. Diabetogenic T cells are therefore already activated before infiltrating the islets.[101] This is supported by a study performed to identify sites where islet-specific lymphocytes are activated, using isolated lymphocytes from spleen, pancreatic lymph nodes, and gut-associated and subcutaneous lymph nodes for transfer into an NOD SCID/SCID recipient. The study showed that most diabetogenic lymphocytes were found in pancreatic lymph nodes, but in young NOD mice (3 weeks of age), such lymphocytes were only found in gut-associated lymph nodes. Diabetogenic T cells might therefore be primed in the gut and the autoimmune response further amplified in the pancreatic lymph nodes.[102] Similar results have been obtained in human studies. One study showed that pancreas-infiltrating lymphocytes adhered specifically to endothelium from gut mucosa and pancreas.[103] Finally, it has been found that glutamate decarboxylase–reactive T cells from patients with T1D expressed α4β7 integrin.[104]

The gut immune system is activated in T1D patients, and the pancreas and the same lymphocytes seem to home to both the pancreas and the gut. It is therefore possible that islet-infiltrating, autoreactive T cells are activated in gut-associated lymphoid tissue.

11. INTAKE OF GLUTEN CHANGES SPECIFIC IMMUNE SYSTEM PARAMETERS

The intake of dietary gluten influences the immune system and studies have elucidated the effect of gluten both on innate cells populations as well as in adaptive immune cell subpopulations.

Gluten intake results in various changes in innate immune parameters. First of all, gluten has been shown to directly stimulate antigen-presenting cells. This was demonstrated in BALB/c mice, in which DCs from bone marrow, when exposed to gluten fragmented by trypsin, induced DC maturation as indicated by increased expression of major histocompatibility complex II and the costimulatory molecules CD86, CD40, and CD54. Upregulation of surface-expressed maturation markers was of similar magnitude to that seen by liposaccharides when 100 μg/mL gluten was compared with 10 ng/mL liposaccharides, thus emphasizes the potency of gluten. In addition, gluten exposure induced cytokine and chemokine production in the DC. The secretion was characterized by high production of the two chemokines MIP-2 and keratinocyte-derived cytokine, which is known to prepare the immune system for stronger reactivity, by attracting leukocytes and by increasing the reactive state of immune cells.[105] Also, gluten fragments obtained by α-chymotrypsin digestion of gluten, stimulated Toll-like receptors 4, 7, 8, and secretion of IFN-α when incubated with bone marrow–derived DC from mice transgenic for HLA-DQ8.[106] A similar result was obtained by using human monocyte-derived DC, in which stimulation increased CD80, CD83, CD86, and HLA-DR expression and increased secretion of IL-6, IL-8, IL-10, TNF-α, and monocyte chemoattractant proteins 1 and 2. In these experiments, gliadin reduced DC endocytosis, but improved the DC capacity to stimulate proliferation of allogeneic T cells.[107] In murine peritoneal macrophages, gluten similarly induces increased expression of pro-inflammatory cytokine genes such TNF-α, IL-12, and IL-15[108] and nitric oxide production[109,110] in an MYD88-dependent manner (myeloid differentiation factor 88 is part of the Toll-like receptor/IL-1R signaling cascade).[108]

At present, neither the gluten epitopes, nor the DC or macrophage receptors involved are known in the stimulation of gluten in antigen-presenting cells, but the magnitude of the signals should encourage clarification of the matter.

Besides having an effect on antigen-presenting cells, gluten intake is also found to induce upregulation of the activating natural killer group 2D receptor (NKG2D) on $DX5^+$ NK cells isolated from spleens of NOD mice and induce proliferation, measured by the expression of CD71. Moreover, intestinal expression of NKG2D ligands was increased in mice receiving gluten and the pancreas had higher insulitis score.[111] Furthermore, gliadin exposure increases direct cytotoxicity and IFN-γ secretion from isolated splenocytes and NK cells toward the pancreatic β-cell line MIN6 cells.[112]

Interestingly, gluten intake does not only have an effect on animals with susceptibility for T1D. Studies have shown that gluten greatly influences the level of different T-cell populations, and their cytokine profile, in normal healthy BALB/c mice. The intake of gluten leads to a decreased proportion of regulatory γδ T cells in all lymphoid compartments studied, and increased the number of Th17 cells, associated with the development of autoimmunity, in pancreas-associated lymph nodes.[113] Furthermore, the gluten-containing diet modified the cytokine pattern in both $Foxp3^-$ T cells and $Foxp3^+$ T cells toward a more inflammatory cytokine profile, with higher levels of IL-17, IL-2, IL-4, and IFN-γ. The GF diet inversely induced an anti-inflammatory cytokine profile with higher proportions of $TGF\text{-}\beta^+$ $Foxp3^-$ T cells in all tested lymphoid tissues and higher IL-10 expression within non-T cells in spleen.[114] In animal models of T1D, the intake of gluten also changes adaptive immune parameters. NOD mice on a wheat-based diet have a Th1 cytokine profile in the small intestine, with a four-fold increase in IFN-γ, TNF-α, and a 10-fold increase in inducible nitric oxide synthase compared with mice fed a semipurified hypoallergenic diet. Both groups of mice showed the same expression of IL-10 and TGF-β. Hence, the Th1/Th2 cytokine balance was shifted toward Th1 in the gut of gluten-exposed mice.[115] The pro-inflammatory effect of wheat has also been demonstrated in young BB rats, as indicated by a pro-inflammatory Th1 bias in the mesenterial lymph nodes. Thus, before insulitis is established, high numbers of $IFN\text{-}\gamma^+$ $CD4^+$ T cells that reacted to wheat protein antigens were found. Further, the study found a reduction in the expression of the Th2 cytokine-specific transcription factor *Gata3*, whereas expression of the Th1 cytokine-specific transcription factor *T-bet* remained unchanged, indicating that lymphocytes in the mesenterial lymph nodes are biased toward a Th1 response because of a Th2 deficit. This study also showed that these changes were seen before the pancreatic inflammation (already 1 week before weaning).[116] Recently, a wheat-containing diet was shown to increase the number of activated $CD4^+$ T cells, DCs ($CD11b^+CD11c^+$), and the number of Th17 cells in the colon of young NOD mice.[117] Thus, as in CD patients, in which gluten induce a Th1 cytokine profile in the gut,[118,119] gluten might act similarly in animal models of T1D.

In addition to the described changes in cytokine expression, a cereal-based diet also induces high IFN-γ and low IL-10 and TGF-β expression in insulitis lesions. This was not found in pancreata of BB rats fed a hydrolyzed casein-based semipurified diet. In these animals, less infiltration, low IFN-γ, and high TGF-β expression were seen. Diet without cereals thus protected BB rats from developing overt disease because they promoted a noninflammatory, pancreatic cytokine pattern.[120]

Dietary gluten is thus found to induce a Th1 cytokine bias in the gut and islet infiltrate of NOD and BB rats, and seems to directly stimulate the innate immune system.

12. GLUTEN IS FOUND IN BLOOD AND COULD AFFECT THE PANCREATIC β CELLS

In addition to affecting several aspects of the immune system, gliadin also directly stimulates the insulin-producing β cells both using enzyme-digested gliadin and a specific 33-mer fragment (residues 57–89 of alpha 2-gliadin), which is known to cause mucosal inflammation in celiac disease. The insulin release was stimulated due to closure of the ATP-sensitive potassium channels, as was demonstrated by patch clamp techniques, in a way similar to the oral antidiabetic drugs sulfonylureas, and in vitro the magnitude was twice as high as the maximal stimulation by glucose alone. Even in the presence of the inhibitory the fatty acid palmitate, gliadin still increased insulin release. This effect on insulin release was most likely the explanation of an observed weight increase in BALB/c mice following intravenous injection of enzyme-digested gliadin.[92] It is widely accepted that β cell stress can contribute to diabetes development as illustrated by reduced incidence in diabetes-prone animal models.[93]

The argument that gliadin can stimulate β cells to increase cellular stress or initiate unspecific inflammation in vivo is obviously only valid if gliadin fragments can enter the blood stream in bigger fragments than is usually considered possible during passage of the digestive system. We therefore examined the fate of the mentioned 33-mer fragment following oral delivery to NOD, BALB/c, and C57BL/6 mice and found that intact 33-mer could be detected (by mass spectroscopy) shortly after ingestion. Many degradation products were also found in pancreatic extract. The 33-mer was subsequently rapidly degraded in the circulation, and radioactive label from the 33-mer was incorporated in other proteins.

Thus, large gliadin fragments are readily absorbed through the intestine, probably facilitated by zonulin secretion, enter the circulation, and may initiate or worsen ongoing diabetes development by afflicting cellular stress to the β cells and by stimulating innate inflammation in the gut and pancreas.

13. CONCLUSION

Different dietary factors have been suspected to play a role in the pathogenesis of T1D, including cow milk protein, short-term breastfeeding, and others. New evidence has in particular pointed to a role for gluten-containing cereals, as gliadin seems to play an important role in the early events of the pathogenesis of T1D, possibly because of the molecule itself, which is resistant to intestinal degradation. During disease development, gluten sensitization of the gastrointestinal tract occurs and could spread to the pancreas area because of similar homing receptors for pancreas and gut. This might facilitate a specific T-cell response.

Application of the present knowledge in humans is under investigation, and early results suggest that GF diet might delay the progress of T1D development.

References

1. Akerblom HK, Vaarala O, Hyoty H, Ilonen J, Knip M. Environmental factors in the etiology of type 1 diabetes. *Am J Med Genet* 2002;**115**:18–29.
2. Virtanen SM, Jaakkola L, Rasanen L, Ylonen K, Aro A, Lounamaa R, et al. Nitrate and nitrite intake and the risk for type 1 diabetes in Finnish children. Childhood Diabetes in Finland Study Group. *Diabet Med* 1994;**11**:656–62.
3. Hypponen E, Laara E, Reunanen A, Jarvelin MR, Virtanen SM. Intake of vitamin D and risk of type 1 diabetes: a birth-cohort study. *Lancet* 2001;**358**:1500–3.
4. Kimpimaki T, Erkkola M, Korhonen S, Kupila A, Virtanen SM, Ilonen J, et al. Short-term exclusive breastfeeding predisposes young children with increased genetic risk of Type I diabetes to progressive beta-cell autoimmunity. *Diabetologia* 2001;**44**:63–9.
5. Virtanen SM, Nevalainen J, Kronberg-Kippila C, Ahonen S, Tapanainen H, Uusitalo L, et al. Food consumption and advanced beta cell autoimmunity in young children with HLA-conferred susceptibility to type 1 diabetes: a nested case-control design. *Am J Clin Nutr* 2012;**95**:471–8.
6. Weber KS, Raab J, Haupt F, Aschemeier B, Wosch A, Ried C, et al. Evaluating the diet of children at increased risk for type 1 diabetes: first results from the TEENDIAB study. *Public Health Nutr* 2014:1–9.
7. Virtanen SM, Rasanen L, Ylonen K, Aro A, Clayton D, Langholz B, et al. Early introduction of dairy products associated with increased risk of IDDM in Finnish children. The Childhood in Diabetes in Finland Study Group. *Diabetes* 1993;**42**:1786–90.
8. Dahlquist G, Savilahti E, Landin-Olsson M. An increased level of antibodies to beta-lactoglobulin is a risk determinant for early-onset type 1 (insulin-dependent) diabetes mellitus independent of islet cell antibodies and early introduction of cow's milk. *Diabetologia* 1992;**35**:980–4.
9. Vaarala O. Is it dietary insulin? *Ann N Y Acad Sci* 2006;**1079**:350–9.
10. Knip M, Akerblom HK, Becker D, Dosch HM, Dupre J, Fraser W, et al. Hydrolyzed infant formula and early beta-cell autoimmunity: a randomized clinical trial. *JAMA* 2014;**311**:2279–87.

11. Emani R, Asghar MN, Toivonen R, Lauren L, Soderstrom M, Toivola DM, et al. Casein hydrolysate diet controls intestinal T cell activation, free radical production and microbial colonisation in NOD mice. *Diabetologia* 2013;**56**:1781–91.
12. Knip M, Virtanen SM, Becker D, Dupre J, Krischer JP, Akerblom HK. Early feeding and risk of type 1 diabetes: experiences from the Trial to Reduce Insulin-dependent diabetes mellitus in the Genetically at Risk (TRIGR). *Am J Clin Nutr* 2011;**94**: 1814S–20S.
13. Virtanen SM, Laara E, Hypponen E, Reijonen H, Rasanen L, Aro A, et al. Cow's milk consumption, HLA-DQB1 genotype, and type 1 diabetes: a nested case-control study of siblings of children with diabetes. Childhood diabetes in Finland study group. *Diabetes* 2000;**49**:912–7.
14. Elliott RB, Harris DP, Hill JP, Bibby NJ, Wasmuth HE. Type I (insulin-dependent) diabetes mellitus and cow milk: casein variant consumption. *Diabetologia* 1999;**42**:292–6.
15. Rosenbauer J, Herzig P, Kaiser P, Giani G. Early nutrition and risk of Type 1 diabetes mellitus—a nationwide case-control study in preschool children. *Exp Clin Endocrinol Diabetes* 2007;**115**:502–8.
16. Merriman TR. Type 1 diabetes, the A1 milk hypothesis and vitamin D deficiency. *Diabetes Res Clin Pract* 2009;**83**:149–56.
17. Antvorskov JC, Josefsen K, Engkilde K, Funda DP, Buschard K. Dietary gluten and the development of type 1 diabetes. *Diabetologia* 2014;**57**:1770–80.
18. Patterson CC, Dahlquist GG, Gyurus E, Green A, Soltesz G. Incidence trends for childhood type 1 diabetes in Europe during 1989–2003 and predicted new cases 2005–20: a multicentre prospective registration study. *Lancet* 2009;**373**:2027–33.
19. Karvonen M, Tuomilehto J, Libman I, LaPorte R. A review of the recent epidemiological data on the worldwide incidence of type 1 (insulin-dependent) diabetes mellitus. World Health Organization DIAMOND Project Group. *Diabetologia* 1993;**36**:883–92.
20. Variation and trends in incidence of childhood diabetes in Europe. EURODIAB ACE Study Group. *Lancet* 2000;**355**:873–6.
21. Tracy S, Drescher KM, Jackson JD, Kim K, Kono K. Enteroviruses, type 1 diabetes and hygiene: a complex relationship. *Rev Med Virol* 2010;**20**:106–16.
22. Wen L, Ley RE, Volchkov PY, Stranges PB, Avanesyan L, Stonebraker AC, et al. Innate immunity and intestinal microbiota in the development of Type 1 diabetes. *Nature* 2008;**455**:1109–13.
23. Strachan DP. Hay fever, hygiene, and household size. *BMJ* 1989; **299**:1259–60.
24. Jabri B, Sollid LM. Tissue-mediated control of immunopathology in coeliac disease. *Nat Rev Immunol* 2009;**9**:858–70.
25. Maki M, Mustalahti K, Kokkonen J, Kulmala P, Haapalahti M, Karttunen T, et al. Prevalence of Celiac disease among children in Finland. *N Engl J Med* 2003;**348**:2517–24.
26. Green PH, Cellier C. Celiac disease. *N Engl J Med* 2007;**357**: 1731–43.
27. West J, Logan RF, Hill PG, Lloyd A, Lewis S, Hubbard R, et al. Seroprevalence, correlates, and characteristics of undetected coeliac disease in England. *Gut* 2003;**52**:960–5.
28. Meloni G, Dore A, Fanciulli G, Tanda F, Bottazzo GF. Subclinical coeliac disease in schoolchildren from northern Sardinia. *Lancet* 1999;**353**:37.
29. Rewers M, Liu E, Simmons J, Redondo MJ, Hoffenberg EJ. Celiac disease associated with type 1 diabetes mellitus. *Endocrinol Metab Clin North Am* 2004;**33**:197–214. xi.
30. Smyth DJ, Plagnol V, Walker NM, Cooper JD, Downes K, Yang JH, et al. Shared and distinct genetic variants in type 1 diabetes and celiac disease. *N Engl J Med* 2008;**359**:2767–77.
31. Plenge RM. Shared genetic risk factors for type 1 diabetes and celiac disease. *N Engl J Med* 2008;**359**:2837–8.
32. Collin P, Kaukinen K, Valimaki M, Salmi J. Endocrinological disorders and celiac disease. *Endocr Rev* 2002;**23**:464–83.
33. Hansen D, Brock-Jacobsen B, Lund E, Bjorn C, Hansen LP, Nielsen C, et al. Clinical benefit of a gluten-free diet in type 1 diabetic children with screening-detected celiac disease: a population-based screening study with 2 years' follow-up. *Diabetes Care* 2006;**29**:2452–6.
34. Shewry P, Casey R. *Seed proteins*. Kluwer Academic Publishers; 2003.
35. Shewry PR, Halford NG. Cereal seed storage proteins: structures, properties and role in grain utilization. *J Exp Bot* 2002;**53**: 947–58.
36. Tatham AS, Shewry PR. Comparative structures and properties of elastic proteins. *Philos Trans R Soc Lond B Biol Sci* 2002;**357**: 229–34.
37. Shewry PR, Halford NG, Belton PS, Tatham AS. The structure and properties of gluten: an elastic protein from wheat grain. *Philos Trans R Soc Lond B Biol Sci* 2002;**357**:133–42.
38. Zubay G. *Biochemistry*. Vm. C. Brown Publishers; 1993.
39. Erickson RH, Kim YS. Digestion and absorption of dietary protein. *Annu Rev Med* 1990;**41**:133–9.
40. Piper JL, Gray GM, Khosla C. Effect of prolyl endopeptidase on digestive-resistant gliadin peptides in vivo. *J Pharmacol Exp Ther* 2004;**311**:213–9.
41. Shan L, Molberg O, Parrot I, Hausch F, Filiz F, Gray GM, et al. Structural basis for gluten intolerance in celiac sprue. *Science* 2002;**297**:2275–9.
42. Vader LW, Stepniak DT, Bunnik EM, Kooy YM, de HW, Drijfhout JW, et al. Characterization of cereal toxicity for celiac disease patients based on protein homology in grains. *Gastroenterology* 2003;**125**:1105–13.
43. Bernardo D, Garrote JA, Nadal I, Leon AJ, Calvo C, Fernandez-Salazar L, et al. Is it true that coeliacs do not digest gliadin? Degradation pattern of gliadin in coeliac disease small intestinal mucosa. *Gut* 2009;**58**:886–7.
44. Helmerhorst EJ, Zamakhchari M, Schuppan D, Oppenheim FG. Discovery of a novel and rich source of gluten-degrading microbial enzymes in the oral cavity. *PLoS One* 2010;**5**:e13264.
45. Schippa S, Iebba V, Barbato M, Di NG, Totino V, Checchi MP, et al. A distinctive 'microbial signature' in celiac pediatric patients. *BMC Microbiol* 2010;**10**:175.
46. Nadal I, Donat E, Ribes-Koninckx C, Calabuig M, Sanz Y. Imbalance in the composition of the duodenal microbiota of children with coeliac disease. *J Med Microbiol* 2007;**56**:1669–74.
47. Hoorfar J, Scott FW, Cloutier HE. Dietary plant materials and development of diabetes in the BB rat. *J Nutr* 1991;**121**:908–16.
48. Coleman DL, Kuzava JE, Leiter EH. Effect of diet on incidence of diabetes in nonobese diabetic mice. *Diabetes* 1990;**39**:432–6.
49. Scott FW. Food-induced type 1 diabetes in the BB rat. *Diabetes Metab Rev* 1996;**12**:341–59.
50. Funda DP, Kaas A, Bock T, Tlaskalova-Hogenova H, Buschard K. Gluten-free diet prevents diabetes in NOD mice. *Diabetes Metab Res Rev* 1999;**15**:323–7.
51. Schmid S, Koczwara K, Schwinghammer S, Lampasona V, Ziegler AG, Bonifacio E. Delayed exposure to wheat and barley proteins reduces diabetes incidence in non-obese diabetic mice. *Clin Immunol* 2004;**111**:108–18.
52. Pastore MR, Bazzigaluppi E, Belloni C, Arcovio C, Bonifacio E, Bosi E. Six months of gluten-free diet do not influence autoantibody titers, but improve insulin secretion in subjects at high risk for type 1 diabetes. *J Clin Endocrinol Metab* 2003;**88**:162–5.
53. Fuchtenbusch M, Ziegler AG, Hummel M. Elimination of dietary gluten and development of type 1 diabetes in high risk subjects. *Rev Diabet Stud* 2004;**1**:39–41.
54. Amin R, Murphy N, Edge J, Ahmed ML, Acerini CL, Dunger DB. A longitudinal study of the effects of a gluten-free diet on glycemic control and weight gain in subjects with type 1 diabetes and celiac disease. *Diabetes Care* 2002;**25**:1117–22.

55. Saadah OI, Zacharin M, O'Callaghan A, Oliver MR, Catto-Smith AG. Effect of gluten-free diet and adherence on growth and diabetic control in diabetics with coeliac disease. *Arch Dis Child* 2004;**89**:871–6.
56. Rami B, Sumnik Z, Schober E, Waldhor T, Battelino T, Bratanic N, et al. Screening detected celiac disease in children with type 1 diabetes mellitus: effect on the clinical course (a case control study). *J Pediatr Gastroenterol Nutr* 2005;**41**:317–21.
57. Westman E, Ambler GR, Royle M, Peat J, Chan A. Children with coeliac disease and insulin dependent diabetes mellitus—growth, diabetes control and dietary intake. *J Pediatr Endocrinol Metab* 1999;**12**:433–42.
58. Sildorf SM, Fredheim S, Svensson J, Buschard K. Remission without insulin therapy on gluten-free diet in a 6-year old boy with type 1 diabetes mellitus. *BMJ Case Rep* 2012:2012.
59. Troncone R, Franzese A, Mazzarella G, Paparo F, Auricchio R, Coto I, et al. Gluten sensitivity in a subset of children with insulin dependent diabetes mellitus. *Am J Gastroenterol* 2003;**98**: 590–5.
60. Auricchio R, Paparo F, Maglio M, Franzese A, Lombardi F, Valerio G, et al. In vitro-deranged intestinal immune response to gliadin in type 1 diabetes. *Diabetes* 2004;**53**:1680–3.
61. Mojibian M, Chakir H, Lefebvre DE, Crookshank JA, Sonier B, Keely E, et al. Diabetes-specific HLA-DR-restricted proinflammatory T-cell response to wheat polypeptides in tissue transglutaminase antibody-negative patients with type 1 diabetes. *Diabetes* 2009;**58**:1789–96.
62. Klemetti P, Savilahti E, Ilonen J, Akerblom HK, Vaarala O. T-cell reactivity to wheat gluten in patients with insulin-dependent diabetes mellitus. *Scand J Immunol* 1998;**47**:48–53.
63. Loit E, Melnyk CW, MacFarlane AJ, Scott FW, Altosaar I. Identification of three wheat globulin genes by screening a *Triticum aestivum* BAC genomic library with cDNA from a diabetes-associated globulin. *BMC Plant Biol* 2009;**9**:93.
64. MacFarlane AJ, Burghardt KM, Kelly J, Simell T, Simell O, Altosaar I, et al. A type 1 diabetes-related protein from wheat (*Triticum aestivum*). cDNA clone of a wheat storage globulin, Glb1, linked to islet damage. *J Biol Chem* 2003;**278**:54–63.
65. Mojibian M, Chakir H, MacFarlane AJ, Lefebvre DE, Webb JR, Touchie C, et al. Immune reactivity to a glb1 homologue in a highly wheat-sensitive patient with type 1 diabetes and celiac disease. *Diabetes Care* 2006;**29**:1108–10.
66. Simpson M, Mojibian M, Barriga K, Scott FW, Fasano A, Rewers M, et al. An exploration of Glo-3A antibody levels in children at increased risk for type 1 diabetes mellitus. *Pediatr Diabetes* 2009;**10**:563–72.
67. Funda DP, Kaas A, Tlaskalova-Hogenova H, Buschard K. Gluten-free but also gluten-enriched (gluten+) diet prevent diabetes in NOD mice; the gluten enigma in type 1 diabetes. *Diabetes Metab Res Rev* 2008;**24**:59–63.
68. Mueller DB, Koczwara K, Mueller AS, Pallauf J, Ziegler AG, Bonifacio E. Influence of early nutritional components on the development of murine autoimmune diabetes. *Ann Nutr Metab* 2009;**54**:208–17.
69. Scott FW, Rowsell P, Wang GS, Burghardt K, Kolb H, Flohe S. Oral exposure to diabetes-promoting food or immunomodulators in neonates alters gut cytokines and diabetes. *Diabetes* 2002;**51**:73–8.
70. Norris JM, Barriga K, Klingensmith G, Hoffman M, Eisenbarth GS, Erlich HA, et al. Timing of initial cereal exposure in infancy and risk of islet autoimmunity. *JAMA* 2003;**290**: 1713–20.
71. Ziegler AG, Schmid S, Huber D, Hummel M, Bonifacio E. Early infant feeding and risk of developing type 1 diabetes-associated autoantibodies. *JAMA* 2003;**290**:1721–8.
72. Atkinson M, Gale EA. Infant diets and type 1 diabetes: too early, too late, or just too complicated? *JAMA* 2003;**290**:1771–2.
73. Strobel S, Mowat AM. Immune responses to dietary antigens: oral tolerance. *Immunol Today* 1998;**19**:173–81.
74. Ivarsson A, Persson LA, Nystrom L, Ascher H, Cavell B, Danielsson L, et al. Epidemic of coeliac disease in Swedish children. *Acta Paediatr* 2000;**89**:165–71.
75. Ivarsson A, Hernell O, Stenlund H, Persson LA. Breast-feeding protects against celiac disease. *Am J Clin Nutr* 2002;**75**:914–21.
76. Jeppesen DL, Hasselbalch H, Lisse IM, Ersboll AK, Engelmann MD. T-lymphocyte subsets, thymic size and breast-feeding in infancy. *Pediatr Allergy Immunol* 2004;**15**:127–32.
77. Hansen AK, Ling F, Kaas A, Funda DP, Farlov H, Buschard K. Diabetes preventive gluten-free diet decreases the number of caecal bacteria in non-obese diabetic mice. *Diabetes Metab Res Rev* 2006;**22**:220–5.
78. Marietta EV, Gomez AM, Yeoman C, Tilahun AY, Clark CR, Luckey DH, et al. Low incidence of spontaneous type 1 diabetes in non-obese diabetic mice raised on gluten-free diets is associated with changes in the intestinal microbiome. *PLoS One* 2013; **8**:e78687.
79. Buschard K, Pedersen C, Hansen SV, Hageman I, Aaen K, Bendtzen K. Anti-diabetogenic effect of fusidic acid in diabetes prone BB rats. *Autoimmunity* 1992;**14**:101–4.
80. Hansen CH, Krych L, Nielsen DS, Vogensen FK, Hansen LH, Sorensen SJ, et al. Early life treatment with vancomycin propagates *Akkermansia muciniphila* and reduces diabetes incidence in the NOD mouse. *Diabetologia* 2012;**55**:2285–94.
81. Sevelsted A, Stokholm J, Bonnelykke K, Bisgaard H. Cesarean section and chronic immune disorders. *Pediatrics* 2015;**135**:e92–8.
82. Cardwell CR, Stene LC, Joner G, Cinek O, Svensson J, Goldacre MJ, et al. Caesarean section is associated with an increased risk of childhood-onset type 1 diabetes mellitus: a meta-analysis of observational studies. *Diabetologia* 2008;**51**: 726–35.
83. Hansen CH, Andersen LS, Krych L, Metzdorff SB, Hasselby JP, Skov S, et al. Mode of delivery shapes gut colonization pattern and modulates regulatory immunity in mice. *J Immunol* 2014; **193**:1213–22.
84. Patrick C, Wang GS, Lefebvre DE, Crookshank JA, Sonier B, Eberhard C, et al. Promotion of autoimmune diabetes by cereal diet in the presence or absence of microbes associated with gut immune activation, regulatory imbalance, and altered cathelicidin antimicrobial peptide. *Diabetes* 2013;**62**:2036–47.
85. Meddings JB, Jarand J, Urbanski SJ, Hardin J, Gall DG. Increased gastrointestinal permeability is an early lesion in the spontaneously diabetic BB rat. *Am J Physiol* 1999;**276**:G951–7.
86. Neu J, Reverte CM, Mackey AD, Liboni K, Tuhacek-Tenace LM, Hatch M, et al. Changes in intestinal morphology and permeability in the biobreeding rat before the onset of type 1 diabetes. *J Pediatr Gastroenterol Nutr* 2005;**40**:589–95.
87. Graham S, Courtois P, Malaisse WJ, Rozing J, Scott FW, Mowat AM. Enteropathy precedes type 1 diabetes in the BB rat. *Gut* 2004;**53**:1437–44.
88. Hardin JA, Donegan L, Woodman RC, Trevenen C, Gall DG. Mucosal inflammation in a genetic model of spontaneous type I diabetes mellitus. *Can J Physiol Pharmacol* 2002;**80**:1064–70.
89. Bosi E, Molteni L, Radaelli MG, Folini L, Fermo I, Bazzigaluppi E, et al. Increased intestinal permeability precedes clinical onset of type 1 diabetes. *Diabetologia* 2006;**49**:2824–7.
90. Secondulfo M, Iafusco D, Carratu R, deMagistris L, Sapone A, Generoso M, et al. Ultrastructural mucosal alterations and increased intestinal permeability in non-celiac, type I diabetic patients. *Dig Liver Dis* 2004;**36**:35–45.
91. Kuitunen M, Saukkonen T, Ilonen J, Akerblom HK, Savilahti E. Intestinal permeability to mannitol and lactulose in children with type 1 diabetes with the HLA-DQB1*02 allele. *Autoimmunity* 2002;**35**:365–8.

92. Fasano A, Not T, Wang W, Uzzau S, Berti I, Tommasini A, et al. Zonulin, a newly discovered modulator of intestinal permeability, and its expression in coeliac disease. *Lancet* 2000;**355**:1518–9.
93. Watts T, Berti I, Sapone A, Gerarduzzi T, Not T, Zielke R, et al. Role of the intestinal tight junction modulator zonulin in the pathogenesis of type I diabetes in BB diabetic-prone rats. *Proc Natl Acad Sci USA* 2005;**102**:2916–21.
94. Visser JT, Lammers K, Hoogendijk A, Boer MW, Brugman S, Beijer-Liefers S, et al. Restoration of impaired intestinal barrier function by the hydrolysed casein diet contributes to the prevention of type 1 diabetes in the diabetes-prone BioBreeding rat. *Diabetologia* 2010;**53**:2621–8.
95. Sapone A, de ML, Pietzak M, Clemente MG, Tripathi A, Cucca F, et al. Zonulin upregulation is associated with increased gut permeability in subjects with type 1 diabetes and their relatives. *Diabetes* 2006;**55**:1443–9.
96. Lammers KM, Lu R, Brownley J, Lu B, Gerard C, Thomas K, et al. Gliadin induces an increase in intestinal permeability and zonulin release by binding to the chemokine receptor CXCR3. *Gastroenterology* 2008;**135**:194–204.
97. Savilahti E, Ormala T, Saukkonen T, Sandini-Pohjavuori U, Kantele JM, Arato A, et al. Jejuna of patients with insulin-dependent diabetes mellitus (IDDM) show signs of immune activation. *Clin Exp Immunol* 1999;**116**:70–7.
98. Westerholm-Ormio M, Vaarala O, Pihkala P, Ilonen J, Savilahti E. Immunologic activity in the small intestinal mucosa of pediatric patients with type 1 diabetes. *Diabetes* 2003;**52**:2287–95.
99. Hanninen A, Salmi M, Simell O, Jalkanen S. Mucosa-associated (beta 7-integrinhigh) lymphocytes accumulate early in the pancreas of NOD mice and show aberrant recirculation behavior. *Diabetes* 1996;**45**:1173–80.
100. Yang XD, Sytwu HK, McDevitt HO, Michie SA. Involvement of beta 7 integrin and mucosal addressin cell adhesion molecule-1 (MAdCAM-1) in the development of diabetes in obese diabetic mice. *Diabetes* 1997;**46**:1542–7.
101. Hanninen A, Jaakkola I, Jalkanen S. Mucosal addressin is required for the development of diabetes in nonobese diabetic mice. *J Immunol* 1998;**160**:6018–25.
102. Jaakkola I, Jalkanen S, Hanninen A. Diabetogenic T cells are primed both in pancreatic and gut-associated lymph nodes in NOD mice. *Eur J Immunol* 2003;**33**:3255–64.
103. Hanninen A, Salmi M, Simell O, Jalkanen S. Endothelial cell-binding properties of lymphocytes infiltrated into human diabetic pancreas. Implications for pathogenesis of IDDM. *Diabetes* 1993;**42**:1656–62.
104. Paronen J, Klemetti P, Kantele JM, Savilahti E, Perheentupa J, Akerblom HK, et al. Glutamate decarboxylase-reactive peripheral blood lymphocytes from patients with IDDM express gut-specific homing receptor alpha4beta7-integrin. *Diabetes* 1997;**46**:583–8.
105. Nikulina M, Habich C, Flohe SB, Scott FW, Kolb H. Wheat gluten causes dendritic cell maturation and chemokine secretion. *J Immunol* 2004;**173**:1925–33.
106. Ciccocioppo R, Rossi M, Pesce I, Ricci G, Millimaggi D, Maurano F, et al. Effects of gliadin stimulation on bone marrow-derived dendritic cells from HLA-DQ8 transgenic MICE. *Dig Liver Dis* 2008;**40**:927–35.
107. Palova-Jelinkova L, Rozkova D, Pecharova B, Bartova J, Sediva A, Tlaskalova-Hogenova H, et al. Gliadin fragments induce phenotypic and functional maturation of human dendritic cells. *J Immunol* 2005;**175**:7038–45.
108. Thomas KE, Sapone A, Fasano A, Vogel SN. Gliadin stimulation of murine macrophage inflammatory gene expression and intestinal permeability are MyD88-dependent: role of the innate immune response in Celiac disease. *J Immunol* 2006;**176**:2512–21.
109. Tuckova L, Flegelova Z, Tlaskalova-Hogenova H, Zidek Z. Activation of macrophages by food antigens: enhancing effect of gluten on nitric oxide and cytokine production. *J Leukoc Biol* 2000;**67**:312–8.
110. Tuckova L, Novotna J, Novak P, Flegelova Z, Kveton T, Jelinkova L, et al. Activation of macrophages by gliadin fragments: isolation and characterization of active peptide. *J Leukoc Biol* 2002;**71**:625–31.
111. Adlercreutz EH, Weile C, Larsen J, Engkilde K, Agardh D, Buschard K, et al. A gluten-free diet lowers NKG2D and ligand expression in BALB/c and non-obese diabetic (NOD) mice. *Clin Exp Immunol* 2014;**177**:391–403.
112. Larsen J, Dall M, Antvorskov JC, Weile C, Engkilde K, Josefsen K, et al. Dietary gluten increases natural killer cell cytotoxicity and cytokine secretion. *Eur J Immunol* 2014;**44**:3056–67.
113. Antvorskov JC, Fundova P, Buschard K, Funda DP. Impact of dietary gluten on regulatory T cells and Th17 cells in BALB/c mice. *PLoS One* 2012;**7**:e33315.
114. Antvorskov JC, Fundova P, Buschard K, Funda DP. Dietary gluten alters the balance of proinflammatory and anti-inflammatory cytokines in T cells of BALB/c mice. *Immunology* 2012.
115. Flohe SB, Wasmuth HE, Kerad JB, Beales PE, Pozzilli P, Elliott RB, et al. A wheat-based, diabetes-promoting diet induces a Th1-type cytokine bias in the gut of NOD mice. *Cytokine* 2003;**21**:149–54.
116. Chakir H, Lefebvre DE, Wang H, Caraher E, Scott FW. Wheat protein-induced proinflammatory T helper 1 bias in mesenteric lymph nodes of young diabetes-prone rats. *Diabetologia* 2005;**48**:1576–84.
117. Alam C, Valkonen S, Palagani V, Jalava J, Eerola E, Hanninen A. Inflammatory tendencies and overproduction of IL-17 in the colon of young NOD mice are counteracted with diet change. *Diabetes* 2010;**59**:2237–46.
118. Nilsen EM, Jahnsen FL, Lundin KE, Johansen FE, Fausa O, Sollid LM, et al. Gluten induces an intestinal cytokine response strongly dominated by interferon gamma in patients with celiac disease. *Gastroenterology* 1998;**115**:551–63.
119. Nilsen EM, Lundin KE, Krajci P, Scott H, Sollid LM, Brandtzaeg P. Gluten specific, HLA-DQ restricted T cells from coeliac mucosa produce cytokines with Th1 or Th0 profile dominated by interferon gamma. *Gut* 1995;**37**:766–76.
120. Scott FW, Cloutier HE, Kleemann R, Woerz-Pagenstert U, Rowsell P, Modler HW, et al. Potential mechanisms by which certain foods promote or inhibit the development of spontaneous diabetes in BB rats: dose, timing, early effect on islet area, and switch in infiltrate from Th1 to Th2 cells. *Diabetes* 1997;**46**:589–98.

S E C T I O N 2

MOLECULAR BIOLOGY OF THE CELL

CHAPTER

6

Oxidative Stress in Diabetes: Molecular Basis for Diet Supplementation

Yonggang Wang[1], *Xiao Miao*[2], *Jian Sun*[1], *Lu Cai*[3]

[1]Cardiovascular Center, The First Hospital of Jilin University, Changchun, China; [2]Department of Ophthalmology, The Second Hospital of Jilin University, Changchun, China; [3]Department of Pediatrics, Wendy L. Novak Diabetes Care Center, University of Louisville, Louisville, KY, USA

1. INTRODUCTION

Diabetes is a major health problem worldwide. Approximately 40% of all diabetic patients develop complications, which mainly affect the cardiovascular system, the retina, the kidneys, and peripheral nerves. Extensive evidence from experimental models of diabetes implicates hyperglycemia as a main pathogenic mechanism in diabetes and its complications.[1,2] Oxidative stress has been linked to both the onset of diabetes and its complications.[3–6] Although several mechanisms underlying diabetic complications have been pursued, all of these pathogenic factors commonly either induce or result from oxidative stress.[6]

2. OXIDATIVE STRESS AND OXIDATION DAMAGE IN DIABETES

Oxidative stress is defined as excess formation or insufficient removal of highly reactive molecules such as reactive oxygen species (ROS) and reactive nitrogen species (RNS).[7] ROS include free radicals such as superoxide ($\bullet O_2^-$), hydroxyl ($\bullet OH$), peroxyl ($\bullet RO_2$), hydroperoxyl (HRO_2^-), and nonradical species such as hydrogen peroxide (H_2O_2) and hydrochlorous acid (HOCl).[8,9] RNS include free radicals like nitric oxide ($\bullet NO$) and nitrogen dioxide ($\bullet NO_2^-$), and nonradicals such as peroxynitrite (ONOO−), nitrous oxide (HNO_2), and alkyl peroxynitrates (RONOO).[8,9] Under physiological conditions, ROS is continuously generated; however, the antioxidant defenses are adequate to prevent ROS-based tissue dysfunction.[10] ROS have also been reported to be involved in cell signaling as well as cellular defense mechanisms.[7] Thus, imbalances between ROS generation and quenching can lead to an increase in oxidative stress and pathological consequences including damage to proteins, lipids, and DNA.[11,12] Production of one ROS or RNS can also lead to the production of others through a radical chain reaction, as summarized in Figure 1.[12]

Direct evidence of oxidative stress in diabetes is based on studies that have focused on measuring oxidative stress markers such as plasma and urinary F2-isoprostane as well as plasma and tissue levels of nitrotyrosine and $\bullet O_2^-$.[13–16] The increased oxidative stress under diabetic conditions is thought to most likely result from hyperglycemia-induced overproduction of ROS and RNS. Hyperglycemia causes increased oxidative stress through the activation of several signaling pathways.[17] These include nonenzymatic, enzymatic, and mitochondrial pathways. Nonenzymatic sources of oxidative stress originate from glucose. Hyperglycemia can directly increase ROS generation. Glucose can also undergo autoxidation and generate $\bullet OH$ radicals.[7,8] Enzymatic sources of augmented ROS production in diabetes include nitrous oxide (NOS), nicotinamide adenine dinucleotide phosphate-oxidase (NOX), and xanthine oxidase.[13,14,18] Studies have shown that there is an enhanced production of $\bullet O_2^-$ in diabetes and this is predominantly mediated by NOX. Furthermore, the NOS-mediated components are increased in patients with diabetes than in patients without diabetes.[14] Hyperglycemia in diabetes is also associated with increased mitochondrial ROS and RNS production.[19] Thus, mitochondria (mitochondrial respiratory chain) are another major nonenzymatic source of ROS production. The electron transport chain generates superoxide radicals, which are inevitable byproducts of complex I and complex III of the respiratory chain. ROS then leads to cellular damage

Molecular Nutrition and Diabetes
http://dx.doi.org/10.1016/B978-0-12-801585-8.00006-3

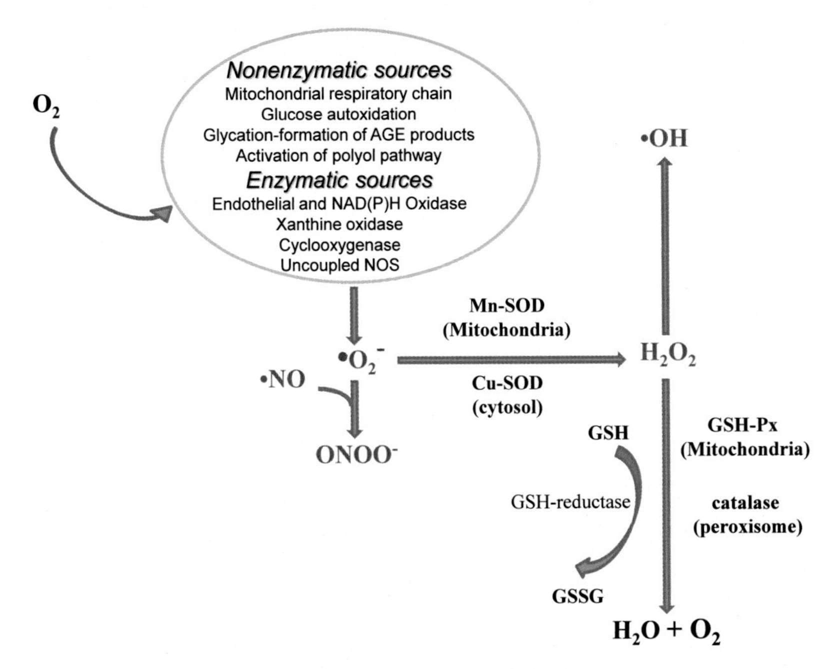

FIGURE 1 **Generation of reactive oxygen and nitrogen species in diabetes.** Highlighted in red are some of the most important reactive oxygen and reactive nitrogen species. Oxygen is converted to $\cdot O_2^-$ via the activation of enzymatic and nonenzymatic pathways, which is then dismutated to hydrogen peroxide by superoxide dismutase (SOD). H_2O_2 can be converted to H_2O by catalase or glutathione peroxidase (GSH-Px) or to •OH after a reaction with Cu or Fe. Glutathione reductase regenerates glutathione (GSH). In addition, $\cdot O_2^-$ reacts rapidly with •NO to form peroxynitrite. Cu, copper; Mn, manganese; NOS, nitrous oxide.

through several mechanisms (oxidation, interference with nitric oxide [NO], and modulation of detrimental intracellular signaling pathways). Therefore, increased ROS causes cellular damage by directly damaging proteins and DNA as well as by inducing apoptosis. Hyperglycemia-induced generation of O_2^- at the mitochondrial level is thought to be the initial trigger for the vicious cycle of oxidative stress in diabetes.[17,19]

Several studies have provided evidence for the pivotal role of oxidative stress in insulin resistance, metabolic syndromes, obesity, and type 1 diabetes (T1D).[20–23] Thus, excessive ROS production is an important trigger for insulin resistance and a relevant factor in the development of diabetes.[24] Production of ROS and RNS occurs in response to extracellular and intracellular stimuli. Extracellular stimuli act through plasma membrane receptors and include tumor necrosis factor-α, hormones, and growth factors, including platelet-derived growth factor, epithelial growth factor, and insulin. ROS and RNS can then react with multiple intracellular components (proteins, lipids, nucleic acids) to generate reversible and irreversible oxidative modifications. They also activate various signaling cascades designated for sensing and responding to "stress," such as the mitogen-activated protein kinase and c-Jun *N*-terminal kinase signaling pathways.[9] Therefore, a causative role of reactive species has been involved in obesity and diabetes.[24,25]

3. OXIDATIVE STRESS AND OXIDATION DAMAGE IN DIABETIC COMPLICATIONS

The occurrence of systemic hyperglycemia in diabetes can ultimately cause long-term damage to multiple organs, leading to severe complications. The microvasculature is a key target of hyperglycemic damage. Damage to small blood vessels can initiate and/or exacerbate systemic complications associated with diabetes. As a result, research has largely focused on the area of peripheral vascular injury. Retinopathy, neuropathy, and nephropathy are all caused, in part, by changes in blood flow or abnormal vessel growth.[26,27] Diabetic cardiovascular injury can impact both the cardiac and peripheral blood vessels. A disorder of the heart muscle in patients with diabetes is called diabetic cardiomyopathy, which can lead to an inability of the heart to circulate blood through the body effectively, a state known as heart failure.[28]

Recent studies have shown that oxidative damage induced by ROS and/or RNS derived from hyperglycemia plays a critical role in diabetic injury in the heart. In rodent models of T1D and type 2 diabetes (T2D), accumulation of triglyceride in the heart is associated with cardiac dysfunction and oxidative stress.[29,30] Expression of antioxidant enzymes and treatment with antioxidant compounds can ameliorate both the triglyceride-associated oxidative stress and

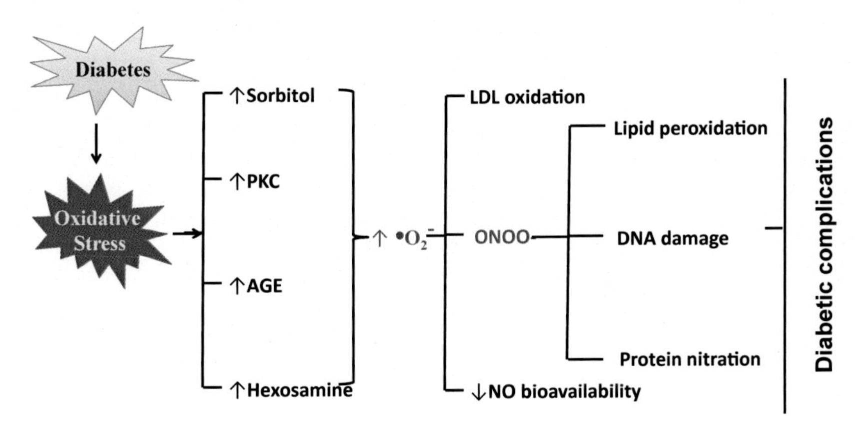

FIGURE 2 **Diabetic oxidative stress and downstream targets.** Hyperglycemia-induced overproduction of mitochondrial reactive oxygen and reactive nitrogen species initiates a vicious cycle of intracellular signaling, involving the stress-sensitive polyol (sorbitol), advanced glycation end products (AGEs), protein kinase C (PKC), and hexosamine pathways. Enhanced production of AGEs and sorbitol exerts a positive feedback on reactive oxygen and reactive nitrogen species synthesis, which is mediated by diabetic complications.

diabetic cardiomyopathy.[29,30] Several mechanisms have been implicated including lipid activation of NOX activity through the downstream production of ceramides, and signaling through pathways that converge on nuclear factor-κB.[31,32] Therefore, through hyperglycemia, hyperlipidemia, hypertension, and possible iron dyshomeostasis, diabetes induces oxidative stress that causes damage to multiple organs, leading to various complications.[33,34] Figure 2 is just one example how diabetes induces oxidative stress that in turn generates superoxide via different pathways, and superoxide further reacts with NO to form peroxynitrate to induce series of detrimental effects, leading the development of diabetic complications.

4. ANTIOXIDANTS IN DIABETES: IMPLICATIONS FOR USE OF BIOACTIVE FOOD COMPONENTS

Given that oxidative and/or nitrosative stress plays a critical role in the development of diabetes and its complications, several antioxidant approaches have been investigated for their potential therapeutic effects on diabetes. Recently, antioxidant therapy has been extensively studied for its potential therapeutic effects on various diabetic complications. Here, we provide some examples that focus on the metallothionein (MT) and nuclear factor erythroid-2-related factor 2 (Nrf2) pathways, which are activated by bioactive food components. These bioactive foods are major targets of antioxidative pathways and can protect against diabetes, based on our own findings and others.

It is generally accepted that food can have health-promoting properties that go beyond its traditional nutritional value. Bioactive compounds are defined as components of food that influence physiological or cellular activities in the animals or humans that consume them. For example, curcumin, sulforaphane (SF), vitamins, and resveratrol are bioactive compounds found in fruits and vegetables that act as antioxidants and shown to be protective against metabolic syndromes such as diabetes. Recently, much attention has been given to bioactive food components that may be beneficial for the prevention of diabetes.

4.1. Curcumin

Turmeric is extracted from the rhizomes of *Curcuma longa*. Turmeric's role as a spice, remedy, and dye has long been recognized. Its use in China and India was recorded as early as 1280 in Marco Polo's travelogues.[35] It is known worldwide by different names in different languages (e.g., Arabic, "Kurkum"; Chinese, "Yu chin").[36] Curcuminoids compose 3–5% of turmeric by weight and include the three bioactive analogs: curcumin (curcumin I), demethoxycurcumin (curcumin II), and bisdemethoxycurcumin (curcumin III).[36–38] Curcumin's occurrence in foods is generally as a food additive in turmeric. In turn, turmeric is often one ingredient in spice blends, mainly curry powder, which generally consists of turmeric, clove, paprika, ginger, cardamom, coriander, cumin, mace, pepper, and cinnamon.[39] Turmeric has been used as a medicinal plant for thousands of years, and there are reported molecular biological targets that may mediate its physiological effects. One important molecular function through which curcumin may act to produce health benefits is its free-radical scavenging and antioxidant characteristics.

Curcumin is composed of two monomers of ferulic acid, and this monomer has been studied for its free-radical scavenging properties. Curcumin has been shown to be a potent scavenger for a variety of ROS, including •OH, nitrogen dioxide radicals, and H_2O_2.[40] Hence, by virtue of its ability to scavenge various types of ROS, curcumin can decrease oxidative damage of proteins, lipids, and DNA.[41–43] In addition, curcumin downregulates NOS expression.[44] Because NO-mediated oxidative stress has been associated with diabetes, downregulating NO production could be beneficial in treating diabetes.[44,45] Studies have shown that curcumin can also prevent streptozotocin (STZ)-induced diabetes in rats as well as diabetic cardiomyopathy.[46,47] We also reported the prevention of diabetic cardiomyopathy by the curcumin analog C66 in an STZ diabetic mouse model through suppression of c-Jun NH_2-terminal kinase-related oxidative stress.[48]

4.2. Sulforaphane

Sulforaphane (4-methylsulfinybutyl isothiocyanate) is a dietary isothiocyanate synthesized from a precursor found in cruciferous vegetables of the genus *Brassica*. These vegetables include cauliflower, broccoli, kale, cole crops, cabbage, collards, Brussels sprouts, mustard, and radish, which have received much attention over the past 20 years because of their antioxidant and other beneficial biological activity. Sulforaphane can be obtained from cruciferous vegetables, such as broccoli, Brussels sprouts, or cabbages. With the increase in the application of natural SF in research, the demands for the high purity compound are increasing. Natural SF can only be obtained through hydrolysis of glucoraphanin by endogenous or exogenous myrosinase. The occurrence and concentration of glucosinolates vary according to the species and cultivar, tissue type, and environmental factors.[49,50] Current research has strongly focused on the therapeutic effects of SF in several pathologies impacted by hyperglycemia that include damage to the brain, kidney, liver, heart, and muscle.

As previously mentioned, SF is found within broccoli and this is the reason why several studies focused on antioxidant effects have exploited dietary ingestion of this cruciferous vegetable. For example, dried broccoli sprouts (200 mg/day) were able to attenuate oxidative stress from the damage.[51] Xue et al. also demonstrated that SF pretreatment can prevent ROS production as well as mitochondrial and biochemical dysfunction induced by hyperglycemia in human microvascular HMEC-1 endothelial cells incubated in low- and high-glucose concentrations, which was associated with increased expression of transketolase.[52] Song et al. also found that SF pretreatment was able to ameliorate STZ-induced islet damage in mice.[53] More specifically, SF pretreatment was able to ameliorate the hyperglycemia and decreased insulin blood concentration by preserving the number of islets producing insulin cells. This protective effect was associated with the prevention of the induction of nuclear factor-κB signaling pathway in islets of these mice. Sulforaphane also suppressed the NOS induction and cyclooxygenase expression in these settings.[53] In support of this, our previous studies also showed that SF had a beneficial effect on vascular and heart complications in T1D and T2D models.[54–56]

4.3. Vitamins

Although it is possible that a single vitamin could act sufficiently as an antioxidant and prevent one or more complications of diabetes, it is more likely that a mixture of vitamins in food determines the total oxidant status of vitamins. Thus, a diet sufficient and balanced in all vitamins and minerals will need to be considered when it comes to treatment avenues. The most benefit would be obtained from vitamin repletion in a patient deficient in vitamins because of inadequate intake, gastrointestinal or urine losses, and pathologies evoking high oxidant stress. Vitamin status should be assessed and levels should be measured. If low, vitamins should be replenished with vegetables and fruit if possible or, if not feasible, with dietary supplements.

Treatment with proper amounts of vitamins and antioxidants is best accomplished with a balanced diet including three servings of vegetables and two servings of fruits. Most individuals do not meet these requirements of fruits and vegetables. Pockets of our population in poor or rural areas of the United States do not have adequate intake of vitamins. Acquired vitamin deficiencies occur with malabsorption secondary to chronic liver, pancreatic, or inflammatory bowel disease, or gastric or small bowel surgery. The impact of vitamin intake may also depend upon the intake and presence of other vitamins and their relative concentrations in the diet.[57] As components of food, vitamins may act differently from when they are given alone. The mixture of foods becomes important and the dose of a vitamin may translate to being different when given in combination with other vitamins or nutrients. A mixture of vitamins and minerals can serve both metabolic and antioxidant functions to allow for normal cell growth, differentiation, and function. This proper mixture could also reduce oxidative stress while allowing for specific biological actions. Interestingly, under some conditions, a vitamin may also have oxidant effects. Whether a vitamin behaves as an oxidant or antioxidant depends upon the mixture of vitamins and the redox state in the blood or in the cell. Giving too much of one vitamin in an altered redox environment may change its effect from an antioxidant to a pro-oxidant.[57] After acting as an antioxidant (donating an electron), a vitamin could also become an oxidant if the net redox state is not corrected.[58] Thus, antioxidants could turn detrimental under these conditions. In other instances, a vitamin could have both oxidant and antioxidant functions. For example, retinoids bind to cysteine-rich domains of protein kinase C (PKC)-α and serve as a redox activator of PKC-α or reversible redox switch in the activation of PKC-α.[59]

The antioxidant properties of a vitamin should also be considered separately from its more specific functions. The blood level and dose required for an antioxidant effect will be different from the dose required for a specific effect. Dose-response curves will be different for an antioxidant versus a specific effect. Thus, proper dosing (amount and frequency) of vitamins must be required to target its specific function. Doses will also be limited by toxicity. Upper tolerable dietary levels and doses are usually based on a side effect of a large dose or blood levels considered higher than normal. Sometimes a specific function can result in changes in the redox state or production of free radicals. For example, retinoids and 1,25-dihydroxy vitamin D3 inhibit nitric oxide synthase, which, in turn, reduce nitric oxide and free radical production in monocytes and vascular smooth muscle cells.[60,61]

4.4. Resveratrol

Resveratrol (*trans*-3,4′,5-trihydroxystilbene) is a nonflavonoid polyphenol, belonging to the stilbenes group and produced naturally in several plants in response to injury or fungal attack.[62] It was first detected in the roots of white hellebore (*Veratrum grandiflorum*) and is found in various foods, such as grapes, berries, red wine, and nuts.[63] Although the molecule exists in two isoforms, *trans*-resveratrol and *cis*-resveratrol, the trans form, which is the preferred steric form in nature, is relatively stable.[64]

Resveratrol has been used as an adjuvant in hypoglycemic therapy. Specifically, the effects of resveratrol (250 mg once daily for 6 months) were investigated in Indian patients with T2D and receiving ongoing oral hypoglycemic treatment (metformin and/or glibenclamide). In this study, hemoglobin A1c was reduced compared to baseline values in the intervention group, and levels of hemoglobin A1c and fasting blood glucose during the study period were significantly improved compared with the control group, suggesting therapeutic effects of resveratrol. However, the authors underline the limitations of their study as it was an open-label trial and only included Indian patients.[65]

Because of its polyphenolic structure, resveratrol plays a prominent role in foods exerting antioxidant activity. The antioxidant properties of resveratrol have been largely studied in vitro. In obese subjects, *trans*-resveratrol did not modify biochemical markers related to oxidative stress, such as total plasma oxidant power, glutathione peroxidase, superoxide dismutase, as well as glucose, protein, and lipid oxidation[66,67]; however, only the expression of a few genes were found to be modified in this setting.[67] In contrast, resveratrol was shown to increase oxidative phosphorylation of gene expression in muscles.[68] Interestingly, coadministration of resveratrol with muscadine polyphenols reduced multiple indices of oxidative stress caused by a high-fat, high-carbohydrate meal.[69] Muscle extracts exhibited significantly reduced levels of reactive oxygen species in mononuclear cells as well as p47^{phox} expression,[69] similar to healthy patients taking *Polygonum cuspidatum* extract had similar results.[70] "Pure" resveratrol could also decrease *ortho*-tyrosine creatine excretion in the urine of subjects with T1D.[71] In experimental models of diabetes, resveratrol treatment was found to reduce ROS production and the incidence of cardiomyocyte death, along with improvement of cardiac function.[72–74]

Studies have shown that resveratrol can exert its actions in multiple ways, which include scavenging ROS and increasing the activity of enzymes that metabolize ROS, such as superoxide dismutase, but also in some instances, decreasing the activity of enzymes that play a role in ROS production.[75–78]

4.5. Metallothionein

MT is a small, cysteine-rich, metal-binding protein that has various biological functions, which include essential metal homeostasis, heavy metal detoxification, and cellular antioxidative defense.[79,80] MT maintains redox status by binding and releasing metals (mainly zinc under physiological conditions). MT isoforms are classified based on their molecular weight, metal bind properties, encoded genes, chromosomes, binding atoms, amino acids environment, etc. Although four isoforms of MTs have been characterized, MT-I and MT-II exist as the major isoforms found in various human and animal organs.[81]

Recent studies have shown that thiolate ligands in MT confer redox activity on zinc clusters, which suggested that MT could control the cellular zinc distribution as a function of the cellular energy state.[82] A recent study has shown that the antioxidant properties of MT can be enhanced in the presence of zinc. The zinc redox−dependent functions of MT are important for regulating physiological processes that depend on zinc as well as pathological processes in which oxidative stress mobilizes zinc. The decrease in zinc availability from mutant MT suggests that the MT is either less reactive toward nitric oxide or it is in an oxidized state that does not bind sufficient amount of zinc.[83] MT has cardioprotective effects in animal models of diabetes

(T1D: STZ-induced and OVE26 transgenic mouse diabetic models and mouse model of obesity-related diabetes) as well as humans with T1D.[84,85] Zinc is an important trace element that is required for more than 300 enzymes and transcription factors as well as important in the heart[86,87] by regulating intracellular zinc, thus, zinc supplementation could provide an alternative approach to intervening with diabetic complications in the clinic in the future.

Previous studies have shown that increasing zinc (via supplementation or genetically enhancing MT synthesis) in the pancreas is sufficient to prevent the development of spontaneous or chemically induced diabetes.[88] Genetic and pharmacological enhancement of MT expression in the heart and kidney were also shown to provide significant protection from diabetes-induced cardiomyopathy and nephropathy.[88] These studies altogether suggest that MT is an adaptive protein that can prevent both diabetes development and diabetic complications.[88]

4.6. Nrf2

The transcription factor, Nrf2, is a member of the cap'n'collar family, which is a master regulator of cellular detoxification responses and redox status.[89,90] Under physiological conditions, Nrf2 is localized in the cytoplasm and associates with its inhibitor kelch-like ECH-associated protein 1 (KEAP1). KEAP1 can function to mediate rapid ubiquitination and subsequent degradation of Nrf2 by the proteasome.[91] However, upon exposure of cells to oxidative stress or electrophilic compounds, Nrf2 is free from KEAP1 and translocates into the nucleus to bind to antioxidant-responsive elements in genes encoding antioxidant enzymes such as nicotinamide adenine dinucleotide phosphate oxidase, quinone oxidoreductase (NQO1), glutathione S-transferase, heme oxygenase-1 (HO1), and γ-glutamylcysteine synthetase. The increased expression of these enzymes plays a key role in mediating cellular detoxification, antioxidation, and anti-inflammatory effects.[91,92]

It is well-established that Nrf2 expression and function is increased in response to oxidative stress in cells in vitro and tissues in vivo. Several studies have shown the induction of ROS and RNS by high levels of glucose in cultured cardiovascular cells, and explored the impact on Nrf2 expression, activation, and its downstream targets in these cells.[52,93] These studies highlighted the protective effect of Nrf2 on oxidative damage induced by high levels of glucose in cultured cells and potentially on the diabetic complications in animal models.

Our recent findings also showed that Nrf2 protein expression was increased in the hearts of mice that were exposed for hyperglycemia for 3 months, but the expression was then significantly downregulated at 6 months.[94] Combined with our earlier studies in which downstream target genes of Nrf2 were increased in the heart of diabetic mice at 2 weeks after STZ-induced hyperglycemia,[93] altogether suggested that Nrf2 may be trying to adaptively remain functional at the early stage of diabetes in order to overcome the diabetic damage. At the late stage of diabetes, however, antioxidant function is further impaired, leading to a decrease in Nrf2 expression. These aforementioned studies imply a preventive function of Nrf2 in diabetes-induced oxidative damage. Upregulation of Nrf2 expression and function by various approaches may provide a preventive effect on diabetes-induced oxidative damage and consequent complications. As described previously, SF is able to induce nuclear translocation of Nrf2 resulting in a significant three- to five-fold increase in the expression of downstream antioxidant genes (transketolase and glutathione reductase).[52] By treating STZ-induced diabetic rats with resveratrol, Palsamy and Subramanian not only validated the protective effects of resveratrol in diabetic nephropathy, but also found that resveratrol could normalize the protein expression of Nrf2, and its downstream genes such as γ-GCS and HO-1 in the diabetic kidney.[95]

5. CONCLUSIONS

A large body of evidence exists to suggest that diabetic patients are prone to significant perturbations at the cellular and molecular level. Diabetes is associated with increased oxidative stress. Increased oxidative stress plays an important role in the pathogenesis and progression of diabetic complications, which are associated with increased morbidity and mortality. Correction of oxidative stress status could be associated with improved survival and a reduction in the diabetic comorbidities. Future studies should be aimed at understanding the potential benefits of combinatorial therapies aimed at reducing oxidative stress and the identification of drugs or natural dietary compounds that may modulate oxidative stress positively.

Acknowledgments

Studies cited from the authors' laboratories were supported in part by a grant from the National Natural Science Foundation of China (No. 81400279 to Y.G. Wang).

References

1. Retnakaran R, Zinman B. Type 1 diabetes, hyperglycaemia, and the heart. *Lancet* 2008;**371**(9626):1790–9.
2. Kugelberg E. Diabetes: regulation of hyperglycaemia—too much can be heart-breaking. *Nat Rev Endocrinol* 2013;**9**(12):690.

3. Baynes JW, Thorpe SR. Role of oxidative stress in diabetic complications: a new perspective on an old paradigm. *Diabetes* 1999;**48**(1): 1–9.
4. Kowluru RA, Engerman RL, Kern TS. Diabetes-induced metabolic abnormalities in myocardium: effect of antioxidant therapy. *Free Radic Res* 2000;**32**(1):67–74.
5. Uemura S, et al. Diabetes mellitus enhances vascular matrix metalloproteinase activity: role of oxidative stress. *Circ Res* 2001;**88**(12): 1291–8.
6. Liu Q, Wang S, Cai L. Diabetic cardiomyopathy and its mechanisms: role of oxidative stress and damage. *J Diabetes Investig* 2014;**5**(6):623–34.
7. Halliwell B. Biochemistry of oxidative stress. *Biochem Soc Trans* 2007;**35**(Pt 5):1147–50.
8. Turko IV, Marcondes S, Murad F. Diabetes-associated nitration of tyrosine and inactivation of succinyl-CoA:3-oxoacid CoA-transferase. *Am J Physiol Heart Circ Physiol* 2001;**281**(6):H2289–94.
9. Evans JL, et al. Oxidative stress and stress-activated signaling pathways: a unifying hypothesis of type 2 diabetes. *Endocr Rev* 2002; **23**(5):599–622.
10. Kaul N, et al. Free radicals and the heart. *J Pharmacol Toxicol Methods* 1993;**30**(2):55–67.
11. Singal PK, et al. Free radicals in health and disease. *Mol Cell Biochem* 1988;**84**(2):121–2.
12. Johansen JS, et al. Oxidative stress and the use of antioxidants in diabetes: linking basic science to clinical practice. *Cardiovasc Diabetol* 2005;**4**(1):5.
13. Guzik TJ, et al. Vascular superoxide production by NAD(P)H oxidase: association with endothelial dysfunction and clinical risk factors. *Circ Res* 2000;**86**(9):E85–90.
14. Guzik TJ, et al. Mechanisms of increased vascular superoxide production in human diabetes mellitus: role of NAD(P)H oxidase and endothelial nitric oxide synthase. *Circulation* 2002;**105**(14):1656–62.
15. Ceriello A, et al. Detection of nitrotyrosine in the diabetic plasma: evidence of oxidative stress. *Diabetologia* 2001;**44**(7):834–8.
16. Pacher P, Beckman JS, Liaudet L. Nitric oxide and peroxynitrite in health and disease. *Physiol Rev* 2007;**87**(1):315–424.
17. Brownlee M. Biochemistry and molecular cell biology of diabetic complications. *Nature* 2001;**414**(6865):813–20.
18. Feoli AM, et al. Xanthine oxidase activity is associated with risk factors for cardiovascular disease and inflammatory and oxidative status markers in metabolic syndrome: effects of a single exercise session. *Oxid Med Cell Longev* 2014;**2014**:587083.
19. Nishikawa T, et al. Normalizing mitochondrial superoxide production blocks three pathways of hyperglycaemic damage. *Nature* 2000;**404**(6779):787–90.
20. Schaffer SW, Jong CJ, Mozaffari M. Role of oxidative stress in diabetes-mediated vascular dysfunction: unifying hypothesis of diabetes revisited. *Vascul Pharmacol* 2012;**57**(5–6):139–49.
21. Campia U, Tesauro M, Cardillo C. Human obesity and endothelium-dependent responsiveness. *Br J Pharmacol* 2012; **165**(3):561–73.
22. Tesauro M, Cardillo C. Obesity, blood vessels and metabolic syndrome. *Acta Physiol (Oxf)* 2011;**203**(1):279–86.
23. Stadler K. Oxidative stress in diabetes. *Adv Exp Med Biol* 2012;**771**: 272–87.
24. Houstis N, Rosen ED, Lander ES. Reactive oxygen species have a causal role in multiple forms of insulin resistance. *Nature* 2006; **440**(7086):944–8.
25. Bashan N, et al. Positive and negative regulation of insulin signaling by reactive oxygen and nitrogen species. *Physiol Rev* 2009;**89**(1):27–71.
26. Francis GS. Diabetic cardiomyopathy: fact or fiction? *Heart* 2001; **85**(3):247–8.
27. Sowers JR, Epstein M, Frohlich ED. Diabetes, hypertension, and cardiovascular disease: an update. *Hypertension* 2001;**37**(4):1053–9.
28. Rubler S, et al. New type of cardiomyopathy associated with diabetic glomerulosclerosis. *Am J Cardiol* 1972;**30**(6):595–602.
29. Cai L, Kang YJ. Oxidative stress and diabetic cardiomyopathy: a brief review. *Cardiovasc Toxicol* 2001;**1**(3):181–93.
30. Zhou YT, et al. Lipotoxic heart disease in obese rats: implications for human obesity. *Proc Natl Acad Sci USA* 2000;**97**(4):1784–9.
31. Zhang DX, Zou AP, Li PL. Ceramide-induced activation of NADPH oxidase and endothelial dysfunction in small coronary arteries. *Am J Physiol Heart Circ Physiol* 2003;**284**(2):H605–12.
32. Inoguchi T, et al. High glucose level and free fatty acid stimulate reactive oxygen species production through protein kinase C–dependent activation of NAD(P)H oxidase in cultured vascular cells. *Diabetes* 2000;**49**(11):1939–45.
33. Wei W, et al. Oxidative stress, diabetes, and diabetic complications. *Hemoglobin* 2009;**33**(5):370–7.
34. Xu YJ, et al. Prevention of diabetes-induced cardiovascular complications upon treatment with antioxidants. *Heart Fail Rev* 2014;**19**(1): 113–21.
35. Aggarwal BB, et al. Curcumin: the Indian solid gold. *Adv Exp Med Biol* 2007;**595**:1–75.
36. Ravindran P, Babu KN, Sivaraman K. *Turmeric: the genus Curcuma*. CRC Press; 2007.
37. Anand P, et al. Biological activities of curcumin and its analogues (Congeners) made by man and Mother Nature. *Biochem Pharmacol* 2008;**76**(11):1590–611.
38. Leela NK, et al. Chemical composition of essential oils of turmeric (*Curcuma longa* L.). *Acta Pharm* 2002;**52**:137–41.
39. Himesh S, et al. Qualitative and quantitative profile of curcumin from ethanolic extract of *Curcuma longa*. *Int Res J Pharm* 2011;**2**: 180–4.
40. Ak T, Gülçin İ. Antioxidant and radical scavenging properties of curcumin. *Chem Biol Interact* 2008;**174**(1):27–37.
41. Singh U, et al. Reactions of reactive oxygen species (ROS) with curcumin analogues: structure-activity relationship. *Free Radic Res* 2011;**45**(3):317–25.
42. Wongcharoen W, Phrommintikul A. The protective role of curcumin in cardiovascular diseases. *Int J Cardiol* 2009;**133**(2):145–51.
43. Khor TO, et al. Pharmacodynamics of curcumin as DNA hypomethylation agent in restoring the expression of Nrf2 via promoter CpGs demethylation. *Biochem Pharmacol* 2011;**82**(9):1073–8.
44. Farhangkhoee H, et al. Differential effects of curcumin on vasoactive factors in the diabetic rat heart. *Nutr Metab (Lond)* 2006;**3**:27.
45. Farhangkhoee H, et al. Vascular endothelial dysfunction in diabetic cardiomyopathy: pathogenesis and potential treatment targets. *Pharmacol Ther* 2006;**111**(2):384–99.
46. Soetikno V, et al. Curcumin prevents diabetic cardiomyopathy in streptozotocin-induced diabetic rats: possible involvement of PKC-MAPK signaling pathway. *Eur J Pharm Sci* 2012;**47**(3):604–14.
47. Yu W, et al. Curcumin alleviates diabetic cardiomyopathy in experimental diabetic rats. *PLoS One* 2012;**7**(12):e52013.
48. Wang Y, et al. Inhibition of JNK by novel curcumin analog C66 prevents diabetic cardiomyopathy with a preservation of cardiac metallothionein expression. *Am J Physiol Endocrinol Metab* 2014;**306**(11): E1239–47.
49. Fahey JW, Zalcmann AT, Talalay P. The chemical diversity and distribution of glucosinolates and isothiocyanates among plants. *Phytochemistry* 2001;**56**(1):5–51.
50. Holst B, Williamson G. A critical review of the bioavailability of glucosinolates and related compounds. *Nat Prod Rep* 2004;**21**(3): 425–47.
51. Wu L, et al. Dietary approach to attenuate oxidative stress, hypertension, and inflammation in the cardiovascular system. *Proc Natl Acad Sci USA* 2004;**101**(18):7094–9.
52. Xue M, et al. Activation of NF-E2-related factor-2 reverses biochemical dysfunction of endothelial cells induced by hyperglycemia linked to vascular disease. *Diabetes* 2008;**57**(10):2809–17.

53. Song MY, et al. Sulforaphane protects against cytokine- and streptozotocin-induced beta-cell damage by suppressing the NF-kappaB pathway. *Toxicol Appl Pharmacol* 2009;**235**(1):57–67.
54. Miao X, et al. Sulforaphane prevention of diabetes-induced aortic damage was associated with the up-regulation of Nrf2 and its down-stream antioxidants. *Nutr Metab (Lond)* 2012;**9**(1):84.
55. Zhang Z, et al. Sulforaphane prevents the development of cardiomyopathy in type 2 diabetic mice probably by reversing oxidative stress-induced inhibition of LKB1/AMPK pathway. *J Mol Cell Cardiol* 2014.
56. Wang Y, et al. Sulforaphane reduction of testicular apoptotic cell death in diabetic mice is associated with the upregulation of Nrf2 expression and function. *Am J Physiol Endocrinol Metab* 2014; **307**(1):E14–23.
57. Olson JA. Benefits and liabilities of vitamin A and carotenoids. *J Nutr* 1996;**126**(4 Suppl.):1208S–12S.
58. Opara EC. Role of oxidative stress in the etiology of type 2 diabetes and the effect of antioxidant supplementation on glycemic control. *J Investig Med* 2004;**52**(1):19–23.
59. Imam A, et al. Retinoids as ligands and coactivators of protein kinase C alpha. *FASEB J* 2001;**15**(1):28–30.
60. Chang JM, et al. 1-alpha,25-Dihydroxyvitamin D3 regulates inducible nitric oxide synthase messenger RNA expression and nitric oxide release in macrophage-like RAW 264.7 cells. *J Lab Clin Med* 2004;**143**(1):14–22.
61. Sirsjo A, et al. Retinoic acid inhibits nitric oxide synthase-2 expression through the retinoic acid receptor-alpha. *Biochem Biophys Res Commun* 2000;**270**(3):846–51.
62. Leiherer A, Mundlein A, Drexel H. Phytochemicals and their impact on adipose tissue inflammation and diabetes. *Vascul Pharmacol* 2013;**58**(1–2):3–20.
63. Burns J, et al. Plant foods and herbal sources of resveratrol. *J Agric Food Chem* 2002;**50**(11):3337–40.
64. Borriello A, et al. Dietary polyphenols: focus on resveratrol, a promising agent in the prevention of cardiovascular diseases and control of glucose homeostasis. *Nutr Metab Cardiovasc Dis* 2010; **20**(8):618–25.
65. Bhatt JK, Thomas S, Nanjan MJ. Resveratrol supplementation improves glycemic control in type 2 diabetes mellitus. *Nutr Res* 2012;**32**(7):537–41.
66. Poulsen MM, et al. High-dose resveratrol supplementation in obese men: an investigator-initiated, randomized, placebo-controlled clinical trial of substrate metabolism, insulin sensitivity, and body composition. *Diabetes* 2013;**62**(4):1186–95.
67. De Groote D, et al. Effect of the intake of resveratrol, resveratrol phosphate, and catechin-rich grape seed extract on markers of oxidative stress and gene expression in adult obese subjects. *Ann Nutr Metab* 2012;**61**(1):15–24.
68. Timmers S, et al. Calorie restriction-like effects of 30 days of resveratrol supplementation on energy metabolism and metabolic profile in obese humans. *Cell Metab* 2011;**14**(5):612–22.
69. Ghanim H, et al. A resveratrol and polyphenol preparation suppresses oxidative and inflammatory stress response to a high-fat, high-carbohydrate meal. *J Clin Endocrinol Metab* 2011;**96**(5): 1409–14.
70. Ghanim H, et al. An antiinflammatory and reactive oxygen species suppressive effects of an extract of *Polygonum cuspidatum* containing resveratrol. *J Clin Endocrinol Metab* 2010;**95**(9):E1–8.
71. Brasnyo P, et al. Resveratrol improves insulin sensitivity, reduces oxidative stress and activates the Akt pathway in type 2 diabetic patients. *Br J Nutr* 2011;**106**(3):383–9.
72. Palsamy P, Subramanian S. Modulatory effects of resveratrol on attenuating the key enzymes activities of carbohydrate metabolism in streptozotocin-nicotinamide-induced diabetic rats. *Chem Biol Interact* 2009;**179**(2–3):356–62.
73. Zhang H, et al. Resveratrol improves left ventricular diastolic relaxation in type 2 diabetes by inhibiting oxidative/nitrative stress: in vivo demonstration with magnetic resonance imaging. *Am J Physiol Heart Circ Physiol* 2010;**299**(4):H985–94.
74. Thirunavukkarasu M, et al. Resveratrol alleviates cardiac dysfunction in streptozotocin-induced diabetes: role of nitric oxide, thioredoxin, and heme oxygenase. *Free Radic Biol Med* 2007;**43**(5):720–9.
75. Li Y, Cao Z, Zhu H. Upregulation of endogenous antioxidants and phase 2 enzymes by the red wine polyphenol, resveratrol in cultured aortic smooth muscle cells leads to cytoprotection against oxidative and electrophilic stress. *Pharmacol Res* 2006;**53**(1):6–15.
76. Riviere C, et al. Inhibitory activity of stilbenes on Alzheimer's beta-amyloid fibrils in vitro. *Bioorg Med Chem* 2007;**15**(2):1160–7.
77. Kohnen S, et al. Resveratrol inhibits the activity of equine neutrophil myeloperoxidase by a direct interaction with the enzyme. *J Agric Food Chem* 2007;**55**(20):8080–7.
78. Juan SH, et al. Mechanism of concentration-dependent induction of heme oxygenase-1 by resveratrol in human aortic smooth muscle cells. *Biochem Pharmacol* 2005;**69**(1):41–8.
79. Cai L, et al. Attenuation by metallothionein of early cardiac cell death via suppression of mitochondrial oxidative stress results in a prevention of diabetic cardiomyopathy. *J Am Coll Cardiol* 2006; **48**(8):1688–97.
80. Coyle P, et al. Metallothionein: the multipurpose protein. *Cell Mol Life Sci* 2002;**59**(4):627–47.
81. Cai L, et al. Metallothionein in radiation exposure: its induction and protective role. *Toxicology* 1999;**132**(2–3):85–98.
82. Higashimoto M, et al. Tissue-dependent preventive effect of metallothionein against DNA damage in dyslipidemic mice under repeated stresses of fasting or restraint. *Life Sci* 2009;**84**(17–18): 569–75.
83. Maret W. Fluorescent probes for the structure and function of metallothionein. *J Chromatogr B Analyt Technol Biomed Life Sci* 2009;**877**(28):3378–83.
84. Giacconi R, et al. +647 A/C and +1245 MT1A polymorphisms in the susceptibility of diabetes mellitus and cardiovascular complications. *Mol Genet Metab* 2008;**94**(1):98–104.
85. Yang L, et al. Polymorphisms in metallothionein-1 and -2 genes associated with the risk of type 2 diabetes mellitus and its complications. *Am J Physiol Endocrinol Metab* 2008;**294**(5):E987–92.
86. Miao X, et al. Zinc homeostasis in the metabolic syndrome and diabetes. *Front Med* 2013;**7**(1):31–52.
87. Li B, et al. The role of zinc in the prevention of diabetic cardiomyopathy and nephropathy. *Toxicol Mech Methods* 2013;**23**(1):27–33.
88. Li X, Cai L, Feng W. Diabetes and metallothionein. *Mini Rev Med Chem* 2007;**7**(7):761–8.
89. Kobayashi M, Yamamoto M. Nrf2-Keap1 regulation of cellular defense mechanisms against electrophiles and reactive oxygen species. *Adv Enzyme Regul* 2006;**46**:113–40.
90. Sykiotis GP, Bohmann D. Stress-activated cap'n'collar transcription factors in aging and human disease. *Sci Signal* 2010;**3**(112):re3.
91. Uruno A, Yagishita Y, Yamamoto M. The Keap1-Nrf2 system and diabetes mellitus. *Arch Biochem Biophys* 2014.
92. de Haan JB. Nrf2 activators as attractive therapeutics for diabetic nephropathy. *Diabetes* 2011;**60**(11):2683–4.
93. He X, et al. Nrf2 is critical in defense against high glucose-induced oxidative damage in cardiomyocytes. *J Mol Cell Cardiol* 2009;**46**(1): 47–58.
94. Bai Y, et al. Prevention by sulforaphane of diabetic cardiomyopathy is associated with up-regulation of Nrf2 expression and transcription activation. *J Mol Cell Cardiol* 2013;**57C**:82–95.
95. Palsamy P, Subramanian S. Resveratrol protects diabetic kidney by attenuating hyperglycemia-mediated oxidative stress and renal inflammatory cytokines via Nrf2-Keap1 signaling. *Biochim Biophys Acta* 2011;**1812**(7):719–31.

CHAPTER

7

Impact of Type 2 Diabetes on Skeletal Muscle Mass and Quality

David Sala[1], *Antonio Zorzano*[2,3,4]

[1]Development, Aging and Regeneration Program (DARe), Sanford-Burnham Medical Research Institute, La Jolla, CA, USA; [2]Institute for Research in Biomedicine (IRB Barcelona), Barcelona, Spain; [3]Departament de Bioquímica i Biologia Molecular, Facultat de Biologia, Universitat de Barcelona, Barcelona, Spain; [4]CIBER de Diabetes y Enfermedades Metabólicas Asociadas (CIBERDEM), Instituto de Salud Carlos III, Madrid, Spain

1. INTRODUCTION

Among metabolic disorders, type 2 diabetes (T2D) is one of the most prevalent. In fact, according to the International Diabetes Federation, around 8% of the total population present with T2D, and its incidence is increasing rapidly, especially in western countries[1]. Given these predictions and the health cost that T2D may represent in the near future, it is necessary and timely to establish effective preventive measures and to develop more efficient treatments. It is therefore crucial to improve our understanding of the causes and consequences of this disease.

T2D is a chronic condition characterized by deficient insulin function (not absence), and it is diagnosed by an increase in circulating glucose levels and glucose intolerance.[2,3] More specifically, peripheral tissues display insulin resistance and do not appropriately sense or respond to the presence of this anabolic hormone. In fact, most current treatments for this pathology have focused on ameliorating insulin resistance in peripheral tissues and do not involve insulin treatment at least in early stages. Thus, during the long onset of insulin resistance and T2D, β cells initially respond by increasing insulin production as a compensatory mechanism.[4] However, with the progression of the disease, insulin secretion is also compromised because of β-cell exhaustion, and individuals may need insulin treatment[4].

Insulin resistance and T2D onset have been associated with various factors of a genetic and environmental nature. In fact, T2D is classified as a polygenic disease with a clear hereditary component, and much research effort has been channeled into identifying the genes associated with it.[5–7] However, this disease is not fully understood.[5–7] In addition, environmental factors play a major role in the development of insulin resistance and T2D, and a strong association with sedentary lifestyle, inappropriate diet, and obesity has been reported.[8–11]

T2D has a huge impact on whole-body glucose, lipid, and protein metabolism. The main tissues affected by insulin resistance that contribute to this metabolic dysregulation are liver, adipose tissue, and skeletal muscle. In fact, the aberrant function of the latter has been described as one of the major contributors to this pathology.[12] Focusing on this tissue, insulin resistance implies deficient glucose uptake and phosphorylation upon nutrient disposal and defective glycogen synthesis, in addition to the altered use of lipids as a result of increased free fatty uptake and deficient oxidation in mitochondria.[12,13] These conditions lead to the accumulation of triglycerides inside myofibers.[13] Overall, this observation indicates that nutrient handling is highly perturbed in skeletal muscle under diabetic conditions. Protein metabolism under T2D has received less attention; however, current knowledge suggests that there are no major changes in skeletal muscle mass in most type 2 diabetic patients in spite of deficient insulin action (except for elderly people or patients that develop diabetes-associated comorbidities).[14–23] However, muscle performance is compromised in these patients,[14,19,24] and there is a reduction in mitochondrial content and a clear impairment of mitochondrial function.[25–28]

Molecular Nutrition and Diabetes
http://dx.doi.org/10.1016/B978-0-12-801585-8.00007-5

In this chapter, we will address the impact of T2D on protein metabolism in skeletal muscle and how this affects muscle mass and performance under these pathological conditions, with a special focus on autophagy and mitochondrial dynamics.

2. REGULATION OF PROTEIN DEGRADATION IN SKELETAL MUSCLE

Research into the regulation of skeletal muscle mass in adult tissue has traditionally focused on the analysis of the pathways that regulate the balance between protein synthesis and protein degradation in myofibers. A net increase in the protein synthesis rate results in larger myofibers and muscle hypertrophy. On the other hand, enhanced protein degradation causes a reduction in myofiber size and muscle atrophy. Of all the proteolytic systems, the two that present a major impact in terms of myofiber size and structure are the ubiquitin-proteasome system (UPS) (Figure 1;) and macroautophagy (hereafter referred to as autophagy). In skeletal muscle, as in most cell types, the UPS is the system that quantitatively degrades more intracellular proteins.[29] In fact, studies performed in rats suggest that around 50% of skeletal muscle protein degradation is via the UPS.[30–32] However, this percentage is variable depending on the muscle analyzed and the chemical compounds used to inhibit proteasome activity.[30–32] The 26S proteasome is one of the basic components of the UPS.[33] It is a complex composed by a central core particle with a cylindrical shape, the 20S proteasome, together with one or two 19S regulatory particles found at one or both sides of this central particle.[34] It is found inside the 20S particle, where protein cleavage takes place, and there are six proteolytic sites with three distinct proteolytic activities: trypsin-like, chymotrypsin-like, and peptidyl-glutamyl peptide-hydrolyzing.[34] However, substrate recognition and translocation inside the 20S proteasome are dependent on 19S particles. These particles contain a number of subunits that unfold and translocate the substrates to be degraded inside the 20S particle by a process that involves the hydrolysis of adenosine triphosphate (ATP).[35]

The recognition of the substrates by the 26S proteasome depends on the previous polyubiquitination of the proteins to be degraded. Ubiquitin is a small polypeptide of 76 amino acids that contains a glycine at the C-terminus.[34] This glycine is crucial for the conjugation of ubiquitin to its target or to other ubiquitins. Ubiquitin contains several internal lysine residues that allow the formation of polyubiquitin chains.[34] Substrate ubiquitination is a multistep process in which three types of enzymes are involved: first, the E1 enzyme or ubiquitin-activating enzyme generates a reactive form of ubiquitin[36]; next, this activated ubiquitin is transferred to the E2 enzymes[34]; and, finally, the E3 enzymes or ubiquitin ligases recognize the substrate to be degraded and

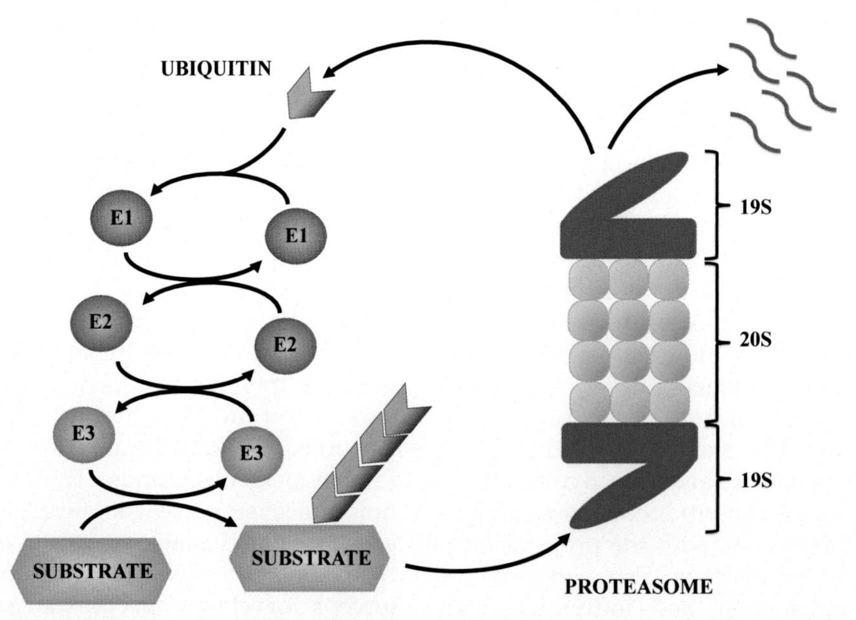

FIGURE 1 **The ubiquitin-proteasome system.** Ubiquitin is conjugated to the proteins that have to be degraded in a three-step process involving three enzymes: first, the E1 enzyme activates ubiquitin; second, this ubiquitin is transferred to the E2 enzyme; and third, the E3 enzyme recognizes the substrate and mediates the conjugation of this ubiquitin to a lysine of this substrate. After that, more ubiquitin molecules are added to the lysine residues of other ubiquitin molecules to generate a polyubiquitin chain. Once this chain is formed, this structure is recognized by the 26S proteasome. The ubiquitin molecules from the chain are recycled while the substrate is linearized and translocated inside the proteasome. Finally, this protein is degraded inside the proteasome and the resulting peptides are released.

promote the transferal of the ubiquitin from the E2 to a lysine of the substrate.[34] Repetition of this multistep process promotes the formation of a polyubiquitin chain that is then recognized by 19S particles and that leads to the degradation of the target.[37] Moreover, 19S particles also contain some subunits with deubiquitinating activity, which allows the removal of ubiquitin molecules from the substrate before it enters the 20S complex.[34] These ubiquitins can be reused in this process.

As previously mentioned, autophagy is also one of the major proteolytic pathways in skeletal muscle (Figure 2). Briefly, autophagy consists of the formation of double-membrane vesicles called autophagosomes.[38] These vesicles can engulf large portions of the cytosol and sequester components of diverse nature, such as protein aggregates, glycogen, and organelles.[38] Once formed, these autophagosomes fuse with the lysosomes that contain the enzymes that will degrade the cargoes previously sequestered.[38] Studies performed using rat skeletal muscles incubated ex vivo and inhibitors of lysosomes suggest that around 25–30% of total protein degradation in this tissue depends on this pathway.[31,32,39] However, this percentage may differ slightly between muscle groups.[31,32,39]

Autophagosome formation is the main step in autophagy. It is a complex process that can be divided into several phases:

1. Autophagosome initiation: One of the main protein complexes involved in this step is the ULK1 complex (formed by ULK1, Atg13, FIP200, and Atg101).[40] This complex is stable regardless of nutritional conditions. It is repressed through direct interaction and phosphorylation by mTORC1 and activated by 5′ adenosine monophosphate-activated protein kinase (AMPK) phosphorylation.[41–44] Thus, under basal conditions with nutrient availability, mTORC1 is active, which causes phosphorylation of the ULK1 complex and its inhibition.[41,43–45] However, upon nutrient scarcity, mTORC1 is inhibited, causing it to dissociate from the ULK1 complex. AMPK is, under these conditions, stimulated and phosphorylates this complex, promoting its activation.[41–45]

   Moreover, the nucleation of the isolation membrane or phagophore also depends on the action of the class III phosphatidylinositol 3-kinase (PI3K) complex. This complex is formed by class III PI3K or Vps34, together with Beclin1, p150, and Atg14L.[46,47] The activity of this complex is regulated by several factors, and the phosphatidylinositol 3-phosphate generated is crucial for the recruitment of proteins involved in autophagosome elongation.[48,49]

   It is still not fully understood how the ULK1 complex and the class III PI3K complex interact to regulate autophagosome formation.

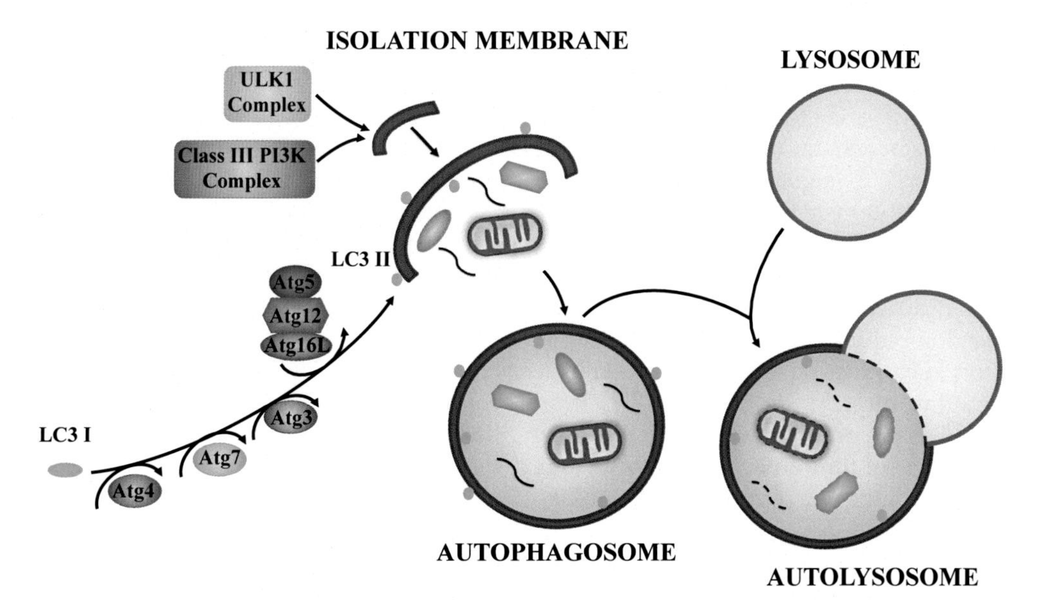

FIGURE 2 **Overview of autophagy.** The nucleation of the isolation membrane depends on the action of two different complexes, the ULK1 complex and the class III phosphatidylinositol 3-kinase (PI3K) complex. Next, autophagosome elongation requires the conjugation of LC3 to the phosphatidylethanolamine, a form called LC3II. LC3 is present in the cell in a soluble cytosolic from named LC3I. LC3I is first cleaved by the protease Atg4 and next an ubiquitin-like conjugation system processes this cleaved form to generate the LC3II. In this system, Atg7 acts like an E1 enzyme, Atg3 like an E2 enzyme, and the complex Atg5/Atg12/Atg16L like an E3 enzyme. This elongation step allows the expansion of the initial membrane, sequestering part of the cytosol. One peculiarity of the autophagosome is that it is a double-membrane vesicle. Upon closure, the external membrane of the autophagosome fuses with the lysosome to form the autolysosome. After fusion, the lysosomal enzymes degrade the proteins and other substrates present in the autolysosome.

2. Autophagosome elongation: The key components for autophagosome elongation are the action of two ubiquitin-like conjugation systems.[48,50] First, Atg12 is activated by Atg7 and conjugated to Atg5 by Atg10.[48,50] The conjugated proteins Atg5/Atg12 then interact with Atg16L to form the final complex.[48,50] On the other hand, LC3 is cleaved by the Atg4 protease, exposing a glycine in this protein. In fact, under basal conditions, LC3 is found mainly in the cytosol as a soluble form named LC3I. Autophagy activation involves LC3 cleavage (form named LC3II), and its conjugation to phosphatidylethanolamine (PE) in a three-step process that involves the action of Atg7, Atg3, and the complex Atg5/Atg12/Atg16L.[48,50]
3. Maturation and degradation of the autophagosome: Upon autophagosome closure, the autophagosome will fuse with a lysosome to form the autolysosome. It is precisely within this cell structure that all the cargoes previously sequestered will be degraded.

Proper regulation of both the UPS and autophagy is crucial for the maintenance of skeletal muscle mass. In fact, pathological activation of these two proteolytic systems in skeletal muscle has been directly related to muscle wasting in various diseases, such as cancer cachexia, insulinopenic diabetes, hepatic cirrhosis, chronic kidney disease, and sepsis.[51–58]

Several signaling pathways affect skeletal muscle mass by regulating protein synthesis and protein degradation (reviewed in Schiaffino et al. and Bonaldo et al.[59,60]). Briefly, one of the most important signaling cascades is the PI3K-Akt-mTOR pathway, which promotes protein synthesis and inhibits autophagy.[59,60] On the other hand, the myostatin (transforming growth factor-β)-ALK4/5-Smad2/3 signaling pathway promotes protein degradation by activating the expression of MuRF1 and atrogin 1, two muscle-specific E3 ubiquitin ligases.[59–62] Some reports suggest that these two pathways interact at different levels in skeletal muscle, although this question requires further study.[59,60] Moreover, inflammatory cytokines also affect the skeletal muscle by promoting protein degradation.[59,60] It is worth drawing attention to nuclear factor-κB, which mediates mainly the effects of tumor necrosis factor-α, and to STAT3, which mediates the effect of interleukin-6.[61,63–66]

Moreover, the UPS and autophagy are coordinately activated in highly catabolic conditions by FoxO transcription factors, especially FoxO3, thus promoting muscle wasting.[67,68] Specifically, FoxO3 induces the expression of genes involved in autophagy, such as LC3, Beclin1, Atg4b, and Bnip3 as well as the expression of MuRF1, atrogin 1, and several proteasome subunits.[67–70] One of the main inhibitors of FoxO transcription factors is Akt, which phosphorylates these transcription factors and causes their translocation from the nucleus (where they serve as activators of transcription) to the cytosol.[67–69,71]

Given the topic of this chapter, we will focus on the effects of insulin on the regulation of muscle mass. Studies performed in muscle cells in culture and using mouse models have revealed that insulin (and other related hormones such as insulin-like growth factor-1) activates the Akt-mTORC1 signaling pathway, thus stimulating protein synthesis and inhibiting protein degradation.[72–75] With respect to protein degradation, insulin causes the repression of several genes involved in the UPS[55,76,77] and inhibits autophagy through mTORC1 activation.[59,60] In fact, in the streptozotocin-induced diabetes mouse model (which lacks insulin) autophagy is activated in skeletal muscle.[78]

Current data about insulin action in human skeletal muscle do not completely recapitulate what has been observed in vitro and in murine models. Thus, protein synthesis in human skeletal muscle does not appear to be affected by insulin.[79–83] However, it is clear that insulin inhibits muscle protein catabolism in this tissue.[79–83]

3. SKELETAL MUSCLE MASS IN INSULIN RESISTANCE AND T2D

From the clinical point of view, most obese, insulin-resistant, and type 2 diabetic patients do not display changes in muscle mass.[15] In fact, there is a positive correlation between body fat content and skeletal muscle mass. Accordingly, increased muscle weight has been reported in obese subjects.[15–17]

Various studies have analyzed skeletal muscle protein metabolism in T2D. Although the results obtained were somewhat controversial, the authors of those studies reached the same conclusion, namely that protein turnover in skeletal muscle is preserved in type 2 diabetic patients.[84–89] Thus, some authors indicated, first, that the increased levels of circulating insulin in type 2 diabetic subjects maintain the protein degradation rate at the same level as that in healthy subjects, and, second, that insulin infusion in diabetic patients causes a similar response as in healthy individuals.[84,85] However, other authors defined what they called insulin resistance of protein metabolism in skeletal muscle.[86,87] Thus, insulin infusion in type 2 diabetic patients does not alter muscle protein anabolism or catabolism, in contrast to what occurs in healthy individuals.[86,87] Thus, as previously proposed by others, differences in the experimental design and methods used to assess protein synthesis and degradation may explain the discrepancies found between studies.[86,88,89]

Overall, T2D is a paradoxical condition because there are no changes in muscle mass or protein turnover in spite of insulin resistance. Given that insulin inhibits protein catabolism, one would expect an increased protein degradation rate and, perhaps, some muscle loss in insulin-resistant subjects. Because this is not the case, it was deduced that there is an adaptive mechanism that allows the maintenance of skeletal muscle mass under this pathological condition. However, this mechanism has barely been studied, and the factors that participate in this adaptation are unclear. One factor is TP53INP2, whose expression is reduced in skeletal muscle of insulin-resistant and type 2 diabetic patients.[78] TP53INP2 is a protein that activates autophagy in skeletal muscle and promotes muscle wasting in a murine model of streptozotocin-induced diabetes.[78] The role of TP53INP2 in skeletal muscle and in diabetes will be further discussed in the following sections.

It is also noteworthy that although T2D per se is not associated with muscle atrophy, the development of associated comorbidities, or the presence of other clinical complications can result in muscle wasting. Some examples are the development of cardiovascular complications or chronic kidney disease.[21,22,90] Moreover, T2D is considered a risk factor for the development of sarcopenia—the muscle atrophy associated with aging.[18,23,91,92] Specifically, elderly individuals with T2D have three times more probability of developing sarcopenia than healthy subjects.[92] In fact, obesity and insulin resistance per se can worsen the compromised muscle performance caused by sarcopenia, a phenomenon that has been called sarcopenic obesity (reviewed in Kob et al.[93]).

4. TP53INP2 AND ITS ROLE IN AUTOPHAGY

As previously commented, the scientific community has made huge efforts to identify and characterize the genes involved in the pathophysiology of T2D. Among these genes, the one coding for TP53INP2 or diabetes and obesity regulated (DOR)[94] appear to be related to this disease. More specifically, *Tp53Inp2* was found to be repressed in skeletal muscle of Zucker Fatty Diabetic (ZFD) rats compared with control animals.[94] TP53INP2 is highly expressed in skeletal muscle and has only one homologous protein, TP53INP1.[78,94,95] The first function reported for TP53INP2 was as a coactivator of various nuclear receptors. Thus, in studies performed in cultured cells, TP53INP2 overexpression with several nuclear receptors, such as TRα1, GR, PPAR-γ, or VDR, increased the activity of these transcription factors upon the presence of the respective ligand.[94,95] Focusing on TRα1, TP53INP2 is required for the proper induction of the myogenic genes regulated by T3 and, accordingly, to ensure correct myogenic differentiation.[94] In this case, TP53INP2 and TRα1 directly interact in the promoter of the genes regulated by triiodothyronine to promote their transcription.[94] In addition, the TP53INP2 ortholog in *Drosophila melanogaster* or *Drosophila* DOR (dDOR) also operates as a nuclear coactivator of the ecdysone receptor.[96] More specifically, dDOR is necessary to maximize the transcriptional activity of this nuclear receptor, thereby indicating that the role of TP53INP2 as coactivator has been conserved during evolution.[96]

In addition, TP53INP2 has also been reported to be an activator of autophagy by promoting autophagosome formation.[78,97,98] Thus, upon autophagy activation, TP53INP2 is localized in the cytoplasm in punctated structures, together with the autophagosome marker LC3.[97,98] However, TP53INP2 is mostly a nuclear protein under basal conditions and it shuttles continuously between the nucleus and the cytoplasm.[97,99] In fact, while shuttling, it promotes the translocation of the nuclear LC3 to the cytosol.[100] TP53INP2 is able to interact directly with LC3 and with other members of the Atg8 protein family, such as GABARAP, GABARAPL1, and GATE16, which are also involved in autophagosome formation.[95,97,98]

Studies performed in cultured cells demonstrated that TP53INP2 overexpression increases protein degradation and the number of autophagosomes in both basal conditions and upon autophagy induction by amino acid starvation.[97,98] Consistently, repression of TP53INP2 showed the opposite phenotype: a reduction in protein degradation and autophagosome number in both conditions.[97,98] Moreover, this function is also conserved in *D. melanogaster*.[97] During the normal development of *Drosophila*, autophagy is highly induced in the larval fat body for pupation.[101,102] However, dDOR knockdown compromises the induction of this process in this organ.[97]

Basal autophagy is a highly selective process and, in fact, several proteins have been described as autophagy receptors (reviewed in Stolz et al.[103]). Thus, these proteins can deliver specific targets for degradation to the nascent autophagosomes. Examples include p62, NBR1, and ALFY, which interact with polyubiquitin chains and LC3 (either directly or indirectly).[104-107] In this regard, TP53INP2 has been proposed as another autophagy receptor for ubiquitinated proteins.[78] This protein interacts with ubiquitin, as shown by coimmunoprecipitation studies performed in cells in culture.[78]

Specifically, TP53INP2 shows preferential binding to K-63 polyubiquitin chain and linked monoubiquitin.[78] Moreover, the induction of the formation of protein aggregates by puromycin treatment causes partial colocalization of TP53INP2 with polyubiquitin-positive dots in muscle cells.[78]

5. TP53INP2 IN SKELETAL MUSCLE AND T2D

TP53INP2 is a negative regulator of skeletal muscle mass, as revealed by results obtained using transgenic mouse models.[78] More specifically, TP53INP2 gain of function in skeletal muscle causes a reduction in muscle weight and myofiber size.[78] In contrast, the ablation of this protein in skeletal muscle increases the weight of various muscle groups and myofiber size.[78] Moreover, TP53INP2 promotes muscle wasting under highly catabolic conditions, such as in diabetes induced by streptozotocin injection. Thus, TP53INP2 gain of function enhances muscle atrophy caused by insulinopenic diabetes, whereas TP53INP2 ablation ameliorates muscle loss under this condition.[78]

The effect of TP53INP2 in skeletal muscle depends on its function as an autophagy activator.[78] In fact, TP53INP2 overexpression reduces LC3I protein content in skeletal muscle under basal conditions and causes a major accumulation of LC3II upon autophagy inhibition by chloroquine treatment.[78] These data indicate that autophagy flux is higher in skeletal muscle upon TP53INP2 gain of function. On the other hand, TP53INP2 ablation increases the amount of both forms of LC3 (LC3I and LC3II) in skeletal muscle under basal conditions.[78] However, upon autophagy blockage, LC3II accumulation is lower in TP53INP2-ablated mice, thereby indicating that autophagy is less active upon TP53INP2 absence in skeletal muscle.[78]

In fact, the increase in autophagy in skeletal muscle as a result of TP53INP2 gain of function causes a 28% increase in the protein degradation rate.[78] This increase is at least partly from the autophagy-dependent degradation of polyubiquitinated proteins.[78] Specifically, autophagy inhibition with chloroquine administration leads to more accumulation of ubiquitinated proteins in skeletal muscles that overexpress TP53INP2 compared with control muscles.[78] This observation further confirms that TP53INP participates in the degradation of polyubiquitinated proteins through autophagy.

Focusing on T2D, TP53INP2 expression in skeletal muscle is repressed under this condition. This repression is observed in classical rodent models of insulin resistance, such as ZFD rats and db/db mice as well as in patients.[78] Thus, type 2 diabetic subjects display a dramatic reduction in *TP53INP2* messenger RNA (mRNA) levels in skeletal muscle.[78] In fact, repression of this protein in human muscle occurs early in the development of insulin resistance. Thus, obese nondiabetic subjects or overweight nondiabetic subjects also display a reduction in TP53INP2 mRNA levels in this muscle type.[78] Early repression of TP53INP2, which is a negative regulator of muscle mass, may contribute to the preservation of skeletal muscle mass seen in most type 2 diabetic patients. It is highly likely that other factors apart from TP53INP2 are involved in this adaptive mechanism to conserve muscle mass under conditions of insulin resistance. This question will require further study in the coming years.

6. SKELETAL MUSCLE QUALITY IN INSULIN RESISTANCE AND T2D

As previously commented, most type 2 diabetic patients do not show changes or even an increase in muscle mass.[15–17] However, they do present muscle weakness.[14,24] These data suggest that, in spite of the preservation of muscle mass, muscle quality, and performance of this tissue is reduced in T2D.[14,108] Muscle quality is used as an index of the functional performance of skeletal muscle and is assessed as specific strength (strength relative to muscle mass). Classically, muscle mass has been considered one of the main factors that determine muscle quality and performance. However, a correlation between these two parameters does not occur in all cases, and muscle quality rather than mass is currently considered a more reliable index of the function of this tissue.[14,108,109]

Reduced muscle quality in T2D has been directly associated with the duration of diabetes and with poor glycemic control.[110,111] Moreover, one of the factors proposed to be involved in the functional decline of skeletal muscle under diabetic conditions is fat infiltration into this tissue.[24,110,112] T2D alters lipid metabolism and causes fat accumulation in tissues other than adipose tissue, such as skeletal muscle.[113] Muscle fat infiltration has been associated with reduced strength in elderly subjects[112] and, in fact, fat infiltration into skeletal muscle is detectable before any change in muscle mass or cross-sectional area occurs.[24] These observations are consistent with the compromised muscle function in diabetes, in spite of the absence of changes in muscle mass.

In addition, the progressive decline in muscle performance under diabetic conditions has also been linked to the development and progression of peripheral neuropathy.[24,114,115] Defective innervation and motor neuron function directly affect muscle performance.

One of the main factors involved in the maintenance of muscle quality is basal autophagy.[41,116] As previously commented, excessive autophagy is detrimental for skeletal muscle mass because it causes exacerbated protein loss. However, the inhibition of autophagy compromises skeletal muscle quality and leads to the development of a myopathy.[41,116] In fact, current data support the notion that defective autophagy is one of the main causes of several myopathies, such as Danon disease and Pompe disease (reviewed in Malicdan et al.[117]). In the first study, which addressed the role of autophagy in the maintenance of muscle mass and quality, the authors used a mouse model with muscle-specific Atg7 ablation.[116] The authors showed that autophagy blockage recapitulates basic features of several muscle myopathies, such as myofiber disorganization, vacuolization, accumulation of aberrant organelles, and reduction in both absolute and specific force.[116] In keeping with these data, autophagy inhibition caused by constitutive mTORC1 activation in skeletal muscle leads to the development of late-onset myopathy with similar morphological features as those observed for Atg7 ablation.[41] Moreover, a reduction in the specific force occurring much earlier than the detection of muscle loss has also been reported.[41] These observations are very similar to what occurs in T2D. Overall, these data indicate that autophagy acts as a quality control mechanism for the maintenance of muscle function.

It is worth mentioning that one of the organelles most affected in T2D in the mitochondrion.[41,116] In this regard, the authors observed the accumulation of unusually large and swollen mitochondria, together with increased oxidative stress in skeletal muscle, upon autophagy inhibition.[41,116] In fact, mitochondrial degradation through autophagy or mitophagy is an essential quality control mechanism to maintain mitochondrial function. Moreover, mitochondrial organization is also critical for the function of the organelles themselves, and mitochondrial dynamics plays a central role in this function (Figure 3). Mitochondrial function is highly compromised in T2D[25,26,118,119]; this will be further discussed in the following section.

7. MITOCHONDRIAL DYNAMICS, MITOPHAGY, AND INSULIN RESISTANCE

Mitochondria are fundamental organelles for proper skeletal muscle function because they are the main energetic support for this tissue by producing ATP through oxidative phosphorylation. T2D and obesity have been associated with altered mitochondrial function in skeletal muscle, understood as a reduction in mitochondrial activity and oxidative capacity.[25,26,118,119] However, it is still not clear whether mitochondrial dysfunction is a cause or a consequence of insulin resistance.[120] Mitochondrial function depends on several parameters, including mitochondrial number, organization, and quality. In this regard, the biogenesis and dynamics of mitochondria and mitophagy play central roles in this function.

Mitochondrial content is decreased in obesity and T2D, as assessed by a reduced amount of mitochondrial DNA and reduced citrate synthase activity.[25,121] In this regard, the expression of several mitochondrial genes is decreased under these pathological conditions, especially those genes related to mitochondrial oxidation and oxidative phosphorylation.[122–125] Moreover, the expression of PG1-α and PG1-β, two classical transcription factors involved in the regulation mitochondrial biogenesis and metabolism, is also reduced.[122,125]

Mitochondria are highly dynamic organelles that are constantly undergoing fusion and fission events inside the cell (Figure 3). Thus, mitochondrial morphology is extremely flexible and can change in order to adapt to different requirements or conditions (reviewed in Chen et al. and Liesa et al.[126,127]). In fact, mitochondrial dynamics is crucial for the regulation of mitochondrial function and maintenance of mitochondrial quality. Mitochondrial fusion favors the maintenance of the former by facilitating the exchange of components such as mitochondrial DNA, proteins (for example, metabolic enzymes), and metabolites.[128,129] On the other hand, mitochondrial fission allows the degradation of damaged mitochondria through mitophagy.[130,131]

Several factors participate in mitochondrial fusion and fission. The former process first involves the fusion of the outer mitochondrial membrane and afterward the fusion of the inner mitochondrial membrane.[132] The main factors regulating fusion are mitofusin 1, mitofusin 2, and OPA1. Mitofusin 1 2, proteins localized in the outer mitochondrial membrane, have guanosine triphosphatase (GTPase) activity, which is essential for their role in mitochondrial fusion.[133,134] Mitofusin 2 expression levels are reduced in skeletal muscle of both Zucker Fatty rats and obese and type 2 diabetic patients.[135,136] These data are consistent with the alterations of mitochondrial dynamics observed in insulin resistance. Such alterations include the reduction in mitochondrial size observed in type 2 diabetic and obese subjects as well as the altered mitochondrial network in skeletal muscle of ZFD rats.[25,135,137]

On the other hand, OPA1, also a GTPase, is localized in the inner mitochondrial membrane. It has a complex regulation as it presents eight distinct splicing isoforms that, when translated, are subjected to proteolysis at various cleavage sites.[138–140] The different forms of OPA1 have a range of effects on mitochondrial fusion.[138–140] Studies performed in cultured cells revealed that OPA1 repression reduces mitochondrial

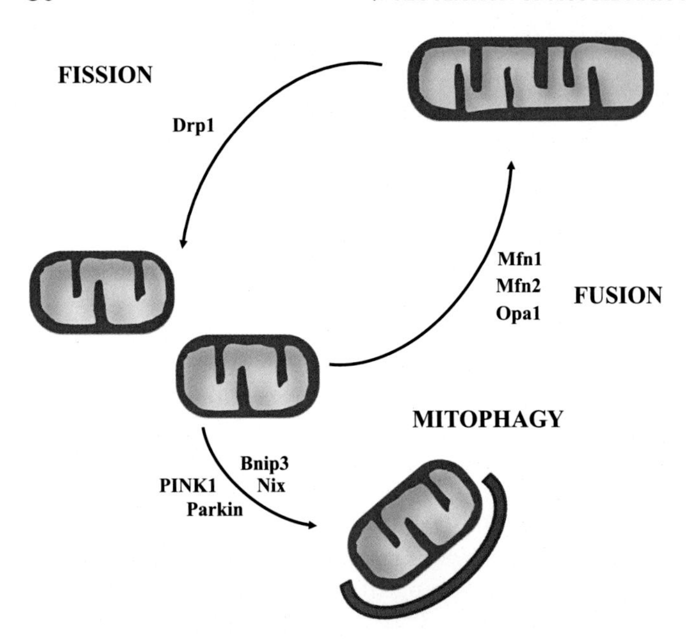

FIGURE 3 **Mitochondrial dynamics and mitophagy.** Mitochondria are dynamic organelles that can undergo processes of fusion and fission in response to different stimuli or to adapt to different conditions. The main players involved in mitochondrial fusion are the mitochondrial proteins Mfn1, Mfn2, and Opa1. The main player in mitochondrial fission is Drp1, a cytosolic protein that is recruited to the mitochondria when necessary by interaction with different mitochondrial proteins. Mitochondrial fission is necessary before mitochondrial degradation through mitophagy. The most studied pathway promoting mitophagy is the PINK1-Parkin pathway. However, other proteins different from this pathway have been also involved in mitophagy, such as Bnip3 or Nix.

membrane potential and respiratory capacity.[141,142] Moreover, the knock-out mouse model of Oma1, one of the metalloproteases that cleaves OPA1, causes dramatic changes in whole-body metabolism and thermogenesis, together with impaired OPA1 cleavage and altered mitochondrial morphology.[143]

Regarding mitochondrial fission, the main regulator of this process is Drp1. Drp1 is a cytosolic GTPase that is recruited to mitochondria by its interaction with various proteins such as Fis1, Mff, MID49, and MID51, which are located in the outer mitochondrial membrane.[144–151] Studies performed using myotubes in culture and obese mouse models suggest that the increase in mitochondrial fission enhances insulin resistance and negatively affects mitochondrial function in skeletal muscle.[152]

Mitochondrial fission is a basic step of the mitochondrial degradation through autophagy or mitophagy. The latter is a crucial quality control mechanism through which proper mitochondrial function is maintained inside the cell.[130,131] To date, the most well-known pathway that triggers mitophagy involves the action of PINK1 and Parkin. Briefly, PINK1 is a mitochondrial kinase with a very high turnover, which causes its rapid degradation after synthesis.[153,154] However, upon mitochondrial depolarization, PINK1 is stabilized and promotes Parkin recruitment to mitochondria.[155–157] Parkin is an E3 ubiquitin ligase that is activated by PINK1 and it ubiquitinates several substrates in the outer mitochondrial membrane, including Mfn2 and Mfn1.[158–160] In addition to inducing the proteasomal degradation of these proteins, this ubiquitination lead to the recruitment of several autophagy receptors that have the capacity to interact with ubiquitin, such as p62 or optineurin, and that will drive mitophagy.[161–163] Other proteins can regulate PINK1-Parkin function. One example is HSP72, a heat shock protein required for the recruitment of Parkin to mitochondria.[164]

Moreover, mitophagy can also be triggered by proteins other than PINK1 and Parkin. Two examples are Bnip3 and NIX.[69,165–168] These two proteins interact with LC3 and promote engulfment of damaged mitochondria by the autophagosome.[69,165–168]

A recent study showed that impaired mitophagy (through HSP72 or Parkin ablation) causes insulin resistance in skeletal muscle.[164] This observation suggests that mitophagy plays a positive role in the maintenance of insulin sensitivity in this tissue and therefore in the preservation of muscle metabolic function.[164,169]

8. CONCLUDING REMARKS

T2D is a pathologic condition that affects whole-body metabolism, including glucose, lipid, and protein metabolism. To find efficient treatments for this disease, it is necessary to gain insight into the mechanisms that cause these metabolic alterations in different tissues and to identify new potential targets. In this chapter, we have focused on skeletal muscle and have summarized current knowledge about the regulation of muscle mass and quality under insulin resistance. Available data indicate that, although muscle mass is not reduced in most cases, muscle quality is highly compromised. Unraveling the evident lack of correlation between muscle mass and quality in T2D will represent a challenge for future research. The protein TP53INP2 may play a critical role in this disconnection. Thus, although TP53INP2 repression may prevent muscle loss under insulin resistance and may compromise muscle quality. In fact, it has recently been reported that enhanced autophagy in skeletal muscle (induced either pharmacologically or by exercise) ameliorates insulin resistance.[170,171] As stated previously, TP53INP2 interacts with ubiquitin and may be responsible for the selective degradation of substrates that are specifically delivered to the autophagosome through polyubiquitination. Examples of these substrates are protein aggregates and some organelles like mitochondria, which are essential for skeletal muscle function. Analysis of these processes will probably involve the identification of new regulators of autophagy and the UPS in skeletal muscle, together with the redefinition of previously identified regulators. Moreover, it will also encompass the study of factors that select the substrates to be degraded under insulin resistance in skeletal muscle.

Acknowledgments

We would like to thank Ms. Tanya Yates for editorial support.

Financial support and sponsorship.

D.S. was a recipient of an FPU fellowship from the "Ministerio de Educación y Cultura," Spain, and currently holds a California Institute for Regenerative Medicine Training grant (TG2-01162). This work was supported by research grants from the Ministry of Economy and Competiveness (SAF2008-03803 and SAF2013-40987R), grants 2009SGR915 and 2014SGR48 from the "Generalitat de Catalunya," CIBERDEM ("Instituto de Salud Carlos III"), INTERREG IV-B-SUDOE-FEDER (DIOMED, SOE1/P1/E178), and DEXLIFE (grant agreement no: 279,228). A.Z. is recipient of an ICREA Acadèmia ("Generalitat de Catalunya").

References

1. http://www.idf.org.
2. Canadian Diabetes Association Clinical Practice Guidelines Expert Committee, Goldenberg R, Punthakee Z. Definition, classification and diagnosis of diabetes, prediabetes and metabolic syndrome. *Can J Diabetes* 2013;**37**(Suppl. 1):S8–11.
3. American Diabetes Association. Diagnosis and classification of diabetes mellitus. *Diabetes Care* 2014;**37**(Suppl. 1):S81–90.
4. Prentki M, Nolan CJ. Islet beta cell failure in type 2 diabetes. *J Clin Invest* 2006;**116**(7):1802–12.
5. Brunetti A, Chiefari E, Foti D. Recent advances in the molecular genetics of type 2 diabetes mellitus. *World J Diabetes* 2014;**5**(2): 128–40.
6. Doria A, Patti ME, Kahn CR. The emerging genetic architecture of type 2 diabetes. *Cell Metab* 2008;**8**(3):186–200.
7. Ahlqvist E, Ahluwalia TS, Groop L. Genetics of type 2 diabetes. *Clin Chem* 2011;**57**(2):241–54.
8. Qiu B, Shi X, Wong ET, et al. NUCKS is a positive transcriptional regulator of insulin signaling. *Cell Rep* 2014;**7**(6):1876–86.
9. Friedman JM. Obesity in the new millennium. *Nature* 2000; **404**(6778):632–4.
10. Kahn BB, Flier JS. Obesity and insulin resistance. *J Clin Invest* 2000; **106**(4):473–81.
11. Owen N, Sparling PB, Healy GN, Dunstan DW, Matthews CE. Sedentary behavior: emerging evidence for a new health risk. *Mayo Clin Proc* 2010;**85**(12):1138–41.
12. DeFronzo RA, Tripathy D. Skeletal muscle insulin resistance is the primary defect in type 2 diabetes. *Diabetes Care* 2009;**32**(Suppl 2): S157–63.
13. Corcoran MP, Lamon-Fava S, Fielding RA. Skeletal muscle lipid deposition and insulin resistance: effect of dietary fatty acids and exercise. *Am J Clin Nutr* 2007;**85**(3):662–77.
14. Park SW, Goodpaster BH, Strotmeyer ES, et al. Decreased muscle strength and quality in older adults with type 2 diabetes: the health, aging, and body composition study. *Diabetes* 2006;**55**(6): 1813–8.
15. Janssen I, Heymsfield SB, Wang ZM, Ross R. Skeletal muscle mass and distribution in 468 men and women aged 18-88 yr. *J Appl Physiol* 2000;**89**(1):81–8.
16. Kanehisa H, Fukunaga T. Association between body mass index and muscularity in healthy older Japanese women and men. *J Physiol Anthropol* 2013;**32**(1):4.
17. Micozzi MS, Harris TM. Age variations in the relation of body mass indices to estimates of body fat and muscle mass. *Am J Phys Anthropol* 1990;**81**(3):375–9.
18. Park SW, Goodpaster BH, Lee JS, et al. Excessive loss of skeletal muscle mass in older adults with type 2 diabetes. *Diabetes Care* 2009;**32**(11):1993–7.
19. Park SW, Goodpaster BH, Strotmeyer ES, et al. Accelerated loss of skeletal muscle strength in older adults with type 2 diabetes: the health, aging, and body composition study. *Diabetes Care* 2007; **30**(6):1507–12.
20. Bell JA, Volpi E, Fujita S, Cadenas JG, Sheffield-Moore M, Rasmussen BB. Skeletal muscle protein anabolic response to increased energy and insulin is preserved in poorly controlled type 2 diabetes. *J Nutr* 2006;**136**(5):1249–55.
21. Doehner W, Erdmann E, Cairns R, et al. Inverse relation of body weight and weight change with mortality and morbidity in patients with type 2 diabetes and cardiovascular co-morbidity: an analysis of the PROactive study population. *Int J Cardiol* 2012; **162**(1):20–6.
22. Pupim LB, Flakoll PJ, Majchrzak KM, Aftab Guy DL, Stenvinkel P, Ikizler TA. Increased muscle protein breakdown in chronic hemodialysis patients with type 2 diabetes mellitus. *Kidney Int* 2005; **68**(4):1857–65.
23. Leenders M, Verdijk LB, van der Hoeven L, et al. Patients with type 2 diabetes show a greater decline in muscle mass, muscle strength, and functional capacity with aging. *J Am Med Dir Assoc* 2013;**14**(8):585–92.
24. Hilton TN, Tuttle LJ, Bohnert KL, Mueller MJ, Sinacore DR. Excessive adipose tissue infiltration in skeletal muscle in individuals with obesity, diabetes mellitus, and peripheral neuropathy: association with performance and function. *Phys Ther* 2008;**88**(11):1336–44.
25. Kelley DE, He J, Menshikova EV, Ritov VB. Dysfunction of mitochondria in human skeletal muscle in type 2 diabetes. *Diabetes* 2002;**51**(10):2944–50.
26. Mogensen M, Sahlin K, Fernstrom M, et al. Mitochondrial respiration is decreased in skeletal muscle of patients with type 2 diabetes. *Diabetes* 2007;**56**(6):1592–9.
27. Petersen KF, Befroy D, Dufour S, et al. Mitochondrial dysfunction in the elderly: possible role in insulin resistance. *Science* 2003; **300**(5622):1140–2.
28. Petersen KF, Dufour S, Befroy D, Garcia R, Shulman GI. Impaired mitochondrial activity in the insulin-resistant offspring of patients with type 2 diabetes. *N Engl J Med* 2004;**350**(7):664–71.
29. Rock KL, Gramm C, Rothstein L, et al. Inhibitors of the proteasome block the degradation of most cell proteins and the generation of peptides presented on MHC class I molecules. *Cell* 1994; **78**(5):761–71.
30. Tawa Jr NE, Odessey R, Goldberg AL. Inhibitors of the proteasome reduce the accelerated proteolysis in atrophying rat skeletal muscles. *J Clin Invest* 1997;**100**(1):197–203.
31. Mitch WE, Medina R, Grieber S, et al. Metabolic acidosis stimulates muscle protein degradation by activating the adenosine triphosphate-dependent pathway involving ubiquitin and proteasomes. *J Clin Invest* 1994;**93**(5):2127–33.
32. Wing SS, Goldberg AL. Glucocorticoids activate the ATP-ubiquitin-dependent proteolytic system in skeletal muscle during fasting. *Am J Physiol* 1993;**264**(4 Pt 1):E668–76.
33. Voges D, Zwickl P, Baumeister W. The 26S proteasome: a molecular machine designed for controlled proteolysis. *Ann Rev Biochem* 1999;**68**:1015–68.

34. Lecker SH, Goldberg AL, Mitch WE. Protein degradation by the ubiquitin-proteasome pathway in normal and disease states. *J Am Soc Nephrol* 2006;**17**(7):1807–19.
35. Benaroudj N, Zwickl P, Seemuller E, Baumeister W, Goldberg AL. ATP hydrolysis by the proteasome regulatory complex PAN serves multiple functions in protein degradation. *Molecular Cell* 2003;**11**(1):69–78.
36. Haas AL, Rose IA. The mechanism of ubiquitin activating enzyme. A kinetic and equilibrium analysis. *J Biol Chem* 1982; **257**(17):10329–37.
37. Jagoe RT, Goldberg AL. What do we really know about the ubiquitin-proteasome pathway in muscle atrophy? *Curr Opin Clin Nutr Metab Care* 2001;**4**(3):183–90.
38. Martinez-Vicente M, Cuervo AM. Autophagy and neurodegeneration: when the cleaning crew goes on strike. *Lancet Neurol* 2007; **6**(4):352–61.
39. Lowell BB, Ruderman NB, Goodman MN. Evidence that lysosomes are not involved in the degradation of myofibrillar proteins in rat skeletal muscle. *Biochem J* 1986;**234**(1):237–40.
40. Mercer CA, Kaliappan A, Dennis PB. A novel, human Atg13 binding protein, Atg101, interacts with ULK1 and is essential for macroautophagy. *Autophagy* 2009;**5**(5):649–62.
41. Castets P, Lin S, Rion N, et al. Sustained activation of mTORC1 in skeletal muscle inhibits constitutive and starvation-induced autophagy and causes a severe, late-onset myopathy. *Cell Metab* 2013; **17**(5):731–44.
42. Egan DF, Shackelford DB, Mihaylova MM, et al. Phosphorylation of ULK1 (hATG1) by AMP-activated protein kinase connects energy sensing to mitophagy. *Science* 2011;**331**(6016):456–61.
43. Kim J, Kundu M, Viollet B, Guan KL. AMPK and mTOR regulate autophagy through direct phosphorylation of Ulk1. *Nat Cell Biol* 2011;**13**(2):132–41.
44. Jung CH, Ro SH, Cao J, Otto NM, Kim DH. mTOR regulation of autophagy. *FEBS Lett* 2010;**584**(7):1287–95.
45. Mizushima N. The role of the Atg1/ULK1 complex in autophagy regulation. *Curr Opin Cell Biol* 2010;**22**(2):132–9.
46. Itakura E, Kishi C, Inoue K, Mizushima N. Beclin 1 forms two distinct phosphatidylinositol 3-kinase complexes with mammalian Atg14 and UVRAG. *Mol Biol Cell* 2008;**19**(12):5360–72.
47. Sun Q, Fan W, Chen K, Ding X, Chen S, Zhong Q. Identification of Barkor as a mammalian autophagy-specific factor for Beclin 1 and class III phosphatidylinositol 3-kinase. *Proc Natl Acad Sci USA* 2008;**105**(49):19211–6.
48. Mehrpour M, Esclatine A, Beau I, Codogno P. Overview of macroautophagy regulation in mammalian cells. *Cell Res* 2010;**20**(7): 748–62.
49. Reidick C, El Magraoui F, Meyer HE, Stenmark H, Platta HW. Regulation of the Tumor-Suppressor Function of the Class III Phosphatidylinositol 3-Kinase Complex by Ubiquitin and SUMO. *Cancers* 2014;**7**(1):1–29.
50. Wing SS, Lecker SH, Jagoe RT. Proteolysis in illness-associated skeletal muscle atrophy: from pathways to networks. *Crit Rev Clin Lab Sci* 2011;**48**(2):49–70.
51. Baracos VE, DeVivo C, Hoyle DH, Goldberg AL. Activation of the ATP-ubiquitin-proteasome pathway in skeletal muscle of cachectic rats bearing a hepatoma. *Am J Physiol* 1995;**268**(5 Pt 1): E996–1006.
52. Penna F, Costamagna D, Pin F, et al. Autophagic degradation contributes to muscle wasting in cancer cachexia. *Am J Pathol* 2013; **182**(4):1367–78.
53. Wang XH, Mitch WE. Mechanisms of muscle wasting in chronic kidney disease. *Nat Rev Nephrol* 2014;**10**(9):504–16.
54. Bailey JL, Zheng B, Hu Z, Price SR, Mitch WE. Chronic kidney disease causes defects in signaling through the insulin receptor substrate/phosphatidylinositol 3-kinase/Akt pathway: implications for muscle atrophy. *J Am Soc Nephrol* 2006;**17**(5):1388–94.
55. Price SR, Bailey JL, Wang X, et al. Muscle wasting in insulinopenic rats results from activation of the ATP-dependent, ubiquitin-proteasome proteolytic pathway by a mechanism including gene transcription. *J Clin Invest* 1996;**98**(8):1703–8.
56. Krause MP, Riddell MC, Hawke TJ. Effects of type 1 diabetes mellitus on skeletal muscle: clinical observations and physiological mechanisms. *Pediatr Diabetes* 2011;**12**(4 Pt 1):345–64.
57. Jakobsen J, Reske-Nielsen E. Diffuse muscle fiber atrophy in newly diagnosed diabetes. *Clin Neuropathol* 1986;**5**(2):73–7.
58. Qiu J, Tsien C, Thapalaya S, et al. Hyperammonemia-mediated autophagy in skeletal muscle contributes to sarcopenia of cirrhosis. *Am J Physiol Endocrinol Metab* 2012;**303**(8):E983–93.
59. Schiaffino S, Dyar KA, Ciciliot S, Blaauw B, Sandri M. Mechanisms regulating skeletal muscle growth and atrophy. *FEBS J* 2013;**280**(17):4294–314.
60. Bonaldo P, Sandri M. Cellular and molecular mechanisms of muscle atrophy. *Dis Model Mech* 2013;**6**(1):25–39.
61. Zhang L, Pan J, Dong Y, et al. Stat3 activation links a C/EBPdelta to myostatin pathway to stimulate loss of muscle mass. *Cell Metab* 2013;**18**(3):368–79.
62. Thomas SS, Mitch WE. Mechanisms stimulating muscle wasting in chronic kidney disease: the roles of the ubiquitin-proteasome system and myostatin. *Clin Exp Nephrol* 2013;**17**(2):174–82.
63. Strassmann G, Fong M, Kenney JS, Jacob CO. Evidence for the involvement of interleukin 6 in experimental cancer cachexia. *J Clin Invest* 1992;**89**(5):1681–4.
64. Bonetto A, Aydogdu T, Kunzevitzky N, et al. STAT3 activation in skeletal muscle links muscle wasting and the acute phase response in cancer cachexia. *PLoS One* 2011;**6**(7):e22538.
65. Peterson JM, Bakkar N, Guttridge DC. NF-kappaB signaling in skeletal muscle health and disease. *Curr Top Dev Biol* 2011;**96**: 85–119.
66. Bonetto A, Aydogdu T, Jin X, et al. JAK/STAT3 pathway inhibition blocks skeletal muscle wasting downstream of IL-6 and in experimental cancer cachexia. *Am J Physiol Endocrinol Metab* 2012;**303**(3):E410–21.
67. Zhao J, Brault JJ, Schild A, et al. FoxO3 coordinately activates protein degradation by the autophagic/lysosomal and proteasomal pathways in atrophying muscle cells. *Cell Metab* 2007;**6**(6):472–83.
68. Stitt TN, Drujan D, Clarke BA, et al. The IGF-1/PI3K/Akt pathway prevents expression of muscle atrophy-induced ubiquitin ligases by inhibiting FOXO transcription factors. *Mol Cell* 2004; **14**(3):395–403.
69. Mammucari C, Milan G, Romanello V, et al. FoxO3 controls autophagy in skeletal muscle in vivo. *Cell Metab* 2007;**6**(6):458–71.
70. Sandri M, Sandri C, Gilbert A, et al. Foxo transcription factors induce the atrophy-related ubiquitin ligase atrogin-1 and cause skeletal muscle atrophy. *Cell* 2004;**117**(3):399–412.
71. Brunet A, Bonni A, Zigmond MJ, et al. Akt promotes cell survival by phosphorylating and inhibiting a Forkhead transcription factor. *Cell* 1999;**96**(6):857–68.
72. Airhart J, Arnold JA, Stirewalt WS, Low RB. Insulin stimulation of protein synthesis in cultured skeletal and cardiac muscle cells. *Am J Physiol* 1982;**243**(1):C81–6.
73. Shen WH, Boyle DW, Wisniowski P, Bade A, Liechty EA. Insulin and IGF-I stimulate the formation of the eukaryotic initiation factor 4F complex and protein synthesis in C2C12 myotubes independent of availability of external amino acids. *J Endocrinol* 2005;**185**(2):275–89.
74. Pain VM, Garlick PJ. Effect of streptozotocin diabetes and insulin treatment on the rate of protein synthesis in tissues of the rat in vivo. *J Biol Chem* 1974;**249**(14):4510–4.
75. Monier S, Le Cam A, Le Marchand-Brustel Y. Insulin and insulin-like growth factor I. Effects on protein synthesis in isolated muscles from lean and goldthioglucose-obese mice. *Diabetes* 1983; **32**(5):392–7.

76. Lecker SH, Jagoe RT, Gilbert A, et al. Multiple types of skeletal muscle atrophy involve a common program of changes in gene expression. *FASEB J* 2004;**18**(1):39–51.
77. Lecker SH, Solomon V, Mitch WE, Goldberg AL. Muscle protein breakdown and the critical role of the ubiquitin-proteasome pathway in normal and disease states. *J Nutr* 1999;**129**(1S Suppl.): 227S–37S.
78. Sala D, Ivanova S, Plana N, et al. Autophagy-regulating TP53INP2 mediates muscle wasting and is repressed in diabetes. *J Clin Invest* 2014;**124**(5):1914–27.
79. Gelfand RA, Barrett EJ. Effect of physiologic hyperinsulinemia on skeletal muscle protein synthesis and breakdown in man. *J Clin Invest* 1987;**80**(1):1–6.
80. Louard RJ, Fryburg DA, Gelfand RA, Barrett EJ. Insulin sensitivity of protein and glucose metabolism in human forearm skeletal muscle. *J Clin Invest* 1992;**90**(6):2348–54.
81. Godil MA, Wilson TA, Garlick PJ, McNurlan MA. Effect of insulin with concurrent amino acid infusion on protein metabolism in rapidly growing pubertal children with type 1 diabetes. *Pediatr Res* 2005;**58**(2):229–34.
82. Nair KS, Ford GC, Ekberg K, Fernqvist-Forbes E, Wahren J. Protein dynamics in whole body and in splanchnic and leg tissues in type I diabetic patients. *J Clin Invest* 1995;**95**(6): 2926–37.
83. Pacy PJ, Bannister PA, Halliday D. Influence of insulin on leucine kinetics in the whole body and across the forearm in post-absorptive insulin dependent diabetic (type 1) patients. *Diabetes Res* 1991;**18**(4):155–62.
84. Tessari P, Kiwanuka E, Coracina A, et al. Insulin in methionine and homocysteine kinetics in healthy humans: plasma vs. intracellular models. *Am J Physiol Endocrinol Metab* 2005;**288**(6): E1270–6.
85. Halvatsiotis P, Short KR, Bigelow M, Nair KS. Synthesis rate of muscle proteins, muscle functions, and amino acid kinetics in type 2 diabetes. *Diabetes* 2002;**51**(8):2395–404.
86. Pereira S, Marliss EB, Morais JA, Chevalier S, Gougeon R. Insulin resistance of protein metabolism in type 2 diabetes. *Diabetes* 2008; **57**(1):56–63.
87. Bassil M, Burgos S, Marliss EB, Morais JA, Chevalier S, Gougeon R. Hyperaminoacidaemia at postprandial levels does not modulate glucose metabolism in type 2 diabetes mellitus. *Diabetologia* 2011;**54**(7):1810–8.
88. Tessari P, Cecchet D, Cosma A, et al. Insulin resistance of amino acid and protein metabolism in type 2 diabetes. *Clin Nutr* 2011; **30**(3):267–72.
89. Bassil MS, Gougeon R. Muscle protein anabolism in type 2 diabetes. *Curr Opin Clin Nutr Metab Care* 2013;**16**(1):83–8.
90. Wang X, Hu Z, Hu J, Du J, Mitch WE. Insulin resistance accelerates muscle protein degradation: Activation of the ubiquitin-proteasome pathway by defects in muscle cell signaling. *Endocrinology* 2006;**147**(9):4160–8.
91. Volpato S, Bianchi L, Lauretani F, et al. Role of muscle mass and muscle quality in the association between diabetes and gait speed. *Diabetes Care* 2012;**35**(8):1672–9.
92. Kim TN, Park MS, Yang SJ, et al. Prevalence and determinant factors of sarcopenia in patients with type 2 diabetes: the Korean Sarcopenic Obesity Study (KSOS). *Diabetes Care* 2010;**33**(7):1497–9.
93. Kob R, Bollheimer LC, Bertsch T, et al. Sarcopenic obesity: molecular clues to a better understanding of its pathogenesis? *Biogerontology* 2015;**16**(1):15–29.
94. Baumgartner BG, Orpinell M, Duran J, et al. Identification of a novel modulator of thyroid hormone receptor-mediated action. *PLoS One* 2007;**2**(11):e1183.
95. Sancho A, Duran J, Garcia-Espana A, et al. DOR/Tp53inp2 and Tp53inp1 constitute a metazoan gene family encoding dual regulators of autophagy and transcription. *PLoS one* 2012;**7**(3):e34034.
96. Francis VA, Zorzano A, Teleman AA. dDOR is an EcR coactivator that forms a feed-forward loop connecting insulin and ecdysone signaling. *Curr Biol* 2010;**20**(20):1799–808.
97. Mauvezin C, Orpinell M, Francis VA, et al. The nuclear cofactor DOR regulates autophagy in mammalian and Drosophila cells. *EMBO Rep* 2010;**11**(1):37–44.
98. Nowak J, Archange C, Tardivel-Lacombe J, et al. The TP53INP2 protein is required for autophagy in mammalian cells. *Mol Biol Cell* 2009;**20**(3):870–81.
99. Mauvezin C, Sancho A, Ivanova S, Palacin M, Zorzano A. DOR undergoes nucleo-cytoplasmic shuttling, which involves passage through the nucleolus. *FEBS Lett* 2012;**586**(19):3179–86.
100. Huang R, Xu Y, Wan W, et al. Deacetylation of Nuclear LC3 Drives Autophagy Initiation under Starvation. *Mol Cell* 2015;**57**(3): 456–66.
101. Penaloza C, Lin L, Lockshin RA, Zakeri Z. Cell death in development: shaping the embryo. *Histochem Cell Biol* 2006;**126**(2):149–58.
102. Neufeld TP, Baehrecke EH. Eating on the fly: function and regulation of autophagy during cell growth, survival and death in Drosophila. *Autophagy* 2008;**4**(5):557–62.
103. Stolz A, Ernst A, Dikic I. Cargo recognition and trafficking in selective autophagy. *Nat Cell Biol* 2014;**16**(6):495–501.
104. Isakson P, Holland P, Simonsen A. The role of ALFY in selective autophagy. *Cell Death and Differ* 2013;**20**(1):12–20.
105. Kirkin V, Lamark T, Sou YS, et al. A role for NBR1 in autophagosomal degradation of ubiquitinated substrates. *Mol Cell* 2009; **33**(4):505–16.
106. Pankiv S, Clausen TH, Lamark T, et al. p62/SQSTM1 binds directly to Atg8/LC3 to facilitate degradation of ubiquitinated protein aggregates by autophagy. *J Biol Chem* 2007;**282**(33): 24131–45.
107. Kraft C, Peter M, Hofmann K. Selective autophagy: ubiquitin-mediated recognition and beyond. *Nat Cell Biol* 2010;**12**(9): 836–41.
108. Goodpaster BH, Park SW, Harris TB, et al. The loss of skeletal muscle strength, mass, and quality in older adults: the health, aging and body composition study. *J Gerontol A Biol Sci Med Sci* 2006; **61**(10):1059–64.
109. Hairi NN, Cumming RG, Naganathan V, et al. Loss of muscle strength, mass (sarcopenia), and quality (specific force) and its relationship with functional limitation and physical disability: the Concord Health and Ageing in Men Project. *J Am Geriatr Soc* 2010;**58**(11):2055–62.
110. Park CW, Kim HW, Ko SH, et al. Accelerated diabetic nephropathy in mice lacking the peroxisome proliferator-activated receptor alpha. *Diabetes* 2006;**55**(4):885–93.
111. De Rekeneire N, Resnick HE, Schwartz AV, et al. Diabetes is associated with subclinical functional limitation in nondisabled older individuals: the Health, Aging, and Body Composition study. *Diabetes Care* 2003;**26**(12):3257–63.
112. Goodpaster BH, Carlson CL, Visser M, et al. Attenuation of skeletal muscle and strength in the elderly: The Health ABC Study. *J Appl Physiol* 2001;**90**(6):2157–65.
113. Schafer AL, Vittinghoff E, Lang TF, et al. Fat infiltration of muscle, diabetes, and clinical fracture risk in older adults. *J Clin Endocrinol Metab* 2010;**95**(11):E368–72.
114. Andersen H, Poulsen PL, Mogensen CE, Jakobsen J. Isokinetic muscle strength in long-term IDDM patients in relation to diabetic complications. *Diabetes* 1996;**45**(4):440–5.
115. Andersen H, Nielsen S, Mogensen CE, Jakobsen J. Muscle strength in type 2 diabetes. *Diabetes* 2004;**53**(6):1543–8.
116. Masiero E, Agatea L, Mammucari C, et al. Autophagy is required to maintain muscle mass. *Cell Metab* 2009;**10**(6):507–15.
117. Malicdan MC, Noguchi S, Nonaka I, Saftig P, Nishino I. Lysosomal myopathies: an excessive build-up in autophagosomes is too much to handle. *Neuromuscul Disord* 2008;**18**(7):521–9.

118. Kelley DE, Goodpaster B, Wing RR, Simoneau JA. Skeletal muscle fatty acid metabolism in association with insulin resistance, obesity, and weight loss. *Am J Physiol* 1999;**277**(6 Pt 1): E1130–41.
119. Kim JY, Hickner RC, Cortright RL, Dohm GL, Houmard JA. Lipid oxidation is reduced in obese human skeletal muscle. *Am J Physiol Endocrinol Metab* 2000;**279**(5):E1039–44.
120. Turner N, Heilbronn LK. Is mitochondrial dysfunction a cause of insulin resistance? *Trends Endocrinol Metab* 2008;**19**(9):324–30.
121. Ritov VB, Menshikova EV, He J, Ferrell RE, Goodpaster BH, Kelley DE. Deficiency of subsarcolemmal mitochondria in obesity and type 2 diabetes. *Diabetes* 2005;**54**(1):8–14.
122. Patti ME, Butte AJ, Crunkhorn S, et al. Coordinated reduction of genes of oxidative metabolism in humans with insulin resistance and diabetes: Potential role of PGC1 and NRF1. *Proc Natl Acad Sci USA* 2003;**100**(14):8466–71.
123. Heilbronn LK, Gan SK, Turner N, Campbell LV, Chisholm DJ. Markers of mitochondrial biogenesis and metabolism are lower in overweight and obese insulin-resistant subjects. *J Clin Endocrinol Metab* 2007;**92**(4):1467–73.
124. Skov V, Glintborg D, Knudsen S, et al. Reduced expression of nuclear-encoded genes involved in mitochondrial oxidative metabolism in skeletal muscle of insulin-resistant women with polycystic ovary syndrome. *Diabetes* 2007;**56**(9):2349–55.
125. Mootha VK, Lindgren CM, Eriksson KF, et al. PGC-1alpha-responsive genes involved in oxidative phosphorylation are coordinately downregulated in human diabetes. *Nat Genet* 2003;**34**(3): 267–73.
126. Chen L, Winger AJ, Knowlton AA. Mitochondrial dynamic changes in health and genetic diseases. *Mol Biol Rep* 2014;**41**(11): 7053–62.
127. Liesa M, Shirihai OS. Mitochondrial dynamics in the regulation of nutrient utilization and energy expenditure. *Cell Metab* 2013;**17**(4): 491–506.
128. Westermann B. Mitochondrial fusion and fission in cell life and death. *Nat Rev Mol Cell Biol* 2010;**11**(12):872–84.
129. Montgomery MK, Turner N. Mitochondrial dysfunction and insulin resistance: an update. *Endocr Connect* 2015;**4**(1):R1–15.
130. Twig G, Elorza A, Molina AJ, et al. Fission and selective fusion govern mitochondrial segregation and elimination by autophagy. *EMBO J* 2008;**27**(2):433–46.
131. Twig G, Hyde B, Shirihai OS. Mitochondrial fusion, fission and autophagy as a quality control axis: the bioenergetic view. *Biochim Biophys Acta* 2008;**1777**(9):1092–7.
132. Malka F, Guillery O, Cifuentes-Diaz C, et al. Separate fusion of outer and inner mitochondrial membranes. *EMBO Rep* 2005; **6**(9):853–9.
133. Chen H, Detmer SA, Ewald AJ, Griffin EE, Fraser SE, Chan DC. Mitofusins Mfn1 and Mfn2 coordinately regulate mitochondrial fusion and are essential for embryonic development. *J Cell Biol* 2003;**160**(2):189–200.
134. Liesa M, Palacin M, Zorzano A. Mitochondrial dynamics in mammalian health and disease. *Physiol Rev* 2009;**89**(3):799–845.
135. Bach D, Pich S, Soriano FX, et al. Mitofusin-2 determines mitochondrial network architecture and mitochondrial metabolism. A novel regulatory mechanism altered in obesity. *J Biol Chem* 2003;**278**(19):17190–7.
136. Bach D, Naon D, Pich S, et al. Expression of Mfn2, the Charcot-Marie-Tooth neuropathy type 2A gene, in human skeletal muscle: effects of type 2 diabetes, obesity, weight loss, and the regulatory role of tumor necrosis factor alpha and interleukin-6. *Diabetes* 2005;**54**(9):2685–93.
137. Toledo FG, Watkins S, Kelley DE. Changes induced by physical activity and weight loss in the morphology of intermyofibrillar mitochondria in obese men and women. *J Clin Endocrinol Metab* 2006;**91**(8):3224–7.
138. Song Z, Chen H, Fiket M, Alexander C, Chan DC. OPA1 processing controls mitochondrial fusion and is regulated by mRNA splicing, membrane potential, and Yme1L. *J Cell Biol* 2007; **178**(5):749–55.
139. Ishihara N, Fujita Y, Oka T, Mihara K. Regulation of mitochondrial morphology through proteolytic cleavage of OPA1. *EMBO J* 2006;**25**(13):2966–77.
140. Zorzano A, Liesa M, Sebastian D, Segales J, Palacin M. Mitochondrial fusion proteins: dual regulators of morphology and metabolism. *Semin Cell Dev Biol* 2010;**21**(6):566–74.
141. Olichon A, Baricault L, Gas N, et al. Loss of OPA1 perturbates the mitochondrial inner membrane structure and integrity, leading to cytochrome c release and apoptosis. *J Biol Chem* 2003;**278**(10): 7743–6.
142. Chen H, Chomyn A, Chan DC. Disruption of fusion results in mitochondrial heterogeneity and dysfunction. *J Biol Chem* 2005; **280**(28):26185–92.
143. Quiros PM, Ramsay AJ, Sala D, et al. Loss of mitochondrial protease OMA1 alters processing of the GTPase OPA1 and causes obesity and defective thermogenesis in mice. *EMBO J* 2012; **31**(9):2117–33.
144. Otera H, Wang C, Cleland MM, et al. Mff is an essential factor for mitochondrial recruitment of Drp1 during mitochondrial fission in mammalian cells. *J Cell Biol* 2010;**191**(6):1141–58.
145. Yoon Y, Krueger EW, Oswald BJ, McNiven MA. The mitochondrial protein hFis1 regulates mitochondrial fission in mammalian cells through an interaction with the dynamin-like protein DLP1. *Mol Cell Biol* 2003;**23**(15):5409–20.
146. Pitts KR, Yoon Y, Krueger EW, McNiven MA. The dynamin-like protein DLP1 is essential for normal distribution and morphology of the endoplasmic reticulum and mitochondria in mammalian cells. *Mol Biol Cell* 1999;**10**(12):4403–17.
147. Smirnova E, Griparic L, Shurland DL, van der Bliek AM. Dynamin-related protein Drp1 is required for mitochondrial division in mammalian cells. *Mol Biol Cell* 2001;**12**(8):2245–56.
148. Mozdy AD, McCaffery JM, Shaw JM. Dnm1p GTPase-mediated mitochondrial fission is a multi-step process requiring the novel integral membrane component Fis1p. *J Cell Biol* 2000;**151**(2):367–80.
149. Gandre-Babbe S, van der Bliek AM. The novel tail-anchored membrane protein Mff controls mitochondrial and peroxisomal fission in mammalian cells. *Mol Biol Cell* 2008;**19**(6):2402–12.
150. Palmer CS, Osellame LD, Laine D, Koutsopoulos OS, Frazier AE, Ryan MT. MiD49 and MiD51, new components of the mitochondrial fission machinery. *EMBO Rep* 2011;**12**(6):565–73.
151. Loson OC, Song Z, Chen H, Chan DC. Fis1, Mff, MiD49, and MiD51 mediate Drp1 recruitment in mitochondrial fission. *Mol Biol Cell* 2013;**24**(5):659–67.
152. Jheng HF, Tsai PJ, Guo SM, et al. Mitochondrial fission contributes to mitochondrial dysfunction and insulin resistance in skeletal muscle. *Mol Cell Biol* 2012;**32**(2):309–19.
153. Valente EM, Abou-Sleiman PM, Caputo V, et al. Hereditary early-onset Parkinson's disease caused by mutations in PINK1. *Science* 2004;**304**(5674):1158–60.
154. Lin W, Kang UJ. Characterization of PINK1 processing, stability, and subcellular localization. *J Neurochem* 2008;**106**(1):464–74.
155. Matsuda N, Sato S, Shiba K, et al. PINK1 stabilized by mitochondrial depolarization recruits Parkin to damaged mitochondria and activates latent Parkin for mitophagy. *J Cell Biol* 2010;**189**(2): 211–21.
156. Jin SM, Lazarou M, Wang C, Kane LA, Narendra DP, Youle RJ. Mitochondrial membrane potential regulates PINK1 import and proteolytic destabilization by PARL. *J Cell Biol* 2010;**191**(5): 933–42.
157. Narendra DP, Jin SM, Tanaka A, et al. PINK1 is selectively stabilized on impaired mitochondria to activate Parkin. *PLoS Biol* 2010; **8**(1):e1000298.

158. Gegg ME, Cooper JM, Chau KY, Rojo M, Schapira AH, Taanman JW. Mitofusin 1 and mitofusin 2 are ubiquitinated in a PINK1/parkin-dependent manner upon induction of mitophagy. *Hum Mol Genet* 2010;**19**(24):4861–70.
159. Poole AC, Thomas RE, Yu S, Vincow ES, Pallanck L. The mitochondrial fusion-promoting factor mitofusin is a substrate of the PINK1/parkin pathway. *PLoS One* 2010;**5**(4):e10054.
160. Sha D, Chin LS, Li L. Phosphorylation of parkin by Parkinson disease-linked kinase PINK1 activates parkin E3 ligase function and NF-kappaB signaling. *Hum Mol Genet* 2010;**19**(2):352–63.
161. Geisler S, Holmstrom KM, Skujat D, et al. PINK1/Parkin-mediated mitophagy is dependent on VDAC1 and p62/SQSTM1. *Nat Cell Biol* 2010;**12**(2):119–31.
162. Huang C, Andres AM, Ratliff EP, Hernandez G, Lee P, Gottlieb RA. Preconditioning involves selective mitophagy mediated by Parkin and p62/SQSTM1. *PLoS One* 2011;**6**(6):e20975.
163. Wong YC, Holzbaur EL. Optineurin is an autophagy receptor for damaged mitochondria in parkin-mediated mitophagy that is disrupted by an ALS-linked mutation. *Proc Natl Acad Sci USA* 2014;**111**(42):E4439–48.
164. Drew BG, Ribas V, Le JA, et al. HSP72 is a mitochondrial stress sensor critical for Parkin action, oxidative metabolism, and insulin sensitivity in skeletal muscle. *Diabetes* 2014;**63**(5):1488–505.
165. Tracy K, Macleod KF. Regulation of mitochondrial integrity, autophagy and cell survival by BNIP3. *Autophagy* 2007;**3**(6):616–9.
166. Bellot G, Garcia-Medina R, Gounon P, et al. Hypoxia-induced autophagy is mediated through hypoxia-inducible factor induction of BNIP3 and BNIP3L via their BH3 domains. *Mol Cell Biol* 2009;**29**(10):2570–81.
167. Zhang H, Bosch-Marce M, Shimoda LA, et al. Mitochondrial autophagy is an HIF-1-dependent adaptive metabolic response to hypoxia. *J Biol Chem* 2008;**283**(16):10892–903.
168. Schweers RL, Zhang J, Randall MS, et al. NIX is required for programmed mitochondrial clearance during reticulocyte maturation. *Proc Natl Acad Sci USA* 2007;**104**(49):19500–5.
169. Henstridge DC, Bruce CR, Drew BG, et al. Activating HSP72 in rodent skeletal muscle increases mitochondrial number and oxidative capacity and decreases insulin resistance. *Diabetes* 2014;**63**(6):1881–94.
170. Shi L, Zhang T, Liang X, et al. Dihydromyricetin improves skeletal muscle insulin resistance by inducing autophagy via the AMPK signaling pathway. *Mol Cell Endocrinol* 2015;**409**:92–102.
171. He C, Bassik MC, Moresi V, et al. Exercise-induced BCL2--regulated autophagy is required for muscle glucose homeostasis. *Nature* 2012;**481**(7382):511–5.

CHAPTER

8

Mechanisms Whereby Whole Grain Cereals Modulate the Prevention of Type 2 Diabetes

Knud Erik Bach Knudsen[1], *Merete Lindberg Hartvigsen*[2], *Mette Skou Hedemann*[1], *Kjeld Hermansen*[2]

[1]**Department of Animal Science, Aarhus University, Tjele, Denmark;** [2]**Department of Endocrinology and Internal Medicine, Aarhus University Hospital, Aarhus, Denmark**

1. INTRODUCTION

Worldwide, 382 million adults (8.3%) are living with diabetes, and the estimate is projected to rise to more than 592 million by 2035.[1] Type 2 diabetes (T2D), which accounts for 90% of all diabetes, has become one of the major causes of premature illness and death, mainly through the increased risk of cardiovascular disease (CVD), which is responsible for up to 80% of these deaths. The prediabetic condition, the metabolic syndrome (MeS), is estimated to affect around 20–25% of the world's adult population.[2] According to the International Diabetes Federation's definition, MeS is characterized by abdominal obesity and a combination of increased blood pressure, dyslipidemia (high triglycerides and low high-density lipoprotein cholesterol), raised fasting plasma glucose, and insulin resistance.[2] People with MeS are twice as likely to die and three times as likely to have a heart attack or stroke compared with people without the syndrome.[2] In addition, people with MeS have a fivefold increased risk of developing T2D.[2] Changes in body weight caused by physical inactivity and unhealthy diet are likely to contribute to the increased rates of T2D.[1]

As a consequence of urbanization and economic growth, many countries have experienced dietary changes favoring a rise in caloric consumption and a decline in overall diet quality.[3] The drastic changes taking place in food production, processing, and distribution systems have enhanced the accessibility of unhealthy foods. Evidence has accumulated from both prospective observational studies and randomized controlled trials (RCTs) that unhealthy diet plays an important role for development of abdominal obesity and T2D.[3,4] One example of this negative transition is the increased use of milling and processing of whole grains (WG) to produce refined products[5,6] such as polished white rice and refined wheat flour. The consequence is reduced content of dietary fiber (DF), micronutrients, and phytochemicals.[5,6]

The mechanisms for WG benefits in relation to human health are not yet fully understood, but most likely involve DF and bioactive components.[5] The main purpose of the present chapter is to discuss mechanisms whereby WG cereals may prevent T2D.

2. WHOLE GRAINS VERSUS REFINED FLOUR

At present, there is no universal accepted definition of WG. However, the definition by the American Association of Cereal Chemists, "intact, ground, cracked or flaked caryopsis [fruit or kernel] of the grain whose principal components, the starchy endosperm, germ and bran, are present in the same relative proportions as they exist in the intact grain," is widely used.[7] Recently, the European-based consortium of scientists and industry working with cereals, the Healthgrain Forum, has proposed an updated definition that essentially is similar but allows for minor losses (up to 2%) of the grain during milling.[8] In both definitions, WG include all or almost all parts of the grain. The main WG cereals consumed worldwide are wheat, rice, and

Molecular Nutrition and Diabetes
http://dx.doi.org/10.1016/B978-0-12-801585-8.00008-7

maize, followed by oats, rye, barley, triticale, millet, and sorghum.[9] Wheat and rye are to a large extent consumed as soft and crisp breads but also as breakfast cereals, whereas the other cereals grains mostly are consumed as breakfast cereals, porridges, or cooked stables.

The cereal grain, irrespective of species, is a complex organ composed of tissues containing cell walls with different properties and composition as illustrated for wheat in Figure 1.[10] The cell walls from the outer part of the kernel have primarily the role of protection and the cell walls in these tissues are consequently thick, hydrophobic, and contain significant amounts of the noncarbohydrate polyphenolic ether lignin.[11] In endosperm tissues that include the aleurone layer, the cell walls are thin and hydrophilic and DF more soluble than those present in the outermost part of the grain. The main DF polysaccharides present in cereals are arabinoxylan (AX), cellulose, and β−glucan, which vary significantly according to the cereal specie but also between different tissues of the grains.[11,12] Rye, triticale, wheat, and corn are rich in AX, whereas barley and oats contain a high level of β−glucan.[11,12] Of importance in a nutritional physiological context is the ability of some of the cell wall polysaccharides—i.e., AX and β−glucan (soluble nonstarch polysaccharide (NSP)), to give rise to viscosity.[11,12] The viscosity is directly related to the fundamental molecular characteristics of the molecules in solution (molecular conformation, molecular weight (M_w), molecular weight distribution) and the concentration of the polymer.[11–13] WG cereals also contain an array of nonnutritive noncarbohydrates constituents predominantly concentrated in the germ and bran fraction.[5,6] Of importance in a nutritional health context is benzoic acid and cinnamic derivatives, lignans, alkylresorcinols (AR), B vitamins, betaine, phytosterols and phytostanols, phytic acids, and avenanthramides.[5,6]

3. META-ANALYSES AND EPIDEMIOLOGICAL STUDIES

3.1 Whole Grains and Metabolic Syndrome

Diets that are rich in WG foods have been linked to a lower prevalence of the MeS in both middle-aged[14,15] and older adults[16] independent of demographic,

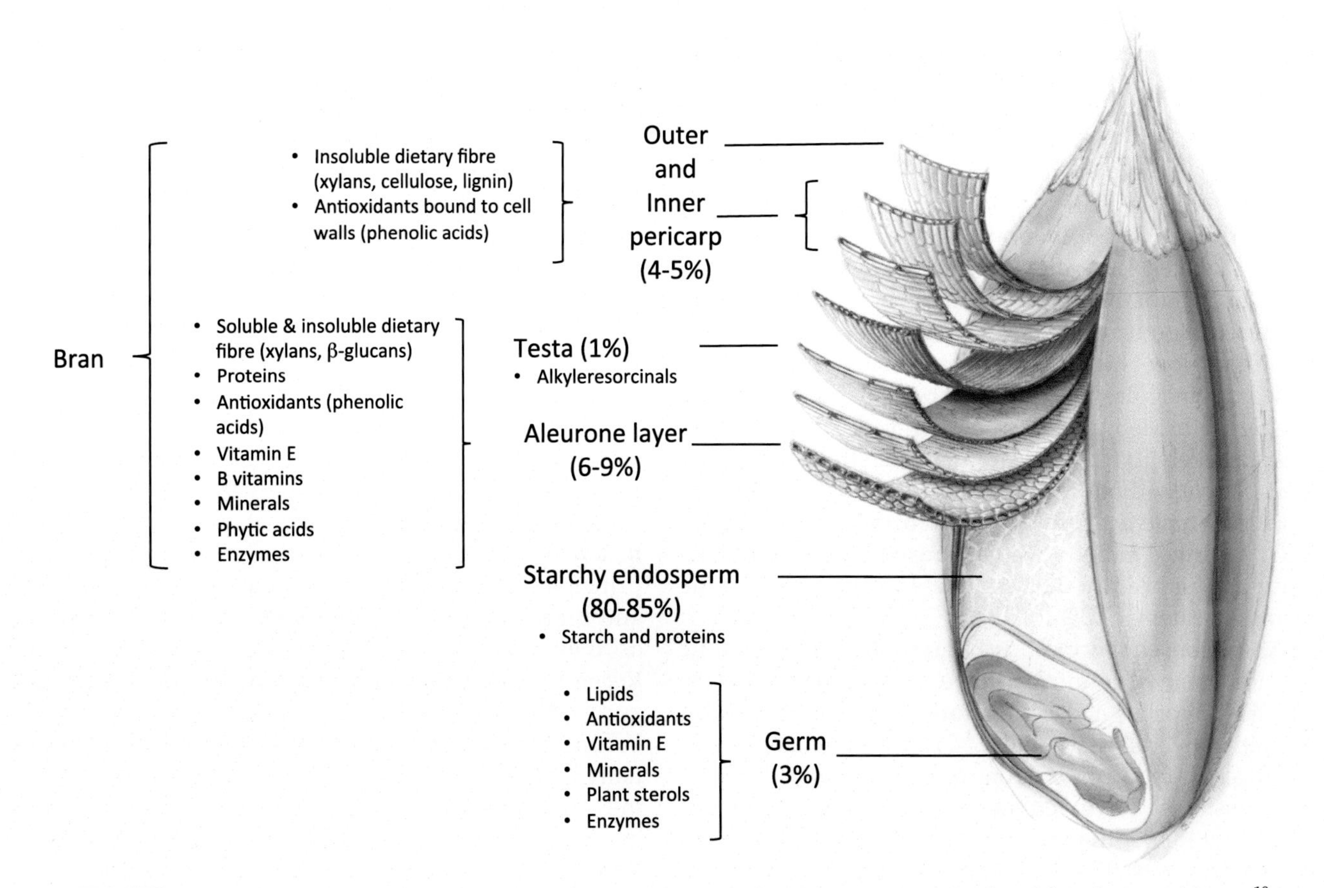

FIGURE 1 Cross-section of a whole grain wheat kernel with location of specific compounds. *Adapted from Surget and Barron.*[10]

lifestyle, and dietary factors.[16] In contrast, intake of refined grain is positively associated with a higher prevalence of MeS.[16] In the Framingham Offspring Cohort, middle-aged men and women who reported consuming three or more servings of WG per day had a 32% lower prevalence of the MeS than did those who reported consuming less servings per day.[17] Similarly, the prevalence of the MeS was 32% lower with higher intakes of WG in Tehranian females.[14] Also in older subjects, it was found that those with a higher intake of WG foods (median intake: 2.9 servings/day) had a lower prevalence of the MeS (odds ratio: 0.46; 95% confidence interval (CI): 0.27–0.79) than did subjects with lower WG intakes (median intake: less than one serving/day).[16] A large body of evidence now supports the association of minimally processed cereal foods as part of a dietary pattern with lower risk of features of the MeS.[18] Processed cereals, in contrast, are associated with increased risk of components of the MeS.[18]

3.2 Whole Grains and T2D

In a Cochrane Review from 2008[19]—based on data up to May 30, 2006—one RCT and eleven prospective cohort studies were identified. The RCT was of low methodological quality. Four of the eleven cohort studies measured cereal fiber intake, three studies measured WG intake, and two studies measured both. Two studies measured the change in WG food intake and one of them also measured change in cereal fiber intake. The incidence of T2D was assessed in nine studies and changes in weight gain in two studies. The prospective studies consistently showed a reduced risk on the development of T2D after high intake of WG foods (27–30%) or cereal fiber (28–37%).[19] The authors concluded that the evidence from only prospective cohort trials was too weak to allow a definite conclusion about the preventive effect of WG foods on the development of T2D. They inquired for properly designed long-term RCT.[19]

Subsequently, several larger studies have reported an inverse association between WG intake and risk of T2D,[20–22] although not all reported this.[23] In the large Women's Health Initiative Observational Study[20] of postmenopausal women, the consumption of WG was inversely associated with incident T2D over a median of 7.9 years follow-up. Women who consumed more than two servings of WG per day had a 43% reduction in risk of incident diabetes compared with women who consumed no WG. The lower risk of diabetes was observed even at relatively low intakes of WG such as one serving per day (hazard ratio, 0.73). The association was robust and dose-related after adjusting for potential confounders.[20] The risk of T2D was slightly reduced after adjusting for dietary magnesium, which is known to be inversely associated with the risk of T2D in a dose–response manner.[24]

Three meta-analyses have been conducted on WG and the risk of T2D.[25–27] The meta-analysis from 2007[26] included six prospective cohort studies. The cohorts included men and women, predominantly white or black populations, from the United States and Finland.[26] Higher WG consumption was only associated with lower fasting and postload plasma glucose concentrations in one cross-sectional study, but not in two other studies.[26] The findings from the cohort studies are consistent with the direct association between WG consumption and insulin sensitivity that has been observed in cross-sectional studies in adolescent and adult US populations.[26] Based on the meta-analysis of six cohort studies including 286,125 participants and 10,944 cases of T2D, a two-serving-per-day increment in WG intake was found to be associated with a 21% decrease in risk of T2D after adjustment for potential confounders and body mass index.[26]

In the meta-analysis by Ye et al.,[27] six cohorts were identified that investigated the relation of WG intake to T2D risk. The overall estimated multivariate-adjusted relative risk (RR) of T2D comparing the highest with the lowest level of intake was RR = 0.74 (i.e., 26% reduction in risk of T2D).[27] They concomitantly identified 11 prospective cohort studies[27] that examined the relation of total DF and/or cereal DF intake to T2D risk. The overall estimate of the multivariable-adjusted RR of T2D comparing the highest and lowest category of DF intake was 0.84 (95% CI: 0.76–0.93) for total DF and 0.87 (95% CI: 0.81–0.94) for total cereal DF.[27] Subgroup analyses in men and women revealed a slightly more protective association among men than among women. Aune et al.[25] identified 16 cohort studies that were included in the analyses of grain intake and T2D risk.[25] Seven studies were from the United States, six were from Europe, two from Asia, and one was from Australia.[25] Ten cohort studies were included in the analysis of total WG intake and T2D risk and included 19,829 cases among 385,868 participants.[25] The summary RR for T2D comparing high versus low intake was 0.74 (95% CI: 0.71–0.78). The summary RR per three servings per day was 0.68 (95% CI: 0.58–0.81) (Figure 2(a)).[25] There was evidence of a nonlinear association between WG intake and T2D risk, with a steeper reduction in risk when increasing intake from low levels, and most of the benefit was observed up to an intake of two servings per day (Figure 2(b)).[25] The summary RR for high versus low intake was 0.82 (95% CI: 0.72–0.94, $n = 4$) for WG bread, 0.66 (95% CI: 0.57–0.77, $n = 3$) for WG cereals, 0.76 (95% CI: 0.69–0.84, $n = 3$) for wheat bran, 0.97

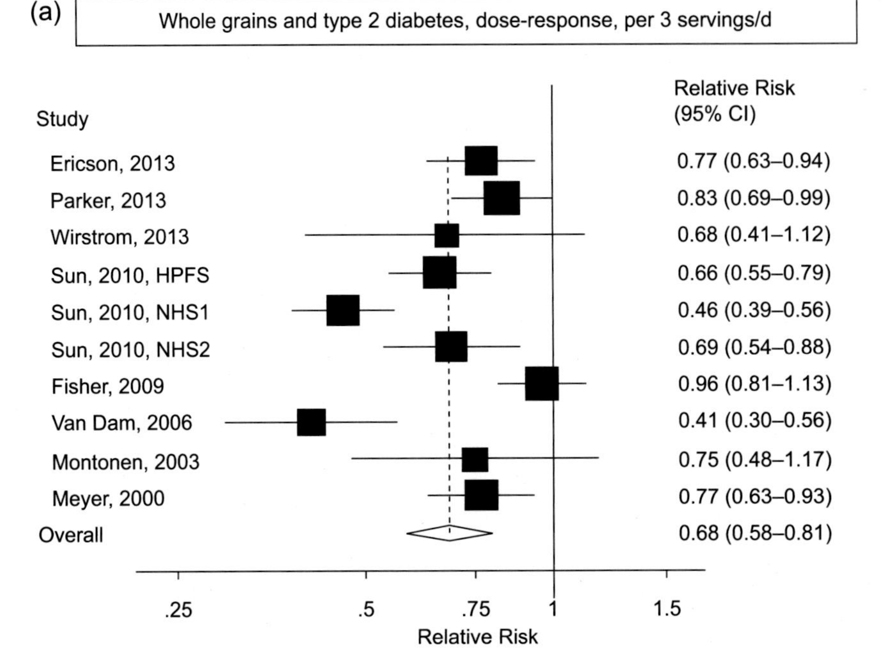

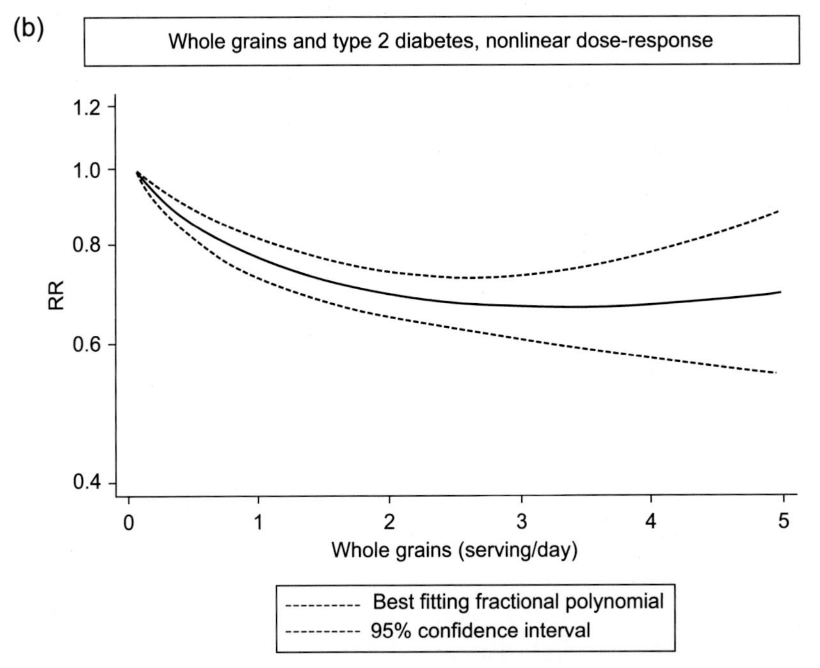

FIGURE 2 Whole grains and risk of T2D: (a) studies and relative risk (RR) and (b) nonlinear dose-response. *Adapted from Aune.*[25]

(95% CI: 0.86–1.10, $n = 3$) for wheat germ, 0.89 (95% CI: 0.81–0.97, $n = 3$) for brown rice, 1.17 (95% CI: 0.93–1.47, $n = 7$) for white rice, and 0.82 (95% CI: 0.56–1.18, $n = 2$) for total cereals.[25] Thus, WG wheat and bran and brown rice had a positive impact on the RR for T2D, whereas wheat germ, white rice, and total cereals had no effect on the RR of T2D.

3.3 Refined Grains and T2D

In many countries, grains are heavily processed and provide a high proportion of total and carbohydrate calories. Carbohydrates from heavily processed cereals are rapidly digested, absorbed, and metabolized with the potential of adverse metabolic effects. Thus, greater intake of white polished rice is associated with an increased risk of T2D, especially in Asian countries where white rice is a stable food.[28] Six studies reported on refined grain intake and T2D[25] included 9545 cases among 258,078 participants. The summary RR for high versus low intake of refined grains was 0.94 (95% CI: 0.82−1.09) and the RR per three servings per day 0.95 (95% CI: 0.88−1.04) (Figure 3). A high load of refined carbohydrates may adversely affect metabolic intermediates potentially by stimulating de novo hepatic lipogenesis, which may increase the risk of MeS and T2D. Systematic reviews and meta-analyses have confirmed the importance of free (added) sugars[29] when consumed ad libitum in contributing to excess body fatness. Other than the need to restrict consumption of free

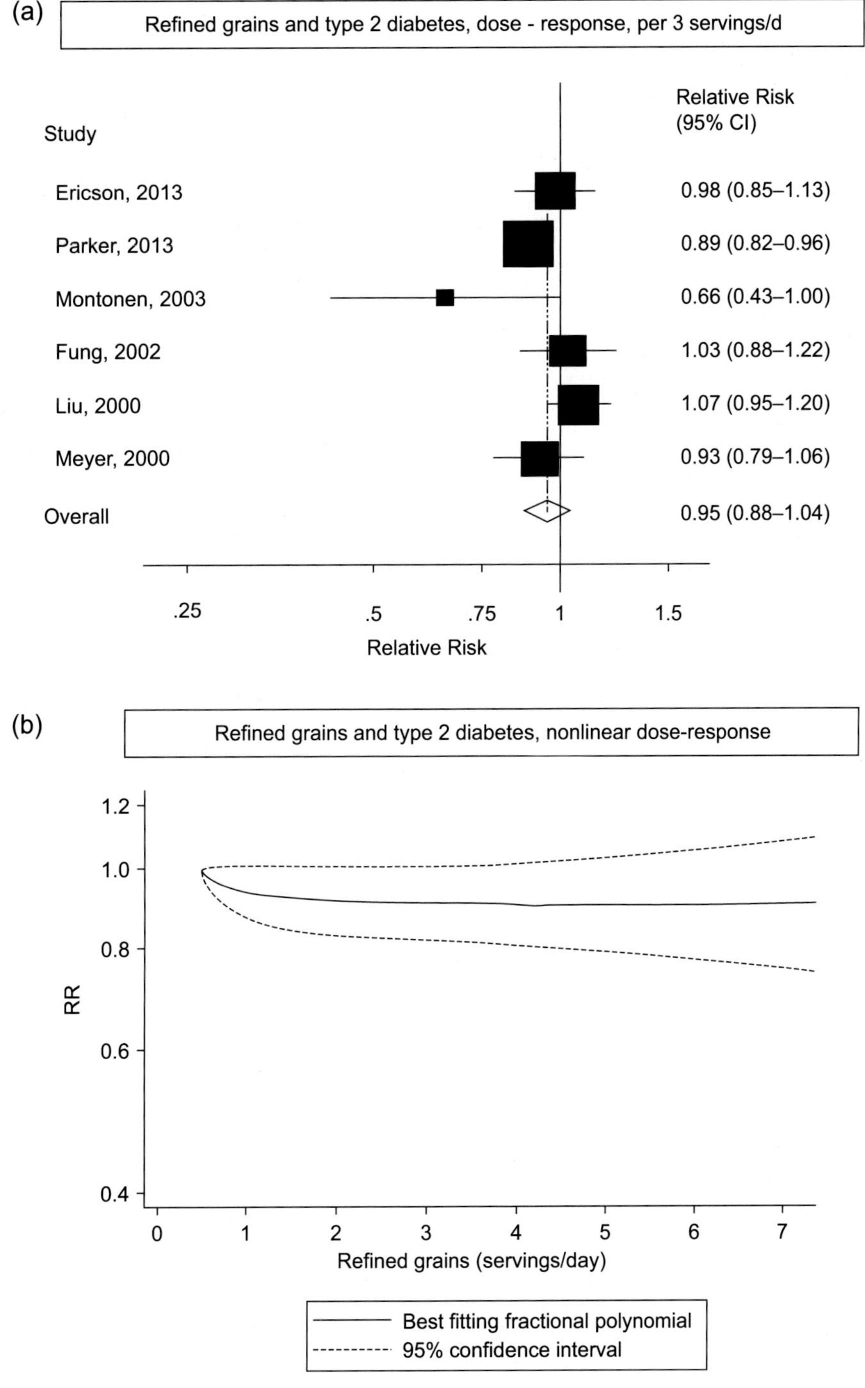

FIGURE 3 Refined grains and risk of type 2 diabetes: (a) studies and relative risk (RR) and (b) nonlinear dose-response. *Adapted from Aune.*[25]

sugars, less attention has been paid to the type of dietary carbohydrate. To lower disease risk and improve vascular health outcomes, it is imperative to replace refined grains with WG.

4. INTERVENTION STUDIES

4.1 Acute Human Studies

Several studies have investigated the effects of WG rye compared with refined wheat on glucose and insulin responses. In these studies, WG rye products repeatedly induced low insulin responses[30–34] compared with refined wheat products, although the glucose responses were similar[35,36] or reduced.[37] The insulin-saving property may, however, depend on the rye varieties,[32,38] the structure of the rye grain,[30] and the presence of bioactives and soluble DF.[32] In addition, the form of food[31] has been proposed to be more important determinants for the low insulin responses than the total DF content. Rye products also induced lower glucose responses, hence preventing hypoglycemia in the late postprandial phase.[33,34,37] Low postprandial insulin responses may reduce the overstimulation of the β cells, thereby protecting the β cells.

In several acute studies, the glucose responses to WG wheat bread did not differ from white refined wheat bread[39–41] and neither did the insulin responses.[41,42] In contrast, the glucose responses[39,40] and insulin responses[39] of refined wheat pasta were significantly lower than for refined wheat bread. The different food structure between bread and pasta has previously been examined, and the latter has consistently been shown to reduce glycemia and insulinemia.[39,43,44] The structural differences between rye and wheat have also been proposed to play a role for the lower postprandial insulin response to rye bread.[30] Interestingly, the glucose responses in type 1 diabetic subjects attaining a constant insulin level via an artificial β cell was unaffected by differences in the extraction rate of wheat white flour (70%) or WG (100%), whereas WG rye bread induced a 25% lower glucose response.[37]

In acute studies, refined wheat products enriched with concentrated AX have been found to reduce the glycemic response,[45] to possess a modest effect on the glucose peak value,[46] or not to affect the acute glycemic response[47] compared with refined wheat products. In combination with whole rye kernels, concentrated AX has been shown to reduce the acute glucose and insulin response.[48]

Oat and barley, which are characterized by a high content of β-glucan, have previously been studied in detail,[49–51] and the European Food Safety Authority has recognized a cause-and-effect relationship between β-glucan and the postprandial glycemic response.[52] The European Food Safety Authority report concluded that 4 g of β-glucan for each 30 g of digestible carbohydrates should be consumed per meal to reduce the postprandial glycemic response.[52] The M_w and solubility of β-glucan are important for the viscosity and thereby the beneficial effect of β-glucan on glucose and insulin responses.[53,54]

The consumption of large amounts of white rice is associated with an increased risk of T2D,[28] but it is noteworthy that acute studies with nonparboiled and parboiled white rice showed that both were low in glycemic index (GI) compared to white bread in T2D patients (GI 50 vs 53).[55] In T2D patients, the glycemic indices to parboiled white rice did not vary depending on the amount of available carbohydrate (GI of 25 and 50 g carbohydrate being 55 and 60, respectively)[56] or gender (GI in men 66 and in women 62, respectively).[57] However, the structure of the starch (amylose and amylopectin ratio) had a pronounced impact on GI; parboiled rice with a high amylose content had a GI of 50 compared with 73 for parboiled rice with a low amylose content ($P < 0.01$).[55] Furthermore, the severity of parboiling also influences GI in T2D subjects. Thus, three meals of cooked polished rice of the same variety being nonparboiled (NP), mildly traditionally parboiled, and severely pressure parboiled were compared with white bread[58] and resulted in GI readings of 55 (NP), 46 (traditionally parboiled), and 39 (pressure parboiled), and lower for pressure parboiled than NP ($P < 0.05$)[58] (Figure 4).

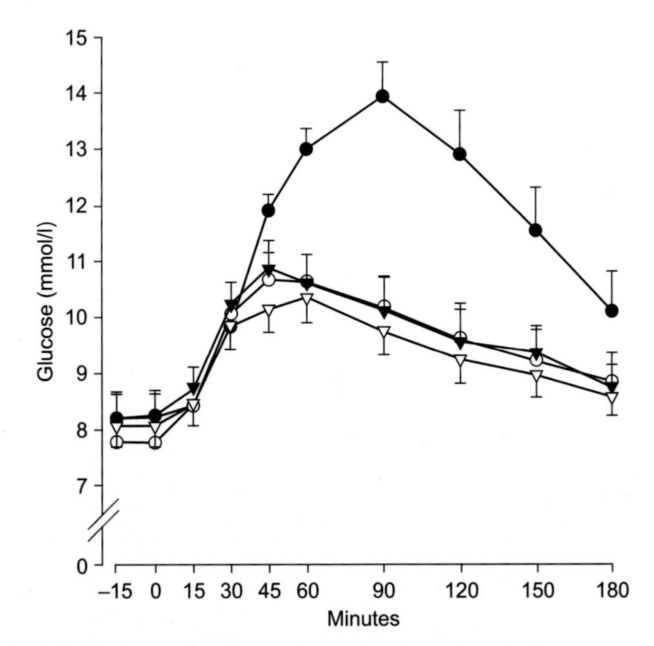

FIGURE 4 Plasma glucose responses (means ± standard error of the mean) in nine type 2 diabetic subjects to four different test meals (50 g available carbohydrates). White bread (●) and cooked rice of the same variety being non-parboiled (○); traditionally parboiled (▼) and pressure parboiled (Ü). *Adapted from Larsen.*[58]

4.2 Long-Term Human Studies

Long-term studies examining the effects of WG intake on glucose and insulin metabolism have, as it is the case in acute studies, provided conflicting findings. The inconsistent results may be ascribed to different study populations, different WG cereals and differences in methodologies used to measure glucose and insulin metabolism. However, also criteria for reporting WG intake have varied widely (e.g., in grams, portions, or servings), making it difficult to precisely explore the relation between WG and grain components in relation to diabetes.[59]

Rye bread and pasta have been studied in subjects with the MeS[60] and postmenopausal, hypercholesterolemic women.[61] In these studies, rye bread as well as rye bread and pasta enhanced the acute insulin response in postmenopausal, hypercholesterolemic women after an 8-week intervention compared with refined wheat bread[61] and early insulin secretion in subjects with the MeS after 12 weeks[60] compared with oat bread, wheat bread, and potatoes, respectively. In both studies, the insulin sensitivity remained unchanged.[60,61] A 6-week WG diet (three bread slices, two crisp bread slices, one portion pasta, and one portion muesli), mainly in milled form, did neither improve insulin sensitivity or inflammatory markers (interleukin-6 and C-reactive protein) compared with a refined diet in overweight adults.[62] Furthermore, in a two-site study conducted in Naples and Kuopio, a 12-week intervention with WG rye and wheat did not change peripheral insulin sensitivity, fasting plasma glucose, fasting insulin, lipids, or inflammatory markers compared with a refined cereal diet in subjects with MeS.[63] However, postprandial plasma insulin and postprandial triglycerides were reduced by the WG rye and wheat diet.[64] In line with the study of Giacco et al.,[63] 12-week administration of a wheat bran high-fiber diet to T2D subjects did not change fasting glucose, hemoglobin A1C, lipids, C-reactive protein, or homocysteine compared with a low-fiber cereal diet.[65] In contrast, 6-week intervention with a WG diet, composed of 80% wheat, reduced fasting plasma insulin levels and improved insulin resistance compared with a refined cereal diet in hyperinsulinemic overweight and obese subjects.[66] Also, improvement of insulin resistance was observed in hyperglycemic obese subjects after 4 weeks of consuming 200 g of a starch-reduced WG wheat product compared with a nutrient dense product with inulin.[67] Furthermore, 4 weeks of intervention with either high-fiber rye or high-fiber wheat decreased postprandial plasma glucose and insulin in overweight men compared with a refined wheat diet[68] and 3 days' consumption of white bread enriched with insoluble oat fiber improved whole-body insulin sensitivity assessed by euglycemic-hyperinsulinemic clamp in overweight and obese women.[69]

4.3 Animal Studies

In porto-arterial catheterized pigs, WG rye bread reduced the insulin secretion 30 min postprandially,[70] whereas a WG rye bread with kernels lowered the glucose flux 15 min postprandially compared with refined wheat bread.[70] Glucose absorption and insulin secretion were also reduced by white bread enriched with concentrated AX at 60 and 30 min postprandially, respectively.[70] In Zucker Diabetic Fatty rats fed the same bread types for 7 weeks, WG rye bread, WG rye bread with kernels, and AX-enriched wheat bread reduced the blood glucose response areas after an oral glucose tolerance test compared with refined wheat bread.[71] The effect of WG rye versus WG wheat was studied for 22 weeks in C57BL/6J mice.[72] The results showed an increased islet insulin release to the WG rye diet compared with the WG wheat diet. The insulin response was also reduced during an intravenous glucose tolerance test for WG rye compared with WG wheat, whereas the glucose response did not differ.[72] In addition, the WG rye diet showed beneficial effects related to insulin signaling, insulin resistance, apoptosis, and inflammation on plasma and/or gene expression levels.[72] In another study, Goto-Kakizaki rats were fed WG diets containing 65% WG flour of barley, oat, wheat, or maize for 5 months.[73] WG wheat, oat, and barley lowered the fasting plasma glucose after 2 months of intervention compared with a basal diet without WG flour. Maize and barley also increased plasma C-peptide concentration more than WG wheat after 2 months. However, all WG diets were ineffective in improving T2D biomarkers when the rats had a more severely deranged glycemic control after 5 months of intervention.[73]

5. MECHANISMS OF ACTION

As is seen from the observational studies, there is convincing evidence for an association between WG intake and T2D risk reduction, whereas the results from the intervention studies with humans and animals are more variable but still in favor of an association. Because WG, in contrast to refined flour, include all botanical constituents of the grain, it is likely that the beneficial effects of WG on T2D is caused by a synergistic action of DF and bioactives present in abundant in the WG. These components will, to a variable extend, be released during the digestion process in the gut, absorbed, and influence the metabolism at the molecular level (Figure 5). Some of the components will be released during the digestion processes in the small intestine, where the physicochemical and physical aspects play a role, whereas others will be released during microbial fermentation in the large intestine.

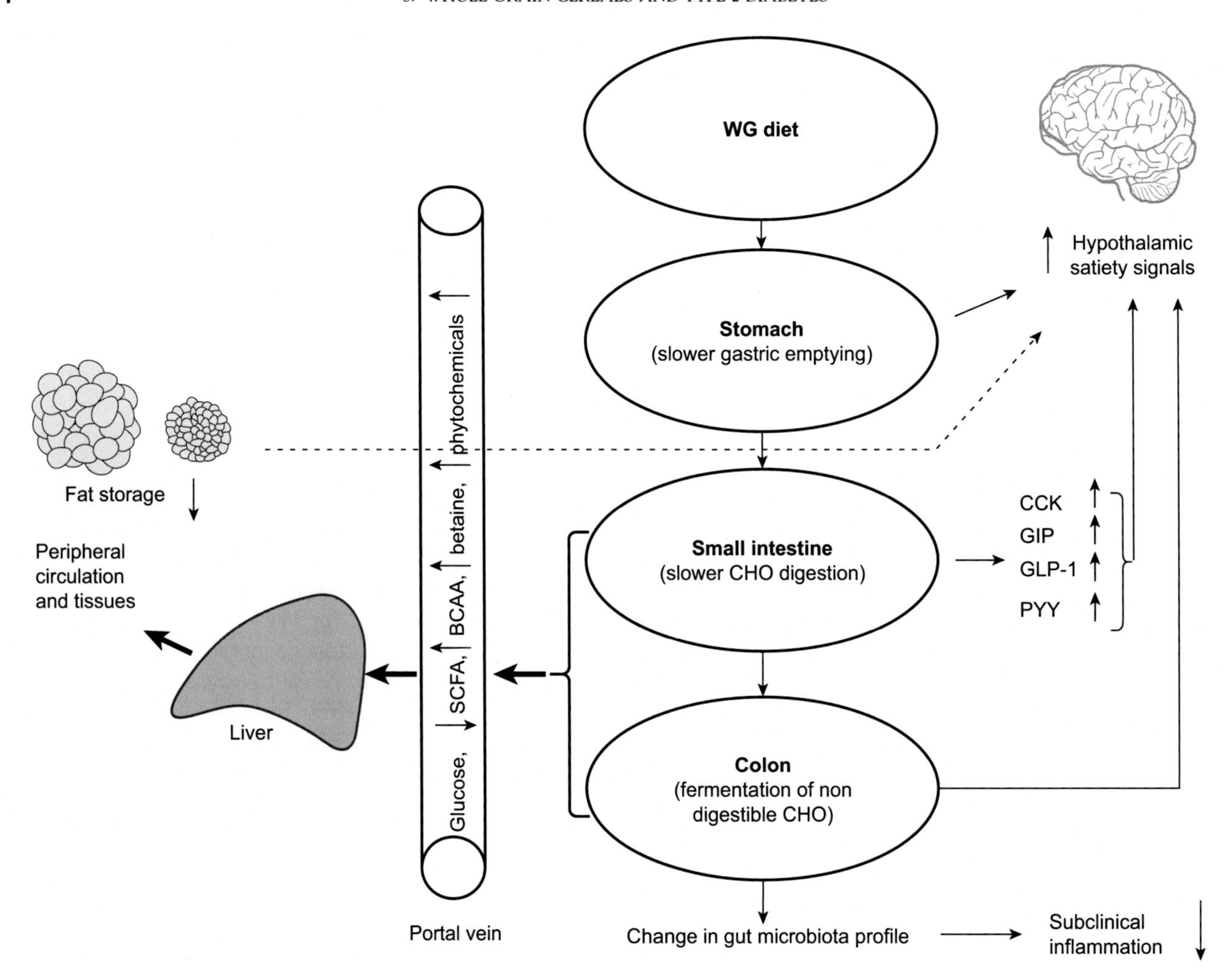

FIGURE 5 Possible mechanisms of action of cereal whole grains. BCAA, branched-chain amino acids; CCK, cholecystokinin; CHO, carbohydrate; FFA, free fatty acids; GIP, gastric inhibitory polypeptide; GLP-1, glucagon-like peptide 1; SCFA, short-chain fatty acids.

5.1 Rate of Glucose Absorption: The Glycemic Index

Differences in the metabolic response to carbohydrates can be classified by the GI,[74] the blood glucose response to a given food compared with a standard (white wheat bread or glucose). Many factors contribute to GI of cereal products including starch structure (amylose, amylopectin),[55] soluble and insoluble DF, food processing, and the presence of intact kernels and other macronutrients. Across different cereal products a positive relationship was found between GI and the content of rapidly available carbohydrate and a negative relationship to the content of slowly available carbohydrate[75]; GI was positively related to the insulin secretion. The postprandial plasma glucose concentration is tightly regulated by homeostatic regulatory systems involving insulin, glucagon, glucagon-like peptide-1 (GLP-1) and glucose-dependent insulinotropic polypeptide. This regulatory system is less exposed to diurnal fluctuations when consuming low-GI compared to high-GI foods. The rapid absorption of glucose following consumption of high-GI foods (e.g., white wheat bread, Figure 4) challenges the homeostatic system and long-time exposure of high-GI foods can initiate a sequence of metabolic events that stimulate hunger, promote fat deposition, and place the pancreatic β cells under increased stress,[76] as illustrated in Figure 6. In addition to the deleterious effects on β-cell function and insulin sensitivity,[77] proinflammatory cytokines are increasingly acknowledged in the pathogenesis of CVD. Benefits on low-grade inflammation may hypothetically be mediated by a range of WG constituents including the DF complex and the associated bioactive components such as methyl donors, antioxidants, and trace minerals.[78–81]

Lowered postprandial glycemia in response to certain WG cereal products, or suppressed insulin responses in the case of rye, might contribute to metabolic benefits of a WG diet,[33,34,82] as may the lower energy density and higher volume of meals rich in WG products because these are contributing factors favoring satiation. In

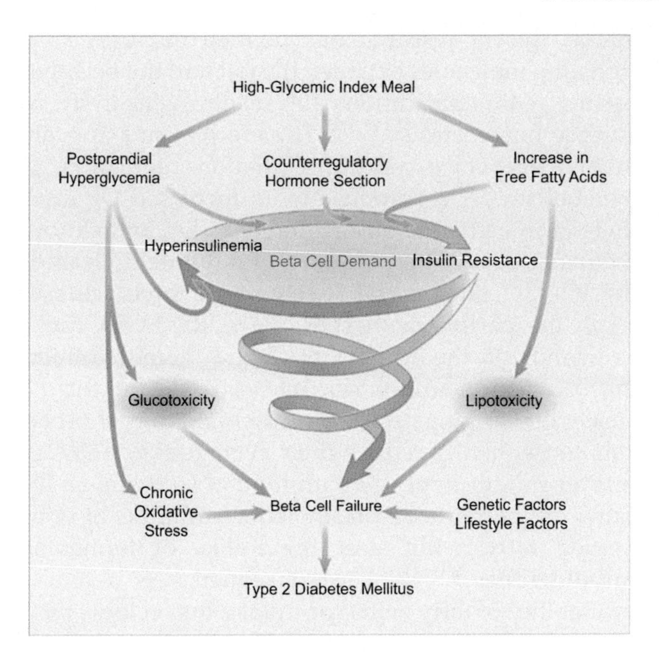

FIGURE 6 The hypothetical model relates a high-glycemic index diet to increased risk for type 2 diabetes. *With permission from Ludwig.*[76]

mice fed a high-GI diet, there was a rapid-onset increase in body fat mass and liver fat, a gene expression profile in liver consistent with elevated lipogenesis, and, after long-term exposure, significantly reduced glucose clearance following a glucose load.[83] Long-term high-GI diet consumption further led to a delayed switch to both carbohydrate and fat oxidation in the postprandial state, indicating reduced metabolic flexibility.[84] In contrast, switching soluble fermentable soluble guar gum with insoluble cereal DF resulted in significantly lower weight gain and improved insulin sensitivity and was further associated with a pattern in liver gene expression consistent with increased fatty acid oxidation.[85] Thus, targeting postprandial hyperglycemia via choice of low-GI foods, which is the case for some WG foods, has a clinically useful effect on glycemic control in T2D.[86]

5.2 The Microbiome

The human intestinal tract contains a unique group of microorganisms (i.e., the microbiota consisting of numerous bacteria, archaea, and viruses). All of these microorganisms generate a biomass of more than 1.5 kg and their combined genomes (i.e., microbiome) exceed the human genome more than 100-fold.[87] Studies, particularly from animals models,[88,89] have demonstrated that the microbiota might be considered as a major player in the development of obesity and, because 80% of patients with T2D are obese, it is logical to evaluate the microbiome in T2D patients. A recent study in which high-throughput sequencing was used on stool samples from Chinese patients with T2D, metagenomic analysis in combination with clinical data showed that patients with T2D exhibited a moderate intestinal dysbiosis characterized especially by a decrease in butyrate-producing *Roseburia intestinalis* and *Faecalibacterium prausnitzii*.[90] It was also found that in the healthy control samples, especially various butyrate-producing bacteria were enriched (e.g., *Clostridiales* sp. SS3/4, *Eubacterium rectale, F. prausnitzii, R. intestinalis*, and others), whereas in T2D patients most metagenomic linkage groups belonged to rather opportunistic pathogens such as *Bacteroides caccae*, various *Clostridiales*, *Escherichia coli*, and others.[90] In a European study using shotgun sequencing on stool samples from postmenopausal female patients with T2D, *R. intestinalis* and *F. prausnitzii* were highly discriminant for T2D.[91] It was also found that when comparing patients with T2D with a group of women with impaired glucose tolerance, an increase in energy metabolism/harvest and fatty acid metabolism could be observed in T2D patients.[91] Although the two human studies have shortcomings, they do suggest that a "gut signature" might exist and that a proinflammatory tone might be initiated in the intestine which could reflect the starting point of low-grade systemic inflammation commonly observed in T2D.[90,91] Lipopolysaccharides, released from unfavorable intestinal bacteria and absorbed into the systemic circulation, may promote a state of metabolic endotoxemia, and act to trigger subclinical inflammation and development of insulin resistance,[92] and potentially obesity.[93]

To the authors' knowledge, intervention studies with WG in T2D patients using high-throughput sequencing technologies to analyze stool samples have not yet been performed. In a study of Costabile et al.,[94] it was found that eating WG wheat breakfast cereal for 21 days significantly increased *Bifidobacterium* spp. numbers in stool samples and resulted in higher Lactobacilli/enterococci ratio compared with an equivalent quantity of wheat bran–based breakfast cereal. Another intervention study with WG versus refined wheat documented significantly higher numbers of *Bifidobacterium* spp. and numerically higher numbers of *Lactobacillus*,[95] whereas the composition of the intestinal microbiota of adults with MeS did not differ depending on whether the subject received a WG fiber-rich rye diet or a refined wheat diet for 12 weeks.[96] A study by Walker et al.[97] further demonstrated that feeding of obese male subjects with diets differing in type and content of DF components (RS_3, NSP from wheat bran and a weight loss diet) for 3 weeks had an impact on fecal microbiota. Thus, fecal microbiota profiles tended to group by individuals more than by diets but with marked changes in the relative abundance of several dominant phylotypes in response to especially increased intake of RS, whereas there was no effect of the NSP from wheat bran.

5.3 Short-Chain Fatty Acids

The colonic bacteria have a far larger repertoire of degrading enzymes and metabolic capabilities than their host.[98] In consequence, dietary changes may affect or interact with the microbial community and the metabolic outcome through several interrelated mechanisms, thereby playing a key function in regulating metabolic pathways in health and disease (Figure 7). Recently, it has been recognized that T2D patients have a lower concentration of butyrate producing *R. intestinalis* and *F. prausnitzii*.[90,91] In this context, it is intriguing that in pigs, a WG rye diet with added enzyme treated wheat bran stimulated the proliferation of *R. intestinalis* and *F. prausnitzii*,[99] butyrate production in the large intestine,[99] and the net portal absorption of butyrate[100] to a larger extent than a diet with equal amounts of DF in the form of RS. A Western-style diet high in refined carbohydrates from sugar and refined wheat flour was used as a reference diet.[99,100] A study of Damen et al.[101] feeding healthy human subjects WG wheat and rye breads reported that the concentration of total short-chain fatty acids (SCFAs) and butyrate in feces was increased after the intervention with the pressure parboiled diet, whereas it was only in connection with xylanase addition to the WG bread that the effect was significant. Similarly, 4 weeks of intervention with high-fiber wheat and high-fiber rye increased fecal butyrate.[68]

An evening meal containing WG barley has also been found to reduce postprandial glucose responses to a standardized breakfast meal. The glucose response was inversely correlated to plasma SCFA concentrations; butyrate,[102,103] acetate,[102] and propionate[104] as well as increased breath hydrogen,[50,104,105] used as markers for colon fermentation. Furthermore, the reduced glucose response was observed concurrently with a decreased concentration of free fatty acids[104] as well as a decreased level of the inflammatory markers interleukin-6[102,103] and tumor necrosis factor-α.[103] In one study, improved insulin sensitivity was also reported.[103]

The SCFA produced in the gut is absorbed by passive diffusion, recovered via monocarboxylic acid transporters, or act as signaling molecules by binding to G protein–coupled receptors (Figure 7).[106] These G protein–coupled receptors are expressed by many cell types including gut epithelia cells, adipocytes, and immune cells. Binding of SCFA to G protein–coupled receptors 41 (Gpr41) (free fatty acid receptor 3) and G protein–coupled receptors 43 (Gpr43) (free fatty acid receptor 2) located on colonic L cells have been found to be involved in the control of anorectic hormones, peptide YY, and GLP-1.[107] These hormones have roles in appetite control, thus providing a potential link between DF-stimulated SCFA production and food intake. Recent research has pointed to SCFA as the signaling molecules between the gut and the peripheral tissues with implications for insulin sensitivity and glucose homeostasis.[103,108] The mechanisms appear to involve adipocyte cell differentiation, regulation, and metabolism.[108] In response to an increased DF intake, there is an increased absorption of SCFA, and although the major part of propionate and butyrate is cleared in the liver,[100] micromolar levels of all three acids will reach the peripheral tissues. Here, the SCFA can act as ligands on the adipocytes[108] and thereby influence the balance of adipokines released and, in turn, increase the adipogenesis and differentiation of fat cells and consequently reduce their average size. SCFA will further inhibit lipolysis within adipose tissue as indicated by the reduced plasma concentrations of nonesterified fatty acids after the intake of fermentable carbohydrates.[109–111] The consequence is a reduced availability of fatty acids for uptake into ectopic fat depots (e.g., liver, pancreas).[108] Overall, the previously mentioned conditions have been linked to improved insulin sensitivity and glucose homeostasis.[103,108,111–113]

5.4 Other Metabolites

The diversity in colonic bacteria degrading enzymes and metabolic capabilities[98] also enable the gut flora to handle a number of the phytochemicals (e.g., phenolic acids,[114] lignans,[115] benzoxazinoids,[116]). These molecules will reach the large intestine in significantly higher quantities when consuming WG compared with refined products. A common feature is that these compounds are converted by the microbiota to an array of other metabolites and that it is the derived metabolites that have the biological function.

5.5 Metabolomics

Metabolomics measures a broad spectrum of metabolites in a biological sample and it is focused on assessing the effect of changes in, for example, environment, health status, or diet.[117] In a plasma or urine sample, the metabolome is composed of endogenous metabolites as well as exogenous compounds originating from the diet and metabolites that originate from the microbial metabolism making it possible to follow changes in the gut microbiota.[117]

The first studies using metabolomics to investigate the global impact of WG were animal (pigs or rats) studies.[118–120] Common to these studies was that betaine was identified as one of the metabolites causing discrimination between a WG diet or a refined diet. These results have subsequently been confirmed in human studies.[121] In a study where WG rye and rye bran were compared

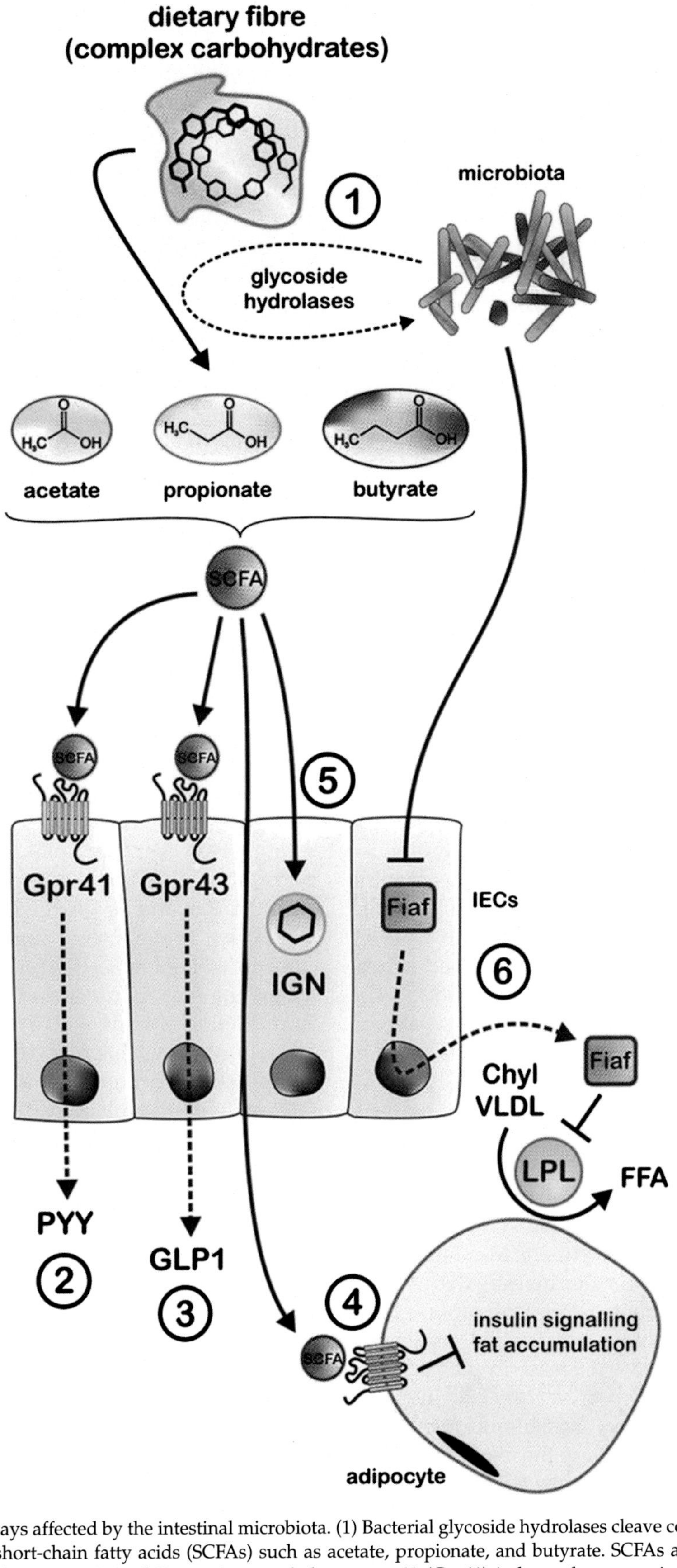

FIGURE 7 Metabolic pathways affected by the intestinal microbiota. (1) Bacterial glycoside hydrolases cleave complex carbohydrates derived from dietary fiber to produce short-chain fatty acids (SCFAs) such as acetate, propionate, and butyrate. SCFAs affect the host's metabolism in several ways. (2) SCFA-dependent activation of G protein–coupled receptor 41 (Gpr41) induces the expression of peptide YY, an intestinal hormone that inhibits gut motility, increases intestinal transit rate, and reduces the harvest of energy from the diet.[47] (3) Engagement of G protein–coupled receptor 43 (Gpr43; and Gpr41) by SCFAs has been shown to trigger the incretin hormone glucagon-like peptide 1 (GLP-1) to increase insulin sensitivity.[46] (4) SCFA-mediated activation of Gpr43 on adipocytes suppresses insulin signaling and inhibits fat accumulation in adipose tissue.[45] (5) The SCFAs butyrate and propionate activate intestinal gluconeogenesis (IGN); butyrate through a cyclic adenosine monophosphate–dependent mechanism, propionate via a gut-brain neuronal circuit involving Gpr41.[48] (6) The intestinal microbiota suppresses the expression of fasting-induced adipose factor (Fiaf), an intestinal epithelial cell (IEC)-derived circulating inhibitor of the lipoprotein lipase (LPL). *With permission from Tilg.*[149]

with refined wheat products, it was found that the level of betaine and its metabolite N,N-dimethylglycine was increased in plasma from subjects eating WG rye and rye bran.[122] The same result was obtained in a study with postmenopausal women who had a high intake of WG rye bread.[123] Increased plasma betaine was also found in healthy subjects eating a WG cereal-rich diet compared to a refined-grain diet[121] and after consumption of wheat aleurone-rich foods.[124] The association between betaine and T2D is not straightforward. Urinary betaine has been assessed as a marker of diabetes in cardiovascular patients,[125] and elevated plasma betaine is a marker of CVD risk in diabetes.[126] A major function of betaine is to act as methyl donor when the enzyme betaine-homocysteine methyl transferase (BHMT) methylates homocysteine to methionine. The activity of BHMT is regulated by insulin and it has been shown in diabetic Zucker Diabetic Fatty rats that the BHMT-activity was increased[127] with a concomitant depletion of the betaine stores.[127] In subjects with the MeS consuming a single test meal of one of four test breads—white wheat bread low in DF, WG rye bread, white wheat bread supplemented with AX, or white wheat bread supplemented with β-glucan—the plasma betaine concentration differed significantly and was highest after consumption of white wheat bread and white wheat bread supplemented with β-glucan.[128] This may be due to increased BHMT activity and hence a rapid conversion of dietary betaine. Various betaine compounds were among the metabolites that differed significantly in a study with participants with features of the MeS assigned to a WG diet or refined wheat breads during a 12-week intervention period.[129] The diverse betaines are most probably derived from altered endogenous metabolism of betaines after consumption of a WG diet (e.g., endogenous catabolism of lysine by intestinal bacteria). The main betaine compounds, glycine-betaine and proline-betaine, did not differ between subjects on WG or refined wheat bread diets, which may be due to rapid metabolism of these compounds. The potential health effects of increased betaine from dietary sources in subjects with the MeS and T2D is complex and remains to be further investigated.

Another group of metabolites, frequently found to be influenced by WG, is amino acids. Several studies on rye-based diets have shown a decrease in plasma level of leucine and isoleucine.[122,123,130,131] Leucine and isoleucine belong to the group of insulinotropic amino acids (i.e., amino acids that increase the insulin response when ingested in combination with carbohydrates). It has been described that there is a strong association of branched chain amino acids and aromatic amino acids with metabolic disease,[132] and they have highly significant association with future diabetes.[133] It has been suggested that WG diets improve insulin sensitivity by suppressing this pathway for increased insulin resistance.[134] Several other amino acids have been found to be affected by WG-enriched diets or high fiber breads including tyrosine, lysine, arginine, tryptophan, phenylalanine, methionine, and glutamic acid.[128,129,135] Special attention has been paid to tryptophan, which is a precursor for the biosynthesis of serotonin, that depresses hunger. In two studies, increased levels of ribitol and ribonic acid, precursors for tryptophan synthesis, has been linked to the higher level of satiety after rye intake.[135,136] Increased urinary secretion of creatinine indicated a change to protein metabolism when consuming WG,[137] further studies are, however, warranted.

The lipidome was investigated in the study by Lankinen et al.[130] They showed that multiple lysophosphatidylcholine (lysoPC) species were decreased following a rye bread and pasta intervention more than when following an oat, wheat bread, and potato intervention.[130] Similar results were obtained in a nontargeted metabolomics study of the acute metabolic response to a single dose of high fiber breads that showed that several phosphatidylcholine and lysoPC species were significantly affected.[128] Both studies were performed in subjects with traits of the MeS. A close association between T2D risk and choline-containing phospholipids has been observed[138,139] and hence a dietary intervention that modifies these species may be potentially interesting. However, because of the huge number of phosphatidylcholine and lysoPC species it may be difficult to target and assess the health potential.

Alkylresorcinols were suggested as markers for WG rye and wheat intake 10 years ago[140]; therefore, it is not surprising that AR are found to be metabolites changing during interventions with WG rye or wheat. In a study on the urinary changes after WG rye consumption compared with consumption of refined wheat bread in subjects with slightly elevated serum cholesterol a sulfate derivative of 3-(3,5-dihydroxyphenyl)-1-propanoic acid was identified as the most discriminative metabolite[141] and is a metabolite of AR; along with this, other metabolites of AR were identified as well. Two glucorinidated ARs were also found as highly discriminative metabolites in plasma from subjects with features of the MeS assigned to either a WG-enriched diet or control diet with refined wheat breads.[129] Information of the bioactivity of AR is scarce and it is difficult to distinguish between effects of AR and other features of WG foods.[142] Recently, it has been demonstrated that the AR C17:0/C21:0 ratio, an indicator of relative WG rye intake, is associated with increased insulin sensitivity in a population with MeS.[143] Concomitantly, AR C17:0/C21:0 ratio was also associated with a favorable blood lipid outcome (e.g., inversely associated with low-density lipoprotein cholesterol concentrations in a population with the MeS).[143,144]

A group of natural compounds recently identified in WG rye and wheat products is benzoxazinoids.[142] A microbial degradation product of 1,3-benzoxazol-2-one was identified in plasma from subjects consuming a high-fiber diet[116] and in volunteers consuming WG for 4 weeks. Several metabolites of benzoxazinoids were found to be explanatory of WG rye bread consumption.[145] Similar results were obtained by Hanhineva et al.,[146] who served test breads as a single meal to healthy participants and observed the postprandial plasma kinetic phytochemical profile. The most discriminant metabolites were two sulfonated phenylacetamides, hydroxy-N-(2-hydroxyphenyl) acetamide, and N-(2-hydroxyphenyl) acetamide, potentially derived from benzoxazinoid metabolites.[146] The studies showed that these compounds are taken up and converted in the body. They have to be taken into account when addressing the bioactivity and potential health effects of WG products; however, no studies have so far been published on their bioactivity.[142]

6. CONCLUSIONS

The inclusion of all parts of the grain in WG is causative for the higher concentration of DF and bioactives in WG than in refined flours. Existing evidence indicates that WG has a beneficial health effect; much of the evidence comes from observational studies that demonstrate an association between WG intake and T2D risk reduction, whereas results from intervention studies are more variable. The macro- and micronutrients along with the bioactive components present in WG synergistically contribute to their beneficial health effects brought about by the impact of WG on glycemia, insulinemia, the microbiome, and cellular metabolism. Recent studies indicate that a "microbiotal gut signature" is present in human obesity and in T2D, and metabolomics have pointed to several metabolites to be altered in consequence of WG consumption. However, the mode of action and their impact on T2D remains to be elucidated. For instance, studies on the association between betaine and T2D are warranted to confirm whether the elevated level of betaine consistently observed in subjects consuming WG holds a health potential. The effect of WG on plasma levels of amino acids and especially the branched chain amino acids represent another important area of future research as branched chain amino acids possesses strong links between early markers for onset of T2D and markers changed during WG interventions. Finally, the vast number of micronutrients and phytochemicals present in WG still represents an important area of future research. In spite of several gaps in our understanding of how WG may prevent T2D, current evidence lends credence to the recommendations to incorporate WG foods into a healthy diet and lifestyle program[27,29] and that individuals at high risk of T2D should consume foods containing WG (one-half of grain intake)[147] and that T2D is treated with foods rich in DF such as WG.[147,148] However, future research needs to report WG intake in grams rather than servings or portions,[59] and to examine the interactions not only among different dietary factors but also between diet and genetic predisposition and between diet and metabolic determinants (e.g., abdominal obesity, hypertension, dyslipidemia). Moreover, it will be important to conduct RCTs of sufficient size and length to further elucidate the role of WG and associated components on risk of the MeS and its features as well as on the risk of T2D.

References

1. International-Diabetes-Federation. *IDF diabetes atlas*. 6th ed. 2013. http://www.idf.org/diabetesatlas.
2. Alberti KG, Zimmet P, Shaw J. Metabolic syndrome—a new world-wide definition. A Consensus Statement from the International Diabetes Federation. *Diabet Med* 2006;**23**(5):469–80.
3. Hu FB. Globalization of diabetes: the role of diet, lifestyle, and genes. *Diabetes Care* 2011;**34**(6):1249–57.
4. Schulze MB, Schulz M, Heidemann C, Schienkiewitz A, Hoffmann K, Boeing H. Fiber and magnesium intake and incidence of type 2 diabetes: a prospective study and meta-analysis. *Arch Intern Med* 2007;**167**(9):956–65.
5. Fardet A. New hypotheses for the health-protective mechanisms of whole-grain cereals: what is beyond fibre? *Nutr Res Rev* 2010; **23**(1):65–134.
6. Hemery Y, Rouau X, Lullien-Pellerin V, Barron C, Abecassis J. Dry processes to develop wheat fractions and products with enhanced nutritional quality. *J Cereal Sci* 2007;**46**(3):327–47.
7. AACC. *AACC members agree on definition of whole grain*. St. Paul (MN): AACC International; 2000.
8. van der Kamp JW, Poutanen K, Seal CJ, Richardson DP. The HEALTHGRAIN definition of "whole grain". *Food Nutr Res* 2014;**58**:1–8. 10.3402/fnr.v58.22100.
9. FAO. World food and agriculture. In: Da Silva JG, editor. *FAO statistical yerabook 2013*. Rome: Food and Agriculture Organization of the United Nations; 2013.
10. Surget A, Barron C. Histologie du grain de blé (histology of the wheat grain). *Ind Cér* 2005;**145**:3–7.
11. Saulnier L, Guillon F, Sado P-E, Rouau X. *Plant cell wall polysaccharides in storage organs: xylans (food application)*. The Netherlands: Elsevier; 2007.
12. Izydorczyk MS, Dexter JE. Barley β-glucans and arabinoxylans: molecular structure, physicochemical properties, and uses in food products—a review. *Food Res Int* 2008;**41**(9):850–68.
13. Wood PJ. Review: oat and rye β-glucan: properties and function. *Cereal Chem J* 2010;**87**(4):315–30.
14. Esmaillzadeh A, Mirmiran P, Azizi F. Whole-grain consumption and the metabolic syndrome: a favorable association in Tehranian adults. *Eur J Clin Nutr* 2004;**59**(3):353–62.
15. McKeown NM, Meigs JB, Liu S, Saltzman E, Wilson PW, Jacques PF. Carbohydrate nutrition, insulin resistance, and the prevalence of the metabolic syndrome in the Framingham Offspring Cohort. *Diabetes Care* 2004;**27**(2):538–46.

16. Sahyoun NR, Jacques PF, Zhang XL, Juan W, McKeown NM. Whole-grain intake is inversely associated with the metabolic syndrome and mortality in older adults. *Am J Clin Nutr* 2006;**83**(1):124–31.
17. McKeown NM. Whole grain intake and insulin sensitivity: evidence from observational studies. *Nutr Rev* 2004;**62**(7 Pt 1):286–91.
18. Baxter AJ, Coyne T, McClintock C. Dietary patterns and metabolic syndrome—a review of epidemiologic evidence. *Asia Pac J Clin Nutr* 2006;**15**(2):134–42.
19. Priebe MG, van Binsbergen JJ, de Vos R, Vonk RJ. Whole grain foods for the prevention of type 2 diabetes mellitus. *Cochrane Database Syst Rev* 2008. 1:CD006061.
20. Parker ED, Liu S, Van Horn L, et al. The association of whole grain consumption with incident type 2 diabetes: the Women's Health Initiative Observational Study. *Ann Epidemiol* 2013;**23**(6):321–7.
21. Sun Q, Spiegelman D, van Dam RM, et al. White rice, brown rice, and risk of type 2 diabetes in US men and women. *Arch Intern Med* 2010;**170**(11):961–9.
22. Wirstrom T, Hilding A, Gu HF, Ostenson C-G, Bjorklund A. Consumption of whole grain reduces risk of deteriorating glucose tolerance, including progression to prediabetes. *Am J Clin Nutr* 2013;**97**(1):179–87.
23. Fisher E, Boeing H, Fritsche A, Doering F, Joost H-G, Schulze MB. Whole-grain consumption and transcription factor-7-like 2 (TCF7L2) rs7903146: gene-diet interaction in modulating type 2 diabetes risk. *Br J Nutr* 2009;**101**(4):478–81.
24. Dong J-Y, Xun P, He K, Qin L-Q. Magnesium intake and risk of type 2 diabetes: meta-analysis of prospective cohort studies. *Diabetes Care* 2011;**34**(9):2116–22.
25. Aune D, Norat T, Romundstad P, Vatten LJ. Whole grain and refined grain consumption and the risk of type 2 diabetes: a systematic review and dose-response meta-analysis of cohort studies. *Eur J Epidemiol* 2013;**28**(11):845–58.
26. de Munter JSL, Hu FB, Spiegelman D, Franz M, van Dam RM. Whole grain, bran, and germ intake and risk of type 2 diabetes: a prospective cohort study and systematic review. *PLoS Med* 2007;**4**(8):e261.
27. Ye EQ, Chacko SA, Chou EL, Kugizaki M, Liu S. Greater whole-grain intake is associated with lower risk of type 2 diabetes, cardiovascular disease, and weight gain. *J Nutr* 2012;**142**(7):1304–13.
28. Hu EA, Pan A, Malik V, Sun Q. White rice consumption and risk of type 2 diabetes: meta-analysis and systematic review. *BMJ* 2012;**344**:e1454.
29. Jonnalagadda SS, Harnack L, Liu RH, et al. Putting the whole grain puzzle together: health benefits associated with whole grains—summary of American Society for Nutrition 2010 Satellite Symposium. *J Nutr* 2011;**141**(5):1011S–22S.
30. Juntunen KS, Laaksonen DE, Autio K, et al. Structural differences between rye and wheat breads but not total fiber content may explain the lower postprandial insulin response to rye bread. *Am J Clin Nutr* 2003;**78**(5):957–64.
31. Juntunen KS, Niskanen LK, Liukkonen KH, Poutanen KS, Holst JJ, Mykkanen HM. Postprandial glucose, insulin, and incretin responses to grain products in healthy subjects. *Am J Clin Nutr* 2002;**75**(2):254–62.
32. Rosén LAH, Ostman EM, Shewry PR, et al. Postprandial glycemia, insulinemia, and satiety responses in healthy subjects after whole grain rye bread made from different rye varieties. 1. *J Agric Food Chem* 2011;**59**(22):12139–48.
33. Rosen LAH, Silva LOB, Andersson UK, Holm C, Ostman EM, Bjorck IME. Endosperm and whole grain rye breads are characterized by low post-prandial insulin response and a beneficial blood glucose profile. *Nutr J* 2009;**8**:42.
34. Rosén LAH, Östman EM, Björck IME. Effects of cereal breakfasts on postprandial glucose, appetite regulation and voluntary energy intake at a subsequent standardized lunch; focusing on rye products. *Nutr J* 2011;**10**(1):7–17.
35. Hlebowicz J, Jonsson JM, Lindstedt S, Bjorgell O, Darwich G, Almer L-O. Effect of commercial rye whole-meal bread on postprandial blood glucose and gastric emptying in healthy subjects. *Nutr J* 2009;**8**:26.
36. Leinonen KS, Poutanen KS, Mykkanen HM. Rye bread decreases serum total and LDL cholesterol in men with moderately elevated serum cholesterol. *J Nutr* 2000;**130**(2):164–70.
37. Rasmussen O, Winther E, Hermansen K. Glycaemic responses to different types of bread in insulin-dependent diabetic subjects (IDDM): studies at constant insulinaemia. *Eur J Clin Nutr* 1991;**45**(2):97–103.
38. Rosen LA, Ostman EM, Bjorck IM. Postprandial glycemia, insulinemia, and satiety responses in healthy subjects after whole grain rye bread made from different rye varieties. 2. *J Agric Food Chem* 2011;**59**(22):12149–54.
39. Jarvi AE, Karlstrom BE, Granfeldt YE, Bjorck IM, Vessby BO, Asp NG. The influence of food structure on postprandial metabolism in patients with non-insulin-dependent diabetes mellitus. *Am J Clin Nutr* 1995;**61**(4):837–42.
40. Kristensen M, Jensen MG, Riboldi G, et al. Wholegrain vs. refined wheat bread and pasta. Effect on postprandial glycemia, appetite, and subsequent ad libitum energy intake in young healthy adults. *Appetite* 2010;**54**(1):163–9.
41. Najjar AM, Parsons PM, Duncan AM, Robinson LE, Yada RY, Graham TE. The acute impact of ingestion of breads of varying composition on blood glucose, insulin and incretins following first and second meals. *Br J Nutr* 2009;**101**(3):391–8.
42. Breen C, Ryan M, Gibney MJ, Corrigan M, O'Shea D. Glycemic, insulinemic, and appetite responses of patients with type 2 diabetes to commonly consumed breads. *Diabetes Educ* 2013;**39**(3):376–86.
43. d'Emden MC, Marwick TH, Dreghorn J, Howlett VL, Cameron DP. Post-prandial glucose and insulin responses to different types of spaghetti and bread. *Diabetes Res Clin Pract* 1987;**3**(4):221–6.
44. Jenkins DJ, Wolever TM, Jenkins AL, Lee R, Wong GS, Josse R. Glycemic response to wheat products: reduced response to pasta but no effect of fiber. *Diabetes Care* 1983;**6**(2):155–9.
45. Lu ZX, Walker KZ, Muir JG, Mascara T, O'Dea K. Arabinoxylan fiber, a byproduct of wheat flour processing, reduces the postprandial glucose response in normoglycemic subjects. *Am J Clin Nutr* 2000;**71**(5):1123–8.
46. Hartvigsen ML, Gregersen S, Laerke HN, Holst JJ, Bach Knudsen KE, Hermansen K. Effects of concentrated arabinoxylan and beta-glucan compared with refined wheat and whole grain rye on glucose and appetite in subjects with the metabolic syndrome: a randomized study. *Eur J Clin Nutr* 2014;**68**(1):84–90.
47. Mohlig M, Koebnick C, Weickert MO, et al. Arabinoxylan-enriched meal increases serum ghrelin levels in healthy humans. *Horm Metab Res* 2005;**37**(5):303–8.
48. Hartvigsen ML, Larke HN, Overgaard A, Holst JJ, Bach Knudsen KE, Hermansen K. Postprandial effects of test meals including concentrated arabinoxylan and whole grain rye in subjects with the metabolic syndrome: a randomised study. *Eur J Clin Nutr* 2014;**68**(5):567–74.
49. Granfeldt Y, Nyberg L, Bjorck I. Muesli with 4 g oat beta-glucans lowers glucose and insulin responses after a bread meal in healthy subjects. *Eur J Clin Nutr* 2008;**62**(5):600–7.
50. Johansson EV, Nilsson AC, Ostman EM, Bjorck IM. Effects of indigestible carbohydrates in barley on glucose metabolism, appetite and voluntary food intake over 16 h in healthy adults. *Nutr J* 2013;**12**:46.

51. Makelainen H, Anttila H, Sihvonen J, et al. The effect of beta-glucan on the glycemic and insulin index. *Eur J Clin Nutr* 2007; **61**(6):779–85.
52. European-Food-Safety-Authority. Scientific Opinion on the substantiation of health claims related to beta-glucans from oats and barley and maintenance of normal blood LDL-cholesterol concentrations (ID 1236, 1299), increase in satiety leading to a reduction in energy intake (ID 851, 852), reduction of post-prandial glycaemic responses (ID 821, 824), and 'digestive function' (ID 850) pursuant to Article 13(1) of Regulation (EC) No 1924/2006. *EFSA J* 2011;**9**:2207–28.
53. Tosh SM. Review of human studies investigating the post-prandial blood-glucose lowering ability of oat and barley food products. *Eur J Clin Nutr* 2013;**67**(4):310–7.
54. Wood PJ, Beer MU, Butler G. Evaluation of role concentration and molecular weight of oat b-glucan in determining effect of viscosity on plasma glucose and insulin following an oral glucose load. *Br J Nutr* 2000;**84**:19–23.
55. Larsen HN, Christensen C, Rasmussen OW, et al. Influence of parboiling and physico-chemical characteristics of rice on the glycaemic index in non-insulin-dependent diabetic subjects. *Eur J Clin Nutr* 1996;**50**(1):22–7.
56. Rasmussen O, Gregersen S, Hermansen K. Influence of the amount of starch on the glycaemic index to rice in non-insulin-dependent diabetic subjects. *Br J Nutr* 1992;**67**(3):371–7.
57. Rasmussen OW, Gregersen S, Dorup J, Hermansen K. Blood glucose and insulin responses to different meals in non-insulin-dependent diabetic subjects of both sexes. *Am J Clin Nutr* 1992; **56**(4):712–5.
58. Larsen HN, Rasmussen OW, Rasmussen PH, et al. Glycaemic index of parboiled rice depends on the severity of processing: study in type 2 diabetic subjects. *Eur J Clin Nutr* 2000;**54**(5):380–5.
59. Ross AB, Kristensen M, Seal CJ, Jacques P, McKeown NM. Recommendations for reporting whole-grain intake in observational and intervention studies. *Am J Clin Nutr* 2015;**101**(5):903–7.
60. Laaksonen DE, Toppinen LK, Juntunen KS, et al. Dietary carbohydrate modification enhances insulin secretion in persons with the metabolic syndrome. *Am J Clin Nutr* 2005;**82**(6):1218–27.
61. Juntunen KS, Laaksonen DE, Poutanen KS, Niskanen LK, Mykkanen HM. High-fiber rye bread and insulin secretion and sensitivity in healthy postmenopausal women. *Am J Clin Nutr* 2003;**77**(2):385–91.
62. Andersson A, Tengblad S, Karlstrom B, et al. Whole-grain foods do not affect insulin sensitivity or markers of lipid peroxidation and inflammation in healthy, moderately overweight subjects. *J Nutr* 2007;**137**(6):1401–7.
63. Giacco R, Lappi J, Costabile G, et al. Effects of rye and whole wheat versus refined cereal foods on metabolic risk factors: a randomised controlled two-centre intervention study. *Clin Nutr* 2013;**32**(6):941–9.
64. Giacco R, Costabile G, Della Pepa G, et al. A whole-grain cereal-based diet lowers postprandial plasma insulin and triglyceride levels in individuals with metabolic syndrome. *Nutr Metab Cardiovasc Dis* 2014;**24**(8):837–44.
65. Jenkins AL, Jenkins DJA, Zdravkovic U, Wursch P, Vuksan V. Depression of the glycemic index by high levels of beta-glucan fiber in two functional foods tested in type 2 diabetes. *Eur J Clin Nutr* 2002;**56**(7):622–8.
66. Pereira MA, Jacobs Jr DR, Pins JJ, et al. Effect of whole grains on insulin sensitivity in overweight hyperinsulinemic adults. *Am J Clin Nutr* 2002;**75**(5):848–55.
67. Rave K, Roggen K, Dellweg S, Heise T, tom Dieck H. Improvement of insulin resistance after diet with a whole-grain based dietary product: results of a randomized, controlled cross-over study in obese subjects with elevated fasting blood glucose. *Br J Nutr* 2007;**98**(5):929–36.
68. McIntosh GH, Noakes M, Royle PJ, Foster PR. Whole-grain rye and wheat foods and markers of bowel health in overweight middle-aged men. *Am J Clin Nutr* 2003;**77**(4):967–74.
69. Weickert MO, Spranger J, Holst JJ, et al. Wheat-fibre-induced changes of postprandial peptide YY and ghrelin responses are not associated with acute alterations of satiety. *Br J Nutr* 2006; **96**(5):795–8.
70. Christensen KL, Hedemann MS, Lærke HN, et al. Concentrated arabinoxylan but not concentrated β-glucan in wheat bread has similar effects on postprandial insulin as whole-grain rye in porto-arterial catheterized pigs. *J Agric Food Chem* 2013;**61**(32): 7760–8.
71. Hartvigsen ML, Jeppesen PB, Laerke HN, Njabe EN, Knudsen KE, Hermansen K. Concentrated arabinoxylan in wheat bread has beneficial effects as rye breads on glucose and changes in gene expressions in insulin-sensitive tissues of Zucker diabetic fatty (ZDF) rats. *J Agric Food Chem* 2013;**61**(21): 5054–63.
72. Andersson U, Rosen L, Ostman E, et al. Metabolic effects of whole grain wheat and whole grain rye in the C57BL/6J mouse. *Nutrition* 2010;**26**(2):230–9.
73. Youn M, Csallany AS, Gallaher DD. Whole grain consumption has a modest effect on the development of diabetes in the Goto-Kakisaki rat. *Br J Nutr* 2012;**107**(2):192–201.
74. Jenkins DJA, Thomas DM, Wolever MS, et al. Glycemic index of foods: a physiological basis for carbohydrate exchange. *Am J Clin Nutr* 1981;**34**:362–6.
75. Englyst KN, Vinoy S, Englyst HN, Lang V. Glycaemic index of cereal products explained by their content of rapidly and slowly available glucose. *Br J Nutr* 2003;**89**(3):329–40.
76. Ludwig DS. The glycemic index: physiological mechanisms relating to obesity, diabetes, and cardiovascular disease. *JAMA* 2002;**287**(18):2414–23.
77. Greenberg AS, McDaniel ML. Identifying the links between obesity, insulin resistance and beta-cell function: potential role of adipocyte-derived cytokines in the pathogenesis of type 2 diabetes. *Eur J Clin Invest* 2002;**32**(Suppl. 3):24–34.
78. Detopoulou P, Panagiotakos DB, Antonopoulou S, Pitsavos C, Stefanadis C. Dietary choline and betaine intakes in relation to concentrations of inflammatory markers in healthy adults: the ATTICA study. *Am J Clin Nutr* 2008;**87**(2):424–30.
79. Price RK, Welch RW, Lee-Manion AM, Bradbury I, Strain JJ. Total phenolics and antioxidant potential in plasma and urine of humans after consumption of wheat bran. *Cereal Chem* 2008;**85**:152–7.
80. Duntas LH. Selenium and inflammation: underlying anti-inflammatory mechanisms. *Horm Metab Res* 2009;**41**(6):443–7.
81. Mateo Anson N, Aura AM, Selinheimo E, et al. Bioprocessing of wheat bran in whole wheat bread increases the bioavailability of phenolic acids in men and exerts antiinflammatory effects ex vivo. *J Nutr* 2011;**141**(1):137–43.
82. Kallio P, Kolehmainen M, Laaksonen DE, et al. Inflammation markers are modulated by responses to diets differing in post-prandial insulin responses in individuals with the metabolic syndrome. *Am J Clin Nutr* 2008;**87**(5):1497–503.
83. Scribner KB, Pawlak DB, Aubin CM, Majzoub JA, Ludwig DS. Long-term effects of dietary glycemic index on adiposity, energy metabolism, and physical activity in mice. *Am J Physiol Endocrinol Metab* 2008;**295**(5):E1126–31.
84. Isken F, Klaus S, Petzke KJ, Loddenkemper C, Pfeiffer AF, Weickert MO. Impairment of fat oxidation under high- vs. low-glycemic index diet occurs before the development of an obese phenotype. *Am J Physiol Endocrinol Metab* 2010;**298**(2):E287–95.
85. Isken F, Klaus S, Osterhoff M, Pfeiffer AF, Weickert MO. Effects of long-term soluble vs. insoluble dietary fiber intake on high-fat diet-induced obesity in C57BL/6J mice. *J Nutr Biochem* 2010; **21**(4):278–84.

86. Hermansen ML, Eriksen NM, Mortensen LS, Holm L, Hermansen K. Can the Glycemic Index (GI) be used as a tool in the prevention and management of Type 2 diabetes? *Rev Diabet Stud* 2006;**3**(2):61–71.
87. Flint HJ, Scott KP, Louis P, Duncan SH. The role of the gut microbiota in nutrition and health. *Nat Rev Gastroenterol Hepatol* 2012; **9**(10):577–89.
88. Backhed F, Ding H, Wang T, et al. The gut microbiota as an environmental factor that regulates fat storage. *Proc Natl Acad Sci USA* 2004;**101**(44):15718–23.
89. Ley RE, Backhed F, Turnbaugh P, Lozupone CA, Knight RD, Gordon JI. Obesity alters gut microbial ecology. *PNAS* 2005; **102**(31):11070–5.
90. Qin J, Li Y, Cai Z, et al. A metagenome-wide association study of gut microbiota in type 2 diabetes. *Nature* 2012;**490**(7418):55–60.
91. Karlsson FH, Tremaroli V, Nookaew I, et al. Gut metagenome in European women with normal, impaired and diabetic glucose control. *Nature* 2013;**498**(7452):99–103.
92. Cani PD, Amar J, Iglesias MA, et al. Metabolic endotoxemia initiates obesity and insulin resistance. *Diabetes* 2007;**56**(7):1761–72.
93. Everard A, Cani PD. Diabetes, obesity and gut microbiota. *Best Pract Res Clin Gastroenterol* 2013;**27**(1):73–83.
94. Costabile A, Klinder A, Fava F, et al. Whole-grain wheat breakfast cereal has a prebiotic effect on the human gut microbiota: a double-blind, placebo-controlled, crossover study. *Br J Nutr* 2008;**99**(1):110–20.
95. Christensen EG, Licht TR, Kristensen M, Bahl MI. Bifidogenic effect of whole-grain wheat during a 12-week energy-restricted dietary intervention in postmenopausal women. *Eur J Clin Nutr* 2013;**67**(12):1316–21.
96. Lappi J, Salojarvi J, Kolehmainen M, et al. Intake of whole-grain and fiber-rich rye bread versus refined wheat bread does not differentiate intestinal microbiota composition in Finnish adults with metabolic syndrome. *J Nutr* 2013;**143**(5):648–55.
97. Walker AW, Ince J, Duncan SH, et al. Dominant and diet-responsive groups of bacteria within the human colonic microbiota. *ISME J* 2011;**5**(2):220–30.
98. Flint HJ, Scott KP, Duncan SH, Louis P, Forano E. Microbial degradation of complex carbohydrates in the gut. *Gut Microbes* 2012; **3**(4):289–306.
99. Nielsen TS, Lærke HN, Theil PK, et al. Diets high in resistant starch and arabinoxylan modulate digestion processes and SCFA pool size in the large intestine and faecal microbial composition in pigs. *Br J Nutr* 2014;**112**(11):1837–49. http://dx.doi.org/10.1017/S000711451400302X.
100. Ingerslev AK, Theil PK, Hedemann MS, Lærke HN, Bach Knudsen KE. Resistant starch and arabinoxylan augment SCFA absorption, but affect postprandial glucose and insulin responses differently. *Br J Nutr* 2014;**111**(09):1564–76.
101. Damen B, Cloetens L, Broekaert WF, et al. Consumption of breads containing in situ-produced arabinoxylan oligosaccharides alters gastrointestinal effects in healthy volunteers. *J Nutr* 2012;**142**(3): 470–7.
102. Nilsson AC, Ostman EM, Knudsen KEB, Holst JJ, Bjorck IME. A cereal-based evening meal rich in indigestible carbohydrates increases plasma butyrate the next morning. *J Nutr* 2010;**140**(11): 1932–6.
103. Priebe MG, Wang H, Weening D, Schepers M, Preston T, Vonk RJ. Factors related to colonic fermentation of nondigestible carbohydrates of a previous evening meal increase tissue glucose uptake and moderate glucose-associated inflammation. *Am J Clin Nutr* 2010;**91**(1):90–7.
104. Nilsson A, Granfeldt Y, Ostman E, Preston T, Bjorck I. Effects of GI and content of indigestible carbohydrates of cereal-based evening meals on glucose tolerance at a subsequent standardised breakfast. *Eur J Clin Nutr* 2006;**60**(9):1092–9.
105. Nilsson AC, Östman EM, Holst JJ, Björck IME. Including indigestible carbohydrates in the evening meal of healthy subjects improves glucose tolerance, lowers inflammatory markers, and increases satiety after a subsequent standardized breakfast. *J Nutr* 2008;**138**(4):732–9.
106. Furness JB, Rivera LR, Cho HJ, Bravo DM, Callaghan B. The gut as a sensory organ. *Nat Rev Gastroenterol Hepatol* 2013;**10**(12):729–40.
107. Sleeth ML, Thompson EL, Ford HE, Zac-Varghese SE, Frost G. Free fatty acid receptor 2 and nutrient sensing: a proposed role for fibre, fermentable carbohydrates and short-chain fatty acids in appetite regulation. *Nutr Res Rev* 2010;**23**(1):135–45.
108. Robertson MD. Metabolic cross talk between the colon and the periphery: implications for insulin sensitivity. *Proc Nutr Soc* 2007;**66**: 351–61.
109. Ferchaud-Roucher V, Pouteau E, Piloquet H, Zair Y, Krempf M. Colonic fermentation from lactulose inhibits lipolysis in overweight subjects. *Am J Physiol Endocrinol Metab* 2005;**289**(4): E716–20.
110. Brighenti F, Benini L, Del Rio D, et al. Colonic fermentation of indigestible carbohydrates contributes to the second-meal effect. *Am J Clin Nutr* 2006;**83**(4):817–22.
111. Robertson MD, Bickerton AS, Dennis AL, Vidal H, Frayn KN. Insulin-sensitizing effects of dietary resistant starch and effects on skeletal muscle and adipose tissue metabolism. *Am J Clin Nutr* 2005;**82**(3):559–67.
112. Gao Z, Yin J, Zhang J, et al. Butyrate improves insulin sensitivity and increases energy expenditure in mice. *Diabetes* 2009;**58**(7): 1509–17.
113. Robertson MD, Currie JM, Morgan LM, Jewell DP, Frayn KN. Prior short-term consumption of resistant starch enhances postprandial insulin sensitivity in healthy subjects. *Diabetologia* 2003; **46**(5):659–65.
114. Norskov NP, Hedemann MS, Theil PK, Fomsgaard IS, Laursen BB, Knudsen KE. Phenolic acids from wheat show different absorption profiles in plasma: a model experiment with catheterized pigs. *J Agric Food Chem* 2013;**61**(37):8842–50.
115. Setchell KD, Lawson AM, Borriello SP, et al. Lignan formation in man–microbial involvement and possible roles in relation to cancer. *Lancet* 1981;**2**(8236):4–7.
116. Johansson-Persson A, Barri T, Ulmius M, Onning G, Dragsted LO. LC-QTOF/MS metabolomic profiles in human plasma after a 5-week high dietary fiber intake. *Anal Bioanal Chem* 2013;**405**(14): 4799–809.
117. Kelli MS, Karnovsky A, Michailidis G, Pennathur S. Metabolomics and diabetes: analytical and computational approaches. *Diabetes* 2015;**64**:718–32.
118. Bertram HC, Bach Knudsen KE, Serena A, et al. NMR-based metabonomic studies reveal changes in the biochemical profile of plasma and urine from pigs fed high-fibre rye bread. *Br J Nutr* 2006;**95**(5):955–62.
119. Bertram HC, Malmendal A, Nielsen NC, et al. NMR-based metabonomics reveals that plasma betaine increases upon intake of high-fiber rye buns in hypercholesterolemic pigs. *Mol Nutr Food Res* 2009;**53**:1055–62.
120. Fardet A, Canlet C, Gottardi G, et al. Whole-grain and refined wheat flours show distinct metabolic profiles in rats as assessed by a 1H NMR-based metabonomic approach. *J Nutr* 2007;**137**(4): 923–9.
121. Ross AB, Bruce SJ, Blondel-Lubrano A, et al. A whole-grain cereal-rich diet increases plasma betaine, and tends to decrease total and LDL-cholesterol compared with a refined-grain diet in healthy subjects. *Br J Nutr* 2011;**105**(10):1492–502.
122. Moazzami AA, Zhang JX, Kamal-Eldin A, et al. Nuclear magnetic resonance-based metabolomics enable detection of the effects of a whole grain rye and rye bran diet on the metabolic profile of plasma in prostate cancer patients. *J Nutr* 2011;**141**(12):2126–32.

123. Moazzami A, Bondia-Pons I, Hanhineva K, et al. Metabolomics reveals the metabolic shifts following an intervention with rye bread in postmenopausal women- a randomized control trial. *Nutr J* 2012;**11**(1):88.
124. Price RK, Keaveney EM, Hamill LL, et al. Consumption of wheat aleurone-rich foods increases fasting plasma betaine and modestly decreases fasting homocysteine and LDL-cholesterol in adults. *J Nutr* 2010;**140**(12):2153–7.
125. Schartum-Hansen H, Ueland PM, Pedersen ER, et al. Assessment of urinary betaine as a marker of diabetes mellitus in cardiovascular patients. *PloS One* 2013;**8**(8):e69454.
126. Lever M, George PM, Slow S, et al. Betaine and trimethylamine-N-oxide as predictors of cardiovascular outcomes show different patterns in diabetes mellitus: an observational study. *PloS One* 2014;**9**(12):19.
127. Wijekoon EP, Brosnan ME, Brosnan JT. Homocysteine metabolism in diabetes. *Biochem Soc Trans* 2007;**35**:1175–9.
128. Nielsen KL, Hartvigsen ML, Hedemann MS, Larke HN, Hermansen K, Bach Knudsen KE. Similar metabolic responses in pigs and humans to breads with different contents and compositions of dietary fibers: a metabolomics study. *Am J Clin Nutr* 2014;**99**(4):941–9.
129. Hanhineva K, Lankinen MA, Pedret A, et al. Nontargeted metabolite profiling discriminates diet-specific biomarkers for consumption of whole grains, fatty fish, and bilberries in a randomized controlled trial. *J Nutr* 2015;**145**(1):7–17.
130. Lankinen M, Schwab U, Gopalacharyulu PV, et al. Dietary carbohydrate modification alters serum metabolic profiles in individuals with the metabolic syndrome. *Nutr Metab Cardiovasc Dis* 2010;**20**(4):249–57.
131. Moazzami AA, Shrestha A, Morrison DA, Poutanen K, Mykkanen H. Metabolomics reveals differences in postprandial responses to breads and fasting metabolic characteristics associated with postprandial insulin demand in postmenopausal women. *J Nutr* 2014;**144**(6):807–14.
132. Newgard Christopher B. Interplay between lipids and branched-chain amino acids in development of insulin resistance. *Cell Metab* 2012;**15**(5):606–14.
133. Wang TJ, Larson MG, Vasan RS, et al. Metabolite profiles and the risk of developing diabetes. *Nat Med* 2011;**17**(4):448–53.
134. Ross AB. Whole grains beyond fibre: what can metabolomics tell us about mechanisms? *Proc Nutr Soc* 2014:1–8. First View.
135. Bondia-Pons I, Nordlund E, Mattila I, et al. Postprandial differences in the plasma metabolome of healthy Finnish subjects after intake of a sourdough fermented endosperm rye bread versus white wheat bread. *Nutr J* 2011;**10**:116.
136. Lankinen M, Schwab U, Seppanen-Laakso T, et al. Metabolomic analysis of plasma metabolites that may mediate effects of rye bread on satiety and weight maintenance in postmenopausal women. *J Nutr* 2011;**141**(1):31–6.
137. Ross AB, Pere-Trépat E, Montoliu I, et al. A whole-grain-rich diet reduces urinary excretion of markers of protein catabolism and gut microbiota metabolism in healthy men after one week. *J Nutr* 2013;**143**(6):766–73.
138. Wang C, Kong HW, Guan YF, et al. Plasma phospholipid metabolic profiling and biomarkers of type 2 diabetes mellitus based on high-performance liquid chromatography/electrospray mass spectrometry and multivariate statistical analysis. *Anal Chem* 2005;**77**(13):4108–16.
139. Floegel A, Stefan N, Yu Z, et al. Identification of serum metabolites associated with risk of type 2 diabetes using a targeted metabolomic approach. *Diabetes* 2013;**62**(2):639–48.
140. Ross AB, Kamal-Eldin A, Aman P. Dietary alkylresorcinols: absorption, bioactivities, and possible use as biomarkers of whole-grain wheat- and rye-rich foods. *Nutr Rev* 2004;**62**(3):81–95.
141. Bondia-Pons I, Barri T, Hanhineva K, et al. UPLC-QTOF/MS metabolic profiling unveils urinary changes in humans after a whole grain rye versus refined wheat bread intervention. *Mol Nutr Food Res* 2013;**57**(3):412–22.
142. Andersson AAM, Dimberg L, Åman P, Landberg R. Recent findings on certain bioactive components in whole grain wheat and rye. *J Cereal Sci* 2014;**59**(3):294–311.
143. Magnusdottir OK, Landberg R, Gunnarsdottir I, et al. Plasma alkylresorcinols C17:0/C21:0 ratio, a biomarker of relative whole-grain rye intake, is associated to insulin sensitivity: a randomized study. *Eur J Clin Nutr* 2014;**68**(4):453–8.
144. Magnusdottir OK, Landberg R, Gunnarsdottir I, et al. Whole grain rye intake, reflected by a biomarker, is associated with favorable blood lipid outcomes in subjects with the metabolic syndrome—a randomized study. *PloS One* 2014;**9**(10):e110827.
145. Beckmann M, Lloyd AJ, Haldar S, Seal C, Brandt K, Draper J. Hydroxylated phenylacetamides derived from bioactive benzoxazinoids are bioavailable in humans after habitual consumption of whole grain sourdough rye bread. *Mol Nutr Food Res* 2013; **57**(10):1859–73.
146. Hanhineva K, Keski-Rahkonen P, Lappi J, et al. The postprandial plasma rye fingerprint includes benzoxazinoid-derived phenylacetamide sulfates. *J Nutr* 2014;**144**(7):1016–22.
147. ADA. Foundations of care: education, nutrition, physical activity, smoking cessation, psychosocial care, and immunization. *Diabetes Care* 2015;**38**:S20–30.
148. Mann JI, De Leeuw I, Hermansen K, et al. Evidence-based nutritional approaches to the treatment and prevention of diabetes mellitus. *Nutr Metab Cardiovasc Dis* 2004;**14**(6):373–94.
149. Tilg H, Moschen AR. Microbiota and diabetes: an evolving relationship. *Gut* 2014;**63**(9):1513–21.

CHAPTER

9

Peroxisome Proliferator-Activated Receptors (PPARs) in Glucose Control

Fausto Chiazza, Massimo Collino

Department of Drug Science and Technology, University of Turin, Turin, Italy

1. PPAR: AN OVERVIEW

Peroxisome proliferator-activated receptors (PPARs) are among the most extensively investigated and characterized nuclear receptors, mainly because they are activated by compounds that are in clinical use for the treatment of hypertriglyceridemia and insulin resistance. There are three PPAR subtypes: α, β/δ, and γ, also named NR1C1, NR1C2, and NR1C3, respectively. PPAR-α is the main pharmacological target of fibrates, which are antihyperlipidemic drugs, whereas thiazolidinediones (TZD), acting via PPAR-γ, influence free fatty acid (FFA) flux and reduce insulin resistance and blood glucose levels. So far, there are no PPAR-β/δ drugs in clinical use.

The three PPAR isoforms are the products of distinct genes. The human PPAR-α gene was mapped on chromosome 22 in the general region 22q12–q13.1, the PPAR-γ gene is located on chromosome 3 at position 3p25, whereas PPAR-β/δ has been assigned to chromosome 6, at position 6p21.1–p21.2. PPARs were originally identified by Isseman and Green[1] after screening the rat liver complementary DNA library with a complementary DNA sequence located in the highly conserved C domain of nuclear hormone receptors (NHRs). The name PPAR is derived from the fact that activation of PPAR-α, the first member of the PPAR family to be cloned, results in peroxisome proliferation in rodent hepatocytes.[2] Activation of neither PPAR-β/δ nor PPAR-γ, however, elicits this response and, interestingly, the phenomenon of peroxisome proliferation does not occur in humans. The molecular basis for this difference between species is not yet clear. PPAR-β/δ was initially reported as PPAR-β in *Xenopus laevis* and NUC1 in humans.[3] Subsequently, a similar transcript was cloned from mice and termed PPAR-δ.[4] Though now recognized as homologs for each other, it was not originally certain whether PPAR-β from *Xenopus* was identical to murine PPAR-δ, hence the terminology PPAR-β/δ.

All members of this superfamily share the typical domain organization of nuclear receptors. The N-terminal A/B domain contains a ligand-independent transactivation function. The C domain is the DNA-binding domain with its typical two zinc finger–like motifs, as previously described for the steroid receptors and the D domain, is the cofactor docking domain. The E/F domain is the ligand binding domain, it contains a ligand-dependent trans-activation function, and is able to interact with transcriptional coactivators such as steroid receptor coactivator-1 and CREB-binding protein.

2. MOLECULAR MECHANISMS OF PPAR ACTIVATION

There are at least three primary mechanisms by which PPARs can regulate biological functions: transcriptional transactivation, transcriptional transrepression, and ligand-independent transrepression.

2.1 Mechanism of Transcriptional Transactivation

PPARs function as heterodimers with their obligatory partner the retinoid X receptor (RXR). Like other NHRs, the PPAR/RXR heterodimer most likely recruits cofactor complexes—either coactivators or co-repressors—that modulate its transcriptional activity.[5] The PPAR/RXR heterodimer then binds to sequence specific PPAR response elements (PPREs), located in the 5′-flanking region of target genes, thereby acting as a transcriptional

Molecular Nutrition and Diabetes
http://dx.doi.org/10.1016/B978-0-12-801585-8.00009-9

regulator.[6] In the absence of a ligand, to prevent PPAR/RXR binding to DNA, high-affinity complexes are formed between the inactive PPAR/RXR heterodimers and co-repressor molecules, such as nuclear receptor co-repressor or silencing mediator for retinoic receptors. In response to ligand binding, PPAR undergoes a conformational change, leading to release of auxiliary proteins and co-repressors and recruitment of coactivators that contain histone acetylase activity. Acetylation of histones by coactivators bound to the ligand-PPAR complex leads to nucleosome remodeling, allowing for recruitment of RNA polymerase II causing target gene transcription. The search for PPAR target genes with identified PPREs has led to the identification of several genes involved in lipid metabolism, oxidative stress, and the inflammatory response, as widely documented in the literature.

2.2 Mechanism of Transcriptional Transrepression

PPARs can also negatively regulate gene expression in a ligand-dependent manner by inhibiting the activities of other transcription factors, such as activated protein-1, nuclear factor-κB (NF-κB), and nuclear factor of activated T cells (ligand-dependent transrepression). In contrast to transcriptional activation, which usually involves the binding of PPARs to specific response elements in the promoter or enhancer regions of target genes, transrepression does not involve binding to typical receptor-specific response elements.[7] Several lines of evidence suggest that PPARs may exert anti-inflammatory effects by negatively regulating the expression of proinflammatory genes. To date, several mechanisms have been suggested to account for this activity, but despite intensive investigation, unifying principles remain to be elucidated.

First, competition for limited amounts of essential, shared transcriptional coactivators may play a role in transrepression. The activated PPAR/RXR heterodimer reduces the availability of coactivators required for gene induction by other transcriptional factors. Thus, without distinct cofactors, transcription factors cannot cause gene expression.

Second, PPAR/RXR complexes may cause a functional inhibition by directly binding to transcription factors, preventing them from inducing gene transcription or inducing the expression of inhibitory proteins, such as the protein inhibitor of kappa B-α, which sequesters the NF-κB subunits in the cytoplasm and consequently reduces their DNA-binding activity.[8]

Third, PPAR/RXR heterodimers may also inhibit phosphorylation and activation of several members of the MAPK family. In general, very little is known about the molecular mechanisms by which PPARs and their ligands modulate kinase activities.

Recent studies have suggested another mechanism based on co-repressor–dependent transrepression by PPARs. PPAR-β/δ controls the inflammatory status of macrophages based on its association with the transcriptional repressor BCL-6.[9] Free BCL-6 suppresses the expression of multiple proinflammatory cytokines and chemokines. PPAR-β/δ, but not PPAR-α and PPAR-γ, exhibits BCL-6 binding ability.[10,11] In the absence of a ligand, PPAR-β/δ sequesters BCL-6 from inflammatory response genes. In contrast, in the presence of a ligand, PPAR-β/δ releases the repressor, which now distributes to NF-κB–dependent promoters and exerts anti-inflammatory effects by repressing transcription from these genes.

2.3 Mechanism of Ligand-Independent Transrepression

PPARs may repress the transcription of direct target genes in the absence of ligands (ligand-independent repression). PPARs bind to response elements in the absence of any ligand and recruit co-repressor complexes that mediate active repression. The co-repressors are capable of fully repressing PPAR-mediated transactivation induced either by ligands or by cAMP-regulated signaling pathways. This suggests co-repressors as general antagonists of the various stimuli inducing PPAR-mediated transactivation. Co-repressors can display different ligand selectivity: the nuclear receptor co-repressor NCoR interacted strongly with the ligand-binding domain of PPAR-β/δ, whereas interactions with the ligand-binding domains of PPAR-γ and PPAR-α were significantly weaker.[12]

3. THE ROLE OF PPARs IN THE CONTROL OF GLUCOSE METABOLISM

3.1 Peroxisome Proliferator-Activated Receptor-α

PPAR-α, highly expressed in the skeletal muscle, brown adipose tissue, and liver, is one of the main regulators of free fatty acid (FFA) oxidation. Fibrates, clinically used as hypolipidemic agents, are known PPAR-α agonists. Interestingly, PPAR-α agonism seems to be involved in the control of other clinical consequences of metabolic derangements. For instance, preclinical studies showed that treatment of PPAR-α null mice with a high-fat diet leads to an extreme increase in body weight, whereas the activation of PPAR-α reduces weight gain.[13–16]

In the past 15 years, evidence has emerged suggesting that PPAR-α is an important regulator of glucose metabolism and insulin sensitivity. PPAR-α regulates glucose synthesis during fasting states by upregulating glycerol-3-phosphate dehydrogenase, glycerol kinase, and glycerol transport proteins.[17] Moreover, PPAR-α regulates gluconeogenesis by stimulating expression of pyruvate dehydrogenase kinase 4, thus promoting the use of pyruvate for gluconeogenesis instead of FFA synthesis.[18] Maida and colleagues have recently suggested an interesting link between PPAR-α activation and metformin, demonstrating that metformin effects on plasma levels of the incretin glucagon-like peptide-1 and expression levels of islet incretin receptor are due to PPAR-α activation.[19] Treatment with PPAR-α agonists, mainly fenofibrate, dramatically improved insulin resistance and glycemic control in type 2 diabetic mice and rat models by reducing adiposity, improving peripheral insulin action and exerting beneficial effects on pancreatic β-cells.[20–22]

PPAR-α activity can be modulated by glucose itself. Hostetler and colleagues concluded that PPAR-α binds glucose and glucose metabolites with high affinity, causing a significantly altered PPAR-α secondary structure.[23]

Recent studies showed that PPAR-α agonism protects pancreatic islets under conditions of lipotoxicity and improves their adaptive response to pathological conditions. For instance, PPAR-α activation stimulates pancreatic islet β-cells, thereby potentiating glucose-stimulated insulin secretion.[24] Moreover, PPAR-α deficiency in a mouse model of obesity-related insulin resistance leads to reduced insulin secretion by pancreatic β cells in response to glucose.[25] In a β-cell line, ectopic overexpression of PPAR-α improved glucose-stimulated insulin secretion in lipotoxic conditions.[26] In a similar manner, activation of PPAR-α prevents FFA-induced insulin resistance and apoptosis in primary human pancreatic islets.[25]

Although these findings identify PPARα as potential pharmacological target for the treatment of insulin resistance and the prevention of type 2 diabetes (T2D), it must be pointed out that most of these data were collected in animals, and if PPAR-α agonists will efficiently prevent T2D development in humans is not certain. Only a few dated clinical trials show an improvement in glucose homeostasis after fibrate treatment.[27–29] In lipoatrophic diabetic patients, the PPAR-α agonist bezafibrate significantly improved glucose intolerance and insulin resistance.[30] However, more recent clinical studies did not reveal any effect of fenofibrate on glucose parameters or on adipocyte and peripheral (muscle) tissue insulin sensitivity.[31,32] Notably, the PPAR-α genotype can affect the age of development of T2D and modulate the progression to T2D in subjects with impaired glucose tolerance. A clinical trial showed that the PPAR-α intron 1 genotype was associated with a more rapid progression to insulin monotherapy, suggesting that PPAR-α influences the rate of development of insulin resistance or progression to β-cell failure.[33,34]

3.2 Peroxisome Proliferator-Activated Receptor-γ

PPAR-γ is most widely expressed in adipose tissue and in immune/inflammatory cells (e.g., monocytes, macrophages). In contrast, its expression in skeletal muscle and liver is lower if compared with the other PPAR isoforms. PPAR-γ is essential for the differentiation and functioning of brown and white adipocytes and promotes the accumulation of lipids in these cells.[35]

TZDs include a class of PPAR-γ agonists that reverse insulin resistance in liver and peripheral tissues through specific PPAR-γ activation. Troglitazone was the first TZD approved for this use, withdrawn from the market in March 2000 following the emergence of serious hepatotoxicity in some patients. Because troglitazone induces CYP3A4, it has been hypothesized that potentially toxic quinones derived from CYP3A4-dependent metabolism could cause liver damage.[36] Rosiglitazone and pioglitazone are the available TZDs. Activation of PPAR-γ by TZDs decreases glycated hemoglobin, fasting and postprandial glucose concentration in blood, and lowers circulating insulin levels in patients with T2D, mainly as a consequence of the improvement in insulin sensitivity. Notably, TZDs not only improve insulin sensitivity but also preserve pancreatic β-cell function, thus reducing the incidence of T2D, as demonstrated by clinical trials on prevention of T2D in high-risk people treated with troglitazone or pioglitazone.[37–39] The protection from diabetes occurred in the presence of improved insulin sensitivity while subjects were taking study medication, whereas there was not significant evidence of sustained effects of PPAR-γ agonists to reduce diabetes incidence when the drugs were no longer administered. Despite the well-characterized patterns of TZDs' beneficial effects in the treatment of insulin resistance, their clinical use is limited, at least in part, by serious, although rare, adverse effects. The first evidence on increased risk of ischemic cardiovascular events has been reported with rosiglitazone in 2007.[40,41] The alterations of fluid and electrolyte induced by rosiglitazone significantly contribute to the appearance of cardiovascular diseases, such as heart failure or myocardial ischemia. Thus, in 2010, several drug agencies, including the Food and Drug Administration and the European Medicines Agency, suspended the marketing of rosiglitazone. In November 2013, after a revision of data from a large long-term clinical trial and the support of an outside expert reevaluation, the US Food and Drug Administration decided to remove the prescribing and dispensing restrictions for rosiglitazone. On the other hand, meta-analysis on pioglitazone clinical trials indicates the possibility of ischemic cardiovascular benefits.[42] Robust evidence also shows that both drugs increase the risk of congestive heart failure and fractures, but whether any meaningful difference exists in the magnitude of risk between the two TZDs is not known.[41,43] Recently, new PPAR-γ agonists, such as partial agonists or selective modulators, have been proposed. For instance, INT131 besylate is a potent non-TZD–selective PPAR-γ modulator that showed a dose-dependent reduction in fasting plasma glucose without evoking fluid retention or weight gain.[44]

PPAR-γ agonism induces beneficial effects against metabolic derangements throughout a complex simultaneous modulation of different intracellular signaling pathways. PPAR-γ activation increases the expression and translocation of the glucose transporters GLUT1 and GLUT4 to the cell membrane, thus increasing glucose uptake in muscle cells and adipocytes, reducing consequently glucose plasma levels.[45–47] In adipocytes, PPAR-γ activation causes adipose remodeling as a result of selective preadipocyte differentiation and apoptosis of older and larger insulin-resistant visceral adipocytes.[48]

In addition, PPAR-γ increases blood levels of adipocytokines, such as adiponectin, which are detectable at low concentrations in the plasma of patients with T2D. The increased adiponectin levels improve insulin sensitivity and FFA oxidation and decrease glucose production in liver.[49,50] Experimental studies have demonstrated that PPAR-γ stimulation enhanced adipocyte FFAs uptake and increased the use of glycerol for triglyceride (TG) production, thus reducing FFA release from adipocytes. The reduced FFA levels in skeletal muscle, liver, and pancreas (which are all insulin-sensitive tissues) lead to an improvement in hepatic glucose production and skeletal muscle glucose utilization.[51] We have recently demonstrated that a selective PPAR-γ agonist, pioglitazone, directly affects the insulin signaling pathway, modulating expression and activity of key insulin signaling molecules, including IRS-2, Akt, and GSK-3β, an Akt substrate.[52] PPAR-γ activation improves transduction of the insulin signal pathway by increasing the expression of intracellular proteins, facilitating phosphorylation of key insulin-signaling molecules.[53] Although TZDs may directly affect the insulin signaling pathway, most recent findings suggest that an important contributor to the insulin-sensitizing efficacy of PPAR-γ ligands involves suppression of local and systemic cytokine production. It is noteworthy that the treatment of nondiabetic obese patients with TZDs reduces circulating levels of cytokines and other proinflammatory markers; this effect was associated with improved insulin sensitivity.[54] We reported that hepatic PPAR-γ activation was associated with reduced expression of suppressor of cytokine signaling-3 (SOCS-3), an inhibitor of cytokine signaling, which has been suggested to be a crucial key link between inflammation and hepatic insulin resistance.[52] SOCS-3 can modulate insulin signaling by direct association with the insulin receptor and promotion of IRS-2 degradation.[55,56] At the same time, SOCS-3 participates in a classical negative feedback loop to modulate cytokine-mediated signaling pathways. Other authors have demonstrated that pioglitazone improves the hepatic regenerative response after partial hepatectomy in obese and diabetic mice by preventing an increase in hepatic SOCS-3 messenger RNA (mRNA).[57] Pioglitazone has also been reported to reduce the overexpression of mRNA for tumor necrosis factor-α and both of its receptors in a mouse model of obesity-linked diabetes.[58] Further in vitro studies have confirmed that PPAR-γ agonists may exert their antidiabetic activities by counteracting the deleterious effects of tumor necrosis factor-α.[59]

3.3 Peroxisome Proliferator-Activated Receptor-β/δ

Contrary to PPAR-α and PPAR-γ, PPAR-β/δ is expressed ubiquitously in the majority of the tissues and has important functions in the skin, gut, placenta, skeletal and heart muscles, adipose tissue, and brain.[60]

Various theories have been proposed to define PPAR-β/δ role and the pathways that this receptor activates. Tanaka and colleagues suggested that PPAR-β/δ activation by the selective agonist GW501516 controls FFA oxidation regulating genes involved in FFA transport, β oxidation, and mitochondrial respiration.[61,62] Oliver and colleagues demonstrated that the same PPAR-β/δ agonist lowered plasma insulin levels in a dose-dependent manner, without side effects on glycemic control, in a primate model of metabolic syndrome.[63] In a similar way, GW501516 improved glucose tolerance and insulin resistance in transgenic ob/ob mice.[61] More recently, the production of mediators of lipid peroxidation has been reported to stimulate the secretion of insulin via a PPAR-β/δ–dependent mechanism.[64] Another mechanism of improved insulin sensitivity involves modulation of the signal transducer and activator of transcription 3 (STAT3), a key regulator of cytokines and insulin pathways. The activation of PPAR-β/δ avoids the interleukin-6–mediated induction of the transcription factor STAT3, thereby contributing to preventing the cytokine-mediated development of insulin resistance.[65] PPAR-β/δ may also reduce high blood glucose concentration by increasing glucose flux through the pentose phosphate pathway and inducing

FA synthesis. The coupling between increased hepatic carbohydrate catabolism and the ability to promote β oxidation in muscle allows PPAR-β/δ to regulate metabolic homeostasis and enhance insulin action by complementary effects in distinct tissues.[66] The role of PPAR-β/δ in the regulation of glucose homeostasis became clearer with the findings that PPAR-β/δ agonists reduce adiposity and improve glucose tolerance and insulin sensitivity in different mouse models of obesity.[61] We have recently demonstrated that the selective PPAR-β/δ agonist GW0742 attenuates the metabolic abnormalities and the renal dysfunction/inflammation evoked by chronic consumption of a high sugar diet.[67] Specifically, we showed that the beneficial effects of the PPAR-β/δ ligand GW0742 were due to its ability to counteract the excessive fructose metabolism in the liver by inhibiting fructokinase, which induces the production of a large amount of TG and FFA. Besides, we documented a reduced renal inflammatory response in animals exposed to drug treatment during dietary manipulation because of a reduced expression of the inflammatory complex NLRP3 inflammasome.

Among the three PPARs, PPAR-β/δ is the most common isoform in skeletal muscle. PPAR-β/δ forces skeletal muscle to burn stored fat. Luquet and colleagues demonstrated that muscle-specific PPAR-β/δ overexpression in mice results in a shift to more oxidative fibers and to the increased expression and activity of proteins implicated in oxidative metabolism. These changes in skeletal muscle composition and function are accompanied by a reduction of body fat mass, mainly because of a large reduction of adipose cell size.[68] Similarly, Schuler and colleagues indicated that mice in which PPAR-β/δ is selectively eliminated in skeletal myocytes showed a fiber-type switching toward less oxidative fibers. The authors demonstrated that this process preceded obesity and diabetes being cause and not a consequence of metabolic derangements.[69] The PPAR-β/δ activation by the ligand GW501516 increases muscle fiber type I, rich in mitochondria, which allow mice to undertake long periods of aerobic activity.[70] Similarly, we recently demonstrated that PPAR-β/δ activation by GW0742 protects the skeletal muscle against the metabolic disorders caused by chronic exposure to high concentration of sugars, by affecting multiple levels of the insulin and inflammatory cascades.[71] We observed that GW0742 reverts the diet-induced increase in the nuclear translocation of NF-kB p65, a transcriptional factor that plays an important role in regulating the transcription of a number of genes, especially those involved in producing mediators of local and systemic inflammation (such as cytokines, chemokines, and cell adhesion molecules). The use of PPAR-β/δ knockout mice, in adipose tissue or skeletal muscle, has shown that activated PPAR-β/δ induces the expression of genes involved in FFA oxidation and in energy expenditure through the induction of uncoupling proteins in brown adipose tissue and in skeletal muscle.[72] Despite these intriguing preclinical findings, data from clinical studies are limited and focused mainly on the effects of PPAR-β/δ agonism on lipid metabolism. In humans, 2-week clinical studies in healthy volunteers[73] and moderately overweight subjects[74] demonstrated that the synthetic PPAR-β/δ agonist GW501516 improves dyslipidemia and glucose metabolism (decreasing plasma insulin), whereas liver fat content was reduced. Similar lipid-lowering effects have been reported when another PPAR-β/δ agonists, MXB-8025, has been tested in obese dyslipidemic patients.[75] More recently, combined PPAR-β/δ and PPAR-α activation has been proposed as an innovative pharmacological strategy to improve insulin sensitivity. A dual PPAR-β/δ and PPAR-α agonist, GFT505, has been demonstrated to improve peripheral and hepatic insulin sensitivity in obese insulin-resistant males.[76] Interestingly, the safety profile of GFT505 in this specific clinical setting was reassuring because GFT505 did not evoke body weight gain or fluid retention, in contrast to what is observed when PPAR-γ agonists are administered to diabetic patients. In addition, there were no significant increases in plasma creatinine or homocysteine levels, which are some classical PPAR-α—related side effects.

4. DIETARY-DERIVED PPAR LIGANDS AS SUPPLEMENTARY STRATEGIES IN GLUCOSE CONTROL

PPARs are specialized receptors to detect FFA-derived signal molecules (Figure 1). Dietary FFAs have been regarded solely as energy sources for some time, but they are now also recognized as important regulators of inflammation (often via the modulation of eicosanoid synthesis). The polyunsaturated fatty acids (PUFAs) are among the most extensively studied natural PPAR agonists. PUFAs have a chain that goes from 18 to 22 carbon atoms, with two or more unsaturated bonds. Some of the most important members of this class are ω3-PUFAs (e.g., α-linolenic acid with C18:3, docosahexaenoic acid with C22:6), and ω6-PUFAs (such as linoleic acid with C18:2 and arachidonic acid with C20:4). PUFAs are abundant in many edible oils, such as oils obtained from walnuts, canola, soy, and flaxseed, but especially in fish oil.

Some studies suggest that PPARs mediate some of the beneficial effects evoked by dietary PUFAs, such as the promotion of β oxidation and the reduction of lipogenesis.[77] A few years ago, Neschen and colleagues demonstrated a strong correlation between adiponectin overproduction and PPAR-γ activation in mice

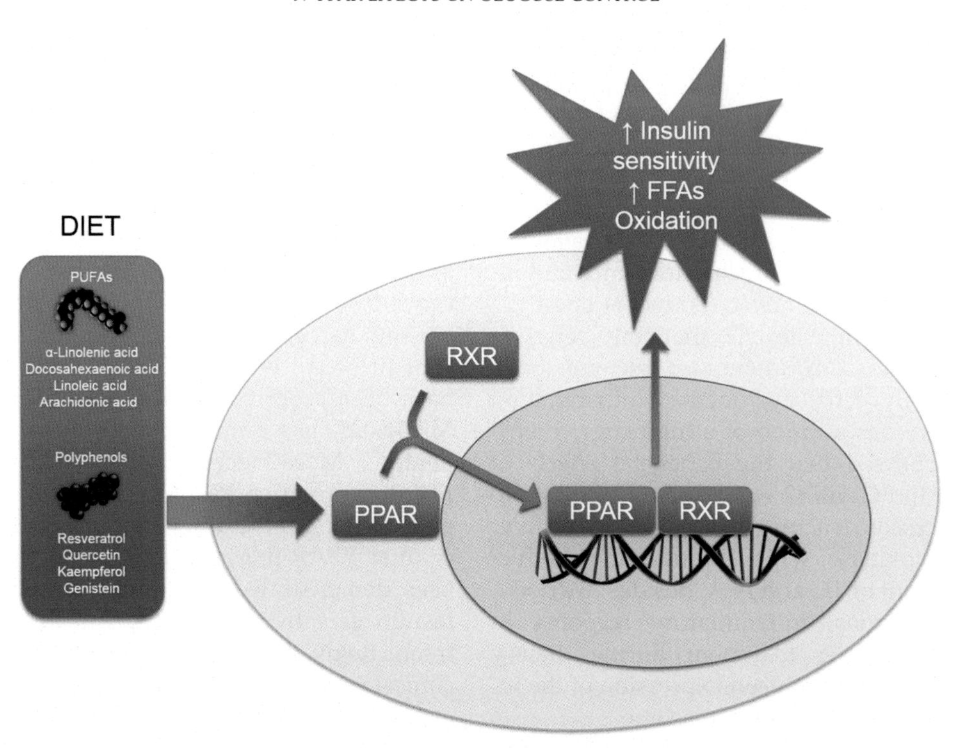

FIGURE 1 Schematic representation of peroxisome proliferator-activated receptor (PPAR) transcriptional activation after binding with polyunsaturated fatty acids or polyphenols originating from the diet. This binding allows the dimerization with retinoid X receptor. The activated transcriptional complex translocates to the nucleus, thus regulating the expression of PPAR target genes. This leads to an improvement in insulin sensitivity and increased free fatty acid oxidation.

chronically exposed to fish oil. Accordingly, no effect of fish oil feeding on adiponectin secretion was recorded in PPAR-α null mice.[78] Not only PUFAs, but also several saturated FFAs, such as stearic acid (C18:0) and myristic acid (C14:0), were found to bind PPAR-α. Optimal binding activity is observed with compounds containing a 16–20 carbon chain length, preferably with several double bonds in the chain.

In vitro studies showed that eicosanoids activate PPARs. PPAR-α is activated by hydroxyeicosatetraenoic acids (with an affinity estimated in the nanomolar-range),[77] leukotriene B4 (the half maximal effective concentration in the micromolar range),[79] and prostacyclin (with an affinity estimated in the micromolar range).[77] Eicosanoids, activating the PPARs, could yield the same benefits as PUFAs, although it is unclear whether they are physiologically relevant ligands for PPARs.[77]

Interestingly, a derivative of the linolenic acid, named 13-oxo-9(Z),11(E),15(Z)-octadecadienoic acid, identified in tomato (*Solanum lycopersicum*), Mandarin orange (*Citrus reticulata*), and bitter gourd (*Momordica charantia*), activated PPAR-γ and induced mRNA expression of PPAR-γ target genes in adipocytes, thus promoting secretion of adiponectin and glucose uptake.[80]

Another important class of dietary-derived ligands of PPARs is the polyphenols (Figure 1). Resveratrol is a natural polyphenol found in grapes, peanuts, and berries. Zhang and colleagues proved that resveratrol markedly reduces the expression of the receptor of advanced glycosylation end-products via PPAR-γ in macrophages.[81] On the other hand, Floyd and colleagues showed that resveratrol inhibits PPAR-γ gene expression in adipocytes, reducing PPAR-γ protein levels and decreasing responsiveness to insulin.[82] Other polyphenols, such as quercetin and kaempferol, are weak partial agonists of PPAR-γ and increase insulin sensitivity and glucose uptake via PPAR-γ agonism.[83] Quercetin has been reported to evoke PPAR-γ activation and induce expression of PPAR-γ target genes in primary human adipocytes.[84]

As demonstrated by Kim et al., polyphenol activity is not limited to PPAR-γ agonism because genistein, the main soy isoflavone, induced the expression of PPAR-α at both mRNA and protein levels in a hepatic cell line.[85]

The search for natural compounds that activate PPARs is not limited to single molecules but takes into account also the screening of whole extracts of natural origin, especially from plants.

Tomato vinegar, a cooking ingredient containing large amounts of phytochemicals, including carotenoids, and polyphenols, has been shown to exert potent antiobesity and anti-insulin resistance effects on high-fat-diet–induced obese mice. These beneficial effects are mediated by PPAR-α upregulation resulting in increased FFA oxidation activity.[86]

Fermented carrot juice, one of the most popular vegetable juices, showed antidiabetic effects on high-fat and low-dose streptozotocin-induced T2D in rats. In diabetic rats, the expression of PPAR-α and PPAR-γ was raised by chronic exposure to the fermented carrot juice and the effects were associated with higher expression levels of proteins involved in lipid and glucose metabolism.[87]

Another interesting natural PPAR agonist is ursolic acid, a pentacyclic triterpenoid present in several traditional medicinal herbs. Jia and colleagues demonstrated that its hypolipidemic and hypoglycemic activities are primarily due to PPAR-α activation and the following induction of hepatic autophagy.[88] Recently, several plants selected from those exhibiting antidiabetic effects were tested for their potential to evoke PPAR agonism. Results showed that extracts of golden root (*Rhodiola rosea*), common elder (*Sambucus nigra*), thyme (*Thymus vulgaris*), and carrot (*Daucus carota*) enhanced glucose uptake in both adipocytes and myotubes by activating PPAR-γ. This molecular mechanism might explain some of the well-known bioactivity of these extracts.[89]

5. CONCLUSIONS

Several studies have convincingly demonstrated that PPARs exert important homeostatic functions in glucose tolerance by affecting expression and activity of several proteins involved in insulin secretion and sensitivity. Besides their "classic" regulatory mechanisms of transcriptional transactivation, which largely depend on direct binding of PPARs to the regulatory regions of target genes involved in the control of lipid and glucose metabolism, PPARs have been demonstrated to exert anti-inflammatory effects via ligand-dependent and ligand-independent transrepression mechanisms. These anti-inflammatory activities lead to reduced synthesis and release of inflammatory markers, which are known to be associated with the development of insulin resistance, thus significantly contribute to counteract the metabolic derangements. Although most preclinical and clinical data focus on the role of the PPAR-γ isoform in the control of glucose metabolism, the evidence presented in this review suggests a key synergic role also for the other two PPAR isoforms, PPAR-α and PPAR-β/δ.

Growing findings show that some foods containing natural compounds, such as PUFAs and polyphenols, are categorized as PPAR agonists. There is a suggestion that PPARs are activated by selective dietary components, many of which also signal through membrane receptors, thereby creating a signaling network between the cell surface and the nucleus, mediated by PPAR agonism. An effective dietary strategy is important in preventing diabetes, managing existing diabetes, and preventing, or at least slowing, the rate of development of diabetes complications. Thus, dietary activation of PPAR by fibers as well as other naturally occurring compounds may represent an efficacious and safe approach for the prevention and amelioration of metabolically related diseases, with limited or no adverse side effects. A better elucidation of the pharmacological interactions between these natural compounds and the PPAR signaling pathway may allow the development of novel dietary and therapeutic strategies for these metabolic diseases that continue to rise in prevalence, with important health and economic implications. Specifically, further efforts in developing effective combination of computational and animal modeling are needed to provide novel avenues for synthesizing and rationalizing existing knowledge on PPAR agonism in a systematic way that accelerates the development of safer and more efficacious nutrition-based therapies. Moreover, identification of specific nutritional approaches aimed to control glucose, lipids, and blood pressure rather than implementing food restrictions can reduce the risk of iatrogenic malnutrition.

References

1. Issemann I, Green S. Activation of a member of the steroid hormone receptor superfamily by peroxisome proliferators. *Nature* October 18, 1990;**347**(6294):645–50.
2. Desvergne B, Wahli W. Peroxisome proliferator-activated receptors: nuclear control of metabolism. *Endocr Rev* October 1999; **20**(5):649–88.
3. Schmidt A, Endo N, Rutledge SJ, Vogel R, Shinar D, Rodan GA. Identification of a new member of the steroid hormone receptor superfamily that is activated by a peroxisome proliferator and fatty acids. *Mol Endocrinol* October 1992;**6**(10):1634–41.
4. Amri EZ, Bonino F, Ailhaud G, Abumrad NA, Grimaldi PA. Cloning of a protein that mediates transcriptional effects of fatty acids in preadipocytes. Homology to peroxisome proliferator-activated receptors. *J Biol Chem* February 3, 1995;**270**(5):2367–71.
5. Shi Y, Hon M, Evans RM. The peroxisome proliferator-activated receptor delta, an integrator of transcriptional repression and nuclear receptor signaling. *Proc Natl Acad Sci USA* March 5, 2002; **99**(5):2613–8.
6. Palmer CN, Hsu MH, Griffin HJ, Johnson EF. Novel sequence determinants in peroxisome proliferator signaling. *J Biol Chem* July 7, 1995;**270**(27):16114–21.
7. Pascual G, Glass CK. Nuclear receptors versus inflammation: mechanisms of transrepression. *Trends Endocrinol Metab* October 2006;**17**(8):321–7.
8. Delerive P, Martin-Nizard F, Chinetti G, Trottein F, Fruchart JC, Najib J, et al. Peroxisome proliferator-activated receptor activators inhibit thrombin-induced endothelin-1 production in human vascular endothelial cells by inhibiting the activator protein-1 signaling pathway. *Circ Res* September 3, 1999;**85**(5):394–402.
9. Lee CH, Chawla A, Urbiztondo N, Liao D, Boisvert WA, Evans RM, et al. Transcriptional repression of atherogenic inflammation: modulation by PPARdelta. *Science* October 17, 2003;**302**(5644):453–7.

10. Barish GD, Atkins AR, Downes M, Olson P, Chong LW, Nelson M, et al. PPARdelta regulates multiple proinflammatory pathways to suppress atherosclerosis. *Proc Natl Acad Sci USA* March 18, 2008; **105**(11):4271–6.
11. Takata Y, Liu J, Yin F, Collins AR, Lyon CJ, Lee CH, et al. PPAR-delta-mediated antiinflammatory mechanisms inhibit angiotensin II-accelerated atherosclerosis. *Proc Natl Acad Sci USA* March 18, 2008;**105**(11):4277–82.
12. Krogsdam AM, Nielsen CA, Neve S, Holst D, Helledie T, Thomsen B, et al. Nuclear receptor corepressor-dependent repression of peroxisome-proliferator-activated receptor delta-mediated transactivation. *Biochem J* April 1, 2002;**363**(Pt 1):157–65.
13. Seedorf U, Assmann G. The role of PPAR alpha in obesity. *Nutr Metab Cardiovasc Dis* June 2001;**11**(3):189–94.
14. Kim BH, Won YS, Kim EY, Yoon M, Nam KT, Oh GT, et al. Phenotype of peroxisome proliferator-activated receptor-alpha (PPAR-alpha) deficient mice on mixed background fed high fat diet. *J Vet Sci* December 2003;**4**(3):239–44.
15. Costet P, Legendre C, More J, Edgar A, Galtier P, Pineau T. Peroxisome proliferator-activated receptor alpha-isoform deficiency leads to progressive dyslipidemia with sexually dimorphic obesity and steatosis. *J Biol Chem* November 6, 1998;**273**(45):29577–85.
16. Vazquez M, Merlos M, Adzet T, Laguna JC. Decreased susceptibility to copper-induced oxidation of rat-lipoproteins after fibrate treatment: influence of fatty acid composition. *Br J Pharmacol* March 1996;**117**(6):1155–62.
17. Patsouris D, Mandard S, Voshol PJ, Escher P, Tan NS, Havekes LM, et al. PPARalpha governs glycerol metabolism. *J Clin Invest* July 2004;**114**(1):94–103.
18. Wu P, Peters JM, Harris RA. Adaptive increase in pyruvate dehydrogenase kinase 4 during starvation is mediated by peroxisome proliferator-activated receptor alpha. *Biochem Biophys Res Commun* September 21, 2001;**287**(2):391–6.
19. Maida A, Lamont BJ, Cao X, Drucker DJ. Metformin regulates the incretin receptor axis via a pathway dependent on peroxisome proliferator-activated receptor-alpha in mice. *Diabetologia* February 2011;**54**(2):339–49.
20. Koh EH, Kim MS, Park JY, Kim HS, Youn JY, Park HS, et al. Peroxisome proliferator-activated receptor (PPAR)-alpha activation prevents diabetes in OLETF rats: comparison with PPAR-gamma activation. *Diabetes* September 2003;**52**(9):2331–7.
21. Aasum E, Belke DD, Severson DL, Riemersma RA, Cooper M, Andreassen M, et al. Cardiac function and metabolism in Type 2 diabetic mice after treatment with BM 17.0744, a novel PPAR-alpha activator. *Am J Physiol Heart Circ Physiol* September 2002;**283**(3):H949–57.
22. Park CW, Zhang Y, Zhang X, Wu J, Chen L, Cha DR, et al. PPAR-alpha agonist fenofibrate improves diabetic nephropathy in db/db mice. *Kidney Int* May 2006;**69**(9):1511–7.
23. Hostetler HA, Huang H, Kier AB, Schroeder F. Glucose directly links to lipid metabolism through high affinity interaction with peroxisome proliferator-activated receptor alpha. *J Biol Chem* January 25, 2008;**283**(4):2246–54.
24. Lefebvre P, Chinetti G, Fruchart JC, Staels B. Sorting out the roles of PPAR alpha in energy metabolism and vascular homeostasis. *J Clin Invest* March 2006;**116**(3):571–80.
25. Lalloyer F, Vandewalle B, Percevault F, Torpier G, Kerr-Conte J, Oosterveer M, et al. Peroxisome proliferator-activated receptor alpha improves pancreatic adaptation to insulin resistance in obese mice and reduces lipotoxicity in human islets. *Diabetes* June 2006; **55**(6):1605–13.
26. Ravnskjaer K, Boergesen M, Rubi B, Larsen JK, Nielsen T, Fridriksson J, et al. Peroxisome proliferator-activated receptor alpha (PPARalpha) potentiates, whereas PPARgamma attenuates, glucose-stimulated insulin secretion in pancreatic beta-cells. *Endocrinology* August 2005;**146**(8):3266–76.
27. Ferrari C, Frezzati S, Romussi M, Bertazzoni A, Testori GP, Antonini S, et al. Effects of short-term clofibrate administration on glucose tolerance and insulin secretion in patients with chemical diabetes or hypertriglyceridemia. *Metabolism* February 1977;**26**(2):129–39.
28. Murakami K, Nambu S, Koh H, Kobayashi M, Shigeta Y. Clofibrate enhances the affinity of insulin receptors in non-insulin dependent diabetes mellitus. *Br J Clin Pharmacol* January 1984;**17**(1):89–91.
29. Kobayashi M, Shigeta Y, Hirata Y, Omori Y, Sakamoto N, Nambu S, et al. Improvement of glucose tolerance in NIDDM by clofibrate. Randomized double-blind study. *Diabetes Care* June 1988;**11**(6): 495–9.
30. Panz VR, Wing JR, Raal FJ, Kedda MA, Joffe BI. Improved glucose tolerance after effective lipid-lowering therapy with bezafibrate in a patient with lipoatrophic diabetes mellitus: a putative role for Randle's cycle in its pathogenesis? *Clin Endocrinol (Oxf)* March 1997;**46**(3):365–8.
31. Anderlova K, Dolezalova R, Housova J, Bosanska L, Haluzikova D, Kremen J, et al. Influence of PPAR-alpha agonist fenofibrate on insulin sensitivity and selected adipose tissue-derived hormones in obese women with type 2 diabetes. *Physiol Res* 2007;**56**(5):579–86.
32. Bajaj M, Suraamornkul S, Hardies LJ, Glass L, Musi N, DeFronzo RA. Effects of peroxisome proliferator-activated receptor (PPAR)-Alpha and PPAR-gamma agonists on glucose and lipid metabolism in patients with type 2 diabetes mellitus. *Diabetologia* August 2007;**50**(8):1723–31.
33. Flavell DM, Ireland H, Stephens JW, Hawe E, Acharya J, Mather H, et al. Peroxisome proliferator-activated receptor alpha gene variation influences age of onset and progression of type 2 diabetes. *Diabetes* February 2005;**54**(2):582–6.
34. Andrulionyte L, Kuulasmaa T, Chiasson JL, Laakso M, STOP-NIDDM Study Group. Single nucleotide polymorphisms of the peroxisome proliferator-activated receptor-alpha gene (PPARA) influence the conversion from impaired glucose tolerance to type 2 diabetes: the STOP-NIDDM trial. *Diabetes* April 2007; **56**(4):1181–6.
35. Tontonoz P, Spiegelman BM. Fat and beyond: the diverse biology of PPARgamma. *Annu Rev Biochem* 2008;**77**:289–312.
36. Yamamoto Y, Yamazaki H, Ikeda T, Watanabe T, Iwabuchi H, Nakajima M, et al. Formation of a novel quinone epoxide metabolite of troglitazone with cytotoxicity to HepG2 cells. *Drug Metab Dispos* February 2002;**30**(2):155–60.
37. Buchanan TA, Xiang AH, Peters RK, Kjos SL, Marroquin A, Goico J, et al. Preservation of pancreatic beta-cell function and prevention of type 2 diabetes by pharmacological treatment of insulin resistance in high-risk hispanic women. *Diabetes* 2002;**51**(9):2796–803.
38. Knowler WC, Hamman RF, Edelstein SL, Barrett-Connor E, Ehrmann DA, Walker EA, Diabetes Prevention Program Research Group, et al. Prevention of type 2 diabetes with troglitazone in the Diabetes Prevention Program. *Diabetes* 2005;**54**(4):1150–6.
39. Xiang AH, Peters RK, Kjos SL, Marroquin A, Goico J, Ochoa C, et al. Effect of pioglitazone on pancreatic beta-cell function and diabetes risk in hispanic women with prior gestational diabetes. *Diabetes* 2006;**55**(2):517–22.
40. Nissen SE, Wolski K. Rosiglitazone revisited: an updated meta-analysis of risk for myocardial infarction and cardiovascular mortality. *Arch Intern Med* July 26, 2010;**170**(14):1191–201.
41. Singh S, Loke YK, Furberg CD. Long-term risk of cardiovascular events with rosiglitazone: a meta-analysis. *JAMA* September 12, 2007;**298**(10):1189–95.
42. Lincoff AM, Wolski K, Nicholls SJ, Nissen SE. Pioglitazone and risk of cardiovascular events in patients with type 2 diabetes mellitus: a meta-analysis of randomized trials. *JAMA* September 12, 2007; **298**(10):1180–8.
43. Loke YK, Singh S, Furberg CD. Long-term use of thiazolidinediones and fractures in type 2 diabetes: a meta-analysis. *CMAJ* January 6, 2009;**180**(1):32–9.

44. Dunn FL, Higgins LS, Fredrickson J, DePaoli AM, I. N. T. study group. Selective modulation of PPARgamma activity can lower plasma glucose without typical thiazolidinedione side-effects in patients with type 2 diabetes. *J Diabetes Complications* May-Jun 2011;**25**(3):151–8.
45. Standaert ML, Kanoh Y, Sajan MP, Bandyopadhyay G, Farese RV. Cbl, IRS-1, and IRS-2 mediate effects of rosiglitazone on PI3K, PKC-lambda, and glucose transport in 3T3/L1 adipocytes. *Endocrinology* May 2002;**143**(5):1705–16.
46. Kramer D, Shapiro R, Adler A, Bush E, Rondinone CM. Insulin-sensitizing effect of rosiglitazone (BRL-49653) by regulation of glucose transporters in muscle and fat of Zucker rats. *Metabolism* November 2001;**50**(11):1294–300.
47. Kim HI, Ahn YH. Role of peroxisome proliferator-activated receptor-gamma in the glucose-sensing apparatus of liver and beta-cells. *Diabetes* February 2004;**53**(Suppl. 1):S60–5.
48. Arner P. The adipocyte in insulin resistance: key molecules and the impact of the thiazolidinediones. *Trends Endocrinol Metab* April 2003;**14**(3):137–45.
49. Yamauchi T, Kamon J, Waki H, Terauchi Y, Kubota N, Hara K, et al. The fat-derived hormone adiponectin reverses insulin resistance associated with both lipoatrophy and obesity. *Nat Med* August 2001;**7**(8):941–6.
50. Pajvani UB, Hawkins M, Combs TP, Rajala MW, Doebber T, Berger JP, et al. Complex distribution, not absolute amount of adiponectin, correlates with thiazolidinedione-mediated improvement in insulin sensitivity. *J Biol Chem* March 26, 2004;**279**(13): 12152–62.
51. Armoni M, Harel C, Bar-Yoseph F, Milo S, Karnieli E. Free fatty acids repress the GLUT4 gene expression in cardiac muscle via novel response elements. *J Biol Chem* October 14, 2005;**280**(41): 34786–95.
52. Collino M, Aragno M, Castiglia S, Miglio G, Tomasinelli C, Boccuzzi G, et al. Pioglitazone improves lipid and insulin levels in overweight rats on a high cholesterol and fructose diet by decreasing hepatic inflammation. *Br J Pharmacol* August 2010; **160**(8):1892–902.
53. Baumann CA, Chokshi N, Saltiel AR, Ribon V. Cloning and characterization of a functional peroxisome proliferator activator receptor-gamma-responsive element in the promoter of the CAP gene. *J Biol Chem* March 31, 2000;**275**(13):9131–5.
54. Samaha FF, Szapary PO, Iqbal N, Williams MM, Bloedon LT, Kochar A, et al. Effects of rosiglitazone on lipids, adipokines, and inflammatory markers in nondiabetic patients with low high-density lipoprotein cholesterol and metabolic syndrome. *Arterioscler Thromb Vasc Biol* March 2006;**26**(3):624–30.
55. Rui L, Yuan M, Frantz D, Shoelson S, White MF. SOCS-1 and SOCS-3 block insulin signaling by ubiquitin-mediated degradation of IRS1 and IRS2. *J Biol Chem* November 1, 2002;**277**(44):42394–8.
56. Farrell GC. Signalling links in the liver: knitting SOCS with fat and inflammation. *J Hepatol* July 2005;**43**(1):193–6.
57. Aoyama T, Ikejima K, Kon K, Okumura K, Arai K, Watanabe S. Pioglitazone promotes survival and prevents hepatic regeneration failure after partial hepatectomy in obese and diabetic KK-A(y) mice. *Hepatology* May 2009;**49**(5):1636–44.
58. Hofmann C, Lorenz K, Braithwaite SS, Colca JR, Palazuk BJ, Hotamisligil GS, et al. Altered gene expression for tumor necrosis factor-alpha and its receptors during drug and dietary modulation of insulin resistance. *Endocrinology* January 1994;**134**(1):264–70.
59. Szalkowski D, White-Carrington S, Berger J, Zhang B. Antidiabetic thiazolidinediones block the inhibitory effect of tumor necrosis factor-alpha on differentiation, insulin-stimulated glucose uptake, and gene expression in 3T3-L1 cells. *Endocrinology* April 1995; **136**(4):1474–81.
60. Berger J, Leibowitz MD, Doebber TW, Elbrecht A, Zhang B, Zhou G, et al. Novel peroxisome proliferator-activated receptor (PPAR) gamma and PPARdelta ligands produce distinct biological effects. *J Biol Chem* March 5, 1999;**274**(10):6718–25.
61. Tanaka T, Yamamoto J, Iwasaki S, Asaba H, Hamura H, Ikeda Y, et al. Activation of peroxisome proliferator-activated receptor delta induces fatty acid beta-oxidation in skeletal muscle and attenuates metabolic syndrome. *Proc Natl Acad Sci USA* December 23, 2003; **100**(26):15924–9.
62. Planavila A, Laguna JC, Vazquez-Carrera M. Nuclear factor-kappaB activation leads to down-regulation of fatty acid oxidation during cardiac hypertrophy. *J Biol Chem* April 29, 2005;**280**(17): 17464–71.
63. Oliver Jr WR, Shenk JL, Snaith MR, Russell CS, Plunket KD, Bodkin NL, et al. A selective peroxisome proliferator-activated receptor delta agonist promotes reverse cholesterol transport. *Proc Natl Acad Sci USA* April 24, 2001;**98**(9):5306–11.
64. Cohen G, Riahi Y, Shamni O, Guichardant M, Chatgilialoglu C, Ferreri C, et al. Role of lipid peroxidation and PPAR-delta in amplifying glucose-stimulated insulin secretion. *Diabetes* November 2011;**60**(11):2830–42.
65. Serrano-Marco L, Rodriguez-Calvo R, El Kochairi I, Palomer X, Michalik L, Wahli W, et al. Activation of peroxisome proliferator-activated receptor-beta/-delta (PPAR-beta/-delta) ameliorates insulin signaling and reduces SOCS3 levels by inhibiting STAT3 in interleukin-6-stimulated adipocytes. *Diabetes* July 2011;**60**(7): 1990–9.
66. Lee CH, Olson P, Hevener A, Mehl I, Chong LW, Olefsky JM, et al. PPARdelta regulates glucose metabolism and insulin sensitivity. *Proc Natl Acad Sci USA* February 28, 2006;**103**(9):3444–9.
67. Collino M, Benetti E, Rogazzo M, Mastrocola R, Yaqoob MM, Aragno M, et al. Reversal of the deleterious effects of chronic dietary HFCS-55 intake by PPAR-delta agonism correlates with impaired NLRP3 inflammasome activation. *Biochem Pharmacol* January 15, 2013;**85**(2):257–64.
68. Luquet S, Lopez-Soriano J, Holst D, Fredenrich A, Melki J, Rassoulzadegan M, et al. Peroxisome proliferator-activated receptor delta controls muscle development and oxidative capability. *FASEB J* December 2003;**17**(15):2299–301.
69. Schuler M, Ali F, Chambon C, Duteil D, Bornert JM, Tardivel A, et al. PGC1alpha expression is controlled in skeletal muscles by PPARbeta, whose ablation results in fiber-type switching, obesity, and type 2 diabetes. *Cell Metab* November 2006;**4**(5):407–14.
70. Smeets PJ, Teunissen BE, Planavila A, de Vogel-van den Bosch H, Willemsen PH, van der Vusse GJ, et al. Inflammatory pathways are activated during cardiomyocyte hypertrophy and attenuated by peroxisome proliferator-activated receptors PPARalpha and PPARdelta. *J Biol Chem* October 24, 2008;**283**(43):29109–18.
71. Benetti E, Mastrocola R, Rogazzo M, Chiazza F, Aragno M, Fantozzi R, et al. High sugar intake and development of skeletal muscle insulin resistance and inflammation in mice: a protective role for PPAR-delta agonism. *Mediators Inflamm* 2013;**2013**:509502.
72. Wang YX, Lee CH, Tiep S, Yu RT, Ham J, Kang H, et al. Peroxisome-proliferator-activated receptor delta activates fat metabolism to prevent obesity. *Cell* April 18, 2003;**113**(2):159–70.
73. Sprecher DL, Massien C, Pearce G, Billin AN, Perlstein I, Willson TM, et al. Triglyceride:high-density lipoprotein cholesterol effects in healthy subjects administered a peroxisome proliferator activated receptor delta agonist. *Arterioscler Thromb Vasc Biol* 2007;**27**(2):359–65.
74. Riserus U, Sprecher D, Johnson T, Olson E, Hirschberg S, Liu A, et al. Activation of peroxisome proliferator-activated receptor (PPAR)delta promotes reversal of multiple metabolic abnormalities, reduces oxidative stress, and increases fatty acid oxidation in moderately obese men. *Diabetes* 2008;**57**(2):332–9.
75. Choi YJ, Roberts BK, Wang X, Geaney JC, Naim S, Wojnoonski K, et al. Effects of the PPAR-delta agonist MBX-8025 on atherogenic dyslipidemia. *Atherosclerosis* 2012;**220**(2):470–6.

76. Cariou B, Hanf R, Lambert-Porcheron S, Zair Y, Sauvinet V, Noel B, et al. Dual peroxisome proliferator-activated receptor alpha/delta agonist GFT505 improves hepatic and peripheral insulin sensitivity in abdominally obese subjects. *Diabetes Care* 2013;**36**(10):2923–30.
77. Forman BM, Chen J, Evans RM. Hypolipidemic drugs, polyunsaturated fatty acids, and eicosanoids are ligands for peroxisome proliferator-activated receptors alpha and delta. *Proc Natl Acad Sci USA* April 29, 1997;**94**(9):4312–7.
78. Neschen S, Morino K, Rossbacher JC, Pongratz RL, Cline GW, Sono S, et al. Fish oil regulates adiponectin secretion by a peroxisome proliferator-activated receptor-gamma-dependent mechanism in mice. *Diabetes* April 2006;**55**(4):924–8.
79. Devchand PR, Keller H, Peters JM, Vazquez M, Gonzalez FJ, Wahli W. The PPARalpha-leukotriene B4 pathway to inflammation control. *Nature* November 7, 1996;**384**(6604):39–43.
80. Takahashi H, Hara H, Goto T, Kamakari K, Wataru N, Mohri S, et al. 13-Oxo-9(Z),11(E),15(Z)-octadecatrienoic acid activates peroxisome proliferator-activated receptor gamma in adipocytes. *Lipids* November 27, 2014.
81. Zhang Y, Luo Z, Ma L, Xu Q, Yang Q, Si L. Resveratrol prevents the impairment of advanced glycosylation end products (AGE) on macrophage lipid homeostasis by suppressing the receptor for AGE via peroxisome proliferator-activated receptor gamma activation. *Int J Mol Med* May 2010;**25**(5):729–34.
82. Floyd ZE, Wang ZQ, Kilroy G, Cefalu WT. Modulation of peroxisome proliferator-activated receptor gamma stability and transcriptional activity in adipocytes by resveratrol. *Metabolism* July 2008;**57**(7 Suppl. 1):S32–8.
83. Fang XK, Gao J, Zhu DN. Kaempferol and quercetin isolated from *Euonymus alatus* improve glucose uptake of 3T3-L1 cells without adipogenesis activity. *Life Sci* March 12, 2008; **82**(11–12):615–22.
84. Chuang CC, Martinez K, Xie G, Kennedy A, Bumrungpert A, Overman A, et al. Quercetin is equally or more effective than resveratrol in attenuating tumor necrosis factor-{alpha}-mediated inflammation and insulin resistance in primary human adipocytes. *Am J Clin Nutr* December 2010;**92**(6):1511–21.
85. Kim S, Shin HJ, Kim SY, Kim JH, Lee YS, Kim DH, et al. Genistein enhances expression of genes involved in fatty acid catabolism through activation of PPARalpha. *Mol Cell Endocrinol* May 31, 2004;**220**(1–2):51–8.
86. Seo KI, Lee J, Choi RY, Lee HI, Lee JH, Jeong YK, et al. Anti-obesity and anti-insulin resistance effects of tomato vinegar beverage in diet-induced obese mice. *Food Funct* July 25, 2014;**5**(7):1579–86.
87. Li C, Ding Q, Nie S, Zhang YS, Xiong T, Xie MY. Carrot juice fermented with *Lactobacillus plantarum* Ncu116 ameliorates type 2 diabetes in rats. *J Agric Food Chem* October 22, 2014.
88. Jia Y, Kim S, Kim J, Kim B, Wu C, Lee JH, et al. Ursolic acid improves lipid and glucose metabolism in high-fat-fed C57bl/6j mice by activating peroxisome proliferator-activated receptor alpha and hepatic autophagy. *Mol Nutr Food Res* November 24, 2014;**59**(2):344–54.
89. El-Houri RB, Kotowska D, Olsen LC, Bhattacharya S, Christensen LP, Grevsen K, et al. Screening for bioactive metabolites in plant extracts modulating glucose uptake and fat accumulation. *Evid Based Complement Altern Med* 2014;**2014**:156398.

CHAPTER

10

High-Fat Diets and β-Cell Dysfunction: Molecular Aspects

Carla Beatriz Collares-Buzato

Department of Biochemistry and Tissue Biology, Institute of Biology, University of Campinas, Campinas, SP, Brazil

1. INTRODUCTION

Type 2 diabetes mellitus (T2D) is a multifactorial metabolic disease characterized by a chronic state of hyperglycemia caused by defects in the production as well as in the peripheral action of insulin.[1–4] T2D is the most frequent form of diabetes, corresponding to 90–95% of cases, and is a rapidly growing global epidemic that already affects more than 300 million people worldwide.[3,4]

The pancreatic β cell plays a central role in T2D pathogenesis, being capable of an adaptive response that involves alteration in its secretory function and in the relative mass of the pancreas.[5–9] To compensate for the resistance of the peripheral tissues to insulin, β cells initially increase biosynthesis and release insulin followed by an incremental increase of β-cell mass by hypertrophy and/or hyperplasia.[6,7,10,11] When the β cells fail to maintain normoglycemia, T2D is triggered. The continuous demand for insulin hypersecretion in association with long-term exposition to hyperglycemia/dyslipidemia and a chronic inflammatory state within the pancreatic islet milieu result in secretory impairment and then β-cell death by apoptosis, which leads to an irreversible body dependence on exogenous insulin.[8,9,12,13]

Among all the genetic and environmental risk factors known to lead to T2D, obesity seems to be the most important factor, being itself another epidemic threat to public health in almost every country of the world.[1,6,14,15] Obesity is a complex interplay of lifestyle and related genes, but is invariably associated with intake of overly high caloric diets rich in carbohydrates and lipids.[16]

Several in vivo and in vitro models have been proposed for investigating the link between obesity and T2D. Among them, the use of animals fed high-fat diets (HFDs) and models of in vitro exposure of β cells to free fatty acids (FFAs) seem to be the two approaches most often used in studies with this goal.[5,16–20] These experimental models are based on the fact that dyslipidemia, in which elevated fatty acid concentration is a component, is a feature of obesity as well as T2D.[4,19,21] Fatty acids may be more readily available in the insulin-resistant state of obesity when insulinemia is insufficient to appropriately suppress lipolysis, particularly if the adipocyte is also resistant to the hormone.[19,21]

A comparison of these HFD studies is sometimes difficult because of the differences in nutritional composition and fat content of the diets combined with differences in animal species, strains, gender, and time frame of exposure to the diet used.[5,17,18,20,22–24] In addition, the relevance of animal-based HFD studies in translating the human obesity/diabetes state has recently been questioned.[16] Despite of all of these considerations, the wealth of information obtained from these works using HFD cannot be denied. They have shed light on our knowledge of the repercussion of high-lipid content exposure in developing and establishing insulin resistance of peripheral tissues/organs (mainly adipose tissue, muscles, and liver), a condition known to occur at a very early phase of T2D.[1,2]

In this chapter, we address a topic that has recently received additional attention: the role of HFD exposure, and particularly of FFAs, in triggering the initial adaptive/compensatory response and later functional failure of β cells during T2D. This subject will be treated in light of the biology of the β cell, focusing on molecular aspects of the β-cell alterations elicited by in vivo and in vitro exposure to a high concentrations of lipids.

Molecular Nutrition and Diabetes
http://dx.doi.org/10.1016/B978-0-12-801585-8.00010-5

2. BIOLOGY OF THE β CELL

The β cell is one of the endocrine cells that forms the parenchyma of the pancreatic islets, which in turn constitute the morphofunctional units of the endocrine pancreas (representing 1–2% of total pancreatic mass). Within the islets, β cells and four other different endocrine cells (namely, α, δ, PP, and ε cells) are arranged in anastomosing cords or plates pervaded by a labyrinthine network of fenestrated capillaries.[25] The islets are demarcated from the surrounding acinar tissue by a thin layer of reticular fibers that extend inward to form a delicate layer around the capillaries (Figure 1(a)).

The arrangement of the endocrine cells, known as the cytoarchitecture, is another important structural feature of the islet and displays species specificity.[20,25–27] In rodents, β cells, the most frequent among all endocrine pancreatic cells, occupy the islet core surrounded by a peripheral mantle formed by the other cell types (Figure 1(b)). Meanwhile, in human islets, β cells occupy preferentially the center of trilaminar plates (that form the structure of the islet cell cords) borded by non-β cells on both sides.[25,28] Both islet structure and cytoarchitecture, which are crucial for the endocrine pancreatic function, depend on the cell junctions and their constitutive proteins that allow the cell–cell recognition/segregation, adhesion, and communication among the endocrine islet cells (Figure 1(c)).[29–31]

Ultrastructurally, β cells typically display characteristics of endocrine cells that secrete protein/polypeptide hormones.[25] These include a nucleus with euchromatin and prominent nucleolus with a well-developed rough endoplasmic reticulum (ER) and Golgi complex (GC) as well as several electrodense secretory granules that store hormone molecules within the cytoplasm until their release when triggered by determined stimulus (Figure 1(d)). The β cell secretes insulin that, along with the other islet hormones (i.e., the α cell–secreted glucagon, the δ cell–secreted somatostatin, the PP cell–secreted pancreatic polypeptide, and the ε cell–secreted ghrelin), regulates carbohydrate, protein, and lipid metabolism. Insulin, once released within the blood circulation, acts on all cells/tissues (except the red blood cells, neurons, and β cells themselves) by interacting with a membrane-specific receptor and triggering an intracellular signaling pathway that culminates with the GLUT-mediated entry of circulating glucose into the cell.[1] This consequently leads to a reduction in blood glucose concentrations.

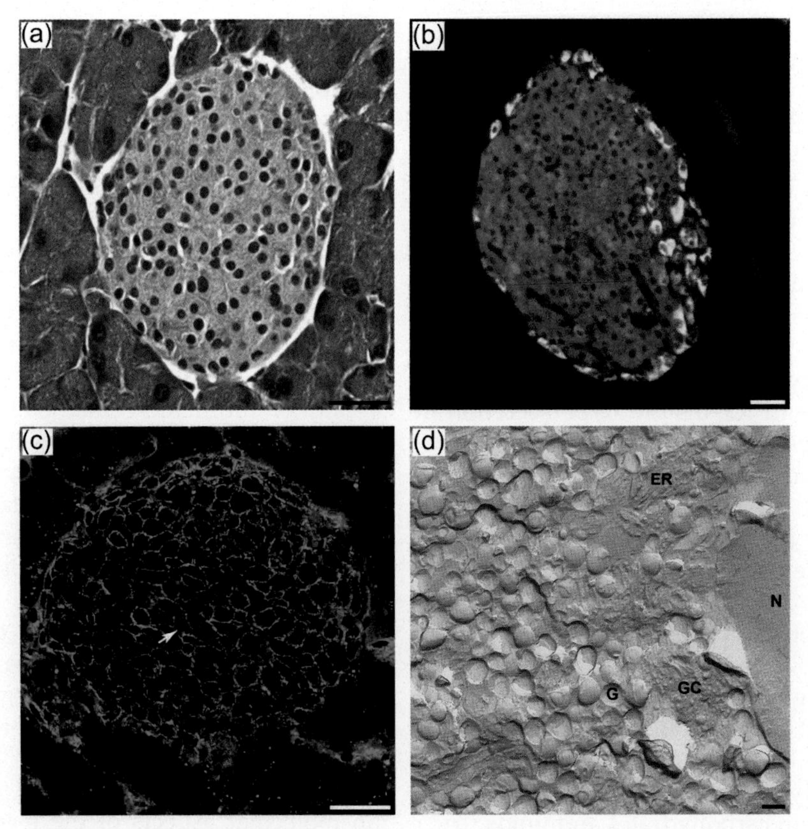

FIGURE 1 Pancreatic islets from mouse stained with hematoxylin-eosin (a), coimmunolabeled for insulin (red) and glucagon (yellow) (b), or immunolabeled for pan-cadherin to display cell–cell contacts (c, arrow). (d) Image of β cell obtained by scanning electron microscopy. ER, rough endoplasmic reticulum; G, insulin granules; GC, Golgi complex; N, nucleus. Bars, 50 μm (a); 25 μm (b,c); 400 nm (d).

Insulin is a polypeptide that is synthesized as part of a larger protein[32,33] (Figure 2). In the rough ER, the messenger RNA (mRNA) transcript is translated into an inactive protein called preproinsulin that is sequentially composed of the signal peptide (S), insulin B domain, C domain flanked by dibasic cleavage sites, and insulin A domain (Figure 2, steps 1a, 2a). The amino-terminal signal sequence of the preproinsulin is required for the precursor hormone to cross the membrane of the ER for posttranslational processing. Upon entering the ER, the preproinsulin signal sequence is proteolytically removed by a signal peptidase to form proinsulin (Figure 2, step 3a), which then folds, forming three disulfide bonds that are conserved among the entire insulin/insulin-like growth factor superfamily (Figure 2, step 4a). Proinsulin forms noncovalently associated homodimers that undergo intracellular transport from the ER to the GC (Figure 2, step 5a) and then into secretory granules, during which proinsulin forms hexamers and is proteolytically processed to C-peptide and mature insulin (Figure 2, step 6a). These products are packaged and stored in the secretory granules (in the presence of zinc) until their release by exocytosis by stimuli.[32]

Several stimuli have been shown to trigger insulin release by β cells; among them, glucose seems to be the major secretagogue.[3,32,33] Glucose is taken up via facilitated diffusion mediated by GLUT2 in rodents and probably GLUT1 in human,[34] that unlike other GLUT subtypes do not require mobilization by insulin (Figure 2, step 1b). Once inside, glucose is first converted to glucose-6-phosphate by glucokinase and then by glycolysis to pyruvate, which enters the mitochondria (Figure 2, step 2b). Metabolism of glucose yields an increased adenosine triphosphate (ATP) level (Figure 2, step 3b), which contributes to closure of the ATP-dependent K^+ channel (known as the K_{ATP} channel) (Figure 2, step 4b). This channel is made up of four central subunits of Kir6.2 (gene name KCNJ11) and four subunits of sulfonylurea receptor (SUR1). The Kir6.2 subunits form the potassium channel itself, whereas the SUR1 membrane proteins surround and influence the pore. Once the K_{ATP} channel is closed, the membrane potential is depolarized, action potential triggered, and L-type voltage-dependent Ca^{2+} channels are opened (Figure 2, step 5b). In response to this rise in intracellular Ca^{2+}, granules fuse with the plasma membrane in a soluble *N*-ethylmaleimide-sensitive factor

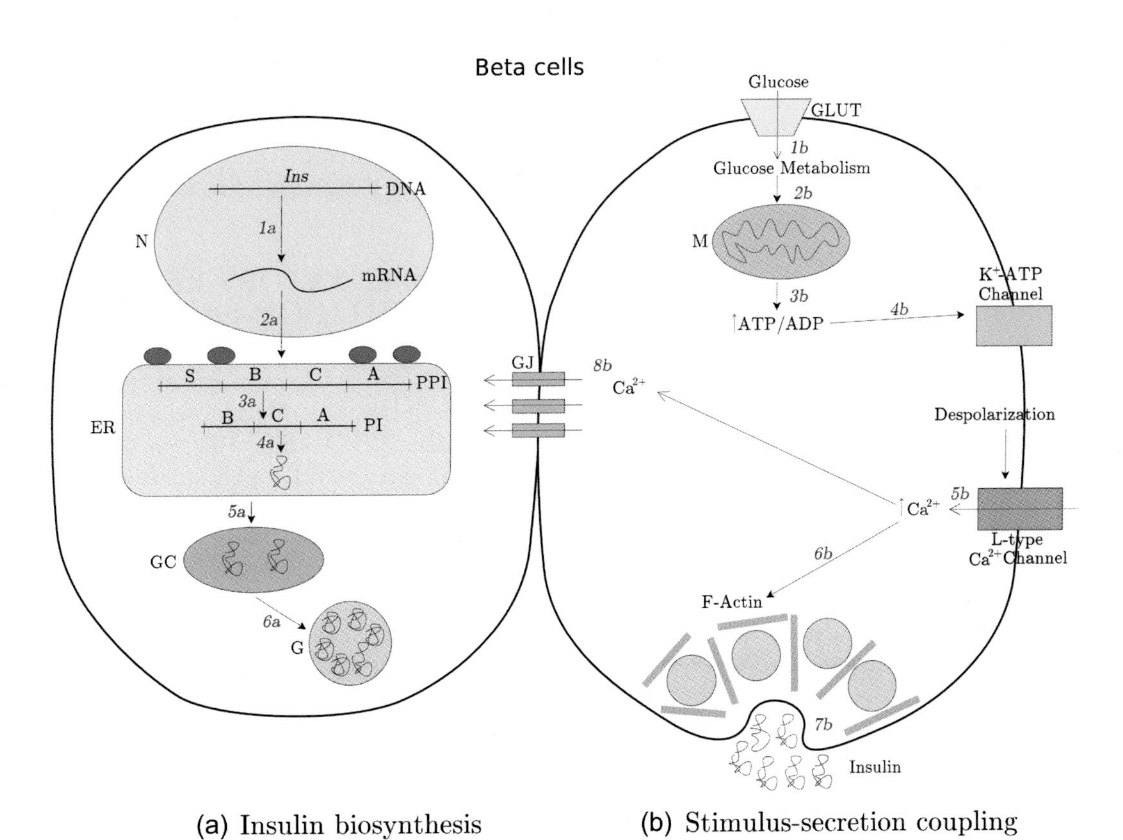

FIGURE 2 Schematic diagram showing the main steps of insulin biosynthesis and secretion. See the text for details. ADP, adenosine diphosphate; ATP, adenosine triphosphate; G, granule; GC, Golgi complex; GJ, gap junction; GLUT, glucose transporter; K^+-ATP, ATP-dependent K^+ channel; mRNA, messenger RNA; PI, proinsulin; PPI, preproinsulin; ER, rough endoplasmic reticulum.

attachment protein receptor (SNARE)-dependent process.[35] This step involves the physical interaction of the vesicular- and target-membrane (v- and t-) SNARE proteins allowing the docking of the insulin-containing granules to the membrane.[35]

The entire process of stimulated insulin secretion is known as the stimulus-secretion coupling. This involves two phases: (1) an initial peak of secretion due to rapid exocytosis, that occurs within 5–10 min of stimulation, known as the first phase, and (2) a subsequent, less vigorous but sustained release, referred to as second-phase secretion, that can last for hours, as long as glucose levels remain high.[32,36] It has been proposed that the first-phase results from the rapid fusion of granules that are predocked at the plasma membrane, also known as "readily releasable pool" of granules.[36] Simultaneously, granules that are deeper within the cell (referred to as the "storage-granule pool") are trafficked to the cell periphery to replenish the "readily releasable pool" at the cell surface, guaranteeing the insulin release during the second phase.

The exocytosis of secretory granules depends on cytoskeleton remodeling, particularly on the cortical actin microfilaments that form an extensive three-dimensional meshwork underneath the plasma membrane.[36] Under basal concentrations of glucose (2–4 mM), cytoskeletal microfilaments seem to function not only as a barrier to block SNARE-complex formation, but also yield tracks for the transportation of secretory granules to the plasma membrane. Glucose-induced Ca^{2+} increase triggers transient reorganization of the cortical actin network (Figure 2, step 6b), mediated by small guanosine triphosphatases (Rho, Rab, and Ras), to allow the granules access to the plasma membrane for subsequent docking, membrane fusion (mediated by interactions between VAMP-2 [a v-SNARE] and syntaxin [a t-SNARE]) and insulin release (Figure 2, step 7b).[32,35,36]

Although insulin biosynthesis and secretion are interrelated processes, the glucose concentration thresholds required to set off insulin release are different from that for biosynthesis. Insulin secretion is triggered by supraliminal concentrations of glucose (>5.6 mM), whereas its biosynthesis is sensitive to changes in glucose within the basal concentration range of this monosaccharide (between 2 and 4 mM).[33,37] Thus, insulin biosynthesis allows a continuous replenishment of insulin granule stores even at normal physiological glucose concentrations. Insulin biosynthesis alone accounts for more than 10% of total protein synthesis in β cell under basal conditions, and this percentage can further increase up to 50% under stimulated conditions.[33,37] As a result of this high demand, proinsulin folding in β cells is very sensitive to changes in the ER environment, and increasing demand for proinsulin synthesis and folding makes the β cell one of the cell types that is most susceptible to ER stress.[38–40] Meanwhile, the glucose metabolism and ATP production are major limiting steps of the glucose-stimulated insulin secretion (GSIS) and, therefore, particularly vulnerable to mitochondrial dysfunction and oxidative stress.[41,42]

Another important feature of the β-cell biology that also plays a role in both processes of insulin biosynthesis and secretion is the degree of β cell–β cell coupling mediated by gap junctions (GJs). These are one of the intercellular junctions and composed of several intercellular channels made of integral proteins belonging to the connexin family.[43] In the case of β cells, these channels are formed of the Cx36 subtype and allow the exchange of ions and small molecules (monosaccharides, amino acids, signaling messengers such as inositol trisphosphate, cyclic adenosine monophosphate/cyclic guanosine monophosphate, and even microRNAs) between adjacent cells so that they are coupled electrically and metabolically.[29,43] During the stimulus-secretion coupling process, gap junctional channels are believed to allow mainly the passage of Ca^{2+} between adjacent cells and therefore to synchronize Ca^{2+} oscillations among coupled β cells in the pancreatic islets (Figure 2, step 8b). This effect optimizes the insulin secretion process by decreasing the functional heterogeneity known to occur within the β-cell population.[29,43] In addition, Cx36 in conjunction with other junctional proteins associated with the other intercellular junctions (e.g., E-cadherin, catenins) are believed to regulate gene expression of insulin and other proteins related to β-cell differentiation and survival/death.[29,31,43–45]

3. COMPENSATORY RESPONSE OF THE β CELL TO HIGH-FAT DIET-INDUCED INSULIN RESISTANCE

The peripheral insulin resistance is a hallmark of the early phase of T2D, known as prediabetes that has been described in both HFD-fed animals and obese human.[1,2] In this state, the insulin action on tissues/organs (mainly in skeletal muscles, adipose tissue, and liver) is impaired.[1,2] Although the molecular mechanisms of insulin resistance are not completely known, it has been suggested that several factors play a role in triggering the process such as: (1) the excess of circulating triglycerides and FFAs; (2) the release of cytokines and inflammatory factors by the inflamed adipose tissue; and (3) the entrance of substances from an altered microbiota into the circulation through a disrupted intestinal barrier.[6,13,46]

As commented previously, the normal β-cell response to insulin resistance is a compensatory insulin hypersecretion designed to maintain normoglycemia.[6,7,10,11]

Hyperinsulinemia associated with lower peripheral sensitivity to insulin (as evaluated by the insulin tolerance test, ITT) and subtle or no change in fasting glycemia have been extensively described in animals exposed to HFD for a short period (days or weeks) and in normoglycemic obese patients.[6,7,20,24,30] Compensation involves enhanced insulin biosynthesis, increased responsiveness of the nutrient-secretion coupling, and/or expansion of β-cell mass (Figure 3; details of these processes are described below).[6,7,10,11,20,24,30]

Pancreatic islets isolated from HFD-fed female and male C57 mice (for 12–16 weeks of diet exposure) have been shown to display a significant increase in insulin release, after static incubation with different glucose concentrations, as well as an enhanced insulin total islet content and higher density of insulin granules in comparison with control animals.[5,47,48] These findings indicate that hyperinsulinemia in obese mice, at early-stage of prediabetes induced by HFD, may be associated with enhanced secretory capacity of their pancreatic β cells. This, in turn, seems to be result of a combination of increased insulin gene transcription/translation and enhanced stimulus-secretion coupling.

A higher insulin mRNA content, as revealed by quantitative polymerase chain reaction, has been described by Gonzalez et al.[48] in prediabetic female mice, in comparison with their controls, and confirmed by our group in male C57 mice exposed to HFD for a shorter (8 weeks) and more prolonged period (8 months)(Maschio DA, Falcão VTFL, Collares-Buzato, CB; unpublished data). Recently, Kanno et al.[47] have shown that the HFD-induced hyperinsulinemia involves also an increase in insulin translation step, as demonstrated by the relatively higher incorporation of the translation marker HAH in proinsulin of isolated islets of HFD-fed mice. In accordance with this idea, Hatanaka et al.[19] reported a general increase in RNA translation in pancreatic islets from mice fed a HFD for 1 week using a different approach: measuring the fraction of RNAs associated with polyribosomes. A similar effect was observed in in vitro conditions by acutely incubating the β-cell lineage, MIN6 cells, with palmitate,[19] which is one the most abundant saturated FFAs found in HFDs and is well-known to be at high blood levels in obese subjects.[4,18,19,21] In addition, looking at the underlying molecular mechanisms, Hatanaka et al.[19] demonstrated that the HFD- and palmitate-induced increase in RNA translation resulted from the activation of the phosphoinositide-3 kinase (PI3K)/Akt/mammalian target of rapamycin (mTOR) signaling pathway, which in turn was dependent upon flux through L-type Ca^{2+} channels.

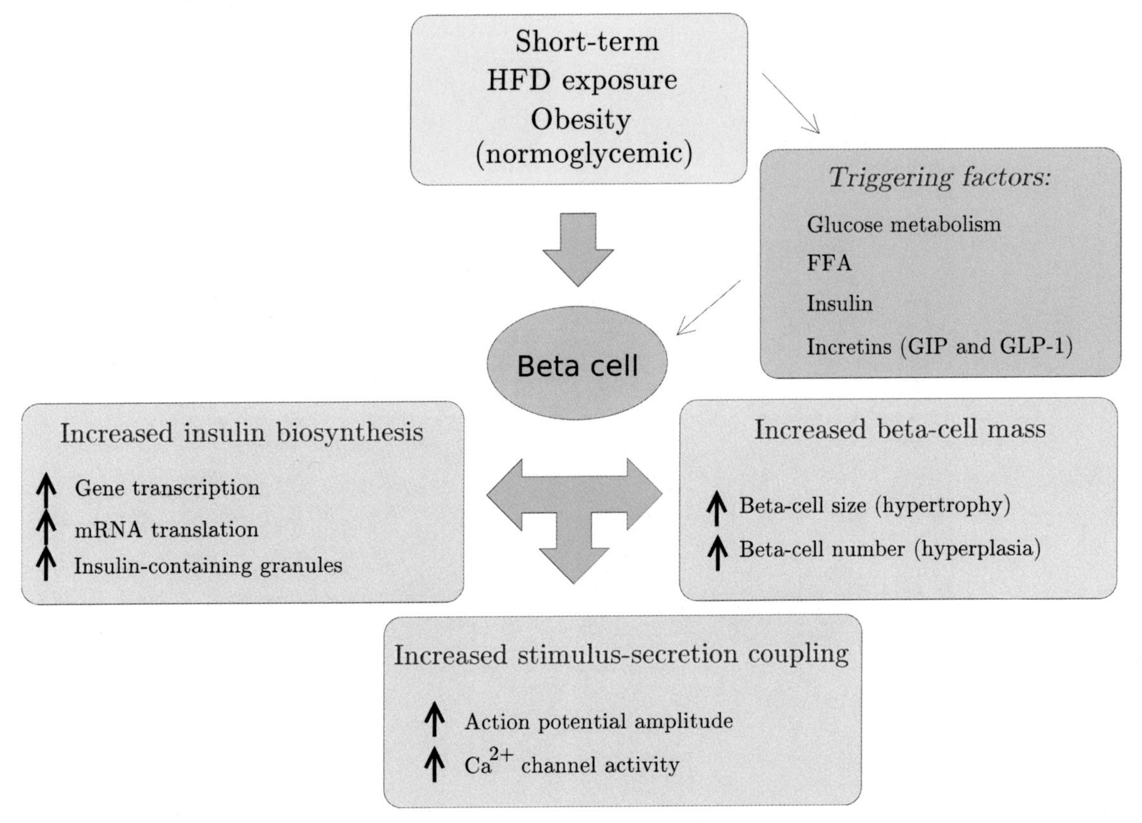

FIGURE 3 β-cell compensation response to short-term high-fat diet (HFD) exposure and obesity. See the text for details. FFA, free fatty acids; GLP-1, glucagon-like protein 1; GIP, glucose-dependent insulinotropic polypeptide; mRNA, messenger RNA.

In addition to insulin biosynthesis, short exposure to a HFD also seems to be associated with activation of some steps of the stimulus-secretion coupling. Gonzalez et al.[48] reported that action potentials during glucose stimulation have larger amplitude in β cells from HFD-fed hyperinsulinemic mice compared with control mice. The HFD-induced activation of insulin secretion may also involve steps upstream of K_{ATP} channels because a markedly increase in mitochondrial metabolism of fuels (other than glucose itself) and in mitochondrial area per β cell has been documented in islets of HFD-treated mice.[49] Gonzalez et al.[48] also showed that the compensatory functional response to HFD involves significantly augmented Ca^{2+} signals (higher amplitude, peak, and area under curve) and cell recruitment in intact islets, in individual β cells within islets, and isolated cells in culture as well as a more efficient exocytosis in medium-sized β cells. In accordance with the idea of Ca^{2+} channel activation during adaptive functional response to HFD-induced insulin resistance, in vitro acute exposure to palmitate or oleate (another FFA) has been shown to increase whole cell Ca^{2+} currents via interaction with L-type Ca^{2+} channel subunits and subsequent channel activation in isolated β cells and other cell lines expressing the same channel subtype.[50,51]

Besides the functional adaptation, numerous studies in both HFD animal models and humans have demonstrated that the hyperinsulinemia resulting from insulin-resistance states is associated also with compensatory structural changes resulting in expansion of the β-cell mass (Figure 4).[7,10,11,20,24,30] A direct correlation between the degree of insulin resistance and the increased mass of β cells was verified in animal model: (1) absence of significant insulin resistance (as assessed by ITT) seen in mice fed a HFD for 30 days resulted in no detectable change in β-cell mass and (2) highly insulin-resistant male fed an identical HFD for 60 days displayed higher increase in β-cell mass as compared with female mice that showed relatively low insulin resistance.[20] A morphometric analysis of the endocrine pancreas of these 60 days HFD-fed mice revealed that the expansion of β-cell mass results from a combination of increased islet size and total number, and enhancement of the proportion of β cell within the islet (Figure 4) [7,10,11,20,24,30]

Regulation of β-cell mass is well known to be determined by the balance of β-cell growth (resulted of hypertrophy, replication, and/or neogenesis) and β-cell loss through apoptosis or necrosis.[5,7,52] In the β-cell mass expansion induced by exposure to HFD in animal models, it seems that both events of hypertrophy and

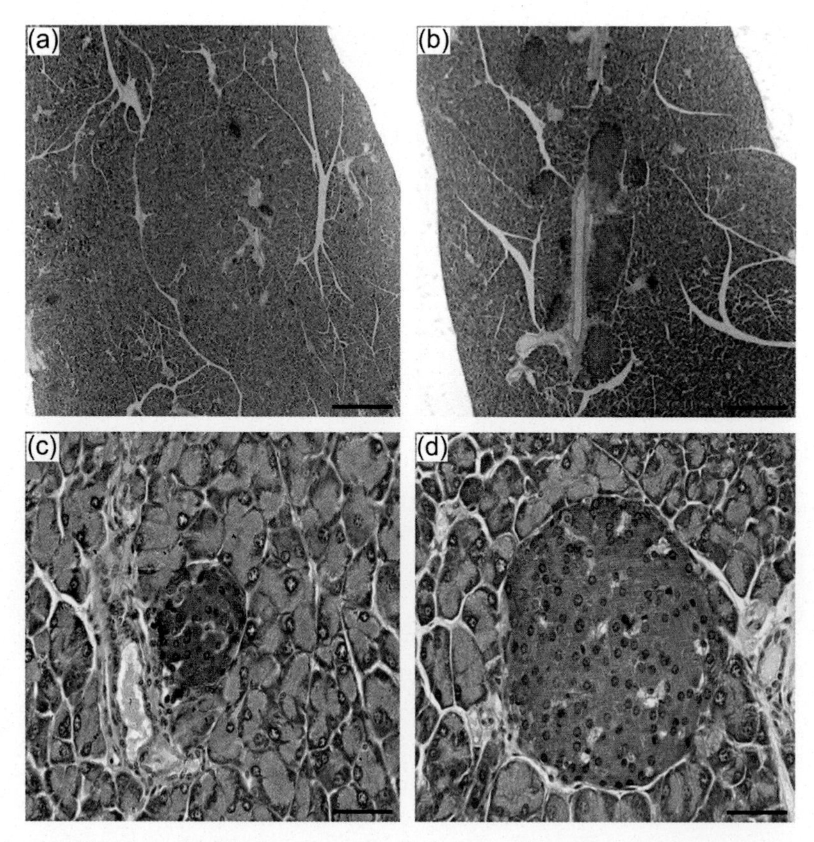

FIGURE 4 Increase in islet size (d) and number (b) after exposure to high-fat diet for 60 days as compared to control mice (a,c). All images are mouse pancreas sections processed by immunoperoxidase for insulin and counterstained with hematoxylin. Bars, 500 μm in (a,b); 50 μm in (c,d).

hyperplasia of this cell type are involved.[5,7,20] Hyperplasia, in turn, resulted from replication of preexisting β cells (as evidenced by immunodetection of cell-cycle markers such as Ki67 and PCNA) and neogenesis from pancreatic ductal cells.[5,20,24,52] Similar processes seem to occur in nondiabetic insulin-resistant obese human subjects, as revealed by morphological studies in pancreatic tissues obtained from autopsy, although the β-cell replication (with lack of evidence for neogenesis) takes place in a lower rate/magnitude as compared with rodents.[7,52–54] Human islets, when transplanted under the kidney capsule of immunodeficient mice fed an HFD for 12 weeks, also exhibited increased β-cell area and increased β-cell:α-cell ratio, although β-cell proliferation was rarely detected in these human islets.[55]

Because of the very limited tools and materials available to study the dynamics of β-cell mass in humans, most of our knowledge about the molecular pathways underlying the process of β-cell mass expansion induced by HFD feeding, comes from studies in rodents and other mammals. Several stimulating factors have been implicated in the compensatory β-cell expansion such as nutrient supply (particularly, glucose, lipids, and FFAs), insulin and other growth factor signaling, and increased levels and sensitivity to incretins (particularly glucagon-like peptide-1, GLP-1, and the glucose-dependent insulinotropic polypeptide, GIP).[7,53,54,56,57] All of these factors can potentially mediate the HFD-induced β-cell expansion because they are expected to be augmented in the prediabetes state.[2,7]

In the context of HFD feeding, of particular note is the evidence of a direct proliferating effect of lipids and FFAs and their indirect effect on β-cell mass expansion mediated by incretins.[53,54,57] FFAs have been postulated to promote compensatory β-cell proliferation only for short times because prolonged treatment with FFAs and lipids has been shown to result in an opposite action leading to β-cell apoptosis (see Section 4 in this chapter). Although the direct proliferating effect of FFAs on human and mice islets is controversial, short-term in vitro elevations in FFA (for few days) increase β-cell proliferation in islets isolated from rats[56,58] and intravenous infusion of a sterile lipid emulsion for up to 96 h into normal rat increases β-cell mass and proliferation.[59,60] Nevertheless, it has been suggested that increased enteric fat supply may also result in β-cell compensation through increased production of the incretins, GLP-1 and GIP, from enteroendocrine cells and within the islet milieu.[7,61] GLP-1 and GIP are released by the L and K cells of intestine, respectively, and their secretion is strongly stimulated by lipid and FFA-enriched diet in animal models.[62] Interestingly, a recent study has indicated that both hormones are also synthesized and secreted by islet α cells at determined conditions,[63] suggesting that these peptides may play a paracrine role. It is well known that GLP-1 and GIP can exert pleiotropic actions on β cells ranging from the stimulation of insulin biosynthesis and secretion to the enhancement of β-cell proliferation and survival.[63,64] Nevertheless, in the case of the HFD-induced β-cell replication, GIP seems to be the main incretin as elegantly demonstrated in a recent work by Mofett et al.[63] These authors showed that HFD-fed C57BL/6 mice exhibited hyperinsulinemia, increased pancreatic GIP level but left GLP-1 unchanged, associated with an increase in β-cell mass and replication. In contrast, GIP receptor knockout mice, when fed an identical HFD, displayed lack of hyperinsulinemic response along with a markedly decreased in the compensatory β-cell mass expansion.[63]

Both GLP-1 and GIP trigger similar intracellular signaling in pancreatic β cell. GLP-1 and GIP receptors are G protein–coupled receptors.[62] Upon ligand binding to their respective receptors, and in the presence of glucose, the α-stimulatory subunit of the G protein dissociates from the G protein complex and activates adenylate cyclase, resulting in the formation of cAMP.[62,65] cAMP activates protein kinase A (PKA), which in turn stimulates the transcription factor cyclic AMP response element–binding protein (CREB). CREB enhances proliferation through simultaneous inhibition of the proliferation inhibitor p27 and activation of cyclins A2 and D1.[62,66] In addition, CREB can promote transcription of IRS-2, triggering the same insulin signaling pathway mediated by IRS-2/PI3K/Akt complex.[67]

Despite of the supportive data obtained from animal models, the possible role of lipid-induced incretin secretion in adaptive β-cell proliferation in human remains unclear because there is no strong evidence for an increased circulating or postprandial concentrations of GPL-1 and GIP in normoglycemic obese and prediabetic individuals.[41,54,68] Interestingly, current therapies for T2D target the incretin pathway, including GLP-1 receptor agonists, and inhibitors of the dipeptidyl peptidase 4 (DPP-4) enzyme that rapidly degrades GLP-1 and GIP.[4,41,62,68] Therefore, much interest remains in understanding the potential of incretin pathways to stimulate β-cell replication.

Because hyperglycemia and hyperinsulinemia are distinguished features of obesity- and HFD-induced insulin resistance, these components have been considered as other mediators of the β-cell structural compensation.[54] In rodents, short-term exposure (for up to 6 days) to mild hyperglycemia through continuous glucose infusion led to increased β-cell hyperplasia, hypertrophy, and/or neogenesis, mediated at least partially by increases in cyclin D2.[69,70] Accordingly, human islets transplanted into immunodeficient mice have increased β-cell replication rates in response to acute glucose infusion.[71] Nevertheless, isolated human islets do not exhibit hyperplasia in response to high glucose alone, but require further inhibition of negative regulators of cell growth to replicate in response to glucose.[72,73]

The mitogenic effect of glucose on β cells seems to depend on its metabolization mediated by glucokinase (GK), as shown in in vivo experiments using genetically manipulated mice displaying haploinsufficiency of β cell–specific glucokinase ($GK^{+/-}$).[74] When fed an HFD, $GK^{+/-}$ mice display decreased β-cell replication and insufficient β-cell hyperplasia associated with reduced expression of IRS-2 compared with wild-type mice that received the same diet.[74] Recently, the effect of various synthetic GK activators, that trigger GK by binding to an allosteric site of the enzyme, have been tested for their capacity to induce β-cell proliferation in vitro.[75–77] The GK agonists, compound A, YH-GKA, GKA50, and LY2121260, increased both cell replication rate and cell numbers when tested at a basal concentration of glucose of 3 mmol/L in INS-1 cells and isolated mouse pancreatic islets. These GK agonists promoted β-cell proliferation in vitro via upregulation of insulin receptor substrate-2 and subsequent activation of Akt/protein kinase B phosphorylation,[75–77] suggesting that glucose may act also through the insulin-mediated proliferating pathway (see the following section). Alternatively, glucose metabolism can lead to an increase in cAMP cell content and consequently activation of the PKA/CREB pathway as well as enhancement of Ca^{2+} influx, triggering the calcium/calmodulin-dependent kinase 4, which also phosphorylates and activates CREB.[67]

The insulin-induced β-cell proliferative response is also complex, involving multiple potential signaling pathways. The classical pathway is triggered by the binding of insulin to its receptor, activation of its tyrosine kinase activity and phosphorylation of the insulin receptor substrate 2 (IRS2).[74,78] Activated IRS2 leads to activation of the PI3K and Akt. Akt inhibits glycogen synthase kinase 3-beta, downregulating the cell-cycle inhibitor p27.[78] Akt is also necessary to phosphorylate the inhibitory transcription factor forkhead box O1 (FOXO1) and promote its exclusion from the nucleus. When FOXO1 inhibition is released, transcription of *PDX1* and other genes promoting cell growth can be expressed.[54]

The importance of the insulin/Akt pathway in the compensatory β-cell expansion has been confirmed in mice knocked out, conditionally or not, homozygous or heterozygous, for components of this pathway. Mice deficient in IRS-2 failed to present compensatory β-cell proliferation when fed a HFD.[74] Conversely, knocking out glycogen synthase kinase 3-beta or p27 genes resulted in massively increased β-cell mass in mice fed a HFD.[79] In addition, insulin can also indirectly modulate β-cell mass expansion in response to obesity via the nuclear hormone receptor peroxisome proliferator-activated receptor gamma (PPAR-γ). Increased PPAR-γ concentrations are induced by insulin and obesity in multiple tissues of mice.[80] Mice with a β-cell–specific PPAR-γ deletion have reduced β-cell mass expansion when fed a HFD.[81] Accordingly, gene knockout of the PPAR-γ2 isoform in an ob/ob background results in failure to enhance β-cell mass in response to obesity when compared with wild-type ob/ob littermates.[82]

4. HIGH-FAT DIET AND β-CELL FAILURE AND DEATH

Although short-time exposition to HFD is associated with functional and structural adaptive response of β cells as described previously, prolonged exposure (ranging from several weeks to months or year) to diets with high lipid content results in general β-cell secretory decline and occasionally in overt T2D in rodents.[5,17,18,24,83,84] Under long-term HFD treatment, β cells display decreased insulin biosynthesis and impaired insulin secretion as well as reduction in β-cell mass as result of cell death mainly by apoptosis (Figure 5).[5,17,18,24,83,84] All of these alterations are due to mitochondrial dysfunction, increased oxidative stress and ER stress, and proinflammatory factor–mediated cell dysfunction/death triggered by chronic exposure to lipids/FFAs and hyperglycemia, among other factors such as the release of adipokines/cytokines and the insulin resistance itself (Figure 5; details of these processes are described below).[41,57,85–88] The insulin resistance, by requiring an enhanced demand for insulin secretion, leads to β-cell functional exhaustion.[12,42]

As compared with animal-based studies, fewer works using postmortem and surgical pancreas specimens yielded crucial data on the pathogenesis of islet β-cell failure in type 2 diabetic human subjects.[89–92] One of these studies reported a 63% and 40% reduction of islet β-cell volume in T2D obese subjects and in those displaying impaired fasting glucose, respectively, compared with weight-matched controls.[89] Accordingly, a study employing Korean human pancreas specimens suggested a role for selective loss of pancreatic β cells in the T2D progression.[90] In addition, another research group showed that islets obtained from T2D cadaveric donors were relatively small, poorly secreted insulin in response to glucose, and did not reverse diabetes upon transplantation into nude mice.[91] Carpentier et al.[93] have demonstrated that a prolonged intravenous infusion of a sterile fat emulsion (containing 20% soybean

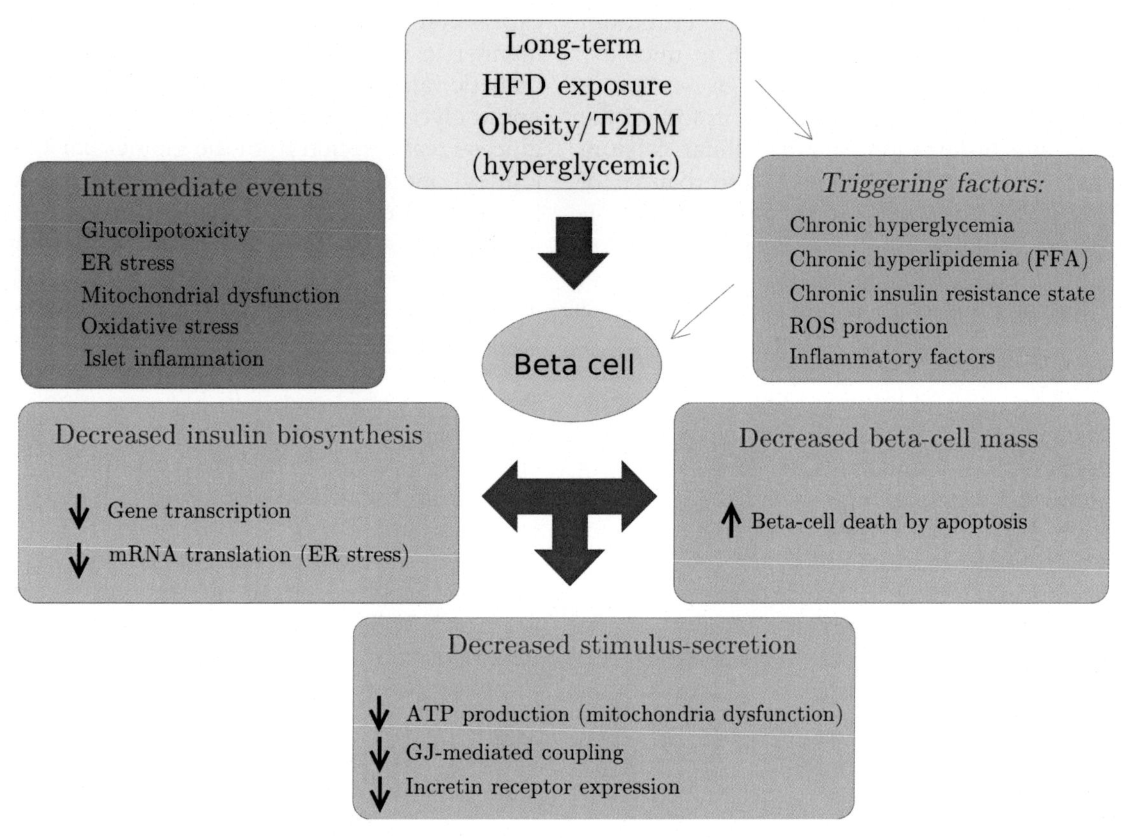

FIGURE 5 Prolonged high-fat diet-induced β-cell failure and death. See text for details. FFA, free fatty acids; ROS, reactive oxygen species; ER, endoplasmic reticulum; GJ, gap junction.

oil and 1.2% egg phospholipids) for 48 h, designed to raise plasma FFA twofold over basal, was associated with a significant reduction in GSIS in obese nondiabetic individuals. All these findings point out the similarities concerning the T2D development between diabetic obese humans and animals after HFD treatment.

It is well-known that chronic exposure to high blood glucose has deleterious effects on β-cell function and population dynamics through multiple mechanisms, which in turn result in aggravation of the hyperglycemia and finally to a vicious circle of continuous functional decline of the β-cell state.[7,15,86,94,95] This condition is aggravated by elevated levels of FFAs that have also a negative effect on β-cell biology and result in accumulation of toxic FFA metabolites (mainly ceramides and peroxides) in the islet cells.[7,15,86,94,95] Several reports (mainly coming from in vitro experiments and animal studies) suggest that the FFA toxicity on β cells depends on several parameters (concentration of unbound FFA, level of saturation, isomer nature, and chain length) and is potentiated by high glucose. The synergistic action of high glucose and saturated FFAs on β-cell deterioration is known as "glucolipotoxicity."[6,94–96] Multiple pathways and mechanisms through which chronic hyperglycemia and dyslipidemia impair β-cell function and cause β-cell apoptosis have been suggested; all them have been shown to negatively affect insulin production, the stimulus-secretion coupling steps and the β-cell survival pathway.

Concerning the insulin secretion process, one of the first alterations noticed in isolated islets of HFD-fed animals was a decreased insulin secretory response, characterized by an increased release of insulin at basal glucose concentration and loss of a pulsatile insulin secretion,[24,30,97] which are defects also seen at the initial phase of T2D in patients.[12,98] As diabetes progresses, the insulin release at stimulatory (supraliminal) concentrations of glucose markedly decreases, indicating a loss of β-cell functioning as "glucose sensor." This alteration has been described in isolated pancreatic islets from animals after several months or years on a HFD[5,84] as well as from obese/diabetic subjects.[91,93,99]

Altered insulin gene expression/protein secretion after a HFD and obesity can be explained at least partially by an impairment of β cell–β cell coupling mediated by GJs. This possibility is reinforced by several experimental and clinical evidences. A direct correlation has been demonstrated between insulin and Cx36 gene expression in islets of normoglycemic and patients T2D.[100] In HFD-fed mice, an increased basal insulin release was paralleled by a reduced GJ Cx36 islet content, smaller size of GJ plaques, and decreased intercellular transfer of a Cx36 channel tracer (ethidium bromide).[24,30] In accordance, loss of Cx36 in mice (by conditionally knocking out the gene) and in β-cell lines (using antisense oligonucleotides) results in loss of the β cell−β cell synchronization of intracellular calcium changes, associated with deleterious disturbances of basal and/or stimulated insulin secretion.[101,102] Interestingly, in vitro studies, in both insulin-producing cells and intact islets, have shown that expression of Cx36 is downregulated through activation of a PKA-dependent pathway after prolonged exposure to supraphysiological glucose concentrations[103] and elevated concentrations of saturated FFAs.[104] This indicates a role for chronic dyslipidemia and hyperglycemia in the reduction of GJ-mediated β cell−β cell coupling and related insulin secretion impairment.

Chronically high glucose itself in synergism or not with FFAs can also affect insulin biosynthesis and other steps of insulin secretion.[41,86,94,95] Chronic hyperglycemia leads to a decreased number of mitochondria and mitochondria morphological abnormalities (i.e. increased volume and outer surface area, reduction of proteins in the inner membrane, and more variably sized mitochondria) in β cells.[41,105] These mitochondrial changes are associated with impaired oxidative phosphorylation, decreased mitochondrial Ca^{2+} capacity, and a decline in ATP generation, which explain in part the hyperglycemia-induced impairment of GSIS.[105,106] Glucose and FFAs (particularly palmitate) can impair insulin biosynthesis by decreasing insulin gene transcription. Reduced insulin mRNA has also been reported in β cells isolated from diabetic human donors.[99] In vitro and in vivo animal studies have shown that the high glucose- and/or palmitate-induced decrease in insulin mRNA level in β cells involves reduced binding of the transcription factors, PDX-1 and Maf-A, to the insulin gene promoter as result of inhibition of the PDX-1 translocation to the nucleus and decrease of Maf-A expression.[86,94,107−109] Ceramide generation has been associated with the FFA-induced suppression of insulin transcription, as well as induction of β-cell death[57] because both events can be prevented by inhibitors of ceramide de novo production.[95] In the case of glucose, its effect on insulin gene expression is at least partially mediated by oxidative stress (see below) because reduction of insulin gene transcription and/or DNA-binding activities of PDX-1 and/or MafA by chronic exposure to high glucose was prevented by antioxidant treatment.[86,94,107]

Besides insulin biosynthesis, chronic exposure to glucose and FFA also decreases insulin secretion, in part, reducing incretin action.[86] Downregulation of GLP-1 receptor expression was observed in T2D subjects as well as in diabetic rodents.[86,110] Furthermore, it has been demonstrated that GLP-1 and GIP receptor expression was decreased in a glucose-dependent manner in islets isolated from 90% pancreatectomized diabetic rats.[86,111] Such a decrease was not observed at normoglycemia condition after phlorizin (that prevents glucose reabsorption from the glomerular filtrate in the kidney), indicating that hyperglycemia per se can lead to the downregulation of incretin receptor expression. In addition, it has been shown that expression of GLP-1 and GIP receptors is downregulated by lipotoxicity, as seen when β-cell lines and islets are exposed to high palmitate concentrations.[112]

ER stress is another mechanism, triggered by chronic exposure to HFD and high FFA and glucose environment that can be also involved in the β-cell failure/apoptosis associated with these conditions.[38,41,85,113] ER stress is an important adaptive cellular signaling (also known as unfolded protein response [UPR]) that aims to balance secretory protein synthesis in the ER against ER protein folding capacity in situation of high secretory activity.[113] UPR involves the attenuation of global protein translation, upregulation of ER chaperones (thereby increasing ER folding capacity), and the degradation of misfolded proteins. The UPR mechanism is complex and requires three main ER stress transducers associated to the ER membrane, namely inositol-requiring enzyme 1 (IRE1), activating transcription factor 6 (ATF6), and protein kinase RNA-like endoplasmic reticulum kinase (PERK). In normal conditions, they are inactive due to binding to the ER chaperone BiP (immunoglobulin heavy chain binding protein). However, in the case of increase in misfolded proteins within ER (from, for instance, protein overload), this interaction is lost and the ER stress transducers are activated. IRE1 activation culminates in increased expression of genes involved in ER expansion, protein folding, and misfolded protein degradation. ATF6 induces ER chaperones, whereas PERK phosphorylates eukaryotic translation initiation factor 2α subunit (eIF2α), thereby inhibiting translation initiation and reducing the arrival of newly synthesized proteins in the ER. In the case of prolonged and excessive ER stress, apoptosis is triggered via at least three pathways: (1) the IRE1-mediated activation of the ER-specific caspase 12 and the CCAAT/enhancer-binding protein homologous protein (CHOP), through c-Jun N-terminal kinase (JNK), (2) loss of the myeloid cell leukemia sequence 1 protein, which is an antiapoptotic member of the BH3 family, because of early translational repression via pancreatic ER kinase/eIF2α phosphorylation, and (3) activation of the ATF4, through phosphorylated eIF2α, resulting also in activation of CHOP pathway.[38,57]

ER stress has been linked to apoptosis in β cells chronically exposed to elevated levels of FFAs,[38,57,114] and ER stress markers have been found to be elevated in pancreatic islets of diabetic patients.[115] Palmitate markedly activates the IRE1, PERK, and ATF6 pathways, whereas the unsaturated oleate leads to milder PERK and IRE1 activation and comparable ATF6 signaling in β-cell lines as well as in fluorescence-activated cell sorting–purified primary rat β cells and human islets.[38] At least in the case of palmitate, an early and sustained depletion of ER Ca^{2+} stores seems to be an important mechanism to trigger ER stress in β cells by impairing protein folding.[38]

Because the main triggering factor of ER stress is ER protein overload, the current idea is that the very same conditions that initially elicited a compensatory increase in insulin biosynthesis and β-cell proliferation (i.e., insulin resistance, acute hyperglycemia, and high lipid/FFA concentration) will lead to, in the long-term, a decline in insulin production/secretion and β-cell apoptosis through ER stress pathways. A link between the two events has elegantly been proved by a recent work by Hatanaka et al.[19] They showed that acute palmitate incubation (24 h) induced increase in RNA translation through activation of the PI3K/Akt/mTOR signaling pathway in the β-cell line, MIN-6. Meanwhile, chronic incubation (>72 h) with this saturated FFA reduced polyribosome-associated RNA (indicative of reduced RNA translation) in this β-cell line through phosphorylation/inactivation of the eIF-2α, consistent with activation of the UPR.[19] When the MIN-6 cells were incubated with a potent inhibitor of mTOR, both the acute effects of palmitate on RNA translation and the chronic effects on the UPR were blocked. The latter effect involved blockage of both events: the eIF-2α phosphorylation and the later increase in expression of UPR genes (i.e., *CHOP, ATF4,* and *BIP*).

Oxidative stress and inflammation are two other major factors thought to contribute to failure/death of β cells in the setting of T2D induced by obesity/HFD.[41,85,86,88] Long-term sustained hyperglycemia and the excessive of FFAs increase the metabolic flux into the mitochondria and induce overproduction of reactive oxygen species (ROS), which lead to chronic oxidative stress. In general, elevated ROS disturbs the integrity and function of cellular proteins (e.g. enzymes, receptors, transport proteins), lipids, and deoxyribonucleic acid (DNA): they degrade polyunsaturated fatty acids of the membranes, induce lipid peroxidation and amino acid oxidation, and damage purine and pyrimidine bases. Because β cells have low levels of antioxidant enzymes (superoxide dismutase, catalase, thioredoxin peroxidase, and glutathione peroxidase), they are relatively vulnerable to free radical damage when exposed to oxidative stress.[116,117]

There are several evidences indicating that oxidative stress is involved in the deterioration of β-cell function found in diabetes. Studies on islets isolated from pancreata of subjects with T2D showed increased markers of oxidative stress (nitrotyrosine and 8-hydroxy-2-deoxyguanosine) and these correlated with the degree of GSIS impairment.[99] Exposure of human and rat islets and HIT β-cell line to high concentrations of glucose also resulted in an increased production of intracellular peroxide and 8-hydroxy-2′-deoxyguanosine.[118,119]

As mentioned previously, oxidative stress is involved in the suppression of insulin gene expression triggered by glucotoxicity (through decreased binding of PDX-1/MafA to insulin gene promoter).[86,94,107–109] It has been suggested that activation of JNK pathway and c-Jun protein mediates the reduction of insulin gene expression by oxidative stress.[107] JNK pathway activation would lead to a decrease in the binding of PDX-1 to the insulin gene promoter because of nucleo-cytoplasmic translocation of PDX-1, therefore resulting in decline in insulin expression. According to this idea, adenoviral overexpression of dominant-negative type JNK1 protected insulin gene expression and secretion in rat islets from oxidative stress induced by hydrogen peroxide by preserving the PDX-1 DNA binding activity.[107] Meanwhile, the c-Jun protein is involved in the oxidative stress–induced decrease of Maf-A expression. This is evidenced by the fact that Maf-A overexpression restored the insulin promoter activity and protein levels, that were suppressed by c-Jun, in isolated β cells of diabetic db/db mice and improved the glycemia control in these animals in vivo.[108,109]

Oxidative stress seems to mediate also the lipotoxicity in β cells. Glucose-stimulated insulin secretion is decreased in MIN-6 cells and rat islets after exposure to FFA for 48 h, which can be obviated by the antioxidant taurine.[114] N-acetylcysteine and Tempol (antioxidant drugs) also prevent the impairment in β-cell function induced by FFA in vivo during hyperglycemic clamping and ex vivo in isolated islets of oleate-treated rats.[114] Although type and cellular localization of ROS in FFA-induced β-cell dysfunction are still unclear, it seems that the membrane lipid peroxidation is important because its products (lipid peroxides) are known to impair insulin secretion.[120] In addition, Koulajian et al.[120] have shown that overexpression of glutathione peroxidase 4, an enzyme that reduces lipid peroxides, induced blockage of FFA-induced increase in ROS and lipid peroxidation in mice and, as a result, these animals were protected from the FFA-induced impairment of β-cell secretory function assessed in vitro, ex vivo, and in vivo.

Recent data indicate that the oxidative stress and ER stress are closely linked inasmuch as cellular ROS can enhance the misfolded protein accumulation in the ER, which amplifies ROS production that, in turn, further enhances ER stress, disrupts insulin biosynthesis, and initiates β-cell death.[41,85,114] Furthermore, a relationship between ER stress/oxidative stress and inflammation of the pancreatic islets in the context of T2D has been suggested, which also culminates in β-cell failure/death.[41,85,114]

T2D has been recently recognized as a chronic inflammation disorder.[8,85,121] Increased cytokine and chemokine levels and immune cell infiltration were observed in pancreatic islets of animals fed a HFD and in islets of type 2 diabetic patients.[122] Although adipose tissue is the main site of inflammation and the chronic release of the hormone-like adipokines by adipocytes, such as leptin, is believed to be involved in β-cell failure/death,[8,41,85,121] the "local" inflammation of the pancreatic islet seems to be equally relevant. Inflammatory signals within the islet can be triggered by ER stress, ROS, and chronically high glucose and/or FFAs, that all converge in signaling pathways involving JNKs and/or the nuclear factor κ-light-chain-enhancer of activated B cells (NF-κB).

Activation of either JNKs or NF-κB pathways in pancreatic β cells has been shown to cause increased expression of proinflammatory molecules, such as interleukin (IL)-1β, IL-8, IL-6, monocyte chemotactic protein-1 (MCP-1), and tumor necrosis factor-α (TNF-α), which have a deleterious effect on β-cell survival and function.[85,123,124] Local release of chemokine and cytokine can also contribute to the inflammatory milieu, attracting host macrophages to the pancreatic β cells, which further increment local inflammation, as seen in islets of T2D patients, HFD-fed mice, and diabetic mice.[122,125]

Among all cytokines, a crucial role seems to be played by IL-1β, a master cytokine that regulates many other proinflammatory cytokines (such as TNFα and IL-6) and chemokines (such as the C-X-C motif ligand 1, IL-8, MCP-1, and macrophage inflammatory protein-α).[121] By enhancing the production of cytokines, cytotoxic factors and chemokines and local attraction of macrophages, IL-1β contributes to β-cell function impairment and apoptosis.[121,122,125] It has been proposed that islet inflammation is mediated by an imbalance of IL-1β and its naturally occurring antagonist, the IL-1 receptor antagonist (IL-1Ra). This contributes to the formation of insulitis by recruiting and activating IL-1β–producing macrophages.[121,122,125] Treatment of obese/diabetic patients or HFD mice with anakinra, the recombinant form of IL-1Ra, or with specific anti–IL-1β antibodies ameliorated glycemia and β-cell function and reduced circulating inflammatory cytokines.[87,122,125–127] Consistent with these findings, the results of microarray analysis of islet cells from patients with T2D linked a group of IL-1β–related genes to reduced insulin secretion.[128]

An integrated model summarizing the interplay between HFD-triggered metabolic stress in β-cell and IL-1β–mediated islet inflammation can be proposed.[8,85,121,125] Increased circulating FFAs bind to their cognate receptors, the Toll-like receptors 2 and 4, leading to NF-kB activation and the production of various proinflammatory chemokines and cytokines, including the proform of IL-1β. Additionally, ROS, generated by oxidative stress resulting from high concentrations of glucose and FFAs, leads to the dissociation of a complex consisting of thioredoxin and thioredoxin-interacting protein (TXNIP) and then to the binding of the liberated TXNIP to NLRP3, which initiates the formation of the inflammasome. Subsequently, pro-IL-1β is processed by the NLRP3-associated caspase-1 and secreted in the extracellular milieu. Both NF-kB and ER stress pathways, triggered by prolonged HFD exposure or high glucose and/or FFA ambiance (mentioned previously), can also promote the activation of the NLRP3 inflammasome through the recruitment of TXNIP in β cells. IL-1β, released through all these signaling pathways, sustains autocrine and paracrine activation of both β cells and macrophages, exacerbating the chronic inflammatory responses in the islets. IL-1β and other cytokines, once interacted with its receptors, induce decrease of insulin gene expression and β-cell apoptosis through activation of JNK pathway.[129–131] Gunnet et al.,[132] investigating molecular mechanisms downstream of the JNK pathway, demonstrated that proinflammatory cytokines (IL-1β, TNF-α, IFN-γ) induce β-cell apoptosis through the canonical mitochondrial pathway.[133] This involves calcineurin-mediated Bad Ser 136 dephosphorylation, Bax activation, increased mitochondrial membrane permeability and consequent cytochrome C release, and activation of caspases-9 and -3 and DNA fragmentation.[132]

5. CONCLUDING REMARKS

In the past decade, research effort has unraveled the complexity of molecular mechanisms underlying β-cell dysfunction in the context of T2D induced by HFD exposure and obesity. This knowledge opened new therapeutic strategies for treating T2D, such as the use of incretin receptor agonists (exendin-4) or DPP-4 inhibitors, orally active chaperones (4-phenylbutyric acid, taurine-conjugated ursodeoxycholic acid), exogenous antioxidants (N-acetylcysteine, vitamin C, α-lipoic acid), and anti-inflammatory drugs (salsalate, anti–TNF-α antibodies, IL-1 antagonists).[4,41,85,121] Continued effort should be directed toward the mechanisms that trigger or underlie the initial compensatory β-cell mass expansion to eventually enable cell therapy of diabetes.

Acknowledgments

The author would like to thank Leandro P. Canuto (Figures 1 and 4) and Luiz E. Buzato (Figures 2, 3, and 5) for helping with the figures of this chapter, and Ricardo B. Oliveira (Figures 1(a, b) and 4 (a–d)) and Junia Santos Silva (Figure 1(c)) for obtaining the images depicted in Figures 1 and 4. CBC-B (CNPq # 307163/2012-1) is the recipient of a Research Fellowship from the Conselho Nacional de Desenvolvimento Científico e Tecnológico (CNPq, Brazil). Financial support for our research group has been provided by FAPESP and CNPq (Brazil).

References

1. Tripathy D, Chavez AO. Defects in insulin secretion and action in the pathogenesis of type 2 diabetes mellitus. *Curr Diab Rep* 2010; **10**:184–91.
2. Ferrannini E, Gastaldelli A, Iozzo P. Pathophysiology of prediabetes. *Med Clin N Am* 2011;**95**:327–39.
3. Ashcroft FM, Rorsman P. Diabetes mellitus and the β-cell: the last ten years. *Cell* 2012;**148**:1160–71.
4. Wu Y, Ding Y, Tanaka Y, Zhang W. Risk factors contributing to type 2 diabetes and recent advances in the treatment and prevention. *Int J Med Sci* 2014;**11**:1185–200.
5. Sone H, Kagawa Y. Pancreatic beta cell senescence contributes to the pathogenesis of type 2 diabetes in high-fat diet-induced diabetic mice. *Diabetologia* 2005;**48**:58–67.
6. Kahn SE, Hull RL, Utzschneider KM. Mechanisms linking obesity to insulin resistance and type 2 diabetes. *Nature* 2006; **14**:840–6.
7. Prentki M, Nolan CJ. Islet beta cell failure in type 2 diabetes. *J Clin Invest* 2006;**116**:1802–12.
8. Hameed I, Masoodi SR, Mir SA, Nabi M, Ghazanfar K, Ganai BA. Type 2 diabetes mellitus: from a metabolic disorder to an inflammatory condition. *World J Diabetes* 2015;**6**(4):598–612.
9. Saisho Y. β-Cell dysfunction: its critical role in prevention and management of type 2 diabetes. *World J Diabetes* 2015;**6**(1):109–24.
10. Lee YC, Nielsen JH. Regulation of beta cell replication. *Mol Cell Endocrinol* 2009;**29**:18–27.
11. Li Q, Lail Z-Q. Recent progress in studies of factors that elicit pancreatic β-cell expansion. *Protein Cell* 2015;**6**(2):81–7.
12. Hollingdal M, Juhl CB, Pincus SM, et al. Failure of physiological plasma glucose excursions to entrain high-frequency pulsatile insulin secretion in type 2 diabetes. *Diabetes* 2000;**49**:1334–40.
13. Shoelson SE, Lee J, Goldfine AB. Inflammation and insulin resistance. *J Clin Invest* 2006;**116**:1793–801.
14. Inadera H. Developmental origins of obesity and type 2 diabetes: molecular aspects and role of chemicals. *Environ Health Prev Med* 2013;**18**:185–97.
15. Al-Goblan AS, Al-Alfi MA, Khan MZ. Mechanism linking diabetes mellitus and obesity. *Diabetes, Metab Syndr Obes* 2014;**7**: 587–91.
16. Lai M, Chandrasekera PC, Barnard ND. You are what you eat, or are you? The challenges of translating high-fat-fed rodents to human obesity and diabetes. *Nutr Diabetes* 2014;**4**:e135.
17. Surwit RS, Kuhn CM, Cochrane C, McCubbin JA, Feinglos MN. Diet-induced type II diabetes in C57BL/6J mice. *Diabetes* 1988; **37**:1163–7.
18. Winzell MS, Ahrén B. The high-fat diet-fed mouse: a model for studying mechanisms and treatment of impaired glucose tolerance and type 2 diabetes. *Diabetes* 2003;**53**:S215–9.
19. Hatanaka M, Maier B, Sims EK, et al. Palmitate induces mRNA translation and increases ER protein load in islet beta-cells via activation of the mammalian target of rapamycin pathway. *Diabetes* 2014;**63**:3404–15.
20. Oliveira RB, Maschio DA, Carvalho CP, Collares-Buzato CB. Influence of gender and time diet exposure on endocrine pancreas remodeling in response to high fat diet-induced metabolic disturbances in mice. *Ann Anat* 2015;**200**:88–97.
21. Arner P, Rydén M. Fatty acids, obesity and insulin resistance. *Obes Facts* 2015;**8**:147–55.
22. Ahren J, Ahren B, Wierup N. Increased β-cell volume in mice fed a high-fat diet A dynamic study over 12 months. *Islets* 2010;**6**: 353–6.
23. Nagao M, Asai A, Inaba W, et al. Characterization of pancreatic islets in two selectively bred mouse lines with different susceptibilities to high-fat diet-induced glucose intolerance. *PLoS One* 2014;**9**(1):e84725.
24. Oliveira RB, Carvalho CP, Polo CC, Dorighello G, Boschero AC, d Oliveira HC, et al. Impaired compensatory beta-cell function and growth in response to high-fat diet in LDL receptor knockout mice. *Int J Exp Pathol* 2014;**95**:296–308.
25. In't Veld P, Marichal M. Microscopic anatomy of the human islet of Langerhans. In: Islam MS, editor. *The islets of Langerhans. Advances in experimental medicine and biology,* vol. 654; 2011. p. 1–19.
26. Carvalho CP, Martins JC, Da Cunha DA, Boschero AC, Collares-Buzato CB. Histomorphology and ultrastructure of pancreatic islet tissue during in vivo maturation of rat pancreas. *Ann Anat* 2006;**188**:221–34.
27. Kim A, Miller K, Junghyo J, Kilimnik G, Wojcik P, Hara M. Islet architecture: a comparative study. *Islets* 2009;**1**(2):129–36.
28. Bosco D, Armanet M, Morel P, et al. Unique arrangement of α- and β-cells in human islets of Langerhans. *Diabetes* 2010;**59**: 1202–10.
29. Jain R, Lammert E. Cell–cell interactions in the endocrine pancreas. *Diabetes Obes Metab* 2009;**11**(Suppl. 4):159–67.
30. Carvalho CP, Oliveira RB, Santos-Silva JC, Boschero AC, Meda P, Collares-Buzato CB. Impaired β-cell-β-cell coupling mediated by Cx36 gap junctions in prediabetic mice. *Am J Physiol Endocrinol Metab* 2012;**303**:144–5.
31. Santos-Silva JC, Carvalho CP, de Oliveira RB, Boschero AC, Collares-Buzato CB. Cell-to-cell contact dependence and junctional protein content are correlated with in vivo maturation of pancreatic beta cells. *Can J Physiol Pharmacol* 2012;**90**:837–50.
32. Fu Z, Gilbert ER, Liu D. Regulation of insulin synthesis and secretion and pancreatic beta-cell dysfunction in diabetes. *Curr Diabetes Rev* 2013;**9**(1):25–53.
33. Suckale J, Solimena M. Pancreas islets in metabolic signaling—focus on the beta-cell. *Front Biosci* 2008;**13**:7156–71.
34. McCulloch LJ, van de Bunt M, Braun M, Frayn KN, Clark A, Gloyn AL. GLUT2 (SLC2A2) is not the principal glucose transporter in human pancreatic beta cells: implications for understanding genetic association signals at this locus. *Mol Genet Metab* 2011;**104**(4):648–53.
35. MacDonald PE. Signal integration at the level of ion channel and exocytotic function in pancreatic β-cell. *Am J Physiol Endocrinol Metab* 2011;**301**:E1065–9.
36. Wang Z, Thurmond DC. Mechanisms of biphasic insulin-granule exocytosis – roles of the cytoskeleton, small GTPases and SNARE proteins. *J Cell Sci* 2009;**122**:893–903.
37. Schuit FC, In't Veld PA, Pipeleers DG. Glucose stimulates proinsulin biosynthesis by a dose-dependent recruitment of pancreatic beta cells (protein synthesis/beta cell heterogeneity/insulin). *Proc Natl Acad Sci USA* 1988;**85**:3865–9.
38. Cnop M, Igoillo-Esteve M, Cunha DA, Ladriere L, Eizirik DL. An update on lipotoxic endoplasmic reticulum stress in pancreatic *β*-cells. *Biochem Soc Trans* 2008;**36**:909–15.
39. Papa FR. Endoplasmic reticulum stress, pancreatic β-cell degeneration, and diabetes. *Cold Spring Harb Perspect Med* 2012;**2**:a007666.
40. Vetere A, Choudhary A, Wagner BK. Targeting the pancreatic β-cell to treat diabetes. *Nat Rev Drug Discov* 2014;**13**(4):278–89.

41. Cernea S, Dobreanu M. Diabetes and beta cell function: from mechanisms to evaluation and clinical implications. *Biochem Medica* 2013;**23**(3):266–80.
42. Liu J, Li J, Li W-J, Wang C-M. The role of uncoupling proteins in diabetes mellitus. *J Diabetes Res* 2013;**2013**:585897–904.
43. Bosco D, Haefliger J-A, Meda P. Connexins: key mediators of endocrine function. *Physiol Rev* 2011;**91**:1393–445.
44. Carvalho CPF, Barbosa HCL, Britan A, Boschero AC, Meda P, Collares-Buzato CB. Beta cell coupling and connexin expression change during the functional maturation of rat pancreatic islets. *Diabetologia* 2010;**53**:1428–37.
45. Collares-Buzato CB, Carvalho CPF, Furtado AG, Boschero AC. Upregulation of the expression of tight and adherens junction-associated proteins during maturation of neonatal pancreatic islets in vitro. *J Mol Histol* 2004;**35**:811–22.
46. De Kort S, Keszthelyi D, Masclee AAM. Etiology and pathophysiology – leaky gut and diabetes mellitus: what is the link? *Obes Rev* 2011;**12**:449–58.
47. Kanno A, Asahara S, Masuda K, et al. Compensatory hyperinsulinemia in high-fat diet-induced obese mice is associated with enhanced insulin translation in islets. *Biochem Biophys Res Commun* 2015;**458**:681–6.
48. Gonzalez A, Merino B, Marroquí L, et al. Insulin hypersecretion in islets from diet-induced hyperinsulinemic obese female mice is associated with several functional adaptations in individual β-cells. *Endocrinology* 2013;**154**:3515–24.
49. Fex M, Nitert MD, Wierup N, Sundler F, Ling C, Mulder H. Enhanced mitochondrial metabolism may account for the adaptation to insulin resistance in islets from C57BL/6J mice fed a high-fat diet. *Diabetologia* 2007;**50**:74–83.
50. Warnotte C, Gilon P, Nenquin M, Henquin JC. Mechanisms of the stimulation of insulin release by saturated fatty acids. A study of palmitate effects in mouse beta-cells. *Diabetes* 1994;**43**(5):703–11.
51. Tian Y, Corkey RF, Yaney GC, Goforth PB, Satin LS, Moitoso de Vargas L. Differential modulation of L-type calcium channel subunits by oleate. *Am J Physiol Endocrinol Metab* 2008;**294**:E1178–86.
52. Montanya E. Insulin resistance compensation: not just a matter of beta-cells? *Diabetes* 2014;**63**:832–4.
53. Sachdeva MM, Stoffers DA. Minireview: meeting the demand for insulin: molecular mechanisms of adaptive postnatal ß-cell mass expansion. *Mol Endocrinol* 2009;**23**:747–58.
54. Linnemann AK, Baan M, Davis DB. Pancreatic β-cell proliferation in obesity. *Adv Nutr* 2014;**5**:278–88.
55. Gargani S, Thévenet J, Yuan JE, et al. Adaptive changes of human islets to an obesogenic environment in the mouse. *Diabetologia* 2013;**56**:350–8.
56. Brelje TC, Bhagroo NV, Stout LE, Sorenson RL. Beneficial effects of lipids and prolactin on insulin secretion and beta-cell proliferation: a role for lipids in the adaptation of islets to pregnancy. *J Endocrinol* 2008;**197**:265–76.
57. Oh YS. Mechanistic insights into pancreatic beta-cell mass regulation by glucose and free fatty acids. *Anat Cell Biol* 2015;**48**:16–24.
58. Milburn Jr JL, Hirose H, Lee YH, et al. Pancreatic beta-cells in obesity. Evidence for induction of functional, morphologic, and metabolic abnormalities by increased long chain fatty acids. *J Biol Chem* 1995;**270**(3):1295–9.
59. Steil GM, Trivedi N, Jonas J-C, et al. Adaptation of beta-cell mass to substrate oversupply: enhanced function with normal gene expression. *Am J Physiol Endocrinol Metab* 2001;**280**:E788–96.
60. Fontés G, Zarrouki B, Hagman DK, et al. Glucolipotoxicity age-dependently impairs beta cell function in rats despite a marked increase in beta cell mass. *Diabetologia* 2010;**53**(11):2369–79.
61. Van Citters GW, Kabir M, Kim SP, et al. Elevated glucagon-like peptide-1-(7–36)-amide, but not glucose, associated with hyperinsulinemic compensation for fat feeding. *J Clin Endocrinol Metab* 2002;**87**:5191–8.
62. Drucker DJ. The biology of incretin hormones. *Cell Metab* 2006;**3**(3):153–65.
63. Moffett RC, Vasu S, Flatt PR. Functional GIP receptors play a major role in islet compensatory response to high fat feeding in mice. *Biochim Biophys Acta* 2015;**1850**:1206–14.
64. McIntosh CH, Widenmaier S, Kim SJ. Glucose-dependent insulinotropic polypeptide (gastric inhibitory polypeptide; GIP). *Vitam Horm* 2009;**80**:409–71.
65. Buteau J. GLP-1 receptor signaling: effects on pancreatic beta-cell proliferation and survival. *Diabetes Metab* 2008;**34**(Suppl. 2):S73–7.
66. Song W-J, Schreiber WE, Zhong E, et al. Exendin-4 stimulation of cyclin A2 in β-cell proliferation. *Diabetes* 2008;**57**:2371–81.
67. Persaud SJ, Liu B, Sampaio HB, Jones PM, Muller DS. Calcium/calmodulin-dependent kinase IV controls glucose-induced Irs2 expression in mouse beta cells via activation of cAMP response element-binding protein. *Diabetologia* 2011;**54**(5):1109–20.
68. Ahrén B, Carr RD, Deacon CF. Incretin hormone secretion over the day. *Vitam Horm* 2010;**84**:203–20.
69. Topp BG, McArthur MD, Finegood DT. Metabolic adaptations to chronic glucose infusion in rats. *Diabetologia* 2004;**47**(9):1602–10.
70. Alonso LC, Yokoe T, Zhang P, et al. Glucose infusion in mice a new model to induce β-cell replication. *Diabetes* 2007;**56**:1792–801.
71. Levitt HE, Cyphert TJ, Pascoe JL, et al. Glucose stimulates human beta cell replication in vivo in islets transplanted into NOD-severe combined immunodeficiency (SCID) mice. *Diabetologia* 2011;**54**(3):572–82.
72. Liu H, Remedi MS, Pappan KL, et al. Glycogen synthase kinase-3 and mammalian target of rapamycin pathways contribute to DNA synthesis, cell cycle progression, and proliferation in human islets. *Diabetes* 2009;**58**:663–72.
73. Rohatgi N, Aly H, Marshall CA, et al. Novel insulin sensitizer modulates nutrient sensing pathways and maintains b- cell phenotype in human islets. *PLoS One* 2013;**8**:e62012.
74. Terauchi Y, Takamoto I, Kubota N, et al. Glucokinase and IRS-2 are required for compensatory β-cell hyperplasia in response to high-fat diet–induced insulin resistance. *J Clin Invest* 2007;**117**(1):246–57.
75. Nakamura A, Terauchi Y, Ohyama S, et al. Impact of small-molecule glucokinase activator on glucose metabolism and β-cell mass. *Endocrinology* 2009;**150**:1147–54.
76. Wei P, Shi M, Barnum S, Cho H, Carlson T, Fraser JD. Effects of glucokinase activators GKA50 and LY2121260 on proliferation and apoptosis in pancreatic INS-1 beta cells. *Diabetologia* 2009;**52**:2142–50.
77. Oh YS, Lee Y-J, Park K, Choi HH, Yoo S, Jun H-S. Treatment with glucokinase activator, YH-GKA, increases cell proliferation and decreases glucotoxic apoptosis in INS-1 cells. *Eur J Pharm Sci* 2014;**51**:137–45.
78. Bernal-Mizrachi E, Kulkarni RN, Scott DK, et al. Human beta-cell proliferation and intracellular signaling part 2: still driving in the dark without a road map. *Diabetes* 2014;**63**(3):819–31.
79. Liu Y, Tanabe K, Baronnier D, et al. Conditional ablation of Gsk-3b in islet beta cells results in expanded mass and resistance to fat feeding-induced diabetes in mice. *Diabetologia* 2010;**53**:2600–10.
80. Vidal-Puig A, Jimenez-Liñan M, Lowell BB, et al. Regulation of PPAR gamma gene expression by nutrition and obesity in rodents. *J Clin Invest* 1996;**97**:2553–61.
81. Rosen ED, Kulkarni RN, Sarraf P, et al. Targeted elimination of peroxisome proliferator activated receptor in beta cells leads to abnormalities in islet mass without compromising glucose homeostasis. *Mol Cell Biol* 2003;**23**:7222–9.

82. Medina-Gomez G, Gray SL, Yetukuri L, et al. PPAR gamma 2 prevents lipotoxicity by controlling adipose tissue expandability and peripheral lipid metabolism. *PLoS Genet* 2007;**3**(4):e64.
83. Shafrir E, Ziv E, Mosthaf L. Nutritionally induced insulin resistance and receptor defect leading to beta-cell failure in animal models. *Ann N Y Acad Sci* 1999;**892**:223–46.
84. Falcão VT, Oliveira RB, Maschio DA, Carvalho CP, Collares-Buzato CB. Cellular distribution and expression of cell adhesion molecules (CAMs) in pancreatic islet cells of obese and diabetic mice. *Mol Biol Cell* 2014;**25**:448 (abstract #P591).
85. Montane J, Cadavez L, Novials A. Stress and the inflammatory process: a major cause of pancreatic cell death in type 2 diabetes. *Diabetes Metab Syndr Obes* 2014;**7**:25–34.
86. Kaneto H, Matsuoka T. Role of pancreatic transcription factors in maintenance of mature β-cell function. *Int J Mol Sci* 2015;**16**: 6281–97.
87. Sauter NS, Thienel C, Plutino Y, et al. Angiotensin II induces interleukin-1β-mediated islet inflammation and beta-cell dysfunction independently of vasoconstrictive effects. *Diabetes* 2015;**64**:1273–83.
88. Tangvarasittichai S. Oxidative stress, insulin resistance, dyslipidemia and type 2 diabetes mellitus. *World J Diabetes* 2015;**6**(3): 456–80.
89. Butler AE, Janson J, Bonner-Weir S, Ritzel R, Rizza RA, Butler PC. Beta-cell deficit and increased beta-cell apoptosis in humans with type 2 diabetes. *Diabetes* 2003;**52**:102–10.
90. Yoon KH, Ko SH, Cho JH, et al. Selective beta-cell loss and alpha-cell expansion in patients with type 2 diabetes mellitus in Korea. *J Clin Endocrinol Metab* 2003;**88**:2300–8.
91. Deng S, Vatamaniuk M, Huang X, et al. Structural and functional abnormalities in the islets isolated from type 2 diabetic subjects. *Diabetes* 2004;**53**:624–32.
92. Gunton JE, Kulkarni RN, Yim S, et al. Loss of ARNT/HIF1beta mediates altered gene expression and pancreatic-islet dysfunction in human type 2 diabetes. *Cell* 2005;**122**:337–49.
93. Carpentier A, Mittelman SD, Bergman RN, Giacca A, Lewis GF. Prolonged elevation of plasma free fatty acids impairs pancreatic β-cell function in obese nondiabetic humans but not in individuals with type 2 diabetes. *Cell* 2000;**49**:399–408.
94. Poitout V. Glucolipotoxicity of the pancreatic beta-cell: myth or reality? *Biochem Soc Trans* 2008;**36**(Pt 5):901–4.
95. Poitout V, Robertson RP. Glucolipotoxicity: fuel excess and β-cell dysfunction. *Endocr Rev* 2008;**29**(3):351–66.
96. Del Prato S. Role of glucotoxicity and lipotoxicity in the pathophysiology of type 2 diabetes mellitus and emerging treatment strategies. *Diabet Med* 2009;**26**:1185–92.
97. Asghar Z, Yau D, Chan F, Leroith D, Chan CB, Wheeler MB. Insulin resistance causes increased beta-cell mass but defective glucose-stimulated insulin secretion in a murine model of type 2 diabetes. *Diabetologia* 2006;**49**:90–9.
98. Schofield CJ, Sutherland C. Disordered insulin secretion in the development of insulin resistance and type 2 diabetes. *Diabet Med* 2012;**29**(8):972–9.
99. Del Guerra S, Lupi R, Marselli L, et al. Functional and molecular defects of pancreatic islets in human type 2 diabetes. *Diabetes* 2005;**54**(3):727–35.
100. Serre-Beinier V, Bosco D, Zulianello L, et al. Cx36 makes channels coupling human pancreatic beta-cells, and correlates with insulin expression. *Hum Mol Genet* 2009;**18**:428–39.
101. Calabrese A, Zang M, Serre-Beinier V, et al. Connexin 36 controls synchronization of Ca^{2+} oscillations and insulin secretion in MIN6 cells. *Diabetes* 2003;**52**:417–24.
102. Ravier MA, Güldenagel M, Charollais A, et al. Loss of connexin36 channels alters beta-cell coupling, islet synchronization of glucose-induced Ca^{2+} and insulin oscillations, and basal insulin release. *Diabetes* 2005;**54**:1798–807.
103. Allagnat F, Martin D, Condorelli DF, Waeber G, Haefliger JA. Glucose represses connexin36 in insulin-secreting cells. *J Cell Sci* 2005;**118**:5335–44.
104. Allagnat F, Alonso F, Martin D, Abderrahmani A, Waeber G, Haefliger JA. ICER-1gamma overexpression drives palmitate-mediated connexin36 down-regulation in insulin-secreting cells. *J Biol Chem* 2008;**283**:5226–34.
105. Ma Z, Wirström T, Borg LA, et al. Diabetes reduces β-cell mitochondria and induces distinct morphological abnormalities, which are reproducible by high glucose in vitro with attendant dysfunction. *Islets* 2012;**4**(3):233–42.
106. Lu H, Koshkin V, Allister EM, Gyulkhandanyan AV, Wheeler MB. Molecular and metabolic evidence for mitochondrial defects associated with β-cell dysfunction in a mouse model of type 2 diabetes. *Diabetes* 2010;**59**:448–59.
107. Kaneto H, Xu G, Fujii N, Kim S, Bonner-Weir S, Weir GC. Involvement of c-Jun N-terminal kinase in oxidative stress-mediated suppression of insulin gene expression. *J Biol Chem* 2002;**277**(33): 30010–8.
108. Matsuoka T, Kaneto H, Miyatsuka T, et al. Regulation of MafA expression in pancreatic β-cells in *db/db* mice with diabetes. *Diabetes* 2010;**59**:1709–20.
109. Matsuoka T, Kaneto H, Kawashima S, et al. Preserving MafA expression in diabetic islet β-cells improves glycemic control in vivo. *J Biol Chem* 2015;**290**(12):7647–57.
110. Shu L, Matveyenko AV, Kerr-Conte J, Cho JH, McIntosh CH, Maedler K. Decreased TCF7L2 protein levels in type 2 diabetes mellitus correlate with downregulation of GIP- and GLP-1 receptors and impaired beta-cell function. *Hum Mol Genet* 2009;**18**: 2388–99.
111. Xu G, Kaneto H, Laybutt DR, et al. Downregulation of GLP-1 and GIP receptor expression by hyperglycemia: possible contribution to the impaired incretin effects in diabetes. *Diabetes* 2007;**56**: 1551–8.
112. Kang ZF, Deng Y, Zhou Y, et al. Pharmacological reduction of NEFA restores the efficacy of incretin-based therapies through GLP-1 receptor signalling in the beta cell in mouse models of diabetes. *Diabetologia* 2013;**56**:423–33.
113. Sun J, Cui J, He Q, Chen Z, Arvan P, Liu M. Proinsulin misfolding and endoplasmic reticulum stress during the development and progression of diabetes. *Mol Asp Med* 2015;**42**:105–18.
114. Giacca A, Xiao C, Oprescu AI, Carpentier AC, Lewis GF. Lipid-induced pancreatic β-cell dysfunction: focus on in vivo studies. *Am J Physiol Endocrinol Metab* 2011;**300**:E255–62.
115. Marchetti P, Bugliani M, Lupi R, et al. The endoplasmic reticulum in pancreatic beta cells of type 2 diabetes patients. *Diabetologia* 2007;**50**:2486–94.
116. Tiedge M, Lortz S, Drinkgern J, Lenzen S. Relation between antioxidant enzyme gene expression and antioxidative defense status of insulin-producing cells. *Diabetes* 1997;**46**(11): 1733–42.
117. Boschero AC, Stoppiglia LF, Collares-Buzato CB, Bosqueiro JR, Delghingaro-Augusto V, Leite A, et al. Expression of a thioredoxin peroxidase in insulin-producing cells. *Diabetes Metab* 2002; **28**(6 Pt 2). 3S25–3S28.
118. Ihara Y, Toyokuni S, Uchida K, et al. Hyperglycemia causes oxidative stress in pancreatic beta-cells of GK rats, a model of type 2 diabetes. *Diabetes* 1999;**48**(4):927–32.
119. Tanaka Y, Tran PO, Harmon J, Robertson RP. A role for glutathione peroxidase in protecting pancreatic beta cells against oxidative stress in a model of glucose toxicity. *Proc Natl Acad Sci USA* 2002;**99**(19):12363–8.
120. Koulajian K, Ivovic A, Ye K, et al. Overexpression of glutathione peroxidase 4 prevents β-cell dysfunction induced by prolonged elevation of lipids in vivo. *Am J Physiol Endocrinol Metab* 2013; **305**:E254–62.

121. Donath MY, Dalmas E, Sauter NS, Boni-Schnetzler M. Inflammation in obesity and diabetes: islet dysfunction and therapeutic opportunity. *Cell Metab* 2013;**17**(6):860–72.
122. Ehses JA, Perren A, Eppler E, et al. Increased number of islet-associated macrophages in type 2 diabetes. *Diabetes* 2007;**56**: 2356–70.
123. Marselli L, Dotta F, Piro S, et al. Th2 cytokines have a partial, direct protective effect on the function and survival of isolated human islets exposed to combined proinflammatory and Th1 cytokines. *J Clin Endocrinol Metab* 2001;**86**(10):4974–8.
124. Zaitseva II, Hultcrantz M, Sharoyko V, Flodstrom-Tullberg M, Zaitsev SV, Berggren PO. Suppressor of cytokine signaling-1 inhibits caspase activation and protects from cytokine-induced beta cell death. *Cell Mol Life Sci* 2009;**66**(23):3787–95.
125. Ehses JA, Meier DT, Wueest S, et al. Toll-like receptor 2-deficient mice are protected from insulin resistance and beta cell dysfunction induced by a high-fat diet. *Diabetologia* 2010;**53**:1795–806.
126. Larsen CM, Faulenbach M, Vaag A, et al. Interleukin-1–Receptor antagonist in type 2 diabetes mellitus. *N Engl J Med* 2007;**356**: 1517–26.
127. van Asseldonk EJ, Stienstra R, Koenen TB, Joosten LA, Netea MG, Tack CJ. Treatment with Anakinra improves disposition index but not insulin sensitivity in nondiabetic subjects with the metabolic syndrome: a randomized, double-blind, placebo-controlled study. *J Clin Endocrinol Metab* 2011;**96**(7):2119–26.
128. Mahdi T, Hanzelmann S, Salehi A, et al. Secreted frizzled-related protein 4 reduces insulin secretion and is overexpressed in type 2 diabetes. *Cell Metab* 2012;**16**:625–33.
129. Ferdaoussi M, Abdelli S, Yang J-Y, et al. Exendin-4 protects β-cells from interleukin-1 beta-induced apoptosis by interfering with the c-Jun NH2-terminal kinase pathway. *Diabetes* 2008; **57**(5):1205–15.
130. Chin-Chance CVT, Newman MV, Aronovitz A, et al. Role of the mitogen-activated protein kinases in cytokine-mediated inhibition of insulin gene expression. *J Invest Med* 2006;**54**(3): 132–42.
131. Abdelli S, Abderrahmani A, Hering BJ, Beckmann JS, Bonny C. The c-Jun N-terminal kinase JNK participates in cytokine- and isolation stress-induced rat pancreatic islet apoptosis. *Diabetologia* 2007;**50**(8):1660–9.
132. Grunnet LG, Aikin R, Tonnesen MF, et al. Proinflammatory cytokines activate the intrinsic apoptotic pathway in β-cells. *Diabetes* 2009;**58**:1807–15.
133. Verma G, Datta M. The critical role of JNK in the ER-mitochondrial crosstalk during apoptotic cell death. *J Cell Physiol* 2012;**227**:1791–5.

CHAPTER

11

Native Fruits, Anthocyanins in Nutraceuticals, and the Insulin Receptor/Insulin Receptor Substrate-1/Akt/Forkhead Box Protein Pathway

Nathalia Romanelli Vicente Dragano[1], *Anne y Castro Marques*[2]

[1]Laboratory of Cell Signaling, Obesity and Comorbidities Research Center (OCRC), Faculty of Medical Sciences (FCM-Unicamp), University of Campinas, Campinas, SP, Brazil; [2]Universidade Federal do Pampa-Campus Itaqui, Itaqui, RS, Brazil

1. ANTHOCYANINS: GENERAL CHARACTERISTICS

Anthocyanins (from Greek anthos = flower and kianos = blue) are secondary metabolites responsible for red, blue, and violet pigmentation in different plant tissues, such as flowers, fruits, leaves, seeds, and roots.[1] The different colors of these compounds occur because their structures suffer interference from various factors, such as temperature, pH, and bonds with other chemical substances.[2]

Because of the chemical structure characteristic, anthocyanins belong to the large class of phenolic compounds and are classified as flavonoids. These compounds have a carbon chain (C_6–C_6–C_6) in which the two aromatic rings are separated by an intermediate heterocyclic ring. They are polyhydroxy or polymetoxy glycosylated molecules, derived from 2-fenilbenzopirilium cation, and feature a rich structural diversity because of: (1) differences in the number of hydroxyl groups in the molecule; (2) degree of methylation of these hydroxyl groups; (3) nature, number, and position of sugar molecules bound to aromatic rings; and (4) nature and number of aliphatic or aromatic acids attached to sugars. Anthocyanins are more common in nature as glycosides aglucons anthocyanidins, which may be acylated with organic acids.[3,4] The anthocyanidins present in higher proportions in different tissues of plants are cyanidin (50%), pelargonidin (12%), peonidin (12%), delphinidin (12%), petunidin (7%), and malvidin (7%).[5] However, anthocyanidins are unstable and convert to anthocyanins through glycosylation.[6] According with He and Giusti,[7] there have been more than 630 anthocyanins identified in nature. Figure 1 illustrates the general structure of some anthocyanins.

In nature, the main functions of anthocyanins are attracting pollinators and to prevent photo-oxidative damage; therefore, its concentration varies in different plant tissues.[8] The biosynthesis of anthocyanins in plants, as well as other flavonoids, occurs from shikimate pathway, phenyl propanoid pathway, and flavonoid pathway, as described by Routray and Orsat.[6] In food, these compounds are available in several fruits such as blueberries, cherries, peaches, grapes, pomegranates, plums as well as many dark greens, such as radish, black beans, red onion, eggplant, purple cabbage, and sweet potato. Dietary intake of anthocyanins is high if compared with other flavonoids, because their large distribution in plant materials. Thus, anthocyanins represent substantial components of nonenergy fraction of the human diet.[7,9] The acceptable daily intake for human was estimated to be 2.5 mg/kg body weight in 1982.[10]

The interests of the scientific community regarding the biochemical aspects and biological effects of anthocyanins increased substantially over the past decade because of large amounts of evidence demonstrating their wide therapeutic potential. The best property of anthocyanins described is their antioxidant activity, which is closely related to other health-promoting effects, such as reduction of cardiovascular disease risk,

Molecular Nutrition and Diabetes
http://dx.doi.org/10.1016/B978-0-12-801585-8.00011-7

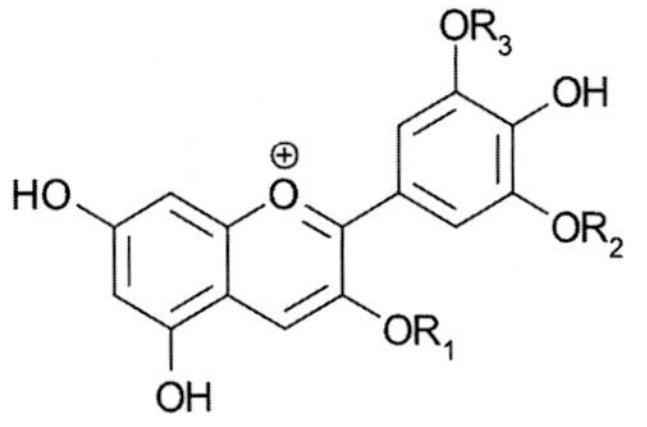

FIGURE 1 General structure of anthocyanins from the skeleton of anthocyanidins (aglycones). R_1: H or glycosidic substituent; R_2: pelargonidin (H), cyanidin (H), delphinidin (OH), peonidin (H), petunidin (OH), or malvinidin (OCH_3); and R_3: pelargonidin (H), cyanidin (OH), delphinidin (OH), peonidin (OCH_3), petunidin (OCH_3), or malvinidin (OCH_3).

prevention and/or inhibition of certain types of cancer, inhibition of platelet aggregation, anti-inflammatory action, modulation of the immune response, and protection from neuronal deficits associated with aging and neurodegenerative diseases.[7,11]

The biological activity of anthocyanins is closely related to its bioavailability. The amount of this compound in foods does not necessarily reflect the amount absorbed and metabolized by the organism.[3,12] After ingestion, anthocyanins are rapidly absorbed in the stomach and small intestine, especially in their intact glycosylated form. After absorption, they enter the systemic circulation after passage through the liver. In this organ, a portion of these compounds may be metabolized by methylation reactions and/or glucuronidation, which are considered two major routes in the metabolism of anthocyanins. In addition to these enzymatic conversions, the pigments can be degraded by intestinal microbiota in sugar molecules, phenolic acids, and aldehydes, and are subsequently absorbed in the colon.[7,13]

However, aspects such as the release of these compounds from food matrix, the extent of absorption in the organism, the influence of structural diversity and training metabolites in bioavailability and the biological efficacy of anthocyanins has not yet been completely clarified.

2. ANTHOCYANIN SOURCES IN FOODS OF PLANT ORIGIN

Anthocyanins, as previously mentioned, are widely distributed in food sources, including vegetable, cereals, flowers, and fruits. Among the vegetables, sources of anthocyanins are radish, black beans, red onion, eggplant, purple cabbage, sweet potato, chili peppers, and purple corn, whereas among the flowers there is the hibiscus.[14–19]

Regarding the fruits with anthocyanins, the most popular sources are blackberries, cherries, strawberries, peaches, grapes, pomegranates, and plums.[20] Currently, the berries (blueberry, blackberry, chokeberry, elderberry, grape, raspberry, strawberries, etc.) have received special attention, as for their versatility cuisine, as much by the amounts of anthocyanins and other bioactive compounds in these foods.[4,21–23] These fruits are characterized by small size, red or blue color, with many seeds, and by accumulating anthocyanins mainly in epicarp. Nowadays, berries are well-known and are consumed in cold climate countries such as the United States, Canada, and Europe.[24]

However, it is useful to highlight the great interest of countries in Latin America, with subtropical and tropical climate, in the study and their production and consumption of these fruits.[25–30] In Table 1 are presented origin, anthocyanins profile, and amounts of anthocyanins about berries and typical food sources with anthocyanins.

In addition to berries, native fruits on different continents have attracted the interest of researchers as potential sources of anthocyanins and other phenolic compounds. See this new world perspective in Table 2, including origin, anthocyanins profile, and amounts of anthocyanins in native fruits.

Acerola or Barbados cherry (*Malphigia* spp.) is a red small fruit known to contain large amounts of vitamin C but is also rich in anthocyanins (cyanidin-3-rhamnoside and pelargonidin-3-rhamnoside) and carotenoids. The fruit consumption is mainly in the form of processed juice[9]; however, Freitas et al.[31] found losses of 86.89% of the initial levels of anthocyanins in acerola-sweetened tropical juice produced by the aseptic (packings carton) process. Therefore, the processes used in the production of acerola pulp and juice can significantly reduce the final concentration of anthocyanins.

Açaí in Brazil (naidi palm, in Colombia; *Euterpe oleraceae* Mart.) is a purple fruit from the Amazon palm tree, and it has a great amount of anthocyanins, especially cyanidin-3-glucoside and cyanidin-3-rutinoside.[32] Gouvea et al.[33] found 35.3 mg per 100 g of cyanidin-3-glucoside and 58.7 mg per 100 g of cyanidin-3-rutinoside in the freeze-dried fruit. In addition to anthocyanins, the açaí also presents important amounts of minerals (6.94%) and phenolic acid as ferulic (10.3 mg per 100 g), caffeic (7.6 mg per 100 g), and *p*-coumaric (2.81 mg per 100 g).[34]

Camu camu (*Myrciaria dubia*) is a native fruit found mainly in the Peruvian Amazon and the Brazilian *cerrado*. As with the acerola fruit, it is also rich in vitamin C and anthocyanins, particularly cyanidin-3-glucoside and delphinidin-3-glucoside. Studies indicate a positive relationship between the degree of fruit ripening and the

TABLE 1 Typical Food Sources of Anthocyanins

Food Source	Origin	Anthocyanins Present	Amount of Anthocyanins	References
Blackberry (*Rubus* spp.)	Asia, Europe, North and South America	Cyanidin-3-glucoside, cyaniding-3-rutinoside, cyaniding-3-xyloside, cyaniding-3-malonyl-glucoside, cyaniding-3-dioxalyl-glucoside, cyaniding-3-arabinoside, pelargonidin-3-glucoside, and peonidin-3-glucoside	67.4–248.0 mg per 100 g	25
Blueberry (*Vaccinium* ssp.)	North America and East Asia	Cyanidin-3-arabinoside, delphinidin-3-arabinoside, delphinidin-3-galactoside, malvidin-3-arabinoside, and malvidin-3-galactoside	40.6–378.3 mg per 100 g (fresh weight)	4,27,29
Chili pepper (*Capsicum* spp.)	South and Central America	Delphinidin-3-*cis*-coumaroylrutinoside-5-glucoside, and delphinidin-3-*trans*-coumaroylrutinoside-5-glucoside	320.0 μg/g (fresh weight)	8,16
Eggplant (*Solanum melongena* L.)	Asia	Delphinidin-3-glucoside, delphinidin-3-rutinoside, delphinidin-3-rutinoside-5-galactoside, and delphinidin-3-rutinoside-5-glucoside	450.0 μg/g (fresh weight)	8,18
Grape (*Vitis vinifera* L.)	Europe and Middle East	Cyanidin-3-glucoside, delphinidin-3-glucoside, malvidin-3-glucoside, peonidin-3-glucoside, and petunidin-3-glucoside	20,000 μg/g fresh weight; 227.0–235.0 mg per 100 g (peel or fruit)	8,20
Purple corn or purple maize (*Zea mays* L.)	Andes (South America)	Cyanidin-3-glucoside, pelargonidin-3-glucoside, and peonidin-3-glucoside	213.6–904.0 mg/kg; 2534.1 μg/g (dry weight)	15,19
Strawberry (*Fragaria* x *ananassa* Duch.)	Europe	Cyanidin-3-glucoside, pelargonidin-3-glucoside, and pelargonidin-3-rutoside	200.0–600.0 mg/kg	21

TABLE 2 Native Fruits Sources of Anthocyanins

Food Source	Origin	Anthocyanins Present	Amount of Anthocyanins	References
Acerola or Barbados cherry (*Malphigia* spp.)	Central America	Cyanidin-3-rhamnoside and pelargonidin-3-rhamnoside	23.0–48.0 mg per 100 g (fresh weight)	9
Açaí or naidi palm (*Euterpe oleraceae* Mart.)	Brazil, Venezuela, Ecuador, Suriname, and Colombia (South America)	Cyanidin-3-glucoside and cyanidin-3-rutinoside	268.5 mg per 100 g (freeze-dried fruit)	33,34
Camu camu (*Myrciaria dubia*)	South America	Cyanidin-3-glucoside and delphinidin-3-glucoside	22.8–86.7 mg per 100 g (peel of ripe fruit)	35,36
Capuli (*Prunus serotina capuli* spp. [Cav] Mc. Vaug Cav)	Tropical America	Cyanidin 3-glucoside and cyanidin-3-rutinoside	2.7 mg/g (crude extract of peel) and 13.4 mg/g (extract rich en monomeric anthocyanin of peel)	37
Ceylon gooseberry (*Dovyalis hebecarpa*)	Sri Lanka (Asia)	Cyanidin-3-glucoside, cyanidin-3-rutinoside, delphinidin-3-glucoside, delphinidin-3-rutinoside, Peonidin-3-rutinoside, and petunidin-3-rutinoside	1617.0–2054.9 mg per 100 g (freeze-dried skin)	38
Chagalapoli (*Ardisia compressa* K.)	Mexico (North America)	Delphinidin-3-galactoside, malvidin-3-galactoside, and petunidin-3-galactoside	796.0 mg per 100 g (fresh weight)	39
Jaboticaba (*Myrciaria* spp.)	Brazil (South America)	Cyanidin-3-glucoside, delphinidin-3-glucoside, and peonidin-3-glucoside	310.0–315.0 mg per 100 g (peel of fruit)	20,42

concentration of anthocyanins and ascorbic acid in the peel.[35,36]

Capuli (*Prunus serotina capuli* spp. [Cav] Mc. Vaug Cav) is a round, purple fruit approximately 1 cm in diameter. The plant, which belongs to the Rosaceae family, grows in several tropical countries, and its fruit is eaten fresh or prepared as juices or homemade jams.[37] Scientific studies are still scarce, especially with regard to the anthocyanin content of the fruit.

Ceylon gooseberry (*Dovyalis hebecarpa*) is a deep red-to-purple exotic berry that is native to Asia. Bochi et al.[38] analyzed the concentration of anthocyanins present in skin and pulp of the fruit produced in Brazil and observed the possible effect of season variation, mainly in delphinidin-3-rutinoside level. The authors found higher levels of this compound in skin samples after the autumn season and higher levels in flesh after months of dry weather and lower temperatures.

Chagalapoli (*Ardisia compressa* K.) is a tropical fruit of deep purple color, with mineral composition similar to common berries (strawberry, blackberry, and blueberry). The fruit also presents a high content of total phenolics (1051.3 mg of gallic acid equivalents per 100 g of fresh weight), among which anthocyanins predominated. Chagalapoli seems to be interesting for the promotion of health because the initial studies showed an antioxidant capacity 40% higher compared with other berries.[39]

Jaboticaba (*Myrciaria* spp.) is a typical Brazilian fruit and generally is eaten raw or used in the preparation of jams and liqueurs. In popular medicine, the peel is used in the treatment of inflammatory conditions, including hemoptysis, asthma, tonsillitis, and diarrhea.[40] The jaboticaba peel is rich in minerals and soluble and insoluble fiber, and also presents relevant amounts of anthocyanins.[41] Reynertson[42] determined the anthocyanin content of the whole fruit, and the values were 433 mg per 100 g and 81 mg per 100 g of dry weight for cyanidin-3-O-β-glucoside and delphinidin-3-O-β-glucoside, respectively. Studies conducted by our research group, with freeze-dried jaboticaba peel powder, revealed the presence of 635.3 mg per 100 g of delphinidin-3-O-β-glucoside os dry weight, and 1964 mg per 100 g of cyanidin-3-O-β-glucoside of dry weight.

The amount and type of anthocyanins in plants are influenced by certain determinants, such as growing conditions, planting time, exposure to ultraviolet light, and harvest method. Therefore, the quantities of anthocyanin among different cultures of the same plant can generate very different results, and it is complex to compare fruits of different crops or different cultures.[20,43]

3. HEALTH EFFECTS OF ANTHOCYANINS

Hippocrates (460 BC–370 BC), considered the father of Western medicine, advocated that the food should be the medicine for the evils that afflict the body. Popular culture, over the centuries, strengthened food's relationship with the cure of diseases, from milder illnesses such as colds and flu, to more severe diseases, such as cancer. In recent years, it has been observed substantial advances in scientific understanding of the bioactive compounds present in food, and its effects on treatment and prevention of various pathologies. Among the phytochemicals, anthocyanins are known to exert positive biological effects.

When the possible biological effects of anthocyanins are discussed, those mainly mentioned in the literature are reduction of oxidative stress,[44] anticarcinogenic,[45] anti-inflammatory,[46,47] and antiatherosclerotic[48] effects; lower insulin resistance[49]; and improvement of visual acuity and cognitive performance.[50,51] Importantly, among the biological effects mentioned previously, one of the first related to anthocyanins was visual improvement. According to Ghosh and Konishi,[52] in World War I, British aviators ate bilberry jam to improve their night vision.

Basically, the biological effects attributed to anthocyanins are related to their antioxidant capacity. Anthocyanins have a direct free radical-scavenging capacity because of the hydrogen donation ability of the flavonoid molecule; they bind to reactive oxygen species (ROS; including free radicals, singlet oxygen, and peroxides), and terminate the chain reaction that is responsible for the oxidative damage.[7] However, antioxidant potential of anthocyanins depends on the chemical structure of the molecule, furthermore may be enhanced by other phytochemicals or vitamins that are also present in fruits.[51,53,54]

ROS are generated to the normal body functions; however, if the oxidants are produced in excess, it can cause damage to cellular structures (for example, nucleic acids, proteins, and lipids) and result in degenerative diseases, such as atherosclerosis and cancer.[7,48] Because of the antioxidant capacity of anthocyanins in vitro, isolated anthocyanins and food rich in these compounds have been tested in vivo as they could be active in the reduction of oxidative stress. Leite et al.[44] fed rats with diet supplemented at 0, 1, 2, and 4% of frieze-dried jaboticaba peel powder; they observed a significant increase in the plasmatic antioxidant potential in groups that received 1% and 2% of jaboticaba peel; however, the group that received 4% of the powder did not show antioxidant effects. Alvarez-Suarez et al.[55] supplemented 23 healthy subjects with 500 g of strawberry per day for 1 month; they observed a significant decreased in serum malondialdehyde, urinary 8-hydroxy-2′-deoxyguanosine, and urinary isoprostanes levels (−31.40, −29.67, and −27.90%, respectively) in addition to an increase in plasma total antioxidant capacity measured by both ferric reducing ability of plasma and oxygen radical absorbance capacity assays and vitamin C levels (+24.97, +41.18, and +41.36%, respectively).

With regard to the cardiovascular effects of anthocyanins, these substances are potentially explanatory for the "French paradox," in which Mediterranean people have low rates of coronary heart disease because the high consumption of red wine, even with a high-fat diet. The cardioprotective effects of anthocyanins include antihypertensive, endothelium protective, antiatherogenic activities, and interact with the estrogenic receptor. As possible causes of cardiovascular complications, these agents have actions, individually or synergistically, on insulin resistance, oxidative stress, and inflammation. In vitro and in vivo studies show that anthocyanins can reduce the oxidative stress involved in the atherosclerotic process; being that several mechanisms may be involved in this process, as the ability of anthocyanins to inhibit the oxidation of low-density lipoprotein (LDL) cholesterol and reduce oxidative injury of vascular endothelial cells.[48,52,54,56–58] Xia et al.[59] showed that anthocyanins, specifically cyanidin-3-glucoside or peonidin-3-glucoside, reduced cholesterol levels in human endothelial cells, resulting in the decrease of tumor necrosis factor (TNF) receptor-associated-2 translocation to lipid recruitment and inhibition of CD40-induced inflammatory signaling pathway. Quin et al.[60] studied the effects of 320 mg/day (four capsules of 80 mg) of anthocyanins (cyanidin and delphinidin majority) extracted from blueberries (*Vaccinium myrtillus*) and black currants (*Ribes nigrum*) in 120 subjects with dyslipidemia. After 12 weeks, the authors observed an increase in high-density lipoprotein cholesterol and a reduction in LDL cholesterol and cholesteryl ester transfer protein activity. Strawberries consumption (500 g per day for 1 month) improved plasma lipids profiles, biomarkers of antioxidant status, antihemolytic defenses, and platelet function in healthy subjects, indicating that intake of this fruit (rich in anthocyanins and vitamin C) may protect against cardiovascular disease.[55]

Inflammation is a complex biological response because of tissue injury. As mentioned previously, there is evidence showing that anthocyanins also exhibit anti-inflammatory properties, which can be explained by different mechanisms: inhibition of activation of the nuclear factor-kappa B (NF-κB); reduction of monocyte chemoattractant protein-1 (MCP-1) plasma concentration; inhibition of the inflammatory response and apoptosis of human endothelial cells induced by CD40 factor; and inhibition of nitric oxide production and expression of inducible nitric oxide synthase.[48,59] Intuyod et al.[61] administered an anthocyanin extract (rich in cyanidin and delphinidin), at 175 and 700 mg/kg body weight, every day for 1 month, to *Opisthorchis viverrini*–infected hamsters, and observed a decrease in inflammatory cells and periductal fibrosis. On the other hand, Basu et al.[62] evaluated antioxidant and anti-inflammatory effects in adults with abdominal obesity and elevated serum lipids supplemented with 50 g of strawberry daily for 12 weeks and, although they observed a reduction in lipid peroxidation, they found no difference in inflammatory markers.

Anticancer activity of anthocyanins has been established largely based on in vitro evidence. Through extensive research, He and Giusti[7] showed that the anticarcinogenic mechanisms of anthocyanins include antimutagenic activity, inhibition of oxidative DNA damage, inhibition of carcinogen activation and induction of phase II enzymes for detoxification, cell-cycle arrest, inhibition of cyclooxygenase-2 enzymes, induction of apoptosis, and antiangiogenesis. In 2012, we investigated the possible antiproliferative (in vitro) and antimutagenic (in vivo) activities of peel jaboticaba extracts (source of delphinidin 3-glucoside and cyanidin 3-glucoside). The polar jaboticaba extract showed antiproliferative effects against a leukemia cell (K-562), and the nonpolar extract was the most active against prostate cancer cell PC-3. A test in mice demonstrated that the polar jaboticaba extract induced no DNA damage, showing no cytotoxic properties on mice bone marrow cells and no mutagenic effects.[45] Bunea et al.[63] also found positive results, indicating that anthocyanins from blueberry could be used as a chemopreventive or adjuvant treatment for metastasis control, whereas inhibited B16-F10 melanoma murine cells proliferation at concentrations higher than 500 μg/mL.

It should be emphasized that animal and human clinical studies on health benefits of anthocyanins are still in the early stages. Furthermore, possibly conflicting results found in the scientific literature about the biological effects of anthocyanins may be the result of the differences in the type of study subjects, methods, concentration, dose, or source of anthocyanins. The most recent researches have investigated the relationship between anthocyanins, insulin resistance, obesity, and diabetes. Given that the incidence of these diseases is increasing all over the world, these topics are extremely important, and therefore will be discussed separately in the next sections.

4. INSULIN SIGNALING PATHWAY

Insulin is the most potent physiological anabolic hormone that is exclusively produced by the β cells of the pancreatic islets of Langerhans.

β cells are able to sense the nutritional state mainly because islets have a dense network of small blood vessels and capillaries surrounding islets are highly fenestrated allowing rapid access to circulating nutrients from the blood stream. Right after a meal, insulin secretion is induced primarily in response to an increase of food breakdown products in blood, such as glucose, amino acids, and fatty acids.[67–69]

The primary role of insulin is to regulate whole-body glucose homeostasis. The main tissues involved in the circulating glucose uptake and in efficient glucose utilization are skeletal muscle, adipose tissue, and liver.[64,65,70–72]

The major metabolic process by which insulin reduces the blood glucose concentration is increasing the rate of glucose uptake, primarily into fat and muscle cells, which is responsible for 85% of whole-body insulin-stimulated glucose uptake. Glucose transport is an essential step in cell metabolism because it controls the rate of glucose utilization. In skeletal muscle and adipose tissue, the clearance of circulating glucose depends on the insulin-stimulated translocation of the glucose transporter glucose transporter 4 (GLUT4) isoform from intracellular pools to the surface cell membrane. Insulin increases the rate of glycolysis by increasing glucose transport and the activities of hexokinase and 6-phosphofructokinase enzymes; it stimulates the glycogen synthesis in muscle, adipose tissue, and liver while reducing the rate of glycogen breakdown in muscle and liver; and reduces hepatic glucose output by decreased gluconeogenesis, which is an important in the maintenance of the glucose blood levels during fasting periods.[69,71–74]

Insulin also has profound effects on both lipid and protein metabolism. Insulin regulates lipid metabolism by decreasing the rate of lipolysis by repressing genes involved in fatty acid oxidation while increasing fatty acid and triacylglycerol synthesis in adipose tissue, liver, and muscle. In addition, insulin is required for the uptake of amino acids into tissues and protein synthesis contributing to its overall anabolic effect and to reduce protein degradation.[66,68,71–73]

In addition to the important insulin actions on the overall metabolism in its classical target organs (muscle, liver, and fat), there are many other important physiological targets of insulin, including the brain, pancreatic β cells, heart, vascular endothelium, kidney, and retina, that help to coordinate and couple metabolic homeostasis under healthy conditions.[75–78]

At the molecular level, insulin signaling is mediated by complex, highly integrated pathway that control systemic nutrient and metabolic homeostasis. Normal insulin signaling comprises a cascade of events initiated through insulin binding to and activation of its transmembrane receptor (insulin receptor, IR) which belongs to a subfamily of receptor tyrosine kinases and consists of two α subunits and two β subunits that are joined by disulfide bonds resulting in a α2β2 heterotetrameric complex. The insulin molecule binds to the extracellular α subunits, transmitting a signal across the plasma membrane that activates the intracellular tyrosine kinase domain of the β subunit. The receptor then undergoes a series of autophosphorylation reactions in which one β subunit phosphorylates its dimer partner on specific tyrosine residues. The activated IR is now able to recruit and phosphorylate multiple tyrosine residues of several intracellular substrates, including the insulin receptor substrate (IRS) proteins. Two members of the IRS family, IRS1 and IRS2, are the key substrates to trigger multiple downstream signaling pathways involved in the glucose transport and hormonal control of metabolism. The tyrosine phosphorylated residues of IRS1/2 generate docking sites for various Src-homology 2 containing proteins such as the p85 regulatory subunit of the class IA phosphatidylinositol 3-kinase (PI3K), small adapter proteins Grb2, and the SHP-2 protein tyrosine phosphatase. Among these, PI3K is a crucial mediator of the metabolic and mitogenic actions of insulin. It is composed of p110 catalytic subunit and its associated regulatory subunit p85, which binds to IRS proteins. Once activated, p110 catalyzes the phosphorylation of its substrate phosphatidylinositol 4, 5-bisphosphate on the 3′ position of the inositol ring to generate the lipid second messenger phosphatidylinositol-3, 4, 5-triphosphate (PIP3), which in turn recruits PH domain–containing proteins that are required for their binding to PIP3 in the plasma membrane activating several serine/threonine phosphorylation cascades. PIP3 main target is PDK1 that activates Akt [also known as protein kinase B)] and atypical protein kinase-C (PKC). Akt plays a central role in the metabolic actions of insulin.

For full Akt activation, phosphorylation at two different residues is required; phosphorylation of Thr308 occurs via PDK1 and phosphorylation of Ser473 is catalyzed by a distinct enzyme, most probably mammalian target of rapamycin (mTOR) complex 2 (defined by the interaction of mTOR and the rapamycin-insensitive companion of mTOR). Activated Akt/PKB phosphorylates multiple substrates relevant to insulin-like signaling including: glucose synthase kinase 3 and protein phosphatase 1 to inhibit glycogen synthesis; AS160 (Akt substrate of 160 kDa) to promote the translocation of insulin-mediated Glut4 from intracellular vesicles to the plasma membrane; tuberous sclerosis complex-1 and -2) to regulate the mTOR1 signaling, protein synthesis and cell growth; forkhead box O (FoxO) transcription factors to control the expression of numerous target genes, playing an essential role in mediating the effects of insulin on diverse physiological functions.[69–78]

The mammalian FoxO transcription factor family is represented by three members—FoxO1, FoxO3a, and FoxO4—all of which, following phosphorylation by Akt, undergo nuclear exclusion to cytoplasm, thereby reducing their transcriptional activities. FoxO proteins are involved in gluconeogenesis, energy metabolism, cellular survival, and proliferation. In the liver, insulin-induced Akt phosphorylation of FoxO1 suppresses the gluconeogenesis by reducing the levels of expression of the two rate-limiting enzymes phosphoenolpyruvate carboxykinase (PEPCK) and glucose-6-phosphatase. Moreover, the inactivation of FoxO1 in the liver reduces the apolipoprotein C III (ApoC-III) expression and the triglyceride metabolism. ApoC-III functions as a potent inhibitor of hepatic lipase, a key enzyme in the hydrolysis of triglycerides in very LDL and chylomicrons (ALTOMONTE 2004). An increase in ApoC-III levels induces the development of hypertriglyceridemia.[79–82]

FoxO proteins also appear to play significant effects in pancreatic β cells, adipocytes, and skeletal muscle. FoxO1 exerts a negative effect on pancreatic duodenal homeobox-1 (PDX-1) promoter activity. PDX-1 is a key regulator for both development (mass expansion) and maintenance of function. The activation of the IR-PI3K-Akt pathway decreases nuclear FoxO1 thus enabling an increase in Pdx-1 expression. Thus, insulin signaling is required to maintain an adequate β-cell mass. FoxO proteins have also been connected in regulating differentiation in adipocytes and muscle cells. In brief, insulin/insulin-like growth factor 1 signaling through Akt leading to nuclear exclusion of FoxO1 induces the terminal differentiation program in these cells. Nuclear FoxO1 promotes the upregulation of p21 cell-cycle inhibitor in preadipocytes and also suppresses peroxisome proliferator-activated receptor-δ (PPARδ) expression at the transcriptional level, which is a main regulator of adipogenesis, by coordinating the expression of many genes involved in the formation of mature adipocyte.[80–83] In muscle, it has been reported that an overexpression of a constitutively nuclear FoxO1 in C2C12 cells inhibits differentiation from myoblasts to myotubes induced by constitutively active Akt, indicating that FoxO proteins are key effectors of Akt-dependent myogenesis.[84]

Among the many molecules involved in the intracellular insulin signal transduction pathway, IRS1/2, Akt, and PI3K (the FOXO transcriptional factors) have attracted particular interest, mainly because their dysfunctions lead to the onset of an insulin resistance state in insulin-target tissues.

5. MOLECULAR MECHANISMS OF INSULIN RESISTANCE

Insulin resistance is a pathophysiological state in which insulin-target cells fail to respond properly to normal levels of circulating insulin. It results in an impairment of insulin-stimulated glucose uptake and glycogen synthesis, unrestrained hepatic gluconeogenesis, reduction in circulating lipids uptake, and disinhibited lipolysis.[85,86]

In the presence of IR, there is an increased metabolic demand for insulin and, thus, higher than ordinary concentrations of insulin are required to maintain normal or near-normal glycemia.[87,88] To this purpose, pancreatic β-cell—enhanced insulin secretion to overcome the reduced efficiency of insulin action, defined as compensatory hyperinsulinemia. However, in individuals destined to develop type 2 diabetes (T2D), the β-cell compensatory response is followed by β-cell failure, in which the pancreas fails to secrete sufficient amounts of insulin and T2D occurs.[75,76,85,89]

The most common underlying cause of IR is associated with overweight and obesity. Insulin resistance is the hallmark of the metabolic syndrome (also called syndrome X or insulin-resistance syndrome), which represents a cluster of different disorders, including fasting hyperglycemia, central adiposity, hypertension, and dyslipidemia (raised triglycerides and lowered high-density lipoprotein cholesterol) that predispose to the development of chronic metabolic diseases including T2D and cardiovascular diseases.[90–93]

Several etiological factors have been proposed in the pathogenesis of obesity-induced insulin resistance. With chronic overfeeding, the ability of white adipose tissue (WAT) to store the excess nutrients as triglycerides is impaired or exceeded resulting in increased concentrations of circulating free fatty acids and abnormal redistribution of lipids to other organs, including the liver and skeletal muscle. Ectopic lipid deposition and accumulation in intracellular content of fatty acid metabolites such as diacylglycerides, fatty acyl- fatty acyl-coenzyme A, and ceramides, have been implicated in the induction of obesity-related IR at these tissues.[91–93]

Originally considered basically as a primary site of storage excess energy, WAT is also considered a dynamic endocrine organ capable of secreting a number of bioactive peptides collectively referred as adipokines, which act locally and on other organs in through autocrine, paracrine, or endocrine function. These peptides include hormones, such as leptin and adiponectin, chemokines such as MCP-1 and interleukin-8 (IL-8), and pro-inflammatory and anti-inflammatory cytokines such as IL-1β/IL-6/TNF-α and IL-10, respectively. They play a central role in the regulation of many physiological processes including insulin sensitivity, metabolism homeostasis, energy balance, blood pressure control, and immune function.[85,94–97]

In states of chronic energy surplus, adipose tissue mass expands through two different mechanisms: hyperplasia, a process in which new adipocytes are formed from preadipocytes (adipogenesis), and hypertrophy, characterized by excessive triglyceride accumulation within adipocytes leading to an increase in cell size. This expansion accelerated by obesity can lead to a myriad of effects, including local hypoxia (deprivation of adequate oxygen supply) at the earliest stages of enlargement, adipocyte cell necrosis and death, and changes in immune cell populations, which together modify adipokine secretory patterns.[94,98,99]

This shift in adipokine secretory pattern from a less inflammatory environment to a predominantly pro-inflammatory profile in adipose tissue typically results in a chronic, low-grade inflammation that is recognized as a unifying mechanism linking obesity to the development of insulin resistance among other comorbidities.

Macrophage recruitment into adipose tissue is a key component of inflammation in obesity. Some of the consequences of adipocyte hypertrophy like hypoxia, cell death, and augmented chemokine secretion (e.g., MCP-1) by adipocytes have been proposed to be initiators of macrophage infiltration and inflammatory response, because adipose tissue macrophages (ATMs) are a major source of pro-inflammatory cytokines.[100,101]

Besides increasing macrophage number, obesity also induces a phenotypic switch in ATMs. Macrophages residing in lean adipose tissue are commonly described as M2 or "alternatively activated" macrophages. They are characterized by high expression of Ym1, arginase1, and anti-inflammatory cytokines, such as IL-10 and IL-1 receptor antagonist, and participate in the blockade of inflammatory responses and in the promotion of tissue repair and angiogenesis. Diet-induced obesity leads to a shift from a predominantly anti-inflammatory M2 macrophage population to an increased proportion of M1 or "classically activated" macrophages that are usually recruited to sites of tissue damage and produce high levels of pro-inflammatory cytokines (e.g., TNF-α, IL-6) and inducible nitric oxide synthase.[100,102,103] This shift in the M2/M1 balance is due to the migration of inflammatory monocytes from the circulation to macrophage clusters and not to the conversion of resident M2 to M1 macrophages in situ.[104]

Activated M1 macrophages recruit additional macrophages establishing a positive feedback loop that further increases ATM content and propagates the chronic inflammatory state. However, macrophage-secreted cytokines exert paracrine effects to activate inflammatory pathways within insulin target tissues, such as liver and skeletal muscle, resulting in systemic insulin resistance.[85,105]

ATMs are not the only immune cell type that can accumulate in obese adipose tissue. Recently studies have shown that T lymphocytes have a central physiological role during the early stages of adipose tissue expansion. More specifically, they revealed that large numbers of $CD8^+$ effector T cells infiltrated obese adipose tissue, whereas the numbers of $CD4^+$ helper and regulatory T cells, which exert anti-inflammatory activity, were diminished. Furthermore, it was demonstrated that T lymphocytes precede the accumulation of macrophage, and that specifically cytotoxic ($CD8^+$) lymphocytes may be key mediators of early adipose tissue inflammation, insulin resistance, and macrophage migration and activation to inflammatory state.[85,106,107]

Saturated fatty acids (SFAs) are also proposed to play an important role for the development of insulin resistance through induction of inflammation. SFAs are released in excess from hypertrophied adipocytes whose circulating levels are often increased in obesity and serve as a naturally occurring ligand for the Toll-like receptor 4. Toll-like receptor 4 is expressed plays a critical role in the innate immune system by activating pro-inflammatory signaling pathways in response to pathogen-associated molecular patterns, particularly lipopolysaccharides, a major cell wall component of gram-negative bacteria.[108]

Many of the more typical pro-inflammatory stimuli in obesity, such as cytokines and SFAs, simultaneously activate inhibitor of nuclear factor-κB kinase β (IKK-β)/NF-κB and c-Jun amino-terminal kinase (JNK) signaling pathways through specific receptor-mediated mechanisms.[105,109–111]

In the resting state, NF-κB exists as a cytoplasmic heterodimeric complex composed mainly of p50 and p65 proteins bound to inhibitory proteins of IκB, which prevents its nuclear translocation and DNA binding. Upon stimulation, IκB protein is phosphorylated and degraded in the proteasome and the NF-κB dimers are subsequently translocated from cytoplasm to nucleus, where it binds on the promoters of various target genes and leading to increased expression of pro-inflammatory cytokines such as TNF-α, IL-6, and IL-1.[85,105,112]

Similarly, the other major intracellular pro-inflammatory pathway involves the JNK1/AP1 system, which is also recognized to respond to cytokines (e.g., TNF. IL-1). JNK can interact with and phosphorylate the N-terminal transactivation domain of c-Jun and additional transcription factors such as JunB, JunD, c-Fos, and ATF, that together with c-Jun, constitute activator protein-1. This leads to transcriptional activation of several inflammatory and stress-responsive genes, which overlap with the set of genes transactivated by NF-κB.[85,113]

In additional to pro-inflammatory cytokines and pattern recognition receptors, cellular stresses can potentiate NF-κB and JNK phosphorylation pathways, including oxidative and endoplasmic reticulum (ER) stress. Obesity overloads the functional capacity of the ER, resulting in activation of the unfolded protein response and this ER stress leads to the accumulation of ROS.[92,105,114] There are many other mechanisms by which obesity induces systemic ROS production and decreases the activity of antioxidant enzymes such as superoxide dismutase, catalase, and glutathione peroxidase in cells and tissues, which in turn promote the secretion of more inflammatory cytokines and chemokines to initiate or amplify inflammatory signals, contributing to insulin resistance.[66,92]

At the molecular level, increased circulating free fatty acid levels, fatty acids derivatives, such as ceramides and diacylglycerides, pro-inflammatory cytokines, and ROS are important negative regulators of insulin signaling by targeting key components of this pathway.

IR/IRS signaling normally occurs through a tyrosine phosphorylation events and is strongly attenuated by counterregulatory serine and/or threonine phosphorylations, creating a state of cellular insulin resistance. This occurs on a number of sites because of several different protein serine/threonine kinases that are activated by inflammatory or stressful stimuli and contribute to inhibition of insulin signaling, including JNK, IKK, and atypical PKC-θ.[115]

IKKβ can affect insulin signaling through both by directly phosphorylating IRS-1 on the inhibitory serine residues and by phosphorylating IκB thus activating NF-κB. This triggers a further vicious cycle of heightened inflammatory responses that feed into the negative regulation of insulin signaling.[92] Another inhibitory effect of IKKβ/NF-κB on insulin signaling is the upregulation of the protein-tyrosine phosphatase 1B expression that dephosphorylates IR and IRS1.[116] Instead, JNK phosphorylates IRS-1 on Ser307 impairing insulin action.[117] A rise in levels of intracellular fatty acid metabolites can activate PKC-θ, which increase Ser307 phosphorylation of IRS-1 and may also activate IKKβ and NF-κB.[105,117–121]

In addition to serine/threonine kinase cascades, other pathways contribute to obesity-induced insulin resistance. For example, suppressor of cytokine signaling (SOCS) proteins, mainly SOCS1 and SOCS3, expression are elevated during obesity, are able to inhibit insulin signaling by three different mechanisms: preventing the association of IRS proteins with activated IR that leads to the blockade of the downstream signals, targeting IRS proteins for degradation and inhibiting the kinase activity of the IR and subsequently decreasing IRS-1 tyrosine phosphorylation.[117,122–125]

6. INSULIN SENSITIZING AND ANTIDIABETIC PROPERTIES OF ANTHOCYANINS

Obesity has reached epidemic proportions globally, leading to a dramatic rise in the incidence of serious chronic diseases, including insulin resistance, T2D, cardiovascular diseases, and certain forms of cancer. The worldwide growth in the prevalence of obesity and its related comorbidities has largely been driven by increased access and consumption of low-cost foods that are low in nutritional value and high in energy combined with declines in physical activity levels, mainly from more mechanized and technologically lifestyles.[126]

Until now, effective and safe pharmacological approaches for the prevention or treatment of obesity remain elusive, and despite both public and private resources being applied to research in the field, the development of antiobesity drugs have been inundated by failures.[127,128]

However, mounting evidence indicates that healthy eating, characterized by high consumption of fruits, vegetables, legumes, nuts, whole grains, and olive oil, can prevent or delay the manifestation of a wide range of pathogenic processes associated with metabolic diseases.

Anthocyanins are among the most abundant polyphenols in fruits, vegetables, and grains. Over the past decade, several animal studies and epidemiological investigations have demonstrated that anthocyanin consumption, either anthocyanin extracts from different plants or pure anthocyanin, can prevent or reverse obesity- and T2D-related disturbances including oxidative stress, chronic inflammation, hyperglycemia, and insulin resistance.[49,129]

In 2003, Tsuda and coworkers[130] published the first report on the preventive effects of anthocyanins against obesity and its related disorders. C57BL/6J mice were fed a high-fat diet (HFD) with or without anthocyanin-rich purple corn color (2 g/kg) for 12 weeks. Anthocyanin supplementation significantly reduced body weight gain and fat accumulation and normalized plasma glucose concentration, hyperinsulinemia, and TNF-α messenger RNA expression, suggesting that dietary anthocyanin may ameliorate insulin resistance in obese mice. In a series of subsequent in vitro studies, these authors showed that isolated rat and human adipocytes treated with typical anthocyanins, cyanidin

3-glucoside or cyanidin, induced significant changes in adipokine expression to a more insulin-sensitive pattern, by upregulating adiponectin and downregulating of plasminogen activator inhibitor-1 and IL-6.[131–133]

In a similar study, mice were initially fed an HFD for 4 weeks and then switched to an HFD containing purified anthocyanins from Cornelian cherries (1 g/kg) for an additional 8 weeks. The anthocyanin-treated mice showed a 24% decrease in weight gain, improved tolerance glucose, and decreased accumulation of lipids in the liver.[134]

Black soybean seed coat anthocyanins and extract are also reported to exhibit an antiobesity effect in HFD-fed mice. The black soybean anthocyanins-added diet (0.037%) suppressed the HFD-induced weight gain and significantly reduced the levels of serum triglyceride and cholesterol, whereas they markedly increased the high-density lipoprotein-cholesterol concentration.[135] Black soybean seed coat extract added in a HFD for 14 weeks suppressed fat accumulation in mesenteric adipose tissue, without affecting food intake, reduced the plasma glucose level, and enhanced insulin sensitivity in obese mice. Moreover, the extract significantly reduced the gene expression of TNF-R, MCP-1, and IL-6 in mesenteric adipose tissue. These findings indicate that this treatment strategy has the potential to ameliorate the inflammatory responses and thus insulin resistance accompanying obesity.[136]

Tart cherries are also a rich fruit source of anthocyanins. Seymour et al.[137] investigated whether anthocyanin-containing whole foods could confer similar effects to concentrated anthocyanin extracts. They observed that an HFD containing whole tart cherry powder (1%) for 90 days reduced fat mass, hyperlipidemia, and IL-6, TNF-α, and NF-κB activity in retroperitoneal adipose tissue of Zucker fatty rats.

Prior and colleagues also compared the effects of anthocyanin extract from blueberries and whole blueberry powder added as a supplement to an HFD in C57BL/6 mice. The intake of anthocyanin-purified extract considerably inhibited weight gain and body fat accumulation, whereas the consumption of whole blueberries did not prevent and may have actually increased obesity.[138]

A similar study conducted by DeFuria et al.[139] showed that whole blueberry powder (4%) supplemented to an HFD for 8 weeks also did not affect body weight gain or adiposity. On the other hand, blueberry intake inhibited WAT inflammation and oxidative stress and protected mice from IR and hyperglycemia induced by HFD.

Jaboticaba fruit, a native Brazilian berry, is also a rich source of anthocyanins. The major anthocyanin in jaboticaba peels consists primarily of cyanidin 3-O-glucoside (~75.6%). Recently, we evaluated the effects of freeze-dried jaboticaba peel powder (FDJPP) on obese Swiss mice. After 4 weeks of HFD, mice received an HFD supplemented with FDJPP (1, 2, or 4%) for an additional 6 weeks. The FDJPP exerted no protective effect on HFD-induced weight gain and glucose intolerance. However, the supplementation was effective in reducing insulin resistance, as evidenced in the insulin tolerance test, and subsequently confirmed by improved signal transduction through the IR/IRS/Akt/FoxO pathway by the attenuation of HFD-induced inflammation in the liver, verified by lower expressions of IL-1β and IL-6 and decreased phosphorylated IkB-a protein levels in all jaboticaba-treated mice. Our results suggest that FDJPP exerts a protective role against obesity-associated insulin resistance.[140]

In a type 2 diabetic mice model (KK-A^y), dietary high-purity anthocyanin or bilberry extract significantly reduced blood glucose concentration and enhanced insulin sensitivity because of the reduction of the adipokine retinol-binding protein 4 expression and via activation of AMP-activated protein kinase, respectively. More specifically, purified anthocyanin upregulated GLUT4, which in turn led to downregulated retinol binding protein 4 expression, which was accompanied by a reduction of the MCP-1 and TNF-α levels in WAT.[141] Dietary bilberry was shown to activate AMP-activated protein kinase in WAT, skeletal muscle, and liver. This activation was accompanied by upregulation of GLUT4 in WAT and skeletal muscle, suppression of hepatic gluconeogenesis by downregulation of phosphoenolpyruvate carboxykinase and glucose-6-phosphatase messenger RNA expression, and reduction of lipid content in the liver, via phosphorylation of ACC and upregulation of PPARα, Acyl-fatty acyl-coenzyme A oxidase, and carnitine palmitoyltransferase-1A, which catalyzes fatty acid oxidation and reduces of lipotoxicity.[142]

7. CONCLUDING REMARKS

In this chapter, we summarized the most important data on anthocyanin-mediated protection against obesity, insulin resistance, and T2D, together with the underlying molecular mechanisms. According to the studies published to date, there is strong evidence suggesting that anthocyanins can improve insulin sensitivity, mainly because of their anti-inflammatory activity in important metabolic tissues. However, a better understanding of the physiological functionality of anthocyanins requires further investigations to effectively help uncover molecular targets for the development of treatment strategies for overweight- and obesity-related diseases.

References

1. Lieberman S. The antioxidant power of purple corn: a research review. *Altern Complement Ther* 2007;**13**:107–10.
2. Bordignon Jr C, Francescatto V, Nienow A, Calvete E, Reginatto F. Influência do pH da solução extrativa no teor de antocianinas em frutos de morango. *Ciênc Tecnol Aliment* 2009;**29**:183–8.
3. Mcghie T, Walton M. The bioavailability and absorption of anthocyanins: towards a better understanding. *Mol Nutr Food Res* 2007; **51**:702–13.
4. Norberto S, Silva S, Meireles M, Faria A, Pintado M, Calhau C. Blueberry anthocyanins in health promotion: a metabolic overview. *J Func Foods* 2013;**5**:1518–28.
5. Kong J, Chia L, Goh N, Chia T, Brouillard R. Analysis and biological activities of anthocyanins. *Phytochemistry* 2003;**64**:923–33.
6. Routray W, Orsat V. Blueberries and their anthocyanins: factors affecting biosynthesis and properties. *Compr Rev Food Sci F* 2010; **10**:303–20.
7. He J, Giusti M. Anthocyanins: natural colorants with health-promoting properties. *Annu Rev Food Sci T* 2010;**1**:163–87.
8. Aza-González C, Núñez-Palenius H, Ochoa-Alejo N. Molecular biology of chili pepper anthocyanin biosynthesis. *J Mex Chem Soc* 2012;**56**:93–8.
9. Brito E, Araújo M, Alves R, Carkeet C, Clevidence B, Novotny J. Anthocyanins present in selected tropical fruits: acerola, jambolão, jussara, and guajiru. *J Agri Food Chem* 2007;**55**:9389–94.
10. Clifford MN. Anthocyanins: nature, occurrence and dietary burden. *J Sci Food Agri* 2000;**80**:1063–72.
11. Talavéra S, Felgines C, Texier O, et al. Anthocyanin metabolism in rats and their distribution to digestive area, kidney, and brain. *J Agri Food Chem* 2005;**53**:3902–8.
12. Biesalski H, Dragsted L, Elmadfa I, et al. Bioactive compounds: definition and assessment of activity. *Nutrition* 2009;**25**:1202–5.
13. Shipp J, Abdel-Aal E-S. Food applications and physiological effects of anthocyanins as functional food ingredients. *T Open Food Sci J* 2010;**4**:7–22.
14. Eibond L, Reynertson K, Luo X, Basile M, Kennelly E. Anthocyanin antioxidants from edible fruits. *Food Chem* 2004;**84**:23–8.
15. Salinas-Moreno Y, Chávez F, Ortiz S, González F. Pigmented maize grains from chiapas, physical characteristics, anthocyanin content and nutraceutical value. *Rev Fitotec Mexi* 2012;**35**:33–41.
16. Aza-González C, Ochoa-Alejo N. Characterization of anthocyanins from fruits of two Mexican chili peppers (*Capsicum annuum* L.). *J Mex Chem Soc* 2012;**56**:149–51.
17. Medina-Carrillo R, Sumaya-Martínez M, Machuca-Sánchez M, Sánchez-Herrera L, Balois-Morales R, Jiménez-Ruiz E. Antioxidant activity of aqueous extracts from dried calyxes of 64 varieties of roselle (*Hibiscus sabdariffa* L.) versus phenolic compounds and total monomeric anthocyanins. *Rev Cie Téc Agr* 2013;**22**:41–4.
18. Arrazola G, Herazo I, Alvis A. Microencapsulación de antocianinas de berenjena (*Solanum melongena* L.) mediante secado por aspersión y evaluación de la estabilidad de su color y capacidad antioxidante. *Inf Tec* 2014;**25**:31–42.
19. Harakotr B, Suriharn B, Tangwongchai R, Scott M, Lertrat K. Anthocyanin, phenolics and antioxidant activity changes in purple waxy corn as affected by traditional cooking. *Food Chem* 2014;**164**: 510–7.
20. Terci D. *Aplicações analíticas e didáticas de antocianinas extraídas de frutas* [Thesis (Doctor)]. Brazil: University of Campinas; 2004.
21. Silva F, Escribano-Bailón M, Alonso J, Rivas-Gonzalo J, Santos-Buelga C. Anthocyanin pigments in strawberry. *LWT - Food Sci Tech* 2007;**40**:374–82.
22. Pantelidis G, Vasilakakis M, Manganaris G, Diamantidis G. Antioxidant capacity, phenol, anthocyanin and ascorbic acid contents in raspberries, blackberries, red currants, gooseberries and Cornelian cherries. *Food Chem* 2007;**102**:777–83.
23. Crecente-Campo J, Nunes-Damaceno M, Romero-Rodríguez M, Vázquez-Odériz M. Color, anthocyanin pigment, ascorbic acid and total phenolic compound determination in organic versus conventional strawberries (*Fragaria* x *ananassa* Duch, cv Selva). *J Food Comp Anal* 2012;**28**:23–30.
24. Fredes C, Montenegro G, Zoffoli J, Santander F, Robert P. Comparison of the total phenolic content, total anthocyanin content and antioxidant activity of polyphenol-rich fruits grown in Chile. *Cie Inv Agr* 2014;**41**:49–60.
25. Ferreira D, Faria A, Grosso C, Mercadante A. Encapsulation of blackberry anthocyanins by thermal gelation of curdlan. *J Bra Chem Soc* 2009;**20**:1908–15.
26. Guerrero J, Ciampi L, Castilla A, et al. Antioxidant capacity, anthocyanins, and total phenols of wild and cultivated berries in Chile. *Chil J Agr Res* 2010;**70**:537–44.
27. Rodrigues E, Poerner N, Rockenbach I, Gonzaga L, Mendes C, Fett R. Phenolic compounds and antioxidant activity of blueberry cultivars grown in Brazil. *Cie Tec Alim* 2011;**31**:911–7.
28. Cantillano R, Ávila J, Peralba M, Pizzolato T, Toralles R. Actividad antioxidante, compuestos fenólicos y ácido ascórbico de frutillas en dos sistemas de producción. *Hort Bra* 2010;**30**:620–6.
29. Yousef G, Brown A, Funakoshi Y, et al. Efficient quantification of the health-relevant anthocyanin and phenolic acid profiles in commercial cultivars and breeding selections of blueberries (*Vaccinium* spp.). *J Agr Food Chem* 2013;**61**:4806–15.
30. Reque P, Steffens R, Jablonski A, Flôres S, Rios A, Jong E. Cold storage of blueberry (*Vaccinium* spp.) fruits and juice: anthocyanin stability and antioxidant activity. *J Food Comp Anal* 2014; **33**:111–6.
31. Freitas C, Maia G, Costa J, Figueiredo R, Sousa P, Fernandes A. Stability of carotenoids, anthocyanins and vitamin C presents in acerola sweetened tropical juice preserved by hot fill and aseptic processes. *Cie Agrotec* 2006;**30**:942–9.
32. Santiago M, Gouvêa A, Godoy R, Neto J, Pacheco S, Rosa J. Adaptação de um método por cromatografia líquida de alta eficiência para determinação de antocianinas em suco de açaí (*Euterpe oleraceae* Mart.). *Comun Técnico Embrapa* 2010; **162**:1–4.
33. Gouvêa A, Araujo M, Schulz D, Pacheco S, Godoy R, Cabral L. Anthocyanins standards (cyanidin-3-O-glucoside and cyanidin-3-O-rutinoside) isolation from freeze-dried açaí (*Euterpe oleraceae* Mart.) by HPLC. *Cie Tec Alim* 2012;**32**:43–6.
34. Rojano B, Vahos I, Arbeláez A, Martínez A, Correa F, Carvajal L. Polyphenols and antioxidant activity of the freeze-dried palm naidi (*Colombian açai*) (*Euterpe oleracea* Mart.). *Rev Fac Nac Agron Medellín* 2011;**64**:6213–20.
35. Villanueva-Tiburcio J, Condezo-Hoyos L, Asquieri E. Antocianinas, ácido ascórbico, polifenoles totales y actividad antioxidante, en la cáscara de camu-camu (*Myrciaria dubia* (H.B.K) McVaugh). *Cie Tec Alim* 2010;**30**:151–60.
36. Gómez J, Rodríguez F, Amaral C, et al. Variación del contenido de vitamina c y antocianinas en *Myrciaria dubia* "camu camu". *Rev la Soc Quim Perú* 2013;**79**:319–30.
37. Hurtado N, Pérez M. Identificación, estabilidad y actividad antioxidante de las antocianinas aisladas de la cáscara del fruto de capulí (*Prunus serotina* spp capuli (Cav) Mc. Vaug Cav). *Inf Tecnol* 2014;**25**:131–40.
38. Bochi V, Godoy H, Giusti M. Anthocyanin and other phenolic compounds in Ceylon gooseberry (*Dovyalis hebecarpa*) fruits. *Food Chem* 2015;**176**:234–43.
39. Joaquín-Cruz E, Dueñas M, García-Cruz L, Salinas-Moreno Y, Santos-Buelga C, García-Salinas C. Anthocyanins and phenolics characterization, chemical composition and antioxidant activity of chagalapoli (*Ardisia compressa* K.) fruit: a tropical source of natural pigments. *Food Res Int* 2015;**70**. http://dx.doi.org/10.1016/j.foodres.2015.01.033.

40. Lima A, Corrêa A, Alves A, Abreu C, Dantas-Barros A. Caracterização química do fruto jaboticaba (*Myrciaria cauliflora* Berg) e de suas frações. *Arch Latinoam Nutr* 2008;**58**:416–21.
41. Dragano N. *Avaliação do efeito da casca de jaboticaba liofilizada sobre o ganho de peso, perfil lipídico e resistência à insulina em camundongos alimentados com dieta hiperlipídica* [Dissertation (Master)]. Brazil: University of Campinas; 2011.
42. Reynertson K. *Phytochemical analysis of bioactive constituents from edible myrtaceae fruits* [Thesis (Graduate)]. New York: University of New York; 2007 [USA].
43. Siriwoharn T, Wrolstad R, Finn C, Pereira C. Influence of cultivar, maturity and sampling on blackberry (*Rubus* L. Hybrids) anthocyanins, polyphenolics, and antioxidant properties. *J Agri Food Chem* 2004;**52**(26):8021–30.
44. Leite A, Malta L, Riccio M, Eberlin M, Pastore G, Maróstica Júnior M. Antioxidant potential of rat plasma by administration of freeze-dried jaboticaba peel (*Myrciaria jaboticaba* Vell Berg). *J Agri Food Chem* 2011;**59**:2277–83.
45. Leite-Legatti A, Batista A, Dragano N, et al. Jaboticaba peel: antioxidant compounds, antiproliferative and antimutagenic activities. *Food Res Int* 2012;**49**:596–603.
46. Kraft T, Dey M, Rogers R, et al. Phytochemical composition and metabolic performance-enhancing activity of dietary berries traditionally used by native North Americans. *J Agri Food Chem* 2008;**56**:654–60.
47. Bowen-Forbes C, Zhang Y, Nair M. Anthocyanin content, antioxidant, anti-inflammatory and anticancer properties of blackberry and raspberry fruits. *J Food Comp Anal* 2010;**23**:554–60.
48. Cardoso L, Leite J, Peluzio M. Efeitos biológicos das antocianinas no processo aterosclerótico. *Ver Col Cie Quí Farm* 2011;**40**:11–38.
49. Sancho RAS, Pastore GM. Evaluation of the effects of anthocyanins in type 2 diabetes. *Food Res Int* 2012;**46**:378–86.
50. Yao Y, Vieira A. Protective activities of Vaccinium antioxidants with potential relevance to mitochondrial dysfunction and neurotoxicity. *Neurotoxicology* 2007;**28**:93–100.
51. Pojer E, Mattivi F, Johnson D, Stockley C. The case for anthocyanin consumption to promote human health: a review. *Compr Rev Food Sci F* 2013;**12**:483–508.
52. Ghosh D, Konishi T. Anthocyanins and anthocyanin-rich extracts: role in diabetes and eye function. *Asia Pac J Clin Nutr* 2007;**16**:200–8.
53. Kähkönen M, Heinonen M. Antioxidant activity of anthocyanins and their aglycons. *J Agri Food Chem* 2003;**51**:628–33.
54. Garzón G. Anthocyanins as natural colorants and bioactive compounds: a review. *Acta Biol Colomb* 2008;**13**:27–36.
55. Alvarez-Suarez J, Giampieri F, Tulipani S, et al. One-month strawberry-rich anthocyanin supplementation ameliorates cardiovascular risk, oxidative stress markers and platelet activation in humans. *J Nutr Biochem* 2014;**25**:289–94.
56. Wang S, Jiao H. Scavenging capacity of berry crops on superoxide radicals, hydrogen peroxide, hydroxyl radicals, and singlet oxygen. *J Agri Food Chem* 2000;**48**:5677–84.
57. Yi L, Chen C, Jin X, Mi M, Ling W, Zhang T. Structural requirements of anthocyanins in relation to inhibition of endothelial injury induced by oxidized low-density lipoprotein and correlation with radical scavenging activity. *FEBS Lett* 2010;**584**:583–90.
58. Pascual-Teresa S. Molecular mechanisms involved in cardiovascular and neuroprotective effects of anthocyanins. *Arch Biochem Biophys* 2014;**559**:68–74.
59. Xia M, Ling W, Zhu H, et al. Anthocyanins prevents from CD40-activated proinflammatory signaling in endothelial cells by regulating cholesterol distribution. *Arterioscler Thromb Vasc Biol* 2007; **27**:519–24.
60. Quin Y, Xia M, Ma J, et al. Anthocyanin supplementation improves serum LDL- and HDL-cholesterol concentrations associated with the inhibition of cholesteryl-ester transfer protein in dyslipidemic subjects. *Am J Clin Nutr* 2009;**90**:485–92.
61. Intuyod K, Priprem A, Limphirat W, et al. Anti-inflammatory and anti-periductal fibrosis effects of an anthocyanin complex in *Opisthorchis viverrini*-infected hamsters. *Food Chem Toxicol* 2014; **74**:206–15.
62. Basu A, Betts N, Nguyen A, Newman E, Fu D, Lyons T. Freeze-dried strawberries lower serum cholesterol and lipid peroxidation in adults with abdominal adiposity and elevated serum lipids. *J Nutr* 2014;**144**(6). http://dx.doi.org/10.3945/jn.113.188169.
63. Bunea A, Rugină D, Sconta Z, et al. Anthocyanin determination in blueberry extracts from various cultivars and their antiproliferative and apoptotic properties in B16-F10 metastatic murine melanoma cells. *Phytochemistry* 2013;**95**:436–44.
64. Pessin JE, Saltiel AR. Signaling pathways in insulin action: molecular targets of insulin resistance. *J Clin Invest* 2000;**106**(2):165–9.
65. Saltiel AR, Pessin JE. Insulin signaling pathways in time and space. *Trends Cell Biol* 2002;**12**(2):65–71.
66. Rains JL, Jain SK. Oxidative stress, insulin signaling, and diabetes. *Free Radic Biol Med* 2011;**50**:567–75.
67. Carvalheira JB, Zecchin HG, Saad MJ. Vias de sinalização da insulina. *Arq Bras Endocrinol Metabol* 2002;**46**(4):419–25.
68. Suckale J, Solimena M. Pancreas islets in metabolic signaling - focus on the beta-cell. *Front Biosci* 2008;**13**:7156–71.
69. Eslam M, Khattab MA, Harrison SA. Insulin resistance and hepatitis C: an evolving story. *Gut* 2011;**60**:1139–51.
70. Saltiel AR, Kahn CR. Insulin signaling and the regulation of glucose and lipid metabolism. *Nature* 2001;**414**:799–806.
71. Samuel VT, Shulman GI. Mechanisms for insulin resistance: common threads and missing links. *Cell* 2012;**148**:852–71.
72. Rask-Madsen C, Kahn CR. Tissue-specific insulin signaling, metabolic syndrome, and cardiovascular disease. *Arterioscler Thromb Vasc Biol* 2012;**32**:2052–9.
73. Newsholme E, Dimitriadis G. Integration of biochemical and physiologic effects of insulin glucose metabolism. *Exp Clin Endocrinol Diabetes* 2001;**109**:122–34.
74. Dimitriadis G, Mitrou P, Lambadiari V, Maratou E, Raptis SA. Insulin effects in muscle and adipose tissue. *Diabetes Res Clin Pract* 2011;**93**:S52–9.
75. Kasuga M. Insulin resistance and pancreatic β cell failure. *J Clin Invest* 2006;**116**:1756–60.
76. Kahn SE, Hull RL, Utzschneider KM. Mechanisms linking obesity to insulin resistance and type 2 diabetes. *Nature* 2006;**444**:840–6.
77. Saine V. Molecular mechanisms of insulin resistance in type 2 diabetes mellitus. *World J Diabetes* 2010;**1**(3):68–75.
78. Altomonte J, Cong L, Harbaran S, Richter A, Xu J, Meseck M, et al. Foxo1 mediates insulin action on apoC-III and triglyceride metabolism. *J Clin Invest* 2004;**114**(10):1493–503.
79. Accili D, Arden KC. FoxOs at the crossroads of cellular metabolism, differentiation, and transformation. *Cell* 2004;**117**(4):421–6.
80. Gross DN, van den Heuvel APJ, Birnbaum MJ. The role of FoxO in the regulation of metabolismo. *Oncogene* 2008;**27**:2320–36.
81. Nakae J, Oki M, Cao Y. The FoxO transcription factors and metabolic regulation. *FEBS Lett* 2008;**582**:54–67.
82. Dowell P, Otto TC, Adi S, Lane MD. Convergence of peroxisome proliferator-activated receptor gamma and Foxo1 signaling pathways. *J Biol Chem* 2003;**278**:45485–91.
83. Armoni M, Harel C, Karni S, et al. FOXO1 represses peroxisome proliferator-activated receptor-gamma1 and -gamma2 gene promoters in primary adipocytes. A novel paradigm to increase insulin sensitivity. *J Biol Chem* 2006;**281**:19881–91.
84. Hribal ML, Nakae J, Kitamura T, Shutter JR, Accili D. Regulation of insulin-like growth factor-dependent myoblast differentiation by Foxo forkhead transcription factors. *J Cell Biol* 2003; **162**:535–41.
85. Olefsky JM, Glass CK. Macrophages, inflammation, and insulin resistance. *Annu Rev Physio* 2010;**72**:219–46.

86. Schenk S, Saberi M, Olefsky JM. Insulin sensitivity: modulation by nutrients and inflammation. *J Clin Investig* 2008;**118**: 2992–3002.
87. Mlinar B, Marc J, Janez A, Pfeifer M. Molecular mechanisms of insulin resistance and associated diseases. *Clin Chim Acta* 2007; **375**:20–35.
88. Sell H, Habich C, Eckel J. Adaptive immunity in obesity and insulin resistance. *Nat Rev Endocrinol* 2012;**8**:709–16.
89. Stumvoll M, Goldstein BJ, van Haeften TW. Type 2 diabetes: principles of pathogenesis and therapy. *Lancet* 2005;**365**:1333–46.
90. Watt MJ, Bruce CR. No need to sweat: is dieting enough to alleviate insulin resistance in obesity? *J Physiol* 2009;**587**:5001–2.
91. Johnson AM, Olefsky JM. The origins and drivers of insulin resistance. *Cell* 2013;**152**(4):673–84.
92. Qatanani M, Lazar MA. Mechanisms of obesity-associated insulin resistance: many choices on the menu. *Genes Dev* 2007;**21**:1443–55.
93. Shulmam GI. Cellular mechanisms of insulin resistance. *J Clin Invest* 2000;**106**:171–6.
94. Kalupahana NS, Moustaid-Moussa N, Claycombe KJ. Immunity as a link between obesity and insulin resistance. *Mol Asp Med* 2012;**33**:26–34.
95. Alexaki VI, Notas G, Pelekanou V, et al. Adipocytes as imune cells: differential expression of TWEAK, BAFF, and APRIL and their receptors (Fn14, BAFF-R, TACI, and BCMA) at different stages of normal and pathological adipose tissue development. *J Immunol* 2009;**183**(9):5948–56.
96. Wang P, Mariman E, Renes J, Keijer J. The secretory function of adipocytes in the physiology of white adipose tissue. *J Cell Physiol* 2008;**216**:3–13.
97. Scherer PE. Adipose tissue: from lipid storage compartment to endocrine organ. *Diabetes* 2006;**55**:1537–45.
98. Sun K, Kusminski CM, Scherer PE. Adipose tissue remodeling and obesity. *J Clin Invest* 2011;**121**:2094–101.
99. Sun K, Scherer PE. *Adipose tissue dysfunction: a multistep process.* Berlin, Germany: Springer-Verlag Berlin Heidelberg; 2010. Research and Perspectives in Endocrine Interactions; 67–75.
100. Maury E, Brichard SM. Adipokine dysregulation, adipose tissue inflammation and metabolic syndrome. *Mol Cell Endocrinol* 2010; **314**:1–16.
101. Weisberg SP, McCann D, Desai M, Rosenbaum M, Leibel RL, Ferrante Jr AW. Obesity is associated with macrophage accumulation in adipose tissue. *J Clin Invest* 2003;**112**:1796–808.
102. Gordon S, Taylor PR. Monocyte and macrophage heterogeneity. *Nat Rev Immunol* 2005;**5**:953–64.
103. Mantovani A, Sica A, Sozzani S, Allavena P, Vecchi A, Massimo L. The chemokine system in diverse forms of macrophage activation and polarization. *Trends Immunol* 2004;**25**:677–86.
104. Lumeng CN, DelProposto JB, Westcott DJ, Saltiel AR. Phenotypic switching of adipose tissue macrophages with obesity is generated by spatiotemporal differences in macrophage subtypes. *Diabetes* 2008;**57**:3239–46.
105. Shoelson SE, Lee J, Goldfine AB. Inflammation and insulin resistance. *J Clin Invest* 2006;**116**(7):1793–801.
106. Winer S, Chan Y, Paltser G, Truong D, Tsui H, et al. Normalization of obesity-associated insulin resistance through immunotherapy. *Nat Med* 2009;**15**:921–9.
107. Travers RL, Motta AC, Betts JA, Bouloumié A, Thompson D. The impact of adiposity on adipose tissue-resident lymphocyte activation in humans. *Int J Obes* 2014:1–8.
108. Akira S, Takeda K. Toll-like receptor signalling. *Nat Ver Immun* 2004;**4**:499–511.
109. Suganami T, Ogawa Y. Adipose tissue macrophages: their role in adipose tissue remodeling. *J Leukoc Biol* 2010;**88**:33–9.
110. Suganami T, Tanimoto-Koyama K, Nishida J, et al. Role of the toll-like receptor 4/NFκB pathway in saturated fatty acid-induced inflammatory changes in the interaction between adipocytes and macrophages. *Arterioscler Thromb Vasc Biol* 2007; **27**:84–91.
111. Lee JY, Sohn KH, Rhee SH, Hwang D. Saturated fatty acids, but not unsaturated fatty acids, induce the expression of cyclooxygenase-2 mediated through toll-like receptor 4. *J Biol Chem* 2001;**276**:16683–9.
112. Takeda K, Akira S. Toll-like receptors in innate immunity. *Int Immunol* 2005;**17**:1–14.
113. Cui J, Zhang M, Zhang Y, Xu Z. JNK pathway: diseases and therapeutic potential. *Acta Pharmacol Sin* 2007;**28**(5):601–8.
114. Wellen KE, Hotamisligil GS. Obesity-induced inflammatory changes in adipose tissue. *J Clin Invest* 2003;**112**:1785–8.
115. Vallerie SN, Hotamisligil JS. The role of JNK proteins in metabolism. *Sci Transl Med* 2010;**2**:60rv65.
116. Tanti J-F, Jager J. Cellular mechanisms of insulin resistance: role of stress-regulated serine kinases and insulin receptor substrates (IRS) serine phosphorylation. *Curr Opin Pharmacol* 2009;**9**(6):753–62.
117. Wellen KE, Hotamisligil GS. Inflammation, stress, and diabetes. *J Clin Invest* 2005;**115**(5):1111–9.
118. Yu C, Chen Y, Cline GW, et al. Mechanism by which fatty acids inhibit insulin activation of insulin receptor substrate-1 (IRS-1)-associated phosphatidylinositol 3-kinase activity in muscle. *J Biol Chem* 2002;**277**:50230–6.
119. Perseghin G, Petersen K, Shulman GI. Cellular mechanism of insulin resistance: potential links with inflammation. *Int J Obes Relat Metab Disord* 2003;**27**:S6–11.
120. Griffin ME, Marcucci MJ, Cline GW, et al. Free fatty acid-induced insulin resistance is associated with activation of protein kinase C theta and alterations in the insulin signaling cascade. *Diabetes* 1999;**48**:1270–4.
121. Itani SI, Ruderman NB, Schmieder F, Boden G. Lipid-induced insulin resistance in human muscle is associated with changes in diacylglycerol, protein kinase C, and IkappaB-alpha. *Diabetes* 2002;**51**:2005–11.
122. Emanuelli B, Peraldi P, Filloux C, et al. SOCS-3 inhibits insulin signaling and is up-regulated in response to tumor necrosis factor-alpha in the adipose tissue of obese mice. *J Biol Chem* 2001;**276**:47944–9.
123. Ueki K, Kondo T, Kahn CR. Suppressor of cytokine signaling 1 (SOCS-1) and SOCS-3 cause insulin resistance through inhibition of tyrosine phosphorylation of insulin receptor substrate proteins by discrete mechanisms. *Mol Cell Biol* 2004;**24**:5434–46.
124. Rui L, Yuan M, Frantz D, Shoelson S, White MF. SOCS-1 and SOCS-3 block insulin signaling by ubiquitin-mediated degradation of IRS1 and IRS2. *J Biol Chem* 2002;**277**:42394–8.
125. Mooney RA, Senn J, Cameron S, et al. Suppressors of cytokine signaling-1 and -6 associate with and inhibit the insulin receptor. A potential mechanism for cytokine-mediated insulin resistance. *J Biol Chem* 2001;**276**:25889–93.
126. Malik VS, Willett WC, Hu FB. Global obesity: trends, risk factors and policy implications. *Nat Rev Endocrinol* 2013;**9**(1):13–27.
127. Dietrich MO, Horvath TL. Limitations in anti-obesity drug development: the critical role of hunger-promoting neurons. *Nat Rev Drug Discov* 2012;**11**(9):675–91.
128. Ioannides-Demos LL, Piccenna L, McNeil J. Pharmacotherapies for obesity: past, current, and future therapies. *J Obes* 2011:179674.
129. Guo H, Ling W. The update of anthocyanins on obesity and type 2 diabetes: experimental evidence and clinical perspectives. *Rev Endocr Metab Disord* 2005. http://dx.doi.org/10.1007/s11154-014-9302-z.

130. Tsuda T, Horio F, Uchida K, et al. Dietary cyanidin 3-O-b-D-glucoside-rich purple corn color prevents obesity and ameliorates hyperglycemia in mice. *J Nutr* 2003;**133**(7):2125–30.
131. Tsuda T, Ueno Y, Aoki H, et al. Anthocyanin enhances adipocytokine secretion and adipocyte-specific gene expression in isolated rat adipocytes. *Biochem Biophys Res Commun* 2004;**316**: 149–57.
132. Tsuda T, Ueno Y, Yoshikawa T, et al. Microarray profiling of gene expression in human adipocytes in response to anthocyanins. *Biochem Pharmacol* 2006;**71**:1184–97.
133. Tsuda T. Regulation of adipocyte function by anthocyanins; possibility of preventing the metabolic syndrome. *J Agric Food Chem* 2008;**56**:642–6.
134. Jayaprakasam B, Olson LK, Schutzki RE, et al. Amelioration of obesity and glucose intolerance in high-fat-fed C57BL/6 mice by anthocyanins and ursolic acid in cornelian cherry (*Cornus mas*). *J Agric Food Chem* 2006;**54**:243–8.
135. Kwon SH, Ahn IS, Kim SO, et al. Anti-obesity and hypolipidemic effects of black soybean anthocyanins. *J Med Food* 2007;**10**:552–6.
136. Kanamoto Y, Yamashita Y, Nanba F, et al. A black soybean seed coat extract prevents obesity and glucose intolerance by up-regulating uncoupling proteins and down-regulating inflammatory cytokines in high-fat diet-fed mice. *J Agric Food Chem* 2010; **59**:8985–93.
137. Seymour EM, Lewis SK, Urcuyo-Llanes DE, et al. Regular tart cherry intake alters abdominal adiposity, adipose gene transcription, and inflammation in obesity-prone rats fed a high fat diet. *J Med Food* 2009;**12**:935–42.
138. Prior RL, Wu X, Gu L, et al. Whole berries versus berry anthocyanins: interactions with dietary fat levels in the C57BL/6J mouse model of obesity. *J Agric Food Chem* 2008;**56**:647–53.
139. DeFuria J, Bennett G, Strissel KJ, et al. Dietary blueberry attenuates whole-body insulin resistance in high fat-fed mice by reducing adipocyte death and its inflammatory sequelae. *J Nutr* 2009;**139**:1510–6.
140. Dragano NRV, Marques AC, Cintra DEC, et al. Freeze-dried jaboticaba peel powder improves insulin sensitivity in high-fat-fed mice. *Brit J Nutr* 2013;**1**:1–9.
141. Sasaki R, Nishimura N, Hoshino H, et al. Cyanidin 3-glucoside ameliorates hyperglycemia and insulin sensitivity due to downregulation of retinol binding protein 4 expression in diabetic mice. *Biochem Pharmacol* 2007;**74**:1619–27.
142. Takikawa M, Inoue S, Horio F, Tsuda T. Dietary anthocyanin-rich bilberry extract ameliorates hyperglycemia and insulin sensitivity via activation of AMPactivated protein kinase in diabetic mice. *J Nutr* 2010;**140**:527–33.

CHAPTER

12

Influence of Dietary Factors on Gut Microbiota: The Role on Insulin Resistance and Diabetes Mellitus

Gemma Xifra, Eduardo Esteve, Wifredo Ricart, José Manuel Fernández-Real

Unit of Diabetes, Endocrinology and Nutrition, Biomedical Research Institute (IDIBGi), Hospital 'Dr. Josep Trueta' of Girona, Girona, Spain; CIBERobn Fisiopatología de la Obesidad y Nutrición, Girona, Spain

1. INTRODUCTION

Human gut microbiota consists of trillions of microorganisms and thousands of bacterial phylotypes that are deeply involved in a different function of host metabolism. The collective genomes of our gut microbes (microbiome) may contain >150 times more genes than our own genome.[1] Eighty to ninety percent of the bacterial phylotypes are members of two phyla: the *Bacteroidetes* (Gram-negative) and the *Firmicutes* (Gram-positive), followed by *Actinobacteria* (Gram-positive) and *Proteobacteria*.[2,3] Each patient has a specific gut microbiota, but a core human gut microbiome is shared among family members, despite different environments.[4] Recent studies described that dietary intervention impact on gut microbial gene richness.[5] Initial studies on gut microbial composition and function were limited by the difficulty to culture all intestinal microbes. The introduction of bacterial genome sequencing and "metagenomic" analysis, has contributed to increase the knowledge about uncultivable microbes, gut microbial functions, its cross-talk with the host, and the potential pathogenic role related to host's diseases.

2. INFLUENCE OF DIETARY FACTORS ON GUT MICROBIOTA

Probably the most important determinant of gut microbiota is the diet. The composition of the diet and the amount of calories consumed are strong modifiers of the microbiota (Figure 1).[5] Germ-free animals (without microbiota) are resistant to high-fat diet–induced obesity and metabolic syndrome.[6] However, genetically identical germ-free animals respond differently to a high-fat diet.[7] Twins, especially monozygotic (MZ) twins, have been reported to have more similar interindividual fecal microbiota than unrelated people.[4] Moreover, genetically identical mice fed a fat-enriched carbohydrate-free diet did not develop insulin resistance uniformly. The subgroup of mice that did show a marked change in insulin sensitivity also presented with distinct gut microbiota.[8]

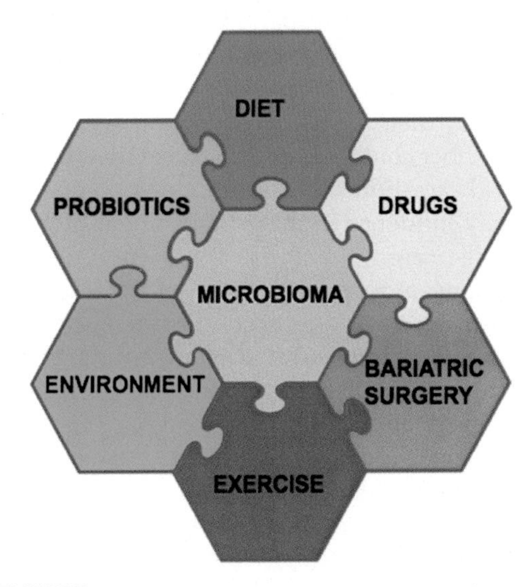

FIGURE 1 Factors capable of modifying microbiomes.

http://dx.doi.org/10.1016/B978-0-12-801585-8.00012-9

These findings have led to the conclusion that the host genotype affects the development of gut microbiota and gut bacterial composition. Nevertheless, concordant, normal weight MZ twins had more similar bacterial populations than the MZ twins discordant for obesity. This suggests the importance of the diet in addition to the genetic background.[9]

The primary activity of gut microbiota is the breakdown of carbohydrates not digested in the ileum to short-chain fatty acids (SCFAs), which are then rapidly absorbed. The principal products of carbohydrate fermentation are SCFAs (acetate, propionate, and butyrate). The amount of energy derived from SCFAs accounts for up to 10% of the total energy requirement of human subjects. Modulation of species will thus affect populations of microbes within microbiota and, thus, modulate energy harvest, storage, and expenditure.

The composition of the diet and the amount of calories consumed are well-known strong modifiers of microbiota.[10] High-fat diets are associated with substantial compositional changes in colonic microbiota at the phylum and genus levels, including reductions in intestinal Gram-negative and Gram-positive bacteria, including levels of bifidobacteria. These changes lead to increased metabolic endotoxemia that induces insulin resistance.[7,11] A study looking at global bacterial populations in vegans, vegetarians, and control subjects found that vegans had decreased levels of *Bacteroides* spp., *Bifidobacterium* spp., *Escherichia coli*, and *Enterobacteriaceae* spp. compared with control patients, with vegans serving as an intermediate population.[12] Recently, in humans, a diet based on whole grains and prebiotics has shown a reduction in phylotypes related to endotoxin-producing opportunistic pathogens such as *Enterobacteriaceae* or *Desulfovibrionaceae*, whereas those related to gut barrier-protecting bacteria of *Bifidobacteriaceae* were increased.[13]

Rats fed a long-term high-fiber diet had higher total bacteria, increased relative abundance of *Bifidobacteria*, and decreased *Firmicutes* compared with rats fed a high-protein diet.[14] Diet modification with increasing protein or fiber content in the maternal diet during pregnancy and lactation modifies gut microbiota of dams, which may influence in the establishment of gut microbiota in offspring.[15]

In humans, a decreased intake of fruits and vegetables is associated with reduced microbial gene richness, suggesting that long-term dietary habits may affect gene richness.[16,17] Dietary intervention improves low gene richness and clinical phenotypes, but seems to be less efficient for inflammatory variables in individuals with lower gene richness. In the weight maintenance phase, the overall tendency was to return to a baseline level, suggesting a transient effect of dietary intervention on gut microbiota.[16] The transitory effect of the diet on microbial population has been described. Short-term consumption of diets composed entirely of animal or plant products can rapidly alter microbial community and diversity.[18] Nevertheless, only the long-term changes on diet seem to define what type of microbial population is present. Microbiota composition is strongly associated with a long-term diet, with *Bacteroides* being associated with diets enriched in animal products and *Prevotella* genus being associated with diets that contained more plant-based foods.[19] Long-term dietary habits could explain the stability of gut enterotypes over time.[20] These long-term changes determine the metabolites produced and the potential impact on the health of the host.

The composition of gut microbiota depends on the age, health, diet, and even geographical location of the individual.[21] At the geographical level, significant differences in the phylogenetic composition and diversity of fecal microbiota were noted between US residents and those from the Malawians and Amerindians.[21] These differences could not be due only to genetic factors. The environment and the typical Westerns diet in US residents have also important roles. In this line, gut microbiota of rural children in Burkina Faso who consumed a plant-rich diet were significantly different from microbiota of children from Italy who consumed a low-fiber diet. African children had lower levels of *Firmicutes* than of *Bacteroidetes*, whereas European children had high levels of *Enterobacteriaceae* (*Shigella* and *Escherichia*).[22] Gut microbiota from Russian cities resembled those of Western countries, which is presumably associated with increased consumption of meat products and processed food. However, the gut microbiota in rural residents of Russia are distinctly enriched in *Firmicutes* and *Actinobacteria*, which is presumably associated with high consumption of starch-rich bread and natural products.[23] In a study of Mongolian rural and urban populations, changes in the seasonal dietary composition in the rural population have been associated with changes in the composition of gut microbiota populations.[24] Long-term dietary intake influences the structure and activity of human intestinal microbiota. These findings imply that the dietary habits formed over a long period may play a key role in the composition of gut microbiota over other variables such as ethnicity, sanitation, and climate.

The microbiota composition is not stable over the whole lifetime of an individual, especially at the extremities of life. Changes of microbiota in the elderly may be attributable to diet, with increased sugar/high-fat foods and diseases.[25] The microbiota, however, may have an active role modulating the response to diet in animals and humans. Early consumption of a high-fiber diet during growth can offer protection from obesity after a high-fat diet.[26] The gut microbiota in obese and lean mice has specific characteristics that are transmittable and confers a different response against high-fat or vegetable diets.[27] In humans, the microbiota composition and abundance of several species such as *Firmicutes* can predict the responsiveness to diet in obese individuals, which could reveal the potential of microbiota signatures for personalized nutrition.[28]

3. IMPACT OF PREBIOTICS, PROBIOTICS, AND EXERCISE ON GUT MICROBIOTA

The beneficial effect of prebiotics on weight, insulin resistance, and metabolic parameters has been widely described.[29] These encouraging results seem to be more robust in animal studies than in humans. Prebiotics may select beneficial bacterial species such as *Bifidobacterium*, reducing bacterial translocation and metabolic endotoxemia. The production of SCFAs as propionate derived from microbial fermentation of dietary fiber prevents body weight gain and reduces intra-abdominal fat accretion in overweight adults by increasing the release of anorectic gut hormones PYY and GLP-1 from gut cells.[30] These results suggest that modulation of the gut microbiota via dietary intervention may enhance the intestinal barrier integrity, reduce circulating antigen load, and ultimately ameliorate the inflammation and metabolically unfavorable phenotypes.

It has been described that not only diet but also exercise can determine gut microbiota. Exercise alters the gut microbiota in mice on both a low- and high-fat diets and normalizes major phylum-level changes of mice on the high-fat diet. These changes of microbiota are different than those seen in their sedentary counterparts.[31]

In summary, dietary habits over a long period may play a key role in the composition of gut microbiota over other variables such as ethnicity, climate, or genetic background. Probably in a given person, different eating habits can generate an environment in which a particular type of flora is established. This flora will confer over time a protection or susceptibility to obesity and insulin resistance induced by high-fat diets.

4. GUT MICROBIOTA INTERACTIONS WITH INSULIN RESISTANCE AND DIABETES

The pathophysiology of obesity, insulin resistance, and diabetes type 2 is complex and involves a combination of genetic and environmental factors. Ingestion of food is a major environmental exposure (Figure 2). Because ingested food is processed through the filter of the intestinal microbial community, it is not surprising that gut microbiota should be considered an environmental factor that modulates host metabolism and is capable to contribute to metabolic diseases.

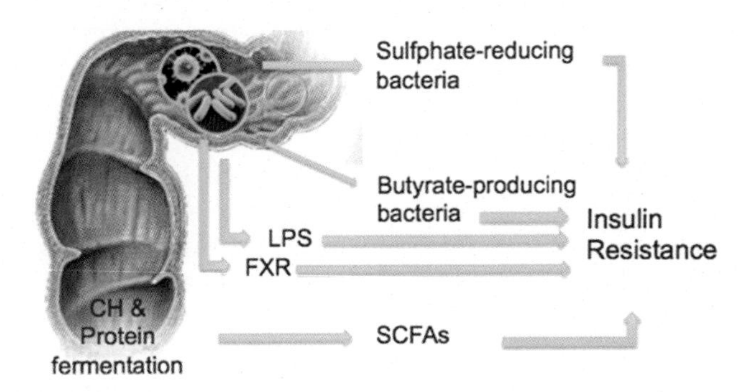

FIGURE 2 Pathophysiological factors linking gut microbiota and insulin resistance. CH, carbohydrates; FXR, nuclear farnesoid X receptor; LPS, gut microbiota-derived lipopolysaccharide; SCFAs, short-chain fatty acids.

Mounting evidence in animal models and humans is accumulating showing that gut microbiota are linked to the onset and development of metabolic disorders, such as insulin resistance and type 2 diabetes (T2D). We summarize here the existing evidence in animal and human models, the pathophysiologic mechanisms underlying this association, how diet interacts with gut microbiota, and how these findings provide novel therapeutic targets for insulin resistance and T2D (Figure 3).

4.1 Pathophysiological Factors Linking Gut Microbiota and Insulin Resistance

Recent studies in mice and humans show that the obese phenotype is associated with differences in the gut microbial composition compared with lean counterparts. In the same line, differences in gut microbiota have been found between patients with insulin resistance and diabetes compared with healthy subjects. In *animal models*, different authors reported that the decreased adiposity in germ-free mice was associated with improved insulin sensitivity and glucose tolerance,[6,32,33] indicating that gut microbiota may directly contribute to altered glucose metabolism. It has been demonstrated that the obesity phenotype is a transmissible trait by showing that fecal microbiota transplantation (from

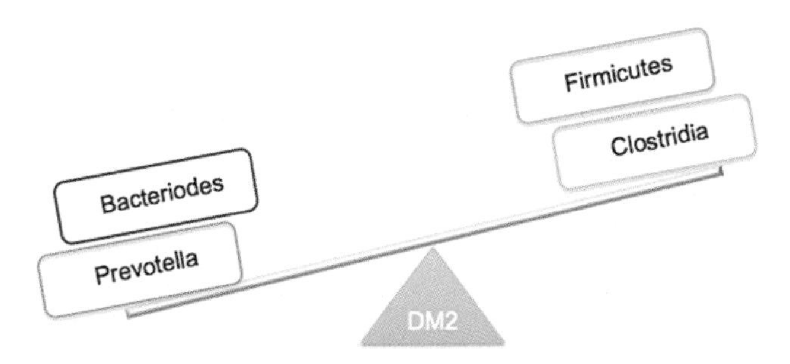

FIGURE 3 Bacterial imbalance observed in diabetes type 2.

conventionally raised mice) into germ-free mice resulted in efficient transmission of the obesity phenotype into recipients, with an increase of body fat up to 60% parallel to insulin resistance.[34]

In humans, several studies provided evidence that gut microbiota are altered in metabolic syndrome[35] as well as in T2D.[36,37] Karlsson and Quin studies reported that subjects with T2D had a lower abundance of butyrate-producing *Clostridiales* and a greater relative abundance of *Clostridiales* that did not produce butyrate. Despite this, the bacterial composition was markedly different in these two cohorts, from Europe and China, demonstrating that the intestinal microbial composition is dramatically impacted by diet and ethnicity. Other studies demonstrated increased levels of *E. coli* and *Proteobacteria* in subjects with T2D. Unfortunately, no information about the possible influence of different drugs (that patients with T2D usually consume) on gut microbiota composition was not considered in these reports.[1,38]

Various mechanisms have been proposed to explain the influence of the microbiota on insulin resistance and T2D, such as metabolic endotoxemia, modifications in the secretion of the incretins, and butyrate production, as summarized in the following section.

4.1.1 Microbial Metabolites

The gut microbiota have profound effects on metabolism of several dietary components and produce numerous metabolites that can be detected in the circulation. One of the most widely studied classes is SCFAs. Intestinal microbes ferment nondigestible carbohydrates to yield energy, leading to the production of SCFAs in the form of acetate (60%), propionate (25%), and butyrate (15%).[39] The concentration of fecal SCFAs is highly dependent on the amount of daily dietary fiber intake. SCFAs are readily absorbed in the plasma of the host via the intestinal epithelium and thus can serve as an energy source, predominantly via metabolism in the liver. The role for intestinal bacteria in the production of SCFA is clearly demonstrated by the observation that germ-free rats and mice are characterized by reduced levels of intestinal SCFAs with a concomitant increase in fecal excretion of nondigestible carbohydrates. Moreover, mice models (ob/ob) and human disease states (e.g., metabolic syndrome), characterized by insulin resistance, also tend to have decreased intestinal SCFA levels with a concomitant reduced excreted energy content in their feces.[1,40]

Different studies reported that patients with T2D exhibited a moderate intestinal dysbiosis characterized especially by a decrease in the number of *Clostridiales* bacteria that produce butyrate (*Roseburia intestinalis* and *Faecalibacterium prausnitzii*[1,40]), supporting the role of butyrate-producing bacteria as regulators of human glucose metabolism, possibly via alterations in intestinal permeability alterations resulting in chronic inflammation.[1,41–43]

Other authors defined gut microbiota-derived lipopolysaccharide (LPS) and metabolic endotoxemia as factors involved in the onset and progression of inflammation and metabolic diseases.[11] LPS is a component of Gram-negative bacteria cell walls and is among the most potent and well-studied inducers of inflammation. It has been reported in animal and human studies that a high-fat diet contributes to a higher abundance of LPS-containing microbiota as well as plasma LPS levels with consequent systemic inflammation.[11,44] Experimental LPS infusion results in hyperglycemia and hyperinsulinemia.[45] Moreover, CD14/Toll-like receptor (TLR)-4 mutant mice, resistant to LPS, were also resistant to high-fat diet–induced metabolic diseases, because the triggering of the inflammatory cascade in the liver and in adipose tissue, was significantly reduced in these animals, suggesting the potential role of CD14 to set the host's insulin sensitivity in physiological conditions.[45]

Sulfate-reducing species such as *Desulfovibrio* seem also important because an increase in their relative abundance is frequently observed in T2D.

4.1.2 Bile Acid Metabolites

Over the past decade, a growing body of evidence has shown that bile acids play an important role in glucose metabolism as signaling molecules and cellular receptor ligands. It has been demonstrated that bile acids inhibit the expression of gluconeogenic genes, such as those encoding phosphoenolpyruvate carboxykinase, fructose-1, 6-biphosphatase-1, and glucose-6-phosphatase in vitro via the nuclear farnesoid X receptor (FXR).[46]

Intestinal microorganisms, such as *Clostridium*, strongly affect bile acid metabolism.[47] Activation of FXR by the gut microbiota reduces the expression levels of most bile acid synthesis enzymes.[48] In turn, bile acids contribute to suppression of bacterial colonization and growth in the gut because of their strong antimicrobial activity. Only microbial populations able to tolerate physiological concentrations of bile acids can survive in the gut. In rat animal models, feeding a diet rich in cholic acid induces phylum-level alterations in the composition of the gut microbiota. Cholic acid feeding also increases significantly the ratio of *Firmicutes* to *Bacteroidetes*, which is similar to the changes induced by high-fat feeding.[49]

The altered gut physiology following bariatric surgery contributes to altered microbial ecology in mice, rats, and humans. Bariatric surgery induces increases in Proteobacteria such as *Escherichia* and *Enterobacter*,[50–54] and has a pronounced effect on the gut microbiota within the first week following surgery, which is stabilized after

5 weeks in mice.[53] This suggests that the altered gut microbiota may contribute to improved host metabolism. Using fecal microbiota transplants, some authors could demonstrate that the altered gut microbiota directly contributed to reduced weight gain.[53] The mechanisms by which bariatric surgery results in rapid improvement of insulin sensitivity <1 week remain unknown. However, bile acid flow alterations could contribute to the improved metabolism following bariatric surgery by altering the gut microbiota and by inducing FXR signaling.

4.1.3 Changes in the Innate Immune System

The majority of obese subjects will eventually develop chronic adipose tissue inflammation leading to production of pro-inflammatory cytokines and subsequent insulin resistance.[55] Recent data have shown that variations in intestinal microbiota were associated with pro-inflammatory changes in adipose gene expression.[51] *F. prausnitzii* is considered the prototypical anti-inflammatory component. Several studies demonstrated that *F. prausnitzii* abundance was lower in patients with T2D,[56] with a lower degree of chronic inflammation. Intestinal microbiota is a source of many pro-inflammatory components—such as LPS or capsule-derived compounds (peptidoglycans, lipoproteins, flagellins)—able to activate innate inflammatory pathways and the development of chronic inflammatory state.[57] Several lines of evidence point toward a direct relation between the intestine and visceral adipose tissue inflammation because the macrophage infiltration in adipose tissue was directly correlated with a pro-inflammatory gene expression profile.[58] Because macrophages are part of the innate immune system, these findings imply that translocation of intestinal bacteria could play a role in the development of chronic inflammatory state. The innate immune system is capable of sensing various types of bacterial components via pattern recognition receptors, such as TLRs.[59,60] It is known that alterations in gut microbiota composition drive activation of TLRs by bacterial LPS and subsequent obesity and insulin resistance.[60] The chronic low-dose LPS exposure in mice is known to result in hepatic insulin resistance, hepatic steatosis, adipose tissue macrophages infiltration, dyslipidemia, fasting hyperglycemia, hyperinsulinemia, and eventually obesity.

4.1.4 Epigenetic Changes

Kumar et al.[61] revealed differences in blood DNA methylation patterns in a cohort of eight healthy pregnant women according to the composition of the dominant phyla. Mothers were selected on the basis of the relative abundance of the dominant phyla (four exhibited a predominance of the Bacteroidetes and Proteobacteria, whereas *Firmicutes* was predominant in the other four women). Deep sequencing of DNA methylomes revealed a clear association between bacterial abundance and epigenetic profiles. The genes with differentially methylated promoters in the group in which *Firmicutes* was dominant were linked to risk of metabolic disease, predominantly cardiovascular disease, and alterations in lipid metabolism, obesity, and the inflammatory response. Despite further longitudinal and in-depth studies are required, the link between gut microbiota and epigenetics could represent another way of interaction between microbiota and metabolic disease.

4.1.5 Drug Influences on Gut Microbiota

It is important to note that different investigations reported differences in the fecal microbiota composition of subjects with T2D, despite that no information about medication was provided in these reports.[1,38] Subjects with T2D who used metformin had increased levels of *Enterobacteriaceae* (i.e., *Escherichia*, *Shigella*, *Klebsiella*, and *Salmonella*) and decreased levels of *Clostridium* and *Eubacterium* compared with those that did not take metformin.[36] Thus, the "true" T2D profile may be masked by metformin or other drugs. For example, metformin increase the levels of *Akkermansia muciniphila*, a mucus-degrading, SCFA-producing bacteria, in the gut.[62] Interestingly, oral administration of *A. muciniphila* protects against obesity and enhances glucose tolerance in mice.[62,63] These findings highlight the importance of drug monitoring and also suggest the possibility that some drugs may act through the modulation of the microbiota. This possibility must be taken into account.

5. GUT MICROBIOTA AND TYPE 1 DIABETES

The gut microbiome are suggested to play a role in the pathogenesis of autoimmune disorders such as type 1 diabetes (T1D). Recent investigations identified an important role of the gut microbiota in the regulation of systemic glucose metabolism and in the modulation of immunologic responses in a mouse model of autoimmune diabetes. The absence of gut microbiota in nonobese diabetic (NOD) mice did not affect overall disease incidence, but resulted in increased insulitis and levels of serum insulin autoantibodies and pro-inflammatory cytokines and worsened glucose metabolism in mice that did not progress to diabetes.[64] Furthermore, these authors observed differences in the serum metabolomics profile between germ-free and conventional NOD mice that resembled the differences between those of newborn infants who later progress to T1D and nonprogressors.[64]

Thus, the metabolomic changes in infants who later progress to T1D may be modulated by the gut microbiota.

In humans, a prospective study with 35 newly diagnosed children with T1D and 35 healthy children revealed decreased abundance of beneficial anaerobic bacteria and a concomitant increase in *Enterobacteriaceae* other than *E. coli* colonization in patients with T1D compared with the control group.[65] A disturbance in the ecological balance of intestinal flora might be a triggering factor in T1D etiology. On the basis of all these results, a role of the gut microbiota as a regulator of diabetic autoimmunity cannot be excluded.

6. FUTURE PERSPECTIVES

It will be essential to determine whether an altered gut microbiota precedes development of insulin resistance and diabetes and to identify the underlying molecular mechanisms. Increased mechanistic insights of how the microbiota modulates metabolic disease in humans may pave the way for identification of innovative microbiota-based diagnosis and/or therapeutics. Despite initial skepticism, fecal microbiota transplantation was recently demonstrated to be a real therapeutic alternative in some diseases such as *Clostridium difficile* infection, being more effective than the usual treatment with vancomycin.[66] In T2D, a trial was performed to investigate whether lean microbiota would improve glucose homeostasis and lipid metabolism in naive males with metabolic syndrome (Fecal Administration to LOSE Insulin Resistance trial).[67] In this double-blind randomized control trial, the authors studied the therapeutic effects of allogenic lean donor feces infusion on insulin resistance in 18 adult male subjects with metabolic syndrome. The subjects were randomized to allogenic (from lean male donors with a body mass index $<23\ kg/m^2$; $n = 9$) or autologous gut microbiota infusion (reinfusion of own collected feces; $n = 9$). Insulin sensitivity and gut microbiota composition was measured before and 6 weeks after gut microbiota infusion. An improvement in peripheral insulin sensitivity was observed in the allogenic infusion group only. Gut microbial diversity was increased significantly after allogenic gut microbiota transfer (from 178 ± 62 to 234 ± 40 species; $P < 0.05$) with no change in the autologous group (from 184 ± 71 to 211 ± 50 [not significant]). Sixteen bacterial groups increased significantly in the allogenic treatment group, including those related to the butyrate-producer *R. intestinalis*, which showed a 2.5-fold increase. However, the effect declined in all subjects over time.

The underlying mechanism for the improvement of insulin resistance remains to be elucidated. However, butyrate-producing bacterial strains were significantly increased in both small intestinal biopsies and fecal samples of metabolic syndrome patients treated with lean donor feces. These authors suggested that butyrate produced bacteria could be the key. They suggest that butyrate-producing bacteria prevent translocation of endotoxin compounds derived from the gut microbiota that have been previously shown to drive insulin resistance.[53]

Intestinal microbiota is thus a future diagnosis and therapeutic target. In other words, "the change of the crew could change the route of the trip."

References

1. Qin J, Li Y, Cai Z, Li S, Zhu J, Zhang F, et al. A metagenome-wide association study of gut microbiota in type 2 diabetes. *Nature* 2012;**490**:55–60.
2. Esteve E, Ricart W, Fernández-Real JM. Gut microbiota interactions with obesity, insulin resistance and type 2 diabetes: did gut microbiote co-evolve with insulin resistance? *Curr Opin Clin Nutr Metab Care* 2011;**14**:483–90.
3. Hill JO. Understanding and addressing the epidemic of obesity: an energy balance perspective. *Endocr Rev* 2006;**27**:750–61.
4. Turnbaugh PJ, Hamady M, Yatsunenko T, Cantarel BL, Duncan A, Ley RE, et al. A core gut microbiome in obese and lean twins. *Nature* 2009;**457**:480–4.
5. Zhang C. Interactions between gut microbiota, host genetics and diet relevant to development of metabolic syndromes in mice. *ISME* 2010;**4**:232–41.
6. Bäckhed F, Manchester JK, Semenkovich CF, Gordon JI. Mechanisms underlying the resistance to diet-induced obesity in germ-free mice. *Proc Natl Acad Sci USA* 2007;**104**:979–84.
7. De La Serre CB, Ellis CL, Lee J, Hartman AL, Rutledge JC, Raybould HE. Propensity to high-fat diet-induced obesity in rats is associated with changes in the gut microbiota and gut inflammation. *Am J Physiol Gastrointest Liver Physiol* 2010;**299**: G440–8.
8. Serino M, Luche E, Gres S, Baylac A, Bergé M, Cenac C, et al. Metabolic adaptation to a high-fat diet is associated with a change in the gut microbiota. *Gut* 2012;**61**:543–53.
9. Simoes CD, Maukonen J, Kaprio J, Rissanen A, Pietiläinen KH, Saarela M, et al. Habitual dietary intake is associated with the stool microbiota composition in monozygotic twins. *J Nutr* 2013;**143**: 417–23.
10. Jeffery IB, O'Toole PW. Diet-microbiota interactions and their implications for healthy living. *Nutrients* 2013;**5**:234–52.
11. Cani PD, Amar J, Iglesias MA, Poggi M, Knauf C, Bastelica D, et al. Metabolic endotoxemia initiates obesity and insulin resistance. *Diabetes* 2007;**56**:1761–72.
12. Zimmer J. A vegan or vegetarian diet substantially alters the human colonic faecal microbiota. *Eur J Clin Nutr* 2012;**66**(1):53–60.
13. Xiao S, Fei N, Pang X, Shen J, Wang L, Zhang B, et al. A gut microbiota-targeted dietary intervention for amelioration of chronic inflammation underlying metabolic syndrome. *FEMS Microbiol Ecol* 2014;**87**:357–67.
14. Saha DC, Reimer RA. Long-term intake of a high prebiotic fiber diet but not high protein reduces metabolic risk after a high fat challenge and uniquely alters gut microbiota and hepatic gene expression. *Nutr Res* 2014;**34**:789–96.
15. Hallam MC, Barile D, Meyrand M, German JB, Reimer RA. Maternal high-protein or high-prebiotic-fiber diets affect maternal milk composition and gut microbiota in rat dams and their offspring. *Obesity* 2014;**22**:2344–51.
16. Cotillard A, Kennedy SP, Kong LC, Prifti E, Pons N, Le Chatelier E, et al. Dietary intervention impact on gut microbial gene richness. *Nature* 2013;**500**:585–8.

17. Kong LC, Holmes BA, Cotillard A, Habi-Rachedi F, Brazeilles R, Gougis S, et al. Dietary patterns differently associate with inflammation and gut microbiota in overweight and obese subjects. *PLoS One* 2014;**9**:e109434.
18. David LA, Maurice CF, Carmody RN, Gootenberg DB, Button JE, Wolfe BE, et al. Diet rapidly and reproducibly alters the human gut microbiome. *Nature* 2014;**23**(505):559–63.
19. Wu GD, Chen J, Hoffmann C, Bittinger K, Chen Y-Y, Keilbaugh SA, et al. Linking long-term dietary patterns with gut microbial enterotypes. *Science* 2011;**334**:105–8.
20. Lim MY, Rho M, Song YM, Lee K, Sung J, Ko G. Stability of gut enterotypes in Korean monozygotic twins and their association with biomarkers and diet. *Sci Rep* 2014;**4**:7348.
21. Yatsunenko T. Human gut microbiome viewed across age and geography. *Nature* 2012;**486**:222–7.
22. De Filippo C, Cavalieri D, Di Paola M, Ramazzotti M, Poullet JB, Massart S, et al. Impact of diet in shaping gut microbiota revealed by a comparative study in children from Europe and rural Africa. *Proc Natl Acad Sci USA* 2010;**107**:14691–6.
23. Tyakht AV, Kostryukova ES, Popenko AS, Belenikin MS, Pavlenko AV, Larin AK, et al. Human gut microbiota community structures in urban and rural populations in Russia. *Nat Commun* 2013;**4**:2469.
24. Zhang J, Guo Z, Qi Lim A, Zheng Y, Koh EY, Ho D, et al. Mongolians core gut microbiota and its correlation with seasonal dietary changes. *Sci Rep* 2014;**4**:5001.
25. Claesson MJ, Jeffery IB, Conde S, Power SE, O'Connor EM, Cusack S, et al. Gut microbiota composition correlates with diet and health in the elderly. *Nature* 2012;**488**:178–84.
26. Maurer AD, Eller LK, Hallam MC, Taylor K, Reimer RA. Consumption of diets high in prebiotic fiber or protein during growth influences the response to a high fat and sucrose diet in adulthood in rats. *Nutr Metab* 2010;**7**:77.
27. Ridaura VK, Faith JJ, Rey FE, Cheng J, Duncan AE, Kau AL, et al. Gut microbiota from twins discordant for obesity modulate metabolism in mice. *Science* 2013;**341**:1241214.
28. Korpela K, Flint HJ, Johnstone AM, Lappi J, Poutanen K, Dewulf E, et al. Gut microbiota signatures predict host and microbiota responses to dietary interventions in obese individuals. *PLoS One* 2014;**9**:e90702.
29. Delzenne NM, Neyrinck AM, Bäckhed F, Cani PD. Targeting gut microbiota in obesity: effects of prebiotics and probiotics. *Nat Rev Endocrinol* 2011;**7**:639–46.
30. Chambers ES, Viardot A, Psichas A, Morrison DJ, Murphy KG, Zac-Varghese SE, et al. Effects of targeted delivery of propionate to the human colon on appetite regulation, body weight maintenance and adiposity in overweight adults. *Gut* 2014;**64**:1744–54 [Epub ahead of print].
31. Evans CC, LePard KJ, Kwak JW, Stancukas MC, Laskowski S, Dougherty J, et al. Exercise prevents weight gain and alters the gut microbiota in a mouse model of high fat diet-induced obesity. *PLoS One* 2014;**9**:e92193.
32. Caesar R, Reigstad CS, Bäckhed HK, Reinhardt C, Ketonen M, Lundén GO, et al. Gut-derived lipopolysaccharide augments adipose macrophage accumulation but is not essential for impaired glucose or insulin tolerance in mice. *Gut* 2012;**61**:1701–7.
33. Rabot S, Membrez M, Bruneau A, Gerard P, Harach T, Moser M, et al. Germ-free C57BL/6J mice are resistant to high-fat-diet-induced insulin resistance and have altered cholesterol metabolism. *FASEB J* 2010;**24**:4948–59.
34. Bäckhed F, Ding H, Wang T, Hooper LV, Koh GY, Nagy A, et al. The gut microbiota as an environmental factor that regulates fat storage. *Proc Natl Acad Sci USA* 2004;**101**:15718–23.
35. Le Chatelier E, Nielsen T, Qin J, Prifti E, Hildebrand F, Falony G, et al. Richness of human gut microbiome correlates with metabolic markers. *Nature* 2013;**500**:541–6.
36. Karlsson FH, Tremaroli V, Nookaew I, Bergström G, Behre CJ, Fagerberg B, et al. Gut metagenome in European women with normal, impaired and diabetic glucose control. *Nature* 2013;**498**: 99–103.
37. Kim MS, Hwang SS, Park EJ, Bae JW. Strict vegetarian diet improves the risk factors associated with metabolic diseases by modulating gut microbiota and reducing intestinal inflammation. *Environ Microbiol Rep* 2013;**5**:765–75.
38. Larsen N, Vogensen FK, van den Berg FW, Nielsen DS, Andreasen AS, Pedersen BK, et al. Gut microbiota in human adults with type 2 diabetes differs from non-diabetic adults. *PLoS One* 2010;**5**:e9085.
39. Turnbaugh PJ, Ley RE, Mahowald MA, Magrini V, Mardis ER, Gordon JI. An obesity-associated gut microbiome with increased capacity for energy harvest. *Nature* 2006;**444**:1027–31.
40. Karlsson F, Tremaroli V, Nielsen J, Bäckhed F. Assessing the human gut microbiota in metabolic diseases. *Diabetes* 2013;**62**: 3341–9.
41. Lewis K, Lutgendorff F, Phan V, Söderholm JD, Sherman PM, McKay DM. Enhanced translocation of bacteria across metabolically stressed epithelia is reduced by butyrate. *Inflamm Bowel Dis* 2010;**16**:1138–48.
42. Kootte RS, Vrieze A, Holleman F, Dallinga-Thie GM, Zoetendal EG, De Vos WM, et al. The therapeutic potential of manipulating gut microbiota in obesity and type 2 diabetes mellitus. *Diab Obes Metab* 2012;**14**:112–20.
43. Karlsson FH, Tremaroli V, Nookaew I, Bergström G, Behre CJ, Fagerberg B, et al. Gut metagenome in European women with normal, impaired and diabetic glucose control. *Nature* 2013;**498**: 99–103.
44. Pussinen PJ, Havulinna AS, Lehto M, Sundvall J, Salomaa V. Endotoxemia is associated with an increased risk of incident diabetes. *Diabetes Care* 2011;**34**:392–7.
45. Cani PD, Delzenne NM. The gut microbiome as therapeutic target. *Pharmacol Ther* 2011;**130**:202–12.
46. Yamagata K, Daitoku H, Shimamoto Y, Matsuzaki H, Hirota K, Ishida J, et al. Bile acids regulate gluconeogenic gene expression via small heterodimer partner-mediated repression of hepatocyte nuclear factor 4 and Foxo1. *J Biol Chem* 2004;**279**:23158–65.
47. Prabha V, Ohri M. Review: bacterial transformations of bile acids. *World J Microbiol Biotechnol* 2006;**22**:191–6.
48. Sayin SI, Wahlström A, Felin J, Jäntti S, Marschall HU, Bamberg K, et al. Gut microbiota regulates bile acid metabolism by reducing the levels of tauro-beta-muricholic acid, a naturally occurring FXR antagonist. *Cell Metab* 2013;**17**:225–35.
49. Islam KB, Fukiya S, Hagio M, Fujii N, Ishizuka S, Ooka T, et al. Bile acid is a host factor that regulates the composition of the cecal microbiota in rats. *Gastroenterology* 2011;**141**:1773–81.
50. Furet JP, Kong LC, Tap J, Poitou C, Basdevant A, Bouillot JL, et al. Differential adaptation of human gut microbiota to bariatric surgery-induced weight loss: links with metabolic and low-grade inflammation markers. *Diabetes* 2010;**59**:3049–57.
51. Kong LC, Tap J, Aron-Wisnewsky J, Pelloux V, Basdevant A, Bouillot JL, et al. Gut microbiota after gastric bypass in human obesity: increased richness and associations of bacterial genera with adipose tissue genes. *Am J Clin Nutr* 2013;**98**:16–24.
52. Li LV, Ashrafian H, Bueter M, Kinross J, Sands C, Le Roux CW, et al. Metabolic surgery profoundly influences gut microbial-host metabolic cross-talk. *Gut* 2011;**60**:1214–23.
53. Liou AP, Paziuk M, Luevano JM, Machineni S, Turnbaugh PJ, Kaplan LM. Conserved shifts in the gut microbiota due to gastric bypass reduce host weight and adiposity. *Sci Transl Med* 2013;**5**: 178ra141.
54. Ryan KK, Tremaroli V, Clemmensen C, Kovatcheva-Datchary PA, Myronovych R, Karns HE, et al. FXR is a molecular target for the effects of vertical sleeve gastrectomy. *Nature* 2014;**8**:183–8.

55. Ortega FJ, Fernández-Real JM. Inflammation in adipose tissue and fatty acid anabolism: when enough is enough! *Horm Metab Res* 2013;**45**:1009–19.
56. Zhang X, Shen D, Fang Z, Jie Z, Qiu X, Zhang C, et al. Human gut microbiota changes reveal the progression of glucose intolerance. *PLoS One* 2013;**8**:e71108.
57. Clarke TB, Davis KM, Lysenko ES, Zhou AY, Yu Y, Weiser JN. Recognition of peptidoglycan from the microbiota by Nod1 enhances systemic innate immunity. *Nat Med* 2010;**16**:228–31.
58. Fernández-Real JM, Pickup JC. Innate immunity, insulin resistance and type 2 diabetes. *Diabetologia* 2012;**55**:273–8.
59. Ghanim H, Abuaysheh S, Sia CL, Korzeniewski K, Chaudhuri A, Fernandez-Real JM, et al. Increase in plasma endotoxin concentrations and the expression of Toll-like receptors and suppressor of cytokine signaling-3 in mononuclear cells after a high-fat, high-carbohydrate meal: implications for insulin resistance. *Diabetes Care* 2009;**32**:2281–7.
60. Sanz Y, Santacruz A, Gauffin P. Gut microbiota in obesity and metabolic disorders. *Proc Nutr Soc* 2010;**69**:434–41.
61. Kumar H, Laiho A, Lundelin K, Ley RE, Isolauri E, Salminen S. Gut microbiota as an epigenetic regulator: Pilot study based on whole-genome methylation analysis. *mBio* 2014;**5**:e02113–4.
62. Shin NR, Lee JC, Lee HY, Kim MS, Whon TW, Lee MS, et al. An increase in the *Akkermansia* spp. population induced by metformin treatment improves glucose homeostasis in diet-induced obese mice. *Gut* 2014;**63**:727–35.
63. Everard A, Belzer C, Geurts L, Ouwerkerk JP, Druart C, Bindels LB, et al. Cross-talk between *Akkermansia muciniphila* and intestinal epithelium controls diet-induced obesity. *Proc Natl Acad Sci USA* 2013;**110**:9066–71.
64. Greiner TU, Hyötyläinen T, Knip M, Bäckhed F, Orešič M. The gut microbiota modulates glycaemic control and serum metabolite profiles in non-obese diabetic mice. *PLoS One* 2014;**9**(11): e110359.
65. Soyucen E, Gulcan A, Aktuglu-Zeybek AC, Onal H, Kiykim E, Aydin A. Differences in the gut microbiota of healthy children and those with type 1 diabetes. *Pediatr Int* 2014;**56**:336–43.
66. van Nood E, Keller JJ, Kuijper EJ, Speelman P. New treatment options for infections with Clostridium difficile. *Ned Tijdschr Geneeskd* 2013;**157**:A6580.
67. Vrieze A, Van Nood E, Holleman F, Salojärvi J, Kootte RS, Bartelsman JF, et al. Transfer of intestinal microbiota from lean donors increases insulin sensitivity in individuals with metabolic syndrome. *Gastroenterology* 2012;**143**:913–6. e7.

CHAPTER

13

Molecular Aspects of Glucose Regulation of Pancreatic β Cells

Rosa Gasa[1,2], *Ramon Gomis*[1,2,3], *Anna Novials*[1,2], *Joan-Marc Servitja*[1,2]

[1]Diabetes and Obesity Research Laboratory, Institut d'Investigations Biomediques August Pi i Sunyer, Barcelona, Spain; [2]Centro de Investigación Biomédica en Red de Diabetes y Enfermedades Metabólicas Asociadas (CIBERDEM), Barcelona, Spain; [3]University of Barcelona, Hospital Clínic, Barcelona, Spain

1. INTRODUCTION

Pancreatic β cells recognize extracellular glucose concentration and secrete insulin and this is the primary mechanism maintaining glucose homeostasis. Glucose-stimulated insulin secretion is tuned over minutes to hours, being modulated by additional factors including nonglucose nutrients, hormones, and neural inputs. In addition to stimulating insulin secretion, glucose controls insulin synthesis by increasing insulin gene expression and by enhancing translation of its messenger RNA (mRNA) and processing of proinsulin to mature insulin. Furthermore, glucose, under some circumstances, acts as a mitogen of β cells. In contrast to these physiological roles, persistent high glucose concentrations can be detrimental for β-cell function, a situation referred to as "glucotoxicity." In this chapter, we attempt to provide a general view of the intracellular pathways regulated by glucose in β cells, focusing our attention in those mechanisms linked to the transcriptional and mitogenic roles of this sugar. Readers are referred to Chapter 3 for specific information on insulin secretion and production. We will also briefly summarize our current knowledge on glucotoxicity, particularly on how this metabolic insult contributes to β-cell damage in diabetes.

2. INTRACELLULAR GLUCOSE SIGNALING

Glucose enters the β cell via glucose transporter (GLUT)2 in rodents (or GLUT1 in humans[1,2]) and is then phosphorylated to glucose-6-phosphate by glucokinase (Gck), a low-affinity hexokinase that, with a Michaelis constant of approximately 8 mM, controls glycolytic flux at the physiological range of this sugar (4–11 mM), thus serving as the "gatekeeper" for metabolic signaling in β cells. Glucose-6-phosphate enters glycolysis to generate adenosine triphosphate (ATP; with concomitant consumption of adenosine diphosphate [ADP] and adenosine monophosphate [AMP]), which closes the potassium inward rectifier/sulfonylurea receptor (Kir6.2/SUR1) complex, which leads to depolarization and opening of voltage-dependent calcium channels, calcium entry, and stimulation of exocytosis. Glucose-dependent effects on β-cell transcription and proliferation are also dependent on intracellular metabolism of glucose and use many of the same signals that link glucose with insulin secretion (Figure 1).

In many instances, glucose-triggered signals act on similar proteins as those used by growth factor–dependent pathways, thereby providing an effective mechanism for combinatorial control of specific β-cell functions.[3,4] This cross-talk complicates the discrimination between direct effects of glucose and indirect/modulatory effects of this hexose on other pathways. A clear example of this scenario is the insulin/insulin-growth factor-1 (IGF-1) pathway. Given that the primary outcome of glucose signaling in β cells is insulin secretion and that β cells are fully equipped with the machinery of the insulin/IGF-1 pathway, it has not been trivial to discern between direct glucose effects from those that require activation of insulin receptors via secreted insulin. However, using β cells that lack the insulin receptor or IRS-2, it was shown that glucose failed to activate the key downstream kinases phosphatidylinositol 3-kinase and Akt kinase,[5] thus demonstrating the requirement of an active insulin-signaling pathway for the glucose

Molecular Nutrition and Diabetes
http://dx.doi.org/10.1016/B978-0-12-801585-8.00013-0

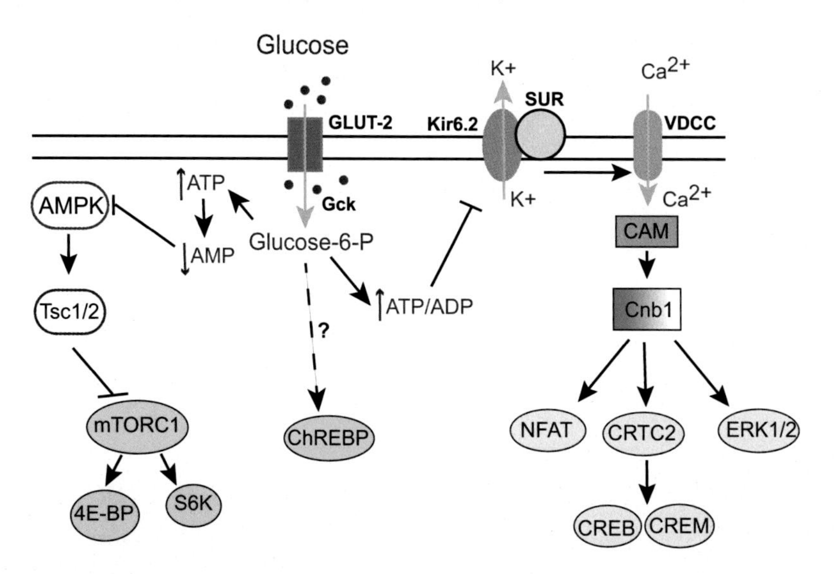

FIGURE 1 Main glucose signaling pathways in the β cell. Briefly, glucose metabolism triggers calcium influx, which binds to calmodulin (CAM) and activates the phosphatase calcineurin (Cnb1), which then activates the nuclear factor of activated T-cell family of transcription factors (NFAT), CRCR2, and extracellular signal-regulated protein kinase 1 and 2 (ERK1/2) families. Metabolism of glucose also activates the transcription factor Carbohydrate response element binding protein (ChREBP) by yet-undefined mechanisms. On the other hand, generation of adenosine triphosphate (ATP) depletes adenosine diphosphate (ADP) and adenosine monophosphate (AMP), which suppresses AMP-activated protein kinase (AMPK), with the resultant activation of mammalian target of rapamycin (mTOR)-dependent signaling.

effects on this pathway. In contrast, other effects of glucose, such as activation of extracellular signal-regulated protein kinases 1 and 2 (ERK1/2), were shown to be insulin-independent.[5]

2.1 The Transcription Factor Carbohydrate Response Element Binding Protein

Carbohydrate response element binding protein (ChREBP) is a glucose-regulated basic helix-loop-helix/leucine-zipper transcription factor. ChREBP and Max-like protein X form heterodimers and bind to carbohydrate response elements in the gene promoters of glucose-responsive target genes.[6,7] Glucose activates ChREBP by promoting its dephosphorylation (via its metabolite xylulose-5-phosphate and activation of protein phosphatase 2A), allowing its nuclear entry and DNA binding.[8] Additionally, ChREBP contains a glucose-sensing module encompassing a low-glucose inhibitory domain and a glucose-response activation conserved element. Glucose responsiveness is mediated by inhibition of the transactivation activity of glucose-response activation conserved element by low-glucose inhibitory domain and relief of the inhibition by glucose.[9] ChREBP is highly expressed in β cells and, among the roles assigned to this factor, is the control of lipogenesis, lipoglutoxicity, and proliferation.[10–13] However, all of this evidence comes from studies in isolated islets and/or β-cell lines, and corroboration in an in vivo setting is still missing. Global knockout for ChREBP exhibits decreased hepatic glycolysis and lipogenesis, modest hyperglycemia, and glucose intolerance, but the β-cell phenotype has not been reported.[14] Recently, a second shortened ChREBP isoform—ChREBPb—that, in contrast to full-length ChREBP—ChREBPa—is constitutively active has been identified and found predominantly in the nucleus.[15] Whether this isoform has a role in β-cell biology will be interesting to investigate.

2.2 Calcium-Dependent Pathways

In β cells, glucose-induced rise in intracellular calcium concentrations activates calmodulin, which then phosphorylates and activates the Ser/Thr phosphatase calcineurin. Mice with a β cell–specific deletion of the calcineurin phosphatase regulatory subunit b1 (Cnb1) develop age-dependent diabetes characterized by decreased β-cell proliferation and mass, reduced pancreatic insulin content and hypoinsulinemia.[16] Further, elimination of calcineurin signaling in endocrine progenitors has revealed that calcineurin is needed for neonatal β-cell functional maturation and growth.[17] Remarkably, this evidence supports the notion that β-cell failure may be the triggering factor of diabetes in patients receiving calcineurin inhibitors (cyclosporine, tacrolimus) as immunosuppressive drugs.[18] Among the calcineurin substrates, three have emerged as possible mediators of the glucose effects: the nuclear factor of activated T-cell family of transcription factors

(NFATs), CREB-regulated transcription coactivator-2, and the ERK1/2.

The NFAT proteins regulate gene transcription to coordinate proliferation, survival, and differentiation of multiple cell types. They are found in phosphorylated form in the cytoplasm and calcineurin-mediated dephosphorylation triggers their translocation to the nucleus, where they bind, among others, promoters of cell cycle activators (cyclins D, c-Myc) and inhibitors (p21, p27). Conditional expression of active NFATc1 in β cell–specific Cnb1 knockouts rescues defective expression of genes linked to proliferation and β-cell function preventing diabetes in these animals,[16] which demonstrates that NFAT factors are important mediators of the glucose/calcineurin pathway.

Glucose also promotes calcineurin-dependent dephosphorylation of TORC2 at Ser-275, which releases cytoplasmic TORC2 from the scaffolding protein 14-3-3, permitting its nuclear entry and association with transcription factors such as CREB and the related proteins CREM and ATF1.[19] CREB-dependent regulation is critical for β-cell proliferation and survival,[20,21] thus linking glucose with these two functions. It is noteworthy that CREB/CREM/ATF1 factors govern cellular responses to the cAMP signaling pathway, hence serving as a common point where glucose-triggered and growth factor–dependent signaling converge to regulate β-cell function.

The ERK1/2 kinases have long been recognized as kinases activated by glucose and other nutrients in β cells.[22,23] This activation requires the metabolism of glucose and is calcium-dependent, likely via calcineurin, although additional mechanisms have not been disregarded.[24] This is in contrast with growth factor–dependent regulation of ERK1/2 in β cells that is thought to be calcium-independent.[25] Although the function of ERK1/2 in cell survival and cell-cycle progression has been extensively studied in multiple cellular contexts, in β cells, ERK1/2 have also been proposed to directly regulate transcription factors that bind the *insulin* gene including Pdx1, NeuroD, or MafA, thus linking these kinases with regulation of *insulin* gene expression.[24]

Apart from calcineurin, Ca^{+2} may also exert its effects via alternative mechanisms. For instance, sorcin, a penta-EF hand calcium-binding protein, is found associated with ChREBP in the cytoplasm. Following glucose stimulation and Ca^{2+} influx, sorcin liberates ChREBP, which is then free to translocate to the nucleus and regulate transcriptional outputs.[26]

2.3 Mammalian Target of Rapamycin Signaling

Mammalian target of rapamycin (mTOR) kinase as part of the mTOR complex mTORC1 is unique in that it integrates signals from nutrients—glucose and amino acids—and mitogens to regulate cellular growth through stimulation of protein synthesis by targeting ribosomal S6 protein kinase 1 and eukaryotic initiation factor 4E-binding protein 1.[27] High glucose levels robustly activate mTOR in an insulin-independent manner in pancreatic islets.[28,29] Numerous studies with its inhibitor, rapamycin, demonstrate that mTORC1 is involved in regulation of β-cell mass and can mediate adaptation of β cells to insulin resistance. Findings using mouse genetic models in which individual upstream regulators of mTOR have been ablated also support a role of this pathway in the regulation of β-cell mass, but their effect on proliferation and the precise mechanisms involved remain unclear.

mTOR is also part of complex mTORC2, which is insensitive to rapamycin and phosphorylates Akt, PKCa, and SGK1. Conditional deletion of Rictor, the mTOR partner in mTORC2, in β cells leads to mild hyperglycemia resulting from decreased β-cell mass, proliferation, and glucose-induced insulin secretion. Islets from these mice exhibit decreased Akt-S473 phosphorylation and increased abundance of FoxO1 and p27 proteins.[30]

2.4 AMP-Activated Protein Kinase Signaling

The AMP-activated protein kinase (AMPK) is one of the best-characterized cellular energy sensors, which is responsible for conserving energy during times of nutrient restriction. AMPK mediates adaptation to decreased nutrients (low ATP/ADP + AMP ratio) by promoting energy production and limiting energy utilization. Activation of AMPK inhibits mTORC1 by acting on its upstream negative regulators TSC1/2. Deletion of the kinase LKB1, which activates AMPK, in adult or developing β cells is associated with reduced AMPK activity and augmented mTORC1 signaling and this leads to enhanced insulin secretion, β-cell proliferation, and hypertrophy.[31–33] Intriguingly, AMPK gain-of-function transgenic models have confirmed the role of this pathway on insulin secretion but failed to detect effects on β-cell proliferation,[31] pointing to additional AMPK-independent pathways downstream of LKB1 involved in β-cell mass regulation. Nonetheless, in support of a relevant role of AMPK in β-cell function, deletion of the catalytic subunits of AMPK in β cells results in impaired glucose tolerance, decreased insulin secretion and, surprisingly, increased β-cell proliferation.[34] Future studies will be needed to delineate the precise mechanisms involved in these discrepant phenotypes.

3. GLUCOSE AS A MITOGENIC SIGNAL FOR β CELLS

To meet the organism demands or recovering from a diabetic injury, adult mice need to increase their β-cell mass. The control of β-cell mass relies on a highly regulated balance involving β-cell proliferation, neogenesis, cell size, and apoptosis. The size of the β-cell pool is dynamically regulated to adapt to insulin demands. β cells need to compensate peripheral insulin resistance and the consequent hyperglycemia by increasing β-cell mass and secrete sufficient insulin levels. Thus, understanding how β-cell mass is regulated and the mechanisms by which natural mitogens drive the expansion of functional β cells is a major focus of the diabetes research community. Different studies in mice showed that the major source of new β cells during adult life and after a pancreatectomy is originated from preexisting β cells rather than pluripotent stem cells.[35,36]

3.1 Glucose Stimulation of β-Cell Proliferation

The effects of glucose on pancreatic β cells are not limited to acute stimulus-induced insulin secretion but also include long-term effects on β-cell physiology. Glucose metabolism has been shown to mediate long-term adaptive responses in β cells, including survival and function. Glucose is also a potent mitogen for β cells, thereby providing a mechanism to match β-cell mass to the organism's demands of this hormone. Glucose has been shown to promote β-cell proliferation in rat and mouse islets in vitro. In cultured mouse islets, the major change in the rate of β-cell replication is observed between 5 and 11 mM,[37] indicating that the modulation of this process takes place within a narrow range of glucose concentrations. In vivo, it has been known for long time that short-term glucose infusion, which induces a moderate increase in blood glucose levels, increases β-cell replication in mice.[38,39] Whether glucose can also induce human β-cell proliferation is still an issue under debate. It has been demonstrated that human β cells do not share the replicative potential of rodent β cells when exposed to the same mitogenic stimuli.[40,41] However, some years ago, it was reported that glucose-induced β-cell replication in vitro when glucose levels raised from 2.8 to 5.6 mM.[42] More recently, human islets transplanted in diabetic mice for a short period also showed increased β-cell replication, yet the highest level of β-cell replication was observed at glucose concentrations lower than 10 mM.[43] These discrepancies could be explained by multiple factors such as the source and the quality of the islet preparation, the period of incubation or the negative effects that glucose could exert on human β cells and that are a confounding factor for the interpretation of the results. For instance, whereas mouse β cells can proliferate in vitro at glucose concentrations as high as 16 mM,[37] human β cells present increased apoptosis when cultured at 11 mM.[41] In conclusion, caution should be taken when translating findings described in rodents to human β cells.

3.2 Glucose as a Systemic Mitogenic Signal for β Cells

Glucose itself has been suggested to be at least one of the systemic signals that allow β cells to sense insulin demands. β cells can sense small changes in glycemia, and β-cell mass is rapidly regulated to adequate the mass of β cells and insulin levels. For instance, transplanting islets to euglycemic mice leads to a reduction of the proliferation of the β cells of the recipient as they detect that less insulin-producing cells are needed to maintain glucose levels.[44] This observation led to the suggestion that the primary signal that drives β-cell replication is the individual β-cell workload,[44] which is defined as the amount of insulin that an individual β cell must secrete to maintain euglycemia: the more glucose is metabolized, the more replication is induced, and therefore more insulin is secreted, allowing an euglycemic state even in incipient stages of insulin resistance. Thus, glucose-induced proliferation can occur in vivo even though glycemia is unaltered.

The glucose proliferative effect is mediated at least in part by the same signals that underlie glucose-induced insulin secretion, that is, glucose metabolism, closure of KATP channels, membrane depolarization, and calcium mobilization.[44] Glucokinase has been proposed as a key player in the regulation of glucose-induced β-cell replication. Homozygous ablation of glucokinase in β cells results in marked suppression of β-cell proliferation despite severe hyperglycemia, whereas activation of this enzyme by pharmacological approaches increases β-cell proliferation, indicating that β-cell mitogenic activity is regulated through direct stimulation of the glycolytic flux.[44] Accordingly, mice haploinsufficient for β-cell glucokinase show decreased β-cell replication and insufficient β-cell hyperplasia when fed a high-fat diet,[45] highlighting a dominant role for glucose metabolism in driving β-cell mass compensation in response to insulin resistance. Remarkably, these mechanisms described for the regulation of β-cell proliferation are likely to be conserved in humans. A gain-of-function mutation of the gene encoding glucokinase in humans has been linked to increased β-cell proliferation and hyperplastic islets.[46] Taken together, these data suggest that glucose acts as a signaling molecule for β-cell replication via its intracellular metabolism, in a cell-autonomous way, similarly to the signals that underlie glucose-stimulated insulin release. Thereby, glucose-induced insulin secretion and β-cell replication are regulated by the same mechanisms.

All these studies point to a systemic role of glucose on β-cell replication, but how the β-cell cycle is controlled in prediabetic settings is now under an intense debate. Strikingly, the role of circulating factors on β-cell proliferation has recently been reported in mouse models of insulin resistance. Circulating factors secreted by the liver have been suggested to promote β-cell proliferation in a liver-specific insulin receptor knockout mouse, a model of insulin resistance.[47] Indeed, it was previously suggested that in this mouse model IGF-1 functions as a liver-derived growth factor to promote compensatory β-cell hyperplasia through the A isoform of the insulin receptor.[48] Recently, betatrophin, a peptide primarily expressed in liver and adipose tissue, was identified as a potential hormone promoting β-cell proliferation in different mouse models of insulin resistance, although serious discrepancies on whether betatrophin plays any major role in β-cell replication have dampened the initial enthusiasm on this molecule in the diabetes research community.[49] Nevertheless, because these circulating factors secreted by peripheral tissues cannot mediate glucose-induced proliferation in isolated islets cultured in vitro, glucose must have an autonomous role on β-cell replication, and probably a joint action of glucose and systemic signals drive adaptive β-cell proliferation. In this context, glucose has been suggested to have a permissive effect on the action of other mitogenic signals.[50] This double regulation would allow a tight regulation of β proliferation, so that β cells only proliferate in response to at least two different cues: an elevated glycolytic flux and increased circulating growth factors for β cells.

3.3 Glucose Regulation of Cell-Cycle Components

Glucose regulates β-cell replication by modulating the levels of cyclins and other components of the cell cycle. A recent mouse islet transcriptome analysis uncovered a cell-cycle gene module as one of the most significantly induced among glucose-induced genes.[37] Cyclin D is responsible for the control of the entry to the cell cycle. Glucose infusion increases the total protein abundance and nuclear localization of cyclin D2 in islets.[39] Consistently, only cyclin D2 protein levels, and not those of cyclin D1, have been shown to be increased by glucose in vitro.[37,51] In line with these observations, cyclin D2 plays a crucial role in postnatal β-cell replication.[36] At the gene level, only slight increases of cyclin D genes by glucose have been reported,[37] suggesting that posttranscriptional regulatory mechanisms influence cyclin D2 expression in β cells. Besides total protein levels, the cytoplasmic-nuclear trafficking of cell cycle molecules is also involved in the cell-cycle progression mediating human β-cell proliferation.[52,53] Although cyclin D2 has also been detected in the nucleus of quiescent β cells,[51] this protein is translocated to the nucleus of glucose-induced proliferative β cells in mice.[39]

3.4 Intracellular and Autocrine/Paracrine Mechanisms of Glucose-Mediated β-Cell Proliferation

An important gap in knowledge is an understanding of how β-cell glycolytic flux translates to progression through the cell cycle. ChREBP has been shown to mediate glucose-stimulated proliferation in rodents and humans β cells, and overexpression of this transcription factor amplifies the response to glucose.[11] However, ChREBP binding to the regulatory regions of cyclin genes has not been observed. A possible link between cyclin genes and ChREBP could be c-Myc, which is a direct activator of cyclin genes and is increased upon overexpression of ChREBP in high glucose. Glucose activation of cyclin genes and proliferation in β cells also requires glucose-stimulated calcium influx,[51] which may lead to cell cycle gene activation by means of calcineurin/NFAT and sorcin/ChREBP. Thus, multiple interrelated signaling pathways activated by glucose metabolism may ultimately result in increased cyclin expression and cell-cycle progression (Figure 2).

However, the fact that control of β-cell proliferation resembles the mechanisms of regulation of insulin secretion raises the question of whether an autocrine/paracrine action of insulin could be central to glucose effects on β cells. Importantly, insulin secretion has been shown to act on an autocrine manner to induce β-cell growth and therefore to mediate at least in part the mitogenic effects of glucose on β cells.[5] It has been demonstrated that insulin secretion[54] and β-cell growth and survival in response to glucose[5] require activation of insulin receptors present in pancreatic β cells. Glucose control of islet β-cell mass and function relies at least in part through the phosphatidylinositol 3-kinase/Akt pathway downstream of insulin signaling and the subsequent inhibition of FoxO1.[55] Remarkably, using genetic studies it has been shown that insulin/FoxO1/Pdx1 signaling is one pathway that is crucial for islet compensatory growth response to insulin resistance[56] (Figure 2).

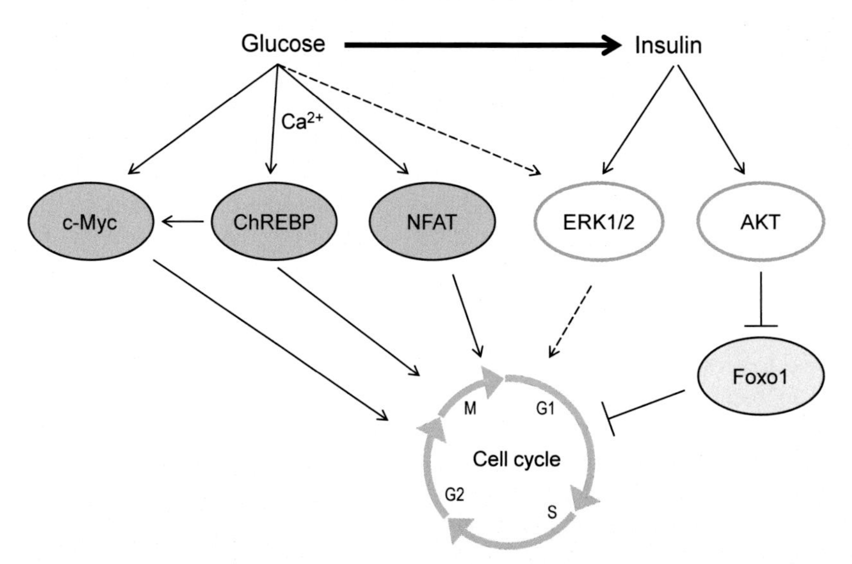

FIGURE 2 Major mechanisms mediating glucose-stimulated β-cell proliferation. Glucose metabolism triggers interrelated intracellular signals that ultimately induce β-cell proliferation. In parallel, secreted insulin activates β-cell insulin receptors through an autocrine/paracrine loop, leading to the activation of mitogenic signaling pathways commonly induced by growth factor receptors.

3.5 Effects of Aging on Glucose-Induced β-Cell Proliferation

An important issue in the field of β-cell proliferation is how far the research performed with young mice can be extrapolated to aged mouse or to human β cells. Incidence of type 2 diabetes (T2D) has been associated with obesity and age. This increased risk of developing diabetes is associated to a gradual increase in insulin resistance in peripheral tissues but also a decline in β-cell function and proliferative capacity during aging.[57] Aged β cells present reduced proliferative capacity and in diabetic individuals; this is further confounded by higher rates of β-cell apoptosis. Comparison of rodent and human islets from old and young donors revealed that β-cell replication is markedly reduced in the adult life.[41,58,59] In fact, under typical circumstances, human β cells are established by young adulthood.[60] For instance, exendin-4 can stimulate β-cell proliferation of islets transplanted into diabetic mice from young human donors and mice, but not those islets belonging to an older human donor.[61] Moreover, β-cell proliferation is severely restricted with advanced age in mice in response to a high-fat diet, short-term exposure to GLP-1, exendin-4, after streptozotocin administration, or partial pancreatectomy.[58,59] In line with these studies, glucose induction of β-cell proliferation is markedly blunted in cultured islets from old mice and rats.[37,62] Glucose has been shown to induce a gene module composed of mitotic genes in mouse pancreatic islets ex vivo. Remarkably, glucose regulation of cyclins D1 and D2 is conserved during aging, indicating that the signaling pathways orchestrated by glucose can efficiently target these cell-cycle regulators,[37] but not glucose induction of mitotic genes. Thus, the inability of glucose to induce the cell cycle in old β cells relies on events downstream of cyclins D. In this context, the expression of *Cdkn2a* is increased during aging because of a concomitant decrease of negative epigenetic marks at the promoter of this gene, which encodes the cell-cycle inhibitors p16 and p19.[63,64] These negative regulators block the interaction of cyclin D with cyclin-dependent kinases, thereby impeding the progression of the cell cycle. However, the deletion of these genes has been shown not to be sufficient to recover the capacity of aged β cells to proliferate in response to glucose,[37] indicating that other players are involved in this age-dependent loss of β-cell replication. The reduced capacity of old β cells to proliferate in settings of increased insulin demands may therefore hamper the adaptive response of β cells, and may be one of the factors accelerating the progression of diabetes in elderly patients.

Other studies, however, highlight the fact that aged β cells still retain the capacity to proliferate. Thus, a compensatory proliferation in response to specific stimuli such as treatment with a glucokinase activator and in response to partial ablation of β cells has been reported.[65] Moreover, β-cell replication is stimulated in islets from aged mice after transplantation into a hyperglycemic recipient, with no difference between the old and the young adult ones.[66] Thus, the effect of aging on β-cell replication and whether old β cells still retain a minimal capacity to proliferate or can reenter the cell cycle in specific settings are currently under an intense debate that warrants further investigation.

4. GLUCOSE SIGNALING AND β-CELL TRANSCRIPTION

Apart from regulating insulin secretion and proliferative responses, glucose controls the expression of genes that are selectively expressed in pancreatic β cells, including those encoding insulin, the glucose transporter GLUT2 (Slc2a2) or islet amyloid polypeptide, among others.[67–69] These transcriptional effects of glucose are pivotal for adaptation of β-cell function to metabolic demands.

4.1 Glucose Regulation of *Insulin* Gene Expression

It has been nearly 50 years since the initial report on glucose regulation of *preproinsulin* mRNA levels.[70] It is now known that glucose increases insulin gene expression through effects in both transcriptional rate and mRNA stability,[71] and that these effects require intracellular metabolism of the sugar. Regulated transcription by glucose is thought to rely on the same *cis*-acting sequences within the proximal promoter region that control cell-type specificity of the *insulin* gene. Thus, mutations in the three conserved insulin enhancer elements E1, A3, and Ripe3b/C1, which are major determinants of β cell–specific expression of the *insulin* gene,[72] have also been shown to lead to impaired glucose-regulated transcription (Figure 3). The mechanisms involved are not fully understood but they are partly mediated by glucose-induced posttranslational modifications of the transcription factors that bind to these motifs, triggering changes in their subcellular localization, DNA binding and/or intrinsic transcriptional activities that ultimately modify *insulin* gene transcription. Moreover, the genes encoding these transcription factors may also be targeted by glucose, thus adding another level of regulation to this process.

The homeodomain transcription factor Pdx1 is mainly expressed in β cells and is a key transactivator not only of the *insulin* gene but also of other β-cell genes such as *Slc2a2, Gck,* or *IAPP.*[73] Pdx1 binds to the core TAAT sequence of the proximal A boxes of the *insulin* promoter. Of these boxes, A3 has been recognized as being the glucose-sensitive element in mouse and human islets.[73] Remarkably, reintroduction of Pdx1 in insulin-producing cells devoid of this factor restored their ability to regulate insulin mRNA levels in response to glucose.[74] Phosphorylation of Pdx1 has been suggested to modulate multiple aspects of PDX-1 activity, including nuclear-cytoplasmic shuttling, DNA binding, transactivation properties, and interactions with transcriptional cofactors.[72,75–77] Several pathways have been implicated in PDX1 phosphorylation including p38/SAPK, glycogen synthase kinase 3 (GSK3), ERK1/2, or Hipk2.[72,77] However, definite demonstration in intact cells is lacking. In general, the precise mechanisms by which glucose affects PDX-1 phosphorylation remains uncertain, as studies using different experimental models have provided inconsistent results.

The basic helix-loop-helix factor NeuroD1/Beta2 is expressed in pancreatic and neuroendocrine cells and is an important regulator of *insulin* gene expression. NeuroD1/Beta2 forms heterodimers with ubiquitously expressed class A basic helix-loop-helix factor proteins such as E12/E47 and binds to E boxes (core sequence CANNTG) in the *insulin* proximal promoter. The rat *insulin I* gene has two E boxes whereas rat insulin II and the human insulin gene only contain one (E1).[72] As for PDX-1, effects of glucose on nuclear translocation and transcriptional capacity of NeuroD1 have been demonstrated in cultured cells. ERK1/2 has been revealed as a possible mediator of the glucose-effects on NeuroD1 transactivation activity and dimer formation.[24,78] In addition to phosphorylation, NeuroD1 can be acetylated[79] and O-linked glycosylated.[80] This latter

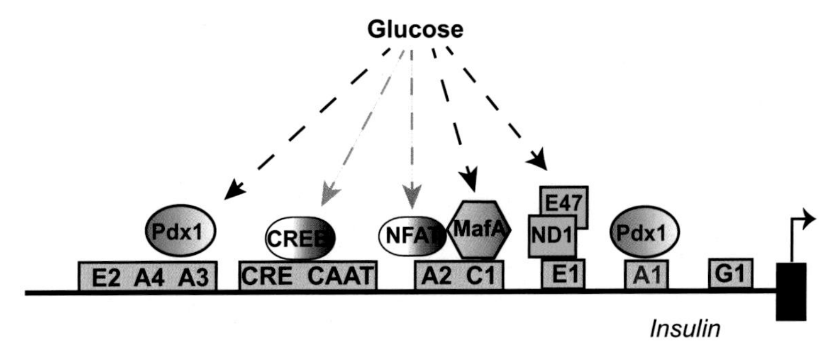

FIGURE 3 Schematic representation of the proximal insulin promoter depicting conserved *cis*-acting elements and their binding factors. Known glucose-responsive elements are shown in red. Pdx1 binds to the A boxes, the heterodimer NeuroD1/E47 (ND1) to the E boxes, and MafA to the C1 element within 400 bp from the transcriptional start site (arrow). In addition to these β cell–specific proteins, the proximal promoter contains conserved motifs that bind ubiquitous factors such as CREB and NFAT. Glucose controls insulin gene transcription by regulating these factors at different levels including expression, cellular sublocalization, DNA binding, transcriptional properties, or interactions with cofactors.

modification is mediated by O-linked GlcNAc transferase in response to high glucose–induced increase in hexosamine biosynthetic pathway flux, and promotes NeuroD1 translocation to the nucleus.[80]

It was not until 2002 that MafA, a member of the Maf family of basic leucine zipper proteins, was identified as the transcription factor binding to the RIPE3b/C1 element of the *insulin* promoter.[81,82] Since then, multiple studies have corroborated the pivotal role of this factor in the regulation of *insulin* gene expression and of glucose-induced insulin secretion by β cells.[83–86] Both the abundance of MafA protein and mRNA are increased by high glucose concentrations in cultured cells,[82,87] consistent with the original finding of glucose-regulated binding of the RIPE3b factor in β-cell nuclear extracts.[88] Glucose metabolism and calcium have been implicated in controlling *MafA* gene expression in β-cell lines,[87,89] but in vivo validation is needed. The turnover and activation potential of MafA are also controlled by multiple phoshorylations[89–92] elicited by several kinases including GSK3, CaMK, p38, or ERK1/2.[22,77] Phosphorylation by GSK3 is needed for proteasomal degradation of MafA,[93] which is thought to allow for rapid inhibition of insulin transcription under low-glucose conditions. Again, most of these studies were performed in vitro. Therefore, additional studies are required to demonstrate that MafA is regulated in vivo by these kinases. It is also unclear at present whether MafA expression and binding occur prior to transcriptional activation of the *insulin* gene in response to glucose.

Finally, the three previously mentioned β cell–specific factors interact with a host of additional proteins including p300, NFAT, CCAAT/enhancer-binding protein beta, c-Jun, and cAMP-response responsive CREB/CREM that also impact regulation of *insulin* gene transcription positively and negatively. It is not surprising that some of these proteins (NFAT, CREB) are mediators of signaling pathways activated by glucose in β cells. It is thus conceivable that effective activation of the *insulin* gene will rely on cooperative interactions among β cell–specific and these ubiquitous proteins. As an example, MafA was identified as the binding partner of NFATc proteins in the insulin promoter.[94] Hence, the specificity of MafA expression could provide a unique mechanism through which the broadly expressed NFATc proteins might regulate the expression of insulin and other β cell–enriched genes and their regulation by nutritional stimuli (Figure 3).

4.2 Characterization of Glucose-Regulated Transcriptional Modules

Rodent islets are usually cultured at 10–11 mM glucose because this is the condition that provides the best outcome for β-cell function and survival.[50,95] Lowering glucose levels down to 2–5 mM or increasing glucose up to 30 mM markedly impairs β-cell function. This tight regulation of β-cell function depending on glucose levels is controlled at the transcriptional level. Thus, glucose is a fundamental signal in the regulation of gene expression of pancreatic islet β cells and, overall, the changes observed are in line with a positive action of glucose on insulin expression and secretion, and on β-cell mass.[37,50,96–98] Global gene expression analyses have revealed dramatic differences in the expression profiles of pancreatic islets cultured at different glucose concentrations. Some genes are induced or repressed unidirectionally between low and high glucose concentrations. Interestingly, other clusters show complex V-shaped or inverse V-shaped profiles, with maximal or minimal expression at 5 or 10 mM glucose.[98]

Apart from the genes encoding β-cell hormones, glucose induces the expression of many other genes that are selectively expressed in pancreatic β cells including key genes for the determination of the β-cell fate such as *Pdx1* and *MafA*. Some of these genes are highly sensitive to glucose, showing modulation already between 5 and 11 mM glucose.[37] Humans have peaks of postprandial glucose (around 11 mM) and phases of lower glucose (around 5 mM); therefore, the expression of these genes are likely to change depending on blood glucose levels during daily fluctuations. Moreover, glucose also induces a cluster of genes shared with neurons and that are involved in cell survival and secretory processes, in line with the glucose induction of a secretory pathway in MIN6 murine pancreatic β cells.[96] This event is likely required to express genes related to insulin secretion according to glucose levels, thereby ensuring that insulin secretion is matched to insulin demands (Figure 4).

Ontology analysis of genes upregulated at low glucose (3 and 5 mM) revealed the enrichment genes related to ribosomes, mitochondrial ribosomes, and oxidative phosphorylation.[37] The transcription factor genes Nrf1 and Myc are induced in the same direction, consistent with the concerted action of c-Myc and NRF1 on the induction of mitochondrial genes and the regulation of mitochondrial biogenesis. On the other hand, glucose represses a cluster of genes related to stress responses,[37,50,95,98] including *Ddit3* (Chop) and *Trib3*. Remarkably, a missense polymorphism in human *TRIB3* that results in greater protein stability has been linked to increased risk of T2D and to impaired insulin exocytosis and β-cell proliferation,[99] in part because of an inhibitory role on the Akt signaling pathway. Ddit3 is a component of the ER stress–mediated apoptosis pathway and has also been reported to increase not only under low glucose but also during long-term exposure to high glucose in the context of

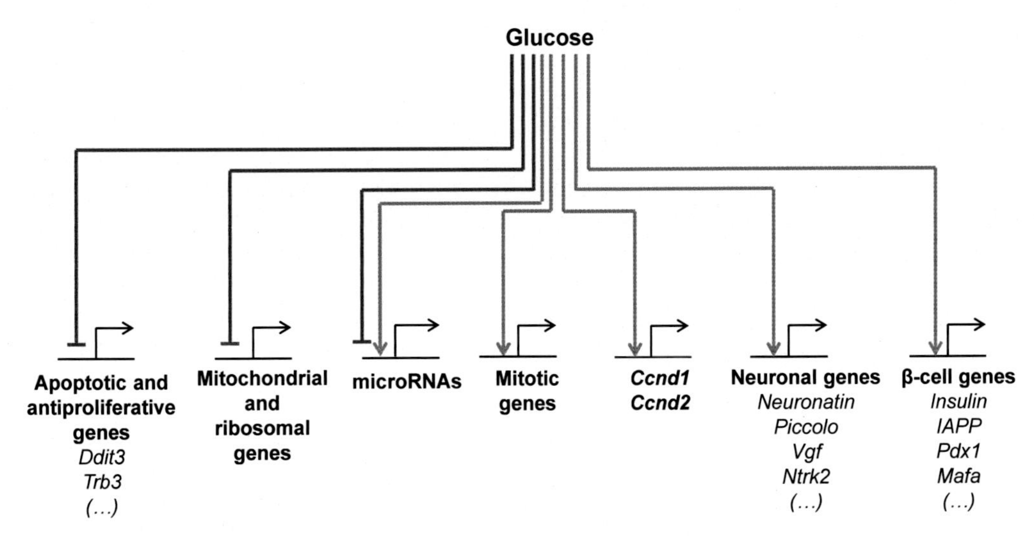

FIGURE 4 Gene modules activated or repressed by stimulatory glucose concentrations in mouse pancreatic β cells. Glucose has a profound impact on the transcriptional program of β cells. Several sets or modules of genes are activated or repressed in mouse pancreatic islets cultured at 11 mM glucose, a stimulatory condition for β cells (see Ref. 37 for more details). The modulation of these gene modules allows β cells to adapt to glucose concentrations by regulating β-cell function, survival, and proliferation. Effects of chronic hyperglycemia are not included in this scheme.

glucotoxicity.[100] Increased levels of Trib3 and Ddit3, among other proteins, may be instrumental for the reduced proliferative rate of β cells at low glucose concentrations.

Among the genes that are induced by glucose, one of the most prominent modules is that composed by cell-cycle genes. The majority of glucose-induced cell-cycle genes are involved in late stages of the cell cycle, namely the progression from G2 to M and the mitotic phase, which are transcriptionally regulated during the cell cycle. Moreover, glucose also induces the expression of genes involved in the G1 phase, such as *Ccnd1* and *Ccnd2*, although only cyclin D2 protein levels are increased by glucose. Importantly, the maximal induction of cell-cycle genes is observed between 5 and 11 mM glucose. This suggests that mild, but sustained, increases in glucose concentrations within this range of concentrations, as occurs in the early stages of diabetes, may induce a proliferative response in β cells. A transcriptional analysis of pancreatic islets from young and old mice cultured at different glucose concentrations revealed that the induction of a cell-cycle gene module is selectively abrogated in aged islets, despite an unexpected conservation of the overall response of islets to glucose. This indicates that the signaling and transcriptional networks are globally maintained in aged islets, except specific steps of the cell cycle. This reduced proliferative capacity of β cells to proliferate in response to glucose might reduce the compensatory response in insulin resistance settings. Indeed, a cell-cycle regulatory module in islets predicts diabetes susceptibility.[101] Studies using models of insulin resistance showed that induction of a cell-cycle module in pancreatic islets correlates with the ability of mice from a diabetes-resistant strain to show increased β-cell replication, which is absent in diabetes-susceptible obese mice.

Therefore, glucose is required to maintain a transcriptional program that maintains fully differentiated and functional β cells. This program involves the induction of genes with a positive effect on β-cell function and mass, in combination with the repression of genes that have antiproliferative and apoptotic effects or is deleterious for β cells. Altogether, this program enables the adaptation of β-cell mass function and mass to increased systemic insulin demands (Figure 4).

4.3 Noncoding RNAs

Gene regulation has been focused for years on the expression of genes encoding proteins. Recent technological advances in the field of genome sequencing have shed new light on the role of noncoding RNAs in gene regulation. Transcriptome analyses have revealed that the vast majority of the genome (at least three-quarters of the human genome) is capable of being transcribed, of which only a small fraction (<2%) is translated into proteins.[102] During the past few years, the interest for noncoding RNAs has emerged and many functions for these genes are being uncovered, including transcriptional functions.

MicroRNAs (miRNAs) are short noncoding RNAs that regulate gene expression by binding to target mRNAs, which leads to reduced protein synthesis or decreased mRNA levels. Hundreds of miRNAs have been identified, and some of them have been shown to play important roles in β-cell function, survival, and proliferation, and to mediate compensatory expansion of the β-cell mass.[103] Several studies have provided evidence that miRNAs are regulated by glucose. For instance, glucose represses the levels of the precursor of miR-375, which is the most abundant miRNA in pancreatic islets. This reduction leads to the concomitant increase of one of its target genes, PDK1, leading to Akt activation and β-cell proliferation.[104] On the other hand, glucose increases the expression of miRNA-30d, which induces insulin transcription factor MafA and insulin production by targeting mitogen-activated protein 4 kinase 4 in pancreatic β cells.[105] These examples illustrate how glucose can regulate the expression of key β-cell transcription factors, insulin synthesis, and β-cell proliferation by modulating the levels of miRNAs.

5. GLUCOTOXICITY

The term "glucotoxicity" refers to the detrimental effects of chronic exposure to supraphysiological glucose concentrations on the phenotype and function of β cells. The association of small increases in plasma glucose levels with dramatic loss of acute glucose-stimulated insulin secretion was first demonstrated in humans nearly 40 years ago.[106] Since then, this finding has been confirmed in several animal models exposed to hyperglycemia including partially pancreatectomized,[107] glucose-infused[108] and neonatal diabetic rats,[109] and in vitro cultured islets.[110,111] Insulin secretion is considered to be the first function compromised by chronic high glucose levels in β cells. Yet, hyperglycemia has also been shown to lead to depletion of insulin stores together with decreased *insulin* gene expression.[110,112] In a similar manner, expression of other highly expressed β cell–specific genes including *Gck* and *Slc2a2*[112,113] and of β-cell identity transcription factors (and upstream regulators of the mentioned genes), such as MafA, Pdx1, or Nkx6.1,[112,114,115] has also been found downregulated by chronic high glucose. Recent studies using genetically manipulated mice that exhibit hyperglycemia have revealed that the reduction in expression of β-cell markers is accompanied by upregulation of embryonic and pluripotent cell markers,[114,116] which has led to suggest that β-cell dedifferentiation may be partly responsible for β-cell dysfunction under high glucose conditions. These are indeed exciting findings that will need to be corroborated in nonmanipulated genetic backgrounds.

Multiple biochemical pathways have been suggested to lead to glucotoxicity, including glucose autoxidation, protein kinase C activation, methylglyoxal formation, glycation, hexosamine metabolism, and sorbitol formation.[117] Common to many of these pathways is the formation of reactive oxygen species that, in excess and over time, can cause chronic oxidative stress, which in turn causes defective *insulin* gene expression and secretion as well as apoptosis. Oxidative excess is particularly hazardous to islets because of their low levels of intrinsic antioxidant defenses.[118,119] One of the mechanisms of action through which hyperglycemia can worsen β-cell function is thought to be decreased levels of the transcription factors Pdx1 and MafA, followed by their reduced binding to the *insulin* promoter.[120–122] Indeed, exposure to antioxidant treatment was shown to protect against loss of Pdx1 and MafA expression,[123,124] and transgenic expression of glutathione peroxidase restored MafA expression in islets of diabetic *db/db* mice.[115] These observations support a role of oxidative stress in downregulation of these transcription factors under high-glucose conditions. One suggested mechanism for reduced expression of *Insulin* and *Pdx1* under chronic hyperglycemia is promoter inactivation mediated by DNA methylation. Indeed, increased DNA methylation of *PDX1* and *INS* in human pancreatic islets from patients with T2D has been reported.[125–127] The finding that short-term exposure to hyperglycemia results in augmented DNA methylation of the insulin promoter in clonal rat β cells supports the concept that glucose itself can trigger epigenetic changes in pancreatic β cells.[125] These epigenetic effects may be important for long-lasting effects of glucose on β cells that could even be transmitted between generations.

Inflammation is another potential factor suggested to be involved in β-cell dysfunction in hyperglycemic settings. High glucose activates the apoptotic gene Fas[41] and, remarkably, interleukin (IL)-1β. β cells themselves were demonstrated to produce IL-1β, which was consistent with the production of IL-1β in pancreatic β cells of T2D patients but not in nondiabetic subjects.[128] Thioredoxin-interacting protein has been suggested to be a link between hyperglycemia and inflammation.[129] Thioredoxin-interacting protein is induced by glucose and is a component of the NLRP3 inflammasome. The activation of this complex leads to caspase-1 activation and subsequent IL-1β maturation and secretion. Although glucose induction of IL-1β has not been reproduced by others,[130] growing evidence indicates that inflammation is central to glucotoxicity in T2D and therefore the IL-1β/NF-κB pathway has emerged as a target to preserve β-cell mass and function in T2D. Indeed, the IL-1 receptor antagonist anakinra has been shown to improve glycemia and β-cell secretory function and reduced markers of systemic inflammation in T2D patients.[131]

6. CONCLUDING REMARKS

Glucose is not only an essential regulator of insulin secretion and production, but it also controls the replicative activity and broad gene expression patterns of β cells. Despite the essential role of glucose for β-cell function, chronic exposure to high glucose levels becomes deleterious for these same functions. Therefore, it is highly important to delineate and define the molecular underpinnings of how glucose transitions from being a physiological regulator to a toxic molecule for β cells. This knowledge is pivotal to understand the pathophysiology of diabetes and may provide new avenues for prevention or amelioration of β-cell dysfunction and loss in diabetes.

References

1. Ferrer J, Benito C, Gomis R. Pancreatic islet GLUT2 glucose transporter mRNA and protein expression in humans with and without NIDDM. *Diabetes* 1995;**44**(12):1369–74.
2. De Vos A, Heimberg H, Quartier E, et al. Human and rat beta cells differ in glucose transporter but not in glucokinase gene expression. *J Clin Invest* 1995;**96**(5):2489–95.
3. Kulkarni RN, Mizrachi EB, Ocana AG, Stewart AF. Human beta-cell proliferation and intracellular signaling: driving in the dark without a road map. *Diabetes* 2012;**61**(9):2205–13.
4. Bernal-Mizrachi E, Kulkarni RN, Scott DK, Mauvais-Jarvis F, Stewart AF, Garcia-Ocana A. Human beta-cell proliferation and intracellular signaling part 2: still driving in the dark without a road map. *Diabetes* 2014;**63**(3):819–31.
5. Assmann A, Ueki K, Winnay JN, Kadowaki T, Kulkarni RN. Glucose effects on beta-cell growth and survival require activation of insulin receptors and insulin receptor substrate 2. *Mol Cell Biol* 2009;**29**(11):3219–28.
6. Filhoulaud G, Guilmeau S, Dentin R, Girard J, Postic C. Novel insights into ChREBP regulation and function. *Trends Endocrinol Metab TEM* 2013;**24**(5):257–68.
7. Lizuka K. Recent progress on the role of ChREBP in glucose and lipid metabolism. *Endocr J* 2013;**60**(5):543–55.
8. Kawaguchi T, Takenoshita M, Kabashima T, Uyeda K. Glucose and cAMP regulate the L-type pyruvate kinase gene by phosphorylation/dephosphorylation of the carbohydrate response element binding protein. *Proc Natl Acad Sci USA* 2001;**98**(24):13710–5.
9. Li MV, Chang B, Imamura M, Poungvarin N, Chan L. Glucose-dependent transcriptional regulation by an evolutionarily conserved glucose-sensing module. *Diabetes* 2006;**55**(5):1179–89.
10. da Silva Xavier G, Rutter GA, Diraison F, Andreolas C, Leclerc I. ChREBP binding to fatty acid synthase and L-type pyruvate kinase genes is stimulated by glucose in pancreatic beta-cells. *J Lipid Res* 2006;**47**(11):2482–91.
11. Metukuri MR, Zhang P, Basantani MK, et al. ChREBP mediates glucose-stimulated pancreatic beta-cell proliferation. *Diabetes* 2012;**61**(8):2004–15.
12. Poungvarin N, Lee JK, Yechoor VK, et al. Carbohydrate response element-binding protein (ChREBP) plays a pivotal role in beta cell glucotoxicity. *Diabetologia* 2012;**55**(6):1783–96.
13. Wang H, Kouri G, Wollheim CB. ER stress and SREBP-1 activation are implicated in beta-cell glucolipotoxicity. *J Cell Sci* 2005; **118**(Pt 17):3905–15.
14. Iizuka K, Bruick RK, Liang G, Horton JD, Uyeda K. Deficiency of carbohydrate response element-binding protein (ChREBP) reduces lipogenesis as well as glycolysis. *Proc Natl Acad Sci USA* 2004;**101**(19):7281–6.
15. Herman MA, Peroni OD, Villoria J, et al. A novel ChREBP isoform in adipose tissue regulates systemic glucose metabolism. *Nature* 2012;**484**(7394):333–8.
16. Heit JJ, Apelqvist AA, Gu X, et al. Calcineurin/NFAT signalling regulates pancreatic beta-cell growth and function. *Nature* 2006; **443**(7109):345–9.
17. Goodyer WR, Gu X, Liu Y, Bottino R, Crabtree GR, Kim SK. Neonatal beta cell development in mice and humans is regulated by calcineurin/NFAT. *Dev Cell* 2012;**23**(1):21–34.
18. Weir MR, Fink JC. Risk for posttransplant diabetes mellitus with current immunosuppressive medications. *Am J Kidney Dis* 1999; **34**(1):1–13.
19. Jansson D, Ng AC, Fu A, Depatie C, Al Azzabi M, Screaton RA. Glucose controls CREB activity in islet cells via regulated phosphorylation of TORC2. *Proc Natl Acad Sci USA* 2008;**105**(29): 10161–6.
20. Hussain MA, Porras DL, Rowe MH, et al. Increased pancreatic beta-cell proliferation mediated by CREB binding protein gene activation. *Mol Cell Biol* 2006;**26**(20):7747–59.
21. Jhala US, Canettieri G, Screaton RA, et al. cAMP promotes pancreatic beta-cell survival via CREB-mediated induction of IRS2. *Genes Dev* 2003;**17**(13):1575–80.
22. Khoo S, Cobb MH. Activation of mitogen-activating protein kinase by glucose is not required for insulin secretion. *Proc Natl Acad Sci USA* 1997;**94**(11):5599–604.
23. Frodin M, Sekine N, Roche E, et al. Glucose, other secretagogues, and nerve growth factor stimulate mitogen-activated protein kinase in the insulin-secreting beta-cell line, INS-1. *J Biol Chem* 1995;**270**(14):7882–9.
24. Lawrence M, Shao C, Duan L, McGlynn K, Cobb MH. The protein kinases ERK1/2 and their roles in pancreatic beta cells. *Acta Physiol* 2008;**192**(1):11–7.
25. Arnette D, Gibson TB, Lawrence MC, et al. Regulation of ERK1 and ERK2 by glucose and peptide hormones in pancreatic beta cells. *J Biol Chem* 2003;**278**(35):32517–25.
26. Noordeen NA, Khera TK, Sun G, et al. Carbohydrate-responsive element-binding protein (ChREBP) is a negative regulator of ARNT/HIF-1beta gene expression in pancreatic islet beta-cells. *Diabetes* 2010;**59**(1):153–60.
27. Foster KG, Fingar DC. Mammalian target of rapamycin (mTOR): conducting the cellular signaling symphony. *J Biol Chem* 2010; **285**(19):14071–7.
28. Xu G, Kwon G, Marshall CA, Lin TA, Lawrence Jr JC, McDaniel ML. Branched-chain amino acids are essential in the regulation of PHAS-I and p70 S6 kinase by pancreatic beta-cells. A possible role in protein translation and mitogenic signaling. *J Biol Chem* 1998;**273**(43):28178–84.
29. McDaniel ML, Marshall CA, Pappan KL, Kwon G. Metabolic and autocrine regulation of the mammalian target of rapamycin by pancreatic beta-cells. *Diabetes* 2002;**51**(10):2877–85.
30. Gu Y, Lindner J, Kumar A, Yuan W, Magnuson MA. Rictor/mTORC2 is essential for maintaining a balance between beta-cell proliferation and cell size. *Diabetes* 2011;**60**(3):827–37.
31. Sun G, Tarasov AI, McGinty JA, et al. LKB1 deletion with the RIP2.Cre transgene modifies pancreatic beta-cell morphology and enhances insulin secretion in vivo. *Am J Physiol Endocrinol Metab* 2010;**298**(6):E1261–73.
32. Granot Z, Swisa A, Magenheim J, et al. LKB1 regulates pancreatic beta cell size, polarity, and function. *Cell Metab* 2009;**10**(4):296–308.
33. Fu A, Ng AC, Depatie C, et al. Loss of Lkb1 in adult beta cells increases beta cell mass and enhances glucose tolerance in mice. *Cell Metab* 2009;**10**(4):285–95.
34. Sun G, Tarasov AI, McGinty J, et al. Ablation of AMP-activated protein kinase alpha1 and alpha2 from mouse pancreatic beta cells and RIP2.Cre neurons suppresses insulin release in vivo. *Diabetologia* 2010;**53**(5):924–36.
35. Dor Y, Brown J, Martinez OI, Melton DA. Adult pancreatic beta-cells are formed by self-duplication rather than stem-cell differentiation. *Nature* 2004;**429**(6987):41–6.

36. Georgia S, Bhushan A. Beta cell replication is the primary mechanism for maintaining postnatal beta cell mass. *J Clin Invest* 2004;**114**(7):963–8.
37. Moreno-Asso A, Castano C, Grilli A, Novials A, Servitja JM. Glucose regulation of a cell cycle gene module is selectively lost in mouse pancreatic islets during ageing. *Diabetologia* 2013;**56**(8):1761–72.
38. Bonner-Weir S, Deery D, Leahy JL, Weir GC. Compensatory growth of pancreatic beta-cells in adult rats after short-term glucose infusion. *Diabetes* 1989;**38**(1):49–53.
39. Alonso LC, Yokoe T, Zhang P, et al. Glucose infusion in mice: a new model to induce beta-cell replication. *Diabetes* 2007;**56**(7):1792–801.
40. Parnaud G, Bosco D, Berney T, et al. Proliferation of sorted human and rat beta cells. *Diabetologia* 2008;**51**(1):91–100.
41. Maedler K, Schumann DM, Schulthess F, et al. Aging correlates with decreased beta-cell proliferative capacity and enhanced sensitivity to apoptosis: a potential role for fas and pancreatic duodenal homeobox-1. *Diabetes* 2006;**55**(9):2455–62.
42. Tyrberg B, Eizirik DL, Hellerstrom C, Pipeleers DG, Andersson A. Human pancreatic beta-cell deoxyribonucleic acid-synthesis in islet grafts decreases with increasing organ donor age but increases in response to glucose stimulation in vitro. *Endocrinology* 1996;**137**(12):5694–9.
43. Levitt HE, Cyphert TJ, Pascoe JL, et al. Glucose stimulates human beta cell replication in vivo in islets transplanted into NOD-severe combined immunodeficiency (SCID) mice. *Diabetologia* 2011;**54**(3):572–82.
44. Porat S, Weinberg-Corem N, Tornovsky-Babaey S, et al. Control of pancreatic beta cell regeneration by glucose metabolism. *Cell Metab* 2011;**13**(4):440–9.
45. Terauchi Y, Takamoto I, Kubota N, et al. Glucokinase and IRS-2 are required for compensatory beta cell hyperplasia in response to high-fat diet-induced insulin resistance. *J Clin Invest* 2007;**117**(1):246–57.
46. Kassem S, Bhandari S, Rodriguez-Bada P, et al. Large islets, beta-cell proliferation, and a glucokinase mutation. *N Engl J Med* 2010;**362**(14):1348–50.
47. El Ouaamari A, Kawamori D, Dirice E, et al. Liver-derived systemic factors drive beta cell hyperplasia in insulin-resistant states. *Cell Rep* 2013;**3**(2):401–10.
48. Escribano O, Guillen C, Nevado C, Gomez-Hernandez A, Kahn CR, Benito M. Beta-cell hyperplasia induced by hepatic insulin resistance: role of a liver-pancreas endocrine axis through insulin receptor A isoform. *Diabetes* 2009;**58**(4):820–8.
49. Kaestner KH. Betatrophin–promises fading and lessons learned. *Cell Metab* 2014;**20**(6):932–3.
50. Martens GA, Pipeleers D. Glucose, regulator of survival and phenotype of pancreatic beta cells. *Vitam Horm* 2009;**80**:507–39.
51. Salpeter SJ, Klochendler A, Weinberg-Corem N, et al. Glucose regulates cyclin D2 expression in quiescent and replicating pancreatic beta-cells through glycolysis and calcium channels. *Endocrinology* 2011;**152**(7):2589–98.
52. Fiaschi-Taesch NM, Kleinberger JW, Salim FG, et al. Human pancreatic beta-cell G1/S molecule cell cycle atlas. *Diabetes* 2013;**62**(7):2450–9.
53. Fiaschi-Taesch NM, Kleinberger JW, Salim FG, et al. Cytoplasmic-nuclear trafficking of G1/S cell cycle molecules and adult human beta-cell replication: a revised model of human beta-cell G1/S control. *Diabetes* 2013;**62**(7):2460–70.
54. Kulkarni RN, Bruning JC, Winnay JN, Postic C, Magnuson MA, Kahn CR. Tissue-specific knockout of the insulin receptor in pancreatic beta cells creates an insulin secretory defect similar to that in type 2 diabetes. *Cell* 1999;**96**(3):329–39.
55. Martinez SC, Cras-Meneur C, Bernal-Mizrachi E, Permutt MA. Glucose regulates Foxo1 through insulin receptor signaling in the pancreatic islet beta-cell. *Diabetes* 2006;**55**(6):1581–91.
56. Okada T, Liew CW, Hu J, et al. Insulin receptors in beta-cells are critical for islet compensatory growth response to insulin resistance. *Proc Natl Acad Sci USA* 2007;**104**(21):8977–82.
57. De Tata V. Age-related impairment of pancreatic beta-cell function: pathophysiological and cellular mechanisms. *Front Endocrinol* 2014;**5**:138.
58. Rankin MM, Kushner JA. Adaptive beta-cell proliferation is severely restricted with advanced age. *Diabetes* 2009;**58**(6):1365–72.
59. Tschen SI, Dhawan S, Gurlo T, Bhushan A. Age-dependent decline in beta-cell proliferation restricts the capacity of beta-cell regeneration in mice. *Diabetes* 2009;**58**(6):1312–20.
60. Perl S, Kushner JA, Buchholz BA, et al. Significant human beta-cell turnover is limited to the first three decades of life as determined by in vivo thymidine analog incorporation and radiocarbon dating. *J Clin Endocrinol Metab* 2010;**95**(10):E234–9.
61. Tian L, Gao J, Weng G, et al. Comparison of exendin-4 on beta-cell replication in mouse and human islet grafts. *Transpl Int* 2011;**24**(8):856–64.
62. Assefa Z, Lavens A, Steyaert C, et al. Glucose regulates rat beta cell number through age-dependent effects on beta cell survival and proliferation. *PloS One* 2014;**9**(1):e85174.
63. Chen H, Gu X, Su IH, et al. Polycomb protein Ezh2 regulates pancreatic beta-cell Ink4a/Arf expression and regeneration in diabetes mellitus. *Genes Dev* 2009;**23**(8):975–85.
64. Dhawan S, Tschen SI, Bhushan A. Bmi-1 regulates the Ink4a/Arf locus to control pancreatic beta-cell proliferation. *Genes Dev* 2009;**23**(8):906–11.
65. Stolovich-Rain M, Hija A, Grimsby J, Glaser B, Dor Y. Pancreatic beta cells in very old mice retain capacity for compensatory proliferation. *J Biol Chem* 2012;**287**(33):27407–14.
66. Chen X, Zhang X, Chen F, Larson CS, Wang LJ, Kaufman DB. Comparative study of regenerative potential of beta cells from young and aged donor mice using a novel islet transplantation model. *Transplantation* 2009;**88**(4):496–503.
67. Gasa R, Gomis R, Casamitjana R, Rivera F, Novials A. Glucose regulation of islet amyloid polypeptide gene expression in rat pancreatic islets. *Am J Physiol* 1997;**272**(4 Pt 1):E543–9.
68. Gasa R, Gomis R, Casamitjana R, Novials A. Signals related to glucose metabolism regulate islet amyloid polypeptide (IAPP) gene expression in human pancreatic islets. *Regul Pept* 1997;**68**(2):99–104.
69. Ferrer J, Gomis R, Fernandez Alvarez J, Casamitjana R, Vilardell E. Signals derived from glucose metabolism are required for glucose regulation of pancreatic islet GLUT2 mRNA and protein. *Diabetes* 1993;**42**(9):1273–80.
70. Jarrett RJ, Keen H, Track N. Glucose and RNA synthesis in mammalian islets of Langerhans. *Nature* 1967;**213**(5076):634–5.
71. Docherty K, Clark AR. Nutrient regulation of insulin gene expression. *FASEB J* 1994;**8**(1):20–7.
72. Melloul D, Marshak S, Cerasi E. Regulation of insulin gene transcription. *Diabetologia* 2002;**45**(3):309–26.
73. Melloul D, Ben-Neriah Y, Cerasi E. Glucose modulates the binding of an islet-specific factor to a conserved sequence within the rat I and the human insulin promoters. *Proc Natl Acad Sci USA* 1993;**90**(9):3865–9.
74. Macfarlane WM, Shepherd RM, Cosgrove KE, James RF, Dunne MJ, Docherty K. Glucose modulation of insulin mRNA levels is dependent on transcription factor PDX-1 and occurs independently of changes in intracellular Ca^{2+}. *Diabetes* 2000;**49**(3):418–23.

75. Macfarlane WM, McKinnon CM, Felton-Edkins ZA, Cragg H, James RF, Docherty K. Glucose stimulates translocation of the homeodomain transcription factor PDX1 from the cytoplasm to the nucleus in pancreatic beta-cells. *J Biol Chem* 1999;**274**(2):1011–6.
76. Elrick LJ, Docherty K. Phosphorylation-dependent nucleocytoplasmic shuttling of pancreatic duodenal homeobox-1. *Diabetes* 2001;**50**(10):2244–52.
77. Andrali SS, Sampley ML, Vanderford NL, Ozcan S. Glucose regulation of insulin gene expression in pancreatic beta-cells. *Biochem J* 2008;**415**(1):1–10.
78. Petersen HV, Jensen JN, Stein R, Serup P. Glucose induced MAPK signalling influences NeuroD1-mediated activation and nuclear localization. *FEBS Lett* 2002;**528**(1–3):241–5.
79. Qiu Y, Guo M, Huang S, Stein R. Acetylation of the BETA2 transcription factor by p300-associated factor is important in insulin gene expression. *J Biol Chem* 2004;**279**(11):9796–802.
80. Andrali SS, Qian Q, Ozcan S. Glucose mediates the translocation of NeuroD1 by O-linked glycosylation. *J Biol Chem* 2007;**282**(21):15589–96.
81. Olbrot M, Rud J, Moss LG, Sharma A. Identification of beta-cell-specific insulin gene transcription factor RIPE3b1 as mammalian MafA. *Proc Natl Acad Sci USA* 2002;**99**(10):6737–42.
82. Kataoka K, Han SI, Shioda S, Hirai M, Nishizawa M, Handa H. MafA is a glucose-regulated and pancreatic beta-cell-specific transcriptional activator for the insulin gene. *J Biol Chem* 2002;**277**(51):49903–10.
83. Zhang C, Moriguchi T, Kajihara M, et al. MafA is a key regulator of glucose-stimulated insulin secretion. *Mol Cell Biol* 2005;**25**(12):4969–76.
84. Kroon E, Martinson LA, Kadoya K, et al. Pancreatic endoderm derived from human embryonic stem cells generates glucose-responsive insulin-secreting cells in vivo. *Nat Biotechnol* 2008;**26**(4):443–52.
85. Matsuoka TA, Zhao L, Artner I, et al. Members of the large Maf transcription family regulate insulin gene transcription in islet beta cells. *Mol Cell Biol* 2003;**23**(17):6049–62.
86. Matsuoka TA, Artner I, Henderson E, Means A, Sander M, Stein R. The MafA transcription factor appears to be responsible for tissue-specific expression of insulin. *Proc Natl Acad Sci USA* 2004;**101**(9):2930–3.
87. Vanderford NL, Andrali SS, Ozcan S. Glucose induces MafA expression in pancreatic beta cell lines via the hexosamine biosynthetic pathway. *J Biol Chem* 2007;**282**(3):1577–84.
88. Sharma A, Stein R. Glucose-induced transcription of the insulin gene is mediated by factors required for beta-cell-type-specific expression. *Mol Cell Biol* 1994;**14**(2):871–9.
89. Vanderford NL, Cantrell JE, Popa GJ, Ozcan S. Multiple kinases regulate mafA expression in the pancreatic beta cell line MIN6. *Arch Biochem Biophys* 2008;**480**(2):138–42.
90. Vanderford NL. Regulation of beta-cell-specific and glucose-dependent MafA expression. *Islets* 2011;**3**(1):35–7.
91. Guo S, Vanderford NL, Stein R. Phosphorylation within the MafA N terminus regulates C-terminal dimerization and DNA binding. *J Biol Chem* 2010;**285**(17):12655–61.
92. Guo S, Burnette R, Zhao L, et al. The stability and transactivation potential of the mammalian MafA transcription factor are regulated by serine 65 phosphorylation. *J Biol Chem* 2009;**284**(2):759–65.
93. Han SI, Aramata S, Yasuda K, Kataoka K. MafA stability in pancreatic beta cells is regulated by glucose and is dependent on its constitutive phosphorylation at multiple sites by glycogen synthase kinase 3. *Mol Cell Biol* 2007;**27**(19):6593–605.
94. Heit JJ. Calcineurin/NFAT signaling in the beta-cell: from diabetes to new therapeutics. *BioEssays* 2007;**29**(10):1011–21.
95. Hoorens A, Van de Casteele M, Kloppel G, Pipeleers D. Glucose promotes survival of rat pancreatic beta cells by activating synthesis of proteins which suppress a constitutive apoptotic program. *J Clin Invest* 1996;**98**(7):1568–74.
96. Webb GC, Akbar MS, Zhao C, Steiner DF. Expression profiling of pancreatic beta cells: glucose regulation of secretory and metabolic pathway genes. *Proc Natl Acad Sci USA* 2000;**97**(11):5773–8.
97. Schuit F, Flamez D, De Vos A, Pipeleers D. Glucose-regulated gene expression maintaining the glucose-responsive state of beta-cells. *Diabetes* 2002;**51**(Suppl. 3):S326–32.
98. Bensellam M, Van Lommel L, Overbergh L, Schuit FC, Jonas JC. Cluster analysis of rat pancreatic islet gene mRNA levels after culture in low-, intermediate- and high-glucose concentrations. *Diabetologia* 2009;**52**(3):463–76.
99. Liew CW, Bochenski J, Kawamori D, et al. The pseudokinase tribbles homolog 3 interacts with ATF4 to negatively regulate insulin exocytosis in human and mouse beta cells. *J Clin Invest* 2010;**120**(8):2876–88.
100. Jonas JC, Bensellam M, Duprez J, Elouil H, Guiot Y, Pascal SM. Glucose regulation of islet stress responses and beta-cell failure in type 2 diabetes. *Diabetes Obes Metab* 2009;**11**(Suppl. 4):65–81.
101. Keller MP, Choi Y, Wang P, et al. A gene expression network model of type 2 diabetes links cell cycle regulation in islets with diabetes susceptibility. *Genome Res* 2008;**18**(5):706–16.
102. Djebali S, Davis CA, Merkel A, et al. Landscape of transcription in human cells. *Nature* 2012;**489**(7414):101–8.
103. Ozcan S. Minireview: microRNA function in pancreatic beta cells. *Mol Endocrinol* 2014;**28**(12):1922–33.
104. El Ouaamari A, Baroukh N, Martens GA, Lebrun P, Pipeleers D, van Obberghen E. miR-375 targets 3′-phosphoinositide-dependent protein kinase-1 and regulates glucose-induced biological responses in pancreatic beta-cells. *Diabetes* 2008;**57**(10):2708–17.
105. Zhao X, Mohan R, Ozcan S, Tang X. MicroRNA-30d induces insulin transcription factor MafA and insulin production by targeting mitogen-activated protein 4 kinase 4 (MAP4K4) in pancreatic beta-cells. *J Biol Chem* 2012;**287**(37):31155–64.
106. Brunzell JD, Robertson RP, Lerner RL, et al. Relationships between fasting plasma glucose levels and insulin secretion during intravenous glucose tolerance tests. *J Clin Endocrinol Metab* 1976;**42**(2):222–9.
107. Bonner-Weir S, Trent DF, Weir GC. Partial pancreatectomy in the rat and subsequent defect in glucose-induced insulin release. *J Clin Invest* 1983;**71**(6):1544–53.
108. Leahy JL, Cooper HE, Deal DA, Weir GC. Chronic hyperglycemia is associated with impaired glucose influence on insulin secretion. A study in normal rats using chronic in vivo glucose infusions. *J Clin Invest* 1986;**77**(3):908–15.
109. Weir GC, Clore ET, Zmachinski CJ, Bonner-Weir S. Islet secretion in a new experimental model for non-insulin-dependent diabetes. *Diabetes* 1981;**30**(7):590–5.
110. Marshak S, Leibowitz G, Bertuzzi F, et al. Impaired beta-cell functions induced by chronic exposure of cultured human pancreatic islets to high glucose. *Diabetes* 1999;**48**(6):1230–6.
111. Eizirik DL, Korbutt GS, Hellerstrom C. Prolonged exposure of human pancreatic islets to high glucose concentrations in vitro impairs the beta-cell function. *J Clin Invest* 1992;**90**(4):1263–8.
112. Jonas JC, Sharma A, Hasenkamp W, et al. Chronic hyperglycemia triggers loss of pancreatic beta cell differentiation in an animal model of diabetes. *J Biol Chem* 1999;**274**(20):14112–21.
113. Laybutt DR, Sharma A, Sgroi DC, Gaudet J, Bonner-Weir S, Weir GC. Genetic regulation of metabolic pathways in beta-cells disrupted by hyperglycemia. *J Biol Chem* 2002;**277**(13):10912–21.
114. Wang Z, York NW, Nichols CG, Remedi MS. Pancreatic beta cell dedifferentiation in diabetes and redifferentiation following insulin therapy. *Cell Metab* 2014;**19**(5):872–82.

115. Guo S, Dai C, Guo M, et al. Inactivation of specific beta cell transcription factors in type 2 diabetes. *J Clin Invest* 2013;**123**(8): 3305–16.
116. Talchai C, Xuan S, Lin HV, Sussel L, Accili D. Pancreatic beta cell dedifferentiation as a mechanism of diabetic beta cell failure. *Cell* 2012;**150**(6):1223–34.
117. Kajimoto Y, Kaneto H. Role of oxidative stress in pancreatic beta-cell dysfunction. *Ann NY Acad Sci* 2004;**1011**:168–76.
118. Robertson RP, Harmon JS. Diabetes, glucose toxicity, and oxidative stress: a case of double jeopardy for the pancreatic islet beta cell. *Free Radic Biol Med* 2006;**41**(2):177–84.
119. Lenzen S. Oxidative stress: the vulnerable beta-cell. *Biochem Soc Trans* 2008;**36**(Pt 3):343–7.
120. Poitout V, Olson LK, Robertson RP. Chronic exposure of betaTC-6 cells to supraphysiologic concentrations of glucose decreases binding of the RIPE3b1 insulin gene transcription activator. *J Clin Invest* 1996;**97**(4):1041–6.
121. Harmon JS, Stein R, Robertson RP. Oxidative stress-mediated, post-translational loss of MafA protein as a contributing mechanism to loss of insulin gene expression in glucotoxic beta cells. *J Biol Chem* 2005;**280**(12):11107–13.
122. Olson LK, Redmon JB, Towle HC, Robertson RP. Chronic exposure of HIT cells to high glucose concentrations paradoxically decreases insulin gene transcription and alters binding of insulin gene regulatory protein. *J Clin investigation* 1993;**92**(1):514–9.
123. Tanaka Y, Gleason CE, Tran PO, Harmon JS, Robertson RP. Prevention of glucose toxicity in HIT-T15 cells and Zucker diabetic fatty rats by antioxidants. *Proc Natl Acad Sci USA* 1999; **96**(19):10857–62.
124. Kaneto H, Kajimoto Y, Miyagawa J, et al. Beneficial effects of antioxidants in diabetes: possible protection of pancreatic beta-cells against glucose toxicity. *Diabetes* 1999;**48**(12):2398–406.
125. Yang BT, Dayeh TA, Kirkpatrick CL, et al. Insulin promoter DNA methylation correlates negatively with insulin gene expression and positively with HbA(1c) levels in human pancreatic islets. *Diabetologia* 2011;**54**(2):360–7.
126. Yang BT, Dayeh TA, Volkov PA, et al. Increased DNA methylation and decreased expression of PDX-1 in pancreatic islets from patients with type 2 diabetes. *Mol Endocrinol* 2012;**26**(7):1203–12.
127. Dayeh T, Volkov P, Salo S, et al. Genome-wide DNA methylation analysis of human pancreatic islets from type 2 diabetic and non-diabetic donors identifies candidate genes that influence insulin secretion. *PLoS Genet* 2014;**10**(3):e1004160.
128. Maedler K, Sergeev P, Ris F, et al. Glucose-induced beta cell production of IL-1beta contributes to glucotoxicity in human pancreatic islets. *J Clin Invest* 2002;**110**(6):851–60.
129. Schroder K, Zhou R, Tschopp J. The NLRP3 inflammasome: a sensor for metabolic danger? *Science (New York, NY)* 2010; **327**(5963):296–300.
130. Welsh N, Cnop M, Kharroubi I, et al. Is there a role for locally produced interleukin-1 in the deleterious effects of high glucose or the type 2 diabetes milieu to human pancreatic islets? *Diabetes* 2005;**54**(11):3238–44.
131. Larsen CM, Faulenbach M, Vaag A, et al. Interleukin-1-receptor antagonist in type 2 diabetes mellitus. *N Engl J Med* 2007; **356**(15):1517–26.

CHAPTER

14

Metals in Diabetes: Zinc Homeostasis in the Metabolic Syndrome and Diabetes

Shudong Wang[1,2], *Gilbert C. Liu*[2], *Kupper A. Wintergerst*[2,3], *Lu Cai*[2,3]

[1]Cardiovascular Center, The First Hospital of Jilin University, Changchun, China; [2]Department of Pediatrics, University of Louisville, Louisville, KY, USA; [3]Wendy L. Novak Diabetes Care Center, University of Louisville, Louisville, KY, USA

1. INTRODUCTION

Diabetes mellitus (DM) is a major cause of morbidity and mortality worldwide. In type 1 diabetes (T1D), there is a lack of insulin production, and in type 2 diabetes (T2D), the resistance to the effect of insulin is predominant. Currently, diabetes treatment involves diet modification, weight reduction, exercise, oral medications, and insulin. Ongoing clinical trials testing various medications to determine their effectiveness in treating the complications of diabetes have met with some success, but there still is much to learn about this disease. Therefore, understanding the mechanisms for the development of metabolic syndrome, such as insulin resistance, and T2D are urgently needed before we can develop an effective approach to prevent and treat these diseases.

Trace elements participate in tissue, cellular, and subcellular functions. These include immune regulation by humoral and cellular mechanisms, nerve conduction, muscle contractions, membrane potential regulations, and mitochondrial activity and enzyme reactions. Dyshomeostasis of trace elements in the body has been identified as a potential pathway to metabolic syndrome and diabetes since the 1980s.[30,65,107] Evidence indicates that micronutrients such as iron and vanadium are higher in T2D. Calcium, magnesium, sodium, chromium, cobalt, iodine, iron, selenium, manganese, and zinc (Zn) seem to be low in T2D. Elements such as potassium and copper have no clear associations with diabetes.[46] Appropriate trace element supplementation might ameliorate some physiological deficiencies associated with diabetes and prevent or retard secondary complications.[16,52] The potential roles of vanadium, chromium, and selenium in diabetes are well reviewed in the literature.[52]

Zn is an essential trace metal that is required for many cell events. Zn is not only a cofactor of numerous enzymes and transcription factors, but it also acts as an intracellular signaling mediator.[11,86] So far, more than 300 catalytically active Zn metalloproteins and more than 2000 Zn-dependent transcription factors have been recognized; therefore, Zn is an integral component of a large variety of proteins and enzymes and participates in a wide variety of metabolic processes including carbohydrate, lipid, protein, and nucleic acid synthesis or degradation. Zn supports a healthy immune system, is needed for wound healing, and helps in maintaining the senses of taste and smell. Zn is also essential for normal growth and development during pregnancy, childhood, and adolescence. Therefore, Zn dyshomeostasis, such as that engendered by Zn deficiency, is associated with various chronic diseases.

The present chapter aims to summarize the information available in the literature, addressing the following issues: (1) Zn and insulin; (2) risk of Zn deficiency for metabolic syndrome and diabetes; (3) effect of diabetes on Zn homeostasis; and (4) improvement of insulin resistance and diabetes with Zn supplementation as well as its possible mechanisms.

2. Zn AND INSULIN

Zn, the second most prevalent trace element in the body, is involved in the structure and function of over 300 enzymes,[86] collectively representing all major biochemical categories and, thus, is essential for normal cell function and metabolism. In our body, 85% of Zn is stored in muscle and bone; 11% is stored in the skin and the liver. In term of organs, the highest

Molecular Nutrition and Diabetes
http://dx.doi.org/10.1016/B978-0-12-801585-8.00014-2

concentrations of Zn are in the prostate, retina, muscle, bone, liver, and kidney.[86,104] In multicellular organisms, virtually all Zn is intracellular. At the cellular level, 30–40% of Zn is located in the nucleus; 50% in the cytoplasm, organelles, and specialized vesicles (for digestive enzymes or hormone storage); and the remainder in the cell membrane. Therefore, Zn in these tissues is not readily mobilized, and there is essentially no free Zn. Zn intake ranges from 107 to 231 mol/d depending on the source, and the Zn requirement for humans is estimated at 15 mg/d. Zn has both catalytic and structural roles in enzymes. In finger motifs, Zn provides a scaffold that organizes protein subdomains for the interaction with either DNA or other proteins. Zn plays critical roles for the function of a number of metalloproteins, inducing members of oxido-reductase, hydrolase ligase, and lyase family and has a coactivating function with copper in superoxide dismutase or phospholipase C. The Zn ion (Zn^{2+}) does not participate in redox reactions, which makes it a stable ion in a biological medium. Tables 1 and 2 summarize key features and associated information related to Zn described in this chapter.

Zn is essential for insulin secretion. Pancreatic β-cells contain large amounts of Zn, and one of the major roles of Zn is the binding of insulin in hexamers, a crystalline structure comprising two Zn ions and six insulin molecules, which are stored in the secretory granules.[38] Part of the Zn ion pool of the β cell is cosecreted with insulin after stimulation with glucose. Immediately after secretion, the hexameric structure dissociates into the active monomer insulin and Zn ions, probably caused by a combination of a rapid decrease in Zn ion pressure and the change in pH from 5.5 to 7.4. An in vitro study suggests that Zn ions cosecreted with insulin during hyperglycemia might contribute to β-cell death by a paracrine mechanism; it has been hypothesized that such activity could link hyperinsulinism with β-cell necrosis and ensuing T2D. The indispensable role of Zn for insulin structure has been recognized in the preparation of insulin analogs to ensure their activity and stabilization.[4]

TABLE 1 Main Features of Zn

- 23rd most abundant element in the earth
- 2nd most common trace element in the body (iron)
- 4th most used metal in the world (iron, aluminum, and copper)
- 70-kg male with 2.3 g Zn in the body
- 75% Zn in skeletal muscle and bone
- In humans, Zn is found in all tissues and tissue fluids
- Antioxidant and MT inducer
- Cofactor for many key molecules in the cell
- Anti-inflammatory effect
- Insulin-like function
- Patients with diabetes and obesity often have zinc deficiency
- Clinically common used for several disorders

These pieces of general information regarding the essentiality were cited from, and detailed information are referred to, several excellent reviews.[86–88]

TABLE 2 General Information

Conditions to Lead to Zn Deficiency

1. Zn deficiency is more common in people in developing countries who eat a high phytate diet
2. Vegetarians are also prone to Zn deficiency
3. Zn absorption takes place in the intestine and thus malabsorption, cirrhosis of the liver, celiac disease, Crohn disease, and chronic diarrhea can all lead to Zn deficiency
4. Too much supplementary iron can interfere with Zn absorption

Food That Are Rich in Zn

1. Meat, eggs, nuts, cheese, oysters, and grains are good sources of Zn
2. Cereal diets rich in phytates decreases the absorption of Zn by binding to Zn

Amount at Which Zn Is Daily Required (the Recommended Daily Allowance of Zn Is)

1. 5 mg in infants
2. 10 mg in children aged 1 to 10 years
3. 15 mg in children older than 11 years
4. 20 mg in adults

Zn ions are hydrophilic and do not cross cell membranes via passive diffusion. As illustrated in Figure 1, Zn homeostasis is thus regulated by several Zn transporters.[60] Metallothionein (MT), a cysteine-rich metal-binding protein, plays a critical role in the intracellular Zn availability since MT functions as both Zn storage for, and donor to, the proteins that need Zn for its functional structure, such as Zn-finger proteins and insulin.[8,10,12] Zn is excreted mostly in feces (12–15 mg/day), with lesser amounts (0.5 mg/day) in urine.

To date, more than 20 Zn transporters have been identified and characterized.[34,59,109] They are classified into two families: the Zn transporter family (ZnT: vertebrate cation diffusion facilitator family proteins, Slc30a family) and the Zip family (Zrt/Irt-like protein, Slc39a family). ZnTs function in Zn efflux from the cytoplasm, while Zips move Zn in the opposite direction. The coordinated action of these two kinds of Zn transporters is essential to the maintenance of Zn homeostasis in the cytoplasm (Figure 1), which play important roles in the regulation of numerous pathways.[35]

A novel member of the ZnT family gene *SLC30A8* (encodes ZnT8) was identified a decade ago.[22] ZnT8 is specifically expressed in the pancreatic β-cells and has been identified as a novel target autoantigen in patients with T1D. Autoantibodies to ZnT8 (ZnT8A) are detected in 50–60% of Japanese patients with acute-onset and 20% with slow-onset autoimmune diabetes. In T2D, a single nucleotide polymorphism in SLC30A8, rs13266634 (Arg325Trp) was reported in several studies.[61,97,112] These studies suggest an important role of ZnT8 in the development of diabetes.

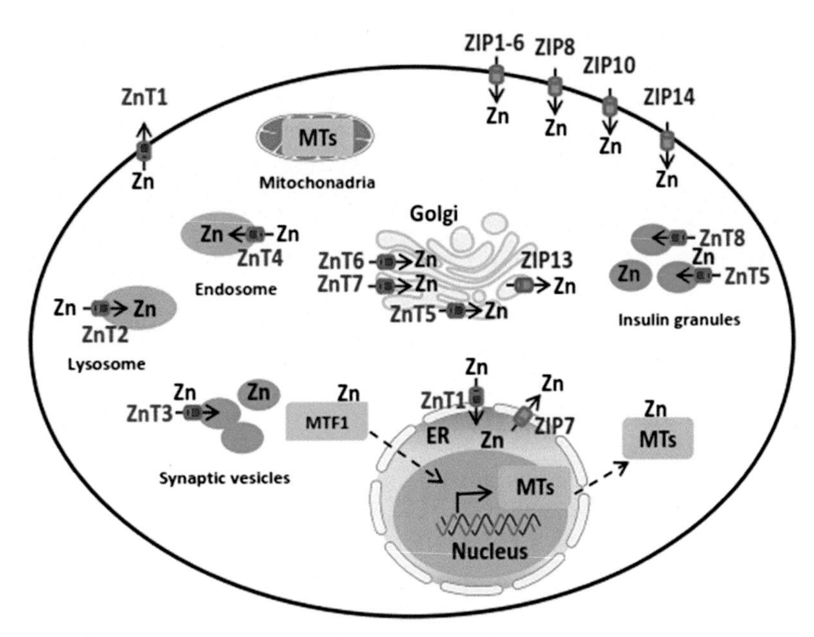

FIGURE 1 Subcellular localization of Zn transporters and MTs. Localization and potential functions of Zn transporters from the Slc39/ZIP (blue) and Slc30/ZnT (red) families, MT, and metal-responsive-element binding transcription factor 1 (MTF1) within the cell. Arrows show the predicted direction of Zn mobilization. ER, endoplasmic reticulum. *The figure was created based on a previous report.*[35]

3. A POTENTIAL RISK OF Zn DEFICIENCY FOR THE METABOLIC SYNDROME AND DIABETES

3.1 Zn Deficiency in Metabolic Syndrome

A study from Park et al. using weaning male Sprague–Dawley rats to investigate the effect of Zn deficiency on glucose tolerance demonstrated the induction of the abnormal glucose metabolism in Zn-deficient rats. They found that abnormal glucose metabolism was not due to altered blood insulin and glucagon levels but rather to peripheral resistance to insulin action.[84] Later, other groups also showed the induction of insulin resistance in the animals fed a Zn-deficient diet.[31,56]

The induction of insulin resistance by Zn deficiency was observed even in the offspring of maternal Zn deficiency. Rats consumed Zn-deficient or control diets ad libitum from age 3 weeks preconception to 21 days postparturition, and then pups were allowed to nurse their original mothers until weaning. Both male and female Zn deficient pups were less sensitive to insulin and glucose stimulation than controls at weeks 5 and 10, suggesting that suboptimal maternal Zn status induces long-term changes in the offspring related to abnormal glucose tolerance.[56]

In addition, Zn-deficient pups significantly increased their body weight compared with controls by days 10 and 20, although there was no difference in birth weight between groups.[56] Associations between Zn deficiency and increased body weight were confirmed, and expanded, by independent studies.[71] For instance, Liu et al. demonstrated that Zn deficiency significantly augmented circulating leptin concentrations and leptin signaling in the liver of obese mice. Zn deficiency also exacerbated macrophage infiltration into adipose tissue in obese mice.[71] These studies with various animal models clearly indicate significant associations between Zn and the induction of insulin resistance, hepatic and adipose inflammation, and increases in body weight.

From human data, a cross-sectional study examined the association between Zn intake and insulin resistance,[99] showing that low levels of Zn intake were inversely associated with central obesity, glucose intolerance, and diabetes in urban subjects but not in rural subjects.[99] In Spanish children, poor Zn nutritional status was found to associate with increased risk of insulin resistance.[83] Another study suggests that Zn deficiency is positively associated with insulin resistance among patients with primary biliary cirrhosis,[46,47] probably via exacerbating iron overload in the liver and inducing hepatic steatosis by facilitating lipid peroxidation.[46]

3.2 Low Zn Is Associated with a High Risk of T2D

Because Zn plays a clear role in the synthesis, storage, and secretion of insulin as well as conformational integrity of insulin in the hexameric form, Zn deficiency may contribute to the pathogenesis of diabetes.[18] In animal studies, Zn deficiency, induced via various Zn chelators, subsequently resulted in diabetes and β-cell destruction in several mammalian species, including rabbits, mice, and hamsters.[36,37] These studies with animal models clearly support the theory that Zn deficiency would be a risk factor for the development of diabetes.

Results from epidemiological studies support the theory that low Zn intake is associated with higher incidence of T2D. A case-control study using a Swedish childhood diabetes registry, with data regarding residence 3 years before the onset of disease, showed that a high concentration of Zn in drinking water was associated with a significant decrease in risk, and low groundwater concentration of Zn was associated with a significant increase in risk.[44] Several other studies consistently support this finding and yielded similar findings for T1D. For instance, Zhao et al. performed a similar study that covered the Cornwall and the former Plymouth Health Authority Regions in southwest England, including 517 children with T1D. They also found that the incidence rate of childhood diabetes is significantly associated with low Zn and magnesium in the drinking water.[115] Other recent studies further confirmed that low Zn in drinking water is associated with the risk of developing T1D during childhood.[5,95] There is one study that does not support the above findings. A Finnish study showed that neither Zn, nor nitrate, nor the urban/rural status of the area had a significant effect on the variation in incidence of T1D during childhood.[80] The authors conjectured that the lack of a significant effect might stem from the aggregated data being too crude to detect it. However, most recent studies reinforce the early observation that T1D is associated with factors that promote Zn deficiency.[32,94]

4. EFFECT OF DIABETES ON Zn HOMEOSTASIS

4.1 Diabetes-Induced Zn Dyshomeostasis

The status of serum Zn in 18 patients with diabetes compared with healthy-age matched controls was examined in the late 1990s,[105] showing that serum Zn concentration was significantly low in diabetic patients compared with controls. In contrast, copper (Cu) levels were not significantly different compared with controls. The significant Zn deficiency in plasma of diabetic patients has been confirmed by multiple studies.[2,3,32,92,94] In diabetic animal models, the low level of serum Zn relative to controls has been reported repeatedly.[30,69] However, these parallel results could not discern whether Zn causes diabetes or diabetes affects Zn homeostasis.

Increasing evidence indicates that diabetes can affect Zn homeostasis in many ways, but it is most likely the hyperglycemia, rather than any primary lesion related to diabetes, that causes the increased urinary loss and decrease in total body Zn. Zn excretion through urine is significantly increased in diabetic patients.[6,25,39,110] The significantly lower plasma Zn levels seen in diabetic patients and animals is dependent on multiple factors, including the type of diabetes, duration of diabetes, and the age of diabetes onset.[2,3,6,25,39,49,92,110]

Quilliot et al. have analyzed the effects of hyperglycemia, malabsorption, and dietary intake on Zn levels in 35 men with alcohol-induced chronic pancreatitis complicated by insulin-treated diabetes, 12 men with chronic pancreatitis without diabetes, 25 men with T1D, and 20 control subjects.[90] Seventeen of the chronic pancreatitis patients had low plasma Zn. None of the type 1 diabetic patients without pancreatitis had low plasma concentrations of Zn. This suggests the perturbations of Zn metabolism are particularly pronounced in subjects with chronic pancreatitis plus diabetes.[90]

Kechrid et al. performed an experiment to study ^{65}Zn turnover under normal and Zn-deficiency conditions in diabetic mice. The study demonstrated that pancreatic Zn was lower in the diabetic mice than the nondiabetic mice and the levels of dietary Zn did not affect the pancreatic levels in either diabetic or nondiabetic mice.[62] However, whole-blood glucose concentration was significantly higher in the mice consuming a low-Zn diet. Rate of ^{65}Zn loss was similar in both diabetic and nondiabetic mice fed with a normal Zn diet, but diabetic mice had a significantly greater whole-body ^{65}Zn loss than nondiabetic mice when they both were fed a low-Zn diet.[62]

In general, these studies suggest that Zn deficiency is strongly associated with insulin resistance, systemic inflammation, increases in body weight, and T1D. Moreover, metabolic syndrome and diabetes also influence Zn homeostasis.

4.2 Effect of Diabetes on Zn Exporter and Importer

Because the ZnT family regulates Zn fluxes into subcellular compartments and β-cells depend on Zn for both insulin crystallization and regulation of cell mass, a study was performed to investigate (1) the effect of glucose and Zn chelation on ZnT gene and protein levels and apoptosis in β-cells and pancreatic islets, (2) the effect of ZnT3 knock-down on insulin secretion in a β-cell line, and (3) the effect of ZnT3 knock-out on glucose metabolism in mice during streptozotocin (STZ)-induced β-cell stress. They found that, in insulinoma-1E

(INS-1E) cells, 2 mmol/L glucose downregulated ZnT3 and upregulated ZnT5 expression, but 16 mmol/L glucose increased ZnT3 and decreased ZnT8 expression, relative to the control (5 mmol/L glucose) group. Zn chelation by diethyldithiocarbamate (DEDTC) lowered INS-1E insulin expression. Like high levels of glucose, Zn depletion also increased ZnT3 and decreased ZnT8 gene expression, whereas the amount of ZnT3 protein in the cells was decreased. The most responsive Zn transporter, ZnT3, was further investigated by immunohistochemistry and Western blotting; 44% knock-down of ZnT-3 by silent, interfering iRNA transfection in INS-1E cells decreased insulin expression and secretion, but STZ-treated mice had higher glucose levels after ZnT3 knock-out, particularly in overt diabetic animals. These data suggest that Zn transporting proteins in β-cells respond to variations in glucose and Zn levels. ZnT3, present in β cells, is upregulated by high levels of glucose and Zn depletion. Knock-down of the ZnT3 gene lowers insulin secretion in vitro and affects in vivo glucose metabolism after STZ treatment.[101]

As mentioned, ZnT8 seems to play an important role in the development of both T1D and T2D.[61,97,112] The expression of ZnT8 in pancreas and adipose tissue was investigated by comparing homozygous db/db mice with heterozygous sibling db/$^{+}$ mice. ZnT8, at both mRNA and protein levels, in the pancreas, epididymal, and visceral fat of db/db mice was found to be reduced. These findings suggest that ZnT8 synthesis in the pancreas and adipose tissue is downregulated in db/db mice, and the reduced ZnT8 production in the pancreas may advance defects in insulin secretion in diabetes.[70]

In a study with 75 patients with T1D or T2D and 75 nondiabetic sex-/age-matched control subjects, serum Zn was found to be significantly lower in diabetic patients compared with controls. Intracellular Zn showed the same tendency. Interestingly, type 2 diabetic patients treated with insulin displayed lower serum Zn compared with those not being treated with insulin. In vitro analyses showed that insulin leads to an increase in intracellular Zn and that insulin signaling was enhanced by elevated intracellular Zn concentrations. These studies suggest that type 1 and type 2 diabetic patients have Zn deficiency, and insulin treatment is associated with significant differences in serum Zn levels among such patients.[53]

4.3 Effect of Diabetes on Zn Exporter and Importer and MT

As illustrated in Figure 1, MT also plays an important role in Zn homeostasis.[8,10,12] The effects of diabetes on tissue MT expression have been explored by Failla et al. since 1981. They examined the MT contents in multiple organs including liver, kidney, intestine, spleen, muscles, and plasma of rats 10 days after STZ treatment to induce diabetes.[28] Only hepatic and renal MT contents were significantly increased in diabetic rats compared with control rats. Furthermore, renal MT mainly bound to Cu, while hepatic MT bound to both Zn and Cu. Later, they extended the study to 21 and 28 days after STZ treatment with same trend: that is, only hepatic and renal MT increased and hepatic MT bound to both Zn and Cu, while renal MT mainly bound to Cu. This also implied that the renal Zn/Cu ratio increased in diabetic rats compared with normal rats.[29] They also demonstrated that the increased Zn and Cu levels in both the liver and kidney were not associated with the increased absorption of Zn and Cu since no difference for their absorption was found between STZ-diabetic rats and control rats.[24]

These studies suggest that hepatic MT synthesis is most likely related to diabetes-related endocrinal imbalance, while renal MT synthesis is more related to increased renal Cu accumulation.[21] Although STZ was found to induce hepatic and renal MT synthesis in a dose-dependent manner,[55] the hepatic and renal MT synthesis in STZ-induced diabetic rats was not predominantly secondary to STZ effect. This is because the hepatic and renal MT synthesis, along with increases in hepatic and renal Zn and Cu, were all preventable by supplementation with insulin.[28] In addition, Zn and Zn-MT were also significantly increased in the livers of ob/ob diabetic mice, as a model of T2D,[64] and *BB* rats, as a model of T1D.[27] Insulin treatment could prevent increased Zn and Cu as well as bound MT,[27] suggesting the increased hepatic MT synthesis is predominantly related to chronic endocrine imbalance. To support the above notion, we have demonstrated that hepatic and renal MT proteins were increased in STZ-induced diabetic rats at 1 and 6 months after diabetes as control rats, along with an increase in both hepatic and renal endothelin-1 (ET-1) mRNA. The increased hepatic MT protein level was associated with decreases in hepatic Cu and Fe, whereas increased renal MT was associated with increases in renal Cu and Fe accumulation. Zn levels were unaltered in both organs in diabetic rats. Treatment with bosentan, a potent orally active dual ET receptor blocker, partially prevented the increase in MT levels in both the liver and kidney. No significant effects of bosentan treatment on nondiabetic rats were observed.[9]

Dietary changes among patients with metabolic syndrome or diabetes likely also affect Zn homeostasis. For instance, phytic acid is known as a major determinant of Zn bioavailability. Studies comparing diabetic patients with matched controls have shown a negative relationship between dietary fiber and Zn serum.

Studies have also demonstrated a low level of the messenger RNA ratio of ZnT1 to Zip1 among diabetic patients with high dietary fiber intake. This suggests that increased dietary fiber intake, not uncommon among diabetic patients, may increase the risk of Zn deficiency.[33]

5. PREVENTION AND/OR IMPROVEMENT OF METABOLIC SYNDROME AND DIABETES BY Zn SUPPLEMENTATION AS WELL AS POSSIBLE MECHANISMS

5.1 Evidence

Zn supplementation potentially preventing alloxan-induced diabetes was reported initially by Tadros et al., but they experimented with multiple minerals including Zn, Mn, Cr, and Co.[102] Consequently Yang and Cherian,[113] followed by others, investigated the prevention of diabetes by Zn supplementation alone. These results are summarized in Table 3. In general, this protective effect of Zn was examined utilizing several fold increases in the concentration of Zn in plasma and pancreas, and controlling for factors related to food intake, body weight gain or tissue copper content. Supplementation of Zn by subcutaneous or intraperitoneal injection, drinking water, and dietary food were all effective in preventing diabetes. Zn supplementation provides a protective effect on diabetes development in multiple diabetic models. Zn supplementation prevents diabetes induced by single dose of STZ or alloxan, as well as preventing diabetes induced by multiple low doses of STZ. Multiple low doses of STZ induce a diabetic model that is mechanistically distinct from that induced by single high dose of STZ.[48,82,100] Furthermore, Zn supplementation prevents diabetes among genetically prodiabetic models such as BB Wistar rat, NOD, and db/db or od/od mouse diabetes.

Leptin is thought to be a lipostatic signal that contributes to body weight regulation. Zn might play an important role in appetite regulation and its administration by stimulating leptin production. To test this hypothesis, a prospective double-blind, randomized, placebo-controlled clinical trial, included 56 normal glucose-tolerant obese women who were treated with 30 mg Zn daily for 4 weeks. Baseline values of both intervention and control groups were similar for age, body mass index (BMI), caloric intake, insulin concentration, insulin resistance, and Zn concentration in diet, plasma, urine, and erythrocytes. After 4 weeks, BMI, fasting glucose, and Zn concentration in plasma and erythrocyte did not change in either group, although Zn concentration in the urine increased in the group with Zn supplementation. Insulin did not change in the placebo group but was significantly decreased in the Zn-supplemented group. Leptin did not change either in the placebo group or Zn group.[73] In contrast to this study, another triple-masked, randomized, placebo-controlled, crossover trial was conducted among 60 obese Iranian children who were randomly assigned to two groups of equal number; one group received 20 mg Zn and the other group received placebo for 8 weeks. A significant decrease was seen for apolipoprotein (Apo)B/ApoA-I ratio, oxidized low-density lipoprotein (LDL), leptin, malondialdehyde (MDA), and total and LDL cholesterol after receiving Zn without significant change after receiving placebo. In addition,

TABLE 3 Evidence for the Preventive Effect of Zn Supplementation on Diabetes

Zn pretreatments	Major mechanisms	Animals and type of diabetes	Outcomes	References
i.p. injection (10 mg/kg)	MT induction	Rats, STZ single dose (75 mg/kg)	++	113
Drinking (20 mmol/L), 8 w	MT (ND)	Ob/ob mice	++++	20
Dietary (1000 ppm), 4 w	MT (ND)	Prodiabetic BB Wistar rats	++++	106
Drinking (25 mmol/L), 1 w	MT induction	C57BL/6 and B6SJL/F_1 mice 5 × 40 mg STZ/kg	++++	82
Dietary (300 ppm), 6 w	MT (ND)	db/db mice	++++	98
Dietary (1000 ppm), 2 w	MT induction	CD-1 mice, ALX (50 mg/kg)	++++	
		STZ (5 × 40 mg/kg)	++++	48
Drinking (25 mmol/L), 1 (12) w	MT (ND)	C57BL/6, ALX (50 mg/kg)	+++	51
Drinking (25 mmol/L), 1 w	Inhibiting nuclear factor-κB and/or AP1	C57BL/6 mice NOD	++++	96
Genetic enhancing MT	Zn-MT	MT-TG mice, STZ (1 × 200 mg/kg)	++++	19

Notes: (MT) ND, no detection of pancreatic MT content; 1 (12) w, 1 week prior to STZ and continued 12 weeks after STZ; inhibiting nuclear factor-κB and/or AP1, through inhibition of pancreatic nuclear factor-κB and/or AP1 activation; ALX, alloxan; NOD, nonobese diabetes.

high-sensitivity C-reactive protein (hs-CRP) and markers of insulin resistance were decreased significantly in the Zn group but increased in the placebo group. Zn supplementation significantly decreased markers of insulin resistance and the mean BMI compared with placebo group.[63] These two studies suggested the different outcomes, probably due to a couple of factors—age difference, Zn dose difference, and population difference—that may involve different nutrition situations.[63,73] However, the latter study clearly indicates the deleterious consequences of childhood obesity and early changes in markers of inflammatory and oxidative stress. Zn supplementation significantly improved these deleterious changes in these obese children with a decrease in plasma leptin level along with other changes.[63]

The effect of Zn supplementation on insulin resistance and components of metabolic syndrome in prepubertal obese children has been investigated using a triple-masked, randomized, placebo-controlled cross-over trial among 60 obese Iranian children. Group 1 participants received 20 mg Zn, and group 2 received placebo on a regular daily basis for 8 weeks. After a 4-week wash-out period (WO, Figure 2), the groups were crossed over. After receiving Zn, in either group 1 or group 2, the mean fasting plasma glucose (FPG), insulin, and homeostasis model assessment for insulin resistance (HOMA-IR) decreased significantly, while BMI, waist circumference, and triglycerides (TG) did not significantly change. After receiving placebo, the mean FPG, insulin, and HOMA-IR increased significantly, while BMI, waist circumference, and TG showed a nonsignificant increase,[45] as summarized in Figure 2.

These results suggest that besides lifestyle modification, Zn supplementation may be considered as a useful and safe additional intervention treatment for improvement of cardiometabolic risk factors related to childhood obesity. However, other studies have found mixed results.[67] Forty Korean obese women aged 19–28 years were recruited for a study of Zn's effects on glucose metabolism. Twenty women were given 30 mg/day of supplemental Zn as Zn gluconate for 8 weeks and the remainder was given placebo. At the beginning of the study, dietary Zn averaged 7.31 mg/day and serum Zn averaged 12.98 μmol/L in the Zn group. Zn supplementation increased serum Zn by 15% and urinary Zn by 56%. HOMA values tended to decrease and insulin sensitivity increased slightly in the study group, but not significantly so. No change for other measurements was seen before and after Zn supplementation in either the Zn or control group.[67]

A meta-analysis evaluated the literature regarding the effect of Zn supplementation on diabetes.[54] The analysis collected 12 studies comparing the effects of Zn supplementation on fasting blood glucose in T2D. The pooled mean difference in fasting blood glucose among subjects receiving Zn supplementation is 18.13 mg/dL lower than that in placebo groups, as illustrated in Figure 3. There were eight studies comparing the effects of Zn supplementation on lipid parameters in patients with T2D. The pooled mean for total cholesterol in the Zn-supplemented group was 32.37 mg/dL lower than that in placebo groups. LDL cholesterol in the Zn-treated group also was also significantly lower (11.19 mg/dL, $p < 0.05$) than that in placebo group (Figure 3). This general conclusion for the beneficial effect of Zn supplementation on glycemic control was again supported by another meta-analysis showing the beneficial effects of Zn supplementation on blood glucose, HbA_{1c}, and TG level, systemic inflammation, and urinary albumin excretion in diabetic nephropathy patients.[16,66,93] Among the studies considered, the vast majority lasted for 6 months or less, indicating the need for longer-duration studies.[93]

Individuals with a common missense variant (rs13266634; R325W) in SLC30A8, a gene encoding a ZnT8 in the β cell, demonstrate a lower early insulin response to glucose and an increased risk of T2D. Maruthur et al. performed a study on the effect of

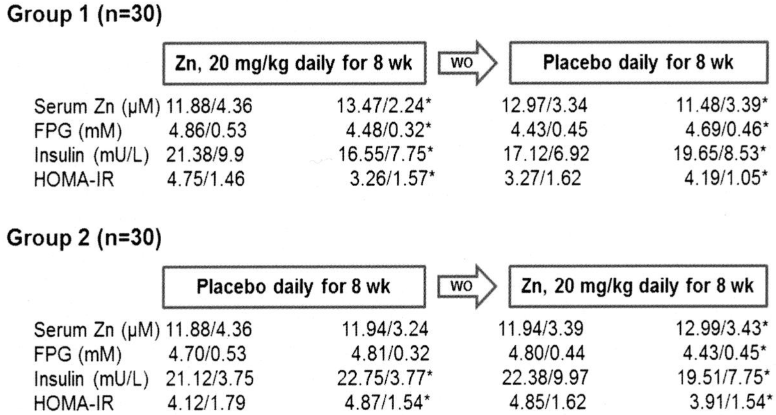

Group 1 (n=30)

		Zn, 20 mg/kg daily for 8 wk	WO	Placebo daily for 8 wk
Serum Zn (μM)	11.88/4.36	13.47/2.24*	12.97/3.34	11.48/3.39*
FPG (mM)	4.86/0.53	4.48/0.32*	4.43/0.45	4.69/0.46*
Insulin (mU/L)	21.38/9.9	16.55/7.75*	17.12/6.92	19.65/8.53*
HOMA-IR	4.75/1.46	3.26/1.57*	3.27/1.62	4.19/1.05*

Group 2 (n=30)

		Placebo daily for 8 wk	WO	Zn, 20 mg/kg daily for 8 wk
Serum Zn (μM)	11.88/4.36	11.94/3.24	11.94/3.39	12.99/3.43*
FPG (mM)	4.70/0.53	4.81/0.32	4.80/0.44	4.43/0.45*
Insulin (mU/L)	21.12/3.75	22.75/3.77*	22.38/9.97	19.51/7.75*
HOMA-IR	4.12/1.79	4.87/1.54*	4.85/1.62	3.91/1.54*

FIGURE 2 Effects of Zn supplementation on insulin resistance and plasma glucose level in obese children. *, $p < 0.05$ versus before receiving either Zn or placebo. *Figure was created based on a published study.*[45]

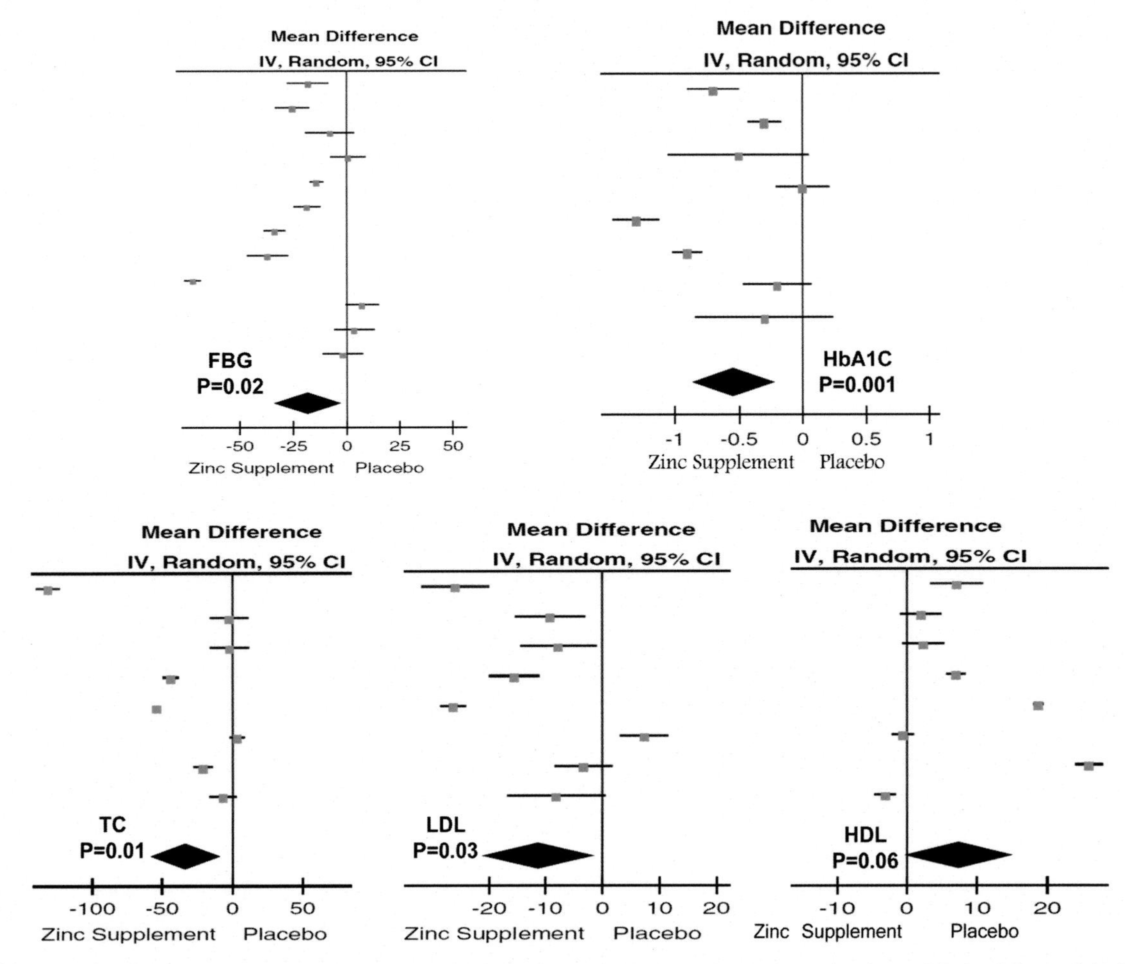

FIGURE 3 Forest plots showing effects of Zn supplementation on FBG (fasting blood glucose), HbA_{1c} (glycated hemoglobin), TC (total cholesterol), LDL (low-density lipoprotein), and HDL (high-density lipoprotein) cholesterol. *The illustration of the pooled mean difference for each measurement was collected from a previous meta-analysis.*[54]

Zn supplementation on insulin secretion specifically examining the interaction between Zn and SLC30A8 genotype among individuals from the Old Order Amish.[74] Maruthur et al.'s investigative team tested the hypothesis that Zn supplementation may improve insulin secretion in a genotype-dependent manner. They demonstrated that Zn supplementation appeared to affect the early insulin response to glucose differentially among those with the rs13266634 genotype and that Zn supplementation could be beneficial for diabetes prevention and/or treatment for individuals based on the presence of the SLC30A8 variation.

5.2 Possible Mechanisms for Zn Improvement of Metabolic Syndrome and Diabetes

5.2.1 Insulin-like Function via Insulin Signaling Pathway

Studies indicate that Zn deficiency significantly decreases the response of tissues to insulin[30] and the metabolic rate, leading to anorexia.[26] Zn has insulin-like effects on cells, including promotion of both lipogenesis and glucose transport; therefore, Zn supplementation to diabetic patients may stimulate tissues to use glucose and maintain more normal lipid metabolism and cellular function. Simon et al. have demonstrated that Zn supplementation enhanced gastrocnemius insulin receptor concentration and tyrosine kinase activity.[98] Fasting serum glucose concentrations were significantly lower in the Zn-treated diabetic group compared with non−Zn-treated diabetic group.

Zn acts with insulin-like function through a direct effect on insulin signaling regulation. Treatment of various cells with Zn significantly increased glucose transport and insulin signaling.[15,40,78,103] The effect of Zn on insulin signaling was associated with stimulation of multiple components including phosphoinositide 3-kinase (PI3K), tyrosine phosphorylation of the insulin receptor β subunit, tyrosine phosphorylation of insulin receptor substrate-1, and serine-473 phosphorylation of Akt. Hence, it appears that Zn can induce an increase in glucose transport into cells and potentiate insulin-induced glucose transport, likely acting through the insulin-signaling pathway.

The insulin-like effects of ionic Zn have been confirmed in isolated rat adipocytes at an early stage of development.[75] Concentrations of Zn between 250 and 1000 μmol/L stimulate 3-*O*-methylglucose transport. These concentrations of Zn also stimulate glucose metabolism to CO_2, glyceride fatty acid, and glyceride glycerol. Selective stimulation of the pentose phosphate cycle have been also observed since a Zn-induced increase in glucose carbon 1 oxidation persists even when glucose transport was blocked with 50 μmol/L cytochalasin B or when transport was no longer rate-limiting for metabolism at high concentrations of glucose.[75]

Interestingly, type 2 diabetic patients treated with insulin display lower serum Zn compared with those not being treated with insulin. In vitro analyses showed that insulin leads to an increase in intracellular Zn and that insulin signaling was enhanced by elevated intracellular Zn concentrations. In conclusion, type 1 and type 2 diabetic patients have Zn deficiency, and Zn supplementation may qualify as a potential treatment adjunct in T2D by promoting insulin signaling, especially in Zn-deficient subjects.[53]

The insulin-like effect of Zn was also reflected by its inactivation of glycogen synthase kinase-3β (GSK-3β), a serine/threonine protein kinase linked with insulin resistance and T2D. Treatment of HEK-293 cells with Zn was reported to enhance glycogen synthase activity and increased the intracellular levels of β-catenin, providing evidence for inhibition of endogenous GSK-3β by Zn. Moreover, Zn ions enhanced glucose uptake 3-fold in isolated mouse adipocytes, an increase similar to activation with saturated concentrations of insulin.[50] Another study found that treatment of cardiac H9c2 cells with $ZnCl_2$ (10 μmol/L) in the presence of Zn ionophore pyrithione for 20 min significantly enhanced GSK-3β phosphorylation at serine-9, indicating that exogenous Zn can inactivate GSK-3β in H9c2 cells. The effect of Zn on GSK-3β activity was blocked by the PI3K inhibitor LY-294,002, but not by the mammalian target of rapamycin (mTOR) inhibitor or the protein kinase C (PKC) inhibitor chelerythrine, implying that PI3K, but not mTOR or PKC, accounts for the action of Zn. In support of this interpretation, Zn induced a significant increase in Akt but not mTOR phosphorylation. Therefore, the PI3K/Akt signaling pathway is responsible for the inactivation of GSK-3β by Zn.[17,68]

In terms of mechanisms by which Zn ions have an insulin-like effect, one possible mechanism may be the particularly sensitive target of Zn ions, protein tyrosine phosphatase 1B (PTP 1B), a key regulator of the phosphorylation state of the insulin receptor. Tyrosine phosphatases seem to be regulated jointly by insulin-induced redox (hydrogen peroxide) signaling, which results in their oxidative inactivation and by their Zn inhibition after oxidative Zn release from other proteins. In diabetes, the significant oxidative stress and associated changes in Zn metabolism modify the cell's response and sensitivity to insulin. Zn deficiency activates stress pathways and may result in a loss of tyrosine phosphatase control, thereby causing insulin resistance.[42] The tight inhibition of protein tyrosine phosphatases by Zn is likely responsible for the known insulinomimetic effects of Zn ions, which increase net phosphorylation of the insulin/insulin-like growth factor (IGF)-1 receptors and activate their signaling cascades. More importantly, not only do extracellular Zn ions affect signal transduction, but growth factors induce cellular Zn fluctuations that are of sufficient magnitude to inhibit protein tyrosine phosphatases.[41]

The tumor suppressor phosphatase and tensin homology deleted on chromosome 10 (PTEN) is a putative negative regulator of the PI3K/Akt pathway. Treatment with Zn resulted in a significant reduction in levels of PTEN protein in a dose- and time-dependent fashion in a human airway epithelial cell line. This effect of Zn was also observed in normal human airway epithelial cells in primary culture and in rat airway epithelium in vivo. Concomitantly, levels of PTEN mRNA were also significantly reduced by Zn exposure. PTEN phosphatase activity evaluated by measuring Akt phosphorylation decreased after Zn treatment. Pretreatment of the cells with a proteasome inhibitor significantly blocked Zn-induced reduction of PTEN protein as well as the increase in Akt phosphorylation, implicating the involvement of proteasome-mediated PTEN degradation.[111]

The important role of Zn in insulin signaling is also supported by findings from a recent study[13] of several structures that induce insulin-like signal transduction to downstream effectors such as the transcription factor FOXO1a and the key gluconeogenic regulatory enzymes phosphoenolpyruvate carboxykinase and glucose 6-phosphatase (G6Pase).[13] Results indicate that β-thujaplicin, DEDTC and its clinically-used dimer disulfiram, induce insulin-like dose-dependent effects on signaling to FOXO1a in a manner that is strictly dependent on the presence of Zn ions. The most potent compound tested on gluconeogenesis is disulfiram, which, in the presence of 10 μmol/L Zn, inhibited both phosphoenolpyruvate carboxykinase and G6Pase with an IC_{50} of 4 μmol/L. These results demonstrate that metal-binding compounds with diverse structures can induce Zn-dependent insulin-like effects on signal transduction and gene expression.[13]

Vardatsikos et al. have systemically summarized the available information for the insulinomimetic and antidiabetic effects of Zn.[108] They state: "The insulin-like properties of Zn have been demonstrated in isolated cells, tissues, and different animal models of T1D and T2D. Zn treatment has been found to improve carbohydrate and lipid metabolism in rodent models of diabetes. In isolated cells, it enhances glucose transport, glycogen and lipid synthesis, and inhibits gluconeogenesis and lipolysis. The molecular mechanism responsible for the insulin-like effects of Zn compounds involves the activation of several key components of the insulin signaling pathways, which include the ERK1/2 and PI3K/Akt pathways. However, the precise molecular mechanisms by which Zn triggers the activation of these pathways remain to be clarified." It clearly indicates the promising insulinomimetic and antidiabetic effects of Zn, but more and mechanistic studies remain required.

5.2.2 *Insulin-like Function via IGF Signaling Pathway*

Zn acts with insulin-like function through IGF regulation. Zn can regulate IGF signaling through maintaining IGF-I and IGF-II in an active form by directly regulating IGF-II binding to IGF-binding proteins (IGFBPs). Zn acts on type 1 IGF receptors (IGF-1R) by preventing secreted IGF-II from binding to IGFBP-3 and IGFBP-5, thus maintaining IGF-II in an "active state" (i.e., readily available for IGF-1R association). In addition, Zn decreases the affinity of IGF-2R. In contrast, Zn increased IGF-I, IGF-II, and R(3)-IGF-I binding to the IGF-1R by increasing ligand-binding affinity. Therefore, a mechanism has been proposed that Zn may alter IGF distribution by increasing IGF-1R binding at the expense of IGF binding to soluble IGFBP-5 and the IGF-2R.[76,77]

5.2.3 *Antioxidant and Anti-inflammation*

Oxidative stress plays a critical role in the development of metabolic syndrome and diabetes. The prevention of suppressed systemic antioxidant capacity either by metabolic syndrome or diabetes may be one reason for the prevention of systemic complications of diabetes by Zn supplementation.[89] Lipid peroxidation is significantly increased and activities of antioxidant enzymes, superoxide dismutase (SOD), catalase and Glutathione (GSH) are significantly decreased in liver of STZ-induced diabetic rats, which suggest that the structural damage to these tissues or complications of diabetes may be due to oxidative stress.[58] Zn functions as a complex antioxidant through participation in SODs and thioredoxins,[23,43] enzymatic and chelator activities, stabilizing cell membranes, and inhibiting lipid peroxidation.[7] It protects tissues subjected to ionizing irradiation and ultraviolet light by induction of MT.[12]

Anderson et al. documented the potential beneficial antioxidant effects generated by combined supplementation of Zn and Cr in people with T2D. These results are particularly important in light of the deleterious consequences of oxidative stress in people with diabetes.[2] Furthermore, similar results were found for Zn supplementation only. Zn supplementation not only improves antioxidant activity, but also increases red blood cell Cu/Zn-SOD and glutathione peroxidase (GPx), which can be significantly decreased in diabetic subjects.[92]

5.2.4 *Improvement of Zn Dyshomeostasis*

The clinical relevance of serum Zn levels less than previously established normal ranges should be further studied since the ratio of Zn to other metals such as Cu or Fe (i.e., Zn/Cu or Zn/Fe) also plays an important role in the pathogenesis of insulin resistance, metabolic syndrome, diabetes, and cardiovascular diseases. In terms of the effects of Zn supplementation on the biokinetics of ^{65}Zn, Pathak et al. found that alloxan-induced diabetic rats showed a significant decrease in both the fast and slow components of biological half-life of ^{65}Zn, which, however, were normalized in the whole body following Zn supplementation. In case of liver Zn, the slow component was brought back to the norm, but the fast component was not increased significantly. The study suggests that the paucity of Zn in the tissues of the diabetic animals was due to decreased retention of tissue Zn as evidenced by increased serum Zn, hyperzincuria, and increased rate of the uptake of ^{65}Zn by the liver. Zn supplementation caused a significant improvement in the retention of Zn in the tissues and is therefore likely to be of benefit in the treatment of diabetes.[85]

Using STZ-induced diabetic rats, Aguilar et al.[1] determined the influence of the duration of diabetes on the Zn/Cu ratio by three time periods: 7, 21, and 60 days after diabetes onset in the insulin target tissues (liver, adipose tissue, and skeletal muscle). They found that there was a significant decrease in the ratio of Zn/Cu in the liver and adipose tissue, but not in the skeletal muscle. This decrease was diabetes dependent and also diabetic duration dependent. Zargar et al. found that serum Zn was not decreased in type 2 diabetic patients, but Cu levels significantly increased, leading to a decrease in the ratio of Zn/Cu.[114] Ripa et al. and Canatan et al. also found that Zn and the Zn/Cu ratio were lower in the patients with dilated cardiac failure or etiology of essential hypertension compared with normal controls.[14,91] The combined findings of these studies suggest the potential risk of decreased Zn/Cu ratio for cardiovascular diseases. However, there was also documentation indicating that Zn/Cu ratio was not lower, but even higher in the diabetic patients. Therefore, the association between the Zn/Cu ratio and diabetes among human subjects is confounded by multiple factors.[1]

Zn supplementation significantly corrects diabetes-induced Zn deficiency.[72] Correcting systemic Zn deficiency may represent a beneficial intervention in instances of diabetes-induced decreases in Zn/Cu ratio. This theory is supported by the observation that diabetes caused Zn deficiency along with Zn-dependent low immune function[79,81] and Zn supplementation can correct plasma Zn levels and Zn/Cu ratio to normal values and reverse a trend of decreased CD4 cells in patients with diabetes.[57]

6. CONCLUSIONS

Zn is an essential trace element, involved in many different cellular processes. A relationship between Zn, pancreatic function, and diabetes was suggested almost 70 years ago. The extensive review of the literature based on the mechanistic studies using different animal models and epidemiological studies with different populations under various conditions, including Zn deficiency and supplementation in both types of diabetes, has clearly indicated the negative association of Zn levels with metabolic syndrome and diabetes, and the overall beneficial effects of Zn supplementation on blood glucose control. The present review suggests that Zn might be a candidate for diabetes prevention and therapy.

7. POTENTIAL CLINICAL IMPLICATION FOR THE MANAGEMENT OF DIABETIC PATIENTS

The earliest known reference to DM may be when Papyrus Ebers noted in 1550 BC that "wheat grains, grits, grapes, and sweet beer" were the foods of choice for those who needed to "eliminate urine which is too plentiful." Medical nutrition therapy continues to evolve, with recent evidence resulting in the following key nutrition recommendations for diabetes treatment and prevention:

- Consistent carbohydrate intake
- Maintaining healthy body weight
- Increasing fiber to meet the Adequate Intake (14 g/1000 kcal) recommended by the Dietary Reference Intakes (DRIs)
- Limiting saturated fats to less than 7% of total calories
- Minimizing *trans* fatty acids
- Lowering cholesterol to less than 200 mg/day
- Reducing sodium intake to less than 2300 mg/day

Suggested food choices consist of fruits, vegetables, whole grains, legumes, low-fat dairy products, lean meats, and unsaturated fats. Individuals at risk for diabetes, or who have diabetes, and consume such diets typically demonstrate reduced insulin requirements, improved glycemic control, lower fasting serum cholesterol and TG values, and weight loss.

The evidence emerging that focuses on zinc and diabetes seems to indicate that this trace element may represent an important nutrient that health care providers need to address during patient education and planning for disease management. Patients with poorly controlled diabetes are susceptible to multiple micronutrient deficiencies. Zinc has potent antioxidant activity. Supplementation may be advisable, although further rigorous trials (e.g., randomized, blinded, placebo controlled) are needed to establish safety, efficacy, dosing, and other therapeutic factors.

Any strategy to modify diet should consider long-term adherence with dietary changes. In most health care systems, dietitians assume primary responsibility for educating individuals on the benefits and use of dietary interventions. Diets must be individualized, with special modifications for patient attributes such as obesity, hyperlipidemia, or physiological states such as pregnancy and lactation.

Acknowledgment

Studies cited here from the authors' laboratories were supported in part by grants from the American Diabetes Association (1-11-BS-17; 1-15-BS-018) and a Starting-Up Fund for Chinese-American Research Institute for Diabetic Complications from Wenzhou Medical College. Dr Liu is supported in part by the Department of Health and Human Services grant number 5R01LM010923-02. Dr Wintergerst is supported in part by the Department of Health and Humans Services, HRSA/MCHB grant number U22MC03963/H46MC24092.

References

1. Aguilar MV, Laborda JM, Martinez-Para MC, Gonzalez MJ, Meseguer I, Bernao A, et al. Effect of diabetes on the tissular Zn/Cu ratio. *J Trace Elem Med Bio* 1998;**12**:155–8.
2. Anderson RA, Roussel AM, Zouari N, Mahjoub S, Matheau JM, Kerkeni A. Potential antioxidant effects of zinc and chromium supplementation in people with type 2 diabetes mellitus. *J Am Coll Nutr* 2001;**20**:212–8.
3. Anetor JI, Senjobi A, Ajose OA, Agbedana EO. Decreased serum magnesium and zinc levels: atherogenic implications in type-2 diabetes mellitus in Nigerians. *Nutr Health* 2002;**16**:291–300.
4. Bakaysa DL, Radziuk J, Havel HA, Brader ML, Li S, Dodd SW, et al. Physicochemical basis for the rapid time-action of LysB28ProB29-insulin: dissociation of a protein-ligand complex. *Protein Sci* 1996;**5**:2521–31.
5. Benson VS, Vanleeuwen JA, Taylor J, Somers GS, McKinney PA, Van Til L. Type 1 diabetes mellitus and components in drinking water and diet: a population-based, case-control study in Prince Edward Island, Canada. *J Am Coll Nutr* 2010;**29**:612–24.
6. Blostein-Fujii A, DiSilvestro RA, Frid D, Katz C, Malarkey W. Short-term zinc supplementation in women with non-insulin-dependent diabetes mellitus: effects on plasma 5′-nucleotidase activities, insulin-like growth factor I concentrations, and lipoprotein oxidation rates in vitro. *Am J Clin Nutr* 1997;**66**:639–42.
7. Bray TM, Bettger WJ. The physiological role of zinc as an antioxidant. *Free Radic Biol Med* 1990;**8**:281–91.

8. Cai L. Metallothionein and cardiomyopathy. In: Zatta P, editor. *Metallothioneins in biochemistry and pathology*, vol. 1. New Jersey: World Scientific; 2008. p. 227–69.
9. Cai L, Chen S, Evans T, Cherian MG, Chakrabarti S. Endothelin-1-mediated alteration of metallothionein and trace metals in the liver and kidneys of chronically diabetic rats. *Int J Exp Diabetes Res* 2002;**3**:193–8.
10. Cai L, Klein JB, Kang YJ. Metallothionein inhibits peroxynitrite-induced DNA and lipoprotein damage. *J Biol Chem* 2000;**275**: 38957–60.
11. Cai L, Li XK, Song Y, Cherian MG. Essentiality, toxicology and chelation therapy of zinc and copper. *Curr Med Chem* 2005;**12**: 2753–63.
12. Cai L, Satoh M, Tohyama C, Cherian MG. Metallothionein in radiation exposure: its induction and protective role. *Toxicology* 1999; **132**:85–98.
13. Cameron AR, Anil S, Sutherland E, Harthill J, Rena G. Zinc-dependent effects of small molecules on the insulin-sensitive transcription factor FOXO1a and gluconeogenic genes. *Metallomics* 2010;**2**:195–203.
14. Canatan H, Bakan I, Akbulut M, Halifeoglu I, Cikim G, Baydas G, et al. Relationship among levels of leptin and zinc, copper, and zinc/copper ratio in plasma of patients with essential hypertension and healthy normotensive subjects. *Biol Trace Elem Res* 2004;**100**:117–23.
15. Canesi L, Betti M, Ciacci C, Gallo G. Insulin-like effect of zinc in mytilus digestive gland cells: modulation of tyrosine kinase-mediated cell signaling. *Gen Comp Endocrinol* 2001;**122**:60–6.
16. Capdor J, Foster M, Petocz P, Samman S. Zinc and glycemic control: a meta-analysis of randomised placebo controlled supplementation trials in humans. *J Trace Elem Med Biol* 2013;**27**: 137–42.
17. Chanoit G, Lee S, Xi J, Zhu M, McIntosh RA, Mueller RA, et al. Exogenous zinc protects cardiac cells from reperfusion injury by targeting mitochondrial permeability transition pore through inactivation of glycogen synthase kinase-3beta. *Am J Physiol Heart Circ Physiol* 2008;**295**:H1227–33.
18. Chausmer AB. Zinc, insulin and diabetes. *J Am Coll Nutr* 1998;**17**: 109–15.
19. Chen H, Carlson EC, Pellet L, Moritz JT, Epstein PN. Overexpression of metallothionein in pancreatic beta-cells reduces streptozotocin-induced DNA damage and diabetes. *Diabetes* 2001;**50**:2040–6.
20. Chen MD, Lin PY, Cheng V, Lin WH. Zinc supplementation aggravates body fat accumulation in genetically obese mice and dietary-obese mice. *Biol Trace Elem Res* 1996;**52**:125–32.
21. Chen ML, Failla ML. Metallothionein metabolism in the liver and kidney of the streptozotocin-diabetic rat. *Comp Biochem Physiol B* 1988;**90**:439–45.
22. Chimienti F, Devergnas S, Favier A, Seve M. Identification and cloning of a beta-cell-specific zinc transporter, ZnT-8, localized into insulin secretory granules. *Diabetes* 2004;**53**:2330–7.
23. Collet JF, D'Souza JC, Jakob U, Bardwell JC. Thioredoxin 2, an oxidative stress-induced protein, contains a high affinity zinc binding site. *J Biol Chem* 2003;**278**:45325–32.
24. Craft NE, Failla ML. Zinc, iron, and copper absorption in the streptozotocin-diabetic rat. *Am J Physiol* 1983;**244**:E122–8.
25. el-Yazigi A, Hannan N, Raines DA. Effect of diabetic state and related disorders on the urinary excretion of magnesium and zinc in patients. *Diabetes Res* 1993;**22**:67–75.
26. Evans SA, Overton JM, Alshingiti A, Levenson CW. Regulation of metabolic rate and substrate utilization by zinc deficiency. *Metabolism* 2004;**53**:727–32.
27. Failla ML, Gardell CY. Influence of spontaneous diabetes on tissue status of zinc, copper, and manganese in the BB Wistar rat. *Proc Soc Exp Biol Med* 1985;**180**:317–22.
28. Failla ML, Kiser RA. Altered tissue content and cytosol distribution of trace metals in experimental diabetes. *J Nutr* 1981;**111**: 1900–9.
29. Failla ML, Kiser RA. Hepatic and renal metabolism of copper and zinc in the diabetic rat. *Am J Physiol* 1983;**244**:E115–21.
30. Faure P, Roussel A, Coudray C, Richard MJ, Halimi S, Favier A. Zinc and insulin sensitivity. *Biol Trace Elem Res* 1992;**32**:305–10.
31. Faure P, Roussel AM, Martinie M, Osman M, Favier A, Halimi S. Insulin sensitivity in zinc-depleted rats: assessment with the euglycaemic hyperinsulinic clamp technique. *Diabetes Metab* 1991;**17**:325–31.
32. Forte G, Bocca B, Peruzzu A, Tolu F, Asara Y, Farace C, et al. Blood metals concentration in type 1 and type 2 diabetics. *Biol Trace Elem Res* 2013;**156**:79–90.
33. Foster M, Karra M, Picone T, Chu A, Hancock DP, Petocz P, et al. Dietary fiber intake increases the risk of zinc deficiency in healthy and diabetic women. *Biol Trace Elem Res* 2012;**149**(2):135–42.
34. Fukada T, Kambe T. Molecular and genetic features of zinc transporters in physiology and pathogenesis. *Metallomics* 2011;**3**:662–74.
35. Fukada T, Yamasaki S, Nishida K, Murakami M, Hirano T. Zinc homeostasis and signaling in health and diseases: zinc signaling. *J Biol Inorg Chem* 2011;**16**:1123–34.
36. Goldberg ED, Eshchenko VA, Bovt VD. Diabetogenic activity of chelators in some mammalian species. *Endocrinologie* 1990;**28**:51–5.
37. Goldberg ED, Eshchenko VA, Bovt VD. The diabetogenic and acidotropic effects of chelators. *Exp Pathol* 1991;**42**:59–64.
38. Goldman J, Carpenter FH. Zinc binding, circular dichroism, and equilibrium sedimentation studies on insulin (bovine) and several of its derivatives. *Biochemistry* 1974;**13**:4566–74.
39. Golik A, Cohen N, Ramot Y, Maor J, Moses R, Weissgarten J, et al. Type II diabetes mellitus, congestive heart failure, and zinc metabolism. *Biol Trace Elem Res* 1993;**39**:171–5.
40. Haase H, Maret W. Intracellular zinc fluctuations modulate protein tyrosine phosphatase activity in insulin/insulin-like growth factor-1 signaling. *Exp Cell Res* 2003;**291**:289–98.
41. Haase H, Maret W. Fluctuations of cellular, available zinc modulate insulin signaling via inhibition of protein tyrosine phosphatases. *J Trace Elem Med Biol* 2005a;**19**:37–42.
42. Haase H, Maret W. Protein tyrosine phosphatases as targets of the combined insulinomimetic effects of zinc and oxidants. *Biometals* 2005b;**18**:333–8.
43. Hagay ZJ, Weiss Y, Zusman I, Peled-Kamar M, Reece EA, Eriksson UJ, et al. Prevention of diabetes-associated embryopathy by overexpression of the free radical scavenger copper zinc superoxide dismutase in transgenic mouse embryos. *Am J Obstet Gynecol* 1995;**173**:1036–41.
44. Haglund B, Ryckenberg K, Selinus O, Dahlquist G. Evidence of a relationship between childhood-onset type I diabetes and low groundwater concentration of zinc. *Diabetes Care* 1996;**19**:873–5.
45. Hashemipour M, Kelishadi R, Shapouri J, Sarrafzadegan N, Amini M, Tavakoli N, et al. Effect of zinc supplementation on insulin resistance and components of the metabolic syndrome in prepubertal obese children. *Hormones (Athens)* 2009;**8**:279–85.
46. Himoto T, Nomura T, Tani J, Miyoshi H, Morishita A, Yoneyama H, et al. Exacerbation of insulin resistance and hepatic steatosis deriving from zinc deficiency in patients with HCV-related chronic liver disease. *Biol Trace Elem Res* 2014; **163**(1–2):81–8.
47. Himoto T, Yoneyama H, Kurokochi K, Inukai M, Masugata H, Goda F, et al. Contribution of zinc deficiency to insulin resistance in patients with primary biliary cirrhosis. *Biol Trace Elem Res* 2011; **144**:133–42.
48. Ho E, Quan N, Tsai YH, Lai W, Bray TM. Dietary zinc supplementation inhibits NFkappaB activation and protects against chemically induced diabetes in CD1 mice. *Exp Biol Med* 2001;**226**: 103–11.

49. Honnorat J, Accominotti M, Broussolle C, Fleuret AC, Vallon JJ, Orgiazzi J. Effects of diabetes type and treatment on zinc status in diabetes mellitus. *Biol Trace Elem Res* 1992;**32**:311–6.
50. Ilouz R, Kaidanovich O, Gurwitz D, Eldar-Finkelman H. Inhibition of glycogen synthase kinase-3beta by bivalent zinc ions: insight into the insulin-mimetic action of zinc. *Biochem Biophys Res Commun* 2002;**295**:102–6.
51. im Walde SS, Dohle C, Schott-Ohly P, Gleichmann H. Molecular target structures in alloxan-induced diabetes in mice. *Life Sci* 2002;**71**:1681–94.
52. Jansen J, Karges W, Rink L. Zinc and diabetes—clinical links and molecular mechanisms. *J Nutr Biochem* 2009;**20**:399–417.
53. Jansen J, Rosenkranz E, Overbeck S, Warmuth S, Mocchegiani E, Giacconi R, et al. Disturbed zinc homeostasis in diabetic patients by in vitro and in vivo analysis of insulinomimetic activity of zinc. *J Nutr Biochem* 2012;**23**(11):1458–66.
54. Jayawardena R, Ranasinghe P, Galappatthy P, Malkanthi R, Constantine G, Katulanda P. Effects of zinc supplementation on diabetes mellitus: a systematic review and meta-analysis. *Diabetol Metab Syndr* 2012;**4**:13.
55. Jin T, Nordberg G, Sehlin J, Vesterberg O. Protection against cadmium-metallothionein nephrotoxicity in streptozotocin-induced diabetic rats: role of increased metallothionein synthesis induced by streptozotocin. *Toxicology* 1996;**106**:55–63.
56. Jou MY, Philipps AF, Lonnerdal B. Maternal zinc deficiency in rats affects growth and glucose metabolism in the offspring by inducing insulin resistance postnatally. *J Nutr* 2010;**140**: 1621–7.
57. Kajanachumpol S, Srisurapanon S, Supanit I, Roongpisuthipong C, Apibal S. Effect of zinc supplementation on zinc status, copper status and cellular immunity in elderly patients with diabetes mellitus. *J Med Assoc Thai* 1995;**78**:344–9.
58. Kakkar R, Mantha SV, Radhi J, Prasad K, Kalra J. Increased oxidative stress in rat liver and pancreas during progression of streptozotocin-induced diabetes. *Clinical Sci* 1998;**94**:623–32.
59. Kambe T. An overview of a wide range of functions of ZnT and Zip zinc transporters in the secretory pathway. *Biosci Biotechnol Biochem* 2011;**75**:1036–43.
60. Kambe T, Yamaguchi-Iwai Y, Sasaki R, Nagao M. Overview of mammalian zinc transporters. *Cell Mol Life Sci* 2004;**61**:49–68.
61. Kawasaki E. ZnT8 and type 1 diabetes. *Endocr J* 2012;**59**:531–7.
62. Kechrid Z, Bouzerna N, Zio MS. Effect of low zinc diet on (65)Zn turnover in non-insulin dependent diabetic mice. *Diabetes Metab* 2001;**27**:580–3.
63. Kelishadi R, Hashemipour M, Adeli K, Tavakoli N, Movahedian-Attar A, Shapouri J, et al. Effect of zinc supplementation on markers of insulin resistance, oxidative stress, and inflammation among prepubescent children with metabolic syndrome. *Metab Syndr Relat Disord* 2010;**8**:505–10.
64. Kennedy ML, Failla ML. Zinc metabolism in genetically obese (ob/ob) mice. *J Nutr* 1987;**117**:886–93.
65. Khan AR, Awan FR. Metals in the pathogenesis of type 2 diabetes. *J Diabetes Metab Disord* 2014;**13**:16.
66. Khan MI, Siddique KU, Ashfaq F, Ali W, Reddy HD, Mishra A. Effect of high-dose zinc supplementation with oral hypoglycemic agents on glycemic control and inflammation in type-2 diabetic nephropathy patients. *J Nat Sci Biol Med* 2013;**4**:336–40.
67. Kim J, Lee S. Effect of zinc supplementation on insulin resistance and metabolic risk factors in obese Korean women. *Nutr Res Pract* 2012;**6**:221–5.
68. Lee S, Chanoit G, McIntosh R, Zvara DA, Xu Z. Molecular mechanism underlying Akt activation in zinc-induced cardioprotection. *Am J Physiol Heart Circ Physiol* 2009;**297**:H569–75.
69. Levine AS, McClain CJ, Handwerger BS, Brown DM, Morley JE. Tissue zinc status of genetically diabetic and streptozotocin-induced diabetic mice. *Am J Clin Nutr* 1983;**37**:382–6.
70. Liu BY, Jiang Y, Lu Z, Li S, Lu D, Chen B. Down-regulation of zinc transporter 8 in the pancreas of db/db mice is rescued by Exendin-4 administration. *Mol Med Rep* 2011;**4**:47–52.
71. Liu MJ, Bao S, Bolin ER, Burris DL, Xu X, Sun Q, et al. Zinc deficiency augments leptin production and exacerbates macrophage infiltration into adipose tissue in mice fed a high-fat diet. *J Nutr* 2013;**143**:1036–45.
72. Maldonado Martin A, Gil Extremera B, Fernandez Soto M, Ruiz Martinez M, Gonzalez Jimenez A, Guijarro Morales A, et al. Zinc levels after intravenous administration of zinc sulphate in insulin-dependent diabetes mellitus patients. *Klinische Wochenschrift* 1991;**69**:640–4.
73. Marreiro DN, Geloneze B, Tambascia MA, Lerario AC, Halpern A, Cozzolino SM. Effect of zinc supplementation on serum leptin levels and insulin resistance of obese women. *Biol Trace Elem Res* 2006;**112**:109–18.
74. Maruthur NM, Clark JM, Fu M, Linda Kao WH, Shuldiner AR. Effect of zinc supplementation on insulin secretion: interaction between zinc and SLC30A8 genotype in Old Order Amish. *Diabetologia* 2014;**58**(2):295–303.
75. May JM, Contoreggi CS. The mechanism of the insulin-like effects of ionic zinc. *J Biol Chem* 1982;**257**:4362–8.
76. McCusker RH, Mateski RL, Novakofski J. Zinc alters the kinetics of IGF-II binding to cell surface receptors and binding proteins. *Endocrine* 2003;**21**:279–88.
77. McCusker RH, Novakofski J. Zinc partitions IGFs from soluble IGF binding proteins (IGFBP)-5, but not soluble IGFBP-4, to myoblast IGF type 1 receptors. *J Endocrinol* 2004;**180**:227–46.
78. Miranda ER, Dey CS. Effect of chromium and zinc on insulin signaling in skeletal muscle cells. *Biol Trace Elem Res* 2004;**101**:19–36.
79. Mocchegiani E, Boemi M, Fumelli P, Fabris N. Zinc-dependent low thymic hormone level in type I diabetes. *Diabetes* 1989;**38**:932–7.
80. Moltchanova E, Rytkonen M, Kousa A, Taskinen O, Tuomilehto J, Karvonen M. Zinc and nitrate in the ground water and the incidence of Type 1 diabetes in Finland. *Diabetes Med* 2004;**21**:256–61.
81. Niewoehner CB, Allen JI, Boosalis M, Levine AS, Morley JE. Role of zinc supplementation in type II diabetes mellitus. *Am J Med* 1986;**81**:63–8.
82. Ohly P, Dohle C, Abel J, Seissler J, Gleichmann H. Zinc sulphate induces metallothionein in pancreatic islets of mice and protects against diabetes induced by multiple low doses of streptozotocin. *Diabetologia* 2000;**43**:1020–30.
83. Ortega RM, Rodriguez-Rodriguez E, Aparicio A, Jimenez AI, Lopez-Sobaler AM, Gonzalez-Rodriguez LG, et al. Poor zinc status is associated with increased risk of insulin resistance in Spanish children. *Br J Nutr* 2012;**107**:398–404.
84. Park JH, Grandjean CJ, Hart MH, Erdman SH, Pour P, Vanderhoof JA. Effect of pure zinc deficiency on glucose tolerance and insulin and glucagon levels. *Am J Physiol* 1986;**251**:E273–8.
85. Pathak A, Sharma V, Kumar S, Dhawan DK. Supplementation of zinc mitigates the altered uptake and turnover of 65Zn in liver and whole body of diabetic rats. *Biometals* 2011;**24**:1027–34.
86. Prasad AS. Zinc: an overview. *Nutrition* 1995;**11**:93–9.
87. Prasad AS. Clinical, immunological, anti-inflammatory and antioxidant roles of zinc. *Exp Gerontol* 2008;**43**:370–7.
88. Prasad AS. Discovery of human zinc deficiency: 50 years later. *J Trace Elem Med Biol* 2012;**26**:66–9.
89. Prasad AS, Bao B, Beck FW, Kucuk O, Sarkar FH. Antioxidant effect of zinc in humans. *Free Radic Biol Med* 2004;**37**:1182–90.
90. Quilliot D, Dousset B, Guerci B, Dubois F, Drouin P, Ziegler O. Evidence that diabetes mellitus favors impaired metabolism of zinc, copper, and selenium in chronic pancreatitis. *Pancreas* 2001;**22**:299–306.
91. Ripa S, Ripa R, Giustiniani S. Are failured cardiomyopathies a zinc-deficit related disease? A study on Zn and Cu in patients with chronic failured dilated and hypertrophic cardiomyopathies. *Minerva Med* 1998;**89**:397–403.

92. Roussel AM, Kerkeni A, Zouari N, Mahjoub S, Matheau JM, Anderson RA. Antioxidant effects of zinc supplementation in Tunisians with type 2 diabetes mellitus. *J Am Coll Nutr* 2003;**22**: 316–21.
93. Ruz M, Carrasco F, Rojas P, Codoceo J, Inostroza J, Basfi-fer K, et al. Zinc as a potential coadjuvant in therapy for type 2 diabetes. *Food Nutr Bull* 2013;**34**:215–21.
94. Salmonowicz B, Krzystek-Korpacka M, Noczynska A. Trace elements, magnesium, and the efficacy of antioxidant systems in children with type 1 diabetes mellitus and in their siblings. *Adv Clin Exp Med* 2014;**23**:259–68.
95. Samuelsson U, Oikarinen S, Hyoty H, Ludvigsson J. Low zinc in drinking water is associated with the risk of type 1 diabetes in children. *Pediatr Diabetes* 2011;**12**:156–64.
96. Schott-Ohly P, Lgssiar A, Partke HJ, Hassan M, Friesen N, Gleichmann H. Prevention of spontaneous and experimentally induced diabetes in mice with zinc sulfate-enriched drinking water is associated with activation and reduction of NF-kappa B and AP-1 in islets, respectively. *Exp Biol Med* 2004;**229**:1177–85.
97. Scotto M, Afonso G, Larger E, Raverdy C, Lemonnier FA, Carel JC, et al. Zinc transporter (ZnT)8(186-194) is an immunodominant $CD8^+$ T cell epitope in HLA-$A2^+$ type 1 diabetic patients. *Diabetologia* 2012;**55**:2026–31.
98. Simon SF, Taylor CG. Dietary zinc supplementation attenuates hyperglycemia in db/db mice. *Exp Biol Med (Maywood)* 2001;**226**: 43–51.
99. Singh RB, Niaz MA, Rastogi SS, Bajaj S, Gaoli Z, Shoumin Z. Current zinc intake and risk of diabetes and coronary artery disease and factors associated with insulin resistance in rural and urban populations of North India. *J Am Coll Nutr* 1998;**17**:564–70.
100. Sitasawad S, Deshpande M, Katdare M, Tirth S, Parab P. Beneficial effect of supplementation with copper sulfate on STZ-diabetic mice (IDDM). *Diabetes Res Clin Pract* 2001;**52**:77–84.
101. Smidt K, Jessen N, Petersen AB, Larsen A, Magnusson N, Jeppesen JB, et al. SLC30A3 responds to glucose- and zinc variations in beta-cells and is critical for insulin production and in vivo glucose-metabolism during beta-cell stress. *PLoS One* 2009;**4**:e5684.
102. Tadros WM, Awadallah R, Doss H, Khalifa K. Protective effect of trace elements (Zn, Mn, Cr, Co) on alloxan-induced diabetes. *Indian J Exp Biol* 1982;**20**:93–4.
103. Tang X, Shay NF. Zinc has an insulin-like effect on glucose transport mediated by phosphoinositol-3-kinase and Akt in 3T3-L1 fibroblasts and adipocytes. *J Nutr* 2001;**131**:1414–20.
104. Tapiero H, Tew KD. Trace elements in human physiology and pathology: zinc and metallothioneins. *Biomed Pharmacother* 2003;**57**: 399–411.
105. Terres-Martos C, Navarro-Alarcon M, Martin-Lagos F, Lopez GdlSH, Perez-Valero V, Lopez-Martinez MC. Serum zinc and copper concentrations and Cu/Zn ratios in patients with hepatopathies or diabetes. *J Trace Elem Med Biol* 1998;**12**:44–9.
106. Tobia MH, Zdanowicz MM, Wingertzahn MA, McHeffey-Atkinson B, Slonim AE, Wapnir RA. The role of dietary zinc in modifying the onset and severity of spontaneous diabetes in the BB Wistar rat. *Mol Genet Metab* 1998;**63**:205–13.
107. Tuvemo T, Gebre-Medhin M. The role of trace elements in juvenile diabetes mellitus. *Pediatrician* 1983;**12**:213–9.
108. Vardatsikos G, Pandey NR, Srivastava AK. Insulino-mimetic and anti-diabetic effects of zinc. *J Inorg Biochem* 2013;**120**:8–17.
109. Wang X, Zhou B. Dietary zinc absorption: A play of Zips and ZnTs in the gut. *IUBMB Life* 2010;**62**:176–82.
110. Williams NR, Rajput-Williams J, West JA, Nigdikar SV, Foote JW, Howard AN. Plasma, granulocyte and mononuclear cell copper and zinc in patients with diabetes mellitus. *Analyst* 1995;**120**: 887–90.
111. Wu W, Wang X, Zhang W, Reed W, Samet JM, Whang YE, et al. Zinc-induced PTEN protein degradation through the proteasome pathway in human airway epithelial cells. *J Biol Chem* 2003;**278**: 28258–63.
112. Xu J, Wang J, Chen B. SLC30A8 (ZnT8) variations and type 2 diabetes in the Chinese Han population. *Genet Mol Res* 2012;**11**: 1592–8.
113. Yang J, Cherian MG. Protective effects of metallothionein on streptozotocin-induced diabetes in rats. *Life Sci* 1994;**55**:43–51.
114. Zargar AH, Shah NA, Masoodi SR, Laway BA, Dar FA, Khan AR, et al. Copper, zinc, and magnesium levels in non-insulin dependent diabetes mellitus. *Postgrad Med J* 1998;**74**:665–8.
115. Zhao HX, Mold MD, Stenhouse EA, Bird SC, Wright DE, Demaine AG, et al. Drinking water composition and childhood-onset Type 1 diabetes mellitus in Devon and Cornwall, England. *Diabetes Med* 2001;**18**:709–17.

CHAPTER

15

Cocoa Flavonoids and Insulin Signaling

Maria Ángeles Martin, Luis Goya, Sonia Ramos

Department of Metabolism and Nutrition, Institute of Food Science and Technology and Nutrition (ICTAN-CSIC), Madrid, Spain

1. INTRODUCTION

Diabetes mellitus is a complex metabolic disorder that is considered a major health problem because of its high frequency and its associated complications causing blindness, kidney failure, heart disease, stroke, and amputations.[1] Type 2 diabetes (T2D) is one of the most common chronic diseases in nearly all countries, and it is continuing to be an increasing global health burden.[2] T2D results from a combination of genetic and acquired factors that impair β-cell function and tissue insulin sensitivity.[3] A hallmark of T2D and a central component in the so-called metabolic syndrome is insulin resistance, which is a major contributor to the sustained hyperglycemia.[3] In this regard, it should be highlighted that current medications are not adequately effective in maintaining long-term glycemic control in a high proportion of patients. In fact, despite the available number of hypoglycemic agents, it is assumed that the most efficient approach to prevent or delay the onset of T2D at the lowest cost is at nutritional level. Therefore, the identification of dietary components as potential antidiabetic agents in the form of functional foods or nutraceuticals has become an essential subject in the current research. This is the case of polyphenols, natural dietary compounds present in fruits and vegetables, which have attracted a great deal of interest because of their potential ability to act as highly effective chemopreventive agents.[4] In addition, the low toxicity and the very few adverse side effects linked to polyphenols consumption give them potential advantages with respect to the traditional chemotherapeutic agents.

The present chapter will focus on the molecular basis of chemopreventive activity of cocoa polyphenolic components related to insulin signaling. First, this chapter summarizes the recent in vitro studies that evaluated the potential beneficial properties of cocoa and their components and the molecular mechanism involved to prevent or delay alterations of the insulin signaling. In the second part of the chapter, investigations into the effect of cocoa in various animal models of diabetes are presented. Finally, current evidence on the link between cocoa and diabetes occurrence, based on interventional and epidemiological studies on humans, is briefly described.

2. PHYSIOLOGY OF INSULIN ACTION

Insulin regulates glucose homeostasis mainly through promotion of glucose uptake and utilization in skeletal muscle, suppression of hepatic glucose production, and inhibition of lipolysis in adipose tissue.[5] Nonetheless, recent research has unraveled a multitude of other biological actions of insulin in alternative targets including vascular endothelium, brain, pancreatic β cell, kidney, gut, and other organs to contribute to metabolic homeostasis.[5]

Under physiological conditions, from the circulation glucose is transported into pancreatic β cells of the islets of Langerhans through glucose transporter (GLUT)-2 and its oxidation leads to insulin secretion, through a mechanism involving closure of adenosine triphosphate (ATP)-sensitive potassium channels, membrane depolarization followed by voltage-dependent calcium influx, and subsequent exocytosis of insulin granules.[3,5] Thus, insulin-mediated signaling lowers blood glucose by (i) enhancing the uptake of glucose through translocation of GLUT-4 vesicles to the plasma membrane in peripheral tissues including skeletal muscle, which is the largest site for disposal of dietary glucose, and adipose tissue; (ii) decreasing in the liver the glucose output rate by increasing the storage of glucose as glycogen (e.g., glycogenesis for storage as glycogen); and (iii) inhibiting lipolysis and promoting lipogenesis in the adipose tissue.[3,5]

Molecular Nutrition and Diabetes
http://dx.doi.org/10.1016/B978-0-12-801585-8.00015-4

At molecular level, insulin actions are initiated by the binding of the hormone to specific cell surface insulin receptors (IRs), which are expressed on nearly every cell in the body. This triggers the insulin signaling network that culminates in the insulin actions in different tissues.[6] Insulin signal transduction pathways promote classic metabolic actions that generally involve autophosphorylation of the insulin receptor tyrosine kinase and then phosphorylation of the insulin receptor substrates (IRS)-1 and -2. Tyrosine phosphorylated IRS-1 binds and activates the lipid kinase phosphatidylinositol 3-kinase (PI3K) that then stimulates a serine kinase cascade (including pyruvate dehydrogenase lipoamide kinase isozyme 1 [PDK-1] and protein kinase B [AKT/PKB]). Activation of this kinase cascade enhances translocation of the insulin-responsive GLUTs from an intracellular compartment to the plasma membrane where GLUTs, particularly GLUT-4, facilitates glucose entry into skeletal muscle and adipose tissue.[7] In the liver, activated AKT inhibits the activity of glycogen synthase kinase-3 (GSK-3) by phosphorylation. Since GSK-3 inhibits glycogen synthase (GS) by phosphorylation, suppression of GSK-3 leads to activation of glycogen synthesis. Protein phosphatase-1 and its associated glycogen targeting subunit have been implicated in dephosphorylation of GS and glycogen phosphorylase (GP) in the liver.[8]

In addition to PI3K-dependent insulin signaling, a mitogen-activated protein kinase (MAPK)-dependent branch of insulin signaling initiates with tyrosine-phosphorylated IRS-1 or adaptor protein Shc binding to the Src homology 2 domain of growth factor receptor-bound protein 2, resulting in activation of the preassociated GTP exchange factor son of sevenless.[6,7] This further activates the small GTP-binding protein Ras, which then triggers a kinase phosphorylation cascade involving Raf, MAPK kinase, and MAPK/extracellular signal–regulated kinase (ERK).[5,6] MAPK-dependent insulin signaling regulates effects of insulin on mitogenesis and cell growth and differentiation.

3. PATHOPHYSIOLOGY OF INSULIN ACTION

T2D is one of the most common chronic diseases in nearly all countries, and it is continuing to be an increasing global health burden.[2] Current medications are not adequately effective in maintaining long-term glycemia control in a high proportion of patients. Therefore, there is an urgent need to continue working on the prevention and control of diabetes.

There are two metabolic hallmarks in T2D: pancreatic β-cell dysfunction and impaired responsiveness to insulin in peripheral tissues. Pancreatic β-cell failure is a metabolic disorder critical in the development of T2D. Decreased viability and dysfunction of β cells would accelerate the diabetic pathogenesis associated with higher mortality. Chronic high glucose exposure would directly increase intracellular generation of reactive oxygen species and deteriorate mitochondrial function to uncouple ATP generation, impairing the glucose-stimulated insulin secretion.[2] On the other hand, the response to insulin by target tissues is damaged in a diabetic condition, which aggravates the ability of insulin to trigger downstream metabolic actions resulting in insulin resistance. In this regard, one of the characteristics of T2D is a misregulation of the insulin pathway in the liver, muscle, and adipose tissue.[3]

In insulin resistance, the earliest defect in the insulin signaling pathway is that insulin receptor autophosphorylation is less responsive to insulin.[9] As a result, downstream cellular insulin action is seriously reduced or impaired. Since the autophosphorylated and activated insulin receptor tyrosine kinase is modulated by protein-tyrosine phosphatases, this superfamily of proteins is a potential contributor to insulin resistance.[10] In the hepatocyte, glycogenolysis and gluconeogenesis are less efficiently suppressed in response to insulin despite adequate circulating glucose levels. This leads to inappropriate glucose release by the liver and results in hyperglycemia which can occur both after meals and during fasting. It addition, insulin resistance also affects the skeletal muscle that is responsible for much of insulin-dependent glucose uptake and utilization and contributes very significantly to the hyperglycemia.

Furthermore, much attention is focused on the adipocyte and the factors that it may release into the circulation to influence insulin action in the liver and skeletal muscle. Multiple cytokines, growth factors, hormones, as well as free fatty acids are actively released into the circulation by adipocytes. Aberrant circulating levels of several of these adipocyte-derived compounds are observed in the obese and type 2 diabetics. Particularly remarkable is the evidence that T2D and obesity have characteristics of an inflammatory condition with increased blood levels or adipocyte production of the proinflammatory cytokines interleukin (IL)-6, IL-1, and tumor necrosis factor (TNF)-α.[3]

4. DIETARY FLAVONOIDS

More than 8000 different compounds have been described, and they can be divided into 10 different general classes based on their chemical structure.[11] The most abundantly occurring polyphenolic group in plants is flavonoids, which account for 60% of dietary polyphenols.

Flavonoids are C6–C3–C6 molecules with two aromatic rings (i.e., rings A and B) linked by three carbons usually arranged as an oxygenated heterocycle (i.e., ring C) (Table 1). Flavonoids are classified according to the degree of oxidation of the ring C into flavanols (also named catechins), flavonols, flavones, isoflavones, flavanones, and anthocyanins. Flavonoids also occur as oligomers and polymers (i.e., tannins), classified as condensed tannins (also known as proanthocyanidins or procyanidins) or hydrolysable tannins.[11] Molecular structure of flavonoids, representative groups, the best-known members of each group, and food sources in which they are present, are listed in Table 1. Most flavonoids are present in nature as *O*-glycosides and other conjugates (catechins are an exception), which contribute to their complexity and the large number of individual molecules that have been identified (more than 5000).

TABLE 1 Basic Chemical Structure and Major Dietary Sources of Commonly Occurring Flavonoids

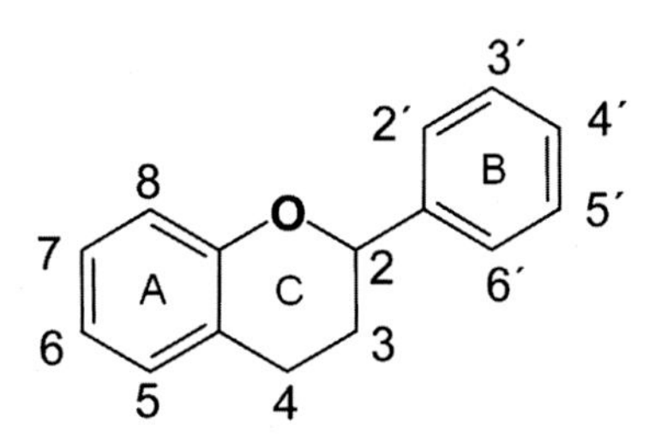

Flavonoid subgroup	Representative flavonoids	Food source
Flavanols (R3=R5=R7=OH)		
R3′=R4′=OH	Epicatechin	Cocoa, beans, tea,
R3′=R4′=OH; R5′=H	Catechin	grape, red wine,
R3=gallate; R3′=R4′=R5′=OH	Epigallocatechin-3-gallate	apple, cherry, apricot, peach
R3′=R4′=R5′=OH	Epigallocatechin	
Flavonols (R3=R5=R7=OH)		
R3′=R4′=OH; R5′=H	Quercetin	Onion, apple,
R3′=R4′=R5′=OH	Myricetin	broccoli, cherry,
R4′=OH; R3′=R5′=H	Kaempferol	tomato, leek,
R3=β-D-glucopyranosyloxide R3′=R4′=OH	Rutin	berries, tea, red wine
Flavones (R5=R7=OH)		
R3′=R4′=OH	Luteolin	Capsicum pepper,
R3′=H; R4′=OH	Apigenin	thyme, celery,
R8=OH	Chrysin	oregano, parsley
R3′=OH; R4′=OCH_3	Diosmetin	
Isoflavones (R7=R4′=OH)		
R5=OH	Genistein	Soya beans,
R5=H	Daidzein	legumes
Flavanones (R5=R7=OH)		
R3′=H; R4′=OH	Naringenin	Orange, lemon,
R3′=OH; R4′=OCH_3	Hesperidin	grapefruit
R3′=R4′=OH	Eriodictyol	
Anthocyanidins (R3=R5=R7=R4′=OH)		
R3′=R5′=H	Pelargonidin	Strawberry, cherry,
R3′=R5′=OH	Delphinidin	rhubarb,
R3′=OH; R5′=H	Cyanidin	aubergine, black
R3′=R5′=OCH_3	Malvidin	grape, red wine

[a] The symbol "=" indicates "equal".

The intake of flavonoids depends on the habits and food preferences, but the daily average in the United Kingdom and the United States has been estimated to range from 20 mg to 1 g.[12] However, it should be considered that the health effects of flavonoids depend on their bioavailability (absorption, distribution, metabolism, and elimination), a factor that is also influenced by their chemical structure.[11] In brief, aglycones can be absorbed in the small intestine, but in food most of phenolic compounds are present as esters, glycosides, or polymers that can either be absorbed in these forms or be hydrolyzed by the intestinal enzymes or the colonic microflora, and consequently the aglycone or the hydrolysis products can be absorbed.[13–15] During absorption, flavonoids are conjugated in the small intestine and later in the liver and other organs[16]: phenolic compounds are methylated, sulfated, or glucuronidated (metabolic detoxification) in order to increase their hydrophilia and facilitate their urinary and biliary excretion.[17] In addition, oligomers and polymers that are not absorbed through the gut barrier could be metabolized by the intestinal microbiota into various phenolic acids of low molecular weight, which are more bioavailable and might be well absorbed through the colon.[16,18] Flavonoids may also be secreted in the bile to the duodenum and be reabsorbed, entering into an enterohepatic cycle and evoking a longer half-life.

Flavonoid amount in the gastrointestinal tract reaches concentrations equal to or frequently greater than micromolar values, with a high presence of oligomers/polymers of flavonoids along with byproducts of colonic microbiota. On the other hand, in the plasma and extracellular fluid, the concentration of polyphenols remains approximately in the micromolar range, with little presence of oligomers/polymers but abundant in products of enterocyte and liver metabolism. In the intracellular environment, polyphenols are probably present at nanomolar or picomolar concentrations, while oligomers/polymers are absent with products of reconjugation still playing an important role.[19] Finally, absorption of flavonoids may be influenced by dosage, vehicle of administration, prior diet, food matrix, gender, and differences in the gut microbial populations.[20]

5. COCOA FLAVONOIDS

Cocoa powder is a rich source of fiber (26–40%), proteins (15–20%), carbohydrates (about 15%), and lipids (10–24%), and it contains minerals and vitamins.[21]

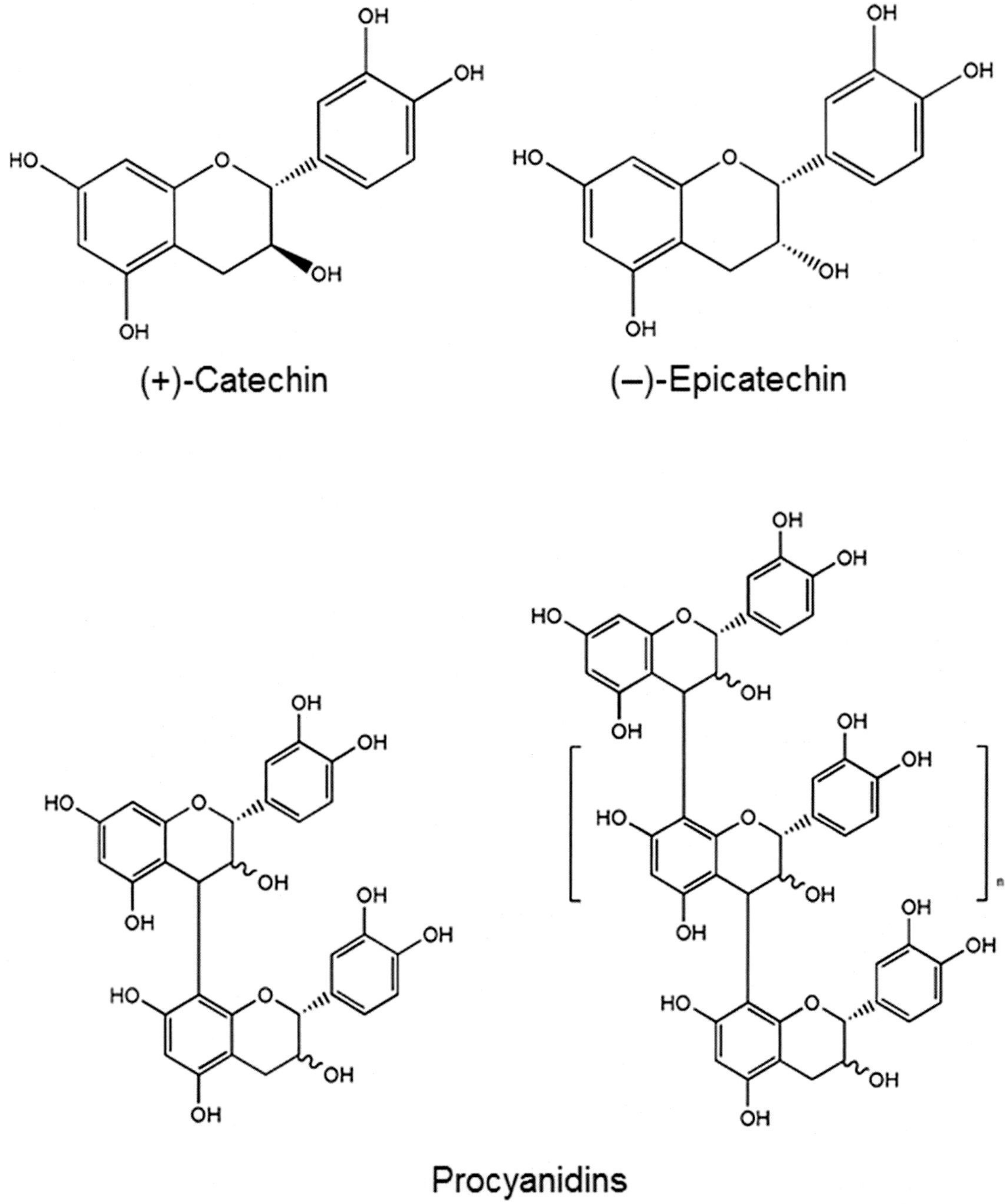

(+)-Catechin

(−)-Epicatechin

Procyanidins

FIGURE 1 Chemical structure of main cocoa flavanols.

Cocoa bean has more than 200 compounds that could be considered as beneficial for humans, but research has mainly been focused on phenolic compounds, which comprise 12–18% of the total weight of dried cocoa nibs.[22] In fact, flavanols are the major compounds in cocoa nibs,[22] and cocoa is the food that possesses the highest flavanol content on a per-weight basis.[23–25] Cocoa mainly contains high amounts of flavanols (−)-epicatechin (EC), (+)-catechin, and their dimers procyanidins B2 and B1 (Figure 1), although other polyphenols such as quercetin, quercetin 3-*O*-glucoside (isoquercitrin), quercetin 3-*O*-arabinose, quercetin 3-*O*-galactoside (hyperoside), naringenin, luteolin, and apigenin are present in lower amounts.[26]

Cocoa and its derivative products are widely consumed worldwide and constitute a larger proportion of the diet of many individuals than green tea, wine, or soybeans.[27,28] In fact, chocolate consumption contributed 2–5 mg of daily catechin intake out of an estimated total of 50 mg per day in a report from the

Netherlands,[27] and in the United States, the average intake of catechins and procyanidins is higher than the estimated intake of other flavonoids.[29]

Health effects derived from cocoa flavonoids depend on their bioavailability, as mentioned earlier.[11] Different studies have proved the absorption of catechin, EC, and dimeric procyanidins after the intake of different cocoa byproducts by animals and humans.[18,30] In particular, monomeric flavonoids are absorbed in the small intestine, and they and their metabolites (methylated, sulfated, and glucuronidated compounds) are rapidly identified in the plasma[13–15] and urine.[17] Accordingly, absorbed flavonoids are widely distributed and can be detected in lymphoid organs, including the thymus, spleen, and mesenteric lymphoid nodes, as well as in the liver and testes at different concentrations.[16] In addition, oligomers and polymers of flavanols that are not absorbed through the gut barrier (large proanthocyanidins appear to be 10- to 100-fold less absorbed[11,31]) could be metabolized by the intestinal microbiota into various phenolic acids of low molecular weight, which are more bioavailable, and might be well absorbed through the colon,[16,18] exerting beneficial biological effects.[32]

6. COCOA FLAVONOIDS AND INSULIN ACTION

Polyphenols have been suggested to ameliorate important hallmarks of T2D, such as insulin signaling alterations. However, the physiological mechanisms involved in the beneficial effects of polyphenols remain poorly understood.

6.1 In Vitro Studies

Cultured cells constitute a useful tool to elucidate the molecular mechanisms of action of cocoa extracts and its polyphenolic compounds. Studies related to the cocoa and its main flavanol on insulin actions are summarized in Table 2.

Cocoa and its flavanols seem to possess an insulin-like activity that could contribute to prevent or ameliorate diabetes and its associated complications and may mediate this effect peripherally. The liver plays a central role in maintaining glucose homeostasis, contributing to preserve insulin sensitivity.[3] In this line, a cocoa phenolic extract (CPE) and its main flavonoid EC exerted an insulinomimetic effect in hepatic HepG2 cultured cells.[33] EC and CPE enhanced tyrosine phosphorylation and total levels of IR, IRS-1, and IRS-2 and activated the PI3K/AKT pathway and 5′-AMP–activated protein kinase (AMPK) at concentrations that are not toxic to hepatic cells and are reachable through the diet (1–10 μmol/L EC and 1–10 μg/mL CPE). These natural substances also modulated hepatic gluconeogenesis and phosphoenolpyruvate carboxykinase expression through AKT and AMPK, and CPE increased GLUT-2 levels.[33] Moreover, EC and CPE (1–10 μmol/L EC and 1–10 μg/mL CPE) alleviated the hepatic insulin resistance induced on HepG2 by treating the cells with a high dose of glucose.[34] Thus, EC and CPE decreased IRS-1 Ser636/639 phosphorylation, enhanced tyrosine phosphorylated and total levels of IR, IRS-1, and IRS-2 and activated the PI3K/AKT pathway, and AMPK. In addition, EC and CPE preserved HepG2 cell functionality by restoring the levels of GLUT-2, increasing glucose uptake, maintaining glycogen synthesis and decreasing glucose production.[34] Accordingly, procyanidin oligomers extracts from cinnamon (5–10 μg/mL), which were rich either in A- or B-type procyanidin oligomers, enhanced the consumption of extracellular glucose in normal and insulin-resistant HepG2 cells, showing that they are able to improve insulin sensitivity.[35]

Skeletal muscle is another therapeutic target for hyperglycemia, as it accounts for approximately 80% of insulin-stimulated glucose uptake in the postprandial state and plays a vital role in maintaining glucose homeostasis.[36] Cacao liquor is a major ingredient of chocolate and cocoa that is abundant in polyphenols, including catechins and their oligomers B-type proanthocyanidins.[37] Cacao liquor procyanidin extract (CLPr) increased glucose uptake in a dose-dependent manner, and showed the maximum effect with 5 μg/mL in L6 myotubes.[38] Additionally, CLPr (5–10 μg/mL) stimulated GLUT-4 translocation in skeletal muscle L6 cells, whereas levels of GLUT-1 and -4 remained unchanged in the plasma membrane.[38] All this suggested that CLPr enhanced the glucose uptake by inducing GLUT-4 translocation in skeletal muscle cultured cells.

In vitro studies suggest that cocoa also has insulinomimetic effects in adipose tissue. Thus, a cocoa polyphenol extract (100–200 μg/mL) did not affect the levels of IR, but significantly inhibited the IR kinase activity by direct binding, without altering total tyrosine phosphorylation of IR or inhibiting its autophosphorylation in 3T3-L1 adipocytes. Moreover, the inhibitory effect of cocoa on the IR kinase activity has been related to a suppression of ERK- and AKT-mediated signaling cascades to facilitate an anti-adipogenic effect.[39] Similarly, EC (50 μmol/L) induced beneficial changes in 3T3-L1 adipocytes, as increased dose- and time-dependently the glucose uptake and GLUT-4 translocation, as well as phosphorylation of protein kinase C (PKC)λ/ξ but not in IR and AKT.[40] However, the PI3K/AKT pathway seemed to play an important role in EC-promoted GLUT-4 translocation, since pharmacological PI3K/AKT inhibitors, such as Wortmannin and

TABLE 2 Cocoa and Cocoa Polyphenols Modulation of Proteins Involved in the Insulin Signaling in Cultured Cells[a]

Cell type	Polyphenol	Treatment	Outcome	References
HEPATIC CELLS				
HepG2	Cocoa phenolic extract (CPE)	1–10 μg/mL, 24 h	↑ p(Tyr)-IR, ↑ IR, ↑ p(Tyr)IRS-1, ↑ IRS-1, ↑ p(Tyr)IRS-2, ↑ IRS-2, ↑ *p*-AKT, ↑ *p*-GSK-3, ↑ *p*-AMPK, ↑ GLUT-2; ↓ *p*-GS, ↓ PEPCK, ↓ glucose production	33
HepG2	Epicatechin (EC)	1–10 μmol/L, 24 h	↑ p(Tyr)-IR, ↑ IR, ↑ p(Tyr)IRS-1, ↑ IRS-1, ↑ p(Tyr)IRS-2, ↑ IRS-2, ↑ *p*-AKT, ↑ *p*-GSK-3, ↑ *p*-AMPK, ↓ *p*-GS, ↓ PEPCK, ↓ glucose production; =GLUT-2	33
HepG2 (insulin-resistant cells)	CPE	1–10 μg/mL, 24 h	↑ p(Tyr)-IR, ↑ IR, ↑ p(Tyr)IRS-1, ↑ IRS-1, ↑ p(Tyr)IRS-2, ↑ IRS-2, ↑ *p*-AKT, ↑ *p*-GSK-3, ↑ *p*-AMPK, ↑ glucose uptake; ↓ p(Ser)-IRS-1; ↓ *p*-GS, ↓ PEPCK, ↓ glucose production; =GLUT-2, =glycogen content	34
HepG2 (insulin-resistant cells)	EC	1–10 μmol/L, 24 h	↑ p(Tyr)-IR, ↑ IR, ↑ p(Tyr)IRS-1, ↑ IRS-1, ↑ p(Tyr)IRS-2, ↑ IRS-2, ↑ *p*-AKT, ↑ *p*-GSK-3, ↑ *p*-AMPK, ↑ glucose uptake; ↓ p(Ser)-IRS-1; ↓ *p*-GS, ↓ PEPCK, ↓ glucose production; =GLUT-2, =glycogen content	34
HepG2	Procyanidin A and B (oligomer)	5, 10 μg/mL, 24 h	↑ consumption extracellular glucose	35
SKELETAL MUSCLE CELLS				
L6	Cocoa liquor procyandin extract	0.05–10 μg/mL, 15 min	↑ glucose uptake, ↑ GLUT-4 translocation; =GLUT-2, =GLUT-1	38
ADIPOCYTE CELLS				
3T3-L1	Cocoa polyphenols	100–200 μg/mL, 4 h	↓ *p*-ERK, ↓p-AKT; =IR	39
3T3-L1	EC	50 μmol/L, 30 min	↑ glucose uptake, ↑ GLUT-4 translocation, ↑ *p*-PKCλ/ε;=*p*-IRβ, =IRβ, *p*-AKT	40
3T3-L1	EC	1–10 μM, 4 h	↓ PTP1B, ↓ IL6, ↓ TNF-α, ↓ *p*-MAPKs, p65- NFκB	41
3T3-L1	Procyanidin	140 mg/L, 1 and 15 h	↑ glucose uptake, ↑ glycogen synthesis, ↑ lipid synthesis	42
3T3-L1	Procyanidin	100 mg/L, 30 min	↑ p(Tyr)-IR, ↑ glucose uptake, ↑ *p*-AKT, ↑ *p*-ERK, ↑ *p*-p38	43
3T3-L1	Procyanidin A (polimers)	0.01–1 mg/mL (0.5–3 h)	↑ IRβ, ↑ GLUT-4 (1 mg/mL, 3 h)	44
PANCREATIC CELLS				
INS-1E	3,4-Dihydroxyphenylacetic acid (DHPAA), and 3-hydroxyphenylpropionic acid (HPPA)	5 μmol/L DHPAA, 20 h 1 μmol/L HPPA, 20 h	↑ insulin secretion, ↑ β-cell survival	45
INS-1E	Procyanidins	25 mg/L, 2 h	↓ insulin, ↓ insulin mRNA, =GK	46
INS-1E	EC, procyanidin B2	0.01–0.32 μmol/L, 90 min	↓ insulin secretion (=insulin secretion for procyanidin B2)	47

The symbol "=" indicates unchanged levels or activity of the parameter.
[a]*The arrow indicates an increase (↑) or decrease (↓) in the levels or activity of the different parameters analyzed.*

LY294002, decreased EC-induced glucose uptake.[40] Accordingly, in 3T3-L1 adipocytes EC (0.5−10 μmol/L) attenuated the TNFα-mediated downregulation of peroxisome proliferator-activated receptor (PPAR)-γ expression and decreased nuclear DNA binding.[41] Moreover, EC also inhibited TNFα-mediated altered transcription of PTP1B, leading both effects to an attenuation of the TNFα-mediated triggering of signaling cascades involved in insulin resistance.[41]

Procyanidin extracts have also demonstrated an insulinomimetic effect in 3T3-L1 adipocytes. A grape seed−derived procyanidin extract (100−140 mg/L) stimulated glycogen and lipid synthesis,[42] as well as activated the IR by interacting with and inducing its phosphorylation, which led to increased glucose uptake mediated by AKT.[43] However, the procyanidin extract phosphorylated proteins of the insulin signaling pathway differently than insulin does, as AKT, ERK, and p38--MAPKs, which were key points for the procyanidin-activated signaling mechanisms. This feature suggested alternative pathways for procyanidins to be effective in an insulin-resistant situation.[43] Similarly, in 3T3-L1 adipocytes procyanidin type-A polymers from cinnamon seemed to play a role in insulin signaling and glucose transport, as it increased IRβ and GLUT-4 protein levels, although it decreased IRβ mRNA and exhibited a biphasic pattern on GLUT-4 mRNA levels.[44]

Cocoa and its phenolic compounds also play an important role to maintain insulin levels as they are able to modulate insulin secretion in β-pancreatic cells. In this regard, the microbial-derived flavonoid metabolites 3,4-dihydroxyphenylacetic acid (5 μmol/L) and 3-hydroxyphenylpropionic acid (1 μmol/L) were able to potentiate glucose-stimulated insulin secretion in a β-cell line INS-1E and in rat pancreatic islets.[45] Moreover, pretreatment of cells with both compounds protected against β-cell dysfunction and death induced by the pro-oxidant *tert*-butylhydroperoxide, through the activation of PKC and ERK pathways.[45] Additionally, procyanidins from grape seed improved β-cell functionality under lipotoxic conditions, as these natural substances reduced insulin mRNA and insulin secretion, although it did not alleviate the oleate-induced upregulation of uncoupling protein (UCP)-2.[46] However, Qa'dan and colleagues[47] have reported that EC and procyanidin B2 at lower concentrations (0.01−0.32 μmol/L) inhibited and did not alter insulin secretion, respectively, in INS-1E cells.

6.2 Animal Studies

In the recent years, there was a growing body of data derived from animal models on the effects of cocoa and cocoa products in ameliorating insulin resistance and improving insulin sensitivity. The more recent in vivo studies with cocoa and their main flavonoids catechin, EC, or procyanidins are provided in Table 3.

In the obese Zucker fatty rats, the ingestion for 7 weeks of a soluble cocoa fiber (SCF)-enriched diet reduced the total cholesterol/high-density lipoprotein cholesterol ratio and the triglyceride levels. In addition, SCF reduced plasma glucose and insulin, and as a consequence the insulin resistance index (homeostasis model assessment of insulin resistance [HOMA-IR]) was also decreased.[48] More recently, an experimental study on male C57BL/6J high-fat−fed obese mice demonstrated that dietary supplementation with 8% cocoa powder for 10 weeks attenuated insulin resistance, as indicated by improved HOMA-IR, and reduced obesity-related inflammation and fatty liver disease. In particular, cocoa supplementation significantly decreased plasma levels of the proinflammatory mediators IL-6 (30.4%) and monocyte chemoattractant protein-1 (25.2%) and increased adiponectin (33.7%) compared with high-fat−fed mice.[49] These effects appeared to be mediated in part by a modulation of dietary fat absorption and inhibition of macrophage infiltration in white adipose tissue. On the contrary, intake of cocoa extract supplemented with polyphenols (2.17 mg EC, 1.52 mg catechin, 0.25 mg dimer, and 0.13 mg trimer/g cocoa extract) and methylxanthines (3.55 mg caffeine and 2.22 mg theobromine/g cocoa extract) for 4 weeks significantly reduced the plasma total cholesterol, triglycerides, and low-density lipoprotein cholesterol of obese-diabetic (*ob/db*) rats while no significant differences were observed in plasma glucose and insulin levels and insulin sensitivity after cocoa extract supplementation.[50]

Concerning the effect of specific cocoa flavanols on insulin action, it has been showed that treatment with a CLPr extract for 13 weeks prevents obesity, hyperglycemia, and insulin resistance in male C57BL/6J High-fat−fed obese mice.[51] Some of the molecular mechanisms involved in CLPr prevention were the activation of the key metabolic regulator AMPK-α, the translocation of GLUT-4, and the upregulation of UCP expressions in skeletal muscle and adipose tissue. In this line, a single dose (10 μg/kg body weight) of tetrameric procyanidin from cacao liquor (Cinnamtannin A2) was able to improve hyperglycemia in male ICR mice by specifically increase the levels of glucagon-like peptide-1 and insulinemia.[52] Cinnamtamin A2 also activates IR and IRS-1 in the soleus muscle of ICR mice, indicating that this cocoa procyanidin may improve hyperglycemia through an incretin-like effect, accompanied by activation of the insulin-signaling pathway. Similarly, Montagut et al.[53] tested the effects of a procyanidin extract from grape seed (25 and 50 mg/kg body weight/day) during 10 or 30 days in a

TABLE 3 Summary of Published Effects of Cocoa and Cocoa Products on Insulin Action in Animal Studies[a]

Animal model	Duration	Treatment	Main outcomes	References
Obese Zucker fatty rats	7 weeks	5% soluble cocoa fiber–enriched diet	↓ glucose, ↓ insulin, ↓ HOMA-IR	48
High-fat–fed obese C57BL/6J mice	10 weeks	8% cocoa powder–enriched diet	=glucose, ↓ insulin ↓ HOMA-IR, ↑ adiponectin, ↓ MCP-1, ↓ IL-6	49
Obese-diabetic (*ob/db*) rats	4 weeks	Cocoa extract polyphenols (600 mg/kg body weight/day)	=glucose, =insulin, =HOMA-IR	50
High-fat–fed obese C57BL/6J mice	13 weeks	0.5% and 0.2% Cacao liquor procyanidin extract –supplemented diet	↓ glucose, ↓ insulin, ↓ HOMA-IR, ↑ *p*-AMPKα, ↑ GLUT-4, ↑ UCP	51
Male ICR mice	A single dose	Cinnamtannin A2 (10 μg/kg body weight)	↓ glucose, ↑ insulin, ↑ GLP-1, ↑ *p*-IR, ↑ *p*-IRS-1	52
Wistar rats with a cafeteria diet	7 weeks	Procyanidin extract (25 and 50 mg/kg body weight/day)	↓ glucose, ↓ insulin, ↓ HOMA-IR	53
Streptozotocin-diabetic Wistar rats	6 weeks	Catechin (20 mg/kg body weight/day)	↓ glucose, =insulin, ↑ GLUT-4	54
High-fructose (HFr)-fed rats	14 weeks	Epicatechin (EC) and proanthocyanidin rich-extract (125 and 250 mg/kg body weight/day)	↓ glucose, ↓ insulin, ↑ GLUT-4, ↑ IRS-1	55
HFr-fed rats	8 weeks	EC (20 mg EC/kg body weight/day)	=glucose, ↓ insulin, ↑ *p*-IR, ↑ *p*-IRS-1, ↑ ERK, ↑AKT ↓ IKK, ↓ JNK, ↓ PKC, ↑ PTP1B	56
High-fat–fed obese C57BL/6J mice	12 weeks	Oligomeric procyanidins (25 mg/kg body weight/day)	↓ glucose, ↓ insulin, ↓ ITT	57

The symbol "=" indicates unchanged levels or activity of the parameter.
[a]*The arrow indicates an increase (↑) or decrease (↓) in the levels or activity of the different parameters analyzed.*

mild model of insulin resistance provoked by feeding healthy female Wistar rats with a cafeteria diet. They demonstrated that procyanidin treatment reduced the fasting plasma insulin levels and improved HOMA-IR index at 10 or 30 days, and this effect was accompanied by downregulation of PPARγ2, GLUT-4, and IRS1 in mesenteric white adipose tissue.

Daisy et al.[54] showed that the flavanol catechin possesses hypoglycemic and insulin mimetic activities. Daily oral administration for 6 weeks of catechin (20 mg/kg body weight/day) to streptozotocin-induced diabetic male Wistar rats markedly increased tissue glycogen and ^{14}C-glucose oxidation without any change in plasma insulin and C-peptide. In particular, catechin restored the altered glucokinase, glucose-6 phosphatase, GS, and GP levels in streptozotocin-induced diabetic rats to near normal. Interestingly, the expression of GLUT-4 mRNA and protein was enhanced in skeletal muscle after catechin treatment indicating the insulin mimetic response of the flavanol. Furthermore, the supplementation of high-fructose (HFr)-fed rats during 14 weeks with an extract rich in EC and proanthocyanidins (125 and 250 mg/kg body weight/day) ameliorated insulin resistance by enhancing the expression in adipocytes of insulin signaling pathway–related proteins, including IRS-1 and GLUT-4.[55] More recently,[56] it has been demonstrated that dietary supplementation with EC (20 mg EC/kg body weight/day) for 8 weeks also mitigates insulin resistance in HFr-fed rats. The in vivo response to insulin by liver and adipose tissues was impaired in the HFr rats as evidenced by decreased activation of components of the insulin signaling cascade (IR, IRS1, AKT, and ERK1/2) and upregulation of negative regulators (PKC, IκB kinase [IKK], JNK, and PTP1B). However, these alterations were partially or totally prevented by EC supplementation. In addition, EC inhibited events that contribute to insulin resistance such as increased expression and activity of NADPH oxidase, activation of redox-sensitive signals (JNK, IKK/nuclear factor-kappa B [NF-κB]), expression of NF-κB–regulated proinflammatory cytokines and chemokines, and some sub-arms of endoplasmic reticulum stress signaling.[56]

Interestingly, the ability of cocoa flavanols to inhibit weight gain and the onset of insulin resistance could be dependent on flavanol polymerization.[57] Dorenkottet al.[57] recently examined the relative activities of cocoa constituents on obesity and insulin resistance in male C57BL/6J mice that were fed a high-fat diet supplemented with either a cocoa flavanol extract or a flavanol fraction enriched with monomeric, oligomeric, or polymeric procyanidins for 12 weeks. Although all three flavanol fractions effectively inhibited hyperinsulinemia, only the oligomer-rich fraction effectively improved glucose tolerance compared to the high-fat diet and inhibited the development of insulin resistance. Altogether, these results indicated that the oligomer-rich fraction was the most effective in preventing weight gain, fat mass, impaired glucose tolerance, and insulin resistance in this model.

6.3 Human Studies

Human studies that have examined the effect of polyphenol-rich cocoa on insulin action have also showed preliminary evidence for an insulin-sensitizing effect of cocoa (Table 4). In a randomized crossover trial,[58] healthy volunteers were given either flavanol-rich dark chocolate (100 g/day) or white chocolate (100 g/day) for 15 days. Ingestion of dark chocolate significantly lowered insulin resistance as indicated by HOMA-IR, and increased insulin sensitivity as indicated by quantitative insulin-sensitivity check index and insulin sensitivity index (ISI), whereas flavonoid-free chocolate was ineffective. Similar results were reported with the same treatment on hypertensive subjects with[59] and without glucose intolerance.[60] Dark chocolate decreased HOMA-IR, increased insulin sensitivity, and increased β-cell function compared to white chocolate. In a longer study, overweight and obese adults that consumed a high-flavanol cocoa (902 mg flavanols/day) for 12 weeks significantly improved insulin sensitivity compared with low-flavanol cocoa.[61] Although it is unclear the mechanism underlying this effect, the improvement in insulin sensitivity and β-cell function directly correlated with flow-mediated dilatation increase, suggesting that the effects of flavonoids on nitric oxide–dependent vascular function and insulin sensitivity might also be linked from a mechanistic point of view.[62] Accordingly, cocoa flavanols may decrease insulin resistance by increasing the nitric oxide bioavailability and decreasing the formation of reactive oxygen and nitrogen species. Nevertheless, in one study, consumption of a flavanol-rich cocoa drink (150 mL twice a day, approximately 900 mg flavanols/day) for 2 weeks improves endothelial function in patients with hypertension without changing insulin sensitivity.[63] Similarly, Mellor et al.[64] showed that the ingestion of a high-polyphenol chocolate providing 16.6 mg/day of ECs during 8 weeks was effective in improving the atherosclerotic cholesterol profile in patients with diabetes without affecting weight, inflammatory markers, insulin resistance, or glycemic control.

Overall, in a systematic review and meta-analysis of some of these randomized controlled trials, Shrime et al.[65] reported that HOMA-IR decreased by 0.94 point (95% CI 0.59 to 1.29; $p < 0.001$) with the consumption of flavonoid-rich cocoa and significantly increased ISI

TABLE 4 Summary of Published Effects of Cocoa on Insulin Action in Human Intervention Studies[a]

Population	Design	Study size	Duration (days)	Control dose (mg/day)	Flavonoid dose (mg/day)	Main outcomes	References
Healthy volunteers	Randomized crossover	30	15	0	500	↓ HOMA-IR, ↑ QUICKI, ↑ ISI	58
Hypertension	Randomized crossover	40	15	0	87.8	↓ HOMA-IR, ↑ QUICKI, ↑ ISI ↑ FMD, ↓ insulin	59
Hypertension glucose intolerant	Randomized crossover	38	15	0	1080	↓ HOMA-IR, ↑ QUICKI, ↑ ISI, ↑ FMD	60
Overweight/obese volunteers	Randomized placebo-controlled	49	84	36	902	↓ HOMA-IR, ↑ FMD	61
Hypertension	Randomized crossover	38	15	28	900	=HOMA-IR, ↑ FMD	63
Diabetic subjects	Randomized crossover	24	56	2	16.6	=HOMA-IR	64
Elderly with MCI	Randomized placebo-controlled	90	56	45	990 520	↓ HOMA-IR, ↑ MCI	67

The symbol "=" indicates unchanged levels or activity of the parameter.
[a]*The arrow indicates an increase (↑) or decrease (↓) in the levels or activity of the different parameters analyzed.*

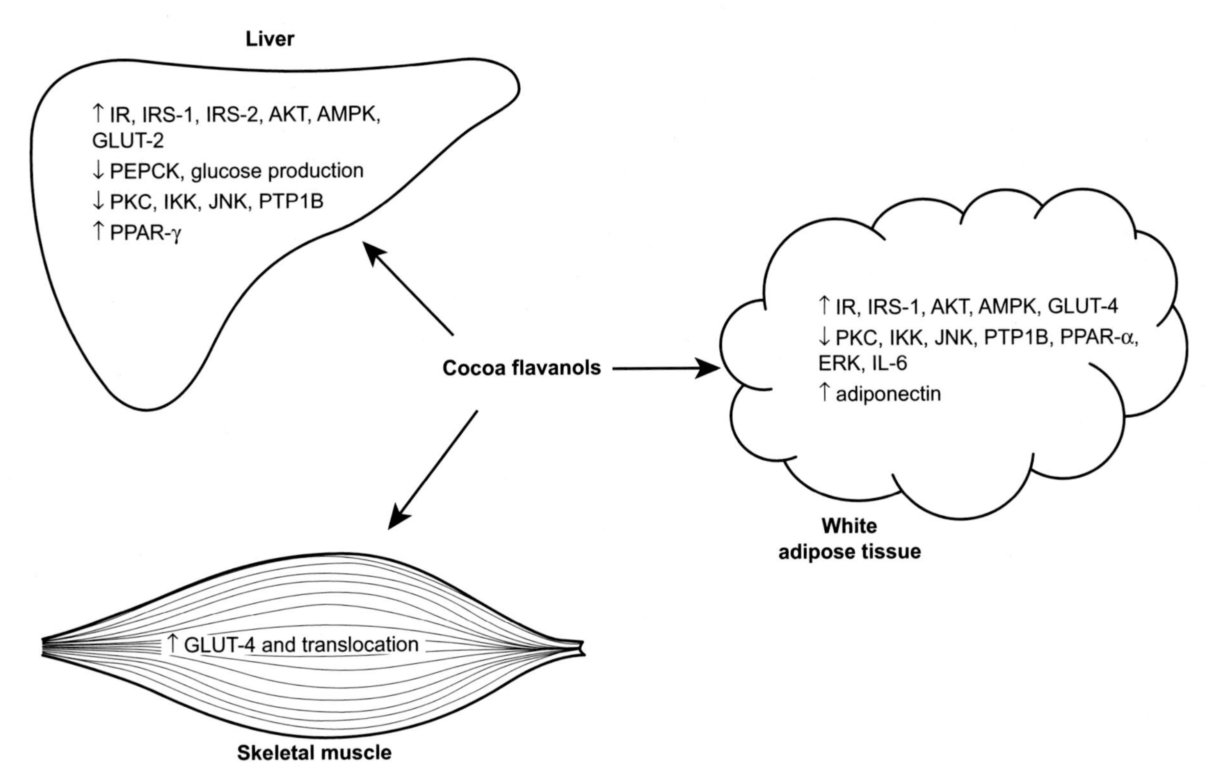

FIGURE 2 General mechanism of action of cocoa and its flavanols.

by 4.95 points (95% CI 2.80 to 7.10; $p < 0.001$). More recently, Hooper et al.,[66] evaluating all the studies, reported that insulin resistance (HOMA-IR: 0.67; 95% CI: 0.98 to −0.36) was improved by chocolate or cocoa because of the significant reductions in serum insulin.

Finally, it is interesting to note that insulin plays an important role in modulating brain function; therefore, alteration in insulin signaling pathway could be a major contributor to cognitive dysfunction. In line with this concept, Desideri et al.[67] showed that cocoa flavanols might be effective in improving cognitive function in elderly volunteers with mild cognitive impairment. In particular, insulin resistance, blood pressure, and lipid peroxidation decreased among volunteers in the high-flavanol (900 mg/daily) and intermediate-flavanol groups (520 mg/daily) in comparison with those assigned to low-flavanol diet (45 mg/daily). Changes of insulin resistance explained about 40% of the effects on improving cognitive function ($p < 0.0001$).

Altogether, despite the scarce number of studies evaluating the effect of cocoa intake on insulin action, the evidence suggests that cocoa may be useful in ameliorating insulin resistance in pathological situations as metabolic syndrome and slowing the progression to T2D. Likewise, dietary inclusion of cocoa flavanols could be one approach to maintain and improve cardiovascular health and cognitive function.

7. CONCLUSIONS

This chapter addresses the potential mechanisms of action related to insulin signaling that have been so far identified for cocoa and its main phenolic compounds, as well as the feasibility that they could occur in vivo. In general terms, those cellular mechanisms include the modulation of glucose transport and key elements in the insulin signal transduction pathway, which leads to an improved glucose tolerance (Figure 2). Similarly, studies performed in animals have demonstrated that cocoa and its main phenolic components would be able to prevent and/or slow down insulin resistance by contributing to the glycemic control through the mentioned mechanisms of action. In addition, several human intervention studies have reported some favorable insulin-sensitizing effects. These studies propose a prominent role for cocoa and its flavanols in the protection afforded by fruits, vegetables, and plant-derived beverages against diseases for which insulin resistance has been implicated as a causal or contributory factor. All these beneficial properties of cocoa and its main phenolic compounds may suggest that they could be a potential chemopreventive tool useful for the nutritional management of diseases such as T2D.

However, widely consumed cocoa products deserve further investigations since the molecular mechanisms of action are not completely characterized and many features remain to be elucidated. Additionally, more extensive, well-controlled clinical trials are needed to fully evaluate the potential of cocoa in terms of optimal dose, insulin-sensitizing effects, and identification of targets modulated by cocoa polyphenols to prevent, delay, or contribute to the treatment of diabetes.

LIST OF ABBREVIATIONS

AKT/PKB protein kinase B
AMPK 5′ AMP-activated protein kinase
ATP adenosine triphosphate
CLPr cacao liquor procyanidin extract
CPE cocoa phenolic extract
DHPAA 3,4-dihydroxyphenylacetic acid
EC (−)-epicatechin
ERK extracellular regulated kinase
FMD flow-mediated dilatation
GLP-1 glucagon-like peptide-1
GLUT glucose transporter
GP glycogen phosphorylase
Grb-2 growth factor receptor-bound protein-2
GS glycogen synthase
GSK-3 glycogen synthase kinase-3
HFr high-fructose-fed rats
HOMA-IR homeostatic model assessment of insulin resistance
HPPA 3-hydroxyphenylpropionic acid
IκB inhibitor of kappa B
IKK IκB kinase
IL interleukin
IR insulin receptor
IRS insulin receptor substrate
ISI insulin-sensitivity index
ITT insulin tolerance test
JNK c-Jun N-terminal Kinase
MAPK mitogen-activated protein kinase
MCI mild cognitive impairment
MCP-1 monocyte chemoattractant protein-1
NF-κB nuclear factor-kappa B
PDK-1 pyruvate dehydrogenase lipoamide kinase isozyme one
PPAR peroxisome proliferator-activated receptor
PEPCK phosphoenolpyruvate carboxykinase
PI3K phosphatidylinositol-3-kinase
PKC protein kinase C
PTP1B phosphatase 1B
QUICKI quantitative insulin sensitivity check index
ROS reactive oxygen species
SCF soluble cocoa fiber
SH2 Src homology 2 domain
Sos son of sevenless
T2D type 2 diabetes
TNF tumor necrosis factor
UCP uncoupling protein
ZF Zucker fatty rats

Acknowledgments

This work was supported by grants AGL2010-17579 and AGL2014-58205-REDC from the Spanish Ministry of Science and Innovation (MICINN).

References

1. Chaturvedi N. The burden of diabetes and its complications: trends and implications for intervention. *Diabetes Res Clin Pract* 2007;**76**:S3–12.
2. Whiting DR, Guariguata L, Weil C, Shaw J. IDF diabetes atlas: global estimates of the prevalence of diabetes for 2011 and 2030. *Diabetes Res Clin Pract* 2011;**94**:311–21.
3. Klover PJ, Mooney RA. Hepatocytes: critical for glucose homeostasis. *Int J Biochem Cell Biol* 2004;**36**:753–8.
4. Babu PVA, Liu D, Gilbert ER. Recent advances in understanding the anti-diabetic actions of dietary flavonoids. *J Nutr Biochem* 2013;**24**:1777–89.
5. Munir KM, Chandrasekaran S, Gao F, Quon MJ. Mechanisms for food polyphenols to ameliorate insulin resistance and endothelial dysfunction: therapeutic implications for diabetes and its cardiovascular complications. *Am J Physiol Endocrinol Metab* 2013;**305**:E679–86.
6. Muniyappa R, Montagnani M, Koh KK, Quon MJ. Cardiovascular actions of insulin. *Endocr Rev* 2007;**28**:463–91.
7. Nystrom FH, Quon MJ. Insulin signalling: metabolic pathways and mechanisms for specificity. *Cell Signal* 1999;**11**:563–74.
8. Yang R, Newgard CB. Hepatic expression of a targeting subunit of protein phosphatase-1 in streptozotocin diabetic rats reverses hyperglycemia and hyperphagia despite depressed glucokinase expression. *J Biol Chem* 2003;**278**:23418–25.
9. Saltiel AR. New perspectives into the molecular pathogenesis and treatment of type 2 diabetes. *Cell* 2001;**104**:517–29.
10. Cheng A, Dube N, Gu F, Tremblay ML. Coordinated action of protein tyrosine phosphatases in insulin signal transduction. *Eur J Biochem* 2002;**269**:1050–9.
11. Manach C, Williamson G, Morand C, Scalbert A, Rémésy C. Bioavailability and bioefficacy of polyphenols in humans. I. Review of 97 bioavailability studies. *Am J Clin Nutr* 2005;**81**(Suppl. 1):230S–42S.
12. Scalbert A, Morand C, Manach C, Rémésy C. Absorption and metabolism of polyphenols in the gut and impact on health. *Biomed Pharmacother* 2002;**56**:276–82.
13. Roura E, Andres-Lacueva C, Jauregui O, et al. Rapid liquid chromatography tandem mass spectrometry assay to quantify plasma (2)-epicatechin metabolites after ingestion of a standard portion of cocoa beverage in humans. *J Agric Food Chem* 2005;**53**:6190–4.
14. Holt RR, Lazarus SA, Sullards MC, et al. Procyanidin dimer B2 [epicatechin-(4b-8)-epicatechin] in human plasma after the consumption of a flavanol-rich cocoa. *Am J Clin Nutr* 2002;**76**:798–804.
15. Baba S, Osakabe N, Yasuda A, et al. Bioavailability of (2)-epicatechin upon intake of chocolate and cocoa in human volunteers. *Free Radic Res* 2000;**33**:635–41.
16. Urpi-Sarda M, Ramiro-Puig E, Khan N, et al. Distribution of epicatechin metabolites in lymphoid tissues and testes of young rats with a cocoa-enriched diet. *Br J Nutr* 2010;**103**:1393–7.
17. Tsang C, Auger C, Mullen W, et al. The absorption, metabolism and excretion of flavan-3-ols and procyanidins following the ingestion of a grape seed extract by rats. *Br J Nutr* 2005;**94**:170–81.
18. Urpi-Sarda M, Monagas M, Khan N, et al. Epicatechin, procyanidins, and phenolic microbial metabolites after cocoa intake in humans and rats. *Anal Bioanal Chem* 2009;**394**:1545–56.
19. Galleano M, Verstraeten SV, Oteiza PI, Fraga CG. Antioxidant actions of flavonoids: thermodynamic and kinetic analysis. *Arch Biochem Biophys* 2010;**501**:23–30.
20. Heim K, Tagliaferro A, Bobilya D. Flavonoid antioxidants: chemistry, metabolism and structure-activity relationships. *J Nutr Biochem* 2002;**13**:572–84.

21. Ramiro-Puig E, Castell M. Cocoa: antioxidant and immunomodulator. *Br J Nutr* 2009;**101**:931–40.
22. Visioli F, Bernaert H, Corti R, et al. Chocolate, lifestyle, and health. *Crit Rev Food Sci Nutr* 2009;**49**:299–312.
23. Rusconi M, Conti A. *Theobroma cacao* L., the Food of the Gods: a scientific approach beyond myths and claims. *Pharmacol Res* 2010;**61**: 5–13.
24. Lee KW, Kim YJ, Lee HJ, Lee CY. Cocoa has more phenolic phytochemicals and a higher antioxidant capacity than teas and red wine. *J Agric Food Chem* 2003;**51**:7292–5.
25. Vinson JA, Proch J, Zubik L. Phenol antioxidant quantity and quality in foods: cocoa, dark chocolate, and milk chocolate. *J Agric Food Chem* 1999;**47**:4821–4.
26. Sánchez-Rabaneda F, Jáuregui O, Casals I, Andrés-Lacueva C, Izquierdo-Pulido M, Lamuela-Raventós RM. Liquid chromatographic/electrospray ionization tandem mass spectrometric study of the phenolic composition of cocoa (*Theobroma cacao*). *J Mass Spectrom* 2003;**38**:35–42.
27. Arts C, Holmann P, Bueno de Mesquita H, Feskens E, Kromhout D. Dietary catechins and epithelial cancer incidence: the Zutphen elderly study. *Int J Cancer* 2001;**92**:298–302.
28. Tabernero M, Serrano J, Saura-Calixto F. The antioxidant capacity of cocoa products: contribution to the Spanish diet. *Int J Food Sci Tech* 2006;**41**(Suppl. 1):28–32.
29. Gu L, Kelm MA, Hammerstone JF, et al. Concentrations of proanthocyanidins in common foods and estimations of normal consumption. *J Nutr* 2004;**134**:613–7.
30. Lamuela-Raventós RM, Romero-Pérez AI, Andrés-Lacueva C, Tornero A. Health effects of cocoa flavonoids. *Food Sci Technol Int* 2005;**11**:159–76.
31. Serra A, Macia A, Romero MP, et al. Bioavailability of procyanidin dimers and trimers and matrix food effects in in vitro and in vivo models. *Br J Nutr* 2010;**103**:944–52.
32. Monagas M, Urpi-Sarda M, Sanchez-Patan F, et al. Insights into the metabolism and microbial biotransformation of dietary flavan-3-ols and the bioactivity of their metabolites. *Food Funct* 2010;**1**: 233–53.
33. Cordero-Herrera I, Martin MA, Bravo L, Goya L, Ramos S. Cocoa flavonoids improve insulin signalling and modulate glucose production via AKT and AMPK in HepG2 cells. *Mol Nutr Food Res* 2013;**57**:974–85.
34. Cordero-Herrera I, Martin MA, Goya L, Ramos S. Cocoa flavonoids attenuate high glucose-induced insulin signalling blockade and modulate glucose uptake and production in human HepG2 cells. *Food Chem Toxicol* 2014;**64**:10–9.
35. Lu Z, Jia Q, Wang R, et al. Hypoglycemic activities of A- and B-type procyanidin oligomer-rich extracts from different cinnamon barks. *Phytomedicine* 2011;**18**:298–302.
36. Saltiel AR, Kahn CR. Insulin signaling and the regulation of glucose and lipid metabolism. *Nature* 2001;**414**:799–806.
37. Natsume M, Osakabe N, Yamagishi M, et al. Analyses of polyphenols in cacao liquor, cocoa, and chocolate by normal-phase and reversed-phase HPLC. *Biosci Biotechnol Biochem* 2000;**64**: 2581–7.
38. Yamashita Y, Okabe M, Natsume M, Ashida H. Cacao liquor procyanidin extract improves glucose tolerance by enhancing GLUT4 translocation and glucose uptake in skeletal muscle. *J Nutr Sci* 2012;**1**(e2):1–9.
39. Min SY, Yang H, Seo SG, et al. Cocoa polyphenols suppress adipogenesis in vitro and obesity in vivo by targeting insulin receptor. *Int J Obes* 2012:1–9.
40. Ueda M, Furuyashiki T, Yamada K, et al. Tea catechins modulate the glucose transport system in 3T3-L1 adipocytes. *Food Funct* 2010;**1**:167–73.
41. Vazquez-Prieto MA, Bettaie A, Haj FG, Fraga CG, Oteiza PI. (-)-Epicatechin prevents TNFa-induced activation of signaling cascades involved in inflammation and insulin sensitivity in 3T3-L1 adipocytes. *Arch Biochem Biophys* 2012;**527**:113–8.
42. Pinent M, Bladé MC, Salvadó MJ, Arola L, Ardevol A. Metabolic fate of glucose on 3T3-L1 adipocytes treated with grape seed-derived procyanidin extract (GSPE). Comparison with the effects of insulin. *J Agric Food Chem* 2005;**53**:5932–5.
43. Montagut G, Onnockx S, Vaqué M, et al. Oligomers of grape-seed procyanidin extract activate the insulin receptor and key targets of the insulin signaling pathway differently from insulin. *J Nutr Biochem* 2010;**21**:476–81.
44. Cao H, Polansky MM, Anderson RA. Cinnamon extract and polyphenols affect the expression of tristetraprolin, insulin receptor, and glucose transporter 4 in mouse 3T3-L1 adipocytes. *Arch Biochem Biophys* 2007;**459**:214–22.
45. Fernández-Millán E, Ramos S, Alvarez C, Bravo L, Goya L, Martín MA. Microbial phenolic metabolites improve glucose-stimulated insulin secretion and protect pancreatic beta cells against *tert*-butylhydroperoxide-induced toxicity via ERKs and PKC pathways. *Food Chem Toxicol* 2014;**66**:245–53.
46. Castell-Auví A, Cedó L, Pallarès V, Blay M, Pinent M, Ardévol A. Grape seed procyanidins improve β-cell functionality under lipotoxic conditions due to their lipid-lowering effect. *J Nutr Biochem* 2013;**24**:948–53.
47. Qa'dan F, Verspohl EJ, Nahrstedt A, Petereit F, Matalka KZ. Cinchonain Ib isolated from *Eriobotrya japonica* induces insulin secretion in vitro and in vivo. *J Ethnopharmacol* 2009;**124**:224–7.
48. Sánchez D, Moulay L, Muguerza B, Quiñones M, Miguel M, Aleixandre A. Effect of a soluble cocoa fiber-enriched diet in Zucker fatty rats. *J Med Food* 2010;**13**:621–8.
49. Gu Y, Yu S, Lambert JD. Dietary cocoa ameliorates obesity-related inflammation in high fat-fed mice. *Eur J Nutr* 2014;**53**:149–58.
50. Jalil AMM, Ismail A, Chong PP, Hamid M, Kamaruddin SHS. Effects of cocoa extract containing polyphenols and methylxanthines on biochemical parameters of obese-diabetic rats. *J Sci Food Agric* 2009;**89**:130–7.
51. Yamashita Y, Okabe M, Natsume M, Ashida H. Prevention mechanisms of glucose intolerance and obesity by cacao liquor procyanidin extract in high-fat diet-fed C57BL/6 mice. *Arch Biochem Biophys* 2012;**527**:95–104.
52. Yamashita Y, Okabe M, Natsume M, Ashida H. Cinnamtannin A2, a tetrameric procyanidin, increases GLP-1 and insulin secretion in mice. *Biosci Biotechnol Biochem* 2013;**77**:888–91.
53. Montagut G, Bladé C, Blay M, et al. Effects of a grape seed procyanidin extract (GSPE) on insulin resistance. *J Nutr Biochem* 2010;**21**: 961–7.
54. Daisy P, Balasubramanian K, Rajalakshmi M, Eliza J, Selvaraj J. Insulin mimetic impact of Catechin isolated from *Cassia fistula* on the glucose oxidation and molecular mechanisms of glucose uptake on streptozotocin-induced diabetic Wistar rats. *Phytomedicine* 2010;**17**: 28–36.
55. Tsai H-Y, Wu L-Y, Hwang LS. Effect of a proanthocyanidin-rich extract from longan flower on markers of metabolic syndrome in fructose-fed rats. *J Agric Food Chem* 2008;**56**:11018–24.
56. Bettaieb A, Vazquez-Prieto MA, Rodriguez-Lanzi C, et al. (-)-Epicatechin mitigates high-fructose-associated insulin resistance by modulating redox signalling and endoplasmic reticulum stress. *Free Rad Biol Med* 2014;**72**:247–56.
57. Dorenkott MR, Griffin LE, Goodrich KM, et al. Oligomeric cocoa procyanidins possess enhanced bioactivity compared to monomeric and polymeric cocoa procyanidins for preventing the development of obesity, insulin resistance, and impaired glucose tolerance during high-fat feeding. *J Agric Food Chem* 2014;**62**: 2216–27.
58. Grassi D, Lippi C, Necozione S, Desideri G, Ferri C. Short-term administration of dark chocolate is followed by a significant increase in insulin sensitivity and a decrease in blood pressure in healthy persons. *Am J Clin Nutr* 2005;**81**:611–4.
59. Grassi D, Necozione S, Lippi C, et al. Cocoa reduces blood pressure and insulin resistance and improves endothelium-dependent vasodilation in hypertensives. *Hypertension* 2005;**46**:398–405.

60. Grassi D, Desideri G, Necozione S, et al. Blood pressure is reduced and insulin sensitivity increased in glucose-intolerant, hypertensive subjects after 15 days of consuming high-polyphenol dark chocolate. *J Nutr Biochem* 2008;**138**:1671–6.
61. Davison K, Coates AM, Buckley JD, Howe PRC. Effect of cocoa flavanols and exercise on cardiometabolic risk factors in overweight and obese subjects. *Int J Obes (Lond)* 2008;**32**: 1289–96.
62. Grassi D, Desideri G, Ferri C. Protective effects of dark chocolate on endothelial function and diabetes. *Curr Opin Clin Nutr Metab Care* 2013;**16**:662–8.
63. Muniyappa R, Hall G, Kolodziej TL, Karne RJ, Crandon SK, Quon MJ. Cocoa consumption for 2 wk enhances insulin-mediated vasodilatation without improving blood pressure or insulin resistance in essential hypertension. *Am J Clin Nutr* 2008;**88**: 1685–96.
64. Mellor DD, Sathyapalan T, Kilpatrick ES, Beckett S, Atkin SL. High cocoa polyphenol-rich chocolate improves HDL cholesterol in type 2 diabetes patients. *Diabet Med* 2010;**27**:1318–21.
65. Shrime MG, Bauer SR, McDonald AC, Chowdhury NH, Coltart C, Ding EL. Flavonoid-rich cocoa consumption affects multiple cardiovascular risk factors in a meta-analysis of short-term studies. *J Nutr* 2011;**141**:1982–8.
66. Hooper L, Kay C, Abdelhamid A, et al. Effects of chocolate, cocoa, and flavan-3-ols on cardiovascular health: a systematic review and meta-analysis of randomized trials. *Am J Clin Nutr* 2012;**95**: 740–51.
67. Desideri G, Kwik-Uribe C, Grassi D, et al. Benefits in cognitive function, blood pressure, and insulin resistance through cocoa flavanol consumption in elderly subjects with mild cognitive impairment: the cocoa, cognition, and aging (CoCoA) study. *Hypertension* 2012;**60**:794–801.

CHAPTER

16

Dietary Proanthocyanidin Modulation of Pancreatic β Cells: Molecular Aspects

Montserrat Pinent, Noemí González-Abuín, Mayte Blay, Anna Ardévol

MoBioFood Research Group, Departament de Bioquímica i Biotecnologia, Universitat Rovira i Virgili, Tarragona, Spain

1. PROANTHOCYANIDINS: A BRIEF DESCRIPTION

This review focuses on the effects of proanthocyanidins (Figure 1), also known as condensed tannins, which are oligomeric and polymeric flavan-3-ols.[1] They are ubiquitous and are the second most abundant natural phenolic after lignin. The flavan-3-ol units are linked mainly through the $C_4 \rightarrow C_8$ bond, but the $C_4 \rightarrow C_6$ bond also exists (both called B type). The flavan-3-ol units can also be doubly linked by an additional ether bond between $C_2 \rightarrow O_7$ (A type). The proanthocyanidins can be described by their degree of polymerization and by their hydroxylation pattern. Based on the latter, there are three common flavan-3-ols: procyanidins, exclusively of (epi)catechin subunits; propelargonidins, containing (epi)afzelechin as subunits; and prodelphinidins, containing (epi) gallocatechin.[1]

Proanthocyanidins are present in the fruit, bark, leaves, and seeds of many plants and plant-derived foods such as green tea, apples, cocoa, chocolate, grapes, apricots, and cherries. They are especially abundant in fruit juices and red wine.[2] Although the average daily intake of flavanols and proanthocyanidins in the context of a Western diet was estimated to lie in the range of 50–100 mg,[3] values in the fifth quintile were reported to be in the range of 135–1050 mg.[4] However, the bioavailability of these compounds is subject to a great deal of controversy.

Monomeric flavan-3-ols are absorbed in the small intestine, where they are extensively metabolized into glucuronide conjugates and go to the systemic circulation.[5] Several studies with animals and humans have indicated that the oligomeric and polymeric flavan-3-ols are not absorbed (reviewed elsewhere[6]). Most of the oligomeric structures pass unaltered to the large intestine, where they are catabolized by the colonic microflora, yielding a diversity of phenolic acids[7,8] that are absorbed into the circulatory system and excreted in urine. Some reports have proven the presence of high levels of conjugated forms of catechin and epicatechin in the plasma of rats acutely treated with grape seed proanthocyanidins. However, other studies have shown very low quantities of nonmodified structures, including proanthocyanidin dimers in the plasma of rats[9–11]; this finding was corroborated in humans after the consumption of a grape seed extract[12] or a flavan-3-ol-rich cocoa.[13] Moreover, studies performed in several animal models have found nonmodified structures together with the conjugated forms distributed in several tissues such as adipose tissue, liver, and brain, although their quantities differed depending on the metabolic situation and the tissue studied.[9,10] As for the controversy over the presence or otherwise of oligomeric structures in biological samples, the results are highly dependent on the constraints of the methodological quantification of these compounds (e.g., equipment, optimal standards), which has only recently been improved.

Proanthocyanidins have been described as highly bioactive compounds against the development of coronary heart diseases.[14–18] They also have been shown to have antigenotoxic effects,[19] improve oxidative or inflammatory states,[20–22] and as antiproliferative agents they have been studied as cancer preventive agents.[23,24] Moreover, several studies have investigated their effects on glucose homeostasis disrupted situations; however, no clear consensus has been reached on these effects.[25]

Molecular Nutrition and Diabetes
http://dx.doi.org/10.1016/B978-0-12-801585-8.00016-6

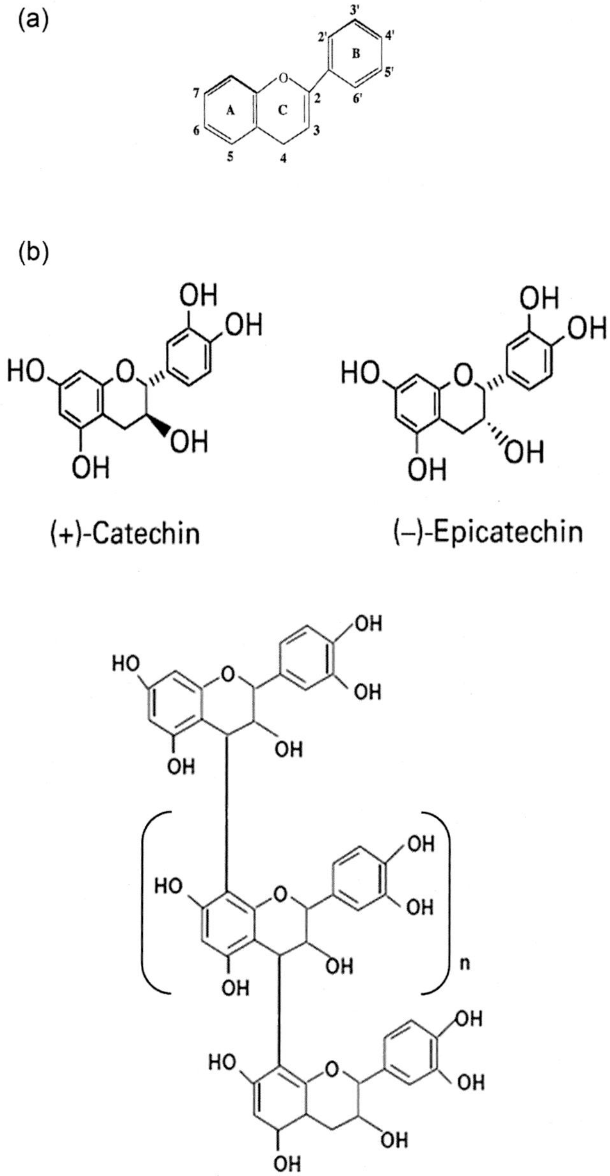

FIGURE 1 **Structure of proanthocyanidins.** (a) Flavonoid nucleus[142] and (b) structure of the monomeric flavan-3-ols (+)-catechin and (−)-epicatechin, and their oligomers called proanthocyanidins.

2. PROANTHOCYANIDINS AND TYPE 1 DIABETES

Type 1 diabetes (T1D) accounts for 10–15% of all forms of diabetes.[26] T1D is caused by an immune-mediated assault on the pancreatic β-cell function. The disease's course could be altered by modulation of the immune system and/or preserving β-cell function before or after clinical disease onset. Some of the approaches alter the immune response include anti-T- or B-cell therapy or anticytokine therapy. Several immunomodulatory agents have been found to slow the rate of β-cell deterioration during the early stages of T1D when compared with a placebo group.

Proanthocyanidin approaches to prevent or to improve T1D are in immunomodulation, anti-inflammatory, and/or antioxidative properties. Most of the studies are based on the administration of streptozotocin or alloxan diabetogenic agents to rats. Both diabetogens lead to β-cell loss, but although in alloxan-induced diabetes reactive oxygen mediates the selective necrosis of β cells, in streptozotocin-induced diabetes DNA alkylation might primarily mediate its toxic action, although the production of reactive oxygen species may also be involved.[27] The antioxidant effects of proanthocyanidins on the pancreas are therefore the main candidate mechanism for ameliorating T1D in these models. We have reviewed all of the studies that include proanthocyanidins and T1D.[25] They include treatments with doses from 10 to 250 mg proanthocyanidin extract/kg bw (body weight). Most of them show improvement of glycemia, and taken together suggest that proanthocyanidins have a short-lived insulin-mimetic effect on internal targets of the organism[28,29] and that these effects are most likely from oligomeric forms[30] and a forced acute dose. Most of these reviewed studies also found an improvement in the general situation because of improved inflammatory status[30] and in diabetic nephropathy.[31]

Regarding the effects on β-cell function, of the studies reviewed in Pinent,[25] only Al-Awwadi[28] analyzed plasma insulin, and found that proanthocyanidins had no significant effect on its levels. More recently, Fu and colleagues[32] showed evidence that dietary supplementation of the monomer epicatechin (0.5% in drinking water) for 30 weeks can prevent the onset of T1D in nondiabetic nonobese diabetic mice, a classical model of autoimmune diabetes. This protective effect is likely due to the epicatechin preventive effect of immune cell infiltration-mediated destruction of pancreatic islets, thereby preserving functional β-cell mass. Also working with nonobese diabetic mice, 0.2% epigallocatechin gallate dissolved in water, administered daily for 14 weeks, normalized the expression of autoimmune markers as PCNA, an early biomarker for exocrine glandular dysfunction, and subsequently decreased PRDX6 antioxidant enzyme levels that could further contribute to oxidative stress. These changes precede inflammatory cell infiltration.[33]

Other approaches, such as directly working on β-cell models, evaluate effects on mechanistic aspects that are common to type 2 diabetes (T2D) and are detailed elsewhere in the chapter.

3. TYPE 2 DIABETES

T2D is the most common type of diabetes.[34] Impaired glucose tolerance (IGT) and T2D result from an interaction between genetic and environmental factors.[35–37] Current evidence suggests a two-step development of T2D (Figure 2). During step 1, the plasma insulin levels are elevated, but β-cell function is clearly impaired. It is therefore important to distinguish between the plasma insulin response and β-cell health.[37] In step 2, IGT advances to T2D because of a progressive decline in β-cell function.[38]

In obesity, the pathological hyperplasia and hypertrophy of adipocytes decrease their capacity to store fat[39]; when their maximal capacity to store lipids is exceeded, circulating fatty acids increase, leading to the ectopic storage of fat in nonadipose tissues such as muscle, liver, and pancreas, impairing their functionality and leading to the beginning of insulin resistance.[40] In the muscle, there is an impaired IRS-1 phosphorylation by the insulin receptor, leading to a defect in insulin signaling and a concomitant decrease in glucose uptake and metabolism.[41] Hepatic insulin resistance leads to increased gluconeogenesis.[42] Additionally, the adipose tissue of obese insulin-resistant subjects is characterized by a progressive infiltration of macrophages, which may increase the secretion of proinflammatory adipokines, leading to a progressive development of insulin resistance.[43]

The defect in insulin secretion in T2D is currently thought to be a combination of two components:

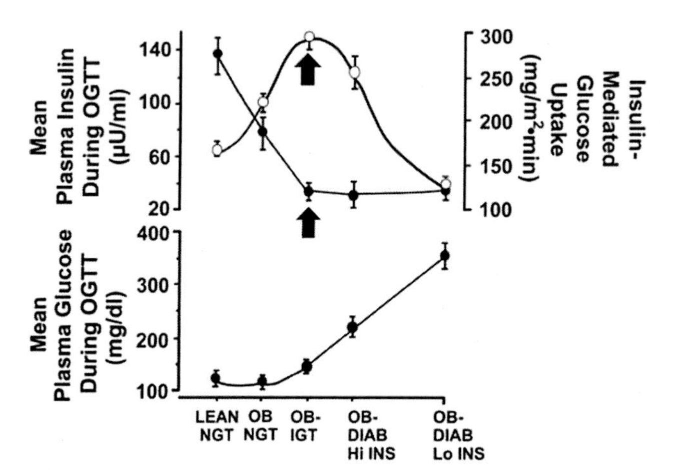

FIGURE 2 Natural history of type 2 diabetes (T2D). Progression form lean normal glucose tolerance (NGT) to obese impaired glucose tolerance (IGT) is linked to increased insulin resistance and compensatory hyperinsulinemia. The development of T2D is associated with a progressive decline in β-cell function, leading to a decrease in insulin secretion, with few changes in insulin resistance. DIAB, diabetes; Hi INS, high insulin; Lo INS, low insulin; OB, obese; OGTT, oral glucose tolerance test. *From Ref. 36.*

intrinsic β-cell dysfunction and reduced functional β-cell mass.[44] T2D patients have disrupted pulsatile insulin secretion, abnormal potentiation of nonglucose secretagogues, and a reduced maximal secretory capacity to glucose and arginine. Also, the interactions of glucagon producing α cells within the islet are altered, leading to increased plasma glucagon levels, which contribute to hyperglycemia.[45] In addition, autopsy studies demonstrated a 63% reduction in the islet mass of T2D patients compared with matched normoglycemic controls. The reduced β-cell mass is not due to the reduced formation of new islets, but is instead caused by increased rates of apoptosis.[46]

The molecular mechanisms underlying β-cell dysfunction and apoptosis in the pathophysiology of T2D remain unclear. However, increasing evidence suggests that in susceptible individuals, hyperglycemia and hyperlipidemia further deteriorate β-cell function by inducing a cascade of processes that are referred to as glucotoxicity and lipotoxicity, respectively.[47] Glucotoxicity refers to the slow and irreversible detrimental effects of chronically elevated glucose levels on β-cell function that are characterized by decreased insulin synthesis from reduced insulin gene expression. The reduction of insulin gene transcription is thought to be secondary to reductions in the transcription or activity of β-cell transcription factors such as PDX1, BETA2/NeuroD, and MafA. Several mechanisms have been proposed to explain this impairment in β-cell function and apoptosis, including endoplasmic reticulum (ER) stress, oxidative stress, mitochondrial dysfunction, and islet inflammation.[44,45] Lipotoxicity refers to the impairment in β-cell function caused by prolonged exposure to pharmacological levels of fatty acids. The mechanisms proposed to explain how fatty acids impair β-cell function include ER stress,[48] oxidative stress,[47,49] and increased ceramide formation, which induces apoptosis and downregulates insulin gene expression.[50] The term glucolipotoxicity has been used to summarize both because the alterations in intracellular lipid partitioning underlying the mechanisms of lipotoxicity are dependent upon elevated glucose levels.[51]

In addition to signal detection and/or signal production, T2D has also been related to disturbances in the incretin system.[52] The decrease in the incretin system in IGT and T2D states is primarily explained by the β-cell resistance to glucose-dependent insulinotropic polypeptide and by the decrease in GLP-1 secretion, which also demonstrates a defect in the intestinal L cells (the GLP-1–secreting cells).[36,38,53] These defects aggravate the hyperglycemic state, although there is little evidence suggesting that they play a major role in the pathogenesis of T2D.[52]

4. PROANTHOCYANIDIN EFFECTS IN GLUCOSE HOMEOSTASIS ON INSULIN RESISTANCE AND ON T2D

Insulin resistance and impaired insulin secretion, which are characteristics of T2D,[36] can be achieved in animals by using different diets, such as high-fructose, high-fat, or cafeteria (high-fat, high-sucrose) diets or by direct genetic disruption.

A review of the studies performed in high-fructose diet-fed rats shows that proanthocyanidins have a clear antihyperglycemic effect in this model.[54–56] Because this effect was more controversial in the case of insulin resistance induced by high-fat diets, we hypothesized that the improvement in glycemia was linked to the effects of proanthocyanidins repressing hepatic lipogenesis,[57] which is a key point in the development of insulin resistance in high-fructose, but not high-fat diet-treated rats.[25] This was further supported by the studies performed by Khanal et al., which reported a tendency for glycemia to decrease after a preventive treatment with cranberry or blueberry proanthocyanidin extracts in this high-fructose model, which was accompanied by an improvement in insulin resistance and insulin sensitivity.[58]

On the other hand, preventive studies in insulin resistant animals induced by a high-fat diet have also reported significant antihyperglycemic effects.[59,60] However, Zhang et al. found no significant changes in glycemia in mice fed a high-fat diet after the daily administration of 80 mg/kg bw of grape seed proanthocyanidins for 6 weeks.[61] Some more recent studies performed on mice fed a high-fat diet showed significant decreases in the glucose and insulin levels after treatments with cacao and black soybean seed coat extracts.[62–64]

Another model for diet-induced insulin resistance is achieved with a cafeteria diet. In this model, studies performed in our group have found that a corrective treatment with 25 mg/kg bw of a grape seed proanthocyanidin extract (GSPE) for 21 or 30 days was able to improve the glycemic state and the insulin resistance,[65,66] which could be linked to its action on peripheral adipose tissue and on β-cell functionality.[66–68] However, the same dose administered for a shorter period (10 days) was unable to correct the increased insulin levels induced by the diet; a higher dose (50 mg/kg bw) was also unable to correct this increase.[65] When administered in a preventive way, concomitantly with the induction of damage with the cafeteria diet, an 8-week GSPE treatment was found to avoid the diet-induced increase on insulinemia after an oral glucose tolerance test. The diet-induced insulin resistance was also prevented after 9 weeks of treatment. However, at the end of the experiment (12 weeks), this effect was lost,[69] suggesting that proanthocyanidins can prevent the diet-induced damage only for a limited period of time. Corrective treatments, therefore, seem to be more effective in situations of diet-induced insulin resistance, although more studies on both the corrective and preventive effects of proanthocyanidins need to be performed to further understand their effectiveness on this model of insulin resistance induced by a cafeteria diet.

The models described here are models of insulin resistance that do not have a disturbed glycemia. The proanthocyanidins' effects on a diabetic situation (e.g., when the basal glucose levels are increased and pancreas functionality is impaired) have also been studied in a high-fat diet-induced rats in which a later stage of T2D is induced with low-dose streptomycin (STZ), causing a mild impairment of insulin secretion.[70–72] In this model, Sundaram et al. reported that several doses of a green tea extract administered orally for 30 days were able to significantly decrease the glucose and increase insulin levels in plasma in a dose-dependent manner. This decrease was related to increased energy consumption and decreased gluconeogenesis in the liver.[70] Ding et al. studied the corrective effect of different doses of proanthocyanidins from grape seeds, and found that a dose higher than that used in most previous studies (500 mg/kg bw) was able to improve glycemia, insulin resistance, and insulin expression in pancreatic β cells, which was linked to decreased β-cell apoptosis and increased hepatic glycogenogenesis.[71,72] Although their effects on β-cell mass cannot be discarded, the antihyperglycemic effect of proanthocyanidins therefore seems to be linked to their insulin mimetic action, which is supported by some studies performed on KK-Ay mice (a model of T2D induced by its genetic background). In this model, Kurimoto et al. reported that a black soybean seed coat extract rich in proanthocyanidins improves glycemia and insulin sensitivity after 6 weeks of treatment, in an effect related to the increased glucose uptake and utilization in muscle and liver and to decreased gluconeogenesis.[73] Furthermore, 600 mg/kg bw of a *Sarcopoterium spinosum* extract administered orally for 13 weeks was found to improve glycemia both in basal- and glucose-stimulated conditions, which was also related to its insulin-like effect.[74]

Studies in db/db mice, a model of obesity induced by the genetic background, have also shown improvements in glucose homeostasis after treatment with 10 mg/kg bw of GSPE for 6 weeks[75] and 40 or 70 mg/kg (approximately) of proanthocyanidins from cacao liquor for 3 weeks.[76] Finally, an oral treatment with 20 mg/kg bw of Provinols for 8 weeks has also been reported as improving glycemia in Zucker fa/fa rats.[77] By contrast, in the same animal model, 35 mg/kg bw of GSPE administered orally did not improve glucose homeostasis.[78] This different effectiveness between experiments could be explained by differences in the stage of T2D (i.e. the impairment on β-cell function). In the last study, although the basal glucose increased in the fa/fa rats, the homeostatic model assessment-β index also increased (129.62 for lean and 746.07 for fa/fa rats, calculated by the published glucose and insulin levels), suggesting that these animals still had not reached β-cell function impairment. However, the other studies do not provide data regarding β-cell functionality, so we cannot reach a clear conclusion on this issue. More studies should be therefore performed to fully confirm the antihyperglycemic effect of proanthocyanidins in this model of obesity.

Taken together, the results show that the antihyperglycemic effect of proanthocyanidins seems to be mediated by different mechanisms, depending on the model of glucose homeostasis disruption. Some mechanisms by which proanthocyanidins might exert their antihyperglycemic effect include their insulin mimetic effect on liver and peripheral tissues, and their action on β-cell function and/or mass. Proanthocyanidins may also have complementary effects on collateral aspects linked to T2D (e.g., lipid accrual, inflammatory profile), as detailed in the next sections.

5. PROANTHOCYANIDIN EFFECTS ON INSULIN SENSING TISSUES

The liver plays a central role in maintaining glucose homeostasis because of its capacity to uptake or produce glucose, depending on the plasma glucose levels, in coordination with peripheral tissues, mainly skeletal muscle and adipose tissue, which are also involved in glucose uptake in situations of elevated circulating levels.[79] Glucose uptake modulation has been identified as an important mechanism for the antihyperglycemic effect of proanthocyanidins, as reviewed previously by Pinent et al.[25] This is in addition to other hepatic enzymes involved in glucose input/output which have been found to be potential targets for proanthocyanidins, such as glucokinase, and glycogen synthase (increased by proanthocyanidins), or glucose-6-phosphatase, phosphoenolpyruvate carboxykinase, and fructose-1,6-biphosphatase (decreased by proanthocyanidins).[61,70,73,80–82] Several in vitro studies have reported an increased glucose uptake in hepatocytes,[61,74,83,84] adipocytes,[29,74] and myotubes,[29,73,74] with adenosine monophosphate–activated protein kinase (AMPK) being a common target in all of these insulin-sensitive cell lines.[65,73,84] An increase in AMPK phosphorylation in hepatic, adipose, and muscle tissues was also found in vivo in high-fat diet-induced insulin-resistant mice[64] and T2D and STZ-induced diabetic models.[73,82]

The proanthocyanidin-induced decrease of hepatic lipogenesis has also been identified as a mechanism for the proanthocyanidins' antihyperglycemic effect in the liver. Quesada et al. reported that chronic treatment with grape seed proanthocyanidins was able to limit lipogenesis and very low-density lipoprotein assembly in a model of cafeteria diet.[85] Similar results were found with chronic treatment with Flavangenol® in TSOD obese diabetic mice and grape seed proanthocyanidins in cafeteria- or high-fructose–fed rats, respectively,[55,86,87] and by other in vitro studies.[11] Peroxisome proliferator-activated receptor (PPAR)-α, a fatty acid oxidative enzyme, was found to be a common target for proanthocyanidins, and their upregulation was closely related to a decrease in hepatic lipogenesis.[75,86] Moreover, other enzymes and transcription factors involved in hepatic lipid homeostasis were suggested as possible targets for proanthocyanidins, including ACO, CPT-1, and ABCA1 (which were upregulated, thus increasing fatty acid oxidation and decreasing cholesterol formation), and FAS and SREBP1/2 (which were downregulated, thus decreasing lipid synthesis and TAG deposition).[75,86,88]

Moreover, the glucose transporter GLUT-4 was found to be upregulated by proanthocyanidins in adipose tissue and muscle[29,56,64,73]; because AMPK activation induces GLUT4 translocation to the plasma membrane, these results support the involvement of AMPK activation on proanthocyanidin-induced glucose uptake. Montagut et al. and Lee et al. also found an increase in AKT phosphorylation in adipocytes, which was related to the increased GLUT4 translocation,[68,89] suggesting an enhancement of the insulin-signaling pathway as an additional mechanism to improve glucose uptake. These effects are complementary to the effects on lipid pathways. Proanthocyanidins were also reported to decrease adipose tissue depots in several models of insulin resistance,[63,64,67,90,91] which could be related to the decrease in PPAR-γ, a key controller of adipogenesis,[9,92] and to the increase in enzymes involved in β-oxidation, such as ACADVL, CPT1β, and PPARα.[90] Other enzymes involved in lipogenesis/lipolysis in adipose tissue, such as DGAT2 or HSL, have also been identified as potential targets for proanthocyanidins, although the results are controversial.[9,90,92] More studies are necessary to understand their actual role on the hypolipidemic effect of procyandins and its relationship with their antihyperglycemic effect.

Finally, obesity has also been found to induce oxidative stress in the mitochondria because of the

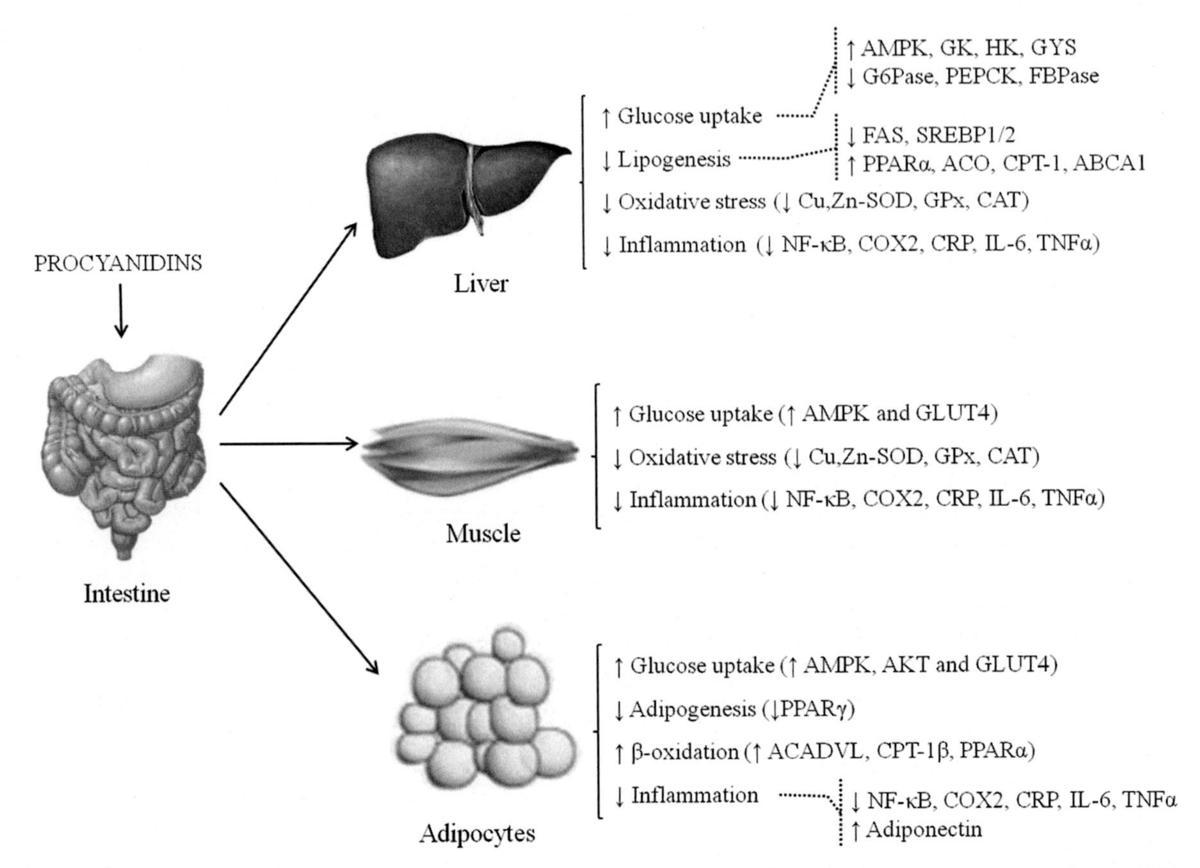

FIGURE 3 **Proanthocyanidin effects in insulin-sensing tissues.** AMPK, adenosine monophosphate–activated protein kinase; CAT, catalase; CRP, C-reactive protein; Cu, copper; GPx, glutathione peroxidase; IL-6, interleukin-6; NF-κB, nuclear factor-κB; PPAR-α, peroxisome proliferator-activated receptor-α; TNF-α, tumor necrosis factor-α; Zn-SOD, superoxide dismutase.

increased fatty acid and glucose metabolism, which in turn activates several inflammatory signaling pathways.[93,94] Several studies performed on insulin-resistant animals have shown that proanthocyanidins are able to decrease the activity of some enzymes involved in the generation of toxic species, such as copper, superoxide dismutase, glutathione peroxidase, and catalase, which was related to an improvement in the oxidative stress status induced by obesity in the liver and muscle.[20,55,75,95,96] In addition, several cytokines that are increased in the obesity-induced proinflammatory state, such as nuclear factor-κB, cyclooxygenase 2, C-reactive protein, interleukin-6, and tumor necrosis factor-α, have been found to be potential targets for proanthocyanidins, and their gene expression was shown to be reduced in liver, adipose tissue, and muscle.[55,60,63,75,97,98] Moreover, the anti-inflammatory cytokine, adiponectin, has been found to be increased in adipose tissue after treatment with grape seed proanthocyanidins,[60,98] further suggesting an improvement in the inflammatory state in insulin-resistant situations. Considering the key role of oxidative stress/inflammation in the development of insulin resistance and T2D, the anti-inflammatory and antioxidant effects exerted by proanthocyanidins could therefore be involved in their antihyperglycemic action.

All these effects in insulin-sensitive tissues are summarized in Figure 3.

6. PROANTHOCYANIDIN EFFECTS ON β-CELL FUNCTIONALITY: CONTROL OF INSULIN PRODUCTION

The location of the pancreas following enteric absorption makes it a direct target for flavonoids, as reviewed by our group,[99] and published data indicate that proanthocyanidins have direct effects on insulin secretion and production and on β-cell apoptosis and proliferation (summarized in Figure 4).

Regarding their effects on insulin secretion and production, under obesogenic situations such as a cafeteria diet for 13 weeks, a corrective treatment for 30 days using a dose of 25 mg GSPE/kg bw reduced insulinemia, counteracted the cafeteria diet-induced increase in insulin levels,[65] which was related to a decrease in insulin production, gene expression and content, and PDX1

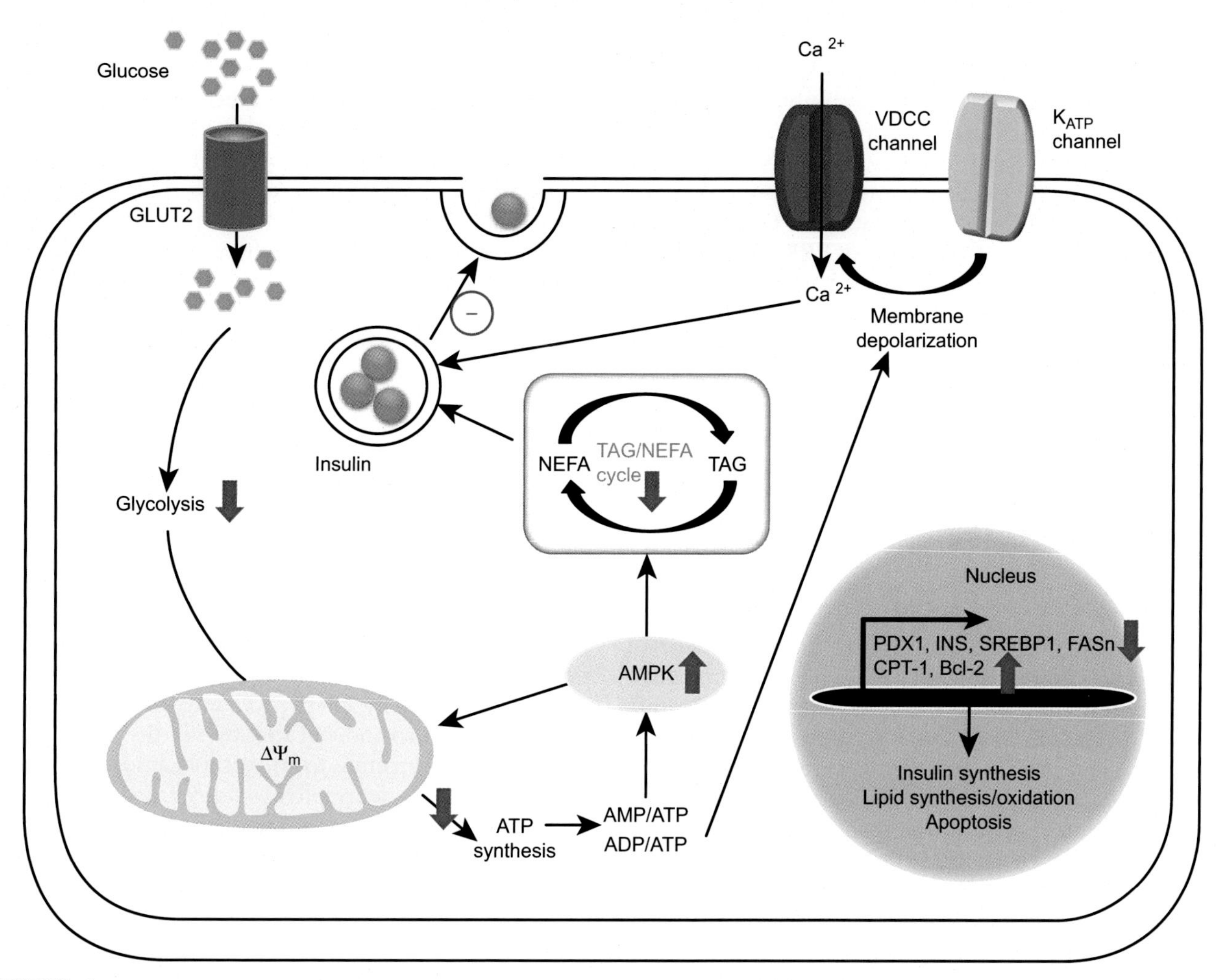

FIGURE 4 Proanthocyanidin effects on pancreatic β cells under obesogenic situations. ADP, adenine diphosphate; AMP, adenosine monophosphate; ATP, adenosine triphosphate; K_{ATP}, ATP-sensitive potassium channel; VDCC, voltage-dependent calcium channel. *From Ref. 143.*

gene expression.[67] A similar situation at the pancreatic level was obtained in Zucker fatty rats treated with 35 mg GSPE/kg bw for 10 weeks, which showed a downregulation of insulin and PDX-1 gene expression. However, these local effects on β cells were not enough to improve plasma hyperinsulinemia in these genetically obese Zucker animals.[78] A proteomic study performed on this model supported the proanthocyanidins effect, limiting insulin synthesis at both the insulin and the carboxypeptidase protein levels[78]; the latter is an enzyme implicated in the conversion of proinsulin to insulin.[100] Proanthocyanidins also downregulated two of the enzymes involved in the glycolytic pathway—pyruvate kinase isozymes M1/M2 and glyceraldehyde-3-phosphate dehydrogenase, and a member of the adenosine triphosphate synthase protein complex, ATP5B.[78]

In disrupted homeostatic situations induced by an excess of lipid substrates (cafeteria diet), proanthocyanidins may also counteract the diet-induced effects by decreasing the lipid accumulation in β cells by upregulating CPT1a, which increases β-oxidation, and downregulating lipid synthesis-related genes, such as FASn and SREBF1.[66,67] Moreover, proanthocyanidin treatment counteracted the cafeteria diet-induced decrease in AMPK protein levels; in the pancreas, AMPK was also defined as a GSPE target.[67]

Several in vitro studies have been performed to further elucidate the effects of proanthocyanidins on insulin production and secretion. Isolated islets from healthy rats treated with GSPE, both acutely and chronically, showed a decreased ability to respond to the glucose stimulus, which was partially explained by the downregulation of insulin and PDX1 gene expression. A similar effect was found when working with the INS-1 β-cell line that was chronically treated with 25 mg/L of GSPE, which showed an effect on both insulin synthesis and insulin secretion. In this last assay, proanthocyanidins impaired glucose-induced mitochondrial hyperpolarization, decreasing adenosine triphosphate synthesis and thus altering cellular membrane potential and limiting the ability of the cells to secrete insulin in response to glucose entry.[101] However, acute treatments with similar and higher doses of *Lotus leaf* extract (25–150 mg/L) showed increased insulin secretion in HIT-M5 β cells and in human isolated islets, which was related to the modulation of the Ca^{2+}-activated protein kinase C–regulated ERK1/2 signaling pathway.[102] Smirin et al. also reported an increased insulin secretion in RIN insulinoma cells and an increased preproinsulin gene expression after an acute treatment with100 mg/L of a *Sarcopoterium spinosum* extract.[74] There is therefore some controversy over the in vitro effects of proanthocyanidins, although these studies analyzed the effects of proanthocyanidins in a nondisturbed situation. In this case of a healthy status, in vivo moderate doses of grape seed proanthocyanidins, such as 15 mg/kg bw for 21 days, increased the insulin plasma levels in healthy rats, but a higher dose of 25 mg/kg bw for 36 days decreased these levels.[101] Similar results were obtained after an acute dose of an *Eriobotrya japonica* water extract, which induced insulin secretion in vitro and decreased insulinemia in vivo. [103] The effects of grape seed proanthocyanidins could be explained by their tendency to reduce insulin biosynthesis through the downregulation of the transcription factor PDX1 and the insulin gene expression, and because of their ability to decrease insulin removal, because the hepatic gene expression of the insulin-degrading enzyme, the primary player in insulin clearance, was found to be decreased at a dose of 15 mg/kg bw.[101]

Another mechanism for proanthocyanidins to modulate insulin secretion may involve microRNA (miRNA) changes. miRNA plays a key role in the regulation of metabolic processes in diabetes,[104] and different proanthocyanidin extracts have been shown to modulate miRNA expression.[105] miR-1249, miR-483, and miR-30c-1* have been reported as downregulated and mir-3544 as upregulated by a 45-day treatment with 25 mg GSPE/kg bw in healthy rats.[106] Despite being briefly described, some of these miRNAs were predicted to control the genes involved in processes related to ion transport and response to stimuli, such as hormones and organic substances.

Finally, the endocrine functionality of pancreatic tissue also depends on the quantity of the β cells, and proanthocyanidins have been found to be good candidates to modulate functionality through their effects on proliferation[92] and apoptosis[24] and because of their antioxidant properties.[107] Additionally, chronic treatments with 10–100 mg/L of GSPE inhibits cell proliferation and increases apoptosis in pancreatic cancer MIA PaCa-2 cells, which is primarily mediated by the downregulation of the antiapoptotic protein Bcl-2 and the depolarization of the mitochondrial membrane.[78] Grape seed proanthocyanidins have been found to enhance the proapoptotic effect of a high-glucose treatment in the INS-1 β-cell line, but they also showed clear antiproliferative effects under high-glucose, high-insulin, and high-palmitate conditions. This study showed that grape seed proanthocyanidins were able to modulate apoptosis and proliferation of β cells under altered, but not basal, conditions.[108] Acute treatments (1 h) with 100–1000 mg/L of *Sarcopoterium spinosum* extract produced an increased proliferation of RINm insulinoma cells, and an increase in cell death occurred at higher doses (10 g/L).[74] In vivo studies in models of insulin resistance support an effect of proanthocyanidins at this level, although the results have not been fully elucidated. GSPE was found to modulate pro- and antiapoptotic markers in the pancreas of female rats fed a cafeteria diet, with different effects depending on the dose of GSPE and the time of treatment.[108] A parallel study in cafeteria-fed male rats showed increased levels of the proapoptotic marker Bax, suggesting a different sensitivity to antiapoptotic properties depending on the animal's gender.[66] Moreover, the previously cited proteomic study in Zucker fa/fa females also showed that two of the main biological processes sensitive to proanthocyanidins effects in pancreatic islets are apoptosis and cell death.[78]

All of the previous studies referred to situations in which the β-cell function/amount was not altered or increased. In an advanced situation, in established T2D, the hyperstimulation of β-cell functionality drives to a loss of β cells.[109] In a model of late-stage T2D (i.e., induced by a high-fat diet plus low STZ), 16 weeks of treatment with 500 mg GSPE/kg bw was found to increase insulin expression and functionality in β cells. This was linked to an amelioration of ER stress and reduction of apoptosis in the pancreas.[71] Additionally, in T1D, in which the loss of β cells is due to an autoimmune assault,[110] Gandhi et al. found that STZ diabetic rats had more β cells after treatment with 400 mg/kg bw of a *Solanumtorvum* extract, which could explain the higher levels of insulin found in the plasma.[81] Finally, Chen et al. found that in genetically obese db/db mice with diabetes, a 4-week treatment with 200 mg/kg bw of two different extracts enriched in proanthocyanidins increased plasma and pancreatic insulin levels, and improved glucose tolerance and insulin sensitivity.[111] However, in this case, it was not stated whether this was linked to an increased number of pancreatic islets. This study supports the theory that when the pancreas is damaged, proanthocyanidins act by increasing insulinemia.

Proanthocyanidins can thus have different effects on β-cell proliferation and apoptosis, depending on the dosage and the duration of the treatment, and on the specific origin of the cell loss. More studies would help further define the exact role of these mechanisms in proanthocyanidin-induced glucose homeostasis modulation.

7. PROANTHOCYANIDIN EFFECTS ON THE INCRETIN SYSTEM

The intestine is the first tissue involved in sensing food components, and because the mechanisms for hormone secretion are very similar in different endocrine cells,[112] enteroendocrine cells have been studied as a possible target for proanthocyanidins. However, few studies suggest that proanthocyanidins might modulate GLP-1, the incretin hormone that is mainly released by the L cells of the distal intestine (ileum and colon).[113] GLP-1 enhances the responsiveness of the β cells to glucose[114] and improves β-cell mass by enhancing proliferation, inhibiting apoptosis, and increasing β-cell differentiation (reviewed previously[115]), which makes it a key hormone for glucose homeostasis regulation.

Torronen et al. reported that a berry puree enriched in proanthocyanidins was able to increase the active GLP-1 levels in healthy humans after simultaneous consumption with sucrose. This increase was accompanied by decreased glucose levels.[116] Cinnamtannnin A2, a tetrameric proanthocyanidin, has also been reported as increasing active GLP-1 secretion and insulin in fasted healthy mice.[117] A chronic treatment with lower doses of GSPE in healthy rats increased the plasma insulin/glucose ratio after an oral glucose tolerance test, but it did not do so when glucose was administered intraperitoneally, suggesting an incretin effect.[118] In fact, an acute treatment of GSPE (1 g/kg bw for 1 h) simultaneous with an oral glucose gavage showed increased total and active GLP-1 levels in healthy rats; this effect was accompanied by an enhancement of the insulin/glucose ratio.[119]

The mechanisms for nutrient-induced GLP-1 secretion by intestinal L cells involve cellular membrane depolarization that leads to calcium entry and calcium-dependent GLP-1 release.[112] Proanthocyanidins have been reported to modulate cellular membrane potential in aortic endothelial RAEC cells[120] and Jurkat cells (T lymphoblasts),[121] and in both liposomes[122] and pancreatic β cells, as previously mentioned.[101] Studies performed on the enteroendocrine cell line STC-1 showed that GSPE also alters the cell membrane potential in a manner that differs from exposures in pancreatic β cells. It was also shown that both alone and in the presence of different types of nutrients (glucose, fatty acids, or amino acids), high doses (5–50 mg/L) of GSPE decreased GLP-1 secretion, whereas low doses (0.05–0.5 mg/L) did not provoke any change, despite the higher membrane depolarization.[123] The direct stimulation of GLP-1 secretion by proanthocyanidins has therefore not been clearly proven.

Another mechanism to modulate active GLP-1 levels is the modulation of dipeptidyl-peptidase 4 (DPP4), the main enzyme that degrades active GLP-1. DPP4 is a serine exopeptidase that cleaves X-proline or X-alanine dipeptides from the N-terminus of polypeptides,[124] and its wide distribution among almost all tissues in the body[125] causes a rapid cleavage of the active GLP-1 almost immediately after its secretion, with a half-life of 1–2 min.[126] In this regard, inhibiting DPP4 has been suggested as a possible mechanism for proanthocyanidins to improve active GLP-1 levels. GSPE and proanthocyanidins from blueberry-blackberry fermented beverages have been found to inhibit DPP4 activity in vitro.[118,127] The effects of GSPE on DPP4 activity were also assessed in several in vivo experimental models; plasma DPP4 activity was unaffected in all cases.[118,128] These in vivo experiments cover a wide range of conditions, including different GSPE doses (acute high doses to chronic lower doses) and different times between GSPE administration and blood sample recovery to assay the DPP4 activity (1, 5, 13, or 17 h after GSPE administration), all of which suggests that although the direct inhibition of plasma DPP4 activity by GSPE cannot be completely ruled out, it is not a primary mechanism that would explain the effects of proanthocyanidin on glucose homeostasis.[118] However, in the acute study mentioned previously, GSPE did decrease the intestinal DPP4 activity; as a result, inhibition of the DPP4 present in the endothelium of the capillaries next to the GLP-1-secreting cells is postulated as a mechanism that is at least partially responsible for the increased glucose-induced GLP-1 secretion.[119] Intestinal DPP4 activity also decreased after a chronic treatment with 25 mg/kg bw of GSPE in healthy animals and in a model of insulin resistance induced by a cafeteria diet.[118] Moreover, Tebib et al. previously reported a significant decrease in intestinal DPP4 activity in healthy rats after chronic treatment with grape seed tannins.[129] Conversely, intestinal DPP4 activity was unaffected by GSPE in Zucker fa/fa rats[118] and increased after a preventive treatment in rats with a diet-induced insulin resistance, thus counteracting the effect provoked by the cafeteria diet.[128] Interestingly, an in vitro approach that mimics the intestinal environment, a coculture of CaCo-2 (enterocytes) and human umbilical vein endothelial cells (endothelial), showed that the GSPE compounds absorbed by the intestinal barrier (i.e., catechin, gallic acid, and B2 dimer) do inhibit endothelial DPP4 activity, suggesting that the effects of proanthocyanidins on active GLP-1 levels might be due to the inhibition of the DPP4 present in the endothelium of the capillaries next to the GLP-1–secreting cells in the intestine.[119] Taken together, these data suggest that the effect of proanthocyanidins on inner intestinal DPP4 activity might be a mechanism that could improve active GLP-1 levels. However, their effects seem to depend on the treatment and the degree of obesity achieved.

Finally, GSPE has been shown to modulate GLP-1 production in a preventive treatment in a model of insulin resistance induced by a cafeteria diet. After 12 weeks, the cafeteria diet-fed animals had lower active GLP-1 levels in plasma, which was linked to a reduction in its production in the colon—an effect that was partially prevented by the treatment with GSPE.[128] These results are similar to those presented by Dao et al., who found that resveratrol, a polyphenolic compound present in red grapes, was able to increase portal active GLP-1 levels due to an increase in its colonic production.[130] Moreover, whether these effects were linked to a direct modulation of GLP-1 gene expression or to a general effect on GLP-1 producing cells was assessed. The results showed that the cafeteria diet led to a decrease in the amount of enteroendocrine cells in colon and that GSPE was able to prevent this.[128] A preliminary hypothesis is that this effect could be due to the modulation of proliferation and/or apoptosis—an effect that has been previously described by our group in pancreatic β cells.[66,99,108] A sensitivity to GSPE was also found in the hypothalamus,[128] a tissue that expresses GLP-1 and its receptor,[131,132] indicating that the hypothalamus is also a potential target for proanthocyanidins.

There is therefore evidence that proanthocyanidins can modulate GLP-1 levels and that this could be achieved through different mechanisms. Whether proanthocyanidins modulate β-cell function and or mass indirectly through GLP-1 modulation remains to be clarified.

8. HUMAN STUDIES

Several clinical studies in humans have been performed to assess the potential antihyperglycemic effect of proanthocyanidins; as we have previously reviewed, they seem to be effective only in situations of glucose homeostatic disruption.[25] In this sense, several clinical studies in T2D subjects have reported that the daily ingestion of 100 mg of Pycnogenol® or 150 mL of muscadine grape products improves glycemic control,[133–135] an effect that is supported by a study in which the daily consumption of 27 g of chocolate (containing 850 mg of flavan-3-ols) for 1 year was found to improve insulinemia and both insulin resistance and sensitivity in T2D women.[136] Moreover, Hsu et al. reported that a daily intake of 1.34 g of green tea proanthocyanidins was able to improve insulinemia and insulin resistance in T2D subjects.[137] However, there are a few studies of T2D subjects who did not show significant changes in their glucose and/or insulin levels after dietary supplementation with a flavanol-rich cacao drink (75, 371, or 960 mg of flavanols/day) or a polyphenol-rich chocolate (16.6 mg of flavanols/day).[138,139] It thus seems that large quantities of flavanols are needed to improve glycemia in humans.

On the other hand, some case-cohort studies have reported that flavan-3-ols consumption is inversely associated with the risk of developing T2D.[140,141] Moreover, Zamora-Ros et al. reported that monomeric flavan-3-ols and proanthocyanidins with a low polymerization degree have a more effective protective role against T2D.[140]

9. CONCLUSIONS

Proanthocyanidins have a clear antihyperglycemic effect in situations of insulin resistance and T2D, which suggests a potential of these compounds as preventive agents against this pathology. In situations of insulin resistance, proanthocyanidins have been found to prevent/counteract the metabolic derangements caused by increased circulating glucose and fatty acids. They act in the gastrointestinal tract by modulating the incretin levels that modify β-cell functionality. A direct effect on pancreas functionality and β-cell mass from the modulation of insulin synthesis and the apoptotic/proliferation rates has also been shown. Furthermore, they act in insulin-sensitive tissues, mainly through their insulinomimetic action, which improves glucose uptake. However, also a decrease in lipid accumulation in peripheral tissues and in the pancreas seems to be a potential mechanism to decrease insulin resistance because of the lower lipotoxicity, which avoids lipid-induced oxidative stress and consequent inflammation. Nevertheless, these mechanisms cannot be confirmed in humans because of the limited number of recent studies; then, more work needs to be done, to clearly identify the amount and the optimal suitability of these compounds to be used as preventive agents in humans.

Acknowledgments

This work has been supported by a grant (AGL2011-23879) from the Spanish government. Montserrat Pinent is a Serra Húnter fellow.

References

1. Gu L, Kelm M, Hammerstone J, et al. Concentrations of proanthocyanidins in common foods and estimations of normal consumption. *J Nutr* 2004;**134**:613–7.
2. Aron PM, Kennedy JA. Flavan-3-ols: nature, occurrence and biological activity. *Mol Nutr Food Res* 2008;**52**:79–104.
3. Schroeter H, Heiss C, Spencer JPE, Keen CL, Lupton JR, Schmitz HH. Recommending flavanols and procyanidins for cardiovascular health: current knowledge and future needs. *Mol Aspects Med* 2010;**31**:546–57.

4. Mink PJ, Scrafford CG, Barraj LM, et al. Flavonoid intake and cardiovascular disease mortality: a prospective study in postmenopausal women. *Am J Clin Nutr* 2007;**85**:895–909.
5. Monagas M, Urpi-Sarda M, Sánchez-Patán F, et al. Insights into the metabolism and microbial biotransformation of dietary flavan-3-ols and the bioactivity of their metabolites. *Food Funct* 2010;**1**:233–53.
6. Crozier A, Del Rio D, Clifford MN. Bioavailability of dietary flavonoids and phenolic compounds. *Mol Aspects Med* 2010;**31**:446–67.
7. Selma MV, Espín JC, Tomás-Barberán FA. Interaction between phenolics and gut microbiota: role in human health. *J Agric Food Chem* 2009;**57**:6485–501.
8. Aura A-M. Microbial metabolism of dietary phenolic compounds in the colon. *Phytochem Rev* 2008;**7**:407–29.
9. Ardévol A, Motilva MJ, Serra A, Blay M, Pinent M. Procyanidins target mesenteric adipose tissue in Wistar lean rats and subcutaneous adipose tissue in Zucker obese rat. *Food Chem* 2013;**141**: 160–6.
10. Arola-Arnal A, Oms-Oliu G, Crescenti A, et al. Distribution of grape seed flavanols and their metabolites in pregnant rats and their fetuses. *Mol Nutr Food Res* 2013;**57**:1741–52.
11. Guerrero L, Margalef M, Pons Z, et al. Serum metabolites of proanthocyanidin-administered rats decrease lipid synthesis in HepG2 cells. *J Nutr Biochem* 2013;**24**:2092–9.
12. Sano A, Tokutake S, Tobe K, Kubota Y, Kikuchi MYJ, Sano A, et al. Procyanidin B1 is detected in human serum after intake of proanthocyanidin-rich grape seed extract. *Biosci Biotechnol Biochem* 2003;**67**:1140–3.
13. Holt RR, Lazarus SA, Sullards MC, et al. Procyanidin dimer B2 [epicatechin-(4β-8)-epicatechin] in human plasma after the consumption of a flavanol-rich cocoa. *Am J Clin Nutr* 2002;**76**: 798–804.
14. Bladé C, Arola L, Salvadó M-J. Hypolipidemic effects of proanthocyanidins and their underlying biochemical and molecular mechanisms. *Mol Nutr Food Res* 2010;**54**:37–59.
15. Baselga-Escudero L, Blade C, Ribas-Latre A, et al. Chronic supplementation of proanthocyanidins reduces postprandial lipemia and liver miR-33a and miR-122 levels in a dose-dependent manner in healthy rats. *J Nutr Biochem* 2014;**25**:151–6.
16. Quiñones M, Guerrero L, Suarez M, et al. Low-molecular procyanidin rich grape seed extract exerts antihypertensive effect in males spontaneously hypertensive rats. *Food Res Int* 2013;**51**: 587–95.
17. Rodriguez-Mateos A, Heiss C, Borges G, Crozier A. Berry (Poly) phenols and cardiovascular health. *J Agric Food Chem* 2013. published online October. DOI:10.1021/jf403757g.
18. Karthikeyan K, Sarala Bai BR, Niranjali Devaraj S. Grape seed proanthocyanidins ameliorates isoproterenol-induced myocardial injury in rats by stabilizing mitochondrial and lysosomal enzymes: an in vivo study. *Life Sci* 2007;**81**:1615–21.
19. Llópiz N, Puiggròs F, Céspedes E, et al. Antigenotoxic effect of grape seed procyanidin extract in Fao cells submitted to oxidative stress. *J Agric Food Chem* 2004;**52**:1083.
20. Fernández-Iglesias A, Pajuelo D, Quesada H, et al. Grape seed proanthocyanidin extract improves the hepatic glutathione metabolism in obese Zucker rats. *Mol Nutr Food Res* 2014;**58**: 727–37.
21. Pallarès V, Fernández-Iglesias A, Cedó L, et al. Grape seed procyanidin extract reduces the endotoxic effects induced by lipopolysaccharide in rats. *Free Radic Biol Med* 2013;**60**:107–14.
22. Martinez-Micaelo N, González-Abuín N, Ardèvol A, Pinent M, Blay M. Procyanidins and inflammation: molecular targets and health implications. *Biofactors* 2012;**38**:257–65.
23. Faria A, Calhau C, deFreitas V, Mateus N, de Freitas V. Procyanidins as antioxidants and tumor cell growth modulators. *J Agric Food Chem* 2006;**54**:2392–7.
24. Mantena SK, Baliga MS, Katiyar SK. Grape seed proanthocyanidins induce apoptosis and inhibit metastasis of highly metastatic breast carcinoma cells. *Carcinogenesis* 2006;**27**:1682–91.
25. Pinent M, Cedó L, Montagut G, Blay M, Ardévol A. Procyanidins improve some disrupted glucose homoeostatic situations: an analysis of doses and treatments according to different animal models. *Crit Rev Food Sci Nutr* 2012;**52**:569–84.
26. Gallagher MP, Goland RS, Greenbaum CJ. Making progress: preserving beta cells in type 1 diabetes. *Ann NY Acad Sci* 2011;**1243**: 119–34.
27. Lenzen S. The mechanisms of alloxan- and streptozotocin-induced diabetes. *Diabetologia* 2008;**51**:216–26.
28. Al-Awwadi N, Azay J, Poucheret P, et al. Antidiabetic activity of red wine polyphenolic extract, ethanol, or both in streptozotocin-treated rats. *J Agric Food Chem* 2004;**52**:1008–16.
29. Pinent M, Blay M, Blade MC, et al. Grape seed-derived procyanidins have an antihyperglycemic effect in streptozotocin-induced diabetic rats and insulinomimetic activity in insulin-sensitive cell lines. *Endocrinology* 2004;**145**:4985–90.
30. Lee Y, Kim Y, Cho E, Yokozawa T. Ameliorative effects of proanthocyanidin on oxidative stress and inflammation in streptozotocin-induced diabetic rats. *J Agric Food Chem* 2007. http://dx.doi.org/10.1021/jf071523u.
31. Li BY, Cheng M, Gao HQ, et al. Back-regulation of six oxidative stress proteins with grape seed proanthocyanidin extracts in rat diabetic nephropathy. *J Cell Biochem* 2008;**104**:668–79.
32. Fu Z, Yuskavage J, Liu D. Dietary flavonol epicatechin prevents the onset of type 1 diabetes in nonobese diabetic mice. *J Agric Food Chem* 2013;**61**:4303–9.
33. Ohno S, Yu H, Dickinson D, et al. Epigallocatechin-3-gallate modulates antioxidant and DNA repair-related proteins in exocrine glands of a primary Sjogren's syndrome mouse model prior to disease onset. *Autoimmunity* 2012;**45**:540–6.
34. Nolan CJ, Damm P, Prentki M. Type 2 diabetes across generations: from pathophysiology to prevention and management. *Lancet* 2011;**378**:169–81.
35. Bi Y, Wang T, Xu M, et al. Advanced research on risk factors of type 2 diabetes. *Diabetes Metab Res Rev* 2012;**28**:32–9.
36. DeFronzo RA. From the triumvirate to the ominous octet: a new paradigm for the treatment of type 2 diabetes mellitus. *Diabetes* 2009;**58**:773–95.
37. DeFronzo RA, Abdul-Ghani MA. Preservation of β-cell function: the key to diabetes prevention. *J Clin Endocrinol Metab* 2011;**96**: 2354–66.
38. Abdul-Ghani MA. Type 2 diabetes and the evolving paradigm in glucose regulation. *Am J Manag Care* 2013;**19**:s43–50.
39. Hansen E, Hajri T, Abumrad NN. Is all fat the same? the role of fat in the pathogenesis of the metabolic syndrome and type 2 diabetes mellitus. *Surgery* 2006;**139**:711–6.
40. Lebovitz HE, Banerji MA. Point: visceral adiposity is causally related to insulin resistance. *Diabetes Care* 2005;**28**:2322–5.
41. Abdul-Ghani MA, DeFronzo RA. Pathogenesis of insulin resistance in skeletal muscle. *J Biomed Biotechnol* 2010;**2010**:476279.
42. DeFronzo RA. Dysfunctional fat cells, lipotoxicity and type 2 diabetes. *Int J Clin Pract Suppl* 2004;**58**:9–21.
43. Goossens GH. The role of adipose tissue dysfunction in the pathogenesis of obesity-related insulin resistance. *Physiol Behav* 2008; **94**:206–18.
44. Bonner-Weir S, O'Brien TD. Islets in type 2 Diabetes: in Honor of Dr Robert C. Turner. *Diabetes* 2008;**57**:2899–904.
45. Salehi M, Aulinger BA, D'Alessio DA. Targeting β-cell mass in type 2 diabetes: promise and limitations of new drugs based on incretins. *Endocr Rev* 2008;**29**:367–79.
46. Ritzel RA, Butler AE, Rizza RA, Veldhuis JD, Butler PC. Relationship between β-cell mass and fasting blood glucose concentration in humans. *Diabetes Care* 2006;**29**:717–8.

47. Van Raalte DH, Diamant M. Glucolipotoxicity and beta cells in type 2 diabetes mellitus: target for durable therapy? *Diabetes Res Clin Pract* 2011;**93**(Suppl. 1):S37–46.
48. Cnop M, Igoillo-Esteve M, Cunha D, Ladrière L, Eizirik D. An update on lipotoxic endoplasmic reticulum stress in pancreatic β-cells. *Biochem Soc Trans* 2008;**36**:909–15.
49. Kim J-W, Yoon K-H. Glucolipotoxicity in pancreatic β-cells. *Diabetes Metab J* 2011;**35**:444–50.
50. Cnop M. Fatty acids and glucolipotoxicity in the pathogenesis of type 2 diabetes. *Biochem Soc Trans* 2008;**36**:348–52.
51. Poitout V, Robertson RP. Glucolipotoxicity: fuel excess and β-cell dysfunction. *Endocr Rev* 2008;**29**:351–66.
52. Meier JJ, Nauck MA. Is the diminished incretin effect in type 2 diabetes just an epi-phenomenon of impaired beta-cell function? *Diabetes* 2010;**59**:1117–25.
53. Holst JJ, Gromada J. Role of incretin hormones in the regulation of insulin secretion in diabetic and nondiabetic humans. *Am J Physiol Endocrinol Metab* 2004;**287**:E199–206.
54. Al-Awwadi NA, Araiz C, Bornet A, et al. Extracts enriched in different polyphenolic families normalize increased cardiac NADPH oxidase expression while having differential effects on insulin resistance, hypertension, and cardiac hypertrophy in high-fructose-fed rats. *J Agric Food Chem* 2005;**53**:151–7.
55. Yokozawa T, Kim HJ, Cho EJ. Gravinol ameliorates high-fructose-induced metabolic syndrome through regulation of lipid metabolism and proinflammatory state in rats. *J Agric Food Chem* 2008;**56**:5026–32.
56. Tsai H-YY, Wu L-YY, Hwang LS. Effect of a proanthocyanidin-rich extract from Longan flower on markers of metabolic syndrome in fructose-fed rats. *J Agric Food Chem* 2008;**56**:11018–24.
57. Baiges I, Palmfeldt J, Bladé C, Gregersen N, Arola L. Lipogenesis is decreased by grape seed proanthocyanidins according to liver proteomics of rats fed a high fat diet. *Mol Cell Proteomics* 2010;**9**: 1499–513.
58. Khanal RC, Rogers TJ, Wilkes SE, Howard LR, Prior RL. Effects of dietary consumption of cranberry powder on metabolic parameters in growing rats fed high fructose diets. *Food Funct* 2010;**1**: 116–23.
59. Décordé K, Teissèdre P-L, Sutra T, Ventura E, Cristol J-P, Rouanet J-M. Chardonnay grape seed procyanidin extract supplementation prevents high-fat diet-induced obesity in hamsters by improving adipokine imbalance and oxidative stress markers. *Mol Nutr Food Res* 2009;**53**:659–66.
60. Terra X, Montagut G, Bustos M, et al. Grape-seed procyanidins prevent low-grade inflammation by modulating cytokine expression in rats fed a high-fat diet. *J Nutr Biochem* 2009;**20**: 210–8.
61. Zhang HJ, Ji BP, Chen G, et al. A combination of grape seed-derived procyanidins and gypenosides alleviates insulin resistance in mice and HepG2 cells. *J Food Sci* 2009;**74**:H1–7.
62. Dorenkott MR, Griffin LE, Goodrich KM, et al. Oligomeric Cocoa procyanidins possess enhanced bioactivity compared to monomeric and polymeric cocoa procyanidins for preventing the development of obesity, insulin resistance, and impaired glucose tolerance during high-fat feeding. *J Agric Food Chem* 2014;**62**: 2216–27.
63. Kanamoto Y, Yamashita Y, Nanba F, et al. A black soybean seed coat extract prevents obesity and glucose intolerance by up-regulating uncoupling proteins and down-regulating inflammatory cytokines in high-fat diet-fed mice. *J Agric Food Chem* 2011; **59**:8985–93.
64. Yamashita Y, Okabe M, Natsume M, Ashida H. Prevention mechanisms of glucose intolerance and obesity by cacao liquor procyanidin extract in high-fat diet-fed C57BL/6 mice. *Arch Biochem Biophys* 2012;**527**:95–104.
65. Montagut G, Bladé C, Blay M, et al. Effects of a grapeseed procyanidin extract (GSPE) on insulin resistance. *J Nutr Biochem* 2010;**21**: 961–7.
66. Cedó L, Castell-Auví A, Pallarès V, Blay M, Ardévol A, Pinent M. Grape seed procyanidin extract improves insulin production but enhances bax protein expression in cafeteria-treated male rats. *Int J Food Sci* 2013;**2013**. 7 pages.
67. Castell-Auví A, Cedó L, Pallarès V, Blay M, Pinent M, Ardévol A. Grape seed procyanidins improve β-cell functionality under lipotoxic conditions due to their lipid-lowering effect. *J Nutr Biochem* 2013;**24**:948–53.
68. Montagut G, Onnockx S, Vaqué M, et al. Oligomers of grape-seed procyanidin extract activate the insulin receptor and key targets of the insulin signaling pathway differently from insulin. *J Nutr Biochem* 2010;**21**:476–81.
69. Gonzalez-Abuin N, González-Abuín N, Martínez-Micaelo N, Blay M, Ardévol A, Pinent M. Grape-seed procyanidins prevent the cafeteria-diet-induced decrease of glucagon-like peptide-1 production. *J Agric Food Chem* 2014;**62**:1066–72.
70. Sundaram R, Naresh R, Shanthi P, Sachdanandam P. Modulatory effect of green tea extract on hepatic key enzymes of glucose metabolism in streptozotocin and high fat diet induced diabetic rats. *Phytomedicine* 2013;**20**:577–84.
71. Ding Y, Zhang Z, Dai X, et al. Grape seed proanthocyanidins ameliorate pancreatic beta-cell dysfunction and death in low-dose streptozotocin- and high-carbohydrate/high-fat diet-induced diabetic rats partially by regulating endoplasmic reticulum stress. *Nutr Metab (Lond)* 2013;**10**:51.
72. Ding Y, Dai X, Jiang Y, et al. Grape seed proanthocyanidin extracts alleviate oxidative stress and ER stress in skeletal muscle of low-dose streptozotocin- and high-carbohydrate/high-fat diet-induced diabetic rats. *Mol Nutr Food Res* 2013;**57**:365–9.
73. Kurimoto Y, Shibayama Y, Inoue S, et al. Black soybean seed coat extract ameliorates hyperglycemia and insulin sensitivity via the activation of AMP-activated protein kinase in diabetic mice. *J Agric Food Chem* 2013;**61**:5558–64.
74. Smirin P, Taler D, Abitbol G, et al. *Sarcopoterium spinosum* extract as an antidiabetic agent: in vitro and in vivo study. *J Ethnopharmacol* 2010;**129**:10–7.
75. Lee YA, Cho EJ, Yokozawa T. Effects of proanthocyanidin preparations on hyperlipidemia and other biomarkers in mouse model of type 2 diabetes. *J Agric Food Chem* 2008;**56**:7781–9.
76. Tomaru M, Takano H, Osakabe N, et al. Dietary supplementation with cacao liquor proanthocyanidins prevents elevation of blood glucose levels in diabetic obese mice. *Nutrition* 2007;**23**:351–5.
77. Agouni A, Lagrue-Lak-Hal A-H, Mostefai HA, et al. Red wine polyphenols prevent metabolic and cardiovascular alterations associated with obesity in Zucker fatty rats (Fa/Fa). *PLoS One* 2009;**4**:e5557.
78. Cedo L, Castell-Auvi A, Pallares V, et al. Pancreatic islet proteome profile in Zucker fatty rats chronically treated with a grape seed procyanidin extract. *Food Chem* 2012;**135**:1948–56.
79. Rosen ED, Spiegelman BM. Adipocytes as regulators of energy balance and glucose homeostasis. *Nature* 2006;**444**:847–53.
80. Fernandez-Larrea J, Montagut G, Bladé C, et al. GSPE has the same effects as insulin on the mRNA levels of the main genes of glucose disposal in the liver of STZ-diabetic animals. *Diab Vasc Dis Res* 2007. March;4 Suppl. 1: S186.
81. Gandhi GR, Ignacimuthu S, Paulraj MG. Solanum torvum Swartz. fruit containing phenolic compounds shows antidiabetic and antioxidant effects in streptozotocin induced diabetic rats. *Food Chem Toxicol* 2011;**49**:2725–33.
82. Huang P-L, Chi C-W, Liu T-Y. Areca nut procyanidins ameliorate streptozocin-induced hyperglycemia by regulating gluconeogenesis. *Food Chem Toxicol* 2013;**55**:137–43.

83. Lu Z, Jia Q, Wang R, et al. Hypoglycemic activities of A- and B-type procyanidin oligomer-rich extracts from different cinnamon barks. *Phytomedicine* 2011;**18**:298–302.
84. Cordero-Herrera I, MÁ M, Goya L, Ramos S. Cocoa flavonoids attenuate high glucose-induced insulin signalling blockade and modulate glucose uptake and production in human HepG2 cells. *Food Chem Toxicol* 2014;**64**:10–9.
85. Quesada H, del Bas JM, Pajuelo D, et al. Grape seed proanthocyanidins correct dyslipidemia associated with a high-fat diet in rats and repress genes controlling lipogenesis and VLDL assembling in liver. *Int J Obes (Lond)* 2009;**33**:1007–12.
86. Shimada T, Tokuhara D, Tsubata M, et al. Flavangenol (pine bark extract) and its major component procyanidin B1 enhance fatty acid oxidation in fat-loaded models. *Eur J Pharmacol* 2012;**677**: 147–53.
87. Baselga-Escudero L, Arola-Arnal A, Pascual-Serrano A, et al. Chronic administration of proanthocyanidins or docosahexaenoic acid reverses the increase of miR-33a and miR-122 in dyslipidemic obese rats. *PLoS One* 2013;**8**:e69817.
88. Baselga-Escudero L, Bladé C, Ribas-Latre A, et al. Grape seed proanthocyanidins repress the hepatic lipid regulators miR-33 and miR-122 in rats. *Mol Nutr Food Res* 2012;**56**:1636–46.
89. Lee H-H, Kim K-J, Lee O-H, Lee B-Y. Effect of pycnogenol on glucose transport in mature 3T3-L1 adipocytes. *Phytother Res* 2010;**24**:1242–9.
90. Caimari A, del Bas JM, Crescenti A, Arola L. Low doses of grape seed procyanidins reduce adiposity and improve the plasma lipid profile in hamsters. *Int J Obes (Lond)* 2013;**37**:576–83.
91. Khanal RC, Howard LR, Wilkes SE, Rogers TJ, Prior RL. Effect of dietary blueberry pomace on selected metabolic factors associated with high fructose feeding in growing Sprague–Dawley rats. *J Med Food* 2012;**15**:802–10.
92. Pinent M, Blade MC, Salvado MJ, et al. Grape-seed derived procyanidins interfere with adipogenesis of 3T3-L1 cells at the onset of differentiation. *Int J Obes Relat Metab Disord* 2005;**29**:934–41.
93. Kolb H, Mandrup-Poulsen T. An immune origin of type 2 diabetes? *Diabetologia* 2005;**48**:1038–50.
94. Houstis N, Lander ESRED, Houstis N, Rosen ED, Lander ES. Reactive oxygen species have a causal role in multiple forms of insulin resistance. *Nature* 2006;**440**:944–8.
95. Pajuelo D, Fernández-Iglesias A, Díaz S, et al. Improvement of mitochondrial function in muscle of genetically obese rats after chronic supplementation with proanthocyanidins. *J Agric Food Chem* 2011;**59**:8491–8.
96. Castrillejo VM, Romero M-M, Esteve M, et al. Antioxidant effects of a grapeseed procyanidin extract and oleoyl-estrone in obese Zucker rats. *Nutrition* 2011;**27**:1172–6.
97. Chacón MR, Ceperuelo-Mallafré V, Maymó-Masip E, et al. Grapeseed procyanidins modulate inflammation on human differentiated adipocytes in vitro. *Cytokine* 2009;**47**:137–42.
98. Terra X, Pallarés V, Ardèvol A, et al. Modulatory effect of grapeseed procyanidins on local and systemic inflammation in diet-induced obesity rats. *J Nutr Biochem* 2011;**22**:380–7.
99. Pinent M, Castell A, Baiges I, Montagut G, Arola L, Ardévol A. Bioactivity of flavonoids on insulin-secreting cell. *Compr Rev Food Sci Food Saf* 2008;**7**:299–308.
100. Hutton JC. Insulin secretory granule biogenesis and the proinsulin-processing endopeptidases. *Diabetologia* 1994;**37**: S48–56.
101. Castell-Auví A, Cedó L, Pallarès V, et al. Procyanidins modify insulinemia by affecting insulin production and degradation. *J Nutr Biochem* 2012;**23**:1565–72.
102. Huang CF, Chen YW, Yang CY, et al. Extract of lotus leaf (*Nelumbo nucifera*) and its active constituent catechin with insulin secretagogue activity. *J Agric Food Chem* 2011;**59**:1087–94.
103. Qa'dan F, Verspohl EJ, Nahrstedt A, Petereit F, Matalka KZ. Cinchonain Ib isolated from *Eriobotrya japonica* induces insulin secretion in vitro and in vivo. *J Ethnopharmacol* 2009;**124**:224–7.
104. Herrera HE, Lockstone J, Taylor M, et al. Global microRNA expression profiles in insulin target tissues in a spontaneous rat model of type 2 diabetes. *Diabetologia* 2010;**53**:1099–109.
105. Arola-Arnal A, Blade C. Proanthocyanidins modulate microRNA expression in human HepG2 cells. *PLoS One* 2011;**6**. http://dx.doi.org/10.1371/journal.pone.0025982.
106. Castell-Auví A, Cedó L, Movassat J, et al. Procyanidins modulate microRNA expression in pancreatic islets. *J Agric Food Chem* 2012; **61**:355–63.
107. Scalbert A, Johnson I, Saltmarsh M. Polyphenols: antioxidants and beyond. *Am J Clin Nutr* 2005;**81**:215S–7S.
108. Cedo L, Castell-Auvi A, Pallares V, et al. Grape seed procyanidin extract modulates proliferation and apoptosis of pancreatic beta-cells. *Food Chem* 2013;**138**:524–30.
109. Bagust A, Beale S. Deteriorating beta-cell function in type 2 diabetes: a long-term model. *QJM* 2003;**96**:281–8.
110. Ichinose K, Kawasaki E, Eguchi K. Recent advancement of understanding pathogenesis of type 1 diabetes and potential relevance to diabetic nephropathy. *Am J Nephrol* 2007;**27**:554–64.
111. Chen L, Sun P, Wang T, et al. Diverse mechanisms of antidiabetic effects of the different procyanidin oligomer types of two different cinnamon species on db/db mice. *J Agric Food Chem* 2012;**60**: 9144–50.
112. Gribble FM. RD Lawrence lecture 2008: targeting GLP-1 release as a potential strategy for the therapy of type 2 diabetes. *Diabet Med* 2008;**25**:889–94.
113. Moran-ramos S, Tovar AR, Torres N. Diet: friend or foe of enteroendocrine cells: how it interacts with enteroendocrine cells. *Adv Nutr* 2012;**3**:8–20.
114. Holz IVGG, Kühtreiber WM, Habener JF. Pancreatic beta-cells are rendered glucose-competent by the insulinotropic hormone glucagon-like peptide-1 (7-37). *Nature* 1993;**361**:362–5.
115. Brubaker PL, Drucker DJ. Minireview: glucagon-like peptides regulate cell proliferation and apoptosis in the pancreas, gut, and central nervous system. *Endocrinology* 2004;**145**:2653–9.
116. Törrönen R, Sarkkinen E, Niskanen T, Tapola N, Kilpi K, Niskanen L. Postprandial glucose, insulin and glucagon-like peptide 1 responses to sucrose ingested with berries in healthy subjects. *Br J Nutr* 2012;**107**:1445–51.
117. Yamashita Y, Okabe M, Natsume M, Ashida H. Cinnamtannin A2, a tetrameric procyanidin, increases GLP-1 and insulin secretion in mice. *Biosci Biotechnol Biochem* 2013. http://dx.doi.org/10.1271/bbb.130095.
118. González-Abuín N, Martínez-Micaelo N, Blay M, et al. Grape seed-derived procyanidins decrease dipeptidyl-peptidase 4 activity and expression. *J Agric Food Chem* 2012;**60**:9055–61.
119. González-Abuín N, Martínez-Micaelo N, Margalef M, et al. A grape seed extract increases active glucagon-like peptide-1 levels after an oral glucose load in rats. *Food Funct* 2014;**5**: 2357–64.
120. Byun E-B, Ishikawa T, Suyama A, et al. A procyanidin trimer, C1, promotes no production in rat aortic endothelial cells via both hyperpolarization and PI3K/Akt pathways. *Eur J Pharmacol* 2012; **692**:52–60.
121. Margina D, Gradinaru D, Manda G, Neagoe I, Ilie M. Membranar effects exerted in vitro by polyphenols - quercetin, epigallocatechin gallate and curcumin—on HUVEC and Jurkat cells, relevant for diabetes mellitus. *Food Chem Toxicol* 2013;**61**:86–93.
122. Verstraeten SV, Hammerstone JF, Keen CL, Fraga CG, Oteiza PI. Antioxidant and membrane effects of procyanidin dimers and trimers isolated from peanut and cocoa. *J Agric Food Chem* 2005;**53**: 5041–8.

123. González-Abuín N, Martínez-Micaelo N, Blay M, Green BD, Pinent M, Ardévol A. Grape-seed procyanidins modulate cellular membrane potential and nutrient-induced GLP-1 secretion in STC-1 cells. *Am J Physiol Cell Physiol* 2014;**306**:C485–92.
124. Matteucci E, Giampietro O. Dipeptidyl peptidase-4 (CD26): knowing the function before inhibiting the enzyme. *Curr Med Chem* 2009;**16**:2943–51.
125. Mentlein R. Dipeptidyl-peptidase IV (CD26)–role in the inactivation of regulatory peptides. *Regul Pept* 1999;**85**:9–24.
126. Hansen L. Glucagon-like peptide-1-(7-36)amide is transformed to glucagon-like peptide-1-(9-36)amide by dipeptidyl peptidase IV in the capillaries Supplying the L Cells of the porcine intestine. *Endocrinology* 1999;**140**:5356–63.
127. Johnson MH, de Mejia EG, Fan J, Lila MA, Yousef GG. Anthocyanins and proanthocyanidins from blueberry-blackberry fermented beverages inhibit markers of inflammation in macrophages and carbohydrate-utilizing enzymes in vitro. *Mol Nutr Food Res* 2013;**57**:1182–97.
128. Gonzalez-Abuin N, Martinez-Micaelo N, Blay M, Ardevol A, Pinent M. Grape-seed procyanidins prevent the cafeteria diet-induced decrease of glucagon-like peptide-1 production. *J Agric Food Chem* 2014;**62**(5):1066–72.
129. Tebib K, Rouanet JM, Besançon P. Effect of grape seed tannins on the activity of some rat intestinal enzyme activities. *Enzyme Protein* 1994;**48**:51–60.
130. Dao T-MA, Waget A, Klopp P, et al. Resveratrol increases glucose induced GLP-1 secretion in mice: a mechanism which contributes to the glycemic control. *PLoS One* 2011;**6**:e20700.
131. Cabou C, Burcelin R. GLP-1, the gut-brain, and brain-periphery axes. *Rev Diabet Stud* 2011;**8**:418–31.
132. Campbell JE, Drucker DJ. Pharmacology, physiology, and mechanisms of incretin hormone action. *Cell Metab* 2013;**17**: 819–37.
133. Liu X, Zhou H-JJ, Rohdewald P. French maritime pine bark extract pycnogenol dose-dependently lowers glucose in type 2 diabetic patients. *Diabetes Care* 2004;**27**:839.
134. Liu X, Wei J, Tan F, et al. Antidiabetic effect of pycnogenol French maritime pine bark extract in patients with diabetes type II. *Life Sci* 2004;**75**:2505–13.
135. Banini AE, Boyd LC, Allen JC, Allen HG, Sauls DL. Muscadine grape products intake, diet and blood constituents of non-diabetic and type 2 diabetic subjects. *Nutrition* 2006;**22**:1137–45.
136. Curtis PJ, Sampson M, Potter J, Dhatariya K, Kroon PA, Cassidy A. Chronic ingestion of flavan-3-ols and isoflavones improves insulin sensitivity and lipoprotein status and attenuates estimated 10-year CVD risk in medicated postmenopausal women with type 2 diabetes A 1-year, double-blind, randomized, controlled trial. *Diabetes Care* 2012;**35**:226–32.
137. Hsu C-H, Liao Y-L, Lin S-C, Tsai T-H, Huang C-J, Chou P. Does supplementation with green tea extract improve insulin resistance in obese type 2 diabetics? A randomized, double-blind, and placebo-controlled clinical trial. *Altern Med Rev* 2011;**16**: 157–63.
138. Balzer J, Rassaf T, Heiss C, et al. Sustained benefits in vascular function through flavanol-containing cocoa in medicated diabetic patients a double-masked, randomized, controlled trial. *J Am Coll Cardiol* 2008;**51**:2141–9.
139. Mellor DD, Sathyapalan T, Kilpatrick ES, Beckett S, Atkin SL. High-cocoa polyphenol-rich chocolate improves HDL cholesterol in type 2 diabetes patients. *Diabet Med* 2010;**27**:1318–21.
140. Zamora-Ros R, Forouhi NG, Sharp SJ, et al. Dietary intakes of individual flavanols and flavonols are inversely associated with incident type 2 diabetes in European populations 1–3. *J Nutr* 2013;**144**:335–43.
141. Jacques PF, Cassidy A, Rogers G, Peterson JJ, Meigs JB, Dwyer JT. Higher dietary flavonol intake is associated with lower incidence of type 2 diabetes. *J Nutr* 2013;**143**:1474–80.
142. Aherne SA, O'Brien NM. Dietary flavonols: chemistry, food content, and metabolism. *Nutrition* 2002;**18**:75–81.
143. Castell Auví A. The effects of grape seed procyanidin extract on insulin synthesis and secretion. http://www.tdx.cat/handle/10803/79133 (accessed 13.11.14).

CHAPTER

17

Dietary Whey Protein and Type 2 Diabetes: Molecular Aspects

Jaime Amaya-Farfan[1], *Carolina S. Moura*[1], *Priscila N. Morato*[1,2], *Pablo C.B. Lollo*[1,2]

[1]Food and Nutrition Department, Protein Resources Laboratory, Faculty of Food Engineering, University of Campinas (UNICAMP), Campinas, São Paulo, Brazil; [2]Faculty of Health Sciences, Federal University of Grande Dourados (UFGD), Dourados, Mato Grosso do Sul, Brazil

1. INTRODUCTION

Even today, too few people realize the extent of damage that excess circulating glucose can cause to the body: glucose is a chemical reagent that attacks and modifies all body proteins. Glucose is a polyolaldehyde and, as such, it readily reacts by a bimolecular mechanism with all primary amines, such as free amino acids, peptides, and whole proteins at body temperature. The reaction, known as glycation, modifies exposed amino groups turning receptors, transporters, and even polypeptide hormones like insulin itself nonfunctional following a structural and stable rearrangement, familiar to those who work with the Maillard reaction.[1] Because the speed of bimolecular reactions depends on the concentration of both reactants and the concentration of proteins in tissues ought to be rather constant, the concentration of glucose becomes the rate-limiting factor. Although the damage on the external domain of membrane proteins can be great, the chaos caused in the cytoplasm by elevated glucose concentrations could be far more damaging should glucose be allowed to flood the inside of the cell, destroying the delicate machinery and never be phosphorylated and metabolized. Biochemically essential for life, glucose is a true tree of good and evil, and hyperglycemia is naturally aberrant to the proper function of the animal body, thus the vital need to keep glycemic levels under control throughout life.

The nutritional importance of dietary proteins, therefore, becomes even more relevant when the body attempts to manage the upsurges of blood glucose, whether in the postprandial or in the fasting stage. Curiously enough, all the interest that has been recently drawn by the health benefits of the milk whey proteins (WPs) gains one more feature to be considered: the known readiness they exhibit to be digested very quickly, like virtually no other food protein. Because WPs render to the bloodstream an abundant array of free amino acids and peptides soon after ingestion,[2] just at the time when gelatinized starches have been also totally converted to glucose during their passage through the intestine, the free α-amino groups may buffer or serve as temporary harbor to the alkylating groups of glucose and spare sensitive functional proteins from glycation, whereas the reaction is still at its reversible stage.

Type 2 diabetes (T2D) shares characteristics of acquired, multifactorial health conditions, affecting more than 347 million people worldwide and accounting for approximately 90% of the two main types of diabetes. If incidence rates remain constant, the number of youth with T2D in the United States is projected to increase by 49% over the next 35 years, whereas the number with type 1 diabetes is expected to climb 23%, according to a study by the Centers for Disease Control and Prevention.[3] The impact of the diet, allied to the difficulty of applying long-range food safety tests to industrialized foods and food combinations, emphasize the need to reevaluate all classical concepts in nutrition under the light of molecular mechanisms.

Fortunately, the revolution of functional foods has facilitated our understanding of the mechanisms of action of bioactive substances because it unlocked the beneficial health properties hidden in traditional foods. Food science and medicine have thus encouraged researchers to study the properties of all food bioactives looking into the possible management of noncommunicable, chronic

Molecular Nutrition and Diabetes
http://dx.doi.org/10.1016/B978-0-12-801585-8.00017-8

diseases such as obesity and T2D. This new vision of research in nutritional science has expanded considerably in the twenty-first century, placing great interest on proteins and, in particular, the WPs as a source of bioactive compounds.

The purpose of the present chapter was therefore to integrate some of the molecular elements outlining isolated mechanisms by which the milk WPs have been shown to counteract the deteriorative biochemical reactions that follow prolonged hyperglycemic states. We have intentionally linked the role of the whey-derived bioactive peptides and amino acids to physical exercise to illustrate the many ways nature has laid down at our body's disposal to overcome the constant metabolic derangements produced by environmental pressures. The connection between the beneficial properties of consuming WPs for better muscle performance and the potential use for the management of glucose energy is not at all coincidental. To ignore this metabolic connection would be equivalent to ignoring the connection between muscle activity and hyper- or hypoglycemia, or even animal life itself.

2. CONSTITUENTS OF THE WP

WP is a complex mixture of the water-soluble proteins of cow's milk, representing about 20% of the total protein. These proteins have been recognized for their high nutritive value, high digestibility, and fast generation of peptides and free amino acids,[4,5] which start appearing in plasma a few minutes after ingestion. The principal components of whey are β-lactoglobulin, α-lactalbumin, serum albumin, lactoferrin, immunoglobulins, lactose, and soluble mineral salts. In terms of relative abundance, β-lactoglobulin and α-lactalbumin are the two major fractions accounting for 50–55% and 20–25% of the total, respectively. Two other significant fractions are the soluble peptide (glycomacropeptide (GMP)) that is cut off the κ-casein at the moment of making the cheese coagulum by proteolysis, and serum albumin, each accounting for 10–15% of the total. Last, lactoferrin and (1% each) lactoperoxidase and others can be identified in bovine milk whey.[4,5]

Whey is commercially found in three forms: a WP concentrate (WPC), WP isolate (WPI), and a more elaborate form that has been enzymatically WP hydrolyzed (WPH). Because the process followed to obtain the WPI eliminates GMP, the WPH will have a more complete array of peptides than those eventually produced from the WPI in the gut. The most common form of whey is the WPC, which exists in two types: one containing 35% and another containing 80% protein. Although the former is practically the dehydrated residue that remains after cheese making, the latter undergoes a microfiltration process for the partial removal of lactose and salts. The WPI in turn is obtained by a more elaborate process, which requires the manipulation of the pH, protein precipitation, and elimination of remaining lactose and salts, thus ending with a ≥90% protein content. One by-product of this process may be the GMP.

3. STUDIES IN SUPPORT OF THE ANTIHYPERGLYCEMIC EFFECT OF WHEY

Many studies have been published on the nutritional and health benefits of the WPs. Of these, we have selected the 15 most recent reports, which include in vitro, animal, and human trials showing some blood glucose-lowering effect (Table 1[6–20]). In synthesis, these studies point out the existence of either proven or potential benefits of WP such as reduction of postprandial blood glucose, stimulation of the release of insulin, and the extension of incretin action by inhibition of the incretin degrading enzyme, dipeptidyl peptidase-4 (DPP4). These studies report consistent results in support of the hypothesis that milk WPs are potential therapeutic or coadjuvant alternatives for prediabetic or diabetic patients. Test trials in humans show that individual needs may have a wide range of variation, but that a most effective dose seems to be between 20 and 50 g/day.

4. WHAT DO EXERCISE AND DIETARY PROTEIN HAVE TO DO WITH HYPERGLYCEMIA?

For years, clinicians have recommended diabetic patients to practice some kind of physical activity. The reason for this is because muscle contraction activates the entire metabolism, including the tendency to approach glucose homeostasis.[21] When considering the role of the diet on glucose homeostasis, it is essential to recognize the functional interaction that exists between liver, skeletal muscle, and the adipose tissue.[22]

Exercise is a valuable strategy for managing diabetes because it is a potent stimulator of glucose uptake by the muscle in an insulin-independent manner and, although the topic of exercise is not central to this chapter, it is instructive to make these considerations for the reader to visualize the connections between physiology, diet, and metabolism that the body automatically makes. Muscle contraction is a molecular trigger for a series of cellular responses that result in the translocation of glucose transporters (GLUTs) to the cell membrane. Among the cellular responses, 5′-adenosine monophosphate-activated protein kinase (AMPK) plays a central role. Because

TABLE 1 Fifteen Most Recent Studies Involving Milk Whey Proteins as Beneficial in the Management of Hyperglycemia and Diabetes

Authors	Aim	Subjects/animals	Main findings
Jakubowicz et al.[6]	Effect of 50 g of WP preload on glucose levels	15 T2D subjects	Consumption of WP shortly before a high-glycemic index breakfast increased the insulin secretion and GLP-1 and reduced postprandial glycemia
Mortensen et al.[7]	Determine the effects of 45 g of four different WP fractions on postprandial lipid and hormone responses: whey isolate, whey hydrolyzate, α-lactalbumin −enhanced whey, and GMP-enhanced whey	12 T2D subjects	Whey isolate and whey hydrolyzate did not differ in their capacity to induce higher insulin levels, and they may be used as potential nutritional protein sources to improve glucose homeostasis in T2D
Ma et al.[8]	Determine the effects of 55 g of WP preload on slow gastric emptying, incretin response, and postprandial glycemia	8 T2D subjects	WP caused a reduction in postprandial glycemia that could be due the stimulation of insulin and incretin hormone secretion and slow gastric emptying
Mortensen et al.[9]	Compare the effects of the 45 g proteins casein, WP, cod, and gluten on postprandial lipid and incretin responses	12 T2D subjects	The glucose response was lower after the whey meal than after the other meals, whereas no significant differences were found in insulin, glucagon, GLP-1, and GIP responses
Pal et al.[10]	Determine the effects of WP supplementation on body composition, lipids, insulin, and glucose in comparison to casein and glucose (control) supplementation	70 overweight/obese subjects	27 g of WP twice a day for 12 weeks decreased insulin and total and LDL cholesterol compared with casein and control.
Akhavan et al.[11]	Identify the mechanism of action of WP beyond insulin on the reduction of postmeal glycemia	10 healthy subjects	A premeal consumption of WP (10 and 20 g) lowered postmeal glycemia by both insulin-dependent and insulin-independent mechanisms
Nilsson et al.[12]	Determine to what extent the insulinotropic properties of WP could be simulated by specific amino acid mixtures	12 healthy subjects	A mixture of leucine, isoleucine, valine, lysine, and threonine resulted in glycemic and insulinemic responses closely mimicking those seen after WP ingestion, but had no effect on GIP or GLP-1
Frid et al.[13]	Evaluate if supplementation of high-glycemic index meals with whey proteins increases insulin secretion and improves blood glucose control in T2D subjects	14 T2D subjects	Addition of whey to meals containing rapidly digested and absorbed carbohydrates stimulated insulin release and reduced postprandial blood glucose after a lunch meal consisting of mashed potatoes and meatballs
Petersen et al.[14]	Determine the glycemic impact of 5, 10 and 20 g of a GILP from WP to a glucose drink	10 healthy subjects	Addition of GILP to an oral glucose bolus reduced blood glucose in a dose-dependent manner
Badr et al.[15]	Determine the effects of camel WP on immune system of diabetic mice	20 diabetic mice	WP supplementation improved the immune response in diabetic mice

Continued

TABLE 1 Fifteen Most Recent Studies Involving Milk Whey Proteins as Beneficial in the Management of Hyperglycemia and Diabetes—cont'd

Authors	Aim	Subjects/animals	Main findings
Shertzer et al.[16]	Determine the influence of WP on systemic energy balance and metabolic changes in mice fed an HF diet	Female C57CL/6J mice	WP consumption led to less hepatosteatosis and insulin resistance, it may be effective in slowing the development of fatty liver disease and T2D
Morato et al.[17]	Determine if WP and WPH could also increase the concentration of the glucose transporters GLUT-1 and GLUT-4 in the PM	48 Wistar Rats (control and exercised)	Consumption of WPH significantly increased the concentrations of GLUT-4 in the PM and glycogen, whereas the GLUT-1 and insulin levels and the health indicators showed no alterations
Morato et al.[18]	Determine which WPH components could modulate translocation of the glucose transporter GLUT-4 to the PM of animal skeletal muscle	47 Wistar rats	Of the WPH components tested, the amino acid L-isoleucine and the peptide L-leucyl-L-isoleucine showed greater efficiency in translocating GLUT-4 to the PM and of increasing glucose capture by skeletal muscle
Belobrajdic et al.[19]	Feeding insulin-resistant rats a high-protein diet (32%) containing WPC would reduce body weight and tissue lipid levels and increase insulin sensitivity more than a diet containing red meat	32 Wistar rats High-fat diets	High-protein diet reduced energy intake and adiposity and that whey protein is more effective than red meat in reducing body weight gain and increasing insulin sensitivity
Konrad et al.[20]	Determine the potential of WPH as the natural inhibitors of DPP4, α-glucosidase, and angiotensin-converting enzyme	In vitro	WPH showed higher inhibitory activities against the three tested and may support postprandial glycemia regulation and blood pressure maintenance, and could be used as functional food ingredients in the diet of patients with T2D

DPP, dipeptidyl peptidase; GILP, glycemic index lowering peptide fraction; GIP, glucose-dependent insulinotropic polypeptide; GLP-1, glucagon-like peptide-1; GLUT, glucose transporter; GMP, glycomacropeptide; LDL, low-density lipoprotein; PM, plasma membrane; T2D, type 2 diabetes; WP, whey protein; WPC, whey protein concentrate; WPH, whey protein hydrolyzate (or prehydrolyzed whey protein).

AMPK can be activated by high concentrations of AMP in the cell, glucose uptake can be further stimulated by hypoxia, an indirect consequence of intense physical exercise (Figure 1(c)). Once AMPK is activated, a cascade of molecular signals will begin to promote the translocation of GLUTs to the myocyte membrane, thus increasing the energy resources of the cell and offsetting the drop in adenosine triphosphate/adenine diphosphate ratio caused by exercise, a negative feedback mechanism.

5. TYPE, AMOUNT, AND FORM OF TAKING THE PROTEIN

Sufficient protein (20–30% of total energy intake, independent of the protein source) is currently considered of upmost importance in the management of T2D.[23] The nature, physicochemical form, and time of feeding have also become relevant and can make a significant difference in the final outcome. In 1998, our research group observed that α-lactalbumin favorably modified glucose homeostasis by increasing muscle glycogen stores in normal exercising rats, provided the protein was given to the animal in the hydrolyzed form. This modification consisted of the animals reaching exhaustion with significantly better levels of blood glucose, greater reserves of blood proteins, high reserves of muscle glycogen, and a high resistance to fatigue.[24] Later, we found that the improved physiological status granted by the hydrolyzed α-lactalbumin was very similar to that afforded by a hydrolyzate of the entire set of WPs.[25] Additionally, it was noticed that the benefits of feeding the prehydrolyzed WPs to the exercising animal included a reduced state of stress, as perceived by the lower activity of intestinal glutaminase.[26] It was evident then that the hydrolyzed

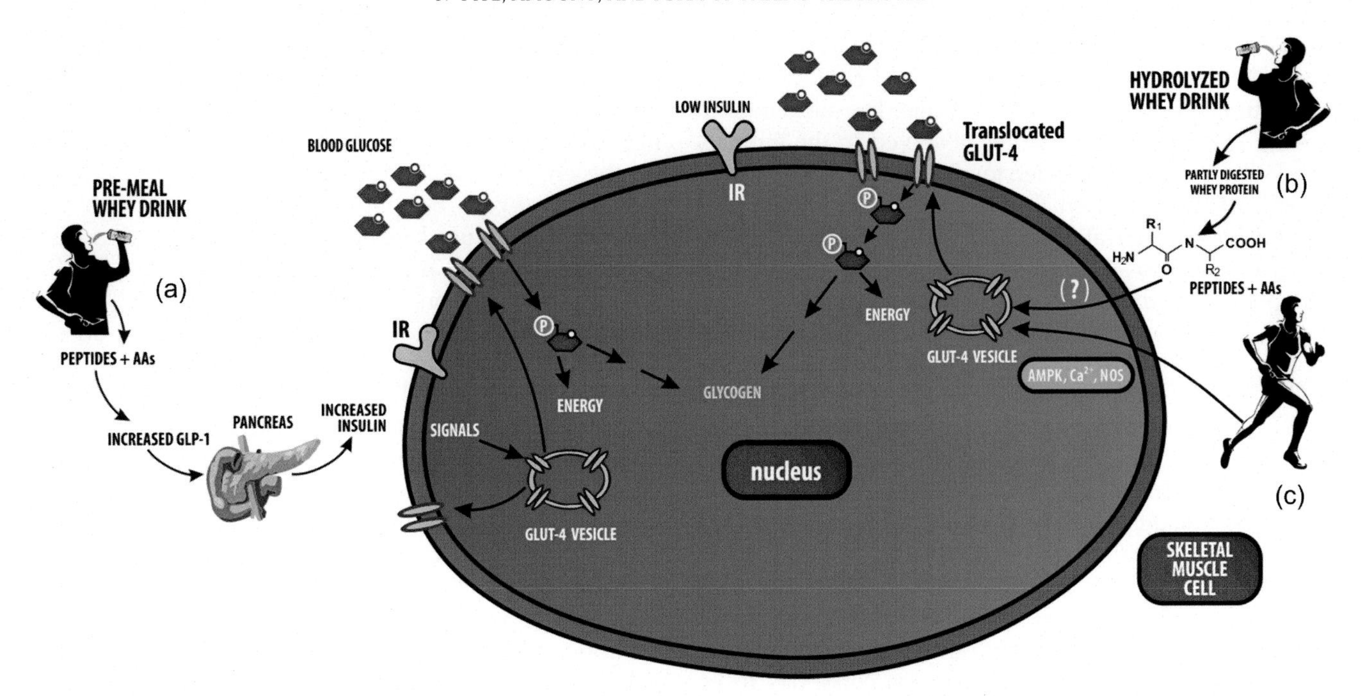

FIGURE 1 Simplified scheme of the two different paths followed by peptides derived from milk whey to achieve the glucose-lowering function. (a) Some tripeptides (like Ile-Pro-Ala) have the ability to inhibit the intestinal proteolytic enzyme dipeptidyl peptidase 4 (DPP4) that has the mission of degrading the incretins (glucagon-like peptide-1 (GLP-1) and glucose-dependent insulinotropic polypeptide-1 (GIP-1)) immediately after use. The incretins' two-fold function is to augment and extend the work of insulin by stimulating the pancreas to release insulin and repressing glucagon formation, thus repressing gluconeogenesis. This is an action proper of the intact whey proteins (WPC) and would be effective in patients with type 1 diabetes (T1D). (b) When the proteins are ingested in the prehydrolyzed form (WPH), dipeptides will trigger the disruption of the glucose transporter (GLUT)-4 vesicle and translocate the transporter to the cell membrane with the net result of lowering blood glucose. (c) This is the case of a normal or diabetic individual when performing physical exercise. The muscle uses up adenosine triphosphate, thus raising the levels of adenosine monophosphate, which will activate 5′-adenosine monophosphate-activated protein kinase (AMPK). Therefore, without the need of a nutritional supplement, but with the intervention of AMPK, calcium ions and nitric oxide synthase (NOS), muscle contraction will produce a similar effect to that of the WPH. In all cases (a–c), the glucose-lowering effect is mediated by translocation of inactive, cytoplasmic GLUTs to the cell membrane. AA, amino acid; IR, insulin receptor; WPC, whey-protein concentrate; WPH, whey-protein hydrolyzate.

whey could be distinguished from a normal whey concentrate because of the superior advantages granted to the animal by the prehydrolyzed WPs. Bearing in mind that the greater physiological value of the hydrolyzate was due to the massive presence of peptides with unknown bioactivities, later we proposed a key part of the mechanism to explain how the hydrolyzate was able to control the glycemic state and result in higher glycogen stores in the exercising rat.[17,18]

It is important to clarify that there are WPHs with different degrees of hydrolysis, or digestion, and that the size as well as the abundance of any particular class of peptide will depend on the extension of digestion, perhaps more so than the type of enzymatic digestion employed in the process. Additionally, a fraction of the peptides formed become substantially less water-soluble than the parent protein, a characteristic feature of the WPs that may determine not only the chances of these peptides being further digested, but also their ability to traverse cell membranes.

An integrated mechanism explaining the glucose-lowering effect of the WPs should involve: (1) modulation of insulin production; (2) translocation of GLUTs from the cytoplasm to the cellular membrane; and (3) modulation of the production, secretion, and degradation of incretins. WPs can control hyperglycemia in at least two ways: the principal one is accomplished by augmenting or extending the activity of incretins glucagon-like peptide-1 (GLP-1) and glucose-dependent insulinotropic polypeptide (GIP), which will in turn stimulate the release of additional insulin.[27,28] Ingestion of a premeal dose of WP is capable of preconditioning the homeostasis to manage a meal of high glycemic index (Figure 1(a)).[6,11] The second way can occur by ingestion of a prehydrolyzed WP supplement, which will mainly stimulate translocation of the GLUT-4 transporter, but without necessarily increasing insulin secretion (Figure 1(b)).[17,18] A diet containing a medium degree-of-hydrolysis (DH) WPH as the sole source of protein was able to promote translocation of GLUT-4 in rats in 9 days[17]; this translocation was shown to occur independently of insulin (i.e., membrane-bound GLUT-4 was increased without insulin levels being any higher in these animals than in those consuming the standard casein diet). This finding, obtained with whey hydrolyzates of about 10% DH, suggested that the WPH could work on target tissues distant from the intestine. The substantially greater muscle glycogen stores observed in the animals fed WPH were then understood to be a direct consequence of the increased flux of glucose into the cell. It should be clarified at this point that glucose is due to enter the cell in a controlled fashion, by an energy-independent manner and as determined by the translocation of a transporter (GLUTs) to the cell membrane, which in turn is commanded by insulin, the presence of glucose in the blood, and the energy demands of the body.[29]

Based on the above findings, we sought to test the ability of some WPH peptides to promote the translocation of GLUT-4 to the cell membrane. Among the peptides tested, L-leucyl-L-isoleucine appeared to display a high ability to stimulate translocation and significantly raise GLUT-4 to levels comparable to those achieved with WPH, thus suggesting a possible hormone-like effect of this peptide on peripheral targets. Current investigation on the role of WPH peptides in glucose homeostasis continues to be an open and promising line of research, which should be extended beyond to other areas of metabolism. Data from the only two studies on this specific topic suggest the possibility that while some peptides act selectively on the muscle cell,[17,18] others act directly on the intestine or on the pancreas modulating insulin secretion or incretin preservation.[30,31] Acting in combination (Figure 1), these two processes could synergistically make the translocation of GLUT-4 to the cell membrane a more effective glucose-lowering process than if the peptides were to act only on one target tissue.

Currently, our understanding is that the amount of GLUT-4 translocated by ingesting a whey hydrolyzate is additive to that produced by exercise alone, but how exactly the bioactive whey peptides or free amino acids trigger the translocation is still unknown. Future studies will decide if peptides derived from a WPH must first act on the AMPK signaling cascade or if they can stimulate a different system to command the translocation of GLUT-4. The results of Morato et al.[18] did show, however, that there was activation of AMPK by the peptide Leu-Ile present in WPH, but these results were still preliminary, and further research will be necessary before we could complete the mechanistic scheme by which this and other peptides contribute to the final glucose-lowering effect of the WPH.

6. WHEY PROTEINS AND THE INCRETINS

Incretins are a kind of protein hormones whose functions include the modulation of glucose metabolism by stimulating the release of insulin by the β cells and, at the same time, inhibiting the release of glucagon by pancreatic α cells.[30,31] The known incretins are GLP-1 and the GIP. Once produced, both GLP-1 and GIP are immediately used and thereafter rapidly inactivated by the enzyme DPP4.[32] Several oligopeptides released from several WPs, but particularly from β-lactoglobulin, have been produced using various types of enzymatic digestion methods and shown to have DPP4—inhibiting activity.[28,33,34] Among them, the tripeptide Ile-Pro-Ala (half maximal inhibitory concentration (IC_{50}) = 3.5 μM) and the pentapeptide Ile-Pro-Ala-Val-Phe (IC_{50} = 49 μM), for example, illustrate the interesting phenomenon that with the passage of time, the pentapeptide will itself degrade to a tripeptide. Nevertheless, as a consequence of this reduction in size, there is a significant structural change that will make the tripeptide 14 times more active than the pentapeptide molecule.[34]

Jakubowicz et al.[6] tested the effects of a preload of WP on glucose homeostasis in 15 patients with T2D. On two separate days, the volunteers consumed either 50 g of WP dissolved in 250 mL of water, or a placebo (250 mL water) followed by a standardized meal containing high-glycemic index carbohydrates. The results showed that, throughout a 3-h postprandial period, blood glucose was 28% lower in those volunteers that took the preload with WP. The authors also noted that the WP preload significantly increased the insulin and C-peptide responses by 105% and 43%, respectively.

7. WHEY PEPTIDES, STRESS, AND THE HEAT-SHOCK PROTEINS

The heat-shock proteins (HSPs) are members of a family of ubiquitous proteins produced by the body in response to injuries of virtually any kind, which tend to alter the structure of functional, rather sensitive protein molecules located outside and inside the cell. It is for this reason that the HSPs are called chaperons or chaperonins and are considered to be part of the natural antioxidant system. The numbers following the three-letter name denote the approximate molecular weight of each member of the family, going from those small HSPs that assist the nascent protein initial folding (e.g., HSP-10) to the larger ones designed to repair damaged quaternary and tertiary structures while the denaturing alterations are still reversible (see Balch et al.[35] for an excellent review on the potential of activation of HSPs for the treatment of chronic diseases). Denaturation caused by heat, oxidation, and forced molecular aggregation can foreseeably be treated by one of the different HSPs, not excluding proteins at the initial stage of glycation. Figure 2 depicts in a simplified manner a process that may start either on the cell membrane or inside the cytoplasm, but that will normally require the

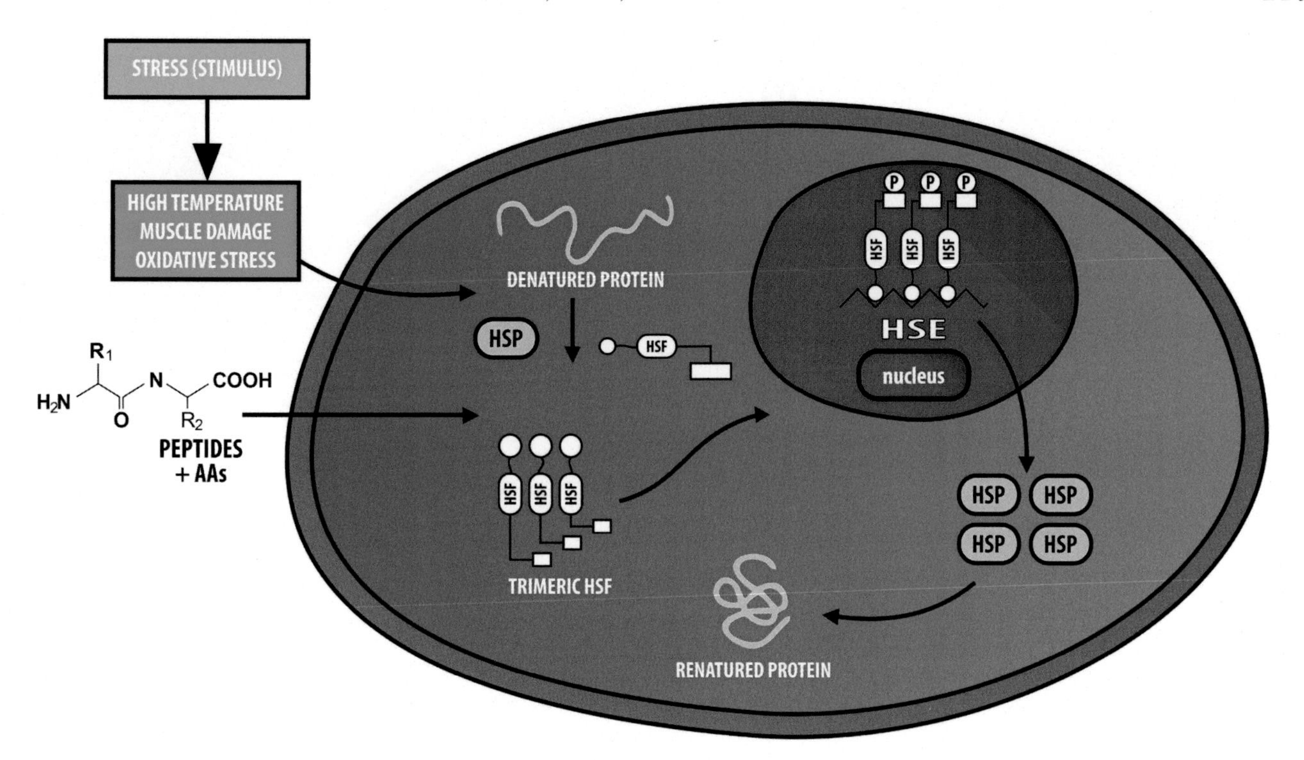

FIGURE 2 The mere presence of protein denaturing factors (all kinds of stress: heat, oxidative, metabolic damage) will trigger the synthesis of heat-shock proteins (HSPs). These causative factors will induce inactive monomers of the heat-shock factor (HSF) to self-assemble into active, phosphorylated trimers that will enter the nucleus. Inside the nucleus, they will bind the heat-shock element (HSE) and command genic DNA to start transcription for the synthesis of new HSP in the cytoplasm. Augmented numbers of HSP molecules will begin repairing the damages by a renaturing process. Because stress of various kinds exacerbates insulin resistance and key peptides and amino acids arising from the whey proteins (WPs) are also capable of eliciting the production of HSP, it is thought that the WPs can contribute to remediate the damaging effects of hyperglycemia in this additional way.

trimerization of inactive heat-shock factors in the cytoplasm, followed by migration into the nucleus. Inside the nucleus, the phosphorylated trimers bind an element essential for commanding transcription.

Based on our earlier observation that rats fed the WPH, instead of the WPC, had a reduced state of stress at the end of a resistance training bout,[26] we compared the WPC with a WPH (~10% DH) in their ability to increase the expression of HSP-70 in nondiabetic rat skeletal muscle and other tissues.[36] In the sedentary animals, all groups showed low HSP expression, but exercise alone increased it to about three times. Dietary WPH, however, produced significantly greater expression of HSP-70 in gastrocnemius, soleus, and lung than the unhydrolyzed WPC, which had a capacity similar to that of casein. Because these were normal rats, it was therefore evident that key peptides present in the hydrolyzate required a bout of resistance exercise, as a stressing effector for the HSP-70 to be expressed.

Oxidative stress can trigger a number of stress pathways of kinases of the serine and threonine family, which in turn have a negative impact on insulin signaling. The manner in which oxidative stress causes reduced glucose uptake by the cell is well known. It begins with a greater expression of such negative biomarkers as nuclear factor-κB (NF-κB), activator protein 1, and hypoxia-inducible factor 1 that in turn trigger overexpression of inflammation interleukins interleukin-6, tumor necrosis factor-α, and monocyte chemoattractant protein-1. Such pro-inflammatory condition will result in the activation of serine/threonine kinases, which themselves will block insulin signaling, automatically stopping GLUT-4 translocation and drastically reducing glucose uptake by the cell.[37]

Stress is therefore expected to contribute to the impairment of glucose uptake, mitochondrial dysfunction, and insulin resistance, as is the case of stress hyperglycemia, which ultimately may lead to diabetes.[37,38] It has also been reported that hyperglycemia is a crucial determinant in the pathogenesis of diabetic nephropathy, and that this disease can differentially modulate the overexpression and phosphorylation of HSP-27, HSP-60, and HSP-70 in the kidney of rats with induced diabetes.[39] Similarly, a study by Reddy et al.[40] showed that in rats with chronic hyperglycemia this condition also induced the upregulation of small HSPs (MKBP, HSPB3, αBC), plus the phosphorylation and translocation of HSP-27 and

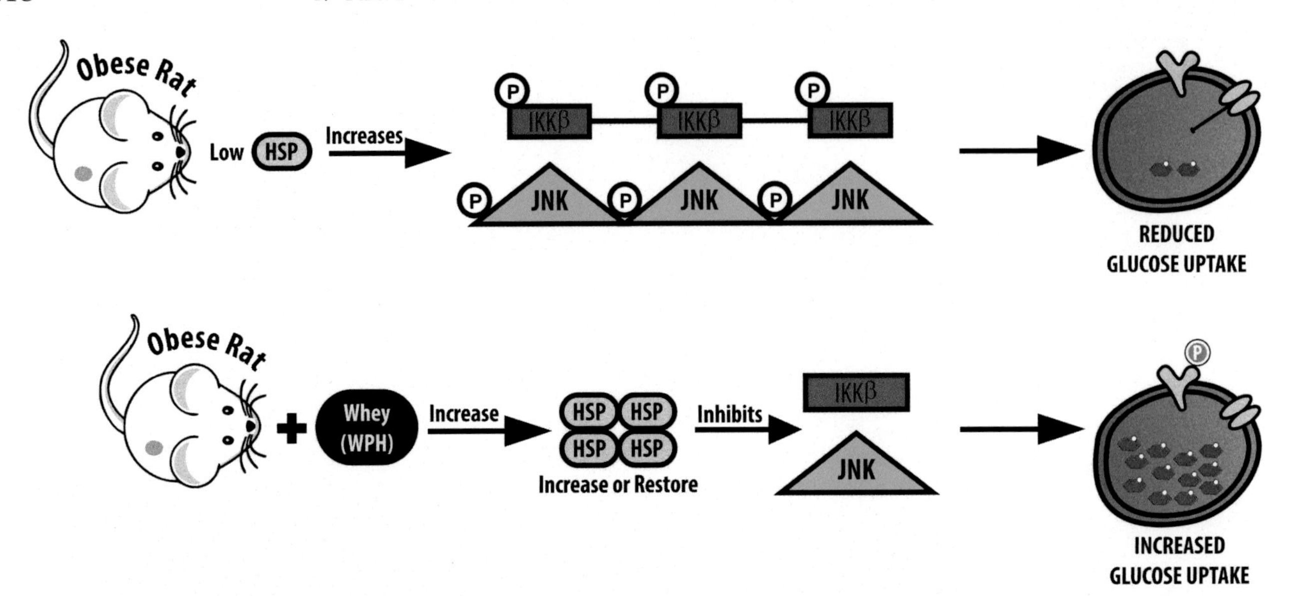

FIGURE 3 How the increase in heat-shock protein (HSP) concentration could improve glucose tolerance in an obese rat. Obesity is akin to low HSP production, a situation that promotes activation (phosphorylation) of kinases that hamper the insulin-signaling cascade. Low levels of HSPs permit phosphorylation of inhibitor KB kinase β (IKKβ) and c-Jun NH_2−terminal kinase (JNK), leading to little glucose transport. Increasing the production of HSPs will mean less phosphorylation of inhibitor kinases and, therefore, more efficient glucose transport will occur. WPH, whey-protein hydrolyzate.

αBC to striated heart muscle. Simar et al.[41] have shown that induction of HSP-72 and HSP-27 by heat stress resulted in the inactivation of stress kinases and lowered levels of IRS-1 serine phosphorylation in monocytes from obese patients. This indicates that metabolic diseases can also affect monocyte metabolism via cellular stress that can be modulated via HSP induction. Although it has been proven that the increased expression of HSPs as provoked by such types of injuries has the objective of remediating the damage, it is not very clear which HSPs renature which proteins to effectively reverse or prevent diabetes.

Several studies have been published reporting the importance to increase HSP expression to ameliorate hyperglycemia, mitochondrial dysfunction, and endoplasmic reticulum alterations in mice and other experimental animals.[37,40,42] Most significant, however, was the study of Chung et al.[43] These authors showed that obese, insulin-resistant patients had low expression of HSP-72 coupled with high phosphorylation of the c-jun amino terminal kinase (JNK). When these individuals were treated with heat shocks in a bathtub, however, the subjects responded by increasing the expression of HSP-72 and inhibiting JNK phosphorylation, thus protecting against hyperglycemia, hyperinsulinemia, glucose intolerance, and insulin resistance. The authors went further and induced HSP-72 expression in mice by two other means: transgenic activation and a hydroxylamine derivative (BGP-15) in vitro adipose tissue. Even if the precise mechanism by which the HSPs protected against insulin resistance has not been fully elucidated, it was clear that HSP-72 acts by inhibiting JNK phosphorylation, which is known to inhibit insulin signal transduction.

Although an increased HSP expression can be produced by hyperglycemic stress, and a whey hydrolyzate is capable of inducing additional expression of HSPs, the exact effect of this overexpression has not been evaluated in hyperglycemic animals or humans. The currently available data, however, strongly suggest that, as in the case of exhausting exercise, the outcome with hyperglycemia could be something analogous. This is to say that the response to some peptides present in a medium DH could magnify the expression of HSPs and will likely contribute to coping with the hyperglycemic state (Figure 3). In the work of De Moura et al.[36] however, the WPH did not have any effect in heart, kidney, or spleen tissues, indicating that there is great selectivity of action. Such selectivity could be determined by various factors, mainly including structure of the peptide and its resulting hydrophobicity. In addition to that, and still depending on the previous findings, resistance to hydrolysis, concentration, and ease of transportation and delivery to target cells should be important.

8. POSSIBLE STRATEGIES FOR A MORE RATIONAL USE OF WHEY PEPTIDES

It is true that not only WPs produce bioactive compounds with possible application in the management of diabetes. Many products of vegetable origin—polyphenols, peptides, or other small molecules—are capable of lowering blood glucose levels, but in view of the various metabolic avenues of both nutritional and medical interest, the WPs tend to become increasingly more attractive than some other sources of bioactives or even pharmaceutical compounds. In this sense, and also considering the trend that world production of milk whey may grow at a pace sufficiently high to meet the increasing demand in a foreseeable future, the synthetic peptide alternative will appear very attractive.

One possible disadvantage from focusing on one specific peptide or on too narrow a selection of peptides as an effective therapeutic strategy can be the automatic exclusion of possible coadjuvant peptides whose functions are still unknown. Nongonierma et al.[44] have evaluated 10 fractions of a hydrolyzate in their capacity to reduce glycemic levels and found that the most potent insulinotropic fraction is the one rich in free amino acids and relatively hydrophilic peptides. Based on the different methods of fractionation, these authors observed that no fraction expressed the power of the whole mixture present in the hydrolyzate, something that is consistent with biological systems, which take advantage of the structure–function relationships for an array of alternate pathways. An example that illustrates the shortcomings of an immediate adoption of selected peptides or free amino acids as a complementary therapy for insulin resistant or T2D patients is the recent identification of whey tripeptides that have shown in vitro anti-inflammatory activity.[45] Using 3T3-F442A murine preadipocytes, these authors found that both Ile-Pro-Pro and Val-Pro-Pro inhibited the expected reduction of adipokine levels and activation of NF-κB pathway, which are pro-inflammatory signs. In addition, the two tripeptides induced beneficial adipogenic cell differentiation, as attested by the lower accumulation of intracellular lipid, the upregulation of peroxisome proliferator-activated receptor-γ and the increase in secretion of adiponectin, a protective lipid hormone. It is interesting to note that these effects are similar to those induced by insulin.

9. CONCLUSIONS

Studies conducted with both animals and humans have shown that intake of the WPs has the benefit of reducing blood glucose levels by mechanisms that can both improve insulin production and sensitivity, and additionally promote the insulin-independent translocation of GLUT-4 to the cell membrane. For the latter, prehydrolyzed WP proved to be far more effective than the intact proteins present in whey. The crude mixture of peptides and free amino acids also raised the expression of muscle HSP-70, consequently reducing overall stress, and substantially increased muscle glycogen stores. The glucose-lowering effect is expected to be dependent on the dose, but todate, has not shown to offer the risk of overshooting. So far, as an alternative to assist in the management of T2D or prediabetic states, food supplements of intact WP (WPC) or hydrolyzates of medium DH may be effective for improving glucose homeostasis. At present, the use of the whole, unhydrolyzed (or intact) WP given as a premeal has been proven to work in type 2 diabetic patients, but the hydrolyzate still remains to be tested in humans.

According to the current state of research, results indicate that doses between 10 and 20 g of WPC, taken 30 min before a meal, can show beneficial effects on glycemic control and should be on the safe side. Furthermore, the use of WPC and/or WPH may have the additional benefits of enhancing the immune system and reducing the levels of stress and inflammation because of their ability to mitigate some of the common complications arising from diabetes, particularly the decreased gene expression of HSPs. Future research in this area should exponentially increase our knowledge on the wide range of metabolic functions that whey peptides can perform in the body, so that more effective food-based therapies for T2D may be developed. For the time being, however, specific selections of WPH peptides or semicrude fractions thereof do not seem to carry either a sufficient wide range of actions or potency to be used as a totally dependable therapy.

Acknowledgments

The authors thank the support of the State of São Paulo Foundation for Research (Fundação de Amparo a pesquisa do Estado de São Paulo, Brazil—FAPESP, Proc 2013/06494-5 and 2013/02862-0). The kind donation of whey material by the Hilmar Ingredients, of Hilmar, CA, USA is also acknowledged.

References

1. Amaya FJ, Lee TC, Chichester CO. Biological inactivation of proteins by the Maillard reaction. Effect of mild heat on the tertiary structure of insulin. *J Agric Food Chem* 1976;**24**:465–7.
2. Hall WL, Millward DJ, Long SJ, et al. Casein and whey exert different effects on plasma amino acid profiles, gastrointestinal hormone secretion and appetite. *Br J Nutr* 2003;**89**:239–48.
3. Imperatore G, Liese AD, Boyle JP, et al. Projections of type 1 and type 2 diabetes burden in the US population aged <20 years through 2050. Dynamic modeling of incidence, mortality, and population growth. *Diabetes Care* 2012;**35**:2515–20.
4. Smithers GW. Whey and whey proteins – from gutter-to-gold. *Int Dairy J* 2008;**18**:695–704.
5. Hulmi JJ, Lockwood CM, Stout JR. Review: effect of protein/essential amino acids and resistance training on skeletal muscle hypertrophy: a case for whey protein. *Nutr Metabol* 2010;**7**:1–11.

6. Jakubowicz D, Froy O, Ahren B, et al. Incretin, insulinotropic and glucose-lowering effects of whey protein pre-load in type 2 diabetes: a randomised clinical trial. *Diabetologia* 2014;**57**:1807–11.
7. Mortensen LS, Holmer-Jensen J, Hartvigsen ML, et al. Effects of different fractions of whey protein on postprandial lipid and hormone responses in type 2 diabetes. *Eur J Clin Nutr* 2013;**66**:799–805.
8. Ma J, Stevens JE, Cukier K, et al. Effects of a protein preload on gastric emptying, glycemia, and gut hormones after a carbohydrate meal in diet-controlled type 2 diabetes. *Diabetes Care* 2009;**32**: 1600–2.
9. Mortensen LS, Hartvigsen ML, Brader LJ, et al. Differential effects of protein quality on postprandial lipemia in response to a fat-rich meal in type-2 diabetes: comparison of whey, casein, gluten, and cod protein. *Am J Clin Nutr* 2009;**90**:41–8.
10. Pal S, Ellis V, Dhaliwal S. Effects of whey protein isolate on body composition, lipids, insulin and glucose in overweight and obese individuals. *Br J Nutr* 2010;**104**:716–23.
11. Akhavan T, Luhovyy BL, Panahi S, et al. Mechanism of action of pre-meal consumption of whey protein on glycemic control in young adults. *J Nutr Biochem* 2014;**25**:36–45.
12. Nilsson M, Holst JJ, Bjorck IM, et al. Metabolic effects of amino acid mixtures and whey protein in healthy subjects: studies using glucose-equivalent drinks. *Am J Clin Nutr* 2007;**85**:996–1004.
13. Frid AH, Nilsson M, Holst JJ. Effect of whey on blood glucose and insulin responses to composite breakfast and lunch meals in type 2 diabetic subjects. *Am J Clin Nutr* 2005;**82**:69–75.
14. Petersen BL, Ward LS, Bastian ED, et al. A whey protein supplement decreases post-prandial glycemia. *Nutr J* 2009;**8**:47. http://www.nutritionj.com/content/8/1/47.
15. Badr G, Mohany M, Metwalli A. Effects of undenatured whey protein supplementation on CXCL12- and CCL21-mediated B and T cell chemotaxis in diabetic mice. *Lipids Health Dis* 2010;**10**:203.
16. Shertzer HG, Woods SE, Krishan M, et al. Dietary whey protein lowers the risk for metabolic disease in mice fed a high-fat diet. *J Nutr* 2011;**141**:582–7.
17. Morato PN, Lollo PCB, Moura CS, et al. Whey protein hydrolysate increases the translocation of GLUT-4 to the plasmatic membrane independent of the insulin response in sedentary, exercised Wistar rats. *PLoS One* 2013;**8**:e71134.
18. Morato PN, Lollo PCB, Moura CS, et al. A dipeptide and an amino acid present in whey protein hydrolysate increase translocation of GLUT-4 to the plasma membrane in Wistar rats. *Food Chem* 2013; **139**:853–9.
19. Belobrajdic DP, McIntosh GH, Owens JA. A high-whey-protein diet reduces body weight gain and alters insulin sensitivity relative to red meat in Wistar rats. *J Nutr* 2004;**134**:1454–8.
20. Konrad B, Anna D, Marek S, et al. The evaluation of dipeptidyl peptidase (DPP)-IV, alpha-glucosidase and angiotensin converting enzyme (ACE) inhibitory activities of whey proteins hydrolyzed with serine protease isolated from Asian pumpkin (*Cucurbita ficifolia*). *Int J Pept Res Ther* 2014;**20**:483–91.
21. Evert AB, Ridell MC. Lifestyle intervention. Nutrition therapy and physical activity. *Med Clin North Am* 2015;**99**:69–85.
22. Kahn SE, Cooper ME, Del Prato S. Pathophysiology and treatment of type 2 diabetes: perspectives on the past, present, and future. *Lancet* 2014;**383**:1068–83.
23. Campbell AP, Rains TM. Dietary protein is important in the practical management of prediabetes and type-2 diabetes. *J Nutr* 2015; **145**:164–9.
24. Tassi EM, Amaya-Farfan J, Azevedo JRM. Hydrolyzed alpha-lactalbumin as a source of protein to the exercising rat. *Nutr Res* 1998;**5**:875–81.
25. Pimenta FVM, Abecia-Soria MI, Auler F, et al. Physical performance of exercising young rats fed hydrolysed whey protein at a sub-optimal level. *Int Dairy J* 2006;**16**:984–91.
26. Nery-Diez AC, Carvalho IR, Amaya-Farfan J, et al. Prolonged ingestion of prehydrolyzed whey protein induces little or no changes in digestive enzymes, but decreases glutaminase activity in exercising rats. *J Med Food* 2010;**13**:992–8.
27. Bowen J, Noakes M, Clifton PM. Appetite hormones and energy intake in obese men after consumption of fructose, glucose and whey protein beverages. *Int J Obes* 2007;**31**:1696–703.
28. Tulipano G, Sibilia V, Caroli AM, et al. Whey proteins as source of dipeptidyl dipeptidase IV (DPP-4) inhibitors. *Peptides* 2011;**32**:835–8.
29. Bryant NJ, Govers R, James ED. Regulated transport of the glucose transporter GLUT4. *Nat Rev* 2002;**3**:267–77.
30. Deacon CF, Ahrén B. Physiology of incretins in health and disease. *Rev Diabet Stud* 2011;**8**:293–306.
31. Jakubowicz D, Froy O. Biochemical and metabolic mechanisms by which dietary whey protein may combat obesity and type 2 diabetes. *J Nutr Biochem* 2013;**24**:1–5.
32. Dicker D. DPP-4 inhibitors: impact on glycemic control and cardiovascular risk factors. *Diabetes Care* 2011;**34**:S276–8.
33. Power O, Hallihan A, Jakeman P. Human insulinotropic response to oral ingestion of native and hydrolysed whey protein. *Amino Acids* 2009;**37**:333–9.
34. Power O, Nongonierma AB, Jakeman P, et al. Food protein hydrolysates as a source of dipeptidyl peptidase IV inhibitory peptides for the management of type-2 diabetes. *Proc Nutr Soc* 2014;**73**:34–46.
35. Balch WE, Morimoto RI, Dillin A, et al. Adapting proteostasis for disease intervention. *Science* 2008;**319**:916–9.
36. De Moura CS, Lollo PCB, Morato PN, et al. Whey protein hydrolysate enhances the exercise-induced heat shock protein (HSP 70) response in rats. *Food Chem* 2013;**136**:1350–7.
37. Rains JL, Jain SK. Oxidative stress, insulin signaling, and diabetes. *Free Rad Biol Med* 2011;**50**:567–75.
38. Dungan KM, Braithwaite SS, Preiser JC. Stress hyperglycemia. *Lancet* 2009;**373**:1798–807.
39. Barutta F, Pinach S, Giunti S, et al. Heat shock protein expression in diabetic nephropathy. *Am J Physiol Renal Physiol* 2008;**295**:F1817–24.
40. Reddy VS, Kumar ChU, Raghu G, et al. Expression and induction of small heat shock proteins in rat heart under chronic hyperglycemic conditions. *Arch Biochem Biophys* 2014;**558**:1–9.
41. Simar NJ, Jacques A, Caillaud C. Heat shock proteins induction reduces stress kinases activation, potentially improving insulin signalling in monocytes from obese subjects. *Cell Stress Chaperones* 2012;**17**:615–21.
42. Lee J-H, Gao J, Kosinski PA, et al. Heat shock protein 90 (HSP90) inhibitors activate the heat shock factor 1 (HSF1) stress response pathway and improve glucose regulation in diabetic mice. *Biochem Biophys Res Com* 2013;**430**:1109–13.
43. Chung J, Nguyen AK, Henstridge DC, et al. HSP72 protects against obesity-induced insulin resistance. *Proc Natl Acad Sci USA* 2008; **105**:1739–44.
44. Nongonierma AB, Gaudel C, Murray BA, et al. Insulinotropic properties of whey protein hydrolysates and impact of peptide fractionation on insulinotropic response. *Int Dairy J* 2013;**32**:163–8.
45. Chakrabarti S, Wu J. Milk-derived tripeptides IPP (Ile-Pro-Pro) and VPP (Val-Pro-Pro) promote adipocyte differentiation and inhibit inflammation in 3T3-F442A cells. *PLoS One* 2015;**10**(2):e0117492. http://dx.doi.org/10.1371/journal.pone.0117492.

CHAPTER

18

Dietary Fatty Acids and C-Reactive Protein

Giovanni Annuzzi, Ettore Griffo, Giuseppina Costabile, Lutgarda Bozzetto

Department of Clinical Medicine and Surgery, Federico II University, Naples, Italy

1. INTRODUCTION

Several studies have shown that low-grade inflammation plays an important role in numerous chronic diseases, including cardiovascular disease (CVD), obesity, and diabetes mellitus.[1–3] The process involves increased production of several inflammatory factors of which the C-reactive protein (CRP) is the most studied. CRP is an acute-phase protein that has been shown to be a marker of systemic inflammation, elevated in response to injury, infection, and other inflammatory stimuli.[4] Its hepatic production is directly related to interleukin-6 (IL-6) stimulation and, unlike other acute-phase reactants, its levels are stable over long periods in the absence of new stimuli.[5] In contrast to many other inflammatory markers, assay techniques for high sensitivity (hs)-CRP are reliable, fully automated, and sensitive, providing a simple clinical tool for the assessment of systemic inflammation.[6]

CRP seems to directly participate in the atherogenic process as suggested by the evidence that it is able to bind to lipids[7] and opsonize native low-density lipoprotein (LDL) for macrophages,[8] and is present in atherosclerotic plaques.[9,10] Systemic inflammation was linked to endothelial dysfunction,[11] and CRP levels were elevated in patients with impaired endothelial vasoreactivity and coronary artery disease.[12,13]

Circulating levels of CRP have been associated directly to a number of cardiovascular risk factors, such as obesity, smoking, blood pressure, serum triglycerides, apolipoprotein B, fasting blood glucose, heart rate, serum fibrinogen and inversely to high-density lipoprotein (HDL)-cholesterol levels, both in children and in adults.[14,15] In 27,939 participants in the Women's Health Study followed for a mean of 8 years, increasing levels of CRP were associated with increased risk of cardiovascular events independently of LDL cholesterol levels, indicating an additive value of CRP to lipid screening in terms of coronary risk prediction.[16] Using widely available high-sensitivity assays, CRP levels of <1, 1–3, and >3 mg/L correspond to low-, moderate-, and high-risk groups for future cardiovascular events (Figure 1).[17]

Because of the overall larger evidence of cardiometabolic relationships with CRP than other serum inflammatory markers, for the purpose of this review we will focus only on CRP concentrations. Although a large panel of serum markers, including cytokines, chemokines, and cell adhesion molecules, is available, investigating only one marker may allow a better, clearer definition of the relations with different outcomes. This is especially true considering that it is difficult to define specific pro- or anti-inflammatory patterns, mainly because the function of a specific cytokine depends on the amount, the target cell, the producing cell, and the sequence of actions of cytokines.[18] Moreover, especially in healthy nonobese populations with low levels of inflammatory markers, the large number of values below detection limits may be a strong limitation for their evaluation.

For the sake of simplicity, aiming at clarifying, only studies in humans are considered in this review, apart from in vitro and animal mechanistic studies. Because of the relatively few data published on this issue concerning patients with diabetes, a relevant part of this chapter refers to data in nondiabetic population.

2. CRP AND DIABETES

Low-grade inflammation may play a pathogenic role in type 2 diabetes (T2D) (Figure 2).[19] Elevated levels of CRP and IL-6 predict insulin resistance and subsequent development of T2D.[20] First evidence came from the Women's Health Study, wherein 188 women who developed diabetes over a 4-year follow-up period were defined as cases and matched by age and fasting status with 362 disease-free controls.[21] The relative risk of

Molecular Nutrition and Diabetes
http://dx.doi.org/10.1016/B978-0-12-801585-8.00018-X

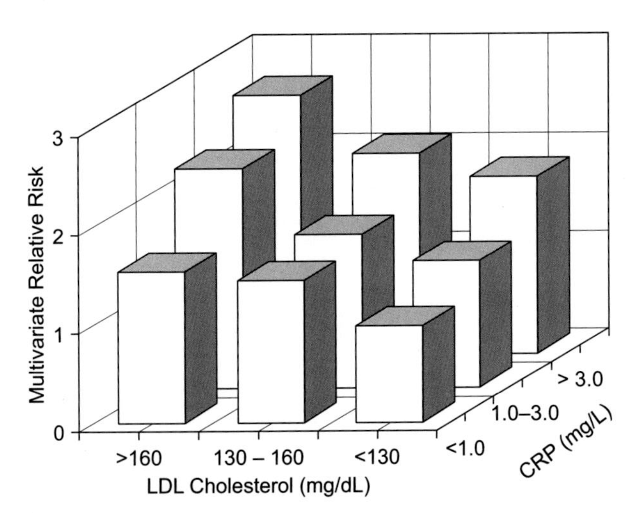

FIGURE 1 Multivariable-adjusted relative risks of cardiovascular disease in the 27,939 participants in the Women's Health Study according to levels of C-reactive protein (CRP) and categories of low-density lipoprotein (LDL) cholesterol. *Figure reproduced from Ridker PM.*[17]

future diabetes in the highest versus lowest quartile of CRP was 15.7 and remained elevated after adjustment for body mass index (BMI), family history of diabetes, smoking, exercise, use of alcohol, and hormone replacement therapy. Elevated levels of CRP have been associated also with development of chronic diabetic complications.[22,23]

Elevation of the concentrations of pro-inflammatory molecules in diabetic patients might be caused by hyperglycemia and oxidative stress. Hyperglycemia seems to activate the immune and macrophage-monocyte systems and to stimulate the production of cytokines and acute-phase proteins. In addition, hyperglycemia induces oxidative stress that is responsible for the activation of the nuclear factor-κB (NF-κB), which increases serum levels of circulating pro-inflammatory cytokines.[24] In small human studies, the anti-inflammatory therapy improved glycemia and β-cell function in T2D patients,[25,26] and reduced CRP in at-risk obese diabetic and nondiabetic subjects.[27]

Abdominal adiposity was associated with elevated CRP independent of BMI in healthy nonobese people.[28] Similarly, in a cohort of patients with T2D ($n = 382$), BMI was associated with CRP and IL-6 levels, and adipose tissue distribution was a specific determinant of systemic inflammation.[29] Moreover, in patients with well-controlled T2D, inflammation was more related to abdominal obesity than glycemic control.[30]

It is controversial whether CRP levels are altered in diabetic patients without obesity. In fact, although higher inflammatory markers were observed in obese T2D subjects but not in nonobese ones,[31] in other studies hs-CRP levels were significantly higher in diabetic patients than in nondiabetic controls and positively associated with diabetes even after adjusting for obesity markers.[32,33]

Lifestyle modifications and drugs in use for the management of diabetes also have additional anti-inflammatory effects. In the Diabetes Prevention Program, the intensive lifestyle intervention including weight reduction, which was able to reduce the incidence of diabetes in individuals with impaired glucose tolerance (IGT), also decreased the levels of CRP by 31%, whereas metformin decreased CRP by only 13%.[34] In the Look AHEAD (Action for Health in Diabetes) study, 1-year lifestyle intervention for weight loss reduced high CRP levels in individuals with T2D.[35] This suggests that lifestyle interventions, even without drug therapy, may decrease insulin resistance

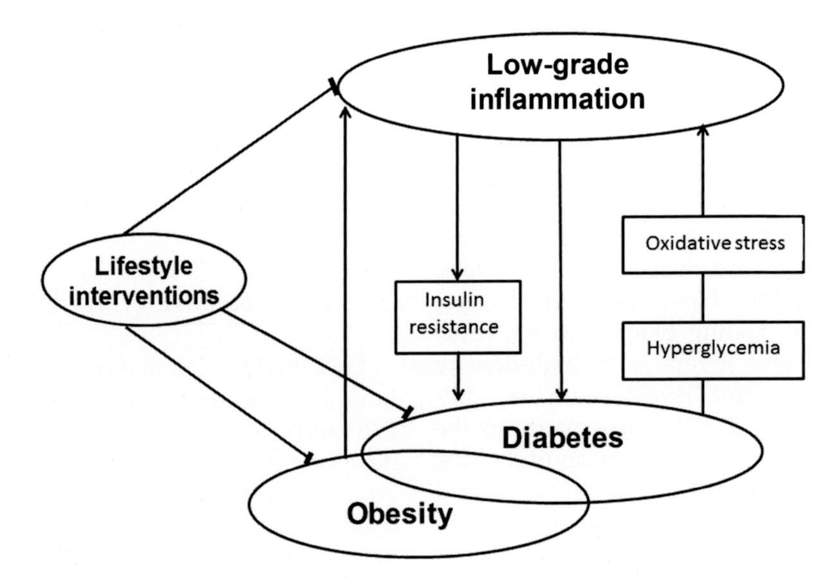

FIGURE 2 Possible role of low-grade inflammation as a pathogenic factor for and a disease mechanism of type 2 diabetes.

and the progression of prediabetes states to diabetes and thus eventually decrease development and progression of diabetes complications also by reducing low-grade inflammation.

3. DIET AND CRP

Association and intervention studies have shown that diet may influence systemic inflammation status, with most evidence coming from weight-losing dietary approaches.

Weight loss diet. Several lifestyle intervention studies have shown a weight loss–mediated improvement of CRP levels. Weight loss reached through an energy-restricted diet rich in plant-derived α-linolenic acid (ALA) significantly reduced CRP levels in 81 overweight/obese patients with metabolic syndrome traits.[36] Similarly, eight men with IGT participating in a weight loss program lost on average 15 kg of body weight and decreased arterial CRP level by 45%.[37] In healthy, moderately obese men, a moderate weight loss (10 kg) significantly improved markers of low-grade systemic inflammation in the fasting state and during the postprandial period.[38] In 83 healthy, obese women placed on very low-fat, energy-restricted diets, after 12 weeks, mean weight loss was 7.9 kg and CRP was significantly decreased by 26%, and a correlation was observed between weight loss and change in CRP.[39]

A systematic review performed in 2007 showed that weight loss was associated with a decline in CRP level across all types of interventions. For each 1 kg of weight loss, the overall mean change in CRP was −0.13 mg/L.[40] The largest changes in CRP level were observed in those surgical intervention studies that demonstrated the most pronounced weight change.

Dietary patterns. At difference with weight loss, studies focusing on quality of diet/dietary patterns have provided some inconsistent results, apart from the positive relation observed between CRP levels and Western-type diets. In fact, CRP concentrations were significantly and positively correlated with the Western diet score, expressing higher intakes of red meats, high-fat dairy products, and refined grains, in a subsample of men ($n = 466$) from the Health Professionals Follow-up Study.[41] Similarly, in 732 women from the Nurses' Health Study I cohort, CRP concentrations were associated positively with the Western pattern characterized by higher intakes of red and processed meats, sweets, desserts, French fries, and refined grains and inversely with the prudent dietary pattern characterized by higher intakes of fruit, vegetables, legumes, fish, poultry, and whole grains.[42] The relations remained significant after adjustment for age, BMI, physical activity, smoking status, and alcohol consumption.

High-fat, low-carbohydrate diets have received considerable attention in recent years, particularly for the short-term beneficial effects reported on cardiovascular risk markers in overweight and obese individuals.[43] However, the results from these studies have been inconsistent, especially regarding inflammation. CRP levels decreased modestly with both a low-carbohydrate or a conventional diet for 6 months in 78 severely obese subjects, including 86% with either diabetes or metabolic syndrome, subjects with baseline level >3 mg/L experiencing a greater decrease in CRP levels on the low-carbohydrate diet, independent of weight loss.[44] In contrast with these results, in a study on overweight/obese women, the composition of the weight loss diet-influenced systemic inflammation because the low-carbohydrate diet increased, whereas the high-carbohydrate diet reduced serum CRP.[45] In line with this observation, Keogh et al.[46] showed that after 8 weeks of weight loss, CRP decreased more with the high-carbohydrate diet than with the low-carbohydrate diet. Some of these discrepancies may be due to the greater weight loss achieved by the high-fat, low-carbohydrate compared with low-fat, high-carbohydrate diets in nonisoenergetic comparisons.[47]

In patients with T2D randomly advised to follow a low-carbohydrate diet or a low-fat diet for 6 months, the effects on weight reduction were similar and CRP levels did not show any significant changes within the groups.[48]

There is wide evidence for the beneficial health effects of the Mediterranean dietary pattern, recently confirmed in a huge randomized trial in persons at high cardiovascular risk, showing that a Mediterranean dietary pattern supplemented with extra virgin olive oil or nuts reduced the incidence of major cardiovascular events by 30%.[49] As for its effects on inflammatory status, a Mediterranean-style diet—higher intakes of fruits, vegetables, nuts, whole grains, and olive oil—showed a decrease in serum hs-CRP with a concomitant greater decrease in body weight in the intervention group compared with the control group in individuals with metabolic syndrome without CVD.[50] On the contrary, Michalsen et al.[51] reported no effects on biomarkers of inflammation with a Mediterranean diet in patients with medically treated coronary artery disease. A recent systematic review and meta-analysis confirmed that a Mediterranean dietary pattern was associated with a significant reduction in CRP levels.[52]

Park et al.[53] evaluated the Alternate Healthy Eating Index (AHEI) and the Alternate Mediterranean Diet Score (aMED) from self-administered 110-item food-frequency questionnaires estimating usual nutrient intake over the past year in 71 males and 80 females, more than 35 years old, and obese (43.7%). CRP was negatively correlated with AHEI and aMED scores.

The percentage of energy from carbohydrate was inversely associated with CRP.

Thus, a Mediterranean-style diet, high in oleic acid, fiber, and antioxidants, may reduce low-grade inflammation, and corresponding coronary events, in middle-aged adults; however, these dietary effects in patients with CVD need further investigation. Moreover, the role of individual dietary factors characterizing the Mediterranean diet in modulating inflammation is not yet defined. In a cross-sectional evaluation of the associations between components of the Mediterranean diet and circulating markers of inflammation in 772 participants in the PREDIMED (Prevención con Dieta Mediterránea) study,[54] after adjusting for age, gender, BMI, diabetes, smoking, use of statins, nonsteroidal anti-inflammatory drugs, and aspirin, a higher consumption of fruits and cereals was associated with lower concentrations of IL-6 but not hs-CRP. The only factor significantly associated with lower concentrations of CRP was a higher intake of dairy products. Moreover, participants with higher adherence to the Mediterranean-type diet did not show significantly lower concentrations of inflammatory markers.

Although a diet rich in whole wheat-fiber products had no effect on inflammation markers,[55] different results were observed by de Mello et al.[56] in overweight and obese individuals at high risk of developing diabetes assigned to either a diet high in fatty fish, bilberry, and whole grain products (Healthy Diet, $n = 36$), or a whole grain–enriched diet (WGED, $n = 34$) or a control diet ($n = 34$). In this study, after completing the 12 week dietary intervention, plasma hs-CRP levels decreased in the Healthy Diet and WGED groups in participants not using statins, also after adjusting for confounding factors, including BMI or insulin sensitivity.

Another dietary component that has recently received particular attention for its possible favorable cardiometabolic effects is polyphenols, the substances that give fruits and vegetables their colors.[57] In vitro data suggest other mechanisms of action for polyphenols beyond antioxidant properties, including anti-inflammatory effects.[58] However, the results of epidemiological and clinical intervention studies on the effects of polyphenol-rich foods on inflammation markers have been contradictory.[59] In a randomized controlled trial in individuals at high risk of T2D and CVD, diets naturally rich in polyphenols significantly reduced plasma triglyceride concentrations, in both fasting and postprandial conditions, improved glucose metabolism and reduced oxidative stress, as detected by an about 20% decrease in urinary 8-isoprostane concentration.[60,61] In this study, the 8-week diet rich in polyphenols (~3 g/day) from different sources induced no significant effects on fasting and postprandial hs-CRP plasma concentrations.[62] It therefore remains unproven that clinical effects of dietary polyphenols are mediated by changes in inflammatory state.

In summary, there are evidences that dietary patterns may influence chronic subclinical inflammation as indicated by changes in circulating CRP concentrations. In particular, the Western diet increases and the Mediterranean diet decreases CRP levels. The role of different dietary components remains controversial.

4. DIETARY FATTY ACIDS AND CRP

An increased dietary intake of saturated fatty acids (SFAs) and *trans* fatty acids (TFAs) is generally considered a risk factor for CVD and metabolic syndrome. The main indication by dietary guidelines for CVD prevention is a reduction in SFA. A partial explanation of the differential effects of dietary fatty acids on CV risk could be a diverse effect on low-grade inflammation. Their dietary effects on CRP levels may be influenced by interactions with CRP genotypes. Nienaber-Rousseau et al.[63] determined 12 single nucleotide polymorphisms (SNPs) in the CRP gene in 2010 black South Africans and observed that an increase in both SFA and monounsaturated fatty acid (MUFA) intake in those homozygous for the polymorphic allele at rs2808630 was associated with a larger increase in CRP. The minor alleles rs3093058 and rs3093062 were associated with significantly higher CRP in the presence of increased triglyceride or cholesterol intakes. A high ratio of n-6/n-3 intake increased inflammation in those harboring the major allele at rs3093058 and rs3093062, whereas the ratio was associated with a greater reduction in inflammation in those harboring the variant allele. Therefore, lipid intake may modulate the pro-inflammatory effects of some SNPs.

4.1 Saturated Fatty Acids

Mechanistic studies. Several mechanisms investigated in in vitro and animal studies might sustain the association between SFA and inflammation, as recently reviewed by Estadella et al.[64] and Santos et al.[65] SFA may deteriorate insulin sensitivity through increased production of cytokines by hypertrophic adipocytes and infiltrating macrophages.[66] Treatment of primary mouse hepatocytes and pancreatic cells with palmitic acid caused sustained JNK activation and insulin resistance and inhibited glucose-induced insulin gene transcription. This effect may be mediated by interference of autocrine insulin signaling through the phosphorylation of insulin receptor substrates 1 and 2.[67] SFA could stimulate inflammatory signaling pathways by a process that involves Toll-like receptor 4[68] and, subsequently, NF-κB, thus increasing the expression of a number of inflammatory genes.[69–72] SFA might also increase macrophages inflammation through their uptake and metabolic processing into ceramide.[73] The effects of SFA on adipokines could occur through their binding and inhibition of the nuclear receptor peroxisome proliferator-activated receptor (PPAR)-γ. It has been shown that dietary SFA decrease the levels of arachidonic acid—the precursor for the PPAR-γ ligands—in adipocyte plasma membrane phospholipids[74] or could be involved in a possible impairment in ligand-dependent activity of PPAR-γ.[75]

Association studies. Evidence of a positive association between diets with a high content of SFA and CRP levels comes from cross-sectional studies. King et al.[76] examined the relation of dietary factors, evaluated by 24-h recall, to levels of hs-CRP in 4900 adult participants in the 1999 to 2000 National Health and Nutrition Examination Survey (NHANES 1999–2000). After controlling for demographic factors, BMI, smoking, alcohol consumption, exercise, and total caloric intake, subjects in the third and fourth highest quartiles of saturated fat consumption had modestly elevated risk of high levels (>3 mg/L) of CRP (third quartile: odds ratio (OR) 1.58, 95% confidence interval (CI) 1.02–2.44; fourth quartile 1.44, 95% CI 0.80–2.58) compared with the lowest quartile.

Food intake was assessed with a validated semiquantitative food frequency questionnaire in 395 noninstitutionalized inhabitants of Porto (Portugal), age range 26–64 years.[77] After adjusting for age, education, regular physical exercise, smoking, and central body fat percentage, hs-CRP was significantly and positively associated with lauric and myristic acids and with SFA/polyunsaturated fatty acid (PUFA) ratio in men, but not in women.

Kalogeropoulos et al.[78] evaluated the associations between dietary (semiquantitative food frequency questionnaire) and plasma (gas chromatography) fatty acids with various inflammation markers in 374 free-living, apparently healthy men and women, randomly selected from the ATTICA (so-named because the participants were from Attica, Greece) study database. They showed that plasma SFA were positively associated with hs-CRP levels. Multiadjusted regression analyses revealed that plasma n-3 and n-6 fatty acids and MUFAs were inversely associated with CRP. At difference with plasma levels, no associations were observed between dietary fatty acids and CRP.

In a cross-sectional population-based study on 264 Swedish men and women aged 70 years, after adjusting for BMI, smoking, physical activity, alcohol consumption, and lipid-lowering therapy, there were no relationships between levels of myristic, palmitic, or stearic acids, measured in serum cholesteryl esters, and hs-CRP.[79] However, the proportion of palmitoleic acid, reflecting high intake of saturated fat, was positively correlated with CRP concentrations.

Some other studies have not confirmed this association. The association between SFA intake and elevated hs-CRP (≥1.0 mg/L) was not significant in a Japanese population of 443 young healthy and lean women with a rather low rate of elevated hs-CRP concentrations (5.6%),[80] and in 335 Italian subjects with multiple metabolic abnormalities.[81]

Causes of discrepancy among these studies could be the different types of populations examined, in terms of either genetic and dietary backgrounds, the use of different cutoff values for defining elevated hs-CRP (i.e., >3.0 mg/L in the NHANES study or ≥1.0 mg/L in the study of young Japanese women) or that the role of dietary factors such as SFA on inflammation could be less evident in dysmetabolic than in healthy subjects.

Intervention studies. The replacement of SFA with MUFA lowers LDL cholesterol concentrations and may improve insulin sensitivity, measured by an intravenous glucose tolerance test, as shown in the KANWU (Kuopio, Aarhus, Naples, Wollongong, Uppsala) study.[82] The effect on insulin sensitivity was not reproduced in the five-center RISCK (Reading, Imperial, Surrey, Cambridge, and Kings) trial[83] that confirmed the decreased total and LDL cholesterol and apolipoprotein B concentrations with the replacement of SFA with MUFA or carbohydrates. In this study, the 24-week intervention on 548 participants was unable to show any effects of the type of fat also on inflammatory markers, including hs-CRP. No effects were also shown by Keogh et al.,[84] who compared a low-fat, high-glycemic load carbohydrate diet with high-SFA, high-MUFA, or high-PUFA diets in a 3-week isocaloric crossover study. The high-SFA diet impaired flow-mediated dilatation compared with all other diets and increased P-selectin compared with MUFA and PUFA, with no significant differences in hs-CRP between diets.

Dairy products. The food source of the fatty acids could be a confounding factor, preventing the effects of specific fatty acids to be disclosed. A good case for it could be the controversy on the effects of dairy products. Although it is generally recommended that animal-derived saturated fats be replaced with plant-derived unsaturated ones, some epidemiological studies have shown an improved cardiometabolic risk with dairy foods. In this respect, it was published very recently that during 14 years of follow-up in 26,930 individuals from the Malmö Diet and Cancer cohort, total intake of high-fat, but not of low-fat, dairy products (regular-fat alternatives) was inversely associated with incident T2D (hazard ratio for highest compared with lowest quintiles: 0.77; 95% CI 0.68–0.87; P for trend <0.001).[85] Intakes of both high-fat meat and low-fat meat were associated with increased risk. Intakes of SFA with 4–10 carbons, lauric acid (12:0), and myristic acid (14:0) were associated with decreased risk of T2D.

In agreement with association studies,[54] positive effects of low-fat dairy products on inflammatory status have been observed also in some small intervention studies, as shown in a blinded, randomized, crossover study in 20 obese/overweight participants comparing dairy-with soy-supplemented eucaloric diets.[86] The 28-day dairy-supplemented diet resulted in lower inflammatory markers including CRP, whereas the soy exerted no significant effect, with no significant differences between overweight and obese subjects. Therefore, this study showed that the effects of dairy products were independent of their fat content because they were observed with low-fat supplements. A recent meta-analysis evaluated the effects on CRP of low- and whole-fat dairy foods.[87] The analysis included six studies with 451 healthy adults randomized to increased dairy food for more than 1 month without additional interventions. An increase in the intake of low- and whole-fat dairy food of 3.6 servings/day was associated with an increase in body weight. For all studies combined, there was no significant change in CRP on a high-dairy diet, also when studies were stratified by duration of dietary intervention, high- or low-fat dairy, and normal or overweight subjects.

In summary, the evidence in the literature of the association between SFA and CRP is mainly sustained by cross-sectional studies, whereas the few intervention studies were generally negative. As also concluded in a systematic review of SFA on inflammation and circulating levels of adipokines, a potential positive association of SFA with hs-CRP but not with adipokines is suggested.[65]

4.2 *Trans* Fatty Acids

TFAs are unsaturated fats with at least one double bond in the *trans* configuration. TFA derived by industrial hydrogenation of vegetable oils (mainly elaidic and linolelaidic acids) have been linked to increased risk for coronary heart diseases, whereas animal-derived TFA (mainly vaccenic acid) have shown beneficial effects in animals while inducing insulin resistance in humans. The consumption of plant-derived TFA is recently increased because they are used for the preparation of fast foods, bakery products, packaged snacks, and margarines. The association of dietary TFA with higher risk of CVD[88] may be mediated by their metabolic effects, including increased incidence of T2D,[89] increased insulin resistance, and worsened plasma lipid profile, but also by their effects on subclinical inflammation.

Mechanisms of action. TFA may influence inflammation through different mechanisms, first of all through changes in the prostaglandin balance, impairing the activity of Δ6 desaturase, the enzyme responsible for the conversion of linoleic acid to arachidonic acid and other n-6 PUFA, therefore changing the phospholipid fatty acid composition at different sites.[64]

Association studies. Whether TFA intake could also act through changes in systemic inflammation was investigated by Mozaffarian et al.[90] in 823 generally healthy women in the Nurses' Health Study I and II. TFA intake was not associated with CRP concentrations overall but was positively associated with CRP in women with higher BMI.

Similarly, in a cross-sectional study of 730 women from the Nurses' Health Study I cohort, free of CVD, cancer, and diabetes, CRP levels were 73% higher among those in the highest quintile of TFA intake, compared with the lowest quintile.[91] TFA intake was positively related to plasma concentration of CRP, in linear regression models after controlling for age, BMI, physical activity, smoking status, alcohol consumption, intake of MUFA, PUFA, SFA, and postmenopausal hormone therapy.

Intervention studies. Intervention trials with TFA have provided somewhat conflicting results.

In a randomized crossover study in 50 healthy men, 8% of fat was replaced across 5-week diets with carbohydrate, oleic acid, TFA, stearic acid, TFA + stearic acid, and SFA as lauric, myristic, and palmitic acids.[92] CRP concentrations were higher after consumption of the TFA diet than after the carbohydrate or oleic acid diets, but were not significantly different from the stearic acid and the saturated diets.

No differences in CRP levels were instead observed between a diet rich in butter with a naturally high content of vaccenic acid, and a concomitantly higher content of MUFA, and a diet rich in butter with a low content of vaccenic acid in a double-blind, randomized, 5-week, parallel intervention study, in 42 healthy young men.[93] Similarly, no significant effect of dietary fat type on CRP levels were observed in moderately hypercholesterolemic subjects provided with each of six diets in randomized order containing as the major source of fat: soybean oil, semiliquid margarine, soft margarine, shortening, traditional stick margarine, or butter.[94]

Although not conclusively, TFA seems to increase circulating CRP levels, likely with a different effect of plant- versus animal-derived TFA. This adds to and could contribute to explain the deleterious cardiometabolic effects associated with the dietary intake of these fatty acids.

Conjugated TFA. Among conjugated TFA, two major isomers of conjugated linoleic acid (CLA), *cis*-9, *trans*-11 and *trans*-10, *cis*-12, show distinctive biological activities.

In a double-blind placebo-controlled trial, 60 men with metabolic syndrome were randomized to receive t10c12 CLA, a CLA mixture, or placebo for 12 weeks.[95] Supplementation with t10c12 CLA markedly increased CRP (110%) compared with placebo, independent of changes in blood glucose and lipids.

Some evidence is available in patients with diabetes. In a randomized, double-blind, placebo-controlled trial, 32 subjects with stable, diet-controlled T2D received CLA (3.0 g/day) or control for 8 weeks. CLA supplementation significantly increased fasting glucose concentrations (6.3%) and reduced insulin sensitivity as measured by homeostasis model assessment, oral glucose insulin sensitivity, and the insulin sensitivity index (composite), but had no effect on CRP.[96] Similar results were observed in a randomized, double-masked crossover study, in 35 postmenopausal, obese women with T2D treated for 16 weeks each with 8 g daily of CLA or safflower oil rich in linoleic acid.[97] Although CLA did not alter measured metabolic parameters, safflower oil decreased HbA1c, increased insulin sensitivity, and HDL cholesterol, and decreased CRP levels.

In summary, the deleterious effects by supplementation with CLA in humans inconsistently relate to CRP levels, and, therefore, further information from studies on effects and mechanisms of specific CLA isomers are required. The adverse effects on insulin and glucose metabolism observed in patients with T2D add further concerns on the consumption of CLA.

4.3 Monounsaturated Fatty Acids

The most common MUFA in daily nutrition is oleic acid, corresponding to ~90% of all MUFAs provided in the diet. Oleic acids/MUFAs are associated with a number of beneficial health effects, mainly related to cardiovascular risk. Discrepancies reported on the association between MUFA and CVD risk could be related to the different sources of MUFA used in most studies. In a Western diet, MUFA is predominantly supplied by foods of animal origin, whereas in south European countries the main source of MUFA is olive oil.[98] In a recent meta-analysis[99] on 32 cohort studies including 841,211 subjects, the comparison of the top versus bottom third of the distribution of a combination of MUFA (of both plant and animal origin), olive oil, oleic acid, and MUFA:SFA ratio resulted in a significant risk reduction for all-cause mortality, cardiovascular mortality, cardiovascular events, and stroke. Following subgroup analyses, significant associations were only found between higher intakes of olive oil and reduced risk of all-cause mortality, cardiovascular events, and stroke, respectively, whereas no significant risk reduction was observed in the MUFA subgroup. The replacement of SFA with MUFA lowers LDL cholesterol concentrations, improves insulin sensitivity,[82] and reduces liver fat content.[100] However, only olive oil seems to be associated with reduced risk. Further research is necessary to evaluate the relation between specific sources of MUFA (i.e., plant vs animal) and cardiovascular risk. Few studies have evaluated the MUFA dietary effects on inflammation, independent of weight reduction.

Intervention studies. As shown in the previously mentioned study by Baer et al.,[92] CRP concentrations were lower after consumption of the oleic acid–enriched diet than after the diets enriched with SFA and TFA. Furthermore, there was no difference in CRP between the oleic acid diet (39% fat) and the standard fat control diet (30% fat), indicating that oleic acid may offset the pro-inflammatory effects of a high-fat diet.

Junker et al.[101] compared in a sequential approach two diets for treatment of 25 nonobese male patients with fasting triglycerides >2.3 mmol/L. The first diet (high fat) was rich in MUFA and marine n-3 PUFA, whereas the second diet (low fat) was rich in complex carbohydrates and dietary fiber. CRP showed a significant decrease after the high-fat period.

On the contrary, no significant differences were observed between MUFA and SFA in the RISCK trial[83] and by Keogh et al.[84]

Equally, no difference in hs-CRP between the two interventions was observed in a sample of normal weight, apparently healthy young adults, supplied with 15 mL/day of either flaxseed oil or olive oil.[102]

There is few evidence in patients with diabetes. A 4-week crossover study in individuals with T2D compared fasting and postprandial CRP response with two dietary approaches recommended for diabetes mellitus and cardiovascular prevention—high-MUFA or high-carbohydrate/high-fiber/low-glycemic index in isocaloric conditions.[103] Fasting hs-CRP plasma concentrations were similar after the two diets (Figure 3). However, hs-CRP postprandial patterns differed between the diets (i.e., hs-CRP concentrations decreased after the MUFA-rich meal, but not after the carbohydrate/fiber meal). Triglyceride-rich lipoprotein (TRL) levels were lower with carbohydrate/fiber than MUFA diet, and correlated with hs-CRP levels only after the carbohydrate/fiber diet. Therefore, the MUFA-rich diet and the carbohydrate/fiber-rich diet seemed to influence hs-CRP levels through different mechanisms (i.e., direct acute postprandial effects by MUFA and TRL-mediated effects by carbohydrate/fiber).

In an acute study,[104] first-degree relatives of T2D patients and control healthy volunteers ingested in

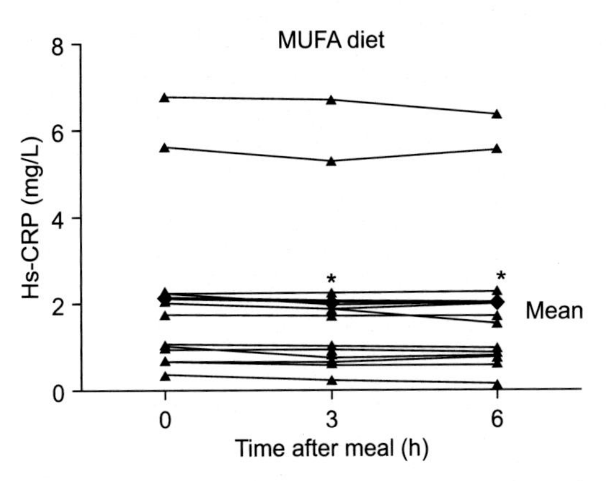

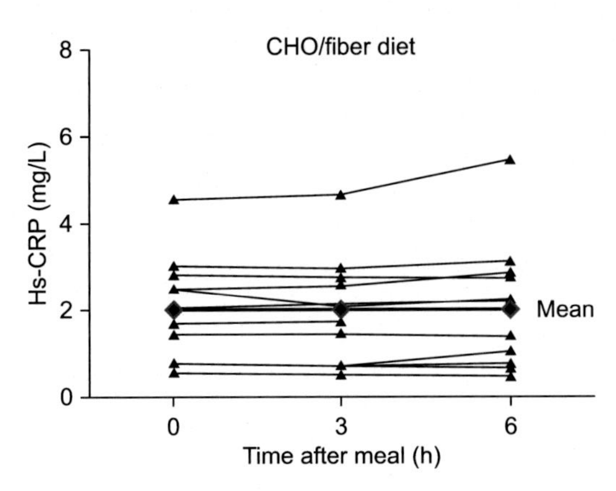

FIGURE 3 High-sensitivity C-reactive protein (hs-CRP) plasma concentrations at fasting and after the test meal at the end of a high-monounsaturated fatty acid (MUFA) or a high-carbohydrate/high-fiber/low-glycemic index [CHO/fiber] diet (*$p < 0.05$, repeated measures analysis of variance). *Figure reproduced from Bozzetto et al.*[102]

randomized order two fat-rich meals, one with 72% energy from MUFA (predominantly oleic acid and palmitoleic acid from macadamia nut oil) and the other with 79% energy from medium- (predominantly lauric acid) and long-chain SFA (predominantly myristic acid) from coconut oil. Mean plasma hs-CRP responses to either meal (adjusted for baseline concentrations, age, gender, BMI, and total body fat percentage) did not differ between relatives and controls. At difference with the previous study, no changes in hs-CRP in response to either meal were detected. The purely acute nature of the latter study and the difference in the source of oleic acid (olive oil vs macadamia nut oil) might explain the different postprandial results.

As a whole, data on MUFA effects on systemic inflammation are scarce and not univocal, mainly because of differences in the food source of the fatty acids and the type of comparator. However, the concept remains, needing further investigations, that an anti-inflammatory effect may be an additional mechanism explaining the relation between dietary patterns characterized by high oleic acid from extra virgin olive oil, such as Mediterranean diets, and a lower incidence of coronary artery disease.

4.4 n-6 PUFA

n-6 PUFAs are generally considered pro-inflammatory. In particular, linoleic acid can be converted into arachidonic acid that is the major substrate for eicosanoids production, which plays an important role in regulating inflammatory and immune responses. Most of the arachidonic acid-derived eicosanoid products are pro-inflammatory, except resolvins and lipoxins, which have anti-inflammatory effects.[105]

Associations between dietary intakes of n-6 and n-3 PUFA and CRP were shown in 843 individuals assessed in 1994–1996 for dietary intakes and 12 years later for CRP concentrations.[106] After adjustment for sociodemographic, lifestyle, anthropometric, and dietary variables, inverse associations were observed between intakes of total n-3 PUFA (ALA, ALA + eicosapentaenoic acid (EPA), EPA + docosapentaenoic acid (DPA), DPA + docosahexaenoic acid (DHA)) and n-6 PUFA (linoleic acid + arachidonic acid) and elevated CRP. These associations were substantial only in individuals with low intakes of vitamin E.

Intervention studies have not shown significant effects of n-6 PUFA on CRP levels. Junker et al.[107] observed no differential effects on CRP levels in 58 healthy students receiving either a 4-week rapeseed oil (high content of MUFA and high n-3/n-6 PUFA ratio), an olive oil (high content of MUFA, low n-3/n-6 PUFA ratio), or a sunflower oil (low content of MUFA, low n-3/n-6 PUFA ratio) diet. The lack of significant effects of n-6 fatty acids has been confirmed in several studies that used different comparators.[84,108,109]

Therefore, both association and intervention studies in humans do not support the pro-inflammatory activity that is generally ascribed to n-6 PUFA.

4.5 n-3 PUFA

n-3 fatty acids from seafood and fish oil seem to affect positively cardiovascular outcomes, particularly in secondary prevention. Epidemiological studies indicate that there is an inverse association also between dietary plant-derived ALA and risk of coronary artery disease.[110,111] Beneficial effects of n-3 PUFA concern blood lipids, blood pressure, vascular reactivity, and antithrombotic activity. The cardioprotective effects might also be mediated by changes in low-grade inflammation. The relationship between dietary n-3 fatty acids and inflammation has been widely investigated with controversial results. Differences may exist between n-3 of plant origin (ALA) and of marine origin (EPA, DHA).

Mechanisms of action. Several in vitro studies have suggested various possible mechanisms explaining n-3 PUFA effects on systemic inflammation.[112,113] As summarized in Table 1, taken from Calder,[114] the possible interlinked mechanisms include altered cell membrane phospholipid fatty acid composition, disruption of lipid rafts, inhibition of activation of the pro-inflammatory transcription factor NF-kB so reducing expression of inflammatory genes, activation of the anti-inflammatory transcription factor NR1C3 (i.e., PPAR-γ), and binding to the GPR120.

4.5.1 α-Linolenic Acid

ALA is the precursor of EPA, with limited conversion of ALA to DHA. An anti-inflammatory activity has been consistently shown in intervention studies. A randomized double-blind placebo-controlled trial in moderately hypercholesterolemic men and women with two other cardiovascular risk factors, compared a margarine enriched in ALA (15% ALA, 46% linolenic acid (LA); $n = 55$) with a margarine enriched in LA (0.3% ALA, 58% LA; $n = 55$). The margarines could be used ad libitum. After 1 and 2 years, ALA users had a significantly lower CRP level than LA users.[108]

TABLE 1 Possible Mechanisms for the Anti-Inflammatory Effects of n-3 PUFA

Anti-inflammatory effect	Likely mechanism involved
Reduced leukocyte chemotaxis	Decreased production of some chemo-attractants (e.g., LTB4) Downregulated expression of receptors for chemo-attractants
Reduced adhesion molecule expression and decreased leukocyte–endothelium interaction	Downregulated expression of adhesion molecule genes (via NF-κB, NR1C3 (e.g., PPAR-γ))
Decreased production of eicosanoids from arachidonic acid	Lowered membrane content of arachidonic acid; inhibition of arachidonic acid metabolism
Decreased production of arachidonic acid containing endocannabinoids	Lowered membrane content of arachidonic acid
Increased production of "weak" eicosanoids from EPA	Increased membrane content of EPA
Increased production of anti-inflammatory EPA and DHA containing endocannabinoids	Increased membrane content of EPA and DHA
Increased production of proresolution resolvins and protectins	Increased membrane content of EPA and DHA; presence of aspirin
Decreased production of inflammatory cytokines	Downregulated expression of inflammatory cytokine genes (via NF-κB, NR1C3 (e.g., PPAR- γ))
Decreased T-cell reactivity	Disruption of membrane rafts (via increased content of EPA and DHA in specific membrane regions)

DHA, docosahexaenoic acid; EPA, eicosapentaenoic acid; NF-κB, nuclear factor-κB; PPAR, peroxisome proliferator-activated receptor.
From Calder PJ.[114]

Similarly, in hypercholesterolemic subjects fed two diets low in saturated fat and high in ALA (6.5% ALA, 10.5% LA) or LA (3.6% ALA, 12.6% LA) compared with an average American diet (0.8% ALA, 7.7% LA), the ALA diet significantly decreased CRP, whereas the LA diet only tended to decrease CRP ($p = 0.08$). Changes in CRP were inversely associated with changes in serum EPA or EPA plus DPA.[115]

Dietary supplementation with ALA (8 g/day, linseed oil, n-6:n-3 ratio 1.3:1, $n = 50$) for 3 months decreased significantly CRP levels in male dyslipidemic patients, whereas LA supplementation (11 g/day, n-6:n-3 ratio 13.2:1, $n = 26$) decreased cholesterol levels with no significant effects on CRP.[109] In this study, it was not possible to exclude the effects of other possible anti-inflammatory components characterizing both background diets such as the high content of walnuts, walnut oil, and flaxseed oil, and the low content of SFA.

The ALA effects on CRP were not confirmed in a large trial in 2425 patients with a history of myocardial infarction, which showed no significant changes in hs-CRP levels after 40-month supplementation with ALA-rich margarin.[116]

Summarizing, these results indicate that dietary supplementation with ALA may reduce CRP levels, although this was not confirmed in a large study in patients with coronary heart disease.

4.5.2 Eicosapentaenoic and Docosahexaenoic Acids

Association studies. Fish consumption was independently associated with lower inflammatory markers in healthy adults in the ATTICA study that enrolled 1514 men and 1528 women in Greece.[117] Around 90% of the participants reported fish consumption at least once a month. Compared with nonfish consumers, those who consumed >300 g of fish per week had on average 33% lower CRP. Significant results were also observed when lower quantities (150–300 g/week) of fish were consumed and associations remained significant after various adjustments. An independent association between greater intake of n-3 PUFA derived from marine products, as measured with a self-administered questionnaire, and a lower prevalence of high CRP concentrations was also observed in an older Japanese population with a diet rich in marine products.[118] In 401 men and 570 women aged ≥70 years, after excluding serum CRP concentrations ≥10.0 mg/L and adjusting for several predictors of inflammation, the OR of high CRP (≥1.0 mg/L) for increasing quartiles of total n-3 PUFA and EPA + DHA were 1.0, 0.72, 0.57, and 0.44 (P for trend <0.01) and 1.0, 0.91, 0.76, and 0.54 (P for trend <0.03), respectively, indicating a beneficial association even at very high intakes of n-3 PUFA.

Therefore, as also shown in the previously mentioned study by Julia et al.[106] and in the Nurses' Health Study I cohort,[91] cross-sectional data indicate that n-3 PUFA intake associates with lower levels of circulating CRP.

Intervention studies. Many studies have evaluated the effects of supplementation with fish oil on CRP. No significant effects of fish oil on CRP levels were observed in healthy individuals,[119–121] healthy moderately obese individuals,[38] dyslipidemic individuals with visceral obesity,[122] and postmenopausal women.[123]

In contrast, a significant decrease in CRP and IL-6 levels was observed in postmenopausal women on hormone replacement therapy after a 5-week supplementation rich in fish oil compared with LA-rich safflower oil.[124] Concordant results were reported in a clinical trial evaluating the effects of intensive lifestyle intervention versus standard care on cardiovascular markers in 128 participants selected regardless of lifestyle/control allocation.[125] After 1 year, subjects in the top tertile of n-3 PUFA intake showed significantly lower serum hs-CRP, TNF-α, and soluble adhesion molecules sICAM-1 and sVCAM-1 than at baseline.

In the few studies performed in patients with diabetes, there were no significant effects of fish oil supplementation on CRP levels (Table 2).[126–130]

Generally, these studies were done with fish-oil capsule supplementation. Results could differ with use of seafood. In fact, different effects between fish oil and whole fish intake were shown during an 8-week intervention trial in 324 subjects from Iceland, Spain, and Ireland randomized to one of four energy-restricted diets: salmon (3 × 150 g/week, 2.1 g n-3 PUFA per day); cod (3 × 150 g/week, 0.3 g n-3 PUFA per day); fish oil capsules (1.3 g n-3 PUFA per day); and control (sunflower oil capsules, no seafood).[131] Taken together for all subjects, there were significant decreases in all inflammation parameters, with greater reductions in CRP during the 8 weeks in the salmon and cod groups. However, according to the multivariate linear model, weight loss (−5.2 ± 3.2 kg) was the only significant predictor of the reduction in hs-CRP. In a study on 13 healthy obese subjects, 4 weeks of herring diet produced a not statistically significant trend toward lower CRP compared with the reference diet (pork and chicken fillets).[132] In a crossover dietary intervention trial, 25 normal to mildly hyperlipidemic subjects were randomly assigned to one of three isoenergetic diets: a walnut diet, a fatty fish diet, or a no nuts or fish control diet.[133] On the walnut diet, the proportion of plasma phospholipid ALA increased 140% and arachidonic acid decreased 7% compared with both the control and fish diets. On the fish diet, the proportion of plasma phospholipid EPA and DHA increased about 200% and 900%, respectively. Both the walnut and fish diets demonstrated no effect on CRP.

We recently investigated this issue in a randomized controlled intervention in isoenergetic conditions, evaluating the effects of a diet naturally rich in marine n-3 PUFA on 73 individuals with high waist circumference and at least one more component of the metabolic syndrome.[62] The main dietary sources of n-3 were salmon, dentex, and anchovies, providing an intake of ~2.5 g per day of EPA + DHA. We observed a significant reduction in fasting and postprandial CRP plasma concentrations after the 8-week intervention with the fish diet.

Several meta-analyses have evaluated the effects of n-3 on inflammatory status, with some controversial results.[134,135] Li et al.[134] included in their meta-analysis 68 RCTs with 4601 subjects. Marine-derived n-3 PUFA supplementation had a significant lowering effect on

TABLE 2 Effects of Fish Oil Supplementation on C-Reactive Protein Concentrations in Randomized Controlled Trials in Individuals with Type 2 Diabetes

	Intervention	Control	Duration (weeks)	Study design	Subjects	CRP change
Mori et al.[126]	EPA/DHA	Olive oil	6	P	T2D and hypertension (n = 17/17/17)	NS
Pooya et al.[127]	Fish oil (3 g)	Sunflower oil	8	P	T2D (n = 40/41)	NS
Wong et al.[129]	Fish oil (4 g)	Olive oil	12	P	T2D (n = 49/48)	NS
Mocking et al.[128]	EPA	Rapeseed/MCT oil	12	P	T2D and depression (n = 12/12)	NS
Malekshahi et al.[130]	Fish oil (3 g)	Sunflower oil	8	P	T2D (n = 42/42)	NS

CRP, C-reactive protein; DHA, docosahexaenoic acid; EPA, eicosapentaenoic acid; MCT, medium-chain triglycerides; NS, not significant; P, parallel; T2D, type 2 diabetes.

TABLE 3 Subgroup Analysis of the Effects of Marine-Derived n-3 PUFA Supplementation on CRP Levels in Healthy Subjects

Subgroup analysis	N	WMD (95% CI)	I_2 (%)	P
Overall	16	−0.18 (−0.28, 0.08)	21.6	0.001
STUDY DESIGN				
Parallel	12	−0.20 (−0.31, −0.09)	24.8	0.000
Crossover	2	−0.01 (−0.30, 0.32)	29.3	0.941
Factorial	2	−0.32 (−0.90, 0.27)	0.0	0.292
MAJOR FATTY ACID IN PLACEBO				
Linoleic acid	2	−0.25 (−0.40, −0.10)	72.9	0.001
Oleic acid	13	−0.12 (−0.26, 0.02)	0.0	0.084
DURATION (WEEK)				
<Median	7	−0.22 (−0.34, −0.11)	26.7	0.000
≥Median	9	−0.09 (−0.26, 0.07)	8.1	0.266
DAILY DOSE OF TOTAL N-3 PUFA (G)				
<Median	8	−0.12 (−0.29, 0.05)	36.2	0.162
≥Median	8	−0.28 (−0.38, −0.17)	0.0	0.000
DAILY DOSE OF EPA (G)				
<Median	8	−0.11 (−0.27, 0.05)	38.5	0.192
≥Median	8	−0.29 (−0.39, −0.18)	0.0	0.000
DAILY DOSE OF DHA (G)				
<Median	8	−0.17 (−0.26, −0.09)	0.0	0.000
≥Median	8	−0.09 (−0.33, 0.15)	38.0	0.478
MEAN AGE (YEAR)				
<Median	8	0.00 (−0.25, 0.25)	0.0	0.995
≥Median	8	−0.21 (−0.31, −0.10)	29.8	0.000
SEX RATIO (MALE/TOTAL SUBJECTS)				
<0.5	6	−0.24 (−0.33, −0.14)	19.2	0.000
≥0.5	9	−0.10 (−0.27, 0.06)	0.0	0.219
BASELINE BMI (LOG-TRANSFORMED)				
<Median	8	−0.07 (−0.24, 0.10)	0.0	0.410
≥Median	8	−0.21 (−0.35, −0.07)	44.5	0.004

BMI, body mass index; CI, confidence interval; CRP, C-reactive protein; DHA, docosahexaenoic acid; EFA, eicosapentaenoic acid; N, number of included studies (or comparisons); $I_2 > 50\%$ indicated significant heterogeneity; PUFA, polyunsaturated fatty acid; WMD, weighted mean difference. Unit for CRP, mg/L (data log-transformed before analysis). No significant differences between subgroups.
Modified from the meta-analysis by Li et al.[134]

CRP in subjects with chronic nonautoimmune disease and healthy subjects. A significant negative linear relationship between duration and effect size of marine-derived n-3 PUFA supplementation on fasting blood levels of TNF-α and IL-6 in subjects with chronic nonautoimmune disease was observed, indicating that longer duration of supplementation could lead to a greater lowering effect. Subgroup analysis showed that the lowering effect of marine-derived n-3 PUFA supplementation on CRP was more evident in the studies when placebo was linoleic acid, and in studies with longer duration, higher daily dose of n-3 PUFA, older age, more females, higher BMI (Table 3). The effect of marine-derived n-3 PUFA from dietary intake was only assessed in subjects with chronic nonautoimmune disease, and a significant lowering effect was observed on IL-6, but not on CRP and TNF-α.

Taken together, the present evidence indicates that the intake of marine-derived n-3 PUFA may have lowering effects on CRP and several mechanisms shown in in vitro and animal studies demonstrate the biological plausibility for these effects. However, although epidemiological studies support the inverse association between n-3 PUFA and CRP levels, intervention studies provided contradictory results. Different factors may explain the lack of concordance, including type of population studied, variations in dosage, treatment length, use of supplement or whole fish, confounding medications, different gender sensibility, small sample size in many studies, or difficulties in disclosing effects occurring within tissues through the evaluation of changes in circulating levels.

Moreover, n-3 PUFA may be effective in healthy subjects preventing chronic inflammatory diseases, whereas not when the disease is already established.

Therefore, available data indicate that the beneficial effects of n-3 PUFA on CVR may be mediated also by their anti-inflammatory properties. However, further investigation is deserved, which should consider all factors potentially influencing the translation of preclinical evidence to clinical results.[136]

5. CONCLUSIONS

There is evidence that dietary changes influence CRP levels and, therefore, may reduce low-grade inflammation. Although no dietary study focused on reducing systemic inflammation has demonstrated it, this might beneficially affect development and course of T2D and CVD.

CRP levels are reduced by healthy dietary patterns, including the Mediterranean diet, compared with Western diets. More controversial is the role of different dietary components. The intake of SFAs and TFAs may increase, whereas n-3 fatty acids, either of plant and marine origin, may decrease CRP levels.

The results of intervention studies have been very conflicting, often not confirming epidemiological data. There are many possible explanations for this inconsistency, including differences in characteristics of the studied cohorts, dietary sources of fatty acids, and type of comparator. Furthermore, generally CRP was not the primary outcome of the intervention studies that were often underpowered to discover a significant effect of treatment or differences between treatments. Moreover, genetic factors may influence the individual response to dietary fat intake and could contribute to explain the observed interindividual heterogeneity.

Available evidence in patients with diabetes does not allow envisaging in these patients, different behaviors or mechanisms of the effects of dietary fatty acids on low-grade inflammation than in the nondiabetic population.

Because of discrepancies between cross-sectional and intervention trials, further "ad hoc studies" are needed aiming at clarifying the role of the confounding factors and exploring possible interactions between genetic and modifiable environmental factors. At the present state of knowledge, the changes in CRP levels induced by dietary fats are not robust enough to modify their current recommended consumptions that are based on their effects on cardiovascular outcomes and other classical CVD risk factor.

References

1. Libby P. Inflammation and cardiovascular disease mechanisms. *Am J Clin Nutr* 2006;**83**(2):456S–60S.
2. Ridker PM. Inflammatory biomarkers and risks of myocardial infarction, stroke, diabetes, and total mortality: implications for longevity. *Nutr Rev* 2007;**65**(12 Pt 2):S253–9.
3. Lee YH, Pratley RE. The evolving role of inflammation in obesity and the metabolic syndrome. *Curr Diab Rep* 2005;**5**(1):70–5.
4. Deodhar SD. C-reactive protein: the best laboratory indicator available for monitoring disease activity. *Cleve Clin J Med* 1989; **56**(2):126–30.
5. Macy EM, Hayes TE, Tracy RP. Variability in the measurement of C-reactive protein in healthy subjects: implications for reference intervals and epidemiological applications. *Clin Chem* 1997;**43**(1): 52–8.
6. Wilkins J, Gallimore JR, Moore EG, Pepys MB. Rapid automated high sensitivity enzyme immunoassay of C-reactive protein. *Clin Chem* 1998;**44**(6 Pt 1):1358–61.
7. de Beer FC, Soutar AK, Baltz ML, Trayner IM, Feinstein A, Pepys MB. Low density lipoprotein and very low density lipoprotein are selectively bound by aggregated C-reactive protein. *J Exp Med* 1982;**156**(1):230–42.
8. Zwaka TP, Hombach V, Torzewski J. C-reactive protein-mediated low density lipoprotein uptake by macrophages: implications for atherosclerosis. *Circulation* 2001;**103**(9):1194–7.
9. Torzewski J, Torzewski M, Bowyer DE, et al. C-reactive protein frequently colocalizes with the terminal complement complex in the intima of early atherosclerotic lesions of human coronary arteries. *Arterioscler Thromb Vasc Biol* 1998;**18**(9):1386–92.
10. Yasojima K, Schwab C, McGeer EG, McGeer PL. Generation of C-reactive protein and complement components in atherosclerotic plaques. *Am J Pathol* 2001;**158**(3):1039–51.
11. Hingorani AD, Cross J, Kharbanda RK, et al. Acute systemic inflammation impairs endothelium-dependent dilatation in humans. *Circulation* 2000;**102**(9):994–9.
12. Fichtlscherer S, Rosenberger G, Walter DH, Breuer S, Dimmeler S, Zeiher AM. Elevated C-reactive protein levels and impaired endothelial vasoreactivity in patients with coronary artery disease. *Circulation* 2000;**102**(9):1000–6.
13. Tomai F, Crea F, Gaspardone A, et al. Unstable angina and elevated C-reactive protein levels predict enhanced vasoreactivity of the culprit lesion. *Circulation* 2001;**104**(13):1471–6.
14. Cook DG, Mendall MA, Whincup PH, et al. C-reactive protein concentration in children: relationship to adiposity and other cardiovascular risk factors. *Atherosclerosis* 2000;**149**(1):139–50.
15. Mendall MA, Patel P, Ballam L, Strachan D, Northfield TC. C reactive protein and its relation to cardiovascular risk factors: a population based cross sectional study. *BMJ* 1996;**312**(7038): 1061–5.
16. Ridker PM, Rifai N, Rose L, Buring JE, Cook NR. Comparison of C-reactive protein and low-density lipoprotein cholesterol levels in the prediction of first cardiovascular events. *N Engl J Med* 2002;**347**(20):1557–65.
17. Ridker PM. Clinical application of C-reactive protein for cardiovascular disease detection and prevention. *Circulation* 2003; **107**(3):363–9.
18. Pot GK, Brouwer IA, Enneman A, Rijkers GT, Kampman E, Geelen A. No effect of fish oil supplementation on serum inflammatory markers and their interrelationships: a randomized controlled trial in healthy, middle-aged individuals. *Eur J Clin Nutr* 2009;**63**(11):1353–9.
19. Donath MY, Shoelson SE. Type 2 diabetes as an inflammatory disease. *Nat Rev Immunol* 2011;**11**(2):98–107.
20. Barzilay JI, Abraham L, Heckbert SR, et al. The relation of markers of inflammation to the development of glucose disorders in the elderly: the Cardiovascular Health Study. *Diabetes* 2001;**50**(10): 2384–9.
21. Pradhan AD, Manson JE, Rifai N, Buring JE, Ridker PM. C-reactive protein, interleukin 6, and risk of developing type 2 diabetes mellitus. *JAMA* 2001;**286**(3):327–34.
22. Kajitani N, Shikata K, Nakamura A, Nakatou T, Hiramatsu M, Makino H. Microinflammation is a common risk factor for progression of nephropathy and atherosclerosis in Japanese patients with type 2 diabetes. *Diabetes Res Clin Pract* 2010;**88**(2):171–6.
23. Pearson TA, Mensah GA, Alexander RW, et al. Markers of inflammation and cardiovascular disease: application to clinical and public health practice: A statement for healthcare professionals from the Centers for Disease Control and Prevention and the American Heart Association. *Circulation* 2003;**107**(3):499–511.
24. Sallam N, Khazaei M, Laher I. Effect of moderate-intensity exercise on plasma C-reactive protein and aortic endothelial function in type 2 diabetic mice. *Mediators Inflamm* 2010;**2010**:149678.
25. Larsen CM, Faulenbach M, Vaag A, et al. Interleukin-1-receptor antagonist in type 2 diabetes mellitus. *N Engl J Med* 2007; **356**(15):1517–26.
26. Goldfine AB, Fonseca V, Jablonski KA, et al. The effects of salsalate on glycemic control in patients with type 2 diabetes: a randomized trial. *Ann Intern Med* 2010;**152**(6):346–57.
27. Goldberg RB. Cytokine and cytokine-like inflammation markers, endothelial dysfunction, and imbalanced coagulation in development of diabetes and its complications. *J Clin Endocrinol Metab* 2009;**94**(9):3171–82.
28. Lapice E, Maione S, Patti L, et al. Abdominal adiposity is associated with elevated C-reactive protein independent of BMI in healthy nonobese people. *Diabetes Care* 2009;**32**(9):1734–6.
29. Sam S, Haffner S, Davidson MH, et al. Relation of abdominal fat depots to systemic markers of inflammation in type 2 diabetes. *Diabetes Care* 2009;**32**(5):932–7.
30. Rytter E, Vessby B, Asgard R, et al. Glycaemic status in relation to oxidative stress and inflammation in well-controlled type 2 diabetes subjects. *Br J Nutr* 2009;**101**(10):1423–6.

31. Hansen D, Dendale P, Beelen M, et al. Plasma adipokine and inflammatory marker concentrations are altered in obese, as opposed to non-obese, type 2 diabetes patients. *Eur J Appl Physiol* 2010;**109**(3):397–404.
32. Thomsen SB, Rathcke CN, Zerahn B, Vestergaard H. Increased levels of the calcification marker matrix Gla Protein and the inflammatory markers YKL-40 and CRP in patients with type 2 diabetes and ischemic heart disease. *Cardiovasc Diabetol* 2010;**9**:86.
33. Mahajan A, Tabassum R, Chavali S, et al. High-sensitivity C-reactive protein levels and type 2 diabetes in urban North Indians. *J Clin Endocrinol Metab* 2009;**94**(6):2123–7.
34. Haffner S, Temprosa M, Crandall J, et al. Intensive lifestyle intervention or metformin on inflammation and coagulation in participants with impaired glucose tolerance. *Diabetes* 2005;**54**(5):1566–72.
35. Belalcazar LM, Reboussin DM, Haffner SM, et al. A 1-year lifestyle intervention for weight loss in individuals with type 2 diabetes reduces high C-reactive protein levels and identifies metabolic predictors of change: from the Look AHEAD (Action for Health in Diabetes) study. *Diabetes Care* 2010;**33**(11):2297–303.
36. Egert S, Baxheinrich A, Lee-Barkey YH, Tschoepe D, Wahrburg U, Stratmann B. Effects of an energy-restricted diet rich in plant-derived alpha-linolenic acid on systemic inflammation and endothelial function in overweight-to-obese patients with metabolic syndrome traits. *Br J Nutr* 2014;**112**(8):1315–22.
37. Corpeleijn E, Saris WH, Jansen EH, Roekaerts PM, Feskens EJ, Blaak EE. Postprandial interleukin-6 release from skeletal muscle in men with impaired glucose tolerance can be reduced by weight loss. *J Clin Endocrinol Metab* 2005;**90**(10):5819–24.
38. Plat J, Jellema A, Ramakers J, Mensink RP. Weight loss, but not fish oil consumption, improves fasting and postprandial serum lipids, markers of endothelial function, and inflammatory signatures in moderately obese men. *J Nutr* 2007;**137**(12):2635–40.
39. Heilbronn LK, Noakes M, Clifton PM. Energy restriction and weight loss on very-low-fat diets reduce C-reactive protein concentrations in obese, healthy women. *Arterioscler Thromb Vasc Biol* 2001;**21**(6):968–70.
40. Selvin E, Paynter NP, Erlinger TP. The effect of weight loss on C-reactive protein: a systematic review. *Arch Intern Med* 2007; **167**(1):31–9.
41. Fung TT, Rimm EB, Spiegelman D, et al. Association between dietary patterns and plasma biomarkers of obesity and cardiovascular disease risk. *Am J Clin Nutr* 2001;**73**(1):61–7.
42. Lopez-Garcia E, Schulze MB, Fung TT, et al. Major dietary patterns are related to plasma concentrations of markers of inflammation and endothelial dysfunction. *Am J Clin Nutr* 2004;**80**(4):1029–35.
43. Santos FL, Esteves SS, da Costa Pereira A, Yancy Jr WS, Nunes JP. Systematic review and meta-analysis of clinical trials of the effects of low carbohydrate diets on cardiovascular risk factors. *Obes Rev* 2012;**13**(11):1048–66.
44. Seshadri P, Iqbal N, Stern L, et al. A randomized study comparing the effects of a low-carbohydrate diet and a conventional diet on lipoprotein subfractions and C-reactive protein levels in patients with severe obesity. *Am J Med* 2004;**117**(6):398–405.
45. Rankin JW, Turpyn AD. Low carbohydrate, high fat diet increases C-reactive protein during weight loss. *J Am Coll Nutr* 2007;**26**(2): 163–9.
46. Keogh JB, Brinkworth GD, Noakes M, Belobrajdic DP, Buckley JD, Clifton PM. Effects of weight loss from a very-low-carbohydrate diet on endothelial function and markers of cardiovascular disease risk in subjects with abdominal obesity. *Am J Clin Nutr* 2008;**87**(3):567–76.
47. Ruth MR, Port AM, Shah M, et al. Consuming a hypocaloric high fat low carbohydrate diet for 12 weeks lowers C-reactive protein, and raises serum adiponectin and high density lipoprotein-cholesterol in obese subjects. *Metabolism* 2013;**62**(12):1779–87.
48. Jonasson L, Guldbrand H, Lundberg AK, Nystrom FH. Advice to follow a low-carbohydrate diet has a favourable impact on low-grade inflammation in type 2 diabetes compared with advice to follow a low-fat diet. *Ann Med* 2014;**46**(3):182–7.
49. Estruch R, Ros E, Salas-Salvado J, et al. Primary prevention of cardiovascular disease with a Mediterranean diet. *N Engl J Med* 2013; **368**(14):1279–90.
50. Esposito K, Marfella R, Ciotola M, et al. Effect of a mediterranean-style diet on endothelial dysfunction and markers of vascular inflammation in the metabolic syndrome: a randomized trial. *JAMA* 2004;**292**(12):1440–6.
51. Michalsen A, Lehmann N, Pithan C, et al. Mediterranean diet has no effect on markers of inflammation and metabolic risk factors in patients with coronary artery disease. *Eur J Clin Nutr* 2006;**60**(4): 478–85.
52. Schwingshackl L, Hoffmann G. Mediterranean dietary pattern, inflammation and endothelial function: a systematic review and meta-analysis of intervention trials. *Nutr Metab Cardiovasc Dis* 2014;**24**(9):929–39.
53. Park KH, Zaichenko L, Peter P, Davis CR, Crowell JA, Mantzoros CS. Diet quality is associated with circulating C-reactive protein but not irisin levels in humans. *Metabolism* 2014;**63**(2):233–41.
54. Salas-Salvado J, Garcia-Arellano A, Estruch R, et al. Components of the Mediterranean-type food pattern and serum inflammatory markers among patients at high risk for cardiovascular disease. *Eur J Clin Nutr* 2008;**62**(5):651–9.
55. Giacco R, Clemente G, Cipriano D, et al. Effects of the regular consumption of wholemeal wheat foods on cardiovascular risk factors in healthy people. *Nutr Metab Cardiovasc Dis* 2010;**20**(3): 186–94.
56. de Mello VD, Schwab U, Kolehmainen M, et al. A diet high in fatty fish, bilberries and wholegrain products improves markers of endothelial function and inflammation in individuals with impaired glucose metabolism in a randomised controlled trial: the Sysdimet study. *Diabetologia* 2011;**54**(11):2755–67.
57. Kishimoto Y, Tani M, Kondo K. Pleiotropic preventive effects of dietary polyphenols in cardiovascular diseases. *Eur J Clin Nutr* 2013;**67**(5):532–5.
58. Kim HS, Quon MJ, Kim JA. New insights into the mechanisms of polyphenols beyond antioxidant properties; lessons from the green tea polyphenol, epigallocatechin 3-gallate. *Redox Biol* 2014; **2**:187–95.
59. Landberg R, Sun Q, Rimm EB, et al. Selected dietary flavonoids are associated with markers of inflammation and endothelial dysfunction in U.S. women. *J Nutr* 2011;**141**(4):618–25.
60. Annuzzi G, Bozzetto L, Costabile G, et al. Diets naturally rich in polyphenols improve fasting and postprandial dyslipidemia and reduce oxidative stress: a randomized controlled trial. *Am J Clin Nutr* 2014;**99**(3):463–71.
61. Bozzetto L, Annuzzi G, Pacini G, et al. Polyphenol-rich diets improve glucose metabolism in people at high cardiometabolic risk: a controlled randomised intervention trial. *Diabetologia* 2015;**58**(7):1551–60.
62. Griffo E, Costabile G, Annuzzi G, et al. Effects of dietary n-3 fatty acid and/or polyphenols on subclinical inflammation in people at high cardiovascular risk: the Etherpaths Project. In: *83rd European Atherosclerosis Society meeting, Glasgow*; 2015.
63. Nienaber-Rousseau C, Swanepoel B, Dolman RC, Pieters M, Conradie KR, Towers GW. Interactions between C-reactive protein genotypes with markers of nutritional status in relation to inflammation. *Nutrients* 2014;**6**(11):5034–50.
64. Estadella D, da Penha Oller do Nascimento CM, Oyama LM, Ribeiro EB, Damaso AR, de Piano A. Lipotoxicity: effects of dietary saturated and transfatty acids. *Mediators Inflamm* 2013;**2013**: 137579.

65. Santos S, Oliveira A, Lopes C. Systematic review of saturated fatty acids on inflammation and circulating levels of adipokines. *Nutr Res* 2013;**33**(9):687–95.
66. Kennedy A, Martinez K, Chuang CC, LaPoint K, McIntosh M. Saturated fatty acid-mediated inflammation and insulin resistance in adipose tissue: mechanisms of action and implications. *J Nutr* 2009;**139**(1):1–4.
67. Solinas G, Naugler W, Galimi F, Lee MS, Karin M. Saturated fatty acids inhibit induction of insulin gene transcription by JNK-mediated phosphorylation of insulin-receptor substrates. *Proc Natl Acad Sci USA* 2006;**103**(44):16454–9.
68. Chait A, Kim F. Saturated fatty acids and inflammation: who pays the toll? *Arterioscler Thromb Vasc Biol* 2010;**30**(4):692–3.
69. Ajuwon KM, Spurlock ME. Palmitate activates the NF-kappaB transcription factor and induces IL-6 and TNFalpha expression in 3T3-L1 adipocytes. *J Nutr* 2005;**135**(8):1841–6.
70. Song MJ, Kim KH, Yoon JM, Kim JB. Activation of Toll-like receptor 4 is associated with insulin resistance in adipocytes. *Biochem Biophys Res Commun* 2006;**346**(3):739–45.
71. Suganami T, Nishida J, Ogawa Y. A paracrine loop between adipocytes and macrophages aggravates inflammatory changes: role of free fatty acids and tumor necrosis factor alpha. *Arterioscler Thromb Vasc Biol* 2005;**25**(10):2062–8.
72. Suganami T, Tanimoto-Koyama K, Nishida J, et al. Role of the Toll-like receptor 4/NF-kappaB pathway in saturated fatty acid-induced inflammatory changes in the interaction between adipocytes and macrophages. *Arterioscler Thromb Vasc Biol* 2007; **27**(1):84–91.
73. Schwartz EA, Zhang WY, Karnik SK, et al. Nutrient modification of the innate immune response: a novel mechanism by which saturated fatty acids greatly amplify monocyte inflammation. *Arterioscler Thromb Vasc Biol* 2010;**30**(4):802–8.
74. Ibrahim A, Natrajan S, Ghafoorunissa R. Dietary trans-fatty acids alter adipocyte plasma membrane fatty acid composition and insulin sensitivity in rats. *Metabolism* 2005;**54**(2):240–6.
75. Saravanan N, Haseeb A, Ehtesham NZ. Ghafoorunissa. Differential effects of dietary saturated and trans-fatty acids on expression of genes associated with insulin sensitivity in rat adipose tissue. *Eur J Endocrinol* 2005;**153**(1):159–65.
76. King DE, Egan BM, Geesey ME. Relation of dietary fat and fiber to elevation of C-reactive protein. *Am J Cardiol* 2003;**92**(11): 1335–9.
77. Santos S, Oliveira A, Casal S, Lopes C. Saturated fatty acids intake in relation to C-reactive protein, adiponectin, and leptin: a population-based study. *Nutrition* 2013;**29**(6):892–7.
78. Kalogeropoulos N, Panagiotakos DB, Pitsavos C, et al. Unsaturated fatty acids are inversely associated and n-6/n-3 ratios are positively related to inflammation and coagulation markers in plasma of apparently healthy adults. *Clin Chim Acta* 2010; **411**(7–8):584–91.
79. Petersson H, Lind L, Hulthe J, Elmgren A, Cederholm T, Riserus U. Relationships between serum fatty acid composition and multiple markers of inflammation and endothelial function in an elderly population. *Atherosclerosis* 2009;**203**(1):298–303.
80. Murakami K, Sasaki S, Takahashi Y, et al. Total n-3 polyunsaturated fatty acid intake is inversely associated with serum C-reactive protein in young Japanese women. *Nutr Res* 2008; **28**(5):309–14.
81. Bo S, Ciccone G, Guidi S, et al. Diet or exercise: what is more effective in preventing or reducing metabolic alterations? *Eur J Endocrinol* 2008;**159**(6):685–91.
82. Vessby B, Uusitupa M, Hermansen K, et al. Substituting dietary saturated for monounsaturated fat impairs insulin sensitivity in healthy men and women: the KANWU Study. *Diabetologia* 2001; **44**(3):312–9.
83. Jebb SA, Lovegrove JA, Griffin BA, et al. Effect of changing the amount and type of fat and carbohydrate on insulin sensitivity and cardiovascular risk: the RISCK (Reading, Imperial, Surrey, Cambridge, and Kings) trial. *Am J Clin Nutr* 2010; **92**(4):748–58.
84. Keogh JB, Grieger JA, Noakes M, Clifton PM. Flow-mediated dilatation is impaired by a high-saturated fat diet but not by a high-carbohydrate diet. *Arterioscler Thromb Vasc Biol* 2005;**25**(6):1274–9.
85. Ericson U, Hellstrand S, Brunkwall L, et al. Food sources of fat may clarify the inconsistent role of dietary fat intake for incidence of type 2 diabetes. *Am J Clin Nutr* 2015;**101**(5):1065–80.
86. Zemel MB, Sun X, Sobhani T, Wilson B. Effects of dairy compared with soy on oxidative and inflammatory stress in overweight and obese subjects. *Am J Clin Nutr* 2010;**91**(1):16–22.
87. Benatar JR, Sidhu K, Stewart RA. Effects of high and low fat dairy food on cardio-metabolic risk factors: a meta-analysis of randomized studies. *PLoS One* 2013;**8**(10):e76480.
88. Kromhout D, Menotti A, Bloemberg B, et al. Dietary saturated and trans fatty acids and cholesterol and 25-year mortality from coronary heart disease: the Seven Countries Study. *Prev Med* 1995; **24**(3):308–15.
89. Salmeron J, Hu FB, Manson JE, et al. Dietary fat intake and risk of type 2 diabetes in women. *Am J Clin Nutr* 2001;**73**(6):1019–26.
90. Mozaffarian D, Pischon T, Hankinson SE, et al. Dietary intake of trans fatty acids and systemic inflammation in women. *Am J Clin Nutr* 2004;**79**(4):606–12.
91. Lopez-Garcia E, Schulze MB, Meigs JB, et al. Consumption of trans fatty acids is related to plasma biomarkers of inflammation and endothelial dysfunction. *J Nutr* 2005;**135**(3):562–6.
92. Baer DJ, Judd JT, Clevidence BA, Tracy RP. Dietary fatty acids affect plasma markers of inflammation in healthy men fed controlled diets: a randomized crossover study. *Am J Clin Nutr* 2004;**79**(6):969–73.
93. Tholstrup T, Raff M, Basu S, Nonboe P, Sejrsen K, Straarup EM. Effects of butter high in ruminant trans and monounsaturated fatty acids on lipoproteins, incorporation of fatty acids into lipid classes, plasma C-reactive protein, oxidative stress, hemostatic variables, and insulin in healthy young men. *Am J Clin Nutr* 2006;**83**(2):237–43.
94. Lichtenstein AH, Erkkila AT, Lamarche B, Schwab US, Jalbert SM, Ausman LM. Influence of hydrogenated fat and butter on CVD risk factors: remnant-like particles, glucose and insulin, blood pressure and C-reactive protein. *Atherosclerosis* 2003;**171**(1): 97–107.
95. Riserus U, Basu S, Jovinge S, Fredrikson GN, Arnlov J, Vessby B. Supplementation with conjugated linoleic acid causes isomer-dependent oxidative stress and elevated C-reactive protein: a potential link to fatty acid-induced insulin resistance. *Circulation* 2002;**106**(15):1925–9.
96. Moloney F, Yeow TP, Mullen A, Nolan JJ, Roche HM. Conjugated linoleic acid supplementation, insulin sensitivity, and lipoprotein metabolism in patients with type 2 diabetes mellitus. *Am J Clin Nutr* 2004;**80**(4):887–95.
97. Asp ML, Collene AL, Norris LE, et al. Time-dependent effects of safflower oil to improve glycemia, inflammation and blood lipids in obese, post-menopausal women with type 2 diabetes: a randomized, double-masked, crossover study. *Clin Nutr* 2011;**30**(4): 443–9.
98. Linseisen J, Welch AA, Ocke M, et al. Dietary fat intake in the European Prospective Investigation into Cancer and Nutrition: results from the 24-h dietary recalls. *Eur J Clin Nutr* 2009;**63**(Suppl. 4):S61–80.
99. Schwingshackl L, Hoffmann G. Monounsaturated fatty acids, olive oil and health status: a systematic review and meta-analysis of cohort studies. *Lipids Health Dis* 2014;**13**:154.

100. Bozzetto L, Prinster A, Annuzzi G, et al. Liver fat is reduced by an isoenergetic MUFA diet in a controlled randomized study in type 2 diabetic patients. *Diabetes Care* 2012;**35**(7):1429–35.
101. Junker R, Pieke B, Schulte H, et al. Changes in hemostasis during treatment of hypertriglyceridemia with a diet rich in monounsaturated and n-3 polyunsaturated fatty acids in comparison with a low-fat diet. *Thromb Res* 2001;**101**(5):355–66.
102. Kontogianni MD, Vlassopoulos A, Gatzieva A, et al. Flaxseed oil does not affect inflammatory markers and lipid profile compared to olive oil, in young, healthy, normal weight adults. *Metabolism* 2013;**62**(5):686–93.
103. Bozzetto L, De Natale C, Di Capua L, et al. The association of hs-CRP with fasting and postprandial plasma lipids in patients with type 2 diabetes is disrupted by dietary monounsaturated fatty acids. *Acta Diabetol* 2013;**50**(2):273–6.
104. Pietraszek A, Gregersen S, Hermansen K. Acute effects of dietary fat on inflammatory markers and gene expression in first-degree relatives of type 2 diabetes patients. *Rev Diabet Stud* 2011;**8**(4): 477–89.
105. Raphael W, Sordillo LM. Dietary polyunsaturated fatty acids and inflammation: the role of phospholipid biosynthesis. *Int J Mol Sci* 2013;**14**(10):21167–88.
106. Julia C, Touvier M, Meunier N, et al. Intakes of PUFAs were inversely associated with plasma C-reactive protein 12 years later in a middle-aged population with vitamin E intake as an effect modifier. *J Nutr* 2013;**143**(11):1760–6.
107. Junker R, Kratz M, Neufeld M, et al. Effects of diets containing olive oil, sunflower oil, or rapeseed oil on the hemostatic system. *Thromb Haemost* 2001;**85**(2):280–6.
108. Bemelmans WJ, Lefrandt JD, Feskens EJ, et al. Increased alpha-linolenic acid intake lowers C-reactive protein, but has no effect on markers of atherosclerosis. *Eur J Clin Nutr* 2004; **58**(7):1083–9.
109. Rallidis LS, Paschos G, Liakos GK, Velissaridou AH, Anastasiadis G, Zampelas A. Dietary alpha-linolenic acid decreases C-reactive protein, serum amyloid A and interleukin-6 in dyslipidaemic patients. *Atherosclerosis* 2003;**167**(2):237–42.
110. Djousse L, Pankow JS, Eckfeldt JH, et al. Relation between dietary linolenic acid and coronary artery disease in the National Heart, Lung, and Blood Institute Family Heart Study. *Am J Clin Nutr* 2001;**74**(5):612–9.
111. Hu FB, Stampfer MJ, Manson JE, et al. Dietary intake of alpha-linolenic acid and risk of fatal ischemic heart disease among women. *Am J Clin Nutr* 1999;**69**(5):890–7.
112. Calder PC. Marine omega-3 fatty acids and inflammatory processes: effects, mechanisms and clinical relevance. *Biochim Biophys Acta* 2015;**1851**(4):469–84.
113. Poudyal H, Panchal SK, Diwan V, Brown L. Omega-3 fatty acids and metabolic syndrome: effects and emerging mechanisms of action. *Prog Lipid Res* 2011;**50**(4):372–87.
114. Calder PC. Omega-3 polyunsaturated fatty acids and inflammatory processes: nutrition or pharmacology? *Br J Clin Pharmacol* 2013;**75**(3):645–62.
115. Zhao G, Etherton TD, Martin KR, West SG, Gillies PJ, Kris-Etherton PM. Dietary alpha-linolenic acid reduces inflammatory and lipid cardiovascular risk factors in hypercholesterolemic men and women. *J Nutr* 2004;**134**(11):2991–7.
116. Hoogeveen EK, Geleijnse JM, Kromhout D, Giltay EJ. No effect of n-3 fatty acids on high-sensitivity C-reactive protein after myocardial infarction: the Alpha Omega Trial. *Eur J Prev Cardiol* 2014; **21**(11):1429–36.
117. Zampelas A, Panagiotakos DB, Pitsavos C, et al. Fish consumption among healthy adults is associated with decreased levels of inflammatory markers related to cardiovascular disease: the ATTICA study. *J Am Coll Cardiol* 2005;**46**(1):120–4.
118. Niu K, Hozawa A, Kuriyama S, et al. Dietary long-chain n-3 fatty acids of marine origin and serum C-reactive protein concentrations are associated in a population with a diet rich in marine products. *Am J Clin Nutr* 2006;**84**(1):223–9.
119. Madsen T, Schmidt EB, Christensen JH. The effect of n-3 fatty acids on C-reactive protein levels in patients with chronic renal failure. *J Ren Nutr* 2007;**17**(4):258–63.
120. Vega-Lopez S, Kaul N, Devaraj S, Cai RY, German B, Jialal I. Supplementation with omega3 polyunsaturated fatty acids and all-rac alpha-tocopherol alone and in combination failed to exert an anti-inflammatory effect in human volunteers. *Metabolism* 2004; **53**(2):236–40.
121. Barcelo-Coblijn G, Murphy EJ, Othman R, Moghadasian MH, Kashour T, Friel JK. Flaxseed oil and fish-oil capsule consumption alters human red blood cell n-3 fatty acid composition: a multiple-dosing trial comparing 2 sources of n-3 fatty acid. *Am J Clin Nutr* 2008;**88**(3):801–9.
122. Chan DC, Watts GF, Barrett PH, Beilin LJ, Mori TA. Effect of atorvastatin and fish oil on plasma high-sensitivity C-reactive protein concentrations in individuals with visceral obesity. *Clin Chem* 2002;**48**(6 Pt 1):877–83.
123. Geelen A, Brouwer IA, Schouten EG, Kluft C, Katan MB, Zock PL. Intake of n-3 fatty acids from fish does not lower serum concentrations of C-reactive protein in healthy subjects. *Eur J Clin Nutr* 2004;**58**(10):1440–2.
124. Ciubotaru I, Lee YS, Wander RC. Dietary fish oil decreases C-reactive protein, interleukin-6, and triacylglycerol to HDL-cholesterol ratio in postmenopausal women on HRT. *J Nutr Biochem* 2003;**14**(9):513–21.
125. Merino J, Sala-Vila A, Kones R, et al. Increasing long-chain n-3 PUFA consumption improves small peripheral artery function in patients at intermediate-high cardiovascular risk. *J Nutr Biochem* 2014;**25**(6):642–6.
126. Mori TA, Woodman RJ, Burke V, Puddey IB, Croft KD, Beilin LJ. Effect of eicosapentaenoic acid and docosahexaenoic acid on oxidative stress and inflammatory markers in treated-hypertensive type 2 diabetic subjects. *Free Radic Biol Med* 2003; **35**(7):772–81.
127. Pooya S, Jalali MD, Jazayery AD, Saedisomeolia A, Eshraghian MR, Toorang F. The efficacy of omega-3 fatty acid supplementation on plasma homocysteine and malondialdehyde levels of type 2 diabetic patients. *Nutr Metab Cardiovasc Dis* 2010;**20**(5):326–31.
128. Mocking RJ, Assies J, Bot M, Jansen EH, Schene AH, Pouwer F. Biological effects of add-on eicosapentaenoic acid supplementation in diabetes mellitus and co-morbid depression: a randomized controlled trial. *PLoS One* 2012;**7**(11):e49431.
129. Wong CY, Yiu KH, Li SW, et al. Fish-oil supplement has neutral effects on vascular and metabolic function but improves renal function in patients with Type 2 diabetes mellitus. *Diabet Med* 2010;**27**(1):54–60.
130. Malekshahi Moghadam A, Saedisomeolia A, Djalali M, Djazayery A, Pooya S, Sojoudi F. Efficacy of omega-3 fatty acid supplementation on serum levels of tumour necrosis factor-alpha, C-reactive protein and interleukin-2 in type 2 diabetes mellitus patients. *Singapore Med J* 2012;**53**(9):615–9.
131. Ramel A, Martinez JA, Kiely M, Bandarra NM, Thorsdottir I. Effects of weight loss and seafood consumption on inflammation parameters in young, overweight and obese European men and women during 8 weeks of energy restriction. *Eur J Clin Nutr* 2010;**64**(9):987–93.
132. Lindqvist H, Langkilde AM, Undeland I, Radendal T, Sandberg AS. Herring (Clupea harengus) supplemented diet influences risk factors for CVD in overweight subjects. *Eur J Clin Nutr* 2007;**61**(9):1106–13.

133. Chiang YL, Haddad E, Rajaram S, Shavlik D, Sabate J. The effect of dietary walnuts compared to fatty fish on eicosanoids, cytokines, soluble endothelial adhesion molecules and lymphocyte subsets: a randomized, controlled crossover trial. *Prostaglandins Leukot Essent Fatty Acids* 2012;**87**(4–5):111–7.
134. Li K, Huang T, Zheng J, Wu K, Li D. Effect of marine-derived n-3 polyunsaturated fatty acids on C-reactive protein, interleukin 6 and tumor necrosis factor alpha: a meta-analysis. *PLoS One* 2014;**9**(2):e88103.
135. Xin W, Wei W, Li X. Effects of fish oil supplementation on inflammatory markers in chronic heart failure: a meta-analysis of randomized controlled trials. *BMC Cardiovasc Disord* 2012;**12**:77.
136. Skulas-Ray AC. Omega-3 fatty acids and inflammation: a perspective on the challenges of evaluating efficacy in clinical research. *Prostaglandins Other Lipid Mediat* 2015;**116–117**:104–11.

CHAPTER

19

Alcoholic Beverage and Insulin Resistance—Mediated Degenerative Diseases of Liver and Brain: Cellular and Molecular Effects

Suzanne M. de la Monte[1,2,3,4,5,6,7], *Susan Huse*[2,7], *Miran Kim*[1,5,6,7]

[1]Department of Medicine, Rhode Island Hospital, Providence, RI, USA; [2]Department of Pathology, Rhode Island Hospital, Providence, RI, USA; [3]Department of Neurology, Rhode Island Hospital, Providence, RI, USA; [4]Department of Neurosurgery, Rhode Island Hospital, Providence, RI, USA; [5]The Liver Research Center, Rhode Island Hospital, Providence, RI, USA; [6]Rhode Island Hospital, Providence, RI, USA; [7]The Warren Alpert Medical School of Brown University, Providence, RI, USA

1. OVERVIEW

1.1 Public Health Problem

After tobacco and obesity, alcohol abuse is the third leading preventable cause of death in the United States (88,000/year).[1] Even more alarming is that when all premature alcohol-related deaths are counted, the actual death toll nearly doubles. Furthermore, heavy drinking exacerbates chronic diseases such as hypertension, diabetes mellitus, and hepatitis, and interferes with the metabolism and therapeutic actions of various medications. Therefore, the long-term consequences of alcohol abuse and addiction are among the costliest of health care problems in the world. Because serious harmful effects of excessive alcohol consumption can occur with either heavy chronic or binge exposures, the National Institutes of Alcohol Abuse and Alcoholism (NIAAA) has established guidelines for upper-limit alcohol intake by adults. For men 21–65 years old, the NIAAA recommends a maximum of 14 standard drinks/week and 4 drinks per day, whereas for women in the same age range and men older than age 65, they recommend upper limits of seven standard drinks per week and three drinks per day. One standard drink equals 14 g of alcohol, which is present in 12 oz. of beer, 5 oz. of wine, 1.5 oz. of 80-proof spirits, 8 oz. of malt liquor, or 3 oz. of fortified wine.

1.2 Alcohol's Toxic/Degenerative Effects Target Liver and Brain

Alcohol is absorbed in the upper gastrointestinal tract by diffusion, quickly distributes to all organs of the body, and is primarily eliminated by oxidation. Ethanol is degraded to acetaldehyde and then to acetate, mainly in the liver. At low ethanol levels, aldehyde dehydrogenase is the main enzyme used for metabolism, whereas at high levels, the microsomal ethanol oxidizing system is recruited into action. Liver and brain are the two major targets of alcohol such that brief, high-level exposures cause cytotoxic injury that is largely reversible, whereas chronic heavy abuse leads to long-lasting injury with functional deficits and structural degeneration. Mechanistically, the earliest adverse effects of alcohol exposure include cellular injury with inflammation, perturbations in membrane lipid composition and function, and impairments in insulin/insulin-like growth factor (IGF) signaling. These effects are linked to multiple pathophysiological processes such as decreased cell survival, growth, and repair, dysregulated lipid metabolism, and deficits in energy metabolism. Downstream consequences including increased oxidative and endoplasmic reticulum (ER) stress, lipotoxicity, mitochondrial dysfunction, and adduct buildup, worsen membrane functions, insulin/IGF signaling, and inflammation, and thereby reinforce or perpetuate the cascade of alcohol-related liver and brain degeneration. In essence, the adverse effects of

Molecular Nutrition and Diabetes
http://dx.doi.org/10.1016/B978-0-12-801585-8.00019-1

alcohol on liver and brain are quite similar, but the consequences differ because of their differential roles in maintaining homeostasis within the organism.

1.3 Indirect Mediators of Liver and Brain Injury and Degeneration

Abundant literature stemming from in vitro experiments clearly documents that alcohol has direct toxic and metabolic effects on hepatocytes and central nervous system (CNS)-derived cells. However, there is also evidence that some adverse effects of alcohol are mediated via indirect pathogenic processes. One example of this is hepatic encephalopathy, which is largely related to astrocyte swelling and neurotoxic injury resulting from ammonia, aromatic amino acids, and toxic lipids. In addition, folate, thiamine, pyridoxine, and zinc deficiencies play critical roles.[2] Correspondingly, meta-analysis of alcohol-induced liver disease showed that nutritional supplementation ameliorates hepatic encephalopathy.[3] Limitations in experimental designs that fail to enable development of both direct and indirect forms of tissue injury limit capacity to draw parallels with human disease or extrapolate pathogenic mechanisms. Recent improvements in the standard alcohol-feeding model, such that bingeing was superimposed on chronic alcohol feeding, dramatically increased acute hepatic injury and inflammation.[4] A second cofactor indirectly contributing to liver and brain degeneration is the nutritional deficiency, chiefly thiamine, but probably other micro- and macro-nutrients as well. A third cofactor is the contribution of other exposures that may damage liver and brain in manners similar to those occurring with alcohol exposure. Tobacco smoke, tobacco-specific nitrosamines, and nitrosamines present in preserved/processed foods cause steatohepatitis and neurodegeneration with features that overlap with those of alcohol. Finally, the concepts that alcohol-related brain injury and degeneration can be mediated by liver disease via a liver–brain axis, and that liver disease can be mediated by alterations in gut microbial flora and increased intestinal permeability with attendant endotoxin-associated injury and inflammation (gut–liver axis), are under intense investigation. This chapter reviews direct and indirect (gut–liver–brain axis) mediators of alcohol-related injury and degeneration in the adult liver and brain.

2. ALCOHOL-RELATED LIVER DISEASE

2.1 Histopathological Correlates of Alcohol-Induced Liver Disease

Alcohol abuse is a leading cause of liver-related morbidity and mortality.[5] Acute alcohol-related liver injury is typically reversible, but in a subset of cases, simple steatosis (fatty liver) progresses to steatohepatitis followed by chronic liver disease, including cirrhosis and liver failure.[6] Acute alcoholic hepatitis (AH)/steatohepatitis is diagnosed using standard clinical criteria together with histopathologic evidence of hepatic steatosis, acute inflammation, necrosis, ballooning degeneration of hepatocytes, disorganization of the lobular architecture, mega-mitochondria, and accumulation of Mallory-Denk bodies[7] (i.e., intracytoplasmic hyaline deposits corresponding to aggregated, misfolded, ubiquitinated proteins).[8] With progression to chronic alcoholic liver disease (ALD), hepatic function deteriorates from multiple interrelated pathophysiological processes, including insulin resistance,[9–11] cytotoxic and lipotoxic injury,[12–14] inflammation,[12] oxidative and ER stress,[15] metabolic and mitochondrial dysfunction,[16] decreased DNA synthesis,[10,17] and increased cell death.[13] Progressive ALD is marked by activation of pro-fibrogenic pathways,[12] setting the stage for eventual development of cirrhosis and finally liver failure.[16] Improved understanding of AH pathogenesis, particularly the role of cofactors, will help guide development of new diagnostic and therapeutic tools for ALD.

2.2 Factors Contributing to ALD

An excellent model of chronic ALD was generated by feeding adult Long Evans (LE) rats Lieber-deCarli ethanol-containing liquid diets (9% v/v) for 5–6 weeks.[10,13,18–20] Compared with Fisher (FS) and Sprague–Dawley (SD) rats, the severity of ALD in LE rats was striking, despite similarly high blood alcohol concentrations.[19] In the LE model, hepatic expression and function of alcohol-metabolizing enzymes were altered in ways that would have caused acetaldehyde to build up. Another distinguishing feature was that LE livers had higher basal levels of oxidative stress, mitochondrial dysfunction, p53 activation, and DNA damage relative to SD and FS rats.[13] This suggests that predisposing endogenous factors can alter host susceptibility to ALD. However, the chronic ALD model lacked several key features of AH and did not match the severity of liver injury observed in humans. In light of the fact that AH usually develops in chronic heavy drinkers who also binge, the chronic LE-ALD model was modified to mimic human clinical scenarios by binge administering ethanol (4 g/kg, 3×/week) during the fourth and fifth weeks of chronic ethanol feeding. Under those circumstances, substantially greater severities of AH occurred compared with bingeing or chronic ethanol feeding alone.[21] The aggregate findings suggest that: (1) susceptibility to chronic ALD is mediated by underlying/intrinsic factors, including genetic; (2) functional biomarkers of susceptibility to ALD and AH are needed; and (3) significant AH is likely mediated by combined effects of chronic liver injury plus superimposed insults that substantially increase metabolic demands (e.g., binge drinking, drug exposures, infection).

2.3 Tobacco Smoking as a Cofactor in ALD

Variability in ALD occurrence/severity suggests that cofactors contribute to its pathogenesis. A very high percentage (~80%) of heavy drinkers/alcoholics also abuse tobacco products, typically by cigarette smoking.[22,23] Although the cocarcinogenic effects of alcohol and cigarette smoke (CS) have been well described,[24–27] particularly with respect to the tobacco-specific nitrosamine, 4-(methylnitrosamino)-1-(3-pyridyl)-1-butanone (NNK) and its metabolites,[24–28] there is now evidence that in low doses, NNK,[29] as with other nitrosamines,[30,31] has toxic-degenerative rather than carcinogenic effects. Correspondingly, both low-dose NNK[29] and chronic CS exposures[32] cause steatohepatitis with hepatic insulin resistance, increased DNA damage, lipid peroxidation, mitochondrial dysfunction, ER stress, increased cytotoxic ceramide generation, and impaired signaling through phosphoinositide-3 kinase (PI3K)-Akt. Furthermore, NNK exacerbates alcohol's adverse effects on the liver.[29] Although CS contains other toxins/carcinogens (e.g., acetaldehyde, acrolein) that generate covalent adducts,[33] these compounds have not been shown to cause hepatic insulin resistance.

2.4 ALD Mechanisms: Hepatic Insulin Resistance (Figures 1 and 2)

Binding of insulin to its cell-surface receptors activates receptor tyrosine kinases that transmit signals downstream through insulin receptor substrate (IRS) proteins,[34] which interact with adaptor molecules including the p85 regulatory subunit of PI3K, which then activates Akt. Signaling through PI3K-Akt: (1) promotes cell survival by inhibiting Bad, Fas-L, and glycogen synthase kinase 3β (GSK-3β); (2) drives cell-cycle progression by inhibiting GSK-3β, thereby releasing cyclin D1 and Myc; (3) inactivates p27 and p130; and (4) promotes growth, metabolism, and protein synthesis via mammalian target of rapamycin and p70S6K.[35,36] Binge and chronic ethanol exposures inhibit insulin and IGF signaling in the liver.[9,10,19,37–39] These effects are partly mediated by reduced tyrosine phosphorylation of insulin and IGF receptors[9,40] because of decreased receptor binding and activation of receptor tyrosine kinases[9,19,40] and increased activation of phosphatases that negatively regulate receptor function.[20,21,41] In addition, ethanol inhibits tyrosine phosphorylation of IRS, blocking downstream signaling through PI3K and Akt.[9,17,19,39,42] Hepatic insulin resistance is likely an early pathophysiologic consequence of ethanol toxicity because it develops shortly after exposure.[43] Furthermore, because ethanol compromises insulin/IGF-regulated hepatocyte growth, regeneration, survival, glucose utilization, energy metabolism and

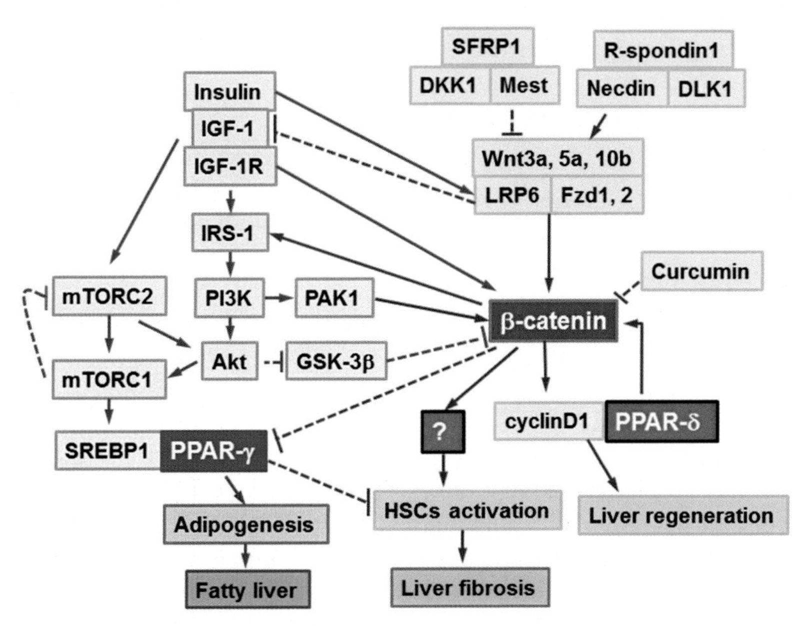

FIGURE 1 Diagram of major components of the insulin and Wnt signaling pathways and the levels at which they cross-talk to promote or inhibit liver growth/regeneration, steatosis, and fibrosis. DKK, Dickkopf; Fzd, frizzled; GSK, glycogen synthase kinase; HSC, hepatic stellate cell; IGF, insulin-like growth factor; IRS, insulin receptor substrate; LRP, low-density lipoprotein receptor-related protein; mTORC, mammalian target of rapamycin complex; PI3K, phosphoinositide-3 kinase; PPAR, peroxisome proliferator-activated receptor; SFRP, secreted frizzled-related protein; SREBP, sterol regulatory element-binding protein.

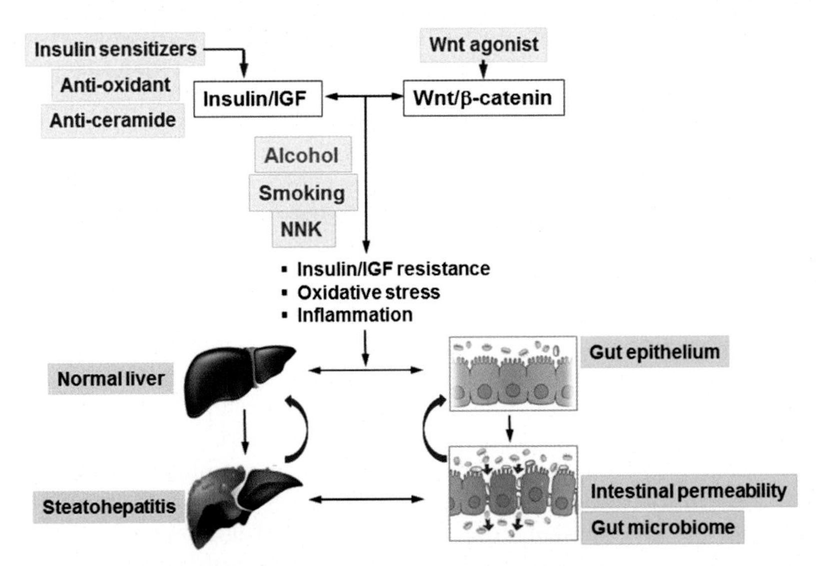

FIGURE 2 Schematic of positive and negative modulators of insulin and Wnt signaling and their effects on liver disease (steatohepatitis), the gut microbiome, and intestinal permeability. IGF, insulin-like growth factor; NNK, 4-(methylnitrosamino)-1-(3-pyridyl)-1-butanone.

protein synthesis,[38,42,44] hepatic insulin/IGF resistance plays a pivotal role in causing acute AH to progress to chronic ALD.

2.5 ALD Mechanisms: Uncoupling of Insulin/IGF Cross-Talk with Wnt (Figures 1 and 2)

Wnt signaling has diverse roles in regulating tissue regeneration and remodeling in the liver and gastrointestinal tract.[45–47] Wnt signaling is activated by Wnt ligand binding to frizzled (Fzd) receptors, leading to stimulation of the canonical β-catenin pathway.[48,49] Resulting stabilization of β-catenin enables its nuclear translocation and accumulation where it interacts with Tcf or Lef transcription factors to upregulate expression of target genes such as cyclin D1 and c-myc,[45] and inhibit activation of hepatic stellate cells (HSCs).[50] The absence of Wnt is permissive to β-catenin phosphorylation within the APC-Axin-GSK-3β destruction complex that targets β-catenin for proteasomal degradation. Chronic ethanol exposure inhibits Wnt signaling via downregulation of multiple Wnt, Fzd, and sFzd isoforms and decreased nuclear β-catenin in liver[51] and brain.[52]

Growing evidence supports the concept that Wnt normally does not act alone, and instead cross-talks with various pathways, including insulin/IGF[53] and via GSK-3β,[54] such that the end-result of impaired insulin/IGF signaling is decreased nuclear β-catenin and downstream target gene expression. Wnt/β-catenin can also positively regulate IGF-1/IRS-1 signaling via Akt-linked metabolic pathways.[55] However, injury and disease can disrupt or uncouple this cross-talk, a phenomenon that may be protective for preserving Wnt functions to combat oxidative stress-mediated tissue injury[56] vis-à-vis inhibition of insulin/IGF/IRS/Akt.[57] Paradoxically, Wnt (presumably unleashed) can promote HSC activation and fibrogenesis.[58] Therefore, uncoupling of insulin/IGF-Wnt with attendant dysregulated activation of Wnt can be detrimental and enable progression of ALD. Correspondingly, evidence suggests that pathological trans-differentiation of HSCs from lipogenic to myofibroblastic phenotypes can be reined in by restoring expression/function of peroxisome proliferator-activated receptor-γ and other lipogenic genes, or treating with Dickkopf-1 coreceptor antagonists.[59] In essence, normalizing liver function in ALD may require support of several linked networks. Because the lists of Wnt ligands/activators and insulin sensitizers continue to grow, in the near future, it may be feasible to restore liver function in chronic ALD by treatment with insulin sensitizers, Wnt agonists, or both.

2.6 ALD Mechanisms: Dysregulated Lipid Metabolism/Lipotoxicity (Figures 1 and 2)

Insulin stimulates lipogenesis and promotes physiological storage of triglycerides in liver.[60] Insulin resistance perturbs lipid homeostasis and promotes lipolysis.[61] Ethanol alters membrane phospholipid levels by activating specific phospholipases; the net result is impaired membrane fluidity[62] and trophic factor signaling.[63] In addition, insulin resistance increases the generation of toxic lipids (i.e., ceramides that further impair insulin signaling[64,65]) by activating pro-inflammatory cytokines and inhibiting PI3K-Akt.[66–69] Hepatic ceramide levels increase with ethanol exposure,[70] because of insulin resistance, inflammation, and activation of acidic and neutral sphingomyelinases.[71] Conceivably, ceramide accumulation in liver plays a key role in the pathogenesis of AH, and mediates AH's progression to ALD because: (1) mice deficient in acidic sphingomyelinase are resistant to ethanol-induced hepatic steatosis[72]; (2) ceramides inhibit adenosine monophosphate–activated protein kinase phosphorylation[73] and thereby promote local hepatocellular injury[74]; and (3) treatment with Myriocin or other inhibitors of ceramide biosynthesis reduce both AH and hepatic insulin resistance in binge or chronic ethanol exposure models.[75,76] In human and experimental livers with AH or ALD, ceramide profiles are substantially altered,[11,44] suggesting that the accumulation of selected ceramide species may be critical to the development of lipotoxic injury and ALD. These concepts are supported by recent imaging mass spectrometry data showing major differences in hepatic ceramide and phospholipid lipid profiles in ALD, and in livers from NNK or chronic CS exposure models.

2.7 ALD Mechanisms: Increased Intestinal Permeability (Figure 2)

In humans, ethanol administration and ALD are associated with increased intestinal permeability[77] and endotoxemia, reflecting impairment of the intestinal epithelial barrier and enhanced bacterial translocation.[78] Likewise, chronic ethanol feeding in rats increases intestinal mucosa permeability.[79] Tight junctions (TJs) regulate paracellular permeability and the apical-basolateral polarity of epithelial cells. TJs are the most apical component of the intercellular junctional complex in epithelial cells and have important roles in establishing cell polarity and functioning as major determinants of epithelial barrier function.[80] Dysregulation of intestinal permeability in various diseases is associated with structural and functional impairments of TJs and displacement of their associated proteins (Occludin, ZO-1, claudin-1, and claudin-3) from microdomains.[81] Inflammation with increased intestinal levels of tumor necrosis factor-α and nuclear factor-κB mediate TJ gene/protein downregulation and dysfunction, and attendant bacterial translocation.[82] Lipopolysaccharide (LPS) also acts directly on intestinal epithelial cells to cause barrier breakdown, increasing gut paracellular permeability and bacterial translocation.[83] Alcohol-induced steatohepatitis has been linked to gut leakiness[84] caused by oxidative stress[85] and cytokine (tumor necrosis factor-α and nuclear factor-κB transcription factor)[86]-mediated disruption of the intestinal barrier and reduced TJs.[87,88] In addition, chronic alcohol exposure enhances pathogenic bacteria-mediated intestinal cell injury, invasion[89] and translocation.[90] These findings helped formulate the hypothesis that ALD is mediated in part via a gut–liver axis.[91] However, alcohol-induced disruption of the intestinal barrier suppresses FOXO4,[86] and Forkhead transcription factors play important roles in mediating insulin, IGF, and other growth factor effects. By extension, alcohol's adverse effects on intestinal epithelial barrier function could be mediated by insulin/IGF resistance. This concept is supported by the recent preliminary findings that insulin/IGF/IRS signaling networks are expressed in intestinal epithelial cells and that their constitutive levels of phosphorylation are reduced by chronic ethanol feeding.

Chronic heavy alcohol consumption causes acetaldehyde, the chief toxic metabolite of alcohol, to accumulate in the colon.[92] In addition to host metabolism, gut microbes also metabolize alcohol to acetaldehyde. Acetaldehyde redistributes TJ proteins and increases TJ protein phosphorylation, thereby disrupting intestinal epithelial TJs and increasing paracellular permeability.[88,92] The colon is where the greatest quantity of bacteria reside and permeability is most affected.[93] Changes in epithelial permeability can be apparent within 1 week of ethanol feeding. Increased permeability allows translocation of bacteria and bacterial products (e.g., LPS) from the gut, leading to portal endotoxemia, followed by hepatic inflammation and injury.[94] Correspondingly, plasma endotoxin levels are higher in current drinkers than controls.[93] The concept that LPS and endotoxemia drive ALD is supported by the findings that mice deficient in TLR-4 or CD-14–mediated LPS signaling are resistant to ALD[95]; germ-free mice are not susceptible to alcohol-induced liver injury[96]; and intestinal sterilization with antibiotics can prevent alcohol-induced liver injury in rats.[97]

2.8 ALD Mechanisms: Altered Gut Microbiota-Dysbiosis

Interest in the role of altered intestinal microbiota grew rapidly with the realization that insulin resistance disorders, including type 2 diabetes, obesity, and nonalcoholic fatty liver disease (NAFLD), were associated with changes or overgrowth of intestinal microbial flora.[98] Furthermore, diet and exercise, which improve insulin responsiveness, also modulate the gut microbiome.[99] Importantly, evidence suggests that intestinal microbes can promote hepatic steatosis and alter host energy balance.[100] Furthermore, increased intestinal permeability with bacterial translocation from gut-barrier dysfunction has been linked to dysbiosis, neurobehavioral abnormalities associated with alcohol dependence,[101] hepatic encephalopathy,[102] spontaneous bacterial peritonitis, and other fatal infections that complicate cirrhosis.[103] Intestinal dysbiosis and bacterial translocation are commonly associated with advanced stages of liver disease, and evidence suggests that bacterial overgrowth and translocation across the epithelial barrier cause liver disease to progress.[104,105] Small intestine bacterial overgrowth is a common form of dysbiosis associated with increased alcohol consumption and marked by excessive quantities of bacteria,[106] correlating with severity of liver disease.[107] Ethanol decreases intestinal motility and thereby favors bacterial proliferation.[106,108] Alterations in relative abundance of common bacterial phyla occur with heavy alcohol abuse. In humans and animal models,[96,103,105,108] there is a decrease in *Firmicutes*, including short-chain fatty acid–producing taxa such as *Ruminococcaceae*, *Lachnospiraceae*, and probiotic *Lactobacilli*, which are associated with epithelial health. In contrast, *Enterobacteriaceae*, *Enterococcus*, *Actinobacteria*, *Fusobacteria*, and *Verrucomicrobia*, primarily gram-negative taxa, are more commonly increased in the dysbiotic gut. Both increases and decreases in the phylum *Bacteroidetes*, have been reported,[93,104,105,108] but this could be explained by an increase in *Prevotellaceae*[105] and compensatory decrease in *Bacteroidaceae*[93]; both are members of *Bacteroidetes*. The ratio of eubiotic to dyspoietic bacteria negatively correlates with plasma endotoxin and liver disease. Controls have the highest ratio (2.05), compensated patients have a ratio of 0.89, decompensated patients have a ratio of 0.66, and inpatients have a ratio of 0.32.[103] *Enterobacteriaceae*, such as *Escherichia coli*, metabolize alcohol[109] and may have a competitive advantage in a high alcohol milieu.[110] Changes in intestinal bile acids can further complicate matters because cholic acid and its deconjugated form, deoxycholic acid, are bacteriocidal and favor eubiotic *Firmicutes* over *Enterobacteriaceae*. Hepatic inflammation decreases bile acid production and gut levels, further altering the eubiotic:dysbiotic bacterial balance.[109,111] Additional evidence that gut dysbiosis exacerbates liver injury is that treatment with probiotic *Lactobacillus rhamnosus*[85] or pretreatment with a high-fiber prebiotic diet[96] ameliorate alcohol-induced liver injury in mice. Increased intestinal permeability combined with bacterial overgrowth and a shift toward more toxigenic bacteria increase bacterial translocation, endotoxemia, and liver damage, and decrease bile acid production, causing a positive feedback loop that facilitates expansion of *Enterobacteriaceae*, and increases endotoxemia and liver injury.[96,98,108,109,111,112]

3. ALCOHOL-RELATED NEURODEGENERATION

3.1 Alcoholic Brain Diseases

Alcohol abuse causes neurobehavioral abnormalities, cognitive dysfunction, dementia, and disability, in large measure from its adverse effects on the central and peripheral nervous systems.[113] Chronic alcoholism impairs cognitive function, and the vast majority (75%) of chronic alcoholics has significant brain damage/degeneration at autopsy. Even after detoxification, more than half continue to have some degree of learning, memory, and executive function deficits. The spectrum of alcohol-related brain degeneration (ARBD) includes disease processes that can be clustered with respect to etiology: (1) acute encephalopathies (e.g., alcohol intoxication and hepatic encephalopathy, myelopathy and cerebral degeneration); (2) nutritional from thiamine deficiency (i.e., Wernicke's encephalopathy or Wernicke-Korsakoff syndrome (WKS)); and (3) neurodegeneration associated with white matter (WM) atrophy, cerebellar, and corticolimbic degeneration.[114] Encephalopathies are mediated by both direct acute toxic effects of alcohol, and indirect consequences of liver disease. The extent to which a liver–brain axis contributes to its development may be related to chronic liver disease with reduced capacity to metabolize/detoxify alcohol.

3.2 Thiamine Deficiency

Thiamine, or vitamin B_1, was the first compound to be recognized as a vitamin and as essential to the human diet. Thiamine deficiency disorders can be serious or even fatal, causing heart failure (beriberi) or neurodegeneration with dementia. Thiamine is required for transketolase, pyruvate dehydrogenase, and α-ketoglutarate dehydrogenase activities, which mediate carbohydrate metabolism and adenosine triphosphate synthesis.[115] Furthermore, thiamine-regulated enzymatic functions are critical for biosynthesis of neurotransmitters, nucleic acids, fatty acids, steroids, and complex carbohydrates. In affluent countries, thiamine deficiency is mainly linked to alcoholism and fad diets. An early study found that up to 80% of alcoholics had some degree of thiamine deficiency. Although the rates of overt thiamine deficiency have declined because of the regular supplementation of packaged commercial foods with vitamins, thiamine deficiency still occurs because of inadequate dietary intake, decreased absorption and conversion of thiamine to its active form, thiamine pyrophosphate, reduced hepatic storage, and impaired utilization by cells.[116] WKS was linked to thiamine deficiency because repletion of thiamine reversed both the symptoms and damage. Moreover, administration of pyrithiamine, which inhibits thiaminokinase, an enzyme needed for thiamine phosphorylation and activation,[117] caused WKS symptoms.[118]

Acute and subacute phases of WKS are associated with encephalopathy, delirium, ophthalmoplegia, incoordination, and confusion, and they are likely to be mediated by metabolic dysfunction. In contrast, chronic phases of WKS are associated with progressive and sustained neurocognitive deficits with psychosis, amnesia (retrograde and anterograde) debilitation, and severe degeneration of corticolimbic WM tracts that are critical to behavior and memory.[114] Despite intensive research, the contributions of thiamine deficiency versus alcohol's toxic effects have not been delineated satisfactorily. Because the neuropathological features of chronic WKS overlap with those of ARBD-mediated WM degeneration, could some aspects of ARBD-associated WM degeneration be caused by the separate or combined effects of thiamine deficiency and alcohol-dependent degeneration? The answer appears to be affirmative in part. Combined exposure studies showed that binge and chronic ethanol administration in the setting of thiamine deficiency caused more substantial progression of dysregulated neural and cognitive functions, impaired neural plasticity associated with reduced GABAergic inhibition and facilitated glutamatergic excitation, hippocampal and cortical cell loss, and reduced neurotrophin protein expression (needed for neuronal survival) compared with each exposure alone. On the other hand, independent and additive effects of thiamine deficiency and alcohol were shown by the findings that repeated bouts of thiamine deficiency caused permanent and severe spatial memory impairments and increased perseverative behavior, whereas alcohol + thiamine deficiency additively reduced WM volume compared with thiamine deficiency alone.[119] The bottom line is that thiamine deficiency should always be regarded as a cofactor mediating alcohol-related neurodegeneration.

3.3 WM Pathology Caused by Chronic Alcohol Exposure

Chronic alcohol abuse causes brain atrophy[120] with selective loss of WM volume.[121] Degrees of WM atrophy correlate with maximum daily and lifetime exposures to alcohol.[113,114,122] Neuroimaging studies showed that the corpus callosum is a vulnerable target and significantly atrophied in alcoholic subjects.[123] Diffusion tensor imaging studies revealed that atrophy of the corpus callosum correlates with disruption of the microstructural integrity of WM.[124] Atrophy of the corpus callosum impairs interhemispheric communication within the cerebrum, compromising exchange of sensory, motor, and cognitive information. Besides the corpus callosum, prefrontal, temporal, and cerebellar WM are notable targets of alcohol-related neurodegeneration.[114] Importantly, WM atrophy and degeneration in alcoholics has been linked to impairments in executive function.[125] Despite the wealth of information about ethanol's adverse effects on WM, the mechanisms of degeneration are poorly understood. Nonetheless, there is ample evidence that the underlying abnormalities are linked to metabolic impairments in oligodendrocyte function, which results in poor myelin maintenance, maturation, and synthesis. Mechanistically, we propose that brain insulin/IGF resistance and lipotoxicity because of aberrant accumulations of ceramides and other sphingolipids, mediate these degenerative and toxic effects on oligodendrocytes.

In both immature and mature brains, WM oligodendrocytes and myelin are major targets of alcohol-mediated toxicity and degeneration.[114,121,126,127] WM atrophy and myelin degeneration are associated with cognitive impairment.[114,128,129] Clinical neuroimaging and postmortem brain studies showed that severity of WM loss or degeneration correlates with lifetime levels of alcohol consumption.[122] Previous studies support the concept that alcohol-associated WM degeneration is mediated by direct neurotoxic effects of alcohol together with impairments in insulin/IGF signaling[44,130–132] needed for myelin synthesis and maintenance.[133–135] Related studies linked alcohol-mediated neurodegeneration to altered sphingolipid (ceramide) levels, profiles, enzymatic activity, and gene expression in the brain,[9,14] and perturbations of ghrelin/leptin circuitry responsible for craving.[136] We hypothesize that accumulation of toxic ceramides and other sphingolipids promote progressive WM degeneration because of its pro-inflammatory, pro-cell death, and insulin resistance effects. Improved means of characterizing the nature and severity of these abnormalities could lead to methods of detecting early indices of neurodegeneration and new strategies to abrogate long-term adverse effects of alcohol on the brain and behavior.

3.4 Ethanol Effects on Oligodendrocyte Function

Oligodendrocytes produce and maintain myelin in the CNS. Myelin insulates and supports CNS axons. Myelin is a specialized membrane that has a very high dry mass of lipids (70–85%) compared with proteins (15–30%). Oligodendrocytes develop from oligodendrocyte precursor cells, which differentiate into immature followed by mature myelin-producing oligodendrocytes. Mature oligodendrocytes express integral membrane proteins including myelin basic protein (MBP), myelin-associated glycoprotein (MAG), myelin oligodendrocyte glycoprotein, and proteolipid protein (PLP).[137] Myelin PLP (30 kDa) is the most abundant protein in CNS myelin, and its spliced variant is DM-20 (26 kDa).[138]

Oligodendrocytes are highly susceptible to the toxic effects of ethanol. Ethanol delays oligodendrocyte maturation, developmental expression of MBP and MAG, and de novo synthesis of myelin.[139,140] In addition, ethanol impairs insulin signaling by reducing insulin/IGF receptor binding[141] and altering oligodendrocyte membrane phospholipid content and membrane fluidity.[142] Inhibition of de novo sphingolipid biosynthesis reduces WM myelination, which leads to cognitive impairment.[143]

3.5 Insulin and IGF Effects in Brain

Insulin and IGF-I play critical roles in regulating and maintaining cognitive and motor functions. Insulin, IGF-I, IGF-2, and their receptors are expressed in CNS neural[144] and glial cells (oligodendroglia) throughout the brain, but mainly in corticolimbic and extrapyramidal motor systems.[34,134,145] Moreover, insulin and IGF-I trophic factor genes are abundantly expressed in the olfactory bulbs, cerebral cortex, cerebellar cortex, hippocampus, thalamus, hypothalamus, brainstem nuclei, spinal cord, and retina.[34,146] Besides myelin production, maintenance, and maturation, insulin and IGFs promote survival of oligodendrocytes.[147] Transgenic and gene depletion models confirmed that aberrant insulin/IGF signaling adversely affects myelination. In transgenic mice, overexpression of IGF-I increases brain size, neuron and oligodendrocyte number, and myelin content, suggesting that IGF-I promotes brain growth and myelination.[148] In contrast, depletion of IGF-I or IGF-I receptor genes or overexpression of IGF-binding proteins impairs brain growth.[149] Because insulin and IGF mediate oligodendrocyte growth, survival, myelin production, energy metabolism, and plasticity,[34,133,150] impairments in insulin/IGF-1 signaling are likely to be important mediators of alcohol-associated cognitive and behavioral deficits linked to WM atrophy and degeneration.

3.6 Chronic Ethanol Abuse Causes Brain Insulin/IGF Deficiency and Resistance

Brain insulin resistance leads to neuronal loss and impaired neurotransmitter function required for plasticity, learning, and memory. In addition, brain insulin resistance impairs oligodendrocyte survival and function, resulting in reduced myelin integrity, and increased generation of ceramides, which further increase brain insulin resistance, neuro-inflammation, oxidative stress, and neurodegeneration. Experimental chronic ethanol feeding leads to loss of neurons and oligodendroglia, increased lipid peroxidation, and DNA damage in the brain.[34,141] In addition, human alcoholic brains have sustained reductions in messenger RNA transcripts encoding insulin and IGF-1 polypeptides and receptors, together with reduced binding to insulin and IGF receptors.[151] Furthermore, chronic gestational exposure to ethanol has dose-dependent adverse effects on brain development with impairments in insulin-stimulated neuronal survival and mitochondrial function.[34,130,141,151,152] In essence, alcohol-induced neurodevelopmental abnormalities are associated with a prominent loss of neurons and oligodendroglia, reduced levels of myelin-associated gene expression, insulin responsiveness, and mitochondrial function, and increased apoptosis, oxidative stress, lipid peroxidation, and DNA damage. In adult brains, the main impact of chronic alcohol abuse is WM atrophy, with impairments in myelin maintenance, production, and maturation. Therefore, ethanol targets oligodendroglia and myelin in both immature and mature brains, and these effects are mediated at least in part by insulin/IGF resistance.

3.7 Ethanol-Induced Brain Insulin/IGF Resistance Is Linked to Altered Membrane Lipid Composition

Previous studies showed that ethanol decreases cholesterol content in rat cerebella and cultured cerebellar neurons, and that chemical depletion of cholesterol impairs insulin signaling, whereas cholesterol repletion just partly restores insulin receptor binding, insulin signaling, and insulin-stimulated glucose uptake in ethanol-exposed cerebella.[141] These findings suggest that ethanol-mediated alterations in membrane lipid composition contribute to insulin/IGF resistance in the brain. However, incomplete restoration of insulin signaling by cholesterol repletion was likely due to ethanol's broad adverse effects on membrane lipid profiles, including depletion or disruption of sphingolipids and phospholipids. Correspondingly, subsequent studies showed that ethanol alters sphingomyelin/ceramide ratios, increasing metabolism of sphingolipids that generate ceramides. Activated ceramide signaling causes CNS glial cell death.[153]

3.8 Role of Ceramides in Neurodegeneration

Chronic alcohol exposure promotes lipolysis and accumulation of toxic lipids, including ceramides in the CNS. Furthermore, exposure of neuronal cells or brains to toxic ceramides impairs insulin/IGF signaling, increases oxidative stress, and causes neurodegeneration with impairments in motor and cognitive functions.[154] Mechanistically, alcohol can disrupt oligodendrocyte function and cause myelin degradation, promoting ceramide accumulation. These findings led to the hypothesis that aberrant ceramide and other toxic sphingolipids in the CNS may be important mediators of neurodegeneration due to inhibition of insulin/IGF-1 signaling, dysregulation of lipid metabolism and myelin maintenance/maturation, increased oxidative stress, and reduced survival mechanisms in oligodendrocytes.

3.9 Potential Roles of Smoking and Tobacco Nitrosamines in ARBD

As noted for ALD, variability in the nature and severity of clinical and pathological features of ARBD suggests that cofactors also contribute to its pathogenesis. Because a very high percentage of heavy drinkers/alcoholics (~80%) also abuse tobacco products, typically smoking cigarettes,[22,23] and both heavy drinking and cigarette smoking adversely affect neurocognitive function[155] and WM structure (neuroimaging),[156] consideration should be given to the concept that tobacco and its toxic metabolites serve as co-factors in ARBDs. Several neuroimaging studies have shown smoking-related brain abnormalities in humans,[157–161] and a recently published meta-analysis revealed smoking-related gray matter loss in the anterior cingulate, prefrontal cortex, and cerebellum[162] (i.e., structures targeted by alcohol). Other neuroimaging studies comparing alcoholics who smoke with those who do not showed that, among smokers, temporal and parietal gray matter structures were relatively reduced,[155,160,163] whereas WM was either unaffected[155] or increased[160] in volume. In a more complex but complete four-way design comparing alcoholic to nonalcoholic smokers and nonsmokers, Durazzo et al.[155,163] found no significant intergroup differences with respect to cortical or WM regions of interest, whereas evidence to the contrary stemmed from a study of abstinent alcoholics in which significant volume loss was detected in the prefrontal cortex, parahippocampal gyrus, temporal pole, amygdala, and globus pallidus of alcoholic smokers compared with nonsmoking controls.[164] Although the aggregate results suggest that chronic smoking causes frontal and temporal lobe atrophy with volume loss in gray matter structures, those studies were probably underpowered because of heterogeneous levels and durations of tobacco smoke exposure and inability to correlate long-term biomarkers of smoking with brain atrophy and cognitive impairment. Moreover, neuroimaging studies and our own recent results provide clear evidence that the WM is targeted by chronic tobacco smoke or tobacco nitrosamine exposures. In essence, additional human and experimental research is needed to address these issues and evaluate the contributions of alcohol and/or tobacco smoke to the pathogenesis of neurodegeneration. In this regard, methodologic approaches beyond measuring gray or WM volumes (e.g., ultrastructure, molecular and biochemical assays) are needed to better characterize early reversible stages of neurodegeneration. Finally, we should determine the agent in tobacco smoke that mediates neurodegeneration and cognitive impairment because that information could be used to develop assays that monitor exposure and early stage effects of neurodegeneration.

Alcohol-tobacco dual effects on carcinogenesis have been well described,[24–27] particularly with respect to the tobacco-specific nitrosamine, NNK, and its metabolites.[24–28] However, in low doses, NNK,[29] like other nitrosamines,[30,31] has toxic-degenerative rather than carcinogenic effects. Previous reports showed that low-dose nitrosamine exposures (e.g., streptozotocin[165–167] and N-nitrosodimethylamine[31,168]) cause degenerative CNS effects mediated by insulin resistance, DNA damage, lipid peroxidation, mitochondrial dysfunction, ER stress, and impairments in PI3K-Akt signaling.[31,168] Follow-up studies revealed: (1) additive and interactive effects of ethanol and submutagenic doses of PLP on brain development in experimental fetal alcohol spectrum disorder[169]; (2) additive and independent adverse effects of ethanol and NNK (submutagenic) on cognitive function, WM structure, insulin/IGF signaling, and myelin-associated gene expression in adolescent brains[132]; and (3) independent adverse effects of chronic tobacco smoking on WM integrity, brain insulin/IGF signaling, and myelin gene expression.[32] The finding that NNK causes brain injury led to the suggestion that smoking can cause cognitive impairment via the toxic effects of NNK.[170]

Recent studies in ethanol-fed (36% Lieber DeCarli × 6 weeks) male LE rats showed that deficits in spatial learning and memory were associated with demyelination and degeneration of myelinated axons in frontal WM,[132] whereas submutagenic NNK (rats) or tobacco smoke (mice) exposures caused mainly axonal loss/degeneration and secondary demyelination. Exploratory matrix-assisted laser desorption/ionization imaging mass spectrometry analysis revealed distinct ethanol- or NNK exposure-linked alterations in WM lipid profiles.[171] To correlate WM ultrastructural pathology with perturbations in myelin lipid biochemical profiles in ethanol (dose effects), tobacco smoke, and NNK exposed rats will help uncover new ways to evaluate WM degeneration, diagnose its underlying causes, and design targeted treatments. NNK effects be studied because we strongly suspect that this toxin and its metabolites are etiopathogenic because they are present in tobacco smoke and can cause neurodegeneration, including WM. The NNK studies will also control for smoking-induced pulmonary injury as a cause of WM degeneration. Our focus on sphingolipids, particularly ceramides, stems from independent reports showing that ceramides, especially those with long-chain fatty acids, can be neurotoxic, impair insulin/IGF signaling, promote lipid peroxidation and DNA damage, and contribute to neurodegeneration.[44,67,154,172–180] Lipotoxicity delays oligodendrocyte maturation and expression of MBP, MAG, and de novo myelin synthesis.[178]

3.10 Liver–Brain Axis of ARBD

Ceramides are lipid-signaling molecules that promote diverse cellular responses including proliferation, motility, plasticity, inflammation, apoptosis, and insulin resistance.[181] Ceramides are generated during biosynthesis or degradation of sphingomyelin.[64,182–185] With disease-associated lipolysis, insulin resistance is initiated by ER stress and mitochondrial dysfunction[74,186] from ceramide activation of pro-inflammatory cytokines and inhibition of insulin signaling through PI3 kinase-Akt.[66–69] In human and experimental models of ALD, hepatic and serum ceramide levels are elevated, and their profiles (molecular forms) are altered, and these responses are mediated in part by changes in messenger RNA expression and enzymatic activity.[11,187]

Many studies have demonstrated direct toxic and degenerative effects of ethanol using in vitro models.[41] Therefore, ethanol that bypasses peripheral detoxification systems can directly cause CNS injury and degeneration by increasing oxidative stress, DNA damage, lipid peroxidation, mitochondrial dysfunction, and perturbing membrane lipid composition, leading to insulin/IGF resistance.[34,141,152] CNS neurons and oligodendrocytes are insulin-responsive.[34,133,145] Insulin and IGF promote their viability, energy metabolism, neurotransmitter synthesis, plasticity, and myelin homeostasis. Metabolic stresses impair oligodendroglial functions, including myelin maintenance.[134,135,150] Because ceramides are generated in brain during myelin turnover and degradation, factors that impair oligodendroglial function would be expected to increase ceramide synthesis.[188,189] Locally increased ceramide production would worsen brain insulin/IGF resistance, neuroinflammation, and oxidative stress.[181,183,185,190,191] The degree to which this scenario mediates alcohol-related neurodegeneration (i.e., WM atrophy) would likely correlate with inefficiency of peripheral (gastrointestinal and hepatic) detoxification systems or binge exposures that overwhelm the network.

Steatohepatitis caused by ALD, NAFLD, or chronic hepatitis C virus infection can be associated with cognitive and neuropsychiatric dysfunction.[128,192,193] Moreover, steatohepatitis produced by chronic high-fat diet feeding and/or low-dose nitrosamine exposures also cause cognitive impairment and neurodegeneration.[31,179,194–196] In those models, steatohepatitis was associated with insulin resistance in both liver and brain, manifested by reduced insulin receptor binding, gene expression, and tyrosine kinase activation; decreased expression of genes required for metabolism and neurotransmitter synthesis in the brain[31,179,194–196]; and increased oxidative stress. Further studies demonstrated that steatohepatitis was associated with upregulated expression of multiple genes that increase ceramide levels in liver, and higher levels of ceramide immunoreactivity in liver and peripheral blood. Furthermore, ethanol-fed rats developed CNS insulin resistance with neurodegeneration and cognitive impairment[9,130] in tandem with steatohepatitis, although over time, ceramide gene dysregulation was measured in both liver and brain as disease progressed.[14]

Mechanistically, toxic lipids, including ceramides, can cross the blood–brain barrier and cause insulin resistance by interfering with critical phosphorylation events[172] and activating pro-inflammatory cytokines.[181,197] We hypothesize that ALD indirectly causes CNS insulin resistance and WM degeneration when the hepatocellular injury is sufficiently severe to enable liver-derived cytotoxic lipids to enter the circulation. Cytotoxic ceramides that penetrate the blood–brain barrier cause CNS insulin resistance, oxidative stress, and pro-inflammatory cytokine activation. These effects could activate a self-reinforcing pathogenic cascade from activation of signaling mechanisms that lead to increased local production and accumulation cytotoxic ceramides within the CNS. In essence, we propose that ARBD is mediated by dual mechanisms: (1) direct neurotoxic/neurodegenerative effects of alcohol and its metabolites, which cause insulin resistance from perturbations in lipid metabolism, reduced membrane integrity, and increased oxidative stress; and (2) secondary injury caused by increased hepatic elaboration of toxic lipids, including ceramides, that cross the blood–brain barrier and cause insulin resistance. This concept suggests new and interesting strategies for monitoring risk for neurodegeneration and preventing cognitive impairment by treating the underlying liver disease.

4. CONCLUDING REMARKS

The implications of the concepts expressed in this review are that alcohol-related chronic degenerative pathology in liver and brain are mediated by interactive pathogenic processes linked to insulin resistance, inflammation, dysregulated lipid metabolism with attendant lipotoxic cellular injury and stress, and mitochondrial dysfunction. Although arguments could be made favoring the pivotal role of any one of these factors, evidence suggests that once the cascade of tissue injury is established, therapeutic interventions must be multipronged and address the collective mediators of disease. Regarding preventive measures, cofactors are seldom given sufficient attention. Both cessation or curtailment of smoking and maintenance of ample nutritional support stand to reduce insulin resistance states throughout the body, including the liver, gut, and brain. The mechanisms by which alcohol, tobacco, and poor nutrition cause pathological shifts in the intestinal microbiome and lead to increased gut permeability require further study because it is quite likely that the resulting bacterial translocation triggers the inflammatory storm that is so characteristic of acute AH, the precursor to chronic progressive ALD. The correlations drawn between steatohepatitis and neurocognitive deficits and neurodegeneration suggest that measures taken to protect the gut and liver would also preserve brain structure and function.

Acknowledgment

Supported by AA-11431, AA-12908, and AA-20587.

References

1. Centers for Disease Control and Prevention. Alcohol-attributable deaths and years of potential life lost among American Indians and Alaska Natives—United States, 2001—2005. *MMWR Morb Mortal Wkly Rep* 2008;**57**(34):938—41.
2. Strohle A, Wolters M, Hahn A. Alcohol intake—a two-edged sword. Part 1: metabolism and pathogenic effects of alcohol. *Med Monatsschr Pharm* 2012;**35**(8):281—92. quiz 93—4.
3. Antar R, Wong P, Ghali P. A meta-analysis of nutritional supplementation for management of hospitalized alcoholic hepatitis. *Can J Gastroenterol* 2012;**26**(7):463—7.
4. Mathews S, Xu M, Wang H, Bertola A, Gao B. Animals models of gastrointestinal and liver diseases. Animal models of alcohol-induced liver disease: pathophysiology, translational relevance, and challenges. *Am J Physiol Gastrointest Liver Physiol* 2014; **306**(10):G819—23.
5. McCullough AJ, O'Shea RS, Dasarathy S. Diagnosis and management of alcoholic liver disease. *J Dig Dis* 2011;**12**(4):257—62.
6. O'Shea RS, Dasarathy S, McCullough AJ. Alcoholic liver disease. *Hepatology* 2010;**51**(1):307—28.
7. Yerian L. Histopathological evaluation of fatty and alcoholic liver diseases. *J Dig Dis* 2011;**12**(1):17—24.
8. Bardag-Gorce F. Effects of ethanol on the proteasome interacting proteins. *World J Gastroenterol* 2010;**16**(11):1349—57.
9. de la Monte SM, Yeon JE, Tong M, et al. Insulin resistance in experimental alcohol-induced liver disease. *J Gastroenterol Hepatol* 2008; **23**(8 Pt 2):e477—86.
10. Pang M, de la Monte SM, Longato L, et al. PPARdelta agonist attenuates alcohol-induced hepatic insulin resistance and improves liver injury and repair. *J Hepatol* 2009;**50**(6):1192—201.
11. Longato L, Ripp K, Setshedi M, et al. Insulin resistance, ceramide accumulation, and endoplasmic reticulum stress in human chronic alcohol-related liver disease. *Oxid Med Cell Longev* 2012; **2012**:479348.
12. Cohen JI, Nagy LE. Pathogenesis of alcoholic liver disease: interactions between parenchymal and non-parenchymal cells. *J Dig Dis* 2011;**12**(1):3—9.
13. Derdak Z, Lang CH, Villegas KA, et al. Activation of p53 enhances apoptosis and insulin resistance in a rat model of alcoholic liver disease. *J Hepatol* 2011;**54**(1):164—72.
14. de la Monte SM, Longato L, Tong M, DeNucci S, Wands JR. The liver-brain axis of alcohol-mediated neurodegeneration: role of toxic lipids. *Int J Environ Res Public Health* 2009;**6**(7):2055—75.
15. Kaplowitz N, Ji C. Unfolding new mechanisms of alcoholic liver disease in the endoplasmic reticulum. *J Gastroenterol Hepatol* 2006;**21**(Suppl. 3):S7—9.
16. Purohit V, Gao B, Song BJ. Molecular mechanisms of alcoholic fatty liver. *Alcohol Clin Exp Res* 2009;**33**(2):191—205.
17. Sasaki Y, Hayashi N, Ito T, Fusamoto H, Kamada T, Wands JR. Influence of ethanol on insulin receptor substrate-1-mediated signal transduction during rat liver regeneration. *Alcohol Alcohol Suppl* 1994;**1**(99):99—106.
18. de la Monte SM, Pang M, Chaudhry R, et al. Peroxisome proliferator-activated receptor agonist treatment of alcohol-induced hepatic insulin resistance. *Hepatol Res* 2011;**41**(4):386—98.
19. Denucci SM, Tong M, Longato L, et al. Rat strain differences in susceptibility to alcohol-induced chronic liver injury and hepatic insulin resistance. *Gastroenterol Res Pract* 2010;**2010**:16.
20. Yeon JE, Califano S, Xu J, Wands JR, De La Monte SM. Potential role of PTEN phosphatase in ethanol-impaired survival signaling in the liver. *Hepatology* 2003;**38**(3):703—14.
21. Ramirez T, Tong M, de la Monte SM. Chronic-binge model of alcoholic hepatitis in long evans rats. *JDAR* 2014;**3**(2014):1—10.
22. Romberger DJ, Grant K. Alcohol consumption and smoking status: the role of smoking cessation. *Biomed Pharmacother* 2004; **58**(2):77—83.
23. Kalman D, Kim S, DiGirolamo G, Smelson D, Ziedonis D. Addressing tobacco use disorder in smokers in early remission from alcohol dependence: the case for integrating smoking cessation services in substance use disorder treatment programs. *Clin Psychol Rev* 2010;**30**(1):12—24.
24. de Boer MF, Sanderson RJ, Damhuis RA, Meeuwis CA, Knegt PP. The effects of alcohol and smoking upon the age, anatomic sites and stage in the development of cancer of the oral cavity and oropharynx in females in the south west Netherlands. *Eur Arch Otorhinolaryngol* 1997;**254**(4):177—9.
25. Duell EJ. Epidemiology and potential mechanisms of tobacco smoking and heavy alcohol consumption in pancreatic cancer. *Mol Carcinog* 2012;**51**(1):40—52.
26. Johnson NW, Warnakulasuriy S, Tavassoli M. Hereditary and environmental risk factors; clinical and laboratory risk matters for head and neck, especially oral, cancer and precancer. *Eur J Cancer* 1996;**5**(1):5—17.
27. Tramacere I, Negri E, Bagnardi V, et al. A meta-analysis of alcohol drinking and oral and pharyngeal cancers. Part 1: overall results and dose-risk relation. *Oral Oncol* 2010;**46**(7):497—503.
28. Go VL, Gukovskaya A, Pandol SJ. Alcohol and pancreatic cancer. *Alcohol* 2005;**35**(3):205—11.
29. Zabala V, Tong M, Yu R, et al. Potential contributions of the tobacco nicotine-derived nitrosamine ketone (NNK) in the pathogenesis of steatohepatitis in a chronic plus binge rat model of alcoholic liver disease. *Alcohol Alcohol* 2015;**50**(2):118—31.
30. Tong M, Longato L, de la Monte SM. Early limited nitrosamine exposures exacerbate high fat diet-mediated type2 diabetes and neurodegeneration. *BMC Endocr Disord* 2010;**10**(1):4.
31. Tong M, Neusner A, Longato L, Lawton M, Wands JR, de la Monte SM. Nitrosamine exposure causes insulin resistance diseases: relevance to type 2 diabetes mellitus, non-alcoholic steatohepatitis, and Alzheimer's disease. *J Alzheimer's Dis* 2009;**17**(4):827—44.
32. Deochand C, Tong M, Agawal AR, Cadenas E, de la Monte SM. Tobacco smoke exposure impairs brain insulin/IGF signaling: potential co-factor role in neurodegeneration. *J Alzhemiers Dis* 2015, in press.

33. Salaspuro M. Acetaldehyde and gastric cancer. *J Dig Dis* 2011; **12**(2):51–9.
34. de la Monte SM, Wands JR. Review of insulin and insulin-like growth factor expression, signaling, and malfunction in the central nervous system: relevance to Alzheimer's disease. *J Alzheimer's Dis* 2005;**7**(1):45–61.
35. Savage DB, Semple RK. Recent insights into fatty liver, metabolic dyslipidaemia and their links to insulin resistance. *Curr Opin Lipidol* 2010;**21**(4):329–36.
36. Verges B. Abnormal hepatic apolipoprotein B metabolism in type 2 diabetes. *Atherosclerosis* 2010;**211**(2):353–60.
37. He J, de la Monte S, Wands JR. Acute ethanol exposure inhibits insulin signaling in the liver. *Hepatology* 2007;**46**(6):1791–800.
38. Ronis MJ, Wands JR, Badger TM, de la Monte SM, Lang CH, Calissendorff J. Alcohol-induced disruption of endocrine signaling. *Alcohol Clin Exp Res* 2007;**31**(8):1269–85.
39. Setshedi M, Longato L, Petersen DR, et al. Limited therapeutic effect of N-acetylcysteine on hepatic insulin resistance in an experimental model of alcohol-induced steatohepatitis. *Alcohol Clin Exp Res* 2011;**35**(12):2139–51.
40. Patel BC, D'Arville C, Iwahashi M, Simon FR. Impairment of hepatic insulin receptors during chronic ethanol administration. *Am J Physiol* 1991;**261**(2 Pt 1):G199–205.
41. Xu J, Yeon JE, Chang H, et al. Ethanol impairs insulin-stimulated neuronal survival in the developing brain: role of PTEN phosphatase. *J Biol Chem* 2003;**278**(29):26929–37.
42. Mohr L, Tanaka S, Wands JR. Ethanol inhibits hepatocyte proliferation in insulin receptor substrate 1 transgenic mice. *Gastroenterology* 1998;**115**(6):1558–65.
43. He J, de la Monte SM, Wands JR. The p85-beta regulatory subunit of PI3K serves as a substrate for PTEN protein phosphatase activity during insulin mediated signaling. *Biochem Biophys Res Commun* 2010;**397**(3):513–9.
44. de la Monte SM. Alcohol-induced liver and brain degeneration: roles of insulin resistance, toxic ceramides, and endoplasmic reticulum stress. In: Sova M, editor. *Alcohol, nutrition, and health consequences*. New York: Humana Press; 2012.
45. Kolligs FT, Bommer G, Goke B. Wnt/beta-catenin/tcf signaling: a critical pathway in gastrointestinal tumorigenesis. *Digestion* 2002; **66**(3):131–44.
46. Krausova M, Korinek V. Wnt signaling in adult intestinal stem cells and cancer. *Cell Signal* 2014;**26**(3):570–9.
47. Nambotin SB, Wands JR, Kim M. Points of therapeutic intervention along the Wnt signaling pathway in hepatocellular carcinoma. *Anticancer Agents Med Chem* 2011;**11**(6):549–59.
48. MacDonald BT, Tamai K, He X. Wnt/beta-catenin signaling: components, mechanisms, and diseases. *Dev Cell* 2009;**17**(1):9–26.
49. Saito-Diaz K, Chen TW, Wang X, et al. The way Wnt works: components and mechanism. *Growth Factors* 2013;**31**(1):1–31.
50. Kordes C, Sawitza I, Haussinger D. Canonical Wnt signaling maintains the quiescent stage of hepatic stellate cells. *Biochem Biophys Res Commun* 2008;**367**(1):116–23.
51. Xu CQ, de la Monte SM, Tong M, Kim M. Chronic ethanol-induced impairment of Wnt/β-catenin signaling is attenuated by PPAR-δ agonist. *Alcohol Clin Exp Res* 2015;**39**(6):969–79.
52. Tong M, Ziplow J, Chen WC, Nguyen QG, Kim C, de la Monte SM. Motor function deficits following chronic prenatal ethanol exposure are linked to impairments in insulin/IGF, notch and Wnt signaling in the cerebellum. *J Diabetes Metab* 2013;**4**(1):238.
53. Whittaker S, Marais R, Zhu AX. The role of signaling pathways in the development and treatment of hepatocellular carcinoma. *Oncogene* 2010;**29**(36):4989–5005.
54. Doble BW, Woodgett JR. GSK-3: tricks of the trade for a multi-tasking kinase. *J Cell Sci* 2003;**116**(Pt 7):1175–86.
55. Ouchi N, Higuchi A, Ohashi K, et al. Sfrp5 is an anti-inflammatory adipokine that modulates metabolic dysfunction in obesity. *Science* 2010;**329**(5990):454–7.
56. Tao GZ, Lehwald N, Jang KY, et al. Wnt/beta-catenin signaling protects mouse liver against oxidative stress-induced apoptosis through the inhibition of forkhead transcription factor FoxO3. *J Biol Chem* 2013;**288**(24):17214–24.
57. Pearl LH, Barford D. Regulation of protein kinases in insulin, growth factor and Wnt signalling. *Curr Opin Struct Biol* 2002; **12**(6):761–7.
58. Ge WS, Wang YJ, Wu JX, Fan JG, Chen YW, Zhu L. Beta-catenin is overexpressed in hepatic fibrosis and blockage of Wnt/beta-catenin signaling inhibits hepatic stellate cell activation. *Mol Med Rep* 2014;**9**(6):2145–51.
59. Miao CG, Yang YY, He X, et al. Wnt signaling in liver fibrosis: progress, challenges and potential directions. *Biochimie* 2013; **95**(12):2326–35.
60. Capeau J. Insulin resistance and steatosis in humans. *Diabetes Metab* 2008;**34**(6 Pt 2):649–57.
61. Kao Y, Youson JH, Holmes JA, Al-Mahrouki A, Sheridan MA. Effects of insulin on lipid metabolism of larvae and metamorphosing landlocked sea lamprey, *Petromyzon marinus*. *Gen Comp Endocrinol* 1999;**114**(3):405–14.
62. French SW. Biochemical basis for alcohol-induced liver injury. *Clin Biochem* 1989;**22**(1):41–9.
63. Hoek JB, Rubin E. Alcohol and membrane-associated signal transduction. *Alcohol Alcohol* 1990;**25**(2–3):143–56.
64. Holland WL, Summers SA. Sphingolipids, insulin resistance, and metabolic disease: new insights from in vivo manipulation of sphingolipid metabolism. *Endocr Rev* 2008;**29**(4):381–402.
65. Langeveld M, Aerts JM. Glycosphingolipids and insulin resistance. *Prog Lipid Res* 2009;**48**(3–4):196–205.
66. Bourbon NA, Sandirasegarane L, Kester M. Ceramide-induced inhibition of Akt is mediated through protein kinase Czeta: implications for growth arrest. *J Biol Chem* 2002;**277**(5):3286–92.
67. Hajduch E, Balendran A, Batty IH, et al. Ceramide impairs the insulin-dependent membrane recruitment of protein kinase B leading to a loss in downstream signalling in L6 skeletal muscle cells. *Diabetologia* 2001;**44**(2):173–83.
68. Nogueira TC, Anhe GF, Carvalho CR, Curi R, Bordin S, Carpinelli AR. Involvement of phosphatidylinositol-3 kinase/AKT/PKCzeta/lambda pathway in the effect of palmitate on glucose-induced insulin secretion. *Pancreas* 2008;**37**(3):309–15.
69. Powell DJ, Hajduch E, Kular G, Hundal HS. Ceramide disables 3-phosphoinositide binding to the pleckstrin homology domain of protein kinase B (PKB)/Akt by a PKCzeta-dependent mechanism. *Mol Cell Biol* 2003;**23**(21):7794–808.
70. Viktorov AV, Yurkiv VA. Effects of ethanol and lipopolysaccharide on the sphingomyelin cycle in rat hepatocytes. *Bull Exp Biol Med* 2008;**146**(6):753–5.
71. Deaciuc IV, Nikolova-Karakashian M, Fortunato F, Lee EY, Hill DB, McClain CJ. Apoptosis and dysregulated ceramide metabolism in a murine model of alcohol-enhanced lipopolysaccharide hepatotoxicity. *Alcohol Clin Exp Res* 2000;**24**(10): 1557–65.
72. Garcia-Ruiz C, Colell A, Mari M, et al. Defective TNF-alpha-mediated hepatocellular apoptosis and liver damage in acidic sphingomyelinase knockout mice. *J Clin Invest* 2003;**111**(2):197–208.
73. Liangpunsakul S, Sozio MS, Shin E, et al. Inhibitory effect of ethanol on AMPK phosphorylation is mediated in part through elevated ceramide levels. *Am J Physiol Gastrointest Liver Physiol* 2010;**298**(6):G1004–12.
74. Anderson N, Borlak J. Molecular mechanisms and therapeutic targets in steatosis and steatohepatitis. *Pharmacol Rev* 2008;**60**(3): 311–57.
75. Lizarazo D, Zabala V, Tong M, Longato L, de la Monte SM. Ceramide inhibitor myriocin restores insulin/insulin growth factor signaling for liver remodeling in experimental alcohol-related steatohepatitis. *J Gastroenterol Hepatol* 2013; **28**(10):1660–8.

76. Tong M, Longato L, Ramirez T, Zabala V, Wands JR, de la Monte SM. Therapeutic reversal of chronic alcohol-related steatohepatitis with the ceramide inhibitor myriocin. *Int J Exp Pathol* 2014;**95**(1):49–63.
77. Elamin E, Masclee A, Troost F, et al. Ethanol impairs intestinal barrier function in humans through mitogen activated protein kinase signaling: a combined in vivo and in vitro approach. *PLoS One* 2014;**9**(9):e107421.
78. Wang L, Llorente C, Hartmann P, Yang AM, Chen P, Schnabl B. Methods to determine intestinal permeability and bacterial translocation during liver disease. *J Immunol Methods* 2015;**421**:44–53.
79. Worthington BS, Meserole L, Syrotuck JA. Effect of daily ethanol ingestion on intestinal permeability to macromolecules. *Am J Dig Dis* 1978;**23**(1):23–32.
80. Takizawa Y, Kishimoto H, Kitazato T, Tomita M, Hayashi M. Changes in protein and mRNA expression levels of claudin family after mucosal lesion by intestinal ischemia/reperfusion. *Int J Pharm* 2012;**426**(1–2):82–9.
81. Li Q, Zhang Q, Wang C, et al. Altered distribution of tight junction proteins after intestinal ischaemia/reperfusion injury in rats. *J Cell Mol Med* 2009;**13**(9B):4061–76.
82. Shen ZY, Zhang J, Song HL, Zheng WP. Bone-marrow mesenchymal stem cells reduce rat intestinal ischemia-reperfusion injury, ZO-1 downregulation and tight junction disruption via a TNF-alpha-regulated mechanism. *World J Gastroenterol* 2013; **19**(23):3583–95.
83. Hanson PJ, Moran AP, Butler K. Paracellular permeability is increased by basal lipopolysaccharide in a primary culture of colonic epithelial cells; an effect prevented by an activator of Toll-like receptor-2. *Innate Immun* 2011;**17**(3):269–82.
84. Pijls KE, Jonkers DM, Elamin EE, Masclee AA, Koek GH. Intestinal epithelial barrier function in liver cirrhosis: an extensive review of the literature. *Liver Int* 2013;**33**(10):1457–69.
85. Wang Y, Liu Y, Sidhu A, Ma Z, McClain C, Feng W. Lactobacillus rhamnosus GG culture supernatant ameliorates acute alcohol-induced intestinal permeability and liver injury. *Am J Physiol Gastrointest Liver Physiol* 2012;**303**(1):G32–41.
86. Chang B, Sang L, Wang Y, Tong J, Wang B. The role of FoxO4 in the relationship between alcohol-induced intestinal barrier dysfunction and liver injury. *Int J Mol Med* 2013;**31**(3):569–76.
87. Tang Y, Forsyth CB, Banan A, Fields JZ, Keshavarzian A. Oats supplementation prevents alcohol-induced gut leakiness in rats by preventing alcohol-induced oxidative tissue damage. *J Pharmacol Exp Ther* 2009;**329**(3):952–8.
88. Elamin E, Jonkers D, Juuti-Uusitalo K, et al. Effects of ethanol and acetaldehyde on tight junction integrity: in vitro study in a three dimensional intestinal epithelial cell culture model. *PLoS One* 2012;**7**(4):e35008.
89. Wood S, Pithadia R, Rehman T, et al. Chronic alcohol exposure renders epithelial cells vulnerable to bacterial infection. *PLoS One* 2013;**8**(1):e54646.
90. Yan AW, Fouts DE, Brandl J, et al. Enteric dysbiosis associated with a mouse model of alcoholic liver disease. *Hepatology* 2011; **53**(1):96–105.
91. Szabo G, Bala S, Petrasek J, Gattu A. Gut-liver axis and sensing microbes. *Dig Dis* 2010;**28**(6):737–44.
92. Dunagan M, Chaudhry K, Samak G, Rao RK. Acetaldehyde disrupts tight junctions in Caco-2 cell monolayers by a protein phosphatase 2A-dependent mechanism. *Am J Physiol Gastrointest Liver Physiol* 2012;**303**(12):G1356–64.
93. Kakiyama G, Hylemon PB, Zhou H, et al. Colonic inflammation and secondary bile acids in alcoholic cirrhosis. *Am J Physiol Gastrointest Liver Physiol* 2014;**306**(11):G929–37.
94. Rao R. Endotoxemia and gut barrier dysfunction in alcoholic liver disease. *Hepatology* 2009;**50**(2):638–44.
95. Uesugi T, Froh M, Arteel GE, Bradford BU, Thurman RG. Toll-like receptor 4 is involved in the mechanism of early alcohol-induced liver injury in mice. *Hepatology* 2001;**34**(1):101–8.
96. Canesso MCC, Lacerda NL, Ferreira CM, et al. Comparing the effects of acute alcohol consumption in germ-free and conventional mice: the role of the gut microbiota. *BMC Microbiol* 2014;**14**:240.
97. Adachi Y, Moore LE, Bradford BU, Gao W, Thurman RG. Antibiotics prevent liver injury in rats following long-term exposure to ethanol. *Gastroenterology* 1995;**108**(1):218–24.
98. Abu-Shanab A, Quigley EM. The role of the gut microbiota in nonalcoholic fatty liver disease. *Nat Rev Gastroenterol Hepatol* 2010;**7**(12):691–701.
99. Kang SS, Jeraldo PR, Kurti A, et al. Diet and exercise orthogonally alter the gut microbiome and reveal independent associations with anxiety and cognition. *Mol Neurodegener* 2014;**9**:36.
100. Luo YH, Zhu WY. The intestinal microbiota and obesity of the host. *Wei Sheng Wu Xue Bao* 2007;**47**(6):1115–8.
101. Leclercq S, Matamoros S, Cani PD, et al. Intestinal permeability, gut-bacterial dysbiosis, and behavioral markers of alcohol-dependence severity. *Proc Natl Acad Sci USA* 2014; **111**(42):E4485–93.
102. Bass NM, Mullen KD, Sanyal A, et al. Rifaximin treatment in hepatic encephalopathy. *N Engl J Med* 2010;**362**(12):1071–81.
103. Bajaj JS, Heuman DM, Hylemon PB, et al. Altered profile of human gut microbiome is associated with cirrhosis and its complications. *J Hepatol* 2014;**60**(5):940–7.
104. Schnabl B, Brenner DA. Interactions between the intestinal microbiome and liver diseases. *Gastroenterology* 2014;**146**(6):1513–24.
105. Chen P, Schnabl B. Host-microbiome interactions in alcoholic liver disease. *Gut Liver* 2014;**8**(3):237–41.
106. Bode C, Bode JC. Effect of alcohol consumption on the gut. *Best Pract Res Clin Gastroenterol* 2003;**17**(4):575–92.
107. Giannelli V, Di Gregorio V, Iebba V, et al. Microbiota and the gut-liver axis: bacterial translocation, inflammation and infection in cirrhosis. *World J Gastroenterol* 2014;**20**(45):16795–810.
108. Fouts DE, Torralba M, Nelson KE, Brenner DA, Schnabl B. Bacterial translocation and changes in the intestinal microbiome in mouse models of liver disease. *J Hepatol* 2012;**56**(6):1283–92.
109. Ridlon JM, Alves JM, Hylemon PB, Bajaj JS. Cirrhowis, bile acids and gut microbiota: unraveling a complex relationship. *Gut Microbes* 2013;**4**(5):382–7.
110. Zhong W, Zhou Z. Alterations of the gut microbiome and metabolome in alcoholic liver disease. *World J Gastrointest Pathophysiol* 2014;**5**(4):514–22.
111. Kakiyama G, Pandak WM, Gillevet PM, et al. Modulation of the fecal bile acid profile by gut microbiota in cirrhosis. *J Hepatol* 2013;**58**(5):949–55.
112. Malaguarnera G, Giordano M, Nunnari G, Bertino G, Malaguarnera M. Gut microbiota in alcoholic liver disease: pathogenetic role and therapeutic perspectives. *World J Gastroenterol* 2014;**20**(44):16639–48.
113. Sutherland GT, Sheahan PJ, Matthews J, et al. The effects of chronic alcoholism on cell proliferation in the human brain. *Exp Neurol* 2013;**247**:9–18.
114. de la Monte SM, Kril JJ. Human alcohol-related neuropathology. *Acta Neuropathol* 2014;**127**(1):71–90.
115. Martin PR, Singleton CK, Hiller-Sturmhofel S. The role of thiamine deficiency in alcoholic brain disease. *Alcohol Res Health* 2003;**27**(2):134–42.
116. Hoyumpa Jr AM. Mechanisms of thiamin deficiency in chronic alcoholism. *Am J Clin Nutr* 1980;**33**(12):2750–61.
117. Koedam JC. The mode of action of pyrithiamine as an inductor of thiamine deficiency. *Biochim Biophys Acta* 1958;**29**(2):333–44.
118. Liu JY, Timm DE, Hurley TD. Pyrithiamine as a substrate for thiamine pyrophosphokinase. *J Biol Chem* 2006;**281**(10):6601–7.

119. Vetreno RP, Hall JM, Savage LM. Alcohol-related amnesia and dementia: animal models have revealed the contributions of different etiological factors on neuropathology, neurochemical dysfunction and cognitive impairment. *Neurobiol Learn Mem* 2011;**96**(4):596–608.
120. Harper C. Neuropathology of brain damage caused by alcohol. *Med J Aust* 1982;**2**(6):276–82.
121. de la Monte SM. Disproportionate atrophy of cerebral white matter in chronic alcoholics. *Arch Neurol* 1988;**45**(9):990–2.
122. Harper C, Dixon G, Sheedy D, Garrick T. Neuropathological alterations in alcoholic brains. Studies arising from the New South Wales Tissue Resource Centre. *Prog Neuropsychopharmacol Biol Psychiatry* 2003;**27**(6):951–61.
123. Pfefferbaum A, Rosenbloom MJ, Adalsteinsson E, Sullivan EV. Diffusion tensor imaging with quantitative fibre tracking in HIV infection and alcoholism comorbidity: synergistic white matter damage. *Brain* 2007;**130**(Pt 1):48–64.
124. Schulte T, Sullivan EV, Muller-Oehring EM, Adalsteinsson E, Pfefferbaum A. Corpus callosal microstructural integrity influences interhemispheric processing: a diffusion tensor imaging study. *Cereb Cortex* 2005;**15**(9):1384–92.
125. Chanraud S, Martelli C, Delain F, et al. Brain morphometry and cognitive performance in detoxified alcohol-dependents with preserved psychosocial functioning. *Neuropsychopharmacology* 2007; **32**(2):429–38.
126. Kril JJ, Halliday GM. Brain shrinkage in alcoholics: a decade on and what have we learned? *Prog Neurobiol* 1999;**58**(4):381–7.
127. Harper CG, Kril JJ, Holloway RL. Brain shrinkage in chronic alcoholics: a pathological study. *Br Med J* 1985;**290**(6467):501–4.
128. Schmidt KS, Gallo JL, Ferri C, et al. The neuropsychological profile of alcohol-related dementia suggests cortical and subcortical pathology. *Dement Geriatr Cogn Disord* 2005;**20**(5):286–91.
129. Jacobus J, Squeglia LM, Bava S, Tapert SF. White matter characterization of adolescent binge drinking with and without co-occurring marijuana use: a 3-year investigation. *Psychiatry Res* 2013;**214**(3): 374–81.
130. Cohen AC, Tong M, Wands JR, de la Monte SM. Insulin and insulin-like growth factor resistance with neurodegeneration in an adult chronic ethanol exposure model. *Alcohol Clin Exp Res* 2007;**31**(9):1558–73.
131. de la Monte SM, Longato L, Tong M, Wands JR. Insulin resistance and neurodegeneration: roles of obesity, type 2 diabetes mellitus and non-alcoholic steatohepatitis. *Curr Opin Investig Drugs* 2009; **10**(10):1049–60.
132. Tong M, Deochand C, Yu R, de la Monte SM. Differential contributions of alcohol and the nicotine-derived nitrosamine ketone (NNK) to insulin and insulin-like growth factor resistance in the adolescent brain. *Alcohol Alcohol* 2015;**50**(6):680–9.
133. Chesik D, De Keyser J, Wilczak N. Insulin-like growth factor system regulates oligodendroglial cell behavior: therapeutic potential in CNS. *J Mol Neurosci* 2008;**35**(1):81–90.
134. Freude S, Leeser U, Muller M, et al. IRS-2 branch of IGF-1 receptor signaling is essential for appropriate timing of myelination. *J Neurochem* 2008;**107**(4):907–17.
135. Gong X, Xie Z, Zuo H. In vivo insulin deficiency as a potential etiology for demyelinating disease. *Med Hypotheses* 2008;**71**(3):399–403.
136. Leggio L, Zywiak WH, Fricchione SR, et al. Intravenous ghrelin administration increases alcohol craving in alcohol-dependent heavy drinkers: a preliminary investigation. *Biol Psychiatry* 2014; **76**(9):734–41.
137. Bordner KA, George ED, Carlyle BC, et al. Functional genomic and proteomic analysis reveals disruption of myelin-related genes and translation in a mouse model of early life neglect. *Front Psychiatry* 2011;**2**:18.
138. Nicklay JJ, Harris GA, Schey KL, Caprioli RM. MALDI imaging and in situ identification of integral membrane proteins from rat brain tissue sections. *Analyt Chem* 2013;**85**(15):7191–6.
139. Chiappelli F, Taylor AN, Espinosa de los Monteros A, de Vellis J. Fetal alcohol delays the developmental expression of myelin basic protein and transferrin in rat primary oligodendrocyte cultures. *Int J Dev Neurosci* 1991;**9**(1):67–75.
140. Gnaedinger JM, Noronha AB, Druse MJ. Myelin gangliosides in developing rats: the influence of maternal ethanol consumption. *J Neurochem* 1984;**42**(5):1281–5.
141. Soscia SJ, Tong M, Xu XJ, et al. Chronic gestational exposure to ethanol causes insulin and IGF resistance and impairs acetylcholine homeostasis in the brain. *Cell Mol Life Sci* 2006;**63**(17):2039–56.
142. Qu W, Zhang B, Wu D, Xiao B. Effects of alcohol on membrane lipid fluidity of astrocytes and oligodendrocytes. *Wei Sheng Yan Jiu* 1999;**28**(3):153–4.
143. Kwon OS, Schmued LC, Slikker Jr W. Fumonisin B1 in developing rats alters brain sphinganine levels and myelination. *Neurotoxicology* 1997;**18**(2):571–9.
144. Hill JM, Lesniak MA, Pert CB, Roth J. Autoradiographic localization of insulin receptors in rat brain: prominence in olfactory and limbic areas. *Neuroscience* 1986;**17**(4):1127–38.
145. Broughton SK, Chen H, Riddle A, et al. Large-scale generation of highly enriched neural stem-cell-derived oligodendroglial cultures: maturation-dependent differences in insulin-like growth factor-mediated signal transduction. *J Neurochem* 2007;**100**(3): 628–38.
146. Unger JW, Livingston JN, Moss AM. Insulin receptors in the central nervous system: localization, signalling mechanisms and functional aspects. *Prog Neurobiol* 1991;**36**(5):343–62.
147. Barres BA, Schmid R, Sendnter M, Raff MC. Multiple extracellular signals are required for long-term oligodendrocyte survival. *Development* 1993;**118**(1):283–95.
148. D'Ercole AJ, Ye P, Gutierrez-Ospina G. Use of transgenic mice for understanding the physiology of insulin-like growth factors. *Horm Res* 1996;**45**(Suppl. 1):5–7.
149. D'Ercole AJ, Ye P, O'Kusky JR. Mutant mouse models of insulin-like growth factor actions in the central nervous system. *Neuropeptides* 2002;**36**(2–3):209–20.
150. Ye P, Kollias G, D'Ercole AJ. Insulin-like growth factor-I ameliorates demyelination induced by tumor necrosis factor-alpha in transgenic mice. *J Neurosci Res* 2007;**85**(4):712–22.
151. de la Monte SM, Tong M, Cohen AC, Sheedy D, Harper C, Wands JR. Insulin and insulin-like growth factor resistance in alcoholic neurodegeneration. *Alcohol Clin Exp Res* 2008;**32**(9):1630–44.
152. de la Monte SM, Wands JR. Chronic gestational exposure to ethanol impairs insulin-stimulated survival and mitochondrial function in cerebellar neurons. *Cell Mol Life Sci* 2002;**59**(5):882–93.
153. Pascual M, Valles SL, Renau-Piqueras J, Guerri C. Ceramide pathways modulate ethanol-induced cell death in astrocytes. *J Neurochem* 2003;**87**(6):1535–45.
154. de la Monte SM, Tong M, Nguyen V, Setshedi M, Longato L, Wands JR. Ceramide-mediated insulin resistance and impairment of cognitive-motor functions. *J Alzheimer's Dis* 2010;**21**(3):967–84.
155. Durazzo TC, Rothlind JC, Cardenas VA, Studholme C, Weiner MW, Meyerhoff DJ. Chronic cigarette smoking and heavy drinking in human immunodeficiency virus: consequences for neurocognition and brain morphology. *Alcohol* 2007;**41**(7):489–501.
156. Wang JJ, Durazzo TC, Gazdzinski S, Yeh PH, Mon A, Meyerhoff DJ. MRSI and DTI: a multimodal approach for improved detection of white matter abnormalities in alcohol and nicotine dependence. *NMR Biomed* 2009;**22**(5):516–22.
157. Almeida OP, Garrido GJ, Lautenschlager NT, Hulse GK, Jamrozik K, Flicker L. Smoking is associated with reduced cortical regional gray matter density in brain regions associated with incipient Alzheimer disease. *Am J Geriatr Psychiatry* 2008;**16**(1):92–8.
158. Fritz HC, Wittfeld K, Schmidt CO, et al. Current smoking and reduced gray matter volume-a voxel-based morphometry study. *Neuropsychopharmacology* 2014;**39**(11):2594–600.

159. Paul RH, Grieve SM, Niaura R, et al. Chronic cigarette smoking and the microstructural integrity of white matter in healthy adults: a diffusion tensor imaging study. *Nicotine Tob Res* 2008;**10**(1):137–47.
160. Gazdzinski S, Durazzo TC, Studholme C, Song E, Banys P, Meyerhoff DJ. Quantitative brain MRI in alcohol dependence: preliminary evidence for effects of concurrent chronic cigarette smoking on regional brain volumes. *Alcohol Clin Exp Res* 2005; **29**(8):1484–95.
161. Durazzo TC, Mattsson N, Weiner MW. Alzheimer's disease neuroimaging I. Smoking and increased Alzheimer's disease risk: a review of potential mechanisms. *Alzheimer's Dement* 2014; **10**(3 Suppl.):S122–45.
162. Pan P, Shi H, Zhong J, et al. Chronic smoking and brain gray matter changes: evidence from meta-analysis of voxel-based morphometry studies. *Neurol Sci* 2013;**34**(6):813–7.
163. Durazzo TC, Rothlind JC, Gazdzinski S, Banys P, Meyerhoff DJ. Chronic smoking is associated with differential neurocognitive recovery in abstinent alcoholic patients: a preliminary investigation. *Alcohol Clin Exp Res* 2007;**31**(7):1114–27.
164. Luhar RB, Sawyer KS, Gravitz Z, Ruiz SM, Oscar-Berman M. Brain volumes and neuropsychological performance are related to current smoking and alcoholism history. *Neuropsychiatr Dis Treat* 2013;**9**:1767–84.
165. Bolzan AD, Bianchi MS. Genotoxicity of streptozotocin. *Mutat Res* 2002;**512**(2–3):121–34.
166. Koulmanda M, Qipo A, Chebrolu S, O'Neil J, Auchincloss H, Smith RN. The effect of low versus high dose of streptozotocin in cynomolgus monkeys (Macaca fascilularis). *Am J Transplant* 2003;**3**(3):267–72.
167. Wang S, Kamat A, Pergola P, Swamy A, Tio F, Cusi K. Metabolic factors in the development of hepatic steatosis and altered mitochondrial gene expression in vivo. *Metab Clin Exp* 2011;**60**(8):1090–9.
168. de la Monte SM, Tong M. Mechanisms of nitrosamine-mediated neurodegeneration: potential relevance to sporadic Alzheimer's disease. *J Alzheimer's Dis* 2009;**17**(4):817–25.
169. Andreani T, Tong M, de la Monte SM. Hotdogs and beer: dietary nitrosamine exposure exacerbates neurodevelopmental effects of ethanol in fetal alcohol spectrum disorder. *JDAR* 2014;**3**(2014):1–9.
170. Ghosh D, Mishra MK, Das S, Kaushik DK, Basu A. Tobacco carcinogen induces microglial activation and subsequent neuronal damage. *J Neurochem* 2009;**110**(3):1070–81.
171. Yalcin EB, Nunez K, Tong M, de la Monte SM. Differential sphingolipid and phospholipid profiles in alcohol and nicotine-derived nitrosamine ketone (NNK) associated white matter degeneration. *Alcohol Clin Exp Res* 2015, in press.
172. Chalfant CE, Kishikawa K, Mumby MC, Kamibayashi C, Bielawska A, Hannun YA. Long chain ceramides activate protein phosphatase-1 and protein phosphatase-2A. Activation is stereospecific and regulated by phosphatidic acid. *J Biol Chem* 1999; **274**(29):20313–7.
173. Seumois G, Fillet M, Gillet L, et al. De novo C16- and C24- -ceramide generation contributes to spontaneous neutrophil apoptosis. *J Leukoc Biol* 2007;**81**(6):1477–86.
174. de la Monte SM. Triangulated mal-signaling in Alzheimer's disease: roles of neurotoxic ceramides, ER stress, and insulin resistance reviewed. *J Alzheimer's Dis* 2012;**30**(Suppl. 2):S231–49.
175. de la Monte SM, Re E, Longato L, Tong M. Dysfunctional pro-ceramide, ER stress, and insulin/IGF signaling networks with progression of Alzheimer's disease. *J Alzheimer's Dis* 2012; **30**(Suppl. 2):S217–29.
176. Tong M, Longato L, Ramirez T, Zabala V, Wands JR, de la Monte SM. Therapeutic reversal of chronic alcohol-related steatohepatitis with the ceramide inhibitor myriocin. *Int J Exp Pathol* 2014;**95**(1):49–63.
177. Hajduch E, Turban S, Le Liepvre X, et al. Targeting of PKCzeta and PKB to caveolin-enriched microdomains represents a crucial step underpinning the disruption in PKB-directed signalling by ceramide. *Biochem J* 2008;**410**(2):369–79.
178. Jana A, Hogan EL, Pahan K. Ceramide and neurodegeneration: susceptibility of neurons and oligodendrocytes to cell damage and death. *J Neurol Sci* 2009;**278**(1–2):5–15.
179. Lyn-Cook Jr LE, Lawton M, Tong M, et al. Hepatic ceramide may mediate brain insulin resistance and neurodegeneration in type 2 diabetes and non-alcoholic steatohepatitis. *J Alzheimer's Dis* 2009; **16**(4):715–29.
180. Tong M, de la Monte SM. Mechanisms of ceramide-mediated neurodegeneration. *J Alzheimer's Dis* 2009;**16**(4):705–14.
181. Summers SA. Ceramides in insulin resistance and lipotoxicity. *Prog Lipid Res* 2006;**45**(1):42–72.
182. Boden G. Ceramide: a contributor to insulin resistance or an innocent bystander? *Diabetologia* 2008;**51**(7):1095–6.
183. Delarue J, Magnan C. Free fatty acids and insulin resistance. *Curr Opin Clin Nutr Metab Care* 2007;**10**(2):142–8.
184. Holland WL, Brozinick JT, Wang LP, et al. Inhibition of ceramide synthesis ameliorates glucocorticoid-, saturated-fat-, and obesity-induced insulin resistance. *Cell Metab* 2007;**5**(3):167–79.
185. Holland WL, Knotts TA, Chavez JA, Wang LP, Hoehn KL, Summers SA. Lipid mediators of insulin resistance. *Nutr Rev* 2007;**65**(6 Pt 2):S39–46.
186. Kaplowitz N, Than TA, Shinohara M, Ji C. Endoplasmic reticulum stress and liver injury. *Semin Liver Dis* 2007;**27**(4):367–77.
187. Ramirez T, Longato L, Tong M, et al. Insulin resistance, ceramide accumulation, and endoplasmic reticulum stress in experimental chronic alcohol-induced steatohepatitis. *Alcohol Alcohol* 2013;**48**(1): 39–52.
188. Blakemore WF, Franklin RJ. Remyelination in experimental models of toxin-induced demyelination. *Curr Top Microbiol Immunol* 2008;**318**:193–212.
189. Soriano JM, Gonzalez L, Catala AI. Mechanism of action of sphingolipids and their metabolites in the toxicity of fumonisin B1. *Prog Lipid Res* 2005;**44**(6):345–56.
190. Teruel T, Hernandez R, Lorenzo M. Ceramide mediates insulin resistance by tumor necrosis factor-alpha in brown adipocytes by maintaining Akt in an inactive dephosphorylated state. *Diabetes* 2001;**50**(11):2563–71.
191. Zierath JR. The path to insulin resistance: paved with ceramides? *Cell Metab* 2007;**5**(3):161–3.
192. Kopelman MD, Thomson AD, Guerrini I, Marshall EJ. The Korsakoff syndrome: clinical aspects, psychology and treatment. *Alcohol Alcohol* 2009;**44**(2):148–54.
193. Weiss JJ, Gorman JM. Psychiatric behavioral aspects of comanagement of hepatitis C virus and HIV. *Curr HIV/AIDS Rep* 2006;**3**(4): 176–81.
194. de la Monte SM, Tong M, Lester-Coll N, Plater Jr M, Wands JR. Therapeutic rescue of neurodegeneration in experimental type 3 diabetes: relevance to Alzheimer's disease. *J Alzheimer's Dis* 2006;**10**(1):89–109.
195. Lester-Coll N, Rivera EJ, Soscia SJ, Doiron K, Wands JR, de la Monte SM. Intracerebral streptozotocin model of type 3 diabetes: relevance to sporadic Alzheimer's disease. *J Alzheimer's Dis* 2006; **9**(1):13–33.
196. Moroz N, Tong M, Longato L, Xu H, de la Monte SM. Limited Alzheimer-type neurodegeneration in experimental obesity and Type 2 diabetes mellitus. *J Alzheimer's Dis* 2008;**15**(1):29–44.
197. Bryan L, Kordula T, Spiegel S, Milstien S. Regulation and functions of sphingosine kinases in the brain. *Biochim Biophys Acta* 2008;**1781**(9):459–66.

SECTION 3

GENETIC MACHINERY AND ITS FUNCTION

CHAPTER

20

Genetic Variants and Risk of Diabetes

Valeriya Lyssenko

Steno Diabetes Center A/S, Gentofte, Denmark; Department of Clinical Sciences, Diabetes and Endocrinology Unit, Lund University Diabetes Center, Lund University, Sweden

1. INTRODUCTION

Diabetes mellitus is steadily increasing, a lifelong, incapacitating disease with a global prevalence of more than 382 million people in 2013, and it is estimated that there will be 592 million diabetic patients in 2035 (http://www.idf.org/diabetesatlas). According to the International Diabetes Federation, as many as 183 million people are unaware that they have diabetes. An early detection of susceptibility markers in individuals at high risk for diabetes development is the prerequisite for successful disease prevention and treatment strategies.

Diabetes is a complex heterogeneous disorder resulting from an interaction between genetic predisposition and environmental factors. Type 2 diabetes (T2D) is the most common type and constitutes approximately 90% of all diabetes cases; the remaining part until now has been ascribed to type 1 diabetes and latent autoimmune diabetes in adults (LADA).[14] This situation, however, is complicated by imprecise classification of T2D, which hinders an adequate diabetes management based on underlying pathophysiology. To get more insights and provide in-depth understanding of disease mechanisms, several risk factors that are consistently associated with T2D development have been identified, including age, sex, obesity, low physical activity, smoking, dietary patterns, ethnicity, family history, history of gestational diabetes, elevated levels of fasting, 2-h or random glucose, increased blood pressure, dyslipidemia, and different drug treatments (diuretics, unselected β-blockers, etc.).[26,34]

There is also ample evidence that T2D has a strong genetic component. The concordance of T2D in monozygotic twins is approximately 70% compared with 20–30% in dizygotic twins.[20] The lifetime risk of developing the disease is about 40% in offspring of one parent with T2D, greater if the mother is affected,[13] the risk approaching 70% if both parents have diabetes. In prospective studies, we and others have demonstrated that first-degree family history is associated with a two-fold increased risk of future T2D.[20,24,26,34] The challenge has been to find genetic markers that explain the excess risk associated with family history of diabetes.

Advances in genotyping and sequencing technology during the past decade have tremendously facilitated rapid progress in the large-scale genetic studies. Genetics have proven to be a successful tool that, in an unbiased way, contributed to our understanding of many common diseases including T2D. Since 2007, genome-wide association studies (GWAS) have identified more than 80 genetic variants that increase the risk of T2D, together explaining around 10% of disease heritability.[39] The main question, however, which continues to be a matter of debate is what explains the remaining proportion of heritability. One of the significant issues relates to our incomplete understanding of constellation patterns of different risk factors that make up specific etiopathological disease subgroups. During the past century, insulin resistance has been postulated as a culprit in the pathogenesis of T2D. Using large prospective part of the Botnia study with longitudinal measurements of insulin secretion, we were the first to demonstrate that, although insulin resistance strongly predicts risk of future T2D, the decline in insulin secretion adjusted for the degree of insulin sensitivity is the key factor contributing to overt diabetes.[24] Recent genetic findings support this notion, demonstrating that many variants identified to date regulate insulin secretion from the pancreatic β cells and not insulin action in target tissues.

This review summarizes current knowledge on genetic variants contributing to T2D pathogenesis, utility of genetics in T2D prediction, and future potential of genetic studies.

Molecular Nutrition and Diabetes
http://dx.doi.org/10.1016/B978-0-12-801585-8.00020-8

2. GENETIC VARIANTS FOR T2D

Early studies on T2D genetics focused on candidate genes using different approaches including case-control, family, and linkage studies identified *PPARG*, *KCNJ11*, transcription factor 7-like 2 (*TCF7L2*), and later *WFS1* associated with risk for T2D, which were also confirmed in the large GWAS.[39] Some recent reports indicate that there are more than 120 genetic loci discovered through GWAS for T2D, glucose, and insulin traits in European and multiethnic cohorts.[36] Puzzling, however, that there is still a limited knowledge on distinct and predominant mechanism by which each of these genetic loci contributes to T2D susceptibility. Given that most of these variants are noncoding, their functional consequences have been challenging to investigate. This review focuses on discovery of genetic loci that have provided new views on the current understanding of the pathogenesis of T2D.

2.1 Transcription Factor 7-Like 2

The number one and the most intriguing T2D gene is the gene encoding for *TCF7L2* (rs7903146), which despite similar effect size of approximately 1.3- to 1.9-fold as reported for other common genetic variants, confers the strongest statistical evidence for association in populations of European and non-European ancestry.[39,51] *TCF7L2* has previously been implicated in the Wnt signaling pathway in different tissues, including skin, intestines, brain, adipose tissue, pancreas, and liver, and was extensively studied in cancer. To some extent, it came as a surprise that it was also the strongest signal for T2D. The enigma of molecular mechanisms by which *TCF7L2* risk variants increase the risk for T2D has still not being fully resolved. One of the main mechanisms involves impaired pancreatic β-cell function, decreased insulin response to glucose and other secretagogues,[5,11,41] impaired proinsulin processing and secretion, and affecting β-cell mass.[54] We have earlier demonstrated deterioration of β-cell function in the *TCF7L2* risk allele carriers over an 8-year follow-up period in individuals who progressed to T2D.[28] Because there is a binding site for *TCF7L2* in the promoter of the preproglucagon, we hypothesized that *TCF7L2* might influence the effect of gut hormone incretins on islets. In our study, carriers of the risk T allele of single nucleotide polymorphism (SNP) rs7903146 showed a weaker response to oral than to intravenous glucose, suggesting a defective enteroinsular axis. These findings have subsequently been replicated by showing impaired glucagon-like peptide (GLP)-1–stimulated insulin secretion in carriers of the risk genotype.[43] The findings that *TCF7L2* risk allele carriers display impaired incretin effects on islets could involve in addition effects on glucagon, which is stimulated by incretin hormone glucose-dependent insulinotropic polypeptide (GIP). Although the concentrations of GIP, GLP-1, and glucagon do not seem to differ between carriers of different *TCF7L2* genotypes,[43] we observed a strong correlation between GIP and glucagon, particularly in carriers of the risk T allele.[28] To provide functional explanation for these observations, we have also shown that expression of *TCF7L2* was five-fold higher in islets from patients with T2D than in islets from nondiabetic cadaver donors, suggesting a link between increased expression of *TCF7L2* mRNA and level of glycemia. The effects of *TCF7L2* on incretin action could to some extent be centrally regulated, given its recently reported expression in the brain[45] and reported interactions of *TCF7L2* variants with dietary patterns on the risk of T2D.[17] We and others have also demonstrated that *TCF7L2* risk alleles were associated with increased hepatic glucose production, even under conditions of similar basal and stimulated insulin levels.[28,35] Taken together, islet dysfunction with increased endogenous glucose production by the liver may result in chronically elevated glucose levels and subsequently contribute to T2D development. Molecular effects and activating and inhibiting properties of different *TCF7L2* isoforms suggest that pancreatic islets is the primary organ for *TCF7L2* diabetogenic effects.[16] Additionally, studies demonstrated the association of *TCF7L2* risk alleles with moderate autoimmunity measured with presence of one or more islet autoantibodies in children[38] and clinically distinct LADA subtype of diabetes in adults.[3,53]

3. GENETIC VARIANTS FOR INSULIN SECRETION AND ACTION

T2D develops in the situation of relative insulin deficiency from inherited or acquired defects when the pancreatic β cells fail to meet demands of increased insulin resistance. The euglycemic-hyperinsulinemic clamp is considered the gold standard method to measure glucose-stimulated insulin secretion and glucose disappearance rate (insulin resistance) during the same test.[31] Early-phase insulin response to glucose during an oral glucose tolerance test (OGTT) strongly correlates with acute insulin response measured with clamp technique[48] and can be assessed by insulinogenic and disposition indices that have been broadly used in the large epidemiological studies.[15] The heritability estimates of OGTT-derived indices of insulin secretion are ranging from 20% to 88%.[2,12,37]

3.1 Melatonin Receptor 1B

Melatonin is a chronobiotic hormone involved in the regulation of circadian rhythms. Insulin secretion was demonstrated to exhibit a circadian rhythm and its perturbations showed to affect glucose concentrations.[44] In a follow-up of our Diabetes Genetic Initiative GWAS for T2D,[8] we performed GWAS for insulin secretion measured with insulinogenic and disposition indices. We identified a common variant in the melatonin receptor 1B (*MTNR1B*) gene (rs10830963) associated with impaired early insulin response to glucose during OGTT.[29] The insulin-reducing allele also conferred a 1.11-fold increased risk for future T2D in two large prospective studies of more than 18,000 individuals, of whom 2201 developed T2D during a mean follow-up period of 23.5 years.[29] Risk allele in the *MTNR1B* gene had a profound effect on impaired early insulin release to both oral (insulinogenic and disposition index) and intravenous (first-phase insulin response) glucose challenge. In addition, the risk allele carriers showed deterioration in insulin secretion over time as compared with non-risk allele carriers.[29] In line with these findings, the Diabetes Prevention Program reported that the association of *MTNR1B* rs10830963 with impaired early insulin release persisted at 1 year despite adjustment for the baseline trait, suggesting a progressive deterioration of the effect at this locus.[55] In our recent largest to-date GWAS meta-analyses for dynamic measurements of insulin secretion during an OGTT in more than 10,000 nondiabetic individuals, *MTNR1B* (rs10830963) was confirmed as the strongest signal for the first-phase insulin secretion.[37] We have also shown that *MTNR1B* mRNA was expressed in human pancreatic islets and more specifically that nondiabetic individuals carrying the risk allele and patients with T2D display increased expression of the receptor.[29] These observations were further confirmed by a large gene-expression analysis of human pancreatic islets.[56] In line with inhibitory effects of exogenously administered melatonin on insulin secretion in rodents,[57] we have shown that melatonin inhibited insulin release in response to glucose in INS-1 rat β cells.[29] As expected from a strong effect of *MTNR1B* (rs10830963) on fasting glucose levels, the risk alleles significantly increased the risk of isolated impaired fasting glucose but not the risk of isolated impaired glucose tolerance[58]; the same SNP was shown to impact the rate of progression from normal fasting glucose to impaired fasting glucose, but not the rate of progression from impaired fasting glucose to T2D.[59] In addition to its effects on insulin secretion and isolated impaired fasting glucose, variants in the *MTNR1B* gene were also reported to be associated with hepatic insulin resistance.[58,60] Notably, the risk carriers of *MTNR1B* were clear outliers when known hyperbolic relationship between insulin secretion and the degree of insulin sensitivity was plotted, showing a strong insulin-resistant phenotype for the given impairment in insulin secretion.[19] An association with insulin sensitivity could involve effects of *MTNR1B* on energy expenditure. In this vein, it has been recently demonstrated that *MTNR1B* variant could modify effects of dietary fat intake on changes in energy expenditure during a 2-year period.[61] Additionally, this could involve the insular–incretin axis as *MTNR1B* variants were shown to be associated with incretin-stimulated insulin secretion.[62] Studies are ongoing to evaluate whether administration of melatonin in carriers of the risk alleles in the *MTNR1B* gene can be unwarranted and may result in an increase in glucose levels.

3.2 Gastric Inhibitory Polypeptide Receptor

Incretins GIP and GLP-1 are gut hormones involved in the regulation of the pancreatic β-cell function by potentiating glucose-stimulated insulin secretion in response to nutrients and by influencing β-cell mass through promoting proliferation and decreasing apoptosis.[18,21] Over the past few years, defects in the enteroinsular axis are emerging as important contributors to the pathogenesis of T2D. Our studies on the physiological effects of T2D susceptibility variants demonstrated that genetic determinants of insulin secretion and T2D were not only restricted to pancreatic β cells, but also included the enteroinsular axis and so far the neglected glucagon hormone secreted from the pancreatic α cells.[19]

In collaboration with the Meta-Analysis of Glucose- and Insulin-related traits Consortium researchers, we identified that the common variant in the gastric inhibitory polypeptide receptor (*GIPR*; rs10423928) gene is associated with impaired glucose tolerance, decreased plasma GIP levels, and reduced GIP-stimulated insulin response.[42] Notably, in our large prospective Malmö Preventive Project, the insulin-reducing effect of *GIPR* alleles was not translated into increased risk of T2D.[25] In line with previous observations in Gipr knock-out mice,[32] we observed that the risk allele carries of *GIPR* rs10423928 had a lean body phenotype as measured with lower body mass index (BMI), waist-to-hip ratio, and lean and fat mass.[25] In the search for a common denominator for these effects, we identified osteopontin (OPN) as a target of GIP. Expression of *OPN* was very abundant in islets, where it was suggested to protect from apoptosis. Carriers of the *GIPR* variant allele also displayed lower expression of *OPN* in human islets.[25] On the contrary, the reduced effects of cytokine *OPN* in adipose tissue were associated with reduced inflammation and thereby better insulin sensitivity in the carriers of *GIPR* variant allele.[1] Recent, large meta-analyses for obesity confirmed the adipogenic effects of the *GIPR/QPCTL* (rs2287019) locus.[23] We also have unpublished observations that, in addition to reported effects of *GIPR* risk alleles in islets and fat, seem to have deleterious effects in the vasculature and bone tissue. These findings support the notion that in addition to the glucose-potentiating effects on insulin and glucagon release in pancreatic islets contribution of incretin hormones to the pathogenesis of T2D might involve a plethora of extrapancreatic effects in different organs including vessels, bone, brain, and kidney.

4. GROWTH FACTOR RECEPTOR-BOUND PROTEIN 10

Across all studies of different ethnicities and populations, the parental family history of diabetes is consistently a strong risk factor for developing T2D in offspring; however, what explains this risk is still unclear. Imprinting is a genomic mechanism predominantly regulated by methylation to control differential gene expression alleles transmitted from the mother or from the father. Hypermethylation of the maternal allele of *KCNQ1* was shown to result in monoallelic activity of the neighboring maternally expressed protein-coding genes in fetal pancreas and biallelically in adult pancreatic islets, suggesting its effects during early pancreas development and thereby contributing to the increased risk of T2D later in life.[22,52] Similarly, maternally expressed *KLF14* only increases the risk when carried on the maternal chromosome, acting as a master *trans*-regulator of adipose tissue gene expression.[46]

In our recent large meta-analyses for dynamic measurements of insulin secretion during OGTT, we identified a variant in the imprinted growth factor receptor-bound protein 10 (*GRB10*) gene to predominantly influencing pancreatic β-cell function in females, but not in males.[37] We observed parent-of-origin effects of *GRB10* on lower fasting plasma glucose and enhanced insulin sensitivity for maternal and elevated glucose and decreased insulin sensitivity for paternal transmissions of the risk alleles. There was a paradoxical effect of risk alleles of *GRB10* on both decreased glucose-stimulated insulin secretion and reduced fasting plasma glucose concentrations. We hypothesized that this could be due to the concomitant effects of *GRB10* on pancreatic glucagon-producing α cells.[9] Indeed, we observed that disruption of *GRB10* by shRNA in human islets resulted in reduction of both insulin and glucagon expression and secretion. The mechanisms by which alleles in the GRB10 gene influence glucose homeostasis seem to involve tissue-specific parent-of-origin effects on DNA methylation and imprinting of *GRB10* in human pancreatic islets. The data further emphasize the need in genetic studies to consider whether risk alleles are inherited from the mother or the father.

5. RARE AND LOW-FREQUENCY VARIANTS

Systematic evaluation of the role of rare and low frequency variants in T2D development is in the early stage; however, there have already been a few successful reports from extremes, family-based studies, or studies in isolates.[4,30,47] As a proof of concept, a study from the Botnia region in Finland of extreme cases with early onset and controls that, despite clustering of risk factors such as age and BMI remain diabetes-free, identified a rare protective variant in the *SLC30A8* gene (R138X, frequency of 0.66%).[10] The identified mutation predicted to result in truncated protein and associated with reduced insulin, proinsulin levels, and random glucose levels. In line with earlier GWAS reports pointing out the important role of cyclin kinases in T2D development, a whole genome sequencing of 2630 Icelanders and imputation into 11,114 Icelandic cases and 267,140 controls followed by testing in Danish and Iranian samples discovered a low-frequency variant in the cyclin D2 gene (*CCND2*, rs76895963, frequency of 1.47%), which despite a concomitant effect on increased BMI was associated with reduced risk of T2D.[47] In the same study, two missense variants in the peptidylglycine alpha-amidating monooxygenase gene (*PAM*, p.Asp563Gly and p.Ser539Trp, frequency of 4.98% and 0.65%, respectively), and in the pancreatic and duodenal homeobox 1 (*PDX1*, p.Gly218Alafs) were associated with increased risk of T2D. Additionally, low-frequency variants in *HNF1A* were identified using large-scale exome sequencing, *TBC1D4* were identified in the Greenland isolates,[33] and solute carriers *SLC16A11* and *SLC16A13* in Mexicans were associated with increased risk for T2D.[4] In contact with other genetic loci influencing insulin secretion, genetic variants in the *TBC1D4* gene seem to increase risk of T2D by influencing insulin resistance.[6,7,33] These early glimpses on the role of rare or low-frequency variants in the pathogenesis of T2D suggest that these protective or deleterious effects with greater success can be detected in extremes or associated with specific familial or ethnic forms of diabetes.

6. GENETIC PREDICTION OF T2D

Identification of large number of novel genetic variants increasing susceptibility to T2D and related traits opened the opportunity to translate this genetic information into clinical practice and possibly improve disease risk prediction. Nevertheless, available data to date do not yet provide convincing evidence to support use of genetic screening for the prediction of T2D.

We and others have assessed the discriminatory ability of the genetic variants individually or combined in cross-sectional and longitudinal studies by grouping individuals based on the number of risk alleles and determining the relative odds of T2D, and by calculating the area under the receiver-operator characteristic curve (AUC). The AUC estimates, using combined information from more than 40 risk variants, ranged from 0.54 to 0.63 in different studies, indicating that genetic factors had limited ability in predicting an individual's risk of the disease. In contrast, the AUC has been considerably larger (from 0.61 to 0.95) for clinical models including different combinations of clinical and laboratory parameters (age, sex, BMI in all models, family history of diabetes, and fasting glucose in most of the models) in predicting the risk of T2D.[27] Although significant, the genetic variants added to the clinical and laboratory parameters, only minimally improved the predictive value at the population level ($P < 0.05$). A recent study using updated genetic risk score consisting of 65 variants reported AUC of 0.60, which confirms previous observations.[49]

There is, however, emerging evidence that genetics can be used to guide choice of therapy. One example of the translational approach based on recruitment by risk genotype has recently been reported for variants in the *ADRA2A* gene.[50] The risk genotypes in *ADRA2A* variant (rs553668) were associated with impaired insulin response to oral and intravenous glucose tolerance tests and increased risk of T2D, effects suggestively mediated by increased expression of *ADRA2A* in human islets.[40] Inhibition of *ADRA2A* by administration of receptor antagonist yohimbine resulted in improved β-cell function in the carriers of risk allele opening possibility of translation of genetic information into tailoring treatment in patients with diabetes.

7. FUTURE DIRECTIONS

The genetic architecture of T2D is now starting to unravel novel mechanisms involved in the disease pathogenesis. Undoubtedly, recent advanced in systematic and unbiased large-scale genotyping and sequencing studies paved the way to a number of novel discoveries and unprecedented findings that otherwise would not be possible. Although the biological mechanisms for many genetic variants are still unknown, in addition to their main effects on insulin-producing β cells, their effects also involve glucagon-producing α cells. Discovery of genes involved in the regulation of insular–incretin axis and circadian rhythms might explain an established in epidemiological studies link between different dietary patterns, sleep disturbances, and the risk for T2D. Ongoing efforts to identify casual variants or rare variants with strong effect size currently suggest that their effects could be family- or ethnicity-specific; therefore, research in different populations is highly warranted. Current efforts of identification of novel genetic variants and pathways involved in the pathogenesis of T2D are focused on large sample sizes and integrative analyses of gene expression, epigenetic, and metabolites, which would benefit from studies addressing effects of new environmental exposures. Finally, better understanding of different diabetes subtypes will allow dissection of genetic architecture of T2D and ultimately pave the way to the precision medicine.

References

1. Ahlqvist E, Osmark P, Kuulasmaa T, Pilgaard K, Omar B, Brons C, et al. Link between GIP and osteopontin in adipose tissue and insulin resistance. *Diabetes* 2013;**62**:2088–94.
2. Almgren P, Lehtovirta M, Isomaa B, Sarelin L, Taskinen MR, Lyssenko V, et al. Heritability and familiality of type 2 diabetes and related quantitative traits in the Botnia study. *Diabetologia* 2011;**54**:2811–9.
3. Cervin C, Lyssenko V, Bakhtadze E, Lindholm E, Nilsson P, Tuomi T, et al. Genetic similarities between latent autoimmune diabetes in adults, type 1 diabetes, and type 2 diabetes. *Diabetes* 2008;**57**:1433–7.
4. SIGMA Type 2 Diabetes Consortium, Williams AL, Jacobs SB, Moreno-Macias H, Huerta-Chagoya A, Churchhouse C, et al. Sequence variants in SLC16A11 are a common risk factor for type 2 diabetes in Mexico. *Nature* 2014;**506**:97–101.
5. Damcott CM, Pollin TI, Reinhart LJ, Ott SH, Shen H, Silver KD, et al. Polymorphisms in the transcription factor 7-like 2 (TCF7L2) gene are associated with type 2 diabetes in the Amish: replication and evidence for a role in both insulin secretion and insulin resistance. *Diabetes* 2006;**55**:2654–9.
6. Dash S, Langenberg C, Fawcett KA, Semple RK, Romeo S, Sharp S, et al. Analysis of TBC1D4 in patients with severe insulin resistance. *Diabetologia* 2010;**53**:1239–42.
7. Dash S, Sano H, Rochford JJ, Semple RK, Yeo G, Hyden CS, et al. A truncation mutation in TBC1D4 in a family with acanthosis nigricans and postprandial hyperinsulinemia. *Proc Natl Acad Sci USA* 2009;**106**:9350–5.
8. Diabetes Genetics Initiative of Broad Institute of Harvard and MIT, Lund University, and Novartis Institutes of BioMedical Research, Saxena R, Voight BF, Lyssenko V, Burtt NP, de Bakker PI, et al. Genome-wide association analysis identifies loci for type 2 diabetes and triglyceride levels. *Science* 2007;**316**:1331–6.
9. Doiron B, Hu W, Norton L, DeFronzo RA. Lentivirus shRNA Grb10 targeting the pancreas induces apoptosis and improved glucose tolerance due to decreased plasma glucagon levels. *Diabetologia* 2012;**55**:719–28.
10. Flannick J, Thorleifsson G, Beer NL, Jacobs SB, Grarup N, Burtt NP, et al. Loss-of-function mutations in SLC30A8 protect against type 2 diabetes. *Nat Genet* 2014;**46**:357–63.
11. Florez JC, Jablonski KA, Bayley N, Pollin TI, de Bakker PI, Shuldiner AR, et al. TCF7L2 polymorphisms and progression to diabetes in the Diabetes Prevention Program. *N Engl J Med* 2006;**355**: 241–50.
12. Gjesing AP, Hornbak M, Allin KH, Ekstrom CT, Urhammer SA, Eiberg H, et al. High heritability and genetic correlation of intravenous glucose- and tolbutamide-induced insulin secretion among non-diabetic family members of type 2 diabetic patients. *Diabetologia* 2014;**57**:1173–81.
13. Groop L, Forsblom C, Lehtovirta M, Tuomi T, Karanko S, Nissen M, et al. Metabolic consequences of a family history of NIDDM (the Botnia study): evidence for sex-specific parental effects. *Diabetes* 1996;**45**:1585–93.

14. Groop L, Pociot F. Genetics of diabetes—are we missing the genes or the disease? *Mol Cell Endocrinol* 2014;**382**:726–39.
15. Hanson RL, Pratley RE, Bogardus C, Narayan KM, Roumain JM, Imperatore G, et al. Evaluation of simple indices of insulin sensitivity and insulin secretion for use in epidemiologic studies. *Am J Epidemiol* 2000;**151**:190–8.
16. Hansson O, Zhou Y, Renstrom E, Osmark P. Molecular function of TCF7L2: consequences of TCF7L2 splicing for molecular function and risk for type 2 diabetes. *Curr Diabetes Rep* 2010;**10**:444–51.
17. Hindy G, Sonestedt E, Ericson U, Jing XJ, Zhou Y, Hansson O, et al. Role of TCF7L2 risk variant and dietary fibre intake on incident type 2 diabetes. *Diabetologia* 2012;**55**:2646–54.
18. Holst JJ, Knop FK, Vilsboll T, Krarup T, Madsbad S. Loss of incretin effect is a specific, important, and early characteristic of type 2 diabetes. *Diabetes Care* 2011;**34**(Suppl. 2):S251–7.
19. Jonsson A, Ladenvall C, Ahluwalia TS, Kravic J, Krus U, Taneera J, et al. Effects of common genetic variants associated with type 2 diabetes and glycemic traits on alpha- and beta-cell function and insulin action in humans. *Diabetes* 2013;**62**:2978–83.
20. Kaprio J, Tuomilehto J, Koskenvuo M, Romanov K, Reunanen A, Eriksson J, et al. Concordance for type 1 (insulin-dependent) and type 2 (non-insulin-dependent) diabetes mellitus in a population-based cohort of twins in Finland. *Diabetologia* 1992;**35**:1060–7.
21. Koehler JA, Baggio LL, Cao X, Abdulla T, Campbell JE, Secher T, et al. Glucagon-like peptide-1 receptor agonists increase pancreatic mass by induction of protein synthesis. *Diabetes* 2015;**64**: 1046–56.
22. Kong A, Steinthorsdottir V, Masson G, Thorleifsson G, Sulem P, Besenbacher S, et al. Parental origin of sequence variants associated with complex diseases. *Nature* 2009;**462**:868–74.
23. Locke AE, Kahali B, Berndt SI, Justice AE, Pers TH, Day FR, et al. Genetic studies of body mass index yield new insights for obesity biology. *Nature* 2015;**518**:197–206.
24. Lyssenko V, Almgren P, Anevski D, Perfekt R, Lahti K, Nissen M, et al. Predictors of and longitudinal changes in insulin sensitivity and secretion preceding onset of type 2 diabetes. *Diabetes* 2005; **54**:166–74.
25. Lyssenko V, Eliasson L, Kotova O, Pilgaard K, Wierup N, Salehi A, et al. Pleiotropic effects of GIP on islet function involve osteopontin. *Diabetes* 2011;**60**:2424–33.
26. Lyssenko V, Jonsson A, Almgren P, Pulizzi N, Isomaa B, Tuomi T, et al. Clinical risk factors, DNA variants, and the development of type 2 diabetes. *N Engl J Med* 2008;**359**:2220–32.
27. Lyssenko V, Laakso M. Genetic screening for the risk of type 2 diabetes: worthless or valuable? *Diabetes Care* 2013;**36**(Suppl. 2): S120–6.
28. Lyssenko V, Lupi R, Marchetti P, Del Guerra S, Orho-Melander M, Almgren P, et al. Mechanisms by which common variants in the TCF7L2 gene increase risk of type 2 diabetes. *J Clin Invest* 2007; **117**:2155–63.
29. Lyssenko V, Nagorny CL, Erdos MR, Wierup N, Jonsson A, Spegel P, et al. Common variant in MTNR1B associated with increased risk of type 2 diabetes and impaired early insulin secretion. *Nat Genet* 2009;**41**:82–8.
30. Mahajan A, Sim X, Ng HJ, Manning A, Rivas MA, Highland HM, et al. Identification and functional characterization of G6PC2 coding variants influencing glycemic traits define an effector transcript at the G6PC2-ABCB11 locus. *PLoS Genet* 2015;**11**: e1004876.
31. Mitrakou A, Vuorinen-Markkola H, Raptis G, Toft I, Mokan M, Strumph P, et al. Simultaneous assessment of insulin secretion and insulin sensitivity using a hyperglycemia clamp. *J Clin Endocrinol Metabol* 1992;**75**:379–82.
32. Miyawaki K, Yamada Y, Ban N, Ihara Y, Tsukiyama K, Zhou H, et al. Inhibition of gastric inhibitory polypeptide signaling prevents obesity. *Nat Med* 2002;**8**:738–42.
33. Moltke I, Grarup N, Jorgensen ME, Bjerregaard P, Treebak JT, Fumagalli M, et al. A common Greenlandic TBC1D4 variant confers muscle insulin resistance and type 2 diabetes. *Nature* 2014; **512**:190–3.
34. Noble D, Mathur R, Dent T, Meads C, Greenhalgh T. Risk models and scores for type 2 diabetes: systematic review. *BMJ* 2011;**343**: d7163.
35. Pilgaard K, Jensen CB, Schou JH, Lyssenko V, Wegner L, Brons C, et al. The T allele of rs7903146 TCF7L2 is associated with impaired insulinotropic action of incretin hormones, reduced 24 h profiles of plasma insulin and glucagon, and increased hepatic glucose production in young healthy men. *Diabetologia* 2009;**52**:1298–307.
36. Prasad RB, Groop L. Genetics of type 2 diabetes-pitfalls and possibilities. *Genes* 2015;**6**:87–123.
37. Prokopenko I, Poon W, Magi R, Prasad BR, Salehi SA, Almgren P, et al. A central role for GRB10 in regulation of islet function in man. *PLoS Genet* 2014;**10**:e1004235.
38. Redondo MJ, Muniz J, Rodriguez LM, Iyer D, Vaziri-Sani F, Haymond MW, et al. Association of TCF7L2 variation with single islet autoantibody expression in children with type 1 diabetes. *BMJ Open Diabetes Res Care* 2014;**2**:e000008.
39. DIAbetes Genetics Replication And Meta-analysis (DIAGRAM) Consortium, Asian Genetic Epidemiology Network Type 2 Diabetes (AGEN-T2D) Consortium, South Asian Type 2 Diabetes (SAT2D) Consortium, Mexican American Type 2 Diabetes (MAT2D) Consortium, Type 2 Diabetes Genetic Exploration by Next-generation sequencing in multi-Ethnic Samples (T2D-GENES) Consortium, Mahajan A, et al. Genome-wide trans-ancestry meta-analysis provides insight into the genetic architecture of type 2 diabetes susceptibility. *Nat Genet* 2014;**46**: 234–44.
40. Rosengren AH, Jokubka R, Tojjar D, Granhall C, Hansson O, Li DQ, et al. Overexpression of alpha2A-adrenergic receptors contributes to type 2 diabetes. *Science* 2010;**327**:217–20.
41. Saxena R, Gianniny L, Burtt NP, Lyssenko V, Giuducci C, Sjogren M, et al. Common single nucleotide polymorphisms in TCF7L2 are reproducibly associated with type 2 diabetes and reduce the insulin response to glucose in nondiabetic individuals. *Diabetes* 2006;**55**:2890–5.
42. Saxena R, Hivert MF, Langenberg C, Tanaka T, Pankow JS, Vollenweider P, et al. Genetic variation in GIPR influences the glucose and insulin responses to an oral glucose challenge. *Nat Genet* 2010;**42**:142–8.
43. Schafer SA, Tschritter O, Machicao F, Thamer C, Stefan N, Gallwitz B, et al. Impaired glucagon-like peptide-1-induced insulin secretion in carriers of transcription factor 7-like 2 (TCF7L2) gene polymorphisms. *Diabetologia* 2007;**50**:2443–50.
44. Scheer FA, Hilton MF, Mantzoros CS, Shea SA. Adverse metabolic and cardiovascular consequences of circadian misalignment. *Proc Natl Acad Sci USA* 2009;**106**:4453–8.
45. Shao W, Wang D, Chiang YT, Ip W, Zhu L, Xu F, et al. The Wnt signaling pathway effector TCF7L2 controls gut and brain proglucagon gene expression and glucose homeostasis. *Diabetes* 2013;**62**: 789–800.
46. Small KS, Hedman AK, Grundberg E, Nica AC, Thorleifsson G, Kong A, et al. Identification of an imprinted master trans regulator at the KLF14 locus related to multiple metabolic phenotypes. *Nat Genet* 2011;**43**:561–4.
47. Steinthorsdottir V, Thorleifsson G, Sulem P, Helgason H, Grarup N, Sigurdsson A, et al. Identification of low-frequency and rare sequence variants associated with elevated or reduced risk of type 2 diabetes. *Nat Genet* 2014;**46**:294–8.
48. Stumvoll M, Mitrakou A, Pimenta W, Jenssen T, Yki-Jarvinen H, Van Haeften T, et al. Use of the oral glucose tolerance test to assess insulin release and insulin sensitivity. *Diabetes Care* 2000;**23**: 295–301.

49. Talmud PJ, Cooper JA, Morris RW, Dudbridge F, Shah T, Engmann J, et al. Sixty-five common genetic variants and prediction of type 2 diabetes. *Diabetes* 2015;**64**:1830–40.
50. Tang Y, Axelsson AS, Spegel P, Andersson LE, Mulder H, Groop LC, et al. Genotype-based treatment of type 2 diabetes with an alpha2A-adrenergic receptor antagonist. *Sci Transl Med* 2014;**6**:257ra139.
51. Tong Y, Lin Y, Zhang Y, Yang J, Zhang Y, Liu H, et al. Association between TCF7L2 gene polymorphisms and susceptibility to type 2 diabetes mellitus: a large Human Genome Epidemiology (HuGE) review and meta-analysis. *BMC Med Genet* 2009;**10**:15.
52. Travers ME, Mackay DJ, Dekker Nitert M, Morris AP, Lindgren CM, Berry A, et al. Insights into the molecular mechanism for type 2 diabetes susceptibility at the KCNQ1 locus from temporal changes in imprinting status in human islets. *Diabetes* 2013;**62**:987–92.
53. Zampetti S, Spoletini M, Petrone A, Capizzi M, Arpi ML, Tiberti C, et al. Association of TCF7L2 gene variants with low GAD autoantibody titre in LADA subjects (NIRAD Study 5). *Diabet Med* 2010; **27**:701–4.
54. Zhou Y, Park SY, Su J, Bailey K, Ottosson-Laakso E, Shcherbina L, et al. TCF7L2 is a master regulator of insulin production and processing. *Hum Mol Genet* 2014;**23**:6419–31.
55. Florez JC, Jablonski KA, McAteer JB, et al. Effects of genetic variants previously associated with fasting glucose and insulin in the Diabetes Prevention Program. *PLoS One* 2012;**7**:e44424.
56. Taneera J, Lang S, Sharma A, et al. A systems genetics approach identifies genes and pathways for type 2 diabetes in human islets. *Cell Metab* 2012;**16**:122–34.
57. Bailey CJ, Atkins TW, Matty AJ. Melatonin inhibition of insulin secretion in the rat and mouse. *Horm Res* 1974;**5**:21–8.
58. Sparsø T, Bonnefond A, Andersson E, et al. G-allele of intronic rs10830963 in MTNR1B confers increased risk of impaired fasting glycemia and type 2 diabetes through an impaired glucose-stimulated insulin release: studies involving 19,605 Europeans. *Diabetes* 2009;**58**:1450–6.
59. Walford GA, Green T, Neale B, et al. Common genetic variants differentially influence the transition from clinically defined states of fasting glucose metabolism. *Diabetologia* 2012;**55**:331–9.
60. Vangipurapu J, Stančáková A, Pihlajamäki J, et al. Association of indices of liver and adipocyte insulin resistance with 19 confirmed susceptibility loci for type 2 diabetes in 6,733 non-diabetic Finnish men. *Diabetologia* 2011;**54**:563–71.
61. Mirzaei K, Xu M, Qi Q, et al. Variants in glucose- and circadian rhythm-related genes affect the response of energy expenditure to weight-loss diets: the POUNDS LOST Trial. *Am J Clin Nutr* 2014;**99**:392–9.
62. Simonis-Bik AM, Nijpels G, van Haeften TW, et al. Gene variants in the novel type 2 diabetes loci CDC123/CAMK1D, THADA, ADAMTS9, BCL11A, and MTNR1B affect different aspects of pancreatic beta-cell function. *Diabetes* 2010;**59**:293–301.

CHAPTER

21

MicroRNA and Diabetes Mellitus

Sofia Salö[1,2], *Julian Geiger*[1], *Anja E. Sørensen*[1,2], *Louise T. Dalgaard*[1]

[1]Department of Science, Systems and Models, Roskilde University, Roskilde, Denmark; [2]Danish Diabetes Academy, Odense, Denmark

1. INTRODUCTION

It was thought of as a special curiosity of the nematode *C. elegans* when the first microRNA (miRNA) was discovered in 1993 by Victor Ambros' and Gary Ruvkuns' groups studying the lin-4–mediated inhibition of the *lin-14* gene.[1,2] The gene *lin-4* encoded small RNA transcripts complementary to the 3′UTR of lin-14, and binding of lin-4 to its 3′UTR would post-transcriptionally downregulate the *lin-14* gene expression. However, when a second nematode miRNA was identified, let-7,[3] it was realized that let-7 was conserved in many species and that miRNAs and miRNA-mediated post-transcriptional regulation of gene expression are ubiquitous mechanisms.[4,5] Now, the term miRNA is defined as a highly conserved, endogenously expressed, noncoding, single-stranded RNA of 21–24 nucleotides.[4,6,7] As of today, more than 2800 miRNAs in *Homo sapiens* are listed with high confidence in the latest release (21.0) of the miRNA database miRBase.[8]

Analyzing sequences outside the protein-coding gene segments through the ENCODE project[9] it has become apparent that parts of the genome previously coined "junk DNA" transcribes noncoding RNA molecules, such as miRNAs.[10] Noncoding RNAs have been demonstrated to influence expression of up to 30% of all protein-coding genes and regulate gene expressions during development and disease.[11] Genetic studies reveal more than 60 loci associated with type 2 diabetes (T2D),[12] for which variants are mostly found in the noncoding segments. Therefore, processes regulating transcription such as miRNA should be further examined.[13] It is likely that miRNA and/or other small regulatory RNA molecules are influencing the development of diabetes.

2. miRNA BIOGENESIS

The position of miRNAs on the genome may be intergenic or outside of gene sequences; they can possess their own promoter or be under transcriptional control of the gene they reside in. Primary miRNA sequences tend to cluster together and members of these clusters are likely transcribed together, but they can also be the sole miRNA of the primary transcript.[14,15] Either way, all miRNAs are transcribed by RNA polymerase II and the initial primary miRNA transcript is processed as conventional mRNA[16] (Figure 1). The ensuing pri-miRNA hairpin transcripts are then cleaved by the enzyme Drosha to form pre-miRNA hairpin structures of about 70 nucleotides.[17] The pre-miRNAs are exported out of the nucleus through the exportin5 complex where the stem-loop structure is cleaved in the cytosol by the enzyme Dicer to form double-stranded RNA duplexes of 21–24 nucleotides. The mature strand is then loaded into the RNA-induced silencing complex (RISC) by Argonaute (Ago) proteins whereby miRNAs direct the complex by complementarity between the miRNA seed sequence and the miRNA recognition element (MRE) in the target 3′UTR of mammalian mRNA.[16] Depending on whether the binding is fully or incompletely complementary, Ago either cleaves the UTR or mediates translational repression, respectively (Figure 1). Due to the rather short seed region of miRNA (6–8 nucleotides), a single miRNA may have hundreds of targets, and, likewise, the mRNA MRE may be targeted by families of miRNAs.[17,18]

Commonly, miRNAs are retained intracellularly but have also been found in the extracellular space, blood, and other body fluids.[19] When in the circulation, miRNAs are suggested to associate with either proteins, microvesicles, and/or lipoprotein complexes.[20–22] Further, studies of miRNAs transported by microvesicles or lipoprotein complexes show that they can be transferred in an active form to recipient cells, thereby possibly contributing in cell-to-cell communication.[20,23,24] A recent study in human pancreatic tissue reveals more than 380 islet and β-cell specific miRNAs.[25]

Molecular Nutrition and Diabetes
http://dx.doi.org/10.1016/B978-0-12-801585-8.00021-X

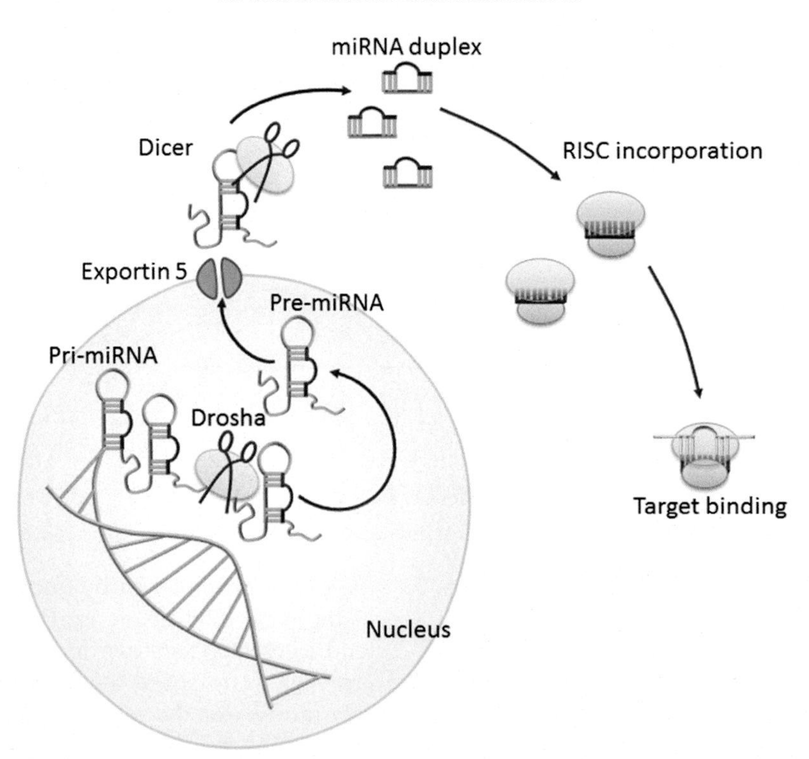

FIGURE 1 MicroRNA (miRNA) biogenesis. miRNA primary transcripts are transcribed by RNA polymerase II and processed in the nucleus by Drosha resulting in precursor miRNAs containing a stem-loop sequence, which is exported to the cytoplasm by Exportin 5. In the cytoplasm Dicer further cleaves the loop from the precursor miRNA yielding a miRNA duplex, which is unwound and the active strand is incorporated in the RNA-induced silencing complex (RISC), where target mRNA interaction takes place.

3. miRNAs ACTING IN β-CELL DEVELOPMENT

The loss of β-cell mass associated with both type 1 diabetes (T1D) and T2D has attracted efforts to conduct alternative routes to islet transplantation, such as that of endogenous pancreatic β-cell regeneration as seen in rodent under both nondiabetic as well as diabetic conditions.[26,27] In humans, proliferation of adult pancreatic β-cells is low to undeterminable under steady-state conditions.[28,29] Finding novel mechanisms involved in β-cell development and regeneration could provide new paths to increase the number of functional β-cells in patients with diabetes.

During mouse embryonic pancreatic development, cell fate is regulated and synchronized by surges of transcription factors of which Neurogenin3 (Neurog3) is a marker of pancreatic endocrine progenitor cells[30] seen largely during the second trimester and not at all in mature islet cells.[31] Following a 70% pancreatectomy performed in adult FVB/NJ mice, pancreatic regeneration was induced. Functional neo-islets appeared within 4 weeks, and it was observed that transcription factors upstream of Neurog3 were expressed in similar patterns during both development and regeneration.[31] However, even though Neurog3 is necessary for β-cell formation during developmental stages, Neurog3 is not expressed during the regenerative phase.[31] Profiling 283 miRNA expression levels of developing and regenerating pancreas revealed high expression of miRNAs targeting Neurog3 (miR-15a, miR-15b, miR-16, and miR-195) during pancreas regeneration (Table 1). Therefore, regenerating pancreas was extracted and digested into single-cell suspension and transfected with antisense miRNAs designed against miR-15a, miR-15b, miR-16, and miR-195 before harvesting. In the transfected cells, expression of Neurog3 and its downstream genes such as NeuroD then reappeared.[31] Thus, miRNA-mediated regulation of Neurog3 expression induces an alternate pathway of regeneration in the adult mouse pancreas.

Another study of miRNA in mouse shows that miR-7 may serve different purposes during development compared with adult pancreas. Distributing antisense miR-7 in vivo through intracardiac delivery to mouse fetal pancreas (e10.5) reveal that the antisense-treated embryos displayed strongly reduced insulin production at e17.5 and extensive apoptosis throughout the pancreatic tissue.[32] In accordance, 2 weeks following birth, the antisense-treated mice exhibited glucose intolerance, lower insulin content, and lower number of insulin-positive cells. Both *Insulin* and *Pdx1* gene expressions were also reduced at this stage.[32] On the contrary,

TABLE 1 MicroRNAs (miRNAs) Regulating β-Cell Development

miRNAs	Known functional effect	Targets	Tissue/cell	References
miR-15a, miR-15b, miR-16, miR-195	Pancreas development, β-cell fate and regeneration	*Neurog3*	Mouse embryo/MIN6	31
miR-7a	Human β-cell proliferation	*p70S6 K, elF4E, Mapkap1, Mknk1, and Mknk2*	Mouse islets	32–34
miR-375	α- and β-cell expansion	*Cav1, Id3, Smarca2, Aifm1, Rasd1, Rgs16, Eef1e1, C1qbp, HuD, Cadm1*	KO mouse islets	78

inhibition of miR-7 in the adult mice promotes β-cell replication.[33,34] By inhibiting miR-7 in mouse MIN6 cells and primary mouse islets, the mTOR pathway is activated.[34] Reduced expression of miR-7a in MIN6 cells lead to gene upregulation of five downstream targets of the mTOR (p70S6 K, elF4E, Mapkap1, Mknk1, and Mknk2), and inhibited miR-7a in adult mouse islets results in increased β-cell proliferation.[34] In humans, miR-7 is highly expressed in both the developing and adult pancreas.[33,35,36] Human dispersed islets transfected with anti−mir-7a undergo a drastic 30-fold increase in proliferation,[34] indicating that miR-7 may serve as a negative regulator of proliferation.

4. miRNAs ACTING ON GLUCOSE-STIMULATED INSULIN SECRETION

The glucose-sensing step in human β-cells begins with glucose entering the cell through a glucose transporter (GLUT1 and/or GLUT2).[37] Following mitochondrial glucose metabolism, the triggering pathway is initiated and executed by the KATP channel consisting of the SUR1 and Kir6.2 subunits.[38] When blood glucose concentration exceeds ~6 mmol/L, closure of KATP channels induces depolarization of the cell membrane sufficient to activate voltage-gated Ca^{2+} and Na^{+} channels,[39] which directly triggers insulin secretion through exocytosis of insulin granules. In MIN6 and INS-1 (832/13) β-cells, miR-124 is glucose regulated and found to target the transcription factor Foxa2, which in turn regulates Kir6.2 and SUR1, thereby altering the Ca^{2+} sensitivity of β-cell insulin secretion.[40]

Insulin from the β-cells is released into the bloodstream by means of exocytosis, the process where insulin-containing vesicles fuse with the plasma membrane. The exocytotic process is largely dependent on the interaction of three SNARE proteins: the association of the vesicle-associated protein VAMP2 with the membrane-bound SNAP25 and Syntaxin-1A.[41−43] Several synaptotagmins, Munc-18, and other proteins such as granuphilin, RIM2, and Noc2 are also involved.[42,44,45] In addition, the vesicular membrane−associated family of Rab GTPases act as molecular switches.[46] Furthermore, numerous effectors of Rabs have been shown to be essential in normal regulation of exocytosis in the pancreatic β-cell, including granuphilin, RIM2, and Noc2 that associate with Rab3a and Rab27a.[45,47] In the mouse insulin-producing cell-line MIN6B1, overexpression of miR-124 increase the levels of SNAP25, whereas the levels of Rab27 and Noc2 decrease, in effect reducing glucose-stimulated insulin secretion (GSIS).[48] In the same study, overexpression of miR-96 lowered the exocytotic capacity by increasing the level of granuphilin and reducing the level of Noc2. Noc2 binding to Rab3 is required for potentiating insulin secretion,[47] whereas granuphilin is a negative modulator of exocytosis.[48] Experiments conducted in the rat cell-line INS-1E show that granuphilin is also indirectly regulated by miR-9; overexpression of miR-9 targeted the granuphilin inhibiting transcription factor Onecut2. In concordance, increased levels of granuphilin associated with reduced insulin secretion.[49] Overexpression of miR-21 and miR-34a in the MIN6 cell-line targets VAMP2 and Rab3a and is, hence, accompanied by lower insulin secretion.[50] The expression of miR-29a in the INS-1E cell-line is glucose dependent and has been shown to target and reduce the expression of Syntaxin-1A, thereby impairing the exocytotic machinery and insulin secretion.[51] Hence, several miRNAs regulate GSIS, and it is likely that their perturbed expression may contribute to the hyperglycemia seen in diabetes (Table 2).

In the presence of glucose, insulin secretion can be further enhanced by second messengers such as the cAMP messenger system potentiated by, for example, glucagon, GLP-1, and GIP. cAMP activates the enzyme PKA, which activates selected targets to enhance insulin secretion.[52,53] The cAMP/PKA pathway has been shown to reduce the miR-375 expression in rodents possibly promoting insulin secretion,[54] via the target myotrophin.[55] Potentiation of insulin secretion also involves a PKA-independent pathway,[56] and glucose has also been shown to decrease miR-375 expression in this manner.[54]

TABLE 2 MicroRNAs (miRNAs) Regulating GSIS and Insulin Transcription

miRNAs	Known functional effect	Targets	Tissue/cell	References
miR-7	Insulin transcription	*Ins1, Ins2*	Mouse islets/MIN6	32
miR-30d	Insulin transcription	(Activator of) *MafA*	MIN6	59,60
miR-204	Insulin transcription	*MafA*	Mouse BS ob/ob islets	58
miR-375	Insulin transcription, insulin secretion	*PDK1, myotrophin*	Mouse islets/INS-1E	55,62
miR-9	Insulin secretion	*Onecut2, granuphilin* (indirectly)	INS-1E	49
miR-21	Insulin secretion	*VAMP2, Rab3a*	MIN6	50
miR-29a	Insulin secretion	*Stx1A*	INS-1E	51,96
miR-34a	Insulin secretion	*VAMP2, Rab3a*	MIN6	50
miR-96	Insulin secretion	*Noc2, granuphilin*	MIN6	48
miR-124a	Insulin secretion	*Foxa2, Rab27a, Noc2, SNAP25*	MIN6 and INS-1832/13	40,48

5. REGULATION OF INSULIN TRANSCRIPTION BY miRNAs

During continuous glucose stimulation, glucose triggers refilling of the insulin synthesis by activating β-cell specific insulin transcription factors such as MafA, PDX1, and NeuroD and miRNAs.[57] In the hyperglycemic diabetic mouse model B6 ob/ob, upregulated miR-204 expression reduces insulin synthesis by targeting and downregulating MafA.[58] Increased expression of miR-9 in MIN6 cells reduces insulin expression levels in a mechanism probably involving the Onecut2 transcription factor.[49] Increased glucose levels in MIN6-cells upregulate miR-30d, which increases insulin transcription through indirect targeting of MafA.[59,60] Further, prolonged glucose treatment in INS-1E cells negatively regulates the levels of mir-375[61] (Table 2). It was found that miR-375 directly targets PDK1, a key player in the PI 3-kinase signaling cascade, thereby decreasing the downstream phosphorylation of PKB and GSK3. Further, miR-375 also inhibited glucose-induced proliferation, probably via the PI3-kinase/PKB cascade.[61] Glucose induces nuclear translocation of the insulin gene transcription factor PDX1 by triggering the phosphorylation of PDX-1 via the PI3-kinase pathway.[62] Thus, the glucose-induced reduction of miR-375 increases PDK1 levels, which activates the PI3-kinase cascade and, hence, stimulates both β-cell proliferation and insulin transcription.

6. β-CELL MASS IN OBESITY AND PREGNANCY

It has been shown that miRNAs play a crucial role in islet formation, where blocked generation of pancreatic miRNAs modeled through a dicer-null mouse resulted in an almost complete loss of insulin-producing cells.[63] In addition, β-cell specific deletion of the Dicer enzyme using the rat insulin promoter 2 (RIP)-Cre transgene results in progressively declining β-cell mass and insulin secretion, and development of diabetes.[64,65] The β-cells in the developing pancreas are highly proliferative; however, as the progenitor cells start producing insulin, the division ceases.[66] Yet, the β-cell mass continues to expand during fetal and postnatal growth,[28,29,67,68] and β-cell mass in the adult human has also been shown to increase in response to insulin resistance during obesity and pregnancy.[28,69–72] Consequently, reduced β-cell mass is associated with both T1D and T2D, and, for example, analyses of pancreatic sections donated from obese patients diagnosed with T2D display decreased β-cell mass compared with obese individuals without diabetes.[69,73–75] Indeed, a microarray comparison of islet miRNA expression levels in the Goto–Kakizaki rat model of T2D with Wistar controls showed a set of upregulated miRNAs enriched in targets within pathways known to be involved in T2D.[76] No "master switch" regulating the human β-cell differentiation versus replication has yet been identified, but evidence from model organisms such as *C. elegans* clearly indicates an important function of miRNAs during the developmental timing of cell fate switches.[1–3]

A microarray performed in islets from pregnant rats at gestational day 14 revealed increased levels of miR-451 and downregulated expression of miR-338-3p (Table 3). When expression was mimicked in INS832/13 and MIN6 cells, miR-338-3p associated with increased proliferation, while miR-451 and miR-338-3p both exerted a protective effect against palmitate and cytokine-induced apoptosis.[77] In the same study, it was found that the maternal hormone 17β-estradiol both increased

TABLE 3 MicroRNAs (miRNAs) Involved in Regulating β-Cell Mass in Obesity and Pregnancy, and β-Cell Failure in Type 2 Diabetes

miRNAs	Known functional effect	Expression change	Tissue/cell	References
miR-375	Islets of *ob/ob* mice	Up	β-Cell proliferation	78
miR-132	Islets of prediabetic *db/db* mice	Up	β-Cell proliferation	82
miR-338-3p	Islets of pregnant rats and islets of prediabetic *db/db* mice and obese mice fed with high-fat diet	Down	β-Cell proliferation/ antiapoptotic	77
miR-338-3p	Cells cultured with estradiol or incretins	Down	β-Cell proliferation/ antiapoptotic	77
miR-451	Islets of pregnant rats, islets of prediabetic *db/db* mice and obese mice fed with high-fat diet	Up	Antiapoptotic	77

miR-451 expression and reduced that of miR-338-3p in INS832/13 cells as well as in rat and human islet cells. Estrogens are known to activate GPR30, increasing cAMP and activating the PKA signaling. Rat islet cells stimulated by cAMP displayed reduced miR-338-3p levels, and the effect of 17β-estradiol on miR-338-3p expression in both rat and human islet cells could be prevented by addition the PKA inhibitor H8978. In accordance, the authors found that GLP-1, known to stimulate cAMP, also downregulated miR-338-3p in obese ob/ob mouse islets. In conclusion, miR-338-3p may act as a regulator of proliferation during both maternal and obesity-related β-cell mass expansion.[77]

Studies utilizing a miR-375 KO mouse model displayed reduced β-cell numbers but increased α-cell mass, in effect potentiating the hyperglycemic state and reduced GSIS.[78] Luciferase assays suggested that miR-375 directly targets several genes (*Cav1, Id3, Smarca2, Aifm1, Rasd1, Rgs16, Eef1e1, C1qbp, HuD, Cadm1*) that are all known to partake in negatively regulating cellular growth and proliferation (Table 3). Furthermore, miR-375 was also upregulated in the obese mouse model (*ob/ob*) with increased β-cell mass.[78] Therefore, miR-375 may serve to maintain a functional β-cell mass by regulating proliferation in obesity.

7. β-CELL FAILURE IN T2D

In one of the few studies comparing miRNA expression in human islets, miR-21, miR-127-3p, and miR-375 expression was higher in islets from glucose intolerant donors than in normoglycemic islets.[79] GSIS and insulin mRNA levels (as a crude estimate for insulin biosynthesis) was measured in the islets from the donors, and miR-122, miR-127-3p, miR-184, and miR-375 in normoglycemic donors correlated positively to insulin mRNA expression. Further, miR-127-3p and miR-184 negatively correlated with GSIS, whereas all these correlations were lost in islets from glucose intolerant donors suggesting a protective effect of these miRNAs against β-cell failure as manifested by glucose intolerance[79] (Table 4).

In a global TaqMan array, islet miRNA expression of individuals with and without T2D was measured revealing a significant increase of miR-187 expression in T2D islets, which inversely correlated with GSIS.[80] Furthermore, mRNA levels of a putative target of miR-187, HIPK3, were reduced in the islets from individuals with T2D. HIPK3 is a known regulator of insulin secretion proposed to regulate two key mediators of glucose responsiveness: PDX1 and GSK3β.[81] Overexpression of miR-187 in primary rat islets and INS-1 cells was found

TABLE 4 MicroRNAs (miRNAs) Involved in Type 2 Diabetes (T2D) β-Cell Failure

miRNAs	Cell types/models	Expression change	Known functional effect	References
miR-21	Islets from individuals with T2D	Up	Not investigated	79
miR-127-3p	Islets from individuals with T2D	Up	Not investigated	79
miR-375	Islets from individuals with T2D	Up	Not investigated	79
miR-187	Islets of individuals with T2D	Up	Glucose-induced insulin secretion	80
miR-210	Islets of diabetic *db/db* mice	Down	Proapoptotic	82

to markedly decrease GSIS and Hipk3 mRNA levels without affecting insulin content or apoptotic index as assessed by TUNEL assay.[80] A luciferase assay performed in the INS-1 cells further confirmed that Hipk3 was a direct target of miR-187 thereby strengthening the suggestion that miR-187 regulate GSIS in human β-cells (Table 4).

One study compared the leptin receptor−deficient *db/db* mouse at 6 weeks of age (when obese, insulin resistant but normoglycemic) with older *db/db* mice having developed diabetes (obese, insulin resistant, and hyperglycemic), thereby attempting to investigate miRNAs of importance for the impending loss of β-cell mass in T2D. In the normoglycemic *db/db* mice miR-184, miR-203, miR-210, and miR-338-3p expression was reduced, whereas miR-132 expression increased.[82] Proliferation was triggered in dispersed β-cells from rat islets upon both miR-184 inactivation and miR-132 overexpression. In contrast, reduction of miR-203 and miR-210 triggered apoptosis in the dispersed rat β-cells, which is in line with the finding that miR-210 was even more reduced in islets from the hyperglycemic db/db mice.[82] In addition, upregulation of miR-199a-3p in the hyperglycemic *db/db* mice was shown to increase rat β-cell apoptosis in rat islet cells in vitro.[82] However, miRNAs involved in human β-cell protection has yet to be identified.

8. miRNAs IN SKELETAL MUSCLE, ADIPOSE TISSUE, AND LIVER

The increasing inability of skeletal muscle, liver, and fat tissue to respond to the hormone insulin is characteristic for the prediabetic and diabetic state. One major risk factor is obesity. A recent study showed that the enzyme heme-oxygenase 1 (HO-1) might be a suitable marker to discriminate "healthy obese" (being insulin sensitive and without comorbidities such as T2D and cardiovascular disease) from the malign forms of obesity, in which the HO-1 is found at higher levels in fat and liver.[83] Interestingly enough this enzyme negatively affects the heme-dependent biogenesis of miRNAs.[84,85] This supports a role for miRNAs during the onset of T2D and insulin resistance and their involvement in late diabetic complications.[86]

Muscle tissue is the largest consumer of glucose in the human body. It accounts for about 75% of the insulin-mediated glucose uptake. Exercise is known to have beneficial effects on T2D and insulin resistance. Whether these effects are mediated by miRNAs is unclear. However, acute exercise resulted in increased levels of miR-1, miR-107, and miR-181.[87] Exercise also reduces miR-23 levels, which is associated with PGC-1α upregulation,[88] and acute exercise increases miR-1 and miR-133a.[89]

Muscle cells express a set of specific miRNAs, termed myomiRs. Known members include miR-1, miR-133, and miR-206. A function in myogenesis, muscle cell differentiation and exercise[88] has been linked to all of them.[90,91] The OH-1 enzyme, generally increased in subjects with the metabolic syndrome, is suspected to have a negative effect on the myomiR expression.[84] In addition, miR-206 seems to play a role in muscle fiber type transition and is predicted to regulate clockwork genes in muscle and heart.[92] A finding of van Rooij et al. indicated that miR-1 and miR-133 might influence electric conductivity in heart muscle linking them to arrhythmic events[93] (Table 5).

Other miRNAs like miR-29a might be relevant for the development of insulin resistance. It has been shown that glucose and saturated fatty acids, like palmitate, enhance the expression of miR-29a.[94−96] Yang et al. report that this miRNA interferes with the insulin-stimulated glucose uptake and impairs insulin-signaling by targeting the 3′UTR of IRS-1.[94] A similar conclusion was drawn by He et al., who placed relevant targets of miR-29a upstream of Akt.[95] Therefore, targeting miR-29a might lead to a potential therapeutic approach for improving insulin signaling.[97]

The miRNAs miR-24 and miR-126 are both downregulated in insulin resistance, although this has been interpreted to represent a consequence rather than a cause.[98] MiR-106, on the other hand, is highly expressed in muscles of diabetic patients. Experiments in cultured C2C12 and HEK293 cells show enhanced expression of miR-106b and miR-494 expression after TNFα treatment.[99,100] Induction of miR-106 under low-grade inflammatory conditions might be a possible link to the mitochondrial dysfunction associated with T2D, because one evaluated target of miR-106 is mitofusin-2, a protein relevant for mitochondrial morphology and metabolism[101,102] (Table 5).

The adipose tissue is another important target of insulin action. The differentiation of adipocytes is a process influenced by several miRNAs,[103] among them miR-27,[104] miR-133,[105] and miR-143.[106] A number of miRNAs (miR-29a,[95] miR-93,[107] miR-103,[105] miR-107,[105] miR-143,[108] miR-221,[109] and miR-320[110]) have been suspected to compromise the insulin sensitivity of adipocytes. Targets of miRNA-caused interference can be located on several levels like (1) insulin signaling mediators,[105,110] (2) adiponectin signaling,[109] or (3) glucose uptake.[95,107] Often the molecular details remain vague, as with miR-143: based on cell culture experiments, the ERK5 protein was one suspected target,[106] although experiments in mice did not support these findings.[111]

In the liver, miRNAs play an important role in posttranscriptional control of cholesterol and lipoprotein metabolism. MiR-33b and miR-33a are co-expressed with SREBF1 and SREBF2, respectively, due to their

TABLE 5 MicroRNAs (miRNAs) in Skeletal Muscle, Adipose Tissue, and Liver Relevant for Diabetes Mellitus

miRNAs	Cell types/models	Expression change	Known functional effect	References
Myomirs: miR-1, miR-133, miR-206	Skeletal muscle	–	Myogenesis	90,91
miR-1, miR-107, miR-133, miR-181	Skeletal muscle	Increased by acute exercise	–	87,89
miR-23a	Skeletal muscle	Reduced by exercise	Associated with PGC-1a upregulation	88
miR-29a	Skeletal muscle	Increased by glucose and palmitate	Impairs insulin-stimulated glucose uptake, IRS-1	94,96
miR-206	Skeletal muscle	–	Fiber type transition	92
miR-106	Skeletal muscle	Increased in diabetes	–	98,99
miR-494	Skeletal muscle	Increased by TNFa	–	99,100
miR-24	Skeletal muscle	Down in insulin resistance	–	98
miR-27	Adipose tissue	–	Adipocyte differentiation	104
miR-29a	Adipose tissue	Increased by glucose	Impairs insulin-stimulated glucose uptake	96
miR-133	Adipose tissue	–	Adipocyte differentiation	105
miR-143	Adipose tissue	–	Adipocyte differentiation, insulin resistance	106,108
miR-93	Adipose tissue	–	Insulin resistance	107
miR-126	Adipose tissue	Down in insulin resistance	–	98
miR-221	Adipose tissue	–	Insulin resistance	109
miR-320	Adipose tissue	–	Insulin resistance	110
miR-103	Adipose tissue/liver	–	Insulin resistance	105
miR-107	Adipose tissue/liver	–	Insulin resistance	105
miR-33	Liver	Altered by cholesterol levels	Controls HDL biogenesis	112,113
miR-122	Liver	Increases with liver cell differentiation	Controls VLDL secretion	116,120
miR-181a	Liver	–	Improves insulin resistance	121
miR-802	Liver	Increased by obesity	Insulin resistance	122

localization in intronic regions of these host genes, and are therefore regulated by the same factors that regulate SREBF1 and -2 mRNA expression.[112,113] MiR-33 has target sites in the 3′UTRs of ATP-binding cassette A1 (ABCA1) and ABCG1, and the endolysosomal transport protein Niemann-Pick C1 (NPC1). ABCA1, ABCG1, and NPC1 all encode proteins involved in cholesterol efflux.[114] Thus, upregulation of miR-33 and SREBFs by low cholesterol levels lead to both an activation of cholesterol synthesis mediated by the SREBFs and a downregulation of cholesterol efflux, mediated by miR-33, thereby controlling HDL-biogenesis in liver. MiR-33 has a number of target genes with roles in cholesterol metabolism, bile acid secretion and fatty acid oxidation,[115,116] and studies of mice or nonhuman primates with lentiviral overexpression or chemical inhibition of miR-33, respectively, confirm these mechanisms.[112,117]

The liver-specific and highly abundant miR-122 constitutes up to 70% of all miRNAs in this tissue and exerts major post-transcriptional control over liver mRNA expression,[118] resulting in modulation of genes involved in cholesterol metabolism via both direct and indirect mechanisms,[119] which is confirmed by studies in liver-specific miR-122 knock-out mice. Therefore, these findings suggest that inhibition of miR-122 results in decreased circulatory lipoprotein levels through reduction of hepatic cholesterol synthesis and VLDL secretion.[116,120] MiR-181a improves hepatocyte insulin sensitivity via downregulation of Sirt1,[121] and the obesity-induced miR-802 impairs hepatic insulin sensitivity via downregulation of HNF-1β[122] (Table 5).

9. miRNAs REGULATED BY NUTRITIONAL STATE AND SPECIFIC INGREDIENTS

Overnutrition characterized by excess intake of carbohydrates and fat contributes to obesity, which is a major risk factor for T2D.[123] However, maternal malnutrition during gestation also has been associated with an increased risk of developing T2D later in life.[124] The mechanism by which malnutrition may impact the expression of certain miRNAs and if dietary compounds also can target the proteins involved in the biogenesis of miRNAs, thus affecting the overall expression of miRNAs, are only sparsely studied. Understanding the influence of nutrition on miRNA expression may provide a unique insight for reducing obesity-linked disorders. To explore the pathophysiological effects of nutrients on miRNA expression, different protocols have been used with different compositions of diets and energy content. Dumortier et al.[125] investigated how progeny of dams fed a low-protein diet affected pancreatic miRNA expression and found that miR-375 and several other miRNAs increased. Expression of miR-375 has been associated with maintenance of glucose homeostasis and α- and β-cell mass and have been observed to be elevated in the *ob/ob* diabetic mouse.[78]

Resveratrol, a polyphenol and a natural plant compound, which can be isolated from some berries, grapes, peanuts, and red wine, has been shown to possess anti-inflammatory, antioxidative,[126,127] and antidiabetic properties[128] and the ability to modulate miRNA expression.[129] Peripheral blood mononuclear cells isolated from male T2D patients consuming 8 mg of resveratrol-enriched grape extract on a daily basis for a year showed an upregulation of miR-21 and miR-181, while miR-155 and miR-34a were downregulated compared with the placebo-treated groups. All of the aforementioned miRNAs have been observed to be involved in inflammation response, and the concomitant observed downregulation of proinflammatory cytokines could in part explain one role of action for resveratrol.[130] Hyperlipidemic mice consuming a high-fat, high-cholesterol diet showed an increased expression of two miRNAs, miR-103 and miR-107 in the liver, which, when fed additionally with plant-derived *Hibiscus sabdariffa* polyphenols, reversed the expression and attenuated metabolic disturbances such as insulin resistance.[131]

Three dietary polyphenols, proanthocyanidin extract (GSPE), cocoa proanthocyanidin (CPE), and pure epigallocatechin gallate (EGCG) were all able to change the miRNA expression profile in human hepatocellular carcinoma (HepG2) cells on stimulation—with miR-30b-3p being repressed by all of the compounds. Target analysis for miR-30b-3p revealed that pathways involved in glucose metabolism, insulin sensitivity, oxidative stress, and inflammation were overrepresented.[132]

TABLE 6 MicroRNA (miRNA) Identified in Serum Samples and Proposed as Biomarkers of Type 2 Diabetes (T2D)

miRNA	Sample type	Expression change in T2D	References
miR-28-3p	Human serum from NGT and T2D and ob/ob mouse serum	Up	137
miR-15a	–	Down	137
miR-20b	–	Down	137
miR-21	–	Down	137
miR-24	–	Down	137
miR-29b	–	Down	137
miR-126	–	Down	137
miR-150	–	Down	137
miR-191	–	Down	137
miR-197	–	Down	137
miR-223	–	Down	137
miR-320	–	Down	137
miR-486	–	Down	137
miR-9	Human serum from NGT, obese, and T2D	Up	136
miR-29a	–	Up	136
miR-30d	–	Up	136
miR-34a	–	Up	136
miR-124	–	Up	136
miR-146a	–	Up	136
miR-375	–	Up	136
miR-503	Human serum from NGT and T2D	Down	143
miR-144	Human serum from NGT and T2D	Up	135
miR-146a	–	Down	135
miR-150	–	Up	135
miR-182	–	Down	135

Human umbilical vein endothelial cells (HUVECs) kept in a mildly elevated glucose and advanced glycation end products (AGEs) environment showed a downregulation of miR-181c, miR-15a, and miR-20b. Addition of calcitriol, the active metabolite of vitamin D and important for Ca^{2+} metabolism, at physiological concentrations, changed the miRNA profile of the three miRNAs which became upregulated, thus having a protective effect against the diabetic-like environment.[133] The physiological relevance of dietary compounds on miRNA expression based on cell culture and animal studies in human context depends largely on exposure concentration as well as the length and time of exposure. Adding to the complexity, the bioavailability of the nutrition after ingestion also as well as the targets needs to be assessed.

10. miRNAs AS CIRCULATING BIOMARKERS

Besides their intracellular function, a large set of miRNAs is released in stable form in body fluids including blood and urine. Variations in the blood miRNA pool are emerging as promising biomarkers of several diseases including diabetes.[134–138] Indeed, circulating miRNAs, including miR-103 and miR-224, have been found in the blood of patients with diabetes.[139] As some miRNA changes are specific to different types of cells or tissues, miRNA measurements in serum or urine may be able to indicate tissue dysfunction. Additionally, it may provide a means to determine those at risk of developing diabetes or diabetic complications, enabling early intervention and subsequently improved outcomes for those with decreased insulin production or increased resistance. For example, miR-375, highly enriched in β-cells, was detectable several weeks prior to onset of diabetes in NOD mice.[140]

In one of the larger studies of miRNA expression in serum of patients with T2D, 80 patients with T2DM and 80 matched controls revealed a defined signature that was also observed in *ob/ob* mice.[138] It was shown that miR-28-3p was upregulated, which was paralleled by downregulated expression of 12 miRNAs: miR-15a, miR-20b, miR-21, miR-29b, miR-126, miR-150, miR-191, miR-197, miR-223, miR-320, and miR-486. Collectively, the signature was able to predict diabetes development in 70% of patients in a 10-year follow-up.[138] In yet another study, seven miRNAs were found to be upregulated in serum (miR-9, miR-29a, miR-30d, miR-34a, miR-124, miR-146a, and miR-375) in patients newly diagnosed with diabetes.[136] However, in the same study, a subgroup of patients with high BMI, impaired glucose tolerance and a family history of diabetes displayed no difference in the expression levels of the seven miRNAs compared with the control group. The implementation

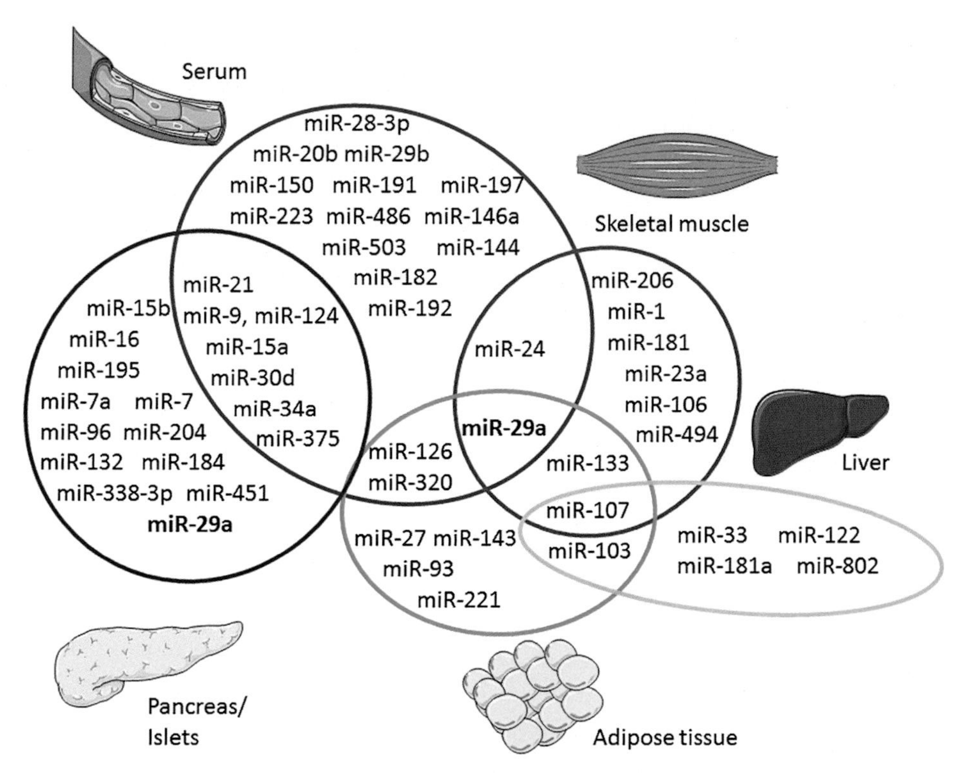

FIGURE 2 Distribution of microRNAs (miRNAs) involved in type 2 diabetes or related phenotypes of the metabolic syndrome organized according to the tissues of observed functional effects or as biomarkers in serum/plasma samples. miRNAs shown are all cited in the text or in Tables 1–6.

of miRNA levels measured in serum has yet to be determined, as it is commonly seen that the expression profiles in serum differ from that of the phenotype-related tissue expression. In recently diagnosed T1D children, a number of miRNAs were upregulated (miR-152, miR-30a-5p, miR-181a, miR-24, miR-148a, miR-210, miR-27a, miR-29a, miR-26a, miR-27b, miR-25, and miR-200a), and miR-25 levels were negatively associated with residual β-cell function.[141] Studies analyzing the profiles of miRNAs in blood to evaluate their roles in the development and progression of T2D have shown disparate results (Table 6), perhaps due to methodological differences, and ethnically or genetically heterogeneous study samples.[135,136,138,142] Of note, very few circulating miRNAs have been identified in more than one study.

11. CONCLUSIONS AND PERSPECTIVES

A large number of miRNAs have been implicated in different facets of the metabolic syndrome and diabetes mellitus (DM), and currently there are no well established unifying sets of miRNAs characterizing the various subphenotypes of hyperglycemia. However, a more defined number of miRNAs appear to affect the function or differentiation of the pancreatic β-cell, whereas miRNAs in skeletal muscle, liver and adipose tissue constitute different and almost nonoverlapping sets of miRNAs (Figure 2). A few miRNAs appear to be consistently increased by hyperglycemia (miR-29a, miR-34a),[136,143,144] and these may form a starting point for defining diabetes-specific miRNA profiles for use as biomarkers. Furthermore, the interactions between dietary components, nutritional state in general and miRNA expression and function are still not well defined. Therefore, a general conclusion is that with many contrasting studies regarding expression regulation and function of miRNAs in different cell types, there is a genuine need for more studies.

LIST OF ABBREVIATIONS

DM Diabetes mellitus
miRNA MicroRNA
RNA Ribonucleic acids

Acknowledgments

This work was supported by grants from the Danish Councils for Independent Research in Health and Disease (DFF–1331-00033), the Danish Diabetes Academy (to S.S. and A.E.S.), and the Novo Nordisk Foundation.

References

1. Lee RC, Feinbaum RL, Ambros V. The *C. elegans* heterochronic gene lin-4 encodes small RNAs with antisense complementarity to lin-14. *Cell* December 3, 1993;**75**(5):843–54.
2. Wightman B, Ha I, Ruvkun G. Posttranscriptional regulation of the heterochronic gene lin-14 by lin-4 mediates temporal pattern formation in *C. elegans*. *Cell* December 3, 1993;**75**(5):855–62.
3. Reinhart BJ, Slack FJ, Basson M, Pasquinelli AE, Bettinger JC, Rougvie AE, et al. The 21-nucleotide let-7 RNA regulates developmental timing in *Caenorhabditis elegans*. *Nature* February 24, 2000; **403**(6772):901–6.
4. Lagos-Quintana M, Rauhut R, Lendeckel W, Tuschl T. Identification of novel genes coding for small expressed RNAs. *Science* October 26, 2001;**294**(5543):853–8.
5. Pasquinelli AE, Reinhart BJ, Slack F, Martindale MQ, Kuroda MI, Maller B, et al. Conservation of the sequence and temporal expression of let-7 heterochronic regulatory RNA. *Nature* November 2, 2000;**408**(6808):86–9.
6. Lau NC, Lim LP, Weinstein EG, Bartel DP. An abundant class of tiny RNAs with probable regulatory roles in *Caenorhabditis elegans*. *Science* October 26, 2001;**294**(5543):858–62.
7. Lee RC, Ambros V. An extensive class of small RNAs in *Caenorhabditis elegans*. *Science* October 26, 2001;**294**(5543):862–4.
8. Kozomara A, Griffiths-Jones S. miRBase: annotating high confidence microRNAs using deep sequencing data. *Nucleic Acids Res* January 2014;**42**(Database issue):D68–73.
9. The ENCODE (ENCyclopedia Of DNA Elements) Project. *Science* October 22, 2004;**306**(5696):636–40.
10. Birney E, Stamatoyannopoulos JA, Dutta A, Guigo R, Gingeras TR, Margulies EH, et al. Identification and analysis of functional elements in 1% of the human genome by the ENCODE pilot project. *Nature* June 14, 2007;**447**(7146):799–816.
11. Hardikar AA, Walker MD, Lynn F. Noncoding RNAs. *Exp Diabetes Res* 2012;**2012**:629249.
12. Hivert MF, Vassy JL, Meigs JB. Susceptibility to type 2 diabetes mellitus—from genes to prevention. *Nat Rev Endocrinol* April 2014;**10**(4):198–205.
13. Lyssenko V, Laakso M. Genetic screening for the risk of type 2 diabetes: worthless or valuable? *Diabetes Care* August 2013; **36**(Suppl. 2):S120–6.
14. Saini HK, Griffiths-Jones S, Enright AJ. Genomic analysis of human microRNA transcripts. *Proc Natl Acad Sci USA* November 6, 2007;**104**(45):17719–24.
15. Parts L, Hedman AK, Keildson S, Knights AJ, Abreu-Goodger C, van de Bunt M, et al. Extent, causes, and consequences of small RNA expression variation in human adipose tissue. *PLoS Genet* 2012;**8**(5):e1002704.
16. Bartel DP. MicroRNAs: genomics, biogenesis, mechanism, and function. *Cell* January 23, 2004;**116**(2):281–97.
17. Olena AF, Patton JG. Genomic organization of microRNAs. *J Cell Physiol* March 2010;**222**(3):540–5.
18. Bartel DP, Chen CZ. Micromanagers of gene expression: the potentially widespread influence of metazoan microRNAs. *Nat Rev Genet* May 2004;**5**(5):396–400.
19. Weber JA, Baxter DH, Zhang S, Huang DY, Huang KH, Lee MJ, et al. The microRNA spectrum in 12 body fluids. *Clin Chem* November 2010;**56**(11):1733–41.
20. Vickers KC, Palmisano BT, Shoucri BM, Shamburek RD, Remaley AT. MicroRNAs are transported in plasma and delivered to recipient cells by high-density lipoproteins. *Nat Cell Biol* April 2011;**13**(4):423–33.
21. Arroyo JD, Chevillet JR, Kroh EM, Ruf IK, Pritchard CC, Gibson DF, et al. Argonaute2 complexes carry a population of circulating microRNAs independent of vesicles in human plasma. *Proc Natl Acad Sci USA* March 22, 2011;**108**(12):5003–8.

22. Gibbings DJ, Ciaudo C, Erhardt M, Voinnet O. Multivesicular bodies associate with components of miRNA effector complexes and modulate miRNA activity. *Nat Cell Biol* September 2009; **11**(9):1143–9.
23. Valadi H, Ekstrom K, Bossios A, Sjostrand M, Lee JJ, Lotvall JO. Exosome-mediated transfer of mRNAs and microRNAs is a novel mechanism of genetic exchange between cells. *Nat Cell Biol* June 2007;**9**(6):654–9.
24. Mathivanan S, Ji H, Simpson RJ. Exosomes: extracellular organelles important in intercellular communication. *J Proteomics* September 10, 2010;**73**(10):1907–20.
25. van de Bunt M, Gaulton KJ, Parts L, Moran I, Johnson PR, Lindgren CM, et al. The miRNA profile of human pancreatic islets and beta-cells and relationship to type 2 diabetes pathogenesis. *PLoS One* 2013;**8**(1):e55272.
26. Bonner-Weir S, Baxter LA, Schuppin GT, Smith FE. A second pathway for regeneration of adult exocrine and endocrine pancreas. A possible recapitulation of embryonic development. *Diabetes* December 1993;**42**(12):1715–20.
27. Hardikar AA, Karandikar MS, Bhonde RR. Effect of partial pancreatectomy on diabetic status in BALB/c mice. *J Endocrinol* August 1999;**162**(2):189–95.
28. Meier JJ, Butler AE, Saisho Y, Monchamp T, Galasso R, Bhushan A, et al. Beta-cell replication is the primary mechanism subserving the postnatal expansion of beta-cell mass in humans. *Diabetes* June 2008;**57**(6):1584–94.
29. Saisho Y, Butler AE, Meier JJ, Monchamp T, Allen-Auerbach M, Rizza RA, et al. Pancreas volumes in humans from birth to age one hundred taking into account sex, obesity, and presence of type-2 diabetes. *Clin Anat* November 2007;**20**(8):933–42.
30. Gu G, Dubauskaite J, Melton DA. Direct evidence for the pancreatic lineage: NGN3+ cells are islet progenitors and are distinct from duct progenitors. *Development* May 2002;**129**(10): 2447–57.
31. Joglekar MV, Parekh VS, Mehta S, Bhonde RR, Hardikar AA. MicroRNA profiling of developing and regenerating pancreas reveal post-transcriptional regulation of neurogenin3. *Dev Biol* November 15, 2007;**311**(2):603–12.
32. Nieto M, Hevia P, Garcia E, Klein D, Alvarez-Cubela S, Bravo-Egana V, et al. Antisense miR-7 impairs insulin expression in developing pancreas and in cultured pancreatic buds. *Cell Transplant* 2012;**21**(8):1761–74.
33. Correa-Medina M, Bravo-Egana V, Rosero S, Ricordi C, Edlund H, Diez J, et al. MicroRNA miR-7 is preferentially expressed in endocrine cells of the developing and adult human pancreas. *Gene Expr Patterns* April 2009;**9**(4):193–9.
34. Wang Y, Liu J, Liu C, Naji A, Stoffers DA. MicroRNA-7 regulates the mTOR pathway and proliferation in adult pancreatic beta-cells. *Diabetes* March 2013;**62**(3):887–95.
35. Rosero S, Bravo-Egana V, Jiang Z, Khuri S, Tsinoremas N, Klein D, et al. MicroRNA signature of the human developing pancreas. *BMC Genomics* September 22, 2010;**11**(1):509.
36. Joglekar MV, Joglekar VM, Hardikar AA. Expression of islet-specific microRNAs during human pancreatic development. *Gene Expr Patterns* February 2009;**9**(2):109–13.
37. De VA, Heimberg H, Quartier E, Huypens P, Bouwens L, Pipeleers D, et al. Human and rat beta cells differ in glucose transporter but not in glucokinase gene expression. *J Clin Invest* November 1995;**96**(5):2489–95.
38. Aittoniemi J, Fotinou C, Craig TJ, de WH, Proks P, Ashcroft FM. Review. SUR1: a unique ATP-binding cassette protein that functions as an ion channel regulator. *Philos Trans R Soc Lond B Biol Sci* January 27, 2009;**364**(1514):257–67.
39. Yang SN, Berggren PO. The role of voltage-gated calcium channels in pancreatic beta-cell physiology and pathophysiology. *Endocr Rev* October 2006;**27**(6):621–76.
40. Baroukh N, Ravier MA, Loder MK, Hill EV, Bounacer A, Scharfmann R, et al. MicroRNA-124a regulates Foxa2 expression and intracellular signaling in pancreatic beta-cell lines. *J Biol Chem* July 6, 2007;**282**(27):19575–88.
41. Ohara-Imaizumi M, Nishiwaki C, Nakamichi Y, Kikuta T, Nagai S, Nagamatsu S. Correlation of syntaxin-1 and SNAP-25 clusters with docking and fusion of insulin granules analysed by total internal reflection fluorescence microscopy. *Diabetologia* December 2004;**47**(12):2200–7.
42. Gandasi NR, Barg S. Contact-induced clustering of syntaxin and munc18 docks secretory granules at the exocytosis site. *Nat Commun* 2014;**5**:3914.
43. Regazzi R, Sadoul K, Meda P, Kelly RB, Halban PA, Wollheim CB. Mutational analysis of VAMP domains implicated in Ca2+-induced insulin exocytosis. *EMBO J* December 16, 1996;**15**(24): 6951–9.
44. Gauthier BR, Wollheim CB. Synaptotagmins bind calcium to release insulin. *Am J Physiol Endocrinol Metab* December 2008; **295**(6):E1279–86.
45. Abderrahmani A, Plaisance V, Lovis P, Regazzi R. Mechanisms controlling the expression of the components of the exocytotic apparatus under physiological and pathological conditions. *Biochem Soc Trans* November 2006;**34**(Pt 5):696–700.
46. Stenmark H. Rab GTPases as coordinators of vesicle traffic. *Nat Rev Mol Cell Biol* August 2009;**10**(8):513–25.
47. Matsumoto M, Miki T, Shibasaki T, Kawaguchi M, Shinozaki H, Nio J, et al. Noc2 is essential in normal regulation of exocytosis in endocrine and exocrine cells. *Proc Natl Acad Sci USA* June 1, 2004;**101**(22):8313–8.
48. Lovis P, Gattesco S, Regazzi R. Regulation of the expression of components of the exocytotic machinery of insulin-secreting cells by microRNAs. *Biol Chem* March 2008;**389**(3):305–12.
49. Plaisance V, Abderrahmani A, Perret-Menoud V, Jacquemin P, Lemaigre F, Regazzi R. MicroRNA-9 controls the expression of Granuphilin/Slp4 and the secretory response of insulin-producing cells. *J Biol Chem* September 15, 2006;**281**(37): 26932–42.
50. Roggli E, Britan A, Gattesco S, Lin-Marq N, Abderrahmani A, Meda P, et al. Involvement of microRNAs in the cytotoxic effects exerted by proinflammatory cytokines on pancreatic beta-cells. *Diabetes* April 2010;**59**(4):978–86.
51. Bagge A, Dahmcke CM, Dalgaard LT. Syntaxin-1a is a direct target of miR-29a in insulin-producing beta-cells. *Horm Metab Res* June 2013;**45**(6):463–6.
52. Ammala C, Ashcroft FM, Rorsman P. Calcium-independent potentiation of insulin release by cyclic AMP in single beta-cells. *Nature* May 27, 1993;**363**(6427):356–8.
53. Renstrom E, Eliasson L, Rorsman P. Protein kinase A-dependent and -independent stimulation of exocytosis by cAMP in mouse pancreatic B-cells. *J Physiol* July 1, 1997;**502**(Pt 1):105–18.
54. Keller DM, Clark EA, Goodman RH. Regulation of microRNA-375 by cAMP in pancreatic beta-cells. *Mol Endocrinol* June 2012; **26**(6):989–99.
55. Poy MN, Eliasson L, Krutzfeldt J, Kuwajima S, Ma X, MacDonald PE, et al. A pancreatic islet-specific microRNA regulates insulin secretion. *Nature* November 11, 2004;**432**(7014): 226–30.
56. Ozaki N, Shibasaki T, Kashima Y, Miki T, Takahashi K, Ueno H, et al. cAMP-GEFII is a direct target of cAMP in regulated exocytosis. *Nat Cell Biol* November 2000;**2**(11):805–11.
57. Andrali SS, Sampley ML, Vanderford NL, Ozcan S. Glucose regulation of insulin gene expression in pancreatic beta-cells. *Biochem J* October 1, 2008;**415**(1):1–10.
58. Xu G, Chen J, Jing G, Shalev A. Thioredoxin-interacting protein regulates insulin transcription through microRNA-204. *Nat Med* September 2013;**19**(9):1141–6.

59. Tang X, Muniappan L, Tang G, Ozcan S. Identification of glucose-regulated miRNAs from pancreatic {beta} cells reveals a role for miR-30d in insulin transcription. *RNA* February 2009;**15**(2):287–93.
60. Zhao X, Mohan R, Ozcan S, Tang X. MicroRNA-30d induces insulin transcription factor MafA and insulin production by targeting mitogen-activated protein 4 kinase 4 (MAP4K4) in pancreatic beta-cells. *J Biol Chem* September 7, 2012;**287**(37):31155–64.
61. El OA, Baroukh N, Martens GA, Lebrun P, Pipeleers D, Van OE. miR-375 targets 3′-phosphoinositide-dependent protein kinase-1 and regulates glucose-induced biological responses in pancreatic beta-cells. *Diabetes* October 2008;**57**(10):2708–17.
62. Rafiq I, da SX, Hooper S, Rutter GA. Glucose-stimulated preproinsulin gene expression and nuclear trans-location of pancreatic duodenum homeobox-1 require activation of phosphatidylinositol 3-kinase but not p38 MAPK/SAPK2. *J Biol Chem* May 26, 2000; **275**(21):15977–84.
63. Lynn FC, Skewes-Cox P, Kosaka Y, McManus MT, Harfe BD, German MS. MicroRNA expression is required for pancreatic islet cell genesis in the mouse. *Diabetes* December 2007;**56**(12):2938–45.
64. Kalis M, Bolmeson C, Esguerra JL, Gupta S, Edlund A, Tormo-Badia N, et al. Beta-cell specific deletion of *Dicer1* leads to defective insulin secretion and diabetes mellitus. *PLoS One* 2011; **6**(12):e29166.
65. Melkman-Zehavi T, Oren R, Kredo-Russo S, Shapira T, Mandelbaum AD, Rivkin N, et al. miRNAs control insulin content in pancreatic beta-cells via downregulation of transcriptional repressors. *EMBO J* March 2, 2011;**30**(5):835–45.
66. Plaisance V, Waeber G, Regazzi R, Abderrahmani A. Role of microRNAs in islet beta-cell compensation and failure during diabetes. *J Diabetes Res* 2014;**2014**:618652.
67. Kassem SA, Ariel I, Thornton PS, Scheimberg I, Glaser B. Beta-cell proliferation and apoptosis in the developing normal human pancreas and in hyperinsulinism of infancy. *Diabetes* August 2000;**49**(8):1325–33.
68. Rahier J, Wallon J, Henquin JC. Cell populations in the endocrine pancreas of human neonates and infants. *Diabetologia* May 1981; **20**(5):540–6.
69. Butler AE, Janson J, Bonner-Weir S, Ritzel R, Rizza RA, Butler PC. Beta-cell deficit and increased beta-cell apoptosis in humans with type 2 diabetes. *Diabetes* January 2003;**52**(1):102–10.
70. Kloppel G, Lohr M, Habich K, Oberholzer M, Heitz PU. Islet pathology and the pathogenesis of type 1 and type 2 diabetes mellitus revisited. *Surv Synth Pathol Res* 1985;**4**(2):110–25.
71. Van Assche FA, Aerts L, de PF. A morphological study of the endocrine pancreas in human pregnancy. *Br J Obstet Gynaecol* November 1978;**85**(11):818–20.
72. Butler AE, Cao-Minh L, Galasso R, Rizza RA, Corradin A, Cobelli C, et al. Adaptive changes in pancreatic beta cell fractional area and beta cell turnover in human pregnancy. *Diabetologia* October 2010;**53**(10):2167–76.
73. Donath MY, Halban PA. Decreased beta-cell mass in diabetes: significance, mechanisms and therapeutic implications. *Diabetologia* March 2004;**47**(3):581–9.
74. Meier JJ, Bhushan A, Butler AE, Rizza RA, Butler PC. Sustained beta cell apoptosis in patients with long-standing type 1 diabetes: indirect evidence for islet regeneration? *Diabetologia* November 2005;**48**(11):2221–8.
75. Gepts W. Pathologic anatomy of the pancreas in juvenile diabetes mellitus. *Diabetes* October 1965;**14**(10):619–33.
76. Esguerra JL, Bolmeson C, Cilio CM, Eliasson L. Differential glucose-regulation of microRNAs in pancreatic islets of non-obese type 2 diabetes model Goto-Kakizaki rat. *PLoS One* 2011; **6**(4):e18613.
77. Jacovetti C, Abderrahmani A, Parnaud G, Jonas JC, Peyot ML, Cornu M, et al. MicroRNAs contribute to compensatory beta cell expansion during pregnancy and obesity. *J Clin Invest* October 1, 2012;**122**(10):3541–51.
78. Poy MN, Hausser J, Trajkovski M, Braun M, Collins S, Rorsman P, et al. miR-375 maintains normal pancreatic alpha- and beta-cell mass. *Proc Natl Acad Sci USA* April 7, 2009; **106**(14):5813–8.
79. Bolmeson C, Esguerra JL, Salehi A, Speidel D, Eliasson L, Cilio CM. Differences in islet-enriched miRNAs in healthy and glucose intolerant human subjects. *Biochem Biophys Res Commun* January 7, 2011;**404**(1):16–22.
80. Locke JM, da SX, Dawe HR, Rutter GA, Harries LW. Increased expression of miR-187 in human islets from individuals with type 2 diabetes is associated with reduced glucose-stimulated insulin secretion. *Diabetologia* January 2014;**57**(1): 122–8.
81. Shojima N, Hara K, Fujita H, Horikoshi M, Takahashi N, Takamoto I, et al. Depletion of homeodomain-interacting protein kinase 3 impairs insulin secretion and glucose tolerance in mice. *Diabetologia* December 2012;**55**(12):3318–30.
82. Nesca V, Guay C, Jacovetti C, Menoud V, Peyot ML, Laybutt DR, et al. Identification of particular groups of microRNAs that positively or negatively impact on beta cell function in obese models of type 2 diabetes. *Diabetologia* October 2013;**56**(10): 2203–12.
83. Jais A, Einwallner E, Sharif O, Gossens K, Lu TT, Soyal SM, et al. Heme oxygenase-1 drives metaflammation and insulin resistance in mouse and man. *Cell* July 3, 2014;**158**(1):25–40.
84. Kozakowska M, Szade K, Dulak J, Jozkowicz A. Role of heme oxygenase-1 in postnatal differentiation of stem cells: a possible cross-talk with microRNAs. *Antioxid Redox Signal* April 10, 2014; **20**(11):1827–50.
85. Quick-Cleveland J, Jacob JP, Weitz SH, Shoffner G, Senturia R, Guo F. The DGCR8 RNA-binding heme domain recognizes primary microRNAs by clamping the hairpin. *Cell Rep* June 26, 2014;**7**(6):1994–2005.
86. McClelland AD, Kantharidis P. MicroRNA in the development of diabetic complications. *Clin Sci (Lond)* January 2014;**126**(2): 95–110.
87. Russell AP, Lamon S, Boon H, Wada S, Guller I, Brown EL, et al. Regulation of miRNAs in human skeletal muscle following acute endurance exercise and short-term endurance training. *J Physiol* September 15, 2013;**591**(Pt 18):4637–53.
88. Safdar A, Abadi A, Akhtar M, Hettinga BP, Tarnopolsky MA. miRNA in the regulation of skeletal muscle adaptation to acute endurance exercise in C57Bl/6J male mice. *PLoS One* 2009;**4**(5): e5610.
89. Nielsen S, Scheele C, Yfanti C, Akerstrom T, Nielsen AR, Pedersen BK, et al. Muscle specific microRNAs are regulated by endurance exercise in human skeletal muscle. *J Physiol* October 15, 2010;**588**(Pt 20):4029–37.
90. Guller I, Russell AP. MicroRNAs in skeletal muscle: their role and regulation in development, disease and function. *J Physiol* November 1, 2010;**588**(Pt 21):4075–87.
91. Koning M, Werker PM, van Luyn MJ, Krenning G, Harmsen MC. A global downregulation of microRNAs occurs in human quiescent satellite cells during myogenesis. *Differentiation* November 2012;**84**(4):314–21.
92. McCarthy JJ. MicroRNA-206: the skeletal muscle-specific myomiR. *Biochim Biophys Acta* November 2008;**1779**(11):682–91.
93. van RE, Liu N, Olson EN. MicroRNAs flex their muscles. *Trends Genet* April 2008;**24**(4):159–66.
94. Yang WM, Jeong HJ, Park SY, Lee W. Induction of miR-29a by saturated fatty acids impairs insulin signaling and glucose uptake through translational repression of IRS-1 in myocytes. *FEBS Lett* June 13, 2014;**588**(13):2170–6.

95. He A, Zhu L, Gupta N, Chang Y, Fang F. Over-expression of miR-29, highly upregulated in diabetic rats, leads to insulin resistance in 3T3-L1 adipocytes. *Mol Endocrinol* November 2007;**21**(11):2785–94.
96. Bagge A, Clausen TR, Larsen S, Ladefoged M, Rosenstierne MW, Larsen L, et al. MicroRNA-29a is up-regulated in beta-cells by glucose and decreases glucose-stimulated insulin secretion. *Biochem Biophys Res Commun* September 21, 2012;**426**(2):266–72. http://dx.doi.org/10.1016/j.bbrc.2012.08.082.
97. Mao Y, Mohan R, Zhang S, Tang X. MicroRNAs as pharmacological targets in diabetes. *Pharmacol Res* September 2013;**75**:37–47.
98. Ferland-McCollough D, Ozanne SE, Siddle K, Willis AE, Bushell M. The involvement of microRNAs in Type 2 diabetes. *Biochem Soc Trans* December 2010;**38**(6):1565–70.
99. Zhang Y, Yang L, Gao YF, Fan ZM, Cai XY, Liu MY, et al. MicroRNA-106b induces mitochondrial dysfunction and insulin resistance in C2C12 myotubes by targeting mitofusin-2. *Mol Cell Endocrinol* December 5, 2013;**381**(1–2):230–40.
100. Lee H, Jee Y, Hong K, Hwang GS, Chun KH. MicroRNA-494, upregulated by tumor necrosis factor-alpha, desensitizes insulin effect in C2C12 muscle cells. *PLoS One* 2013;**8**(12):e83471.
101. Chen Y, Dorn GW. PINK1-phosphorylated mitofusin 2 is a Parkin receptor for culling damaged mitochondria. *Science* April 26, 2013; **340**(6131):471–5.
102. Zorzano A, Hernandez-Alvarez MI, Palacin M, Mingrone G. Alterations in the mitochondrial regulatory pathways constituted by the nuclear co-factors PGC-1alpha or PGC-1beta and mitofusin 2 in skeletal muscle in type 2 diabetes. *Biochim Biophys Acta* June 2010;**1797**(6–7):1028–33.
103. Hilton C, Neville MJ, Karpe F. MicroRNAs in adipose tissue: their role in adipogenesis and obesity. *Int J Obes (Lond)* March 2013; **37**(3):325–32.
104. Lin Q, Gao Z, Alarcon RM, Ye J, Yun Z. A role of miR-27 in the regulation of adipogenesis. *FEBS J* April 2009;**276**(8):2348–58.
105. Trajkovski M, Hausser J, Soutschek J, Bhat B, Akin A, Zavolan M, et al. MicroRNAs 103 and 107 regulate insulin sensitivity. *Nature* June 30, 2011;**474**(7353):649–53.
106. Esau C, Kang X, Peralta E, Hanson E, Marcusson EG, Ravichandran LV, et al. MicroRNA-143 regulates adipocyte differentiation. *J Biol Chem* December 10, 2004;**279**(50):52361–5.
107. Chen YH, Heneidi S, Lee JM, Layman LC, Stepp DW, Gamboa GM, et al. miRNA-93 inhibits GLUT4 and is overexpressed in adipose tissue of polycystic ovary syndrome patients and women with insulin resistance. *Diabetes* July 2013;**62**(7):2278–86.
108. Jordan SD, Kruger M, Willmes DM, Redemann N, Wunderlich FT, Bronneke HS, et al. Obesity-induced overexpression of miRNA-143 inhibits insulin-stimulated AKT activation and impairs glucose metabolism. *Nat Cell Biol* April 2011;**13**(4):434–46.
109. Meerson A, Traurig M, Ossowski V, Fleming JM, Mullins M, Baier LJ. Human adipose microRNA-221 is upregulated in obesity and affects fat metabolism downstream of leptin and TNF-alpha. *Diabetologia* September 2013;**56**(9):1971–9.
110. Ling HY, Ou HS, Feng SD, Zhang XY, Tuo QH, Chen LX, et al. Changes in microRNA (miR) profile and effects of miR-320 in insulin-resistant 3T3-L1 adipocytes. *Clin Exp Pharmacol Physiol* September 2009;**36**(9):e32–9.
111. Takanabe R, Ono K, Abe Y, Takaya T, Horie T, Wada H, et al. Upregulated expression of microRNA-143 in association with obesity in adipose tissue of mice fed high-fat diet. *Biochem Biophys Res Commun* November 28, 2008;**376**(4):728–32.
112. Rayner KJ, Suarez Y, Davalos A, Parathath S, Fitzgerald ML, Tamehiro N, et al. miR-33 contributes to the regulation of cholesterol homeostasis. *Science* June 18, 2010;**328**(5985):1570–3.
113. Davalos A, Goedeke L, Smibert P, Ramirez CM, Warrier NP, Andreo U, et al. miR-33a/b contribute to the regulation of fatty acid metabolism and insulin signaling. *Proc Natl Acad Sci USA* May 31, 2011;**108**(22):9232–7.
114. Rayner KJ, Moore KJ. MicroRNA control of high-density lipoprotein metabolism and function. *Circ Res* January 3, 2014;**114**(1):183–92.
115. Rayner KJ, Sheedy FJ, Esau CC, Hussain FN, Temel RE, Parathath S, et al. Antagonism of miR-33 in mice promotes reverse cholesterol transport and regression of atherosclerosis. *J Clin Invest* July 2011;**121**(7):2921–31.
116. Aranda JF, Madrigal-Matute J, Rotllan N, Fernandez-Hernando C. MicroRNA modulation of lipid metabolism and oxidative stress in cardiometabolic diseases. *Free Radic Biol Med* September 2013; **64**:31–9.
117. Rayner KJ, Esau CC, Hussain FN, McDaniel AL, Marshall SM, van Gils JM, et al. Inhibition of miR-33a/b in non-human primates raises plasma HDL and lowers VLDL triglycerides. *Nature* October 20, 2011;**478**(7369):404–7.
118. Lagos-Quintana M, Rauhut R, Yalcin A, Meyer J, Lendeckel W, Tuschl T. Identification of tissue-specific MicroRNAs from mouse. *Curr Biol* April 30, 2002;**12**(9):735–9.
119. Elmen J, Lindow M, Silahtaroglu A, Bak M, Christensen M, Lind-Thomsen A, et al. Antagonism of microRNA-122 in mice by systemically administered LNA-antimiR leads to up-regulation of a large set of predicted target mRNAs in the liver. *Nucleic Acids Res* March 2008;**36**(4):1153–62.
120. Tsai WC, Hsu SD, Hsu CS, Lai TC, Chen SJ, Shen R, et al. MicroRNA-122 plays a critical role in liver homeostasis and hepatocarcinogenesis. *J Clin Invest* August 1, 2012;**122**(8):2884–97.
121. Zhou B, Li C, Qi W, Zhang Y, Zhang F, Wu JX, et al. Downregulation of miR-181a upregulates sirtuin-1 (SIRT1) and improves hepatic insulin sensitivity. *Diabetologia* July 2012;**55**(7):2032–43.
122. Kornfeld JW, Baitzel C, Konner AC, Nicholls HT, Vogt MC, Herrmanns K, et al. Obesity-induced overexpression of miR-802 impairs glucose metabolism through silencing of Hnf1b. *Nature* February 7, 2013;**494**(7435):111–5.
123. Eckel RH, Kahn SE, Ferrannini E, Goldfine AB, Nathan DM, Schwartz MW, et al. Obesity and type 2 diabetes: what can be unified and what needs to be individualized? *Diabetes Care* June 2011; **34**(6):1424–30.
124. Saenger P, Czernichow P, Hughes I, Reiter EO. Small for gestational age: short stature and beyond. *Endocr Rev* April 2007; **28**(2):219–51.
125. Dumortier O, Hinault C, Gautier N, Patouraux S, Casamento V, Van OE. Maternal protein restriction leads to pancreatic failure in offspring: role of misexpressed microRNA-375. *Diabetes* October 2014;**63**(10):3416–27.
126. de la Lastra CA, Villegas I. Resveratrol as an antioxidant and pro-oxidant agent: mechanisms and clinical implications. *Biochem Soc Trans* November 2007;**35**(Pt 5):1156–60.
127. Vang O. What is new for resveratrol? Is a new set of recommendations necessary? *Ann N Y Acad Sci* July 2013;**1290**:1–11.
128. Szkudelska K, Szkudelski T. Resveratrol, obesity and diabetes. *Eur J Pharmacol* June 10, 2010;**635**(1–3):1–8.
129. Tili E, Michaille JJ. Resveratrol, microRNAs, inflammation, and cancer. *J Nucleic Acids* 2011;**2011**:102431.
130. Tome-Carneiro J, Larrosa M, Gonzalez-Sarrias A, Tomas-Barberan FA, Garcia-Conesa MT, Espin JC. Resveratrol and clinical trials: the crossroad from in vitro studies to human evidence. *Curr Pharm Des* 2013;**19**(34):6064–93.
131. Joven J, Espinel E, Rull A, Aragones G, Rodriguez-Gallego E, Camps J, et al. Plant-derived polyphenols regulate expression of miRNA paralogs miR-103/107 and miR-122 and prevent diet-induced fatty liver disease in hyperlipidemic mice. *Biochim Biophys Acta* July 2012;**1820**(7):894–9.
132. Arola-Arnal A, Blade C. Proanthocyanidins modulate microRNA expression in human HepG2 cells. *PLoS One* 2011;**6**(10):e25982.

133. Zitman-Gal T, Green J, Pasmanik-Chor M, Golan E, Bernheim J, Benchetrit S. Vitamin D manipulates miR-181c, miR-20b and miR-15a in human umbilical vein endothelial cells exposed to a diabetic-like environment. *Cardiovasc Diabetol* January 7, 2014;**13**:8.
134. Guay C, Jacovetti C, Nesca V, Motterle A, Tugay K, Regazzi R. Emerging roles of non-coding RNAs in pancreatic beta-cell function and dysfunction. *Diabetes Obes Metab* October 2012;**14**(Suppl. 3): 12–21.
135. Karolina DS, Armugam A, Tavintharan S, Wong MT, Lim SC, Sum CF, et al. MicroRNA 144 impairs insulin signaling by inhibiting the expression of insulin receptor substrate 1 in type 2 diabetes mellitus. *PLoS One* 2011;**6**(8):e22839.
136. Kong L, Zhu J, Han W, Jiang X, Xu M, Zhao Y, et al. Significance of serum microRNAs in pre-diabetes and newly diagnosed type 2 diabetes: a clinical study. *Acta Diabetol* September 21, 2010.
137. Pescador N, Perez-Barba M, Ibarra JM, Corbaton A, Martinez-Larrad MT, Serrano-Rios M. Serum circulating microRNA profiling for identification of potential type 2 diabetes and obesity biomarkers. *PLoS One* 2013;**8**(10):e77251.
138. Zampetaki A, Kiechl S, Drozdov I, Willeit P, Mayr U, Prokopi M, et al. Plasma microRNA profiling reveals loss of endothelial miR-126 and other microRNAs in type 2 diabetes. *Circ Res* September 17, 2010;**107**(6):810–7.
139. Bonner C, Nyhan KC, Bacon S, Kyithar MP, Schmid J, Concannon CG, et al. Identification of circulating microRNAs in HNF1A-MODY carriers. *Diabetologia* August 2013;**56**(8):1743–51.
140. Erener S, Mojibian M, Fox JK, Denroche HC, Kieffer TJ. Circulating miR-375 as a biomarker of beta-cell death and diabetes in mice. *Endocrinology* February 2013;**154**(2):603–8.
141. Nielsen LB, Wang C, Sorensen K, Bang-Berthelsen CH, Hansen L, Andersen ML, et al. Circulating levels of microRNA from children with newly diagnosed type 1 diabetes and healthy controls: evidence that miR-25 associates to residual beta-cell function and glycaemic control during disease progression. *Exp Diabetes Res* 2012;**2012**:896362.
142. Karolina DS, Tavintharan S, Armugam A, Sepramaniam S, Pek SL, Wong MT, et al. Circulating miRNA profiles in patients with metabolic syndrome. *J Clin Endocrinol Metab* December 2012;**97**(12):E2271–6.
143. Suarez Y, Fernandez-Hernando C. New insights into microRNA-29 regulation: a new key player in cardiovascular disease. *J Mol Cell Cardiol* March 2012;**52**(3):584–6.
144. Chen X, Ba Y, Ma L, Cai X, Yin Y, Wang K, et al. Characterization of microRNAs in serum: a novel class of biomarkers for diagnosis of cancer and other diseases. *Cell Res* October 2008;**18**(10): 997–1006.

CHAPTER

22

Diabetes Mellitus and Intestinal Niemann-Pick C1–Like 1 Gene Expression

Pooja Malhotra[1], *Ravinder K. Gill*[1], *Pradeep K. Dudeja*[1,2], *Waddah A. Alrefai*[1,2]

[1]Division of Gastroenterology and Hepatology, Department of Medicine, University of Illinois at Chicago, Chicago, IL, USA; [2]The Jesse Brown VA Medical Center, Chicago, IL, USA

1. CHOLESTEROL HOMEOSTASIS

Diabetes mellitus is associated with increased risk for serious complications such as atherosclerosis and cardiovascular diseases. The development of these detrimental conditions needs to be carefully monitored in diabetic patients. Previous studies have shown that lowering the levels of plasma cholesterol significantly reduces the risk of cardiovascular diseases associated with diabetes mellitus even in diabetic patients with normal levels of plasma cholesterol.[1] Accordingly, the current guidelines of the American College of Cardiology and the American Heart Association (ACC/AHA) recommend decreasing blood cholesterol to levels less than 70 mg/dL in patients with high risk for cardiovascular diseases including diabetic patients.[2] Notwithstanding current advances in therapeutic options, decreasing plasma cholesterol to meet these stringent low levels remains challenging. Therefore, there is a dire need to define more potential therapeutic targets by expanding our knowledge about cholesterol metabolism and how different organ systems such as the liver and the intestine coordinate their cellular mechanisms to control body cholesterol homeostasis.

Cholesterol is a vital component for living cells and accounts for 20–25% of total lipids in plasma membranes (PMs).[3] Also, cholesterol plays a number of essential physiological functions in the body. For example, cholesterol is the precursor of bile acids and steroid hormones such as androgens, estrogens, and glucocorticoids. Cholesterol metabolites also play important roles in membrane trafficking as well as signaling events.[4] Given the essential physiological roles of cholesterol and the fact that excess of cholesterol leads to deleterious effects, efficient mechanisms have evolved in mammals to secure sufficient supplies and to maintain physiological levels. In humans, a balance between input and output pathways maintains a pool of ~120–130 g of total cholesterol in the body.[5] These pathways include cholesterol synthesis, cholesterol elimination, and intestinal cholesterol absorption, which are discussed below with a major emphasis on mechanisms of intestinal cholesterol absorption.

1.1 Cholesterol Synthesis

Cholesterol synthesis takes place in all cell types. However, the liver represents the main site for de novo cholesterol synthesis contributing to approximately 80% of total cholesterol synthesis in mammals. Cholesterol is generated from acetyl-CoA via a complex multistep process in which the 3-hydroxy-methylglutaryl-CoA reductase (HMG-CoA-R) mediates the rate-limiting step catalyzing the conversion of HMG-CoA to mevalonic acid.[6] Insulin was shown to reduce the expression of HMG-CoA-R in primary isolated rat hepatocytes.[7] Conversely, HMG-CoA-R expression and activity in the liver and intestine were shown to be increased in rats with diabetes mellitus.[8] These observations suggested that increased synthesis of cholesterol is a contributing factor to hypercholesterolemia associated with diabetes mellitus. This also provided the rationale for inhibiting HMG-CoA-R to lower plasma cholesterol in diabetic patients. In fact, the inhibitors of HMG-CoA-R, statins are widely used cholesterol-lowering drugs for the treatment of primary

Molecular Nutrition and Diabetes
http://dx.doi.org/10.1016/B978-0-12-801585-8.00022-1

hypercholesterolemia and to lower cholesterol in high-risk patients such as those with diabetes mellitus. Although statins effectively reduce plasma cholesterol, a significant portion of patients exhibits resistance to their cholesterol-lowering effects and fails to reach low target levels of cholesterol set by the ACC/AHA guidelines for high-risk patients. Moreover, some patients are intolerant to statin treatment due to adverse effects including myopathy.[9] These hurdles further compel the need for better therapeutic modalities to lower the levels of plasma cholesterol.

1.2 Elimination of Cholesterol from the Body

The human body eliminates approximately 1 g of cholesterol per day, of which the conversion of cholesterol to bile acids in the liver accounts for ~500 mg, and the disposal of free cholesterol in the feces represents the other 500 mg of eliminated cholesterol.[10]

The hepatic biosynthesis of bile acids from cholesterol is mediated by complex series of reactions of which cholesterol 7α-hydroxylase (CYP7A1) mediates the rate-limiting step.[11] After their synthesis, bile acids are secreted via the bile into the intestine and undergo efficient reabsorption by which ~95% is returned back to the liver to establish the enterohepatic circulation of bile acids. Only ~500 mg are secreted in the feces accounting for the loss of 500 mg of cholesterol.[10] Increased levels of bile acids fluxing through the hepatocytes have a negative feedback effect on CYP7A1 activity, resulting in suppression of hepatic bile acid biosynthesis. In cases of an increase in the absorption, large amount of bile acids circulate between the liver and the intestine, leading to a suppression of bile acid biosynthesis in the liver and an elevation of the levels of plasma cholesterol.[11] Indeed, previous studies in humans and in animal models of diabetes mellitus suggested that the expansion of the circulating pool of bile acids in diabetes mellitus may contribute to associated hypercholesterolemia.[12] On the other hand, blocking the absorption of bile acids reduces the suppression of hepatic CYP7A1 and promotes cholesterol conversion to bile acids and, hence, decreases plasma cholesterol.[11] Recent studies showed that inhibitors of bile acid absorption improve the associated dyslipidemia and control the hyperglycemia in models of diabetes mellitus.[13] Indeed, the observation that the inhibition of bile acid absorption represents a potential therapeutic option to lower cholesterol in diabetic patients signifies the important roles of the intestine in controlling cholesterol homeostasis and how the intestinal processes offer potential targets for future therapy.

Besides the hepatic conversion of cholesterol to bile acids, the secretion of free cholesterol into the bile is also an important pathway for cholesterol elimination from the body. Biliary secretion of cholesterol is part of the reverse cholesterol transport system that transfers cholesterol from the peripheral tissues via high-density lipoprotein (HDL) to the liver for elimination. HDL cholesterol is taken up by the scavenger receptor type B1 (SR-B1) expressed on the sinusoidal membrane of the hepatocytes, and cholesterol is then secreted into the bile via the action of the heterodimer of ABCG5/G8 transporters expressed on the canalicular membrane of the hepatocytes.[14,15] Many factors appear to be essential for biliary cholesterol secretion such as the presence of phospholipids in the bile. Indeed, studies in *mdr2*-knockout mice, in which phospholipid secretion to bile is impaired, provided evidence for the importance of phospholipid for biliary cholesterol secretion.[16]

Emerging evidence indicates that the hepatobiliary route is not the only way to eliminate cholesterol via feces. Studies in animal models and humans demonstrated that plasma cholesterol could be excreted by a direct secretion into intestinal lumen across the intestinal epithelium.[14,17–19] This novel pathway known as *T*rans-*I*ntestinal *C*holesterol *E*xcretion (TICE) contributes to ~30% of the total fecal sterol in mice.[20,21] The molecular mechanisms underlying TICE are yet to be elucidated. Evidence suggests that the cholesterol secreted via TICE is not provided by HDL and may be derived from the very low-density lipoprotein (VLDL) particles (Figure 1).[22] Also, the nature of intestinal receptors and transporters mediating the flux of cholesterol across the intestinal epithelium is not clearly defined.[22] Nevertheless, this pathway signifies the importance of the intestine in controlling body cholesterol homeostasis and provides a novel therapeutic target that could be activated to reduce plasma cholesterol.

2. INTESTINAL CHOLESTEROL ABSORPTION

In addition to the biliary and intestinal excretion of cholesterol that amount to about 800–1000 mg/day, dietary cholesterol (Western-type diet) contributes ~300–500 mg to luminal cholesterol. Further, cholesterol derived from the sloughing intestinal epithelial cells provides an additional 300 mg.[23,24] Therefore, ~1400 mg of cholesterol is presented in the intestinal lumen every day. However, the intestine absorbs ~50% of total luminal cholesterol regardless of its input source and only ~500 mg is lost in the feces.[23,25] The absorbed cholesterol reaches the liver and then is secreted into the circulation as part of the VLDL and low-density lipoprotein (LDL). Intestinal cholesterol absorption represents an important process directly involved in controlling plasma levels of cholesterol because of its major contribution to plasma LDL

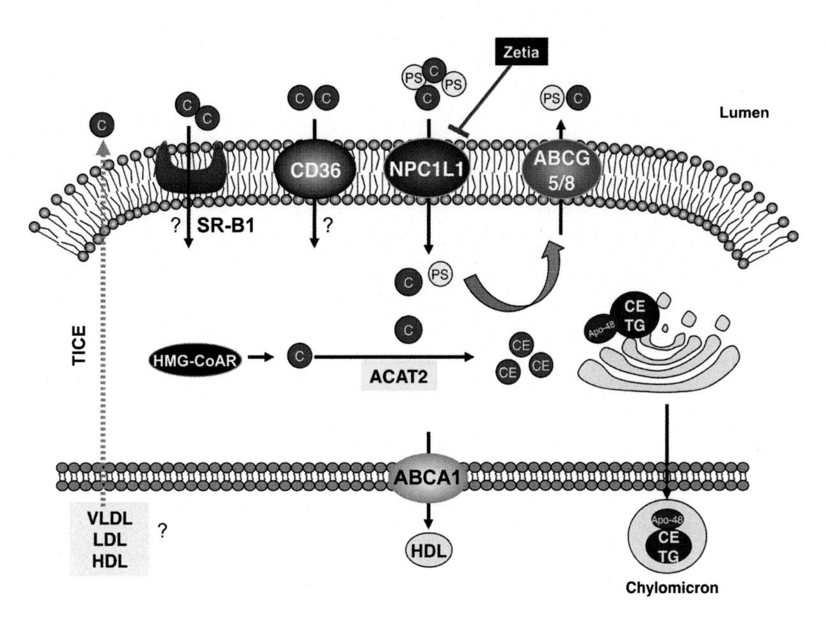

FIGURE 1 Mechanisms of cholesterol transport in intestinal epithelial cells. The intestinal epithelial cells acquire cholesterol from the lumen or from endogenous synthesis in which the HMG-CoA reductase mediates the rate-limiting step. Several sterol transport proteins involved in cholesterol transport are expressed on the apical membrane and the basolateral membrane of intestinal epithelial cells. NPC1L1 plays a central role in cholesterol absorption as evident by the fact that cholesterol absorption is decreased in NPC1L1 knockout mice and that NPC1L1 is the target for the cholesterol absorption inhibitor ezetimibe. CD36 and SR-B1 may mediate cholesterol uptake, however, eliminating their expression in mice had no effects on cholesterol absorption. ABCG5 and ABCG8 form a heterodimer that mediates the efflux of plant sterols (PS) and cholesterol (but to a lesser extent) back to the intestinal lumen. *Trans*-Epithelial Cholesterol Efflux (TICE) contributes to cholesterol secretion into the intestinal lumen but the mechanisms and the source of cholesterol are not yet fully defined. The ABCA1 on the baslolateral membrane mediates the efflux of cholesterol from the plasma membrane of the enterocytes to high-density lipoprotein (HDL) particles.

cholesterol.[23] The following sections focus on reviewing our current knowledge regarding intestinal cholesterol absorption with a focus on the intestinal cholesterol transporter NPC1L1.

2.1 Mechanisms of Intestinal Cholesterol Absorption

The efficiency of intestinal absorption of cholesterol exhibits a wide range of interindividual variations between 29% and 80% of the total luminal cholesterol.[26,27] Furthermore, studies provided evidence for a positive correlation between the efficiency of cholesterol absorption and the levels of plasma cholesterol.[23] Indeed, the process of intestinal cholesterol absorption depends on a number of factors including cellular epithelial and luminal factors. Cholesterol is virtually insoluble in aqueous phase of the intestinal lumen. Luminal cholesterol is emulsified by bile acids and phospholipids to form micelles, which facilitate its delivery to the apical membrane of the intestinal epithelial cells.[28] Free cholesterol is transported across the luminal membrane of the enterocytes and is then esterified by the action of Acyl-CoA cholesterol acyl transferase 2 (ACAT2). Esterified cholesterol is then packaged along with triglycerides and apolipoprotein B48 (ApoB48) into chylomicrons that are then secreted into the lymph to complete the absorption process. Chylomicrons lose their fatty acids in the tissues through hydrolysis of their triglyceride cargo by the action of the lipoprotein lipase producing lipoprotein particles rich with cholesterol known as chylomicron remnants.[29] The liver receives chylomicron remnants and their content of cholesterol is either esterified and stored in the liver or resecreted into the circulation via the VLDL particles.[10] Therefore, intestinal cholesterol absorption provides an important source of cholesterol for all types of cells in the body and offers a physiological means to reduce the energy demand required by the complex process of the de novo synthesis of cholesterol.

Various transport proteins, as illustrated in Figure 1, are present on the apical as well as the basolateral membrane of the enterocytes and have been shown to be involved in the transport of cholesterol. The contribution of these transport proteins to intestinal cholesterol absorption was gleaned out from the studies in knockout mice. Since the first step in cholesterol absorption involves an influx across the luminal membranes of enterocytes, initial studies focused on identifying the luminal membrane transporters responsible for cholesterol absorption. Altmann and colleagues using a bioinformatic approach were successful in identifying the Niemann-Pick type C1–like 1 (NPC1L1) transporter as a candidate protein to mediate the uptake of luminal cholesterol into the enterocytes. Their seminal work showed that knocking out *NPC1L1* expression in mice results in ~70% reduction in intestinal cholesterol absorption.[30] In fact, their studies were the first to demonstrate that a single transport protein is pivotal for the complex, multifactorial process of intestinal cholesterol absorption.

Besides NPC1L1 on the apical membrane of the enterocytes, other sterol transporters are also expressed including the ABCG5 and ABCG8 as well SR-B1. The ABCG5 and G8 are sterol efflux transporters that are localized to luminal membrane of enterocytes and also, as mentioned earlier, in the canalicular membrane of hepatocytes. In the intestine, ABCG5 and G8 form a heterodimer that mediates the efflux of plant sterols (PSs) and, to a lesser extent, cholesterol from the enterocytes into the lumen[31,32] (Figure 1). Mutations in *ABCG5 or G8* result in sitosterolemia, a condition in which the levels of PSs in the blood are elevated.[32] Similarly, studies in mice showed that deletion of *ABCG5* and *G8* results in ~30-fold increase in plasma sitosterols levels.[33] Absorption of PSs upregulates the expression of ABCG5/G8 leading to an increase in cholesterol efflux, thereby reducing cholesterol absorption in mice.[34] These observations may demonstrate the molecular basis explaining one possible mechanism underlying the cholesterol-lowering effects of PSs. However, this notion was challenged by another study showing that the cholesterol lowering effects of PSs were independent of ABCG5/G8 transporters. .[35]

The scavenger receptor class B type 1 (SR-B1) is a receptor of HDL particles in the liver and is found on the apical membrane of the enterocytes. Such a pattern of expression for SR-B1 in the intestine suggested a potential role in cholesterol absorption.[36–38] However, the knockdown of *SR-B1* in mice does not reduce intestinal cholesterol absorption and the exact function of intestinal SR-B1 remains elusive.[39] Recently, a study in intestinal epithelial Caco2 cells showed that SR-B1 may act as a sensor for luminal micelles to trigger intracellular signaling pathways and facilitate fat absorption.[40] Also, the scavenger receptor CD36 that mediates fatty acid absorption is also located on the apical membrane of the enterocytes. Evidence was provided showing that CD36 could mediate cholesterol uptake[41] despite the fact that CD36 knockout mice showed no significant changes in cholesterol absorption.[29,42,43]

Other sterol transporters such as ABCA1 are expressed on the basolateral membranes of enterocytes and may play a role in cholesterol absorption. ABCA1 is a member of ABC family of transporters expressed in the macrophages, liver, and intestine. This transport protein functions to efflux the cholesterol from the cells to nascent HDL particles containing apolipoprotein A-1 acceptor (ApoA1).[44] In Tangier disease, that results from mutations in ABCA1, cholesterol was shown to be accumulated in the intestine due to lack of cholesterol efflux from intestine to ApoA1–containing particles.[29,45,46] However, intestine-specific knockout of *ABCA1* in mice showed no effect on intestinal absorption suggesting a minimal involvement in cholesterol absorption.[47] Indeed, more detailed studies in intestinal epithelial Caco2 cells showed that the cholesterol transported from the enterocytes by the intestinal ABCA1 transporter to nascent HDL is directly derived from the apical PM regardless to its origin whether from the intestinal luminal or from the de novo synthesized cholesterol in the enterocytes.[48]

Despite the presence of multiple sterol transporters in the enterocytes, only the lack of NPC1L1 remarkably reduced cholesterol absorption. NPC1L1 then became the target of intensive investigations that aimed to discern the mechanisms by which NPC1L1 mediates cholesterol absorption and how the function of this protein could be targeted to lower the levels of plasma cholesterol. In fact, a number of elegant studies demonstrated that NPC1L1 is indeed the molecular target of ezetimibe, the well-known inhibitor of intestinal cholesterol absorption.[49,50]

2.2 Inhibition of Cholesterol Absorption

Ezetimibe was the first new approved drug by the US Food and Drug Administration (FDA) to block intestinal cholesterol absorption and to lower plasma cholesterol.[51,52] In humans, treatment with ezetimibe causes a 54% reduction in cholesterol absorption leading to ~15–20% decrease in plasma cholesterol levels.[53,54] Studies showed that ezetimibe improves the postprandial hyperlipidemia and its associated endothelial dysfunction.[55] The efficiency of cholesterol-lowering effect, however, may be hindered by the activation of compensatory pathways such as synthesis of cholesterol, especially in the liver. Combination of ezetimibe with statin to inhibit cholesterol synthesis was thought to be the regimen of choice to reduce plasma cholesterol more effectively compared with treatment with statin or ezetimbe alone.[50,56–60]

The potential effects of the combination therapy, however, have recently been questioned in light of recent reports arguing for the lack of additive effect of ezetimibe treatment in patients previously treated with statin.[61] Also, clinical trials addressing the efficacy of treatment with ezetimibe in reducing the risk of cardiovascular disease were discouraging. For example, the Ezetimibe and Simvastatin in Hypercholesterolemia Enhances Atherosclerosis Regression (ENHANCE) randomized clinical trial examined the outcome after 2-year treatment with ezetimibe and simvastain together or simvastatin alone in 720 patients with heterozygous familial hypercholesterolemia to check their effects on carotid intima-media thickness. Despite a significant reduction in LDL cholesterol levels in both groups of patients, there was no significant difference in carotid intima-media thickness.[62] However, this clinical trial had a number of limitations such as duration of trial, study population, subject number, and methodology used to assess the effect on atherosclerosis and primary end points.[63] Large, long-term clinical trials of ezetimibe such as Improved Reduction of Outcomes: Vytorin Efficacy International Trial (Improve-IT) and Study of heart and renal protection (SHARP) investigated the effect of ezetimibe on morbidity and mortality. While the results of SHARP trial demonstrate that combination of ezetimibe and simvastatin drugs is able to decrease cholesterol in chronic kidney diseases and that ezetimibe is safe to use, the SHARP study also showed that the monotherapy with statin or ezetimibe had the same efficacy in decreasing plasma cholesterol compared with combination therapy.[64] The preliminary findings of the Improve-IT clinical trial provided evidence for a modest benefit of ezetimibe and demonstrated that the combination of ezetimibe and statin decreases LDL cholesterol by an average of 17 mg/dL and reduces the risk of cardiovascular events compared with statin alone.[65]

It should be noted that very recent studies conducted in large number of human subjects demonstrated that naturally occurring inactivating mutations in *NPC1L1* that disrupt its function are associated with lower plasma LDL cholesterol and a significant decrease in the risk of coronary heart disease.[66] These studies clearly show that the failure to achieve the expected outcome by treatment with ezetimibe reported in a number of recent studies may be attributed to factors related to the pharmacokinetic characteristics of ezetimibe or the possible off-target effects. Recently, a number of other inhibitors of NPC1L1 are being investigated. In this regard, spiroimidazolidinone is another inhibitor of NPC1L1 that has been developed by virtual screening and shown to decrease cholesterol absorption by 67% in a mouse model.[67] MD-0727 is also introduced as a new cholesterol absorption inhibitor and is shown to successfully decrease LDL cholesterol in clinical trials.[68]

The fact that inactivating mutations in *NPC1L1* are associated with reduced risk of cardiovascular disease further underscores the essential role of NPC1L1 in controlling cholesterol homeostasis. It is evident, therefore, that understanding how NPC1L1 transports luminal cholesterol and delineating the regulatory pathways controlling its expression may unravel additional targets for better therapeutic modalities to lower plasma cholesterol and decrease the risk of cardiovascular diseases. The following sections will provide a review of the current knowledge about NPC1L1 structure–function relationship, its regulation under normal conditions, and what is known about the deregulation of NPC1L1 in diseases such as diabetes mellitus.

3. INTESTINAL NPC1L1 CHOLESTEROL TRANSPORTER

3.1 Structure–Function Relationship of NPC1L1

NPC1L1 shares 51% similarity and 42% identity with NPC1, an intracellular cholesterol transporter of which mutations cause Niemann-Pick type C1 disease, a lysosomal cholesterol storage disorder.[69,70] Human *NPC1L1* gene resides on chromosome 7, spans ~29 kb, and has 20 exons and codes for a protein of ~1359 amino acids.[30,69,70] In humans, NPC1L1 is expressed in both the intestine and liver, whereas its expression in mice is restricted to the intestine.[30,69,71] NPC1L1 is also found in the gallbladder and stomach of humans, but its functional roles in these organs are not yet understood.[30,72] With respect to cellular localization, NPC1L1 is expressed on the apical surface of enterocytes and its expression is restricted to canalicular membrane of hepatocytes.[73–75] Like its homolog NPC1, the NPC1L1 protein has a secretion signal, N-linked glycosylation sites located within extracellular loops of the protein, 13 transmembrane domains, 3 large extracellular loops, several small cytoplasmic loops, and a C-terminal cytoplasmic tail.[30,73,76] Two conserved domains were identified in the NPC1L1: sterol sensing domain (SSD) and N-terminal domain (NTD).[77,78] SSD resides within the transmembrane domains of NPC1L1, and this region is conserved in other proteins involved in cholesterol metabolism such as SCAP, HMG-CoA-R, and NPC1.[69,78–80] The crystal structure of the NTD of NPC1L1 has been recently demonstrated, which showed that this domain has a sterol binding pocket.[78,81]

Recent studies showed that ezetimibe blocks the internalization of NPC1L1, thus inhibiting cholesterol absorption.[82] Further, Weingless and his group[83] mapped ezetimibe binding site to the second extracellular loop of NPC1L1 protein and showed by various in vitro binding assays that ezetimibe inhibits NPC1L1 through direct binding.[50] Another in vivo study in mouse small intestine also confirmed that ezetimibe blocks the internalization of NPC1L1 and cholesterol absorption.[49]

3.2 Intestinal NPC1L1-Mediated Cholesterol Transport

Until recently, not much information was available regarding the molecular mechanism by which cholesterol is absorbed via intestinal NPC1L1. Studies of Zhang et al.[84] showed that the transport of cholesterol from the apical PM into the cell depends on the binding of cholesterol to the sterol sensing pocket of the NTD of NPC1L1. This binding causes a conformational change in NPC1L1 protein, which facilitates the transfer of cholesterol to the lipid raft protein flotillin that is present in the membrane and forms NPC1L1–flotillin–cholesterol membrane microdomains (NFC). This complex is then internalized by clathrin-mediated AP2 pathway, which is the first step in cholesterol uptake and transported to endocytic recycling compartment (ERC).[85] Mutations in the L216 residue of NTD inhibit cholesterol uptake, suggesting the importance of NTD in cholesterol transport. The NTD domain of NPC1L1 specifically binds cholesterol but not PSs, indicating that it mediates selective uptake of cholesterol.[84] A recent study identified the YVNXXF motif in the cytoplasmic terminal of NPC1L1, which functions as an endocytic peptide signal. This motif has been suggested to be recognized by Numb, a clathrin adapter, and appears to be required for endocytosis of NPC1L1. Also, intestine-specific knockout of Numb protein in mice showed a decrease in dietary cholesterol absorption as well as plasma cholesterol levels further confirming that NPC1L1–Numb interaction is important for the internalization of NPC1L1 and cholesterol absorption.[86] Interestingly, cholesterol levels in the lumen modulate trafficking of NPC1L1 from membrane to intracellular compartments. Using a mouse model, Xie et al.[49] showed that in the absence of cholesterol, NPC1L1 resides on the brush border membrane of enterocytes. The same studies also showed that cholesterol administration to the lumen causes an increase in NPC1L1-positive vesicles in the subapical site under brush border membrane where it partially co-localizes with endosomal marker Rab 11. Overall, NPC1L1 mediates cholesterol uptake by transporting cholesterol-enriched vesicles from the apical membrane to the ERC. How cholesterol is further transported from ERC to endoplasmic reticulum for esterification is unknown.

Besides its role in intestinal cholesterol absorption, NPC1L1 also functions as a hepatic cholesterol transporter. The overexpression of human NPC1L1 in mice liver was shown to reduce biliary cholesterol and to increase plasma cholesterol, providing evidence for a role of NPC1L1 in biliary cholesterol reabsorption.[87] Also, treatment with ezetimibe reverses these effects, suggesting that hepatic NPC1L1 could be targeted by ezetimibe in humans to treat hepatic steatosis.[81,88]

4. TRANSCRIPTIONAL REGULATION OF NPC1L1

Owing to the importance of NPC1L1 in cholesterol absorption, studies over the past decade have rigorously investigated the transcriptional regulation of NPC1L1 and showed that *NPC1L1* gene expression is regulated by different transcription factors as presented in the following sections (Table 1).

4.1 Sterol Response Element-Binding Proteins

Sterol response element-binding proteins (SREBPs) belong to the basic-helix-loop-helix-lucine-zipper family of transcription factors, which control the expression of genes involved in cholesterol and fatty acid synthesis and thereby act as master regulator of

TABLE 1 Transcription Factors and Regulation of NPC1L1 Expression

Transcription factor	Model system		NPC1L1 expression	References
SREBP2	Cell line	Caco2	↑	97–99
		HUH7	↑	
	Human	Duodenal biopsy	↑	100
HNF4α + SREBP2	Cell line	HepG2	↑	104
HNF1α	Cell line	HuH7	↑	99
HNF4α	Human	Duodenal biopsy	↑	100
PPARα	Cell line	HepG2	↑	107
	Mouse	Small intestine	↓	108
PPARδ	Cell line	Caco2	↓	114
	Rat	Cholangiocytes	↑	109
LXR	Cell line	Caco2	↓	116,117
	Mouse	Duodenum	↓	116

cholesterol and lipid homeostasis.[89] In the inactive precursor form, SREBPs are located in the endoplasmic reticulum membrane and on activation undergoes cleavage process to release the nuclear SREBPs that influence gene transcription. To date, three isoforms of SREBPs have been identified: SREBP2, SREBP1c, and SREBP1a.[90,91] Studies using various transgenic mice models have enabled identification of the different functional roles played by these transcription factors. For example, SREBP1c has been shown to be involved in fatty acid synthesis, and SREBP2 is involved in cholesterol synthesis. The role of SREBP1a is implicated in both of these pathways.[89,92–95] A mutation was identified in the regulatory domain of human SREBP2 that disrupts its cleavage, resulting in decreased expression of LDLR and increased plasma cholesterol, indicating the importance of this transcription factor in cholesterol metabolism.[96] In this regard, studies in our laboratory[97] showed that SREBP2 mediates the effects of cholesterol on intestinal NPC1L1 expression and promoter activity. We have shown that NPC1L1 expression and promoter activity in intestinal epithelial Caco2 cells were increased under cholesterol depletion conditions and decreased by treatment of intestinal epithelial cells with 25-hydroxycholesterol. We have further identified two *cis* elements spanning the regions 35/−26 bp and −657/−648 bp in human NPC1L1 promoter that bind to SREBP2. Studies from our laboratory further showed that SREBP2 mediates the modulation of NPC1L1 regulation by curcumin, a South Asian spice that is well known to reduce intestinal cholesterol absorption. Our studies showed that the overexpression of SREBP2 alleviated the inhibitory effect of curcumin on intestinal NPC1L1 expression.[98]

The role of SREBP2 in the regulation of NPC1L1 was further confirmed in heptocytes. Chromatin immunoprecipitation assay (ChIP) confirmed that SREBP2 binds to both sterol regulatory elements in NPC1L1 promoter from samples derived from human liver.[99]

Recent study in human duodenal biopsies from hyperlipidemic men illustrated the link between NPC1L1 expression and SREBP2. Findings from this study suggest that administration of atorvastatin (an inhibitor of HMG-CoA reductase) to hyperlipidemic subjects inhibited cholesterol synthesis and increased the expression of intestinal NPC1L1 and SREBP2 mRNA. This study demonstrated a positive correlation between the expression of SREBP2 and NPC1L1 in the human intestine.[100] It should be noted that the SREBP2 pathway is disrupted in the kidney of diabetic rats leading to renal cholesterol accumulation.[101] It remains interesting to determine if changes in intestinal SREBP2 occur in diabetes mellitus to alter NPC1L1 expression and cholesterol absorption.

4.2 Hepatocytes Nuclear Factor

Hepatocytes nuclear factor (HNF) 4α, an orphan member of the nuclear receptor family, is an important regulator of lipid metabolism expressed widely in various tissues including the liver, kidney, pancreas, and intestine. Knockdown of HNF4α in mouse liver results in increased hepatic cholesterol accumulation and decreased serum cholesterol and triglycerides levels, suggesting that HNF4α is a major regulator of genes involved in hepatic lipid metabolism.[102] Previously, in hepatocytes, the interaction of HNF4α with SREBP2 was shown to regulate the expression of genes involved in cholesterol metabolism.[103] Further studies implicated the role of HNF4α with SREBP2 in regulating cholesterol absorption. For instance, studies in HepG2 cells showed that HNF4α is crucial for the transcriptional regulation of NPC1L1 since co-transfection of HNF4α and SREBP2 significantly increased the NPC1L1 promoter activity, whereas HNF4α alone had no effect. HNF4α exerts its effect on NPC1L1 promoter activity by directly binding to its putative binding sites located between −209 and −197 and between −52 and −40 of NPC1L1 promoter. These effects were further confirmed by attenuating the expression of HNF4α, which resulted in a significant decrease in NPC1L1 mRNA expression.[104] HNF4α is also a positive regulator of HNF1α transcription factor, and studies by Pramfalk et al.[99] in HuH7 cells showed that HNF1α can directly regulate the NPC1L1 promoter activity as well as mRNA expression. Also, these studies confirmed the presence of important binding site of HNF1α (−158/−144) in the NPC1L1 promoter, the mutation of which completely abolished the regulatory effect of HNF1α on NPC1L1. Further, an in vivo study in human liver also confirmed the binding of HNF1α with NPC1L1 demonstrating that both HNFs transcriptionally regulate NPC1L1 expression and thereby act as modulator of lipid homeostasis.[99] The majority of HNF-related studies have been done in hepatic cell lines and liver with limited studies available in intestine. Only a recent study using human duodenal biopsy samples showed a positive correlation between HNF4α and NPC1L1 expression.[100]

4.3 Peroxisome Proliferator-Activated Receptor

Peroxisome proliferator-activated receptors (PPARs) are nuclear receptors that regulate the expression of genes by binding to PPRE (PPAR response elements) present within the promoter region of targeted genes. A number of PPAR isoforms have been described including PPARα, PPARβ, PPAR δ, and PPARγ that are encoded by different genes.[105] Upon activation, PPARs form heterodimer with RXR and this complex then interact with cofactors to stimulate gene transcription.[106] PPARs also act as a cholesterol sensor and thereby regulates lipid metabolism.[105] PPARα is highly expressed in the liver and intestine, and studies in HepG2 cells showed that NPC1L1 expression is transcriptionally regulated by PPARα. These studies identified PPARα responsive elements (PPRE) located between −846 and −834 nucleotide of human NPC1L1 promoter and showed that PPARα induces NPC1L1 expression by directly binding to PPRE. Also, PPARγ cofactor 1-α (PGC1-α) has a stimulatory effect on PPARα/RXRα-mediated transactivation of human NPC1L1 promoter.[107] In contrast, studies in mice[108] showed that PPARα is a negative regulator of NPC1L1 as treatment with the PPARα agonist fenofibrate decreased NPC1L1 expression in wild-type mice but not in PPARα knockout mice, suggesting that species differences in PPARα-mediated regulation of NPC1L1 may exist. It should be noted that PPRE were not identified in mouse NPC1L1 promoter.

In a recent study in murine cholangiocytes, PPARδ agonist (GW501516) was shown to induce NPC1L1 mRNA and protein expressions. Similar results were obtained by overexpression of PPARδ, whereas attenuating PPARδ expression decreased NPC1L1. A putative binding site for PPARδ was identified at −142 bp upstream from the transcription start site of the NPC1L1 promoter, and ChIP assays confirmed the binding. Moreover, mutation of this binding site ameliorated the effect of PPARδ on NPC1L1 expression.[109]

Docosahexaenoic acid (DHA) is a fish oil–derived fatty acid known to decrease cholesterol absorption as well as plasma cholesterol in both humans and animals.[110–113] DHA was shown to decrease the expression of NPC1L1 mRNA and protein concomitant with a decrease in the cholesterol trafficking to the endoplasmic reticulum in intestinal Caco2 cells.[114] Additionally, DHA-fed hamsters demonstrated a decreased expression of NPC1L1 in duodenum and jejunum. In these studies, PPARδ agonist decreased NPC1L1 expression similar to the effects of DHA.[114]

4.4 Liver X Receptor

Liver X receptor (LXR) α and LXRβ belong to nuclear receptor family which when activated form a heterodimer with RXR (retinoid X receptor). This heterodimer complex binds to the promoter region of its targeted genes and increases their transcription. LXRα and LXRβ also act as cholesterol sensor and thereby regulate cholesterol homeostasis.[115] Data from both in vitro (Caco2 cells) and in vivo studies showed that treatment with LXR agonist downregulates NPC1L1 expression, suggesting that NPC1L1 is a novel target of LXR.[116,117] Further, bioinformatic analysis identified two potential binding sites for LXR in the NPC1L1 promoter region, suggesting a direct regulation of NPC1L1 by LXR.[117] However, limited information exists regarding the transcriptional regulation of NPC1L1 by LXR. More studies are needed to explore the detailed mechanism of LXR-mediated transcriptional regulation of NPC1L1 both in vitro and in vivo.

5. NPC1L1 AND DISEASES

Since NPC1L1 is a major player in intestinal cholesterol absorption, the implication of increased NPC1L1 has been associated with various diseases discussed in the following sections.

5.1 Atherosclerosis and Coronary Heart Diseases

As there is a positive correlation between blood cholesterol levels and incidence of atherosclerotic events, elegant studies explored the roles of NPC1L1 in $ApoE^{-/-}$ knockout mice, an excellent animal model for atherosclerosis.[118] These studies showed that ezetimibe reduces plasma cholesterol and inhibits atherogenesis in $ApoE^{-/-}$ mice, accompanied with a decrease in cholesterol in the VLDL and LDL fractions and an increase in cholesterol in the HDL fractions. Ezetimibe treatment also reduced aortic atherosclerotic lesions surface area in these, mice suggesting that reducing cholesterol absorption inhibits the development of atherosclerosis. These observations were further confirmed in *NPC1L1/ApoE* double knockout mice, demonstrating a significant reduction in atherosclerosis and confirming the beneficial effects of suppressing NPC1L1 in the protection from the development of atherosclerosis.[119]

Also, a recent human study recruited samples from large group of individuals has identified 15 naturally occurring inactivating mutations in NPC1L1. Statistical analysis showed that the inactivating mutations are associated with a reduction in plasma LDL cholesterol and a significant decrease in the risk of coronary heart disease.[66] These novel findings support the notion that the inhibition of NPC1L1 reduces the risk for cardiovascular diseases associated with diseases such as diabetes mellitus.

5.2 HCV Infection

A novel role of NPC1L1 as an antiviral target was recently demonstrated. NPC1L1 transporter was recognized as a critical factor in hepatitis C virus

(HCV) entry in HuH7 cells (a human hepatoma cell line). Further, the expression of NPC1L1 was shown to be necessary for HCV infection since its inhibition by silencing or by blocking antibody as well as ezetimibe inhibited the HCV uptake in a cell culture model in a cholesterol-dependent manner. The inhibition of NPC1L1 caused a delayed HCV infection in mouse model supporting the novel role of NPC1L1 in viral pathogenesis.[120]

6. NPC1L1 AND DIABETES

6.1 Intestinal Cholesterol Absorption and Diabetes Mellitus

Multiple studies have previously attempted to determine if cholesterol absorption is altered in individuals with either type 1 diabetes (T1D) or type 2 diabetes (T2D). There was some degree of discrepancy between the results of these studies that could be attributed to many factors related to the fact that the direct measurement of cholesterol absorption in humans is difficult, time consuming, and expensive.[121] T2D patients exhibit postprandial dyslipidemia[122] and increased levels of ApoB48–containing chylomicrons in the bloodstream, which is considered to be atherogeneic.[123] However, it appears that an increase in intestinal cholesterol absorption may not occur in patients with T2D.[1,124] On the other hand, studies indicate that cholesterol absorption may be increased in patients with T1D. In this regard, Miettinen and colleagues[125] investigated the surrogate markers for cholesterol synthesis and absorption in patients with T1D and compared the results with T2D patients. The efficiency of cholesterol absorption was assessed by measuring the levels of PSs such as cholestanol, campesterol, and sitosterol, and the rate of cholesterol synthesis was evaluated using cholesterol precursors such as cholestenol, desmosterol, and lathosterol as surrogate markers. Results from their findings suggested an increase in the cholesterol absorption and a decrease in cholesterol synthesis in T1D patients compared with T2D. Despite the fact that PSs as surrogate markers may not be altered in T2D, the roles of deregulated intestinal processes in the pathophysiology of associated cholesterol-related disorders cannot be ruled out.

It should be noted that the observed postprandial increase in ApoB48 particles in diabetes mellitus may reflect an increase in chylomicrons synthesis. Both the availability of triglycerides and cholesterol is required in the intestine for chylomicrons synthesis.[29] Indeed, studies showed that treatment with simvastatin and ezetimibe together showed a 50% reductions in postprandial ApoB48 chylomicrons in subjects with T2D and hypercholesterolemia.[126] Ezetimibe treatment was also shown to decrease cholesterol content of the chylomicrons and ratio of triglyceride to cholesterol in chylomicrons, suggesting that the inhibition of cholesterol absorption may correct the dyslipidemia associated with diabetes mellitus.[126]

6.2 NPC1L1 Expression and Diabetes Mellitus

Previous study by Lally et al.[127] showed an increase in mRNA expression of NPC1L1 and MTTP (microsomal triglyceride transport protein, involved in chylomicron synthesis) and a decrease in ABCG5 and G8 mRNA expression in diabetic patients compared with nondiabetic controls, suggesting that the processes of cholesterol absorption and excretion are dysregulated in diabetes mellitus. Also, these studies demonstrated a positive correlation between NPC1L1 and MTTP expression and a negative correlation between NPC1L1 and ABCG5 mRNA expression in diabetic patients. These findings suggest that increased cholesterol absorption and decreased cholesterol efflux results in increased chylomicron cholesterol.

The role of intestinal and hepatic NPC1L1 was further explored using streptozotocin-induced diabetes mellitus in rats.[128] The results of this study showed that the levels of triglycerides, cholesterol, ApoB48, and ApoB100 were significantly higher in chylomicrons and VLDL fractions of diabetic rats. The expression of NPC1L1 and MTTP were increased and the expression of ABCG5 and G8 were decreased in both intestine and liver of diabetic rats. A positive correlation between intestinal NPC1L1 mRNA levels and chylomicron cholesterol and hepatic NPC1L1 mRNA levels and VLDL cholesterol was observed. These findings suggested that an increase in both hepatic and intestinal NPC1L1 gene expression is responsible for abnormal levels of VLDL and chylomicrons in diabetes mellitus. It is interesting to mention that recent studies demonstrated that ezetimibe treatment improved glycemic control and increased the secretion of intestinal glucagon-like peptide-1.[129] These findings imply a relationship between cholesterol absorption and insulin sensitivity. In fact, a study showed that NPC1L1 knockout mice were shown to have increased insulin sensitivity compared with wild-type mice, providing evidence for the link between NPC1L1-mediated pathway and insulin sensitivity.[130]

The effects of alterations in the levels of glucose on NPC1L1 were also investigated. Recent studies utilizing intestinal Caco2 cells showed that incubation with high glucose (25 mmol/L) significantly increased ezetimibe-sensitive cholesterol uptake compared with low glucose (5 mmol/L).[131] Treatment with structurally related compounds such as sorbitol and mannitol had no effect on cholesterol uptake, proving that the increase in cholesterol uptake is specific to glucose. The studies demonstrated an increase in NPC1L1 and CD36 mRNA expression. Expression of SR-B1 was decreased in response to high glucose, with no significant differences in the expression of ABCA1 and ABCG8. Incubation with high levels of glucose was associated with a reduction in the activity of HMG-CoA-R and an increase in the expression of various transcription factors such as LXRα, LXRβ, PPARβ, and PPARδ. The observed changes in these transcription factors may be involved in altering the expression of cholesterol transporters and cholesterol uptake.

Recent studies from our laboratory showed that D-glucose directly regulates *NPC1L1* gene expression by transcriptional mechanisms.[132] Our studies made a novel observation that the removal of glucose from culture medium of Caco2 cells resulted in a decrease in NPC1L1 mRNA, protein expression, and function (Figure 2). On the other hand, replenishment of glucose to the cells after its depletion led to an increase in NPC1L1 expression in a dose-dependent manner. The removal or addition of glucose in the intestinal cells was associated with the changes in NPC1L1 promoter activity, suggesting the involvement of transcriptional regulation. The stimulatory effects of glucose on NPC1L1 expression were dependent on its cellular metabolism as equal concentration of nonmetabolizable form of glucose, 3-*o*-methyl-glucopyranose (OMG) showed no effect on NPC1L1 promoter activity. Once the glucose is metabolized, it activates various pathways such as glycolytic pathway, hexose monophosphate shunt pathway (HMP), or hexosamine biosynthetic pathway (HBP). Metabolites of the HMP pathway such as xylulose-5-phosphate stimulate protein phosphatase 2A (PP2A) that in turn activates certain transcription factors such as carbohydrate response element-binding protein. Interestingly, the inhibition of PP2A by the pharmacological inhibitor okadaic acid blocked glucose-mediated increase in NPC1L1 promoter activity. Okadaic acid also resulted in a decrease in the basal activity of NPC1L1 promoter in the presence of glucose, suggesting that glucose stimulates basal NPC1L1 expression in a PP2A--dependent manner (Figure 2). The effect of addition or removal of glucose on NPC1L1 expression was shown to be mediated by *cis* elements located in a region between −291 and +56 of the NPC1L1 promoter. The effects of glucose on NPC1L1 promoter activity are independent of SREBP2. Similar to the findings in cell culture system, overnight fasting in mice decreased NPC1L1 expression in the small intestine. These observations suggested that depletion of ATP during fasting decreases NPC1L1 expression similar to effects of glucose withdrawal on NPC1L1 expression in the cell culture system. Collectively, these studies illustrated a novel link between glucose metabolism and cholesterol absorption and suggest that hyperglycemia may cause an increase in NPC1L1 expression seen in diabetic patients as well as in diabetic animal models.

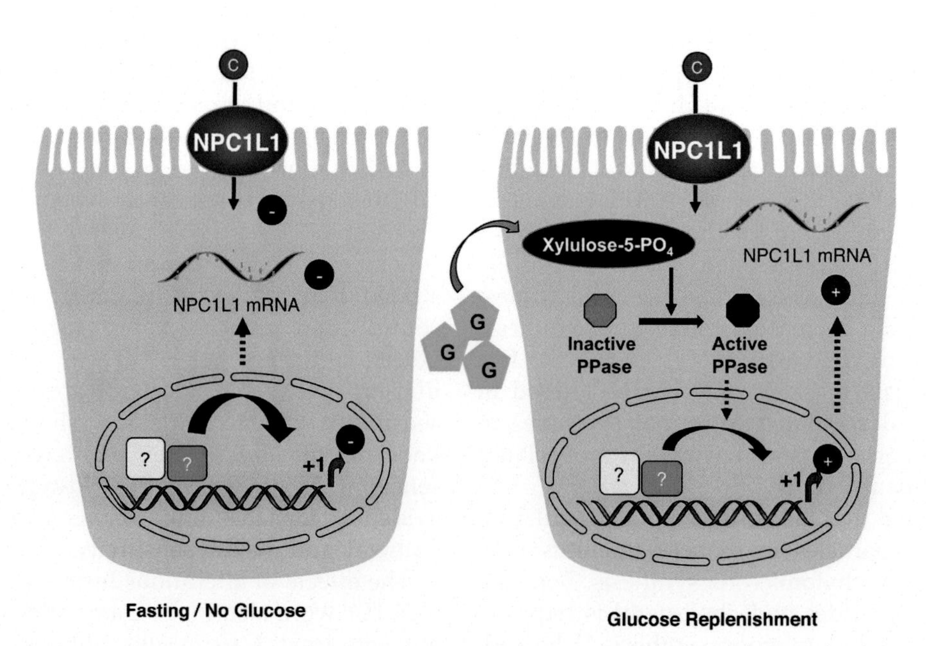

FIGURE 2 Effects of glucose on NPC1L1 expression in intestinal epithelial cells. The depletion of glucose or fasting in mice decreases the expression and the promoter activity of NPC1L1. The replenishment of glucose after depletion stimulates NPC1L1 promoter activity in protein phosphatase-dependent pathway. The increase in the promoter activity results in an increase in NPC1L1 mRNA and protein expression as well as cholesterol uptake.

7. CONCLUSION

Intestinal cholesterol absorption plays a central role in controlling cholesterol levels in mammals. Multiple sterol transport proteins are expressed in intestinal epithelial cells including SR-B1, CD36, and NPC1L1. However, studies in knockout mice established the indispensible role for NPC1L1 in mediating the process of cholesterol absorption. Also, targeting NPC1L1 function represents the basis for blocking cholesterol absorption by the widely used cholesterol-lowering drug ezetimibe. Inactivating mutations in NPC1L1 are associated with lower levels of plasma cholesterol and reduced risk for cardiovascular diseases. Therefore, NPC1L1 appears to be a prime target for therapeutic intervention to lower plasma cholesterol in high-risk patients such as patients with diabetes mellitus. This notion is further supported by the observations that NPC1L1 expression is increased in diabetic patients and in animal models of diabetes mellitus. Also, the lack of NPC1L1 expression in knockout mice increased insulin sensitivity linking NPC1L1-mediated pathway to insulin function and glucose metabolism. There was, however, a lacuna in understanding of how NPC1L1 is increased in diabetes mellitus until few recent studies showing that high concentrations of glucose induce NPC1L1 expression and promoter activity. Given the central role of NPC1L1 in the pathophysiology of complications associated with diabetes mellitus, it is imperative to determine the exact mechanisms involved in the stimulation of its expression in diabetes mellitus. Attempts to design more efficient treatment to lower plasma cholesterol will benefit from future studies to dissect the roles of transcription factors such as SREBP2, LXR, and PPARs in the increase of NPC1L1 associated with diabetes mellitus.

Acknowledgment

Studies in the author's laboratories are funded by grants from the Veterans Administration: BX000152 (WAA) and BX002011 (PKD) and grants from the National Institutes of Health: DK71596 (WAA), DK98170 (RKG), and DK54016, DK81858, DK992441 (PKD).

References

1. Gleeson A, Owens D, Collins P, Johnson A, Tomkin GH. The relationship between cholesterol absorption and intestinal cholesterol synthesis in the diabetic rat model. *Int J Exp Diabetes Res* 2000;**1**(3): 203–10.
2. Stone NJ, Robinson JG, Lichtenstein AH, Bairey Merz CN, Blum CB, Eckel RH, et al. 2013 ACC/AHA guideline on the treatment of blood cholesterol to reduce atherosclerotic cardiovascular risk in adults: a report of the American College of Cardiology/ American Heart Association Task Force on Practice Guidelines. *Circulation* 2014;**129**(25 Suppl. 2):S1–45.
3. Ikonen E. Cellular cholesterol trafficking and compartmentalization. *Nat Rev Mol Cell Biol* 2008;**9**(2):125–38.
4. Goedeke L, Fernandez-Hernando C. Regulation of cholesterol homeostasis. *Cell Mol Life Sci* 2012;**69**(6):915–30.
5. Cook RP. *Cholesterol: chemistry, biochemistry and pathology.* New York: Academic Press; 1958.
6. Tomkin GH. Dyslipidaemia–hepatic and intestinal cross-talk. *Atheroscler Suppl* 2010;**11**(1):5–9.
7. Devery RA, O'Meara N, Collins PB, Johnson AH, Scott L, Tomkin GH. A comparative study of the rate-limiting enzymes of cholesterol synthesis, esterification and catabolism in the rat and rabbit. *Comp Biochem Physiol B* 1987;**87**(4):697–702.
8. Lally S, Owens D, Tomkin GH. The different effect of pioglitazone compared with insulin on expression of hepatic and intestinal genes regulating post-prandial lipoproteins in diabetes. *Atherosclerosis* 2007;**193**(2):343–51.
9. Reiner Z. Resistance and intolerance to statins. *Nutr Metab Cardiovasc Dis* 2014;**24**(10):1057–66.
10. Nair SVG, Wang Y. A new perspective on the development of cholesterol-lowering products. *Intech* 2013.
11. Alrefai WA, Gill RK. Bile acid transporters: structure, function, regulation and pathophysiological implications. *Pharm Res* 2007; **24**(10):1803–23.
12. Bennion LJ, Grundy SM. Effects of diabetes mellitus on cholesterol metabolism in man. *N Engl J Med* 1977;**296**(24):1365–71.
13. Chen L, Yao X, Young A, McNulty J, Anderson D, Liu Y, et al. Inhibition of apical sodium-dependent bile acid transporter as a novel treatment for diabetes. *Am J Physiol Endocrinol Metab* 2012; **302**(1):E68–76.
14. van der Wulp MY, Verkade HJ, Groen AK. Regulation of cholesterol homeostasis. *Mol Cell Endocrinol* 2013;**368**(1–2):1–16.
15. Dikkers A, Tietge UJ. Biliary cholesterol secretion: more than a simple ABC. *World J Gastroenterol* 2010;**16**(47):5936–45.
16. Oude Elferink RP, Ottenhoff R, van Wijland M, Smit JJ, Schinkel AH, Groen AK. Regulation of biliary lipid secretion by mdr2 P-glycoprotein in the mouse. *J Clin Invest* 1995;**95**(1): 31–8.
17. Brufau G, Groen AK, Kuipers F. Reverse cholesterol transport revisited: contribution of biliary versus intestinal cholesterol excretion. *Arterioscler Thromb Vasc Biol* 2011;**31**(8):1726–33.
18. Temel RE, Brown JM. A new framework for reverse cholesterol transport: non-biliary contributions to reverse cholesterol transport. *World J Gastroenterol* 2010;**16**(47):5946–52.
19. DenBesten L, Reyna RH, Connor WE, Stegink LD. The different effects on the serum lipids and fecal steroids of high carbohydrate diets given orally or intravenously. *J Clin Invest* 1973;**52**(6): 1384–93.
20. Tietge UJ, Groen AK. Role the TICE?: advancing the concept of transintestinal cholesterol excretion. *Arterioscler Thromb Vasc Biol* 2013;**33**(7):1452–3.
21. van der Veen JN, van Dijk TH, Vrins CL, van Meer H, Havinga R, Bijsterveld K, et al. Activation of the liver X receptor stimulates trans-intestinal excretion of plasma cholesterol. *J Biol Chem* 2009; **284**(29):19211–9.
22. Marshall SM, Kelley KL, Davis MA, Wilson MD, McDaniel AL, Lee RG, et al. Reduction of VLDL secretion decreases cholesterol excretion in Niemann-Pick C1-like 1 hepatic transgenic mice. *PLoS One* 2014;**9**(1):e84418.
23. Wang DQ. Regulation of intestinal cholesterol absorption. *Annu Rev Physiol* 2007;**69**:221–48.
24. Grundy SM, Metzger AL. A physiological method for estimation of hepatic secretion of biliary lipids in man. *Gastroenterology* 1972; **62**(6):1200–17.
25. Sudhop T, Lutjohann D, Kodal A, Igel M, Tribble DL, Shah S, et al. Inhibition of intestinal cholesterol absorption by ezetimibe in humans. *Circulation* 2002;**106**(15):1943–8.

26. Bosner MS, Lange LG, Stenson WF, Ostlund Jr RE. Percent cholesterol absorption in normal women and men quantified with dual stable isotopic tracers and negative ion mass spectrometry. *J Lipid Res* 1999;**40**(2):302–8.
27. Sehayek E, Nath C, Heinemann T, McGee M, Seidman CE, Samuel P, et al. U-shape relationship between change in dietary cholesterol absorption and plasma lipoprotein responsiveness and evidence for extreme interindividual variation in dietary cholesterol absorption in humans. *J Lipid Res* 1998;**39**(12):2415–22.
28. Iqbal J, Hussain MM. Evidence for multiple complementary pathways for efficient cholesterol absorption in mice. *J Lipid Res* 2005;**46**(7):1491–501.
29. Abumrad NA, Davidson NO. Role of the gut in lipid homeostasis. *Physiol Rev* 2012;**92**(3):1061–85.
30. Altmann SW, Davis Jr HR, Zhu LJ, Yao X, Hoos LM, Tetzloff G, et al. Niemann-Pick C1 like 1 protein is critical for intestinal cholesterol absorption. *Science* 2004;**303**(5661):1201–4.
31. Graf GA, Yu L, Li WP, Gerard R, Tuma PL, Cohen JC, et al. ABCG5 and ABCG8 are obligate heterodimers for protein trafficking and biliary cholesterol excretion. *J Biol Chem* 2003;**278**(48):48275–82.
32. Klett EL, Lee MH, Adams DB, Chavin KD, Patel SB. Localization of ABCG5 and ABCG8 proteins in human liver, gall bladder and intestine. *BMC Gastroenterol* 2004;**4**:21.
33. Yu L, Hammer RE, Li-Hawkins J, Von Bergmann K, Lutjohann D, Cohen JC, et al. Disruption of Abcg5 and Abcg8 in mice reveals their crucial role in biliary cholesterol secretion. *Proc Natl Acad Sci USA* 2002;**99**(25):16237–42.
34. Repa JJ, Berge KE, Pomajzl C, Richardson JA, Hobbs H, Mangelsdorf DJ. Regulation of ATP-binding cassette sterol transporters ABCG5 and ABCG8 by the liver X receptors alpha and beta. *J Biol Chem* 2002;**277**(21):18793–800.
35. Plosch T, Kruit JK, Bloks VW, Huijkman NC, Havinga R, Duchateau GS, et al. Reduction of cholesterol absorption by dietary plant sterols and stanols in mice is independent of the Abcg5/8 transporter. *J Nutr* 2006;**136**(8):2135–40.
36. Hauser H, Dyer JH, Nandy A, Vega MA, Werder M, Bieliauskaite E, et al. Identification of a receptor mediating absorption of dietary cholesterol in the intestine. *Biochemistry* 1998;**37**(51):17843–50.
37. Cai SF, Kirby RJ, Howles PN, Hui DY. Differentiation-dependent expression and localization of the class B type I scavenger receptor in intestine. *J Lipid Res* 2001;**42**(6):902–9.
38. Lobo MV, Huerta L, Ruiz-Velasco N, Teixeiro E, de la Cueva P, Celdran A, et al. Localization of the lipid receptors CD36 and CLA-1/SR-BI in the human gastrointestinal tract: towards the identification of receptors mediating the intestinal absorption of dietary lipids. *J Histochem Cytochem* 2001;**49**(10):1253–60.
39. Mardones P, Quinones V, Amigo L, Moreno M, Miquel JF, Schwarz M, et al. Hepatic cholesterol and bile acid metabolism and intestinal cholesterol absorption in scavenger receptor class B type I-deficient mice. *J Lipid Res* 2001;**42**(2):170–80.
40. Beaslas O, Cueille C, Delers F, Chateau D, Chambaz J, Rousset M, et al. Sensing of dietary lipids by enterocytes: a new role for SR-BI/CLA-1. *PLoS One* 2009;**4**(1):e4278.
41. Nassir F, Wilson B, Han X, Gross RW, Abumrad NA. CD36 is important for fatty acid and cholesterol uptake by the proximal but not distal intestine. *J Biol Chem* 2007;**282**(27):19493–501.
42. Nauli AM, Nassir F, Zheng S, Yang Q, Lo CM, Vonlehmden SB, et al. CD36 is important for chylomicron formation and secretion and may mediate cholesterol uptake in the proximal intestine. *Gastroenterology* 2006;**131**(4):1197–207.
43. Nguyen DV, Drover VA, Knopfel M, Dhanasekaran P, Hauser H, Phillips MC. Influence of class B scavenger receptors on cholesterol flux across the brush border membrane and intestinal absorption. *J Lipid Res* 2009;**50**(11):2235–44.
44. Oram JF, Lawn RM. ABCA1. The gatekeeper for eliminating excess tissue cholesterol. *J Lipid Res* 2001;**42**(8):1173–9.
45. Hoffman HN, Fredrickson DS. Tangier disease (familial high density lipoprotein deficiency). Clinical and genetic features in two adults. *Am J Med* 1965;**39**(4):582–93.
46. Kang MH, Singaraja R, Hayden MR. Adenosine-triphosphate-binding cassette transporter-1 trafficking and function. *Trends Cardiovasc Med* 2010;**20**(2):41–9.
47. Brunham LR, Kruit JK, Iqbal J, Fievet C, Timmins JM, Pape TD, et al. Intestinal ABCA1 directly contributes to HDL biogenesis in vivo. *J Clin Invest* 2006;**116**(4):1052–62.
48. Field FJ, Watt K, Mathur SN. Origins of intestinal ABCA1-mediated HDL-cholesterol. *J Lipid Res* 2008;**49**(12):2605–19.
49. Xie C, Zhou ZS, Li N, Bian Y, Wang YJ, Wang LJ, et al. Ezetimibe blocks the internalization of NPC1L1 and cholesterol in mouse small intestine. *J Lipid Res* 2012;**53**(10):2092–101.
50. Garcia-Calvo M, Lisnock J, Bull HG, Hawes BE, Burnett DA, Braun MP, et al. The target of ezetimibe is Niemann-Pick C1-Like 1 (NPC1L1). *Proc Natl Acad Sci USA* 2005;**102**(23):8132–7.
51. Clader JW. The discovery of ezetimibe: a view from outside the receptor. *J Med Chem* 2004;**47**(1):1–9.
52. Davis Jr HR, Zhu LJ, Hoos LM, Tetzloff G, Maguire M, Liu J, et al. Niemann-Pick C1 Like 1 (NPC1L1) is the intestinal phytosterol and cholesterol transporter and a key modulator of whole-body cholesterol homeostasis. *J Biol Chem* 2004;**279**(32):33586–92.
53. Bays HE, Moore PB, Drehobl MA, Rosenblatt S, Toth PD, Dujovne CA, et al. Effectiveness and tolerability of ezetimibe in patients with primary hypercholesterolemia: pooled analysis of two phase II studies. *Clin Ther* 2001;**23**(8):1209–30.
54. Knopp RH, Gitter H, Truitt T, Bays H, Manion CV, Lipka LJ, et al. Effects of ezetimibe, a new cholesterol absorption inhibitor, on plasma lipids in patients with primary hypercholesterolemia. *Eur Heart J* 2003;**24**(8):729–41.
55. Yunoki K, Nakamura K, Miyoshi T, Enko K, Kohno K, Morita H, et al. Ezetimibe improves postprandial hyperlipemia and its induced endothelial dysfunction. *Atherosclerosis* 2011;**217**(2):486–91.
56. Gagne C, Bays HE, Weiss SR, Mata P, Quinto K, Melino M, et al. Efficacy and safety of ezetimibe added to ongoing statin therapy for treatment of patients with primary hypercholesterolemia. *Am J Cardiol* 2002;**90**(10):1084–91.
57. Kerzner B, Corbelli J, Sharp S, Lipka LJ, Melani L, LeBeaut A, et al. Efficacy and safety of ezetimibe coadministered with lovastatin in primary hypercholesterolemia. *Am J Cardiol* 2003;**91**(4):418–24.
58. Melani L, Mills R, Hassman D, Lipetz R, Lipka L, LeBeaut A, et al. Efficacy and safety of ezetimibe coadministered with pravastatin in patients with primary hypercholesterolemia: a prospective, randomized, double-blind trial. *Eur Heart J* 2003;**24**(8):717–28.
59. Davidson MH, McGarry T, Bettis R, Melani L, Lipka LJ, LeBeaut AP, et al. Ezetimibe coadministered with simvastatin in patients with primary hypercholesterolemia. *J Am Coll Cardiol* 2002;**40**(12):2125–34.
60. Ballantyne CM, Houri J, Notarbartolo A, Melani L, Lipka LJ, Suresh R, et al. Effect of ezetimibe coadministered with atorvastatin in 628 patients with primary hypercholesterolemia: a prospective, randomized, double-blind trial. *Circulation* 2003;**107**(19):2409–15.
61. West AM, Anderson JD, Meyer CH, Epstein FH, Wang H, Hagspiel KD, et al. The effect of ezetimibe on peripheral arterial atherosclerosis depends upon statin use at baseline. *Atherosclerosis* 2011;**218**(1):156–62.
62. Kastelein JJ, Akdim F, Stroes ES, Zwinderman AH, Bots ML, Stalenhoef AF, et al. Simvastatin with or without ezetimibe in familial hypercholesterolemia. *N Engl J Med* 2008;**358**(14):1431–43.
63. Park SW. Intestinal and hepatic Niemann-Pick c1-like 1. *Diabetes Metab J* 2013;**37**(4):240–8.

64. Baigent C, Landray MJ, Reith C, Emberson J, Wheeler DC, Tomson C, et al. The effects of lowering LDL cholesterol with simvastatin plus ezetimibe in patients with chronic kidney disease (Study of Heart and Renal Protection): a randomised placebo-controlled trial. *Lancet* 2011;**377**(9784):2181–92.
65. Corbin LJ, Timpson NJ. Using inactivating mutations to provide insight into drug action. *Genome Med* 2015;**7**(1):7.
66. Stitziel NO, Won HH, Morrison AC. Inactivating mutations in NPC1L1 and protection from coronary heart disease. *N Engl J Med* 2014;**371**(22):2072–82.
67. Howell KL, DeVita RJ, Garcia-Calvo M, Meurer RD, Lisnock J, Bull HG, et al. Spiroimidazolidinone NPC1L1 inhibitors. Part 2: structure-activity studies and in vivo efficacy. *Bioorg Med Chem Lett* 2010;**20**(23):6929–32.
68. Bayes M. Gateways to clinical trials. *Methods Find Exp Clin Pharmacol* 2007;**29**(2):153–73.
69. Davies JP, Levy B, Ioannou YA. Evidence for a Niemann-pick C (NPC) gene family: identification and characterization of NPC1L1. *Genomics* 2000;**65**(2):137–45.
70. Jia L, Betters JL, Yu L. Niemann-pick C1-Like 1 (NPC1L1) protein in intestinal and hepatic cholesterol transport. *Annu Rev Physiol* 2011;**73**:239–59.
71. Davies JP, Scott C, Oishi K, Liapis A, Ioannou YA. Inactivation of NPC1L1 causes multiple lipid transport defects and protects against diet-induced hypercholesterolemia. *J Biol Chem* 2005; **280**(13):12710–20.
72. Xie P, Zhu H, Jia L, Ma Y, Tang W, Wang Y, et al. Genetic demonstration of intestinal NPC1L1 as a major determinant of hepatic cholesterol and blood atherogenic lipoprotein levels. *Atherosclerosis* 2014;**237**(2):609–17.
73. Davis Jr HR, Altmann SW. Niemann-Pick C1 Like 1 (NPC1L1) an intestinal sterol transporter. *Biochim Biophys Acta* 2009;**1791**(7): 679–83.
74. Yu L, Bharadwaj S, Brown JM, Ma Y, Du W, Davis MA, et al. Cholesterol-regulated translocation of NPC1L1 to the cell surface facilitates free cholesterol uptake. *J Biol Chem* 2006;**281**(10):6616–24.
75. Sane AT, Sinnett D, Delvin E, Bendayan M, Marcil V, Menard D, et al. Localization and role of NPC1L1 in cholesterol absorption in human intestine. *J Lipid Res* 2006;**47**(10):2112–20.
76. Wang J, Chu BB, Ge L, Li BL, Yan Y, Song BL. Membrane topology of human NPC1L1, a key protein in enterohepatic cholesterol absorption. *J Lipid Res* 2009;**50**(8):1653–62.
77. Kuwabara PE, Labouesse M. The sterol-sensing domain: multiple families, a unique role? *Trends Genet* 2002;**18**(4):193–201.
78. Kwon HJ, Palnitkar M, Deisenhofer J. The structure of the NPC1L1 N-terminal domain in a closed conformation. *PLoS One* 2011;**6**(4):e18722.
79. Cooper MK, Wassif CA, Krakowiak PA, Taipale J, Gong R, Kelley RI, et al. A defective response to Hedgehog signaling in disorders of cholesterol biosynthesis. *Nat Genet* 2003;**33**(4):508–13.
80. Betters JL, Yu L. NPC1L1 and cholesterol transport. *FEBS Lett* 2010;**584**(13):2740–7.
81. Wang LJ, Song BL. Niemann-Pick C1-like 1 and cholesterol uptake. *Biochim Biophys Acta* 2012;**1821**(7):964–72.
82. Ge L, Wang J, Qi W, Miao HH, Cao J, Qu YX, et al. The cholesterol absorption inhibitor ezetimibe acts by blocking the sterol-induced internalization of NPC1L1. *Cell Metab* 2008;**7**(6):508–19.
83. Weinglass AB, Kohler M, Schulte U, Liu J, Nketiah EO, Thomas A, et al. Extracellular loop C of NPC1L1 is important for binding to ezetimibe. *Proc Natl Acad Sci USA* 2008;**105**(32): 11140–5.
84. Zhang JH, Ge L, Qi W, Zhang L, Miao HH, Li BL, et al. The N-terminal domain of NPC1L1 protein binds cholesterol and plays essential roles in cholesterol uptake. *J Biol Chem* 2011; **286**(28):25088–97.
85. Ge L, Qi W, Wang LJ, Miao HH, Qu YX, Li BL, et al. Flotillins play an essential role in Niemann-Pick C1-like 1-mediated cholesterol uptake. *Proc Natl Acad Sci USA* 2011;**108**(2):551–6.
86. Li PS, Fu ZY, Zhang YY, Zhang JH, Xu CQ, Ma YT, et al. The clathrin adaptor Numb regulates intestinal cholesterol absorption through dynamic interaction with NPC1L1. *Nat Med* 2014;**20**(1): 80–6.
87. Temel RE, Tang W, Ma Y, Rudel LL, Willingham MC, Ioannou YA, et al. Hepatic Niemann-Pick C1-like 1 regulates biliary cholesterol concentration and is a target of ezetimibe. *J Clin Invest* 2007;**117**(7): 1968–78.
88. Tang W, Jia L, Ma Y, Xie P, Haywood J, Dawson PA, et al. Ezetimibe restores biliary cholesterol excretion in mice expressing Niemann-Pick C1-like 1 only in liver. *Biochim Biophys Acta* 2011; **1811**(9):549–55.
89. Horton JD, Shimomura I, Ikemoto S, Bashmakov Y, Hammer RE. Overexpression of sterol regulatory element-binding protein-1a in mouse adipose tissue produces adipocyte hypertrophy, increased fatty acid secretion, and fatty liver. *J Biol Chem* 2003;**278**(38): 36652–60.
90. Hua X, Sakai J, Ho YK, Goldstein JL, Brown MS. Hairpin orientation of sterol regulatory element-binding protein-2 in cell membranes as determined by protease protection. *J Biol Chem* 1995; **270**(49):29422–7.
91. Eberle D, Hegarty B, Bossard P, Ferre P, Foufelle F. SREBP transcription factors: master regulators of lipid homeostasis. *Biochimie* 2004;**86**(11):839–48.
92. Shimano H, Horton JD, Hammer RE, Shimomura I, Brown MS, Goldstein JL. Overproduction of cholesterol and fatty acids causes massive liver enlargement in transgenic mice expressing truncated SREBP-1a. *J Clin Invest* 1996;**98**(7):1575–84.
93. Horton JD, Shah NA, Warrington JA, Anderson NN, Park SW, Brown MS, et al. Combined analysis of oligonucleotide microarray data from transgenic and knockout mice identifies direct SREBP target genes. *Proc Natl Acad Sci USA* 2003;**100**(21): 12027–32.
94. Shimano H, Horton JD, Shimomura I, Hammer RE, Brown MS, Goldstein JL. Isoform 1c of sterol regulatory element binding protein is less active than isoform 1a in livers of transgenic mice and in cultured cells. *J Clin Invest* 1997;**99**(5):846–54.
95. Horton JD, Shimomura I, Brown MS, Hammer RE, Goldstein JL, Shimano H. Activation of cholesterol synthesis in preference to fatty acid synthesis in liver and adipose tissue of transgenic mice overproducing sterol regulatory element-binding protein-2. *J Clin Invest* 1998;**101**(11):2331–9.
96. Miserez AR, Muller PY, Barella L, Barella S, Staehelin HB, Leitersdorf E, et al. Sterol-regulatory element-binding protein (SREBP)-2 contributes to polygenic hypercholesterolaemia. *Atherosclerosis* 2002;**164**(1):15–26.
97. Alrefai WA, Annaba F, Sarwar Z, Dwivedi A, Saksena S, Singla A, et al. Modulation of human Niemann-Pick C1-like 1 gene expression by sterol: role of sterol regulatory element binding protein 2. *Am J Physiol Gastrointest Liver Physiol* 2007;**292**(1): G369–76.
98. Kumar P, Malhotra P, Ma K, Singla A, Hedroug O, Saksena S, et al. SREBP2 mediates the modulation of intestinal NPC1L1 expression by curcumin. *Am J Physiol Gastrointest Liver Physiol* 2011; **301**(1):G148–55.
99. Pramfalk C, Jiang ZY, Cai Q, Hu H, Zhang SD, Han TQ, et al. HNF1alpha and SREBP2 are important regulators of NPC1L1 in human liver. *J Lipid Res* 2010;**51**(6):1354–62.
100. Tremblay AJ, Lamarche B, Lemelin V, Hoos L, Benjannet S, Seidah NG, et al. Atorvastatin increases intestinal expression of NPC1L1 in hyperlipidemic men. *J Lipid Res* 2011;**52**(3): 558–65.

101. Sun H, Yuan Y, Sun ZL. Cholesterol contributes to diabetic nephropathy through SCAP-SREBP-2 pathway. *Int J Endocrinol* 2013;**2013**:592576.
102. Hayhurst GP, Lee YH, Lambert G, Ward JM, Gonzalez FJ. Hepatocyte nuclear factor 4alpha (nuclear receptor 2A1) is essential for maintenance of hepatic gene expression and lipid homeostasis. *Mol Cell Biol* 2001;**21**(4):1393–403.
103. Misawa K, Horiba T, Arimura N, Hirano Y, Inoue J, Emoto N, et al. Sterol regulatory element-binding protein-2 interacts with hepatocyte nuclear factor-4 to enhance sterol isomerase gene expression in hepatocytes. *J Biol Chem* 2003;**278**(38):36176–82.
104. Iwayanagi Y, Takada T, Suzuki H. HNF4alpha is a crucial modulator of the cholesterol-dependent regulation of NPC1L1. *Pharm Res* 2008;**25**(5):1134–41.
105. Berger J, Moller DE. The mechanisms of action of PPARs. *Annu Rev Med* 2002;**53**:409–35.
106. Miyata KS, McCaw SE, Marcus SL, Rachubinski RA, Capone JP. The peroxisome proliferator-activated receptor interacts with the retinoid X receptor in vivo. *Gene* 1994;**148**(2):327–30.
107. Iwayanagi Y, Takada T, Tomura F, Yamanashi Y, Terada T, Inui K, et al. Human NPC1L1 expression is positively regulated by PPARalpha. *Pharm Res* 2011;**28**(2):405–12.
108. Valasek MA, Clarke SL, Repa JJ. Fenofibrate reduces intestinal cholesterol absorption via PPARalpha-dependent modulation of NPC1L1 expression in mouse. *J Lipid Res* 2007;**48**(12):2725–35.
109. Xia X, Jung D, Webb P, Zhang A, Zhang B, Li L, et al. Liver X receptor beta and peroxisome proliferator-activated receptor delta regulate cholesterol transport in murine cholangiocytes. *Hepatology* 2012;**56**(6):2288–96.
110. Castillo M, Amalik F, Linares A, Garcia-Peregrin E. Dietary fish oil reduces cholesterol and arachidonic acid levels in chick plasma and very low density lipoprotein. *Mol Cell Biochem* 1999;**200**(1–2): 59–67.
111. Chen IS, Hotta SS, Ikeda I, Cassidy MM, Sheppard AJ, Vahouny GV. Digestion, absorption and effects on cholesterol absorption of menhaden oil, fish oil concentrate and corn oil by rats. *J Nutr* 1987;**117**(10):1676–80.
112. Higuchi T, Shirai N, Suzuki H. Reduction in plasma glucose after lipid changes in mice fed fish oil, docosahexaenoic acid, and eicosapentaenoic acid diets. *Ann Nutr Metab* 2006;**50**(2):147–54.
113. Parks JS, Crouse JR. Reduction of cholesterol absorption by dietary oleinate and fish oil in African green monkeys. *J Lipid Res* 1992;**33**(4):559–68.
114. Mathur SN, Watt KR, Field FJ. Regulation of intestinal NPC1L1 expression by dietary fish oil and docosahexaenoic acid. *J Lipid Res* 2007;**48**(2):395–404.
115. Millatt LJ, Bocher V, Fruchart JC, Staels B. Liver X receptors and the control of cholesterol homeostasis: potential therapeutic targets for the treatment of atherosclerosis. *Biochim Biophys Acta* 2003;**1631**(2):107–18.
116. Duval C, Touche V, Tailleux A, Fruchart JC, Fievet C, Clavey V, et al. Niemann-Pick C1 like 1 gene expression is down-regulated by LXR activators in the intestine. *Biochem Biophys Res Commun* 2006;**340**(4): 1259–63.
117. Alvaro A, Rosales R, Masana L, Vallve JC. Polyunsaturated fatty acids down-regulate in vitro expression of the key intestinal cholesterol absorption protein NPC1L1: no effect of monounsaturated nor saturated fatty acids. *J Nutr Biochem* 2010;**21**(6):518–25.
118. Davis Jr HR, Compton DS, Hoos L, Tetzloff G. Ezetimibe, a potent cholesterol absorption inhibitor, inhibits the development of atherosclerosis in ApoE knockout mice. *Arterioscler Thromb Vasc Biol* 2001;**21**(12):2032–8.
119. Davis Jr HR, Hoos LM, Tetzloff G, Maguire M, Zhu LJ, Graziano MP, et al. Deficiency of Niemann-Pick C1 like 1 prevents atherosclerosis in ApoE-/- mice. *Arterioscler Thromb Vasc Biol* 2007;**27**(4):841–9.
120. Sainz Jr B, Barretto N, Martin DN, Hiraga N, Imamura M, Hussain S, et al. Identification of the Niemann-Pick C1-like 1 cholesterol absorption receptor as a new hepatitis C virus entry factor. *Nat Med* 2012;**18**(2):281–5.
121. Matthan NR, Lichtenstein AH. Approaches to measuring cholesterol absorption in humans. *Atherosclerosis* 2004;**174**(2): 197–205.
122. Iovine C, Vaccaro O, Gentile A, Romano G, Pisanti F, Riccardi G, et al. Post-prandial triglyceride profile in a population-based sample of Type 2 diabetic patients. *Diabetologia* 2004;**47**(1):19–22.
123. Tomkin GH, Owens D. The chylomicron: relationship to atherosclerosis. *Int J Vasc Med* 2012;**2012**:784536.
124. Gylling H, Miettinen TA. Cholesterol absorption, synthesis, and LDL metabolism in NIDDM. *Diabetes Care* 1997;**20**(1):90–5.
125. Miettinen TA, Gylling H, Tuominen J, Simonen P, Koivisto V. Low synthesis and high absorption of cholesterol characterize type 1 diabetes. *Diabetes Care* 2004;**27**(1):53–8.
126. Arca M. Alterations of intestinal lipoprotein metabolism in diabetes mellitus and metabolic syndrome. *Atheroscler Suppl* 2015; **17**:12–6.
127. Lally S, Tan CY, Owens D, Tomkin GH. Messenger RNA levels of genes involved in dysregulation of postprandial lipoproteins in type 2 diabetes: the role of Niemann-Pick C1-like 1, ATP-binding cassette, transporters G5 and G8, and of microsomal triglyceride transfer protein. *Diabetologia* 2006;**49**(5): 1008–16.
128. Lally S, Owens D, Tomkin GH. Genes that affect cholesterol synthesis, cholesterol absorption, and chylomicron assembly: the relationship between the liver and intestine in control and streptozotosin diabetic rats. *Metabolism* 2007;**56**(3):430–8.
129. Chang E, Kim L, Choi JM, Park SE, Rhee EJ, Lee WY, et al. Ezetimibe stimulates intestinal glucagon-like peptide 1 secretion via the MEK/ERK pathway rather than dipeptidyl peptidase 4 inhibition. *Metabolism* 2015;**64**(5):633–41.
130. Jia L, Ma Y, Rong S, Betters JL, Xie P, Chung S, et al. Niemann-Pick C1-like 1 deletion in mice prevents high-fat diet-induced fatty liver by reducing lipogenesis. *J Lipid Res* 2010;**51**(11): 3135–44.
131. Ravid Z, Bendayan M, Delvin E, Sane AT, Elchebly M, Lafond J, et al. Modulation of intestinal cholesterol absorption by high glucose levels: impact on cholesterol transporters, regulatory enzymes, and transcription factors. *Am J Physiol Gastrointest Liver Physiol* 2008;**295**(5):G873–85.
132. Malhotra P, Boddy CS, Soni V, Saksena S, Dudeja PK, Gill RK, et al. D-Glucose modulates intestinal Niemann-Pick C1-Like 1 (NPC1L1) gene expression via transcriptional regulation. *Am J Physiol Gastrointest Liver Physiol* 2013;**304**(2):G203–10.

CHAPTER

23

Dietary Long Chain Omega-3 Polyunsaturated Fatty Acids and Inflammatory Gene Expression in Type 2 Diabetes

Jency Thomas[1], *Manohar L. Garg*[2]

[1]Department of Human Biosciences, LaTrobe University, Victoria, Australia; [2]Nutraceuticals Research Group, School of Biomedical & Sciences and Pharmacy, University of Newcastle, Newcastle, NSW, Australia

1. INTRODUCTION

Currently, the approach to mitigate and control type 2 diabetes (T2D) is by supplementation with naturally occurring compounds and by a change in lifestyle. T2D manifests because of compounded factors including genetic predisposition, lack of physical activity, and poor nutritional habits, resulting in development of obesity, dyslipidemia, and hypertension that ultimately accelerates the rate of mortality and morbidity because of this disease condition.[1] Complications range from coronary heart disease and other cardiovascular diseases, retinal damage, renal disorders, and neurodegenerative disorders.

The most common type of diabetes is T2D, affecting 85–90% of all people with diabetes.[2] Although there is no single cause for T2D, there are well-established risk factors. One of the major risk factor is obesity or more specifically abdominal or visceral adiposity. A dysfunctional adiposity phenotype is characterized by excess visceral fat, insulin resistance (IR),[3] and inflammatory and adipokine dysregulation.[4] Uncontrolled hyperglycemia leads to poor cardiovascular outcomes, causing high rates of mortality. Pathological mechanisms that are thought to eventually lead to vascular complications in diabetes includes oxidative stress, formation of excess reactive oxygen species,[5] endothelial dysfunction,[6] activation of the coagulation cascade, and formation of advanced glycation end-products (AGEs). Endothelial dysfunction of the cerebral blood vessels is a common feature of vascular-related disorders, including Alzheimer's disease.[7] Longitudinal studies demonstrate that adults with T2D have increased levels of plasma free fatty acids (FFAs).[8] FFAs are metabolic intermediates that are obtained through diet or synthesized endogenously.[9] An FFA overload has been shown to be associated with inflammatory response and IR.[10] Multiple mechanisms underlie progression of complications in diabetes, including glucotoxicity, lipotoxicity, oxidative stress, and endoplasmic reticulum (ER) stress.[11,12] The relative contribution of each of these mechanisms remains unclear; however, all these mechanisms are linked to inflammatory responses.

2. INFLAMMATION IN T2D

Accumulating evidence suggests that acute phase inflammatory response induced by cytokines is associated with complications in T2D. Pro-inflammatory cytokine, tumor necrosis factor-α (TNF-α), and interleukins, such as IL-6 and IL1-β, potentially damage micro- and microvascular circulation and interfere with insulin signaling ultimately leading to IR.[13] These cytokines have been shown to be strong predictors of development of T2D.[14–16] Epidemiological studies involving diabetic patients suggest an elevation of leukocyte counts,[14] C-reactive protein, and fibrinogen, which are considered the hallmarks of inflammation. The leukocytes in diabetic patients are suggested to generate reactive oxygen species,[17] which creates a pro-oxidant status.

Molecular Nutrition and Diabetes
http://dx.doi.org/10.1016/B978-0-12-801585-8.00023-3

Hotamisligil et al. demonstrated the role of TNF-α in the pathophysiology of T2D in a rodent model almost 20 years ago.[18] Association of TNF-α with IR were confirmed in humans in later studies.[19,20] Cytokines such as TNF-α and IL-6 have been shown to activate stress-induced kinases such as c-Jun NH_2-terminal kinase and Janus kinase (JAK)-signal transducer and activator of transcription (STAT) signaling cascade,[21] which inhibit insulin signaling by phosphorylating the STAT3. STAT3 transfers to the nucleus to induce inflammatory molecules such as C-reactive protein, haptoglobulin[19,22] suppressor of cytokine signaling 3, which inhibit hepatic insulin signaling.[23] Previous studies on type 2 diabetic subjects indicate that lower levels of blood glucose are accompanied by reduced inflammatory markers,[24] thus suggesting a direct correlation between inflammation and hyperglycemia because of diabetes. Morohoshi et al. reported that high blood glucose levels in healthy volunteers stimulated the production of IL-6 from monocytes in vitro.[25] Hyperglycemia resulting from T2D also induces oxidative stress, causing oxidation of arachidonic acid, yielding pro-inflammatory molecules (i.e., the eicosanoids, leukotriene, and prostaglandins).[26] Inflammatory and oxidative events have been shown to interact concertedly in progression of deleterious effects of T2D.[27] Oxidative stress established within the cells because of hyperglycemia contributes to the production of AGEs through lipid glyco-oxidation and lipid peroxidation because the two processes are mutually dependent.[28] In streptozotocin-induced diabetic animal model, AGEs are shown to interact with their receptors and activate nuclear factor-κB,[29] promoting the expression of pro-inflammatory cytokines, thus causing further cell damage. Emerging evidence suggest that complications of diabetes are caused by ER stress-induced oxidative stress and by the deposition of AGEs.[30] Conversely, AGEs have also been shown to reduce insulin sensitivity.[31]

3. INFLAMMATORY GENE EXPRESSION IN T2D

Loss of insulin sensitivity in tissues from excess lipids worsens hyperglycemia in T2D.[32,33] However, the specific mechanisms by which lipids cause IR is unclear, although it has been suggested that lipid overload in adipocytes causes release of pro-inflammatory cytokines, via the activation of Toll-like receptors 4 (TLR4) by the fetuin-A lipid complex. These inflammatory cytokines causes an adipose dysfunction and leads to IR. Activation of TLR4 and IL-1β in T2D promotes cyclooxygenase-2 (COX-2) gene expression. Sheu et al. and Kellogg et al. observed upregulation of COX-2 and prostaglandin E_2 in myocardial and umbilical vein cells treated with high glucose concentrations.[34,35] COX-2 increases the production of prostanoids (prostaglandins and thromboxanes from arachidonic acid involved in inflammatory processes). COX-2 messenger RNA expression is mediated by wide range of inflammatory mediators through numerous intracellular pathways.[36] Recently, a novel role for sphingosine kinase 1 has been identified in inflammation and immune cell trafficking in metabolic disorder associated with obesity and IR.[37,38] Using an animal model, Wang et al. demonstrated that pharmacological inhibition of sphingosine kinase 1 results in upregulation of IL-10 and adiponectin, increased insulin sensitivity, improved glucose homeostasis, and reduced TNF-α, IL-6, monocyte chemotactic protein 1, and suppressor of cytokine signaling 3[39] levels. Other than adipocytes, the other tissue that has a major role in the progression of T2D is skeletal muscle, which has the largest store of glucose.[40,41] Accumulation of lipid in the skeletal muscle leads to the activation of protein kinase C (PKC)ε primarily responsible for insulin sensitivity.[42,43] Gogoi et al. reported that saturated fatty acids cause impairment of insulin signaling by phosphorylation of PKCε. Phosphorylated PKCε further phosphorylates HMGA1, which in turn interacts with positively charged histones, allowing its retention time in heterochromatin region and thus inhibiting an IR promoter and causing an adverse effect on IRβ gene expression.[42] One of the other pathways through which inflammation causes deleterious effects in T2D is by the activation of p38 mitogen-activated protein kinase in both skeletal muscle and adipocytes.[44,45] One of the other mechanism by which T2D causes inflammation is by degradation of glucagon-like peptide 1 (GLP1) by dipeptidyl 1 peptidase 4. GLP1 is a 30-amino acid peptide secreted mainly in the distal gut by the L cells and that functions to naturally reduce blood glucose levels by stimulating insulin secretion. GLP1 receptor agonists are modified forms of GLP1. The expression of GLP1 receptors has been found to be in various organs, such as liver, brain, islet, and adipose tissue cells of animals. GLP1 receptor agonists also aid in reducing blood glucose levels, assist in weight loss, and are commonly used in treatment of T2D. Interestingly GLP1 receptor agonists have been suggested to be anti-inflammatory by reducing cytokines such as IL-6 and TNF-α, and also by improving inflammatory pathways such as by inhibiting IκB kinase beta/nuclear factor-κβ and the c-Jun NH 2-terminal kinase (JNK pathway).

4. LONG CHAIN OMEGA-3 POLYUNSATURATED FATTY ACIDS ON INFLAMMATION AND T2D

Omega-3 PUFA have been shown to have beneficial effects on metabolic and inflammatory processes in diabetes (Figure 1). It is interesting to note that although an overload of FFA can cause IR (discussed in the Section 1), much of the beneficial effects of omega-3 PUFAs have been attributed to two main long chain fatty acid specific receptors FFA1 and FFA4.[46] Both FFA1 and FFA4 have been suggested to have potential beneficial effects on metabolic functions.[46,47] Interestingly, these FFAs have been suggested to be downregulated in the islet cells of type 2 diabetic patients.[48] In a transgenic diabetic mouse model, overexpression of FFA1 improves insulin secretion, glucose tolerance, and prevents the development of hyperglycemia.[49] FFA4 is activated by omega-3 PUFA and is found in liver, adipocytes, brain, heart, and ectoendocrine cells, which are the specialized endocrine cells of the gastrointestinal tract and pancreas. FFA4 activation in these tissues has been shown to decrease pro-inflammatory gene expression and increase the expression of anti-inflammatory genes in M1 and M2 macrophages.[50] It has been suggested that beneficial effects of FFA4 are mediated through the recruitment of β-arrestin 2, which inhibits lipopolysaccharide-induced TNF-α and blocks TLR4.

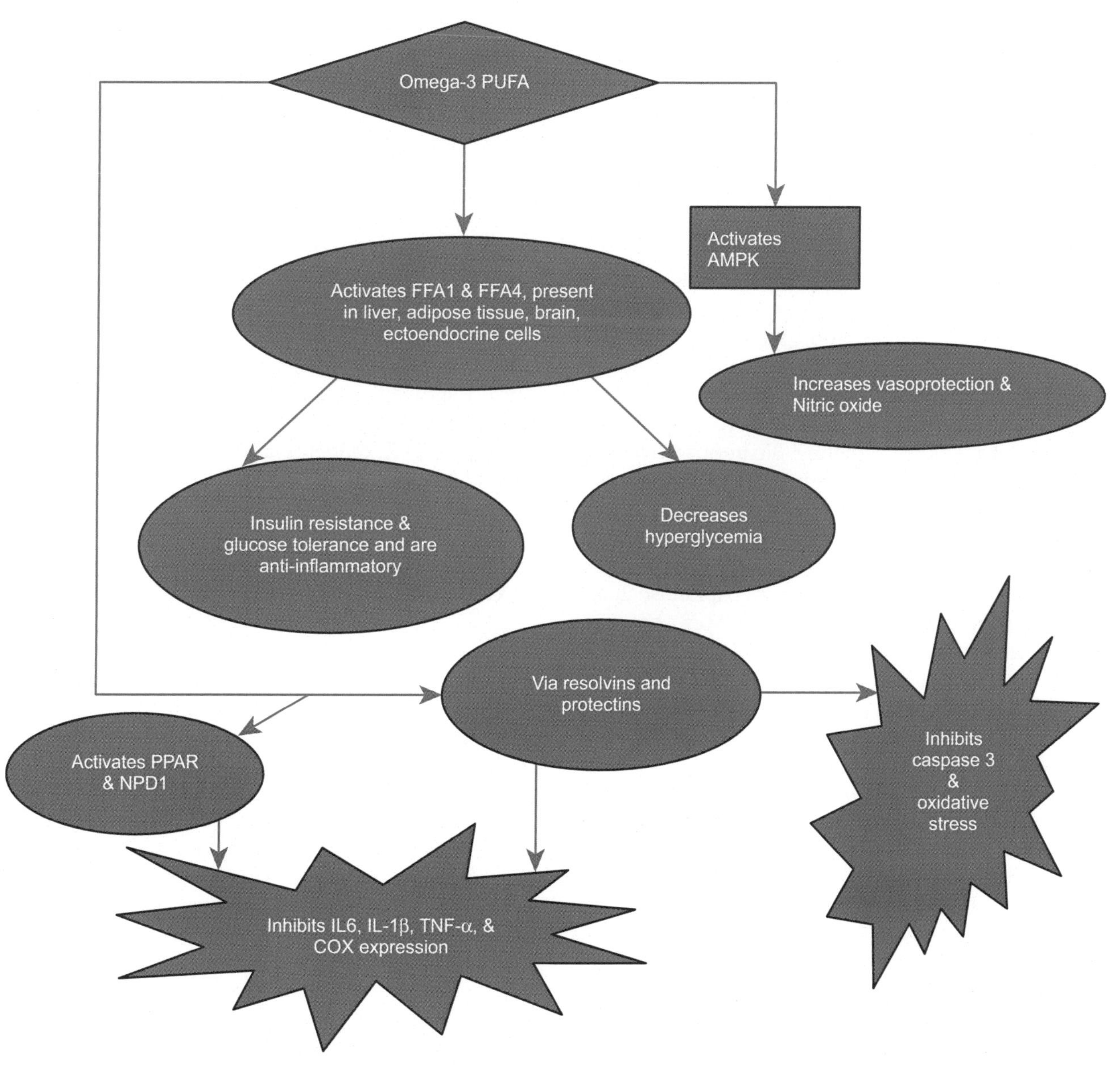

FIGURE 1 Multiple mechanisms through which omega-3 PUFA exerts its beneficial effects. AMPK, adenosine monophosphate–activated protein kinase; COX, cyclooxygenase; FFA, free fatty acid; IL, interleukin; PUFA, polyunsaturated fatty acid; TNF, tumor necrosis factor.

These anti-inflammatory effects of FFA4 extends to the hypothalamus and aid in reducing diet-induced inflammation and adiposity,[50–54] as seen in metabolic disorders including diabetes. In T2D, the content of omega-3 PUFAs is decreased, mainly because of mitochondrial and peroxisomal β-oxidation.[55] Increased ratio of omega-6 to omega-3 fatty acid causes an increase in arachidonic acid in the adipose tissue of obese human subjects, which makes their adipose tissue vulnerable to inflammation.[56,57] Supplementation with omega-3 PUFAs for 26 weeks has been shown to downregulate gene expression related to inflammatory and atherogenic pathways.[58] In metabolic disturbances such as T2D, cytokines are increased through a number of pathways in skeletal muscles including nuclear factor-kappa B, JNK, and the JAK-STAT.[40] Eicosapentaenoic acid (EPA) has been shown to cause activation of adenosine monophosphate–activated protein kinase (AMPK), thereby providing vasoprotection without altering serum lipids. It has been suggested that in T2D repression of pro-inflammatory IL-6 in liver tissues occurs via AMPK, mainly by comprising a catalytic α subunit and regulatory β and ϒ subunits, all of which act in pro-inflammatory cascades.[59] Activated AMPK provides metabolic balance via glucose uptake and enhanced lipid oxidation, decreased lipolysis and lipogenesis in white adipose tissue, and improves overall systemic insulin sensitivity and glucose tolerance.[60] In addition to providing metabolic balance, AMPK is emerging as playing a crucial role in anti-inflammatory processes by inhibiting IL-6–stimulated suppressor of cytokine signaling 3in mouse liver and human hepatocarcinoma cell line. Recently, Cansby et al. also demonstrated that AMPK mediates suppression of IL-6–induced phosphorylation of downstream components of the signaling cascade JAK1, SH2-domain containing protein tyrosine phosphatase 2, and STAT3.[61] Furthermore, activation of AMPK has been shown to increase nitric oxide activity, and increases fatty acid oxidation. Low levels of nitric oxide have been reported in diabetic patients and in animal model of diabetes.[62] Wu et al. demonstrated that AMPK activation releases nitric oxide because of upregulation in uncoupling protein 2 via EPA in the endothelial cells, in vitro and in vivo.[63] Interestingly, all classes of fatty acids or their metabolites upregulate UPC-2 messenger RNA in adipose cells.[64,65] Another mechanism suggested is that EPA activates proliferator-activated receptor-γ (PPAR-γ) because omega-3 is suggested to be endogenous ligands of PPARs,[66] thereby decreasing pro-inflammatory IL6 and TNF-α and improving lipid metabolism.[67] In diabetic conditions, activation of peroxisome PPAR-γ stimulates the differentiation of preadipocytes to adipocytes, which in turn decreases IR by increasing insulin receptors, activation of PPARs, apart from having anti-inflammatory effects in diabetes, has been suggested to prevent induction of diabetes.[68]

In the past decade, two new lipid mediators produced from omega-3 PUFAs have been studied, namely resolvins, which are produced from EPA and docosahexaenoic acid (DHA), and protectins, produced mainly from DHA. Both resolvins and protectins exert anti-inflammatory properties by preventing infiltration of neutrophils into the site of inflammation and inhibit production of pro-inflammatory cytokines such as IL-1β and TNF-α.[69–71] Resolvin D1 has also been shown to have an insulin-sensitizing property.[72] Pathogenesis of metabolic disorders such as obesity and T2D is affected by stress on ER. Jung et al. demonstrated that pretreatment with resolvin D1 attenuated ER stress-induced apoptosis and decreased caspase 3 activity, which is a critical mediator of apoptosis,[73] mainly by inhibiting JNK expression.[72] A recent study in a murine model exposed to hypoxia suggests that treatment with resolvin significantly reduced decreased oxidative stress by increasing glutathione production and decreased tissue inflammation.[74] In addition, resolvins have also been recently shown to impact functional outcomes on traumatic sleep and microglial activation following traumatic brain injury in rodents.[75] Although a vast amount of evidence from animal studies[67,76,77] demonstrates the potential of omega-3 PUFAs to lower inflammatory gene expression, there are very few studies demonstrating the same in humans; few studies conducted showed no positive results.[78] Therefore, future studies could focus on optimal dose of omega-3 on insulin responses in diabetes and use a larger sample size in clinical trials.

5. N-3 POLYUNSATURATED FATTY ACIDS ON NEUROINFLAMMATION IN DIABETES

Neuroinflammation in diabetes has been associated with the progression of neurodegenerative and neuropsychiatric disorders.[79] A vast amount of evidence suggests that, in diabetes, damage to glial cells produces excessive cytokines, which in turn activates the inflammatory pathways thus causing neuroinflammation.[80] Dietary intake of DHA has been greatly emphasized, especially with regard to neurodevelopment and brain health particularly during prenatal brain development.[81] DHA is incorporated in large amounts into the fetal brain through fatty acid transport protein 4,[82] which is made of integral plasma membrane protein and is known to facilitate bidirectional long chain fatty acid movements.[83] DHA that is transferred from the maternal circulation to fetal brain plays crucial role in brain growth especially with regard to synaptogenesis.[84,85] DHA is selectively enriched in neuronal tissues, especially in neuronal and synaptic membranes, oligodendrocytes, and also subcellular particles such as myelin, and nerve endings.[86–88] Experimental studies conducted in our laboratory on streptozotocin-induced diabetic animal models showed an increase in expression of pro-inflammatory TNF-α, IL-6, and decreased expression of insulin-like growth factor *Igf II* in the hippocampus (unpublished data), a critical region of brain for consolidation of memory. Types 1 and 2 diabetes are associated with neuroinflammation memory deficits, and the hippocampus is particularly vulnerable, with structural and functional alterations being caused by this disease processes.[89–91] Supplementation with 50 mg/kg/day of DHA for 6 weeks showed significant downregulation of these pro-inflammatory genes (unpublished data). Genome-wide association studies conducted in our laboratory also showed that diabetes had deleterious effects on synaptic plasticity and neurogenesis.[76] A decrease in DHA, as seen in aging or any other disease condition, such as diabetes, could affect some of the prominent neurotransmitters like glutamate, found in all parts of the brain; this is a major excitatory neurotransmitter and is involved in the integrity of brain function such as learning/memory development and aging.[81,88,92] In neurodegenerative conditions such as Alzheimer's disease, DHA has been shown to be significantly depleted.[86] Dietary supplementation of DHA has been shown to increase the levels of brain-derived neurotrophic factor, which plays a crucial role in modulating synaptic plasticity for hippocampal dependent cognitive functions.[93] Akbar et al. suggest that because DHA is highly enriched in neuronal membranes, and has also been shown to facilitate the activation of Akt via an increase in phosphatidylserine, Akt signaling is a critical pathway in neuronal survival.[94] Activation of Akt can thus cause an increase in brain-derived neurotrophic factor, which further strengthens synaptic plasticity and cell survival. Alternatively, Calon et al. suggest that a diet rich in DHA activates Ca^{2+}/calmodulin-dependent protein kinase, whose signaling cascade is critical for learning and memory, plays a crucial role in induction and maintenance of long-term potentiation in hippocampus.[95,96]

In recent years, our understanding about the role of DHA has further advanced, in that DHA has been shown to be a crucial regulator of neural gene expression.[97] Previous studies have shown that DHA acts as an endogenous ligand for retinoic acid receptors (RAR), retinoid X receptors (RXR) and PPARs, which are the transcription factors involved in many cellular processes such as learning and memory.[98] RAR and RXR have been shown to decrease with age and are associated with age-related memory deficits. Dyall et al. suggest that a reversal in the decrease of RAR and RXR by DHA could alleviate the memory deficits.[98] In agreement, Greiner et al.[99] also suggest that deficiency of DHA as seen in neurodegenerative conditions such as aging and Alzheimer's disease strongly correlates with learning and memory deficits. In patients diagnosed with Alzheimer's disease, DHA was found to be significantly lower in blood plasma and brain, which could be from lower dietary intake of omega-3 fatty acid and could be from increased oxidation of PUFAs.[100] Clinical and preclinical evidences suggest that diet enriched with DHA showed reduced amyloid formation in dementia associated with AD.[101]

In brain, DHA has been shown to exert its beneficial effects also through its derivate neuroprotectin D1 (NPD1). NPD1 has been shown to enhance the expression of anti-apoptotic Bcl2, Bcl-xl, and downregulate pro-apoptotic such as Bax.[102] Interestingly, both DHA and NPD1 have been shown to downregulate pro-inflammatory cytokines such as IL-1β, TNF-α, and COX, which have otherwise been shown to be upregulated in neurodegenerative conditions.[103]

6. CONCLUSION

Considerable progress has been made in the past decade on the beneficial effects of omega-3 PUFA in T2D, indicating that most of the clinical trials conducted yield minimal to nil effect in preventing the deleterious effects of diabetes.[104–107] Furthermore, gene expression studies on the omega-3 PUFA and T2D in humans are very few and indicate no positive effects either[78]; few other studies conducted indicate adverse effects.[108,109] Concerns that n-3 PUFAs may cause lipid peroxidation have also been attributed to the negative or nil effect of omega-3.[110] However, data available at present on the effects of n-3 PUFAs on lipid peroxidation and oxidative stress are contradictory.[111] The clinical literature on the effects of n-3 PUFAs on inflammation are inconclusive and have reported mixed effects. These inconsistencies are related mostly to small sample sizes, dose, and duration of supplementation, EPA/DHA ratio, sex differences, and background diet. Carefully planned intervention studies involving dose-dependent, longer duration, and adequate sample size keeping in mind the gender differences are warranted.

Future direction on omega-3 PUFAs could focus on the following:

1. Setting up an optimum dosage of omega-3 on insulin responses in diabetes.
2. Acquiring sufficient sample size in clinical trials.
3. Examine the effect of gene expression in different tissues before and after supplementation.

References

1. Contreras-Shannon V, Heart DL, Paredes RM, Navaira E, Catano G, Maffi SK, et al. Clozapine-induced mitochondria alterations and inflammation in brain and insulin-responsive cells. *PLoS One* 2013;**8**(3):e59012.
2. Diabetes Australia. https://www.diabetesaustralia.com.au/type-2-diabetes; 2013.
3. Neeland IJ, Turer AT, Ayers CR, Powell-Wiley TM, Vega GL, Farzaneh-Far R, et al. Dysfunctional adiposity and the risk of prediabetes and type 2 diabetes in obese adults. *JAMA* September 19, 2012;**308**(11):1150–9.
4. Boyko EJ, Fujimoto WY, Leonetti DL, Newell-Morris L. Visceral adiposity and risk of type 2 diabetes: a prospective study among Japanese Americans. *Diabetes Care* April 2000;**23**(4):465–71.
5. Baynes JW, Thorpe SR. Role of oxidative stress in diabetic complications: a new perspective on an old paradigm. *Diabetes* January 1999;**48**(1):1–9.
6. Idris I, Gray S, Donnelly R. Protein kinase C activation: isozyme-specific effects on metabolism and cardiovascular complications in diabetes. *Diabetologia* June 2001;**44**(6):659–73.
7. Miller AA, Budzyn K, Sobey CG. Vascular dysfunction in cerebrovascular disease: mechanisms and therapeutic intervention. *Clin Sci (Lond)* 2010;**119**(1):1–17.
8. Salgin B, Ong KK, Thankamony A, Emmett P, Wareham NJ, Dunger DB. Higher fasting plasma free fatty acid levels are associated with lower insulin secretion in children and adults and a higher incidence of type 2 diabetes. *J Clin Endocrinol Metab* September 2012;**97**(9):3302–9.
9. Legrand-Poels S, Esser N, L'Homme L, Scheen A, Paquot N, Piette J. Free fatty acids as modulators of the NLRP3 inflammasome in obesity/type 2 diabetes. *Biochem Pharmacol* August 28, 2014;**92**(1):131–41.
10. Einstein FH, Huffman DM, Fishman S, Jerschow E, Heo HJ, Atzmon G, et al. Aging per se increases the susceptibility to free fatty acid-induced insulin resistance. *J Gerontol* August 2010; **65**(8):800–8.
11. Gao Y, Zhang J, Li G, Xu H, Yi Y, Wu Q, et al. Protection of vascular endothelial cells from high glucose-induced cytotoxicity by emodin. *Biochem Pharmacol* March 1, 2015;**94**(1):39–45.
12. Agrawal NK, Kant S. Targeting inflammation in diabetes: newer therapeutic options. *World J Diabetes* October 15, 2014;**5**(5):697–710.
13. Bullo M, Garcia-Lorda P, Salas-Salvado J. Plasma soluble tumor necrosis factor alpha receptors and leptin levels in normal-weight and obese women: effect of adiposity and diabetes. *European J Endocrinol* March 2002;**146**(3):325–31.
14. Schmidt MI, Duncan BB, Sharrett AR, Lindberg G, Savage PJ, Offenbacher S, et al. Markers of inflammation and prediction of diabetes mellitus in adults (Atherosclerosis Risk in Communities study): a cohort study. *Lancet* May 15, 1999;**353**(9165):1649–52.
15. Duncan BB, Schmidt MI, Offenbacher S, Wu KK, Savage PJ, Heiss G. Factor VIII and other hemostasis variables are related to incident diabetes in adults. The Atherosclerosis Risk in Communities (ARIC) Study. *Diabetes Care* May 1999;**22**(5):767–72.
16. Spranger J, Kroke A, Mohlig M, Hoffmann K, Bergmann MM, Ristow M, et al. Inflammatory cytokines and the risk to develop type 2 diabetes: results of the prospective population-based European Prospective Investigation into Cancer and Nutrition (EPIC)-Potsdam Study. *Diabetes* March 2003;**52**(3):812–7.
17. Lipinski B. Pathophysiology of oxidative stress in diabetes mellitus. *J Diabetes Complications* 2001 Jul–Aug;**15**(4):203–10.
18. Hotamisligil GS, Shargill NS, Spiegelman BM. Adipose expression of tumor necrosis factor-alpha: direct role in obesity-linked insulin resistance. *Science* January 1, 1993;**259**(5091):87–91.
19. Pickup JC. Inflammation and activated innate immunity in the pathogenesis of type 2 diabetes. *Diabetes Care* March 2004;**27**(3): 813–23.
20. Gower BA, Goss AM. A lower-carbohydrate, higher-fat diet reduces abdominal and intermuscular fat and increases insulin sensitivity in adults at risk of type 2 diabetes. *J Nutr* 2015;**145**(1): 177S–83S.
21. McGillicuddy FC, Chiquoine EH, Hinkle CC, Kim RJ, Shah R, Roche HM, et al. Interferon gamma attenuates insulin signaling, lipid storage, and differentiation in human adipocytes via activation of the JAK/STAT pathway. *J Biol Chem* November 13, 2009; **284**(46):31936–44.
22. Alonzi T, Maritano D, Gorgoni B, Rizzuto G, Libert C, Poli V. Essential role of STAT3 in the control of the acute-phase response as revealed by inducible gene inactivation [correction of activation] in the liver. *Mol Cell Biol* March 2001;**21**(5):1621–32.
23. Ramadoss P, Unger-Smith NE, Lam FS, Hollenberg AN. STAT3 targets the regulatory regions of gluconeogenic genes in vivo. *Mol Endocrinol (Baltimore, Md)* June 2009;**23**(6):827–37.
24. Ceriello A, Mercuri F, Fabbro D, Giacomello R, Stel G, Taboga C, et al. Effect of intensive glycaemic control on fibrinogen plasma concentrations in patients with Type II diabetes mellitus. Relation with beta-fibrinogen genotype. *Diabetologia* November 1998; **41**(11):1270–3.
25. Morohoshi M, Fujisawa K, Uchimura I, Numano F. Glucose-dependent interleukin 6 and tumor necrosis factor production by human peripheral blood monocytes in vitro. *Diabetes* July 1996;**45**(7):954–9.
26. Simopoulos AP. Omega-3 fatty acids in inflammation and autoimmune diseases. *J Am Coll Nutr* December 2002;**21**(6):495–505.
27. Almeida-Pititto B, Almada Filho Cde M, Cendoroglo MS. Cognitive deficit: another complication of diabetes mellitus? *Arq Bras Endocrinol Metabol* October 2008;**52**(7):1076–83.
28. Reddy VP, Zhu X, Perry G, Smith MA. Oxidative stress in diabetes and Alzheimer's disease. *J Alzheimers Dis* April 2009;**16**(4):763–74.
29. Tobon-Velasco JC, Cuevas E, Torres-Ramos MA, Santamaria A. Receptor for AGEs (RAGE) as mediator of NF-kB pathway activation in neuroinflammation and oxidative stress. *CNS Neurol Disord Drug Targets* August 6, 2014.
30. Sandu O, Song K, Cai W, Zheng F, Uribarri J, Vlassara H. Insulin resistance and type 2 diabetes in high-fat-fed mice are linked to high glycotoxin intake. *Diabetes* August 2005;**54**(8):2314–9.
31. Tokita Y, Hirayama Y, Sekikawa A, Kotake H, Toyota T, Miyazawa T, et al. Fructose ingestion enhances atherosclerosis and deposition of advanced glycated end-products in cholesterol-fed rabbits. *J Atheroscler Thromb* 2005;**12**(5):260–7.
32. Bhattacharya S, Dey D, Roy SS. Molecular mechanism of insulin resistance. *J Biosci* March 2007;**32**(2):405–13.
33. Boden G, Chen X, Ruiz J, White JV, Rossetti L. Mechanisms of fatty acid-induced inhibition of glucose uptake. *J Clin Invest* June 1994; **93**(6):2438–46.
34. Sheu ML, Ho FM, Yang RS, Chao KF, Lin WW, Lin-Shiau SY, et al. High glucose induces human endothelial cell apoptosis through a phosphoinositide 3-kinase-regulated cyclooxygenase-2 pathway. *Arterioscler Thromb Vasc Biol* March 2005;**25**(3):539–45.
35. Kellogg AP, Converso K, Wiggin T, Stevens M, Pop-Busui R. Effects of cyclooxygenase-2 gene inactivation on cardiac autonomic and left ventricular function in experimental diabetes. *Am J Physiol Heart Circ Physiol* February 2009;**296**(2):H453–61.
36. Hinz B, Brune K. Cyclooxygenase-2–10 years later. *J Pharmacol Exp Therap* February 2002;**300**(2):367–75.
37. Maceyka M, Harikumar KB, Milstien S, Spiegel S. Sphingosine-1-phosphate signaling and its role in disease. *Trends Cell Biol* January 2012;**22**(1):50–60.

38. Taha TA, Hannun YA, Obeid LM. Sphingosine kinase: biochemical and cellular regulation and role in disease. *J Biochem Mol Biol* March 31, 2006;**39**(2):113–31.
39. Wang J, Badeanlou L, Bielawski J, Ciaraldi TP, Samad F. Sphingosine kinase 1 regulates adipose proinflammatory responses and insulin resistance. *Am J Physiol Endocrinol Metab* April 1, 2014; **306**(7):E756–68.
40. DeFronzo RA, Ferrannini E, Sato Y, Felig P, Wahren J. Synergistic interaction between exercise and insulin on peripheral glucose uptake. *J Clin Invest* December 1981;**68**(6):1468–74.
41. Shulman GI, Rothman DL, Jue T, Stein P, DeFronzo RA, Shulman RG. Quantitation of muscle glycogen synthesis in normal subjects and subjects with non-insulin-dependent diabetes by 13C nuclear magnetic resonance spectroscopy. *N Engl J Med* January 25, 1990;**322**(4):223–8.
42. Gogoi B, Chatterjee P, Mukherjee S, Buragohain AK, Bhattacharya S, Dasgupta S. A polyphenol rescues lipid induced insulin resistance in skeletal muscle cells and adipocytes. *Biochem Biophys Res Commun* September 26, 2014;**452**(3):382–8.
43. Dey D, Basu D, Roy SS, Bandyopadhyay A, Bhattacharya S. Involvement of novel PKC isoforms in FFA induced defects in insulin signaling. *Mol Cell Endocrinol* February 26, 2006;**246**(1–2): 60–4.
44. Carlson CJ, Koterski S, Sciotti RJ, Poccard GB, Rondinone CM. Enhanced basal activation of mitogen-activated protein kinases in adipocytes from type 2 diabetes: potential role of p38 in the downregulation of GLUT4 expression. *Diabetes* March 2003; **52**(3):634–41.
45. Lee J, Sun C, Zhou Y, Lee J, Gokalp D, Herrema H, et al. p38 MAPK-mediated regulation of Xbp1s is crucial for glucose homeostasis. *Nat Med* October 2011;**17**(10):1251–60.
46. Watterson KR, Hudson BD, Ulven T, Milligan G. Treatment of type 2 diabetes by free fatty acid receptor agonists. *Front Endocrinol* 2014;**5**:137.
47. Offermanns S. Free fatty acid (FFA) and hydroxy carboxylic acid (HCA) receptors. *Annu Rev Pharmacol Toxicol* 2014;**54**:407–34.
48. Del Guerra S, Bugliani M, D'Aleo V, Del Prato S, Boggi U, Mosca F, et al. G-protein-coupled receptor 40 (GPR40) expression and its regulation in human pancreatic islets: the role of type 2 diabetes and fatty acids. *Nutr Metab Cardiovasc Dis* January 2010;**20**(1): 22–5.
49. Nagasumi K, Esaki R, Iwachidow K, Yasuhara Y, Ogi K, Tanaka H, et al. Overexpression of GPR40 in pancreatic beta-cells augments glucose-stimulated insulin secretion and improves glucose tolerance in normal and diabetic mice. *Diabetes* May 2009;**58**(5): 1067–76.
50. Li X, Yu Y, Funk CD. Cyclooxygenase-2 induction in macrophages is modulated by docosahexaenoic acid via interactions with free fatty acid receptor 4 (FFA4). *FASEB J* December 2013;**27**(12): 4987–97.
51. Oh DY, Talukdar S, Bae EJ, Imamura T, Morinaga H, Fan W, et al. GPR120 is an omega-3 fatty acid receptor mediating potent anti-inflammatory and insulin-sensitizing effects. *Cell* September 3, 2010;**142**(5):687–98.
52. Halder S, Kumar S, Sharma R. The therapeutic potential of GPR120: a patent review. *Expert Opin Ther Pat* December 2013; **23**(12):1581–90.
53. Cintra DE, Ropelle ER, Moraes JC, Pauli JR, Morari J, Souza CT, et al. Unsaturated fatty acids revert diet-induced hypothalamic inflammation in obesity. *PLoS One* 2012;**7**(1):e30571.
54. Wellhauser L, Belsham DD. Activation of the omega-3 fatty acid receptor GPR120 mediates anti-inflammatory actions in immortalized hypothalamic neurons. *J Neuroinflammation* 2014;**11**:60.
55. Hou L, Lian K, Yao M, Shi Y, Lu X, Fang L, et al. Reduction of n-3 PUFAs, specifically DHA and EPA, and enhancement of peroxisomal beta-oxidation in type 2 diabetic rat heart. *Cardiovasc Diabetol* 2012;**11**:126.
56. van Dijk SJ, Feskens EJ, Bos MB, Hoelen DW, Heijligenberg R, Bromhaar MG, et al. A saturated fatty acid-rich diet induces an obesity-linked proinflammatory gene expression profile in adipose tissue of subjects at risk of metabolic syndrome. *Am J Clin Nutr* December 2009;**90**(6):1656–64.
57. Pietilainen KH, Rog T, Seppanen-Laakso T, Virtue S, Gopalacharyulu P, Tang J, et al. Association of lipidome remodeling in the adipocyte membrane with acquired obesity in humans. *PLoS Biology* June 2011;**9**(6):e1000623.
58. Bouwens M, van de Rest O, Dellschaft N, Bromhaar MG, de Groot LC, Geleijnse JM, et al. Fish-oil supplementation induces antiinflammatory gene expression profiles in human blood mononuclear cells. *Am J Clin Nutr* August 2009;**90**(2):415–24.
59. Cansby E, Nerstedt A, Amrutkar M, Duran EN, Smith U, Mahlapuu M. Partial hepatic resistance to IL-6-induced inflammation develops in type 2 diabetic mice, while the anti-inflammatory effect of AMPK is maintained. *Mol Cell Endocrinol* August 5, 2014;**393**(1–2):143–51.
60. Long YC, Zierath JR. AMP-activated protein kinase signaling in metabolic regulation. *J Clin Invest* July 2006;**116**(7): 1776–83.
61. Nerstedt A, Johansson A, Andersson CX, Cansby E, Smith U, Mahlapuu M. AMP-activated protein kinase inhibits IL-6-stimulated inflammatory response in human liver cells by suppressing phosphorylation of signal transducer and activator of transcription 3 (STAT3). *Diabetologia* November 2010; **53**(11):2406–16.
62. Suresh Y, Das UN. Long-chain polyunsaturated fatty acids and chemically induced diabetes mellitus: effect of omega-6 fatty acids. *Nutrition (Burbank, Los Angeles County, Calif)* February 2003;**19**(2):93–114.
63. Wang S, Zhang C, Zhang M, Liang B, Zhu H, Lee J, et al. Activation of AMP-activated protein kinase alpha2 by nicotine instigates formation of abdominal aortic aneurysms in mice in vivo. *Nat Med* June 2012;**18**(6):902–10.
64. Reilly JM, Thompson MP. Dietary fatty acids Up-regulate the expression of UCP2 in 3T3-L1 preadipocytes. *Biochem Biophys Res Commun* November 2, 2000;**277**(3):541–5.
65. Aubert J, Champigny O, Saint-Marc P, Negrel R, Collins S, Ricquier D, et al. Up-regulation of UCP-2 gene expression by PPAR agonists in preadipose and adipose cells. *Biochem Biophys Res Commun* September 18, 1997;**238**(2):606–11.
66. Su CG, Wen X, Bailey ST, Jiang W, Rangwala SM, Keilbaugh SA, et al. A novel therapy for colitis utilizing PPAR-gamma ligands to inhibit the epithelial inflammatory response. *J Clin Invest* August 1999;**104**(4):383–9.
67. Figueras M, Olivan M, Busquets S, Lopez-Soriano FJ, Argiles JM. Effects of eicosapentaenoic acid (EPA) treatment on insulin sensitivity in an animal model of diabetes: improvement of the inflammatory status. *Obesity (Silver Spring, Md)* February 2010;**19**(2): 362–9.
68. Krishna Mohan I, Das UN. Prevention of chemically induced diabetes mellitus in experimental animals by polyunsaturated fatty acids. *Nutrition (Burbank, Los Angeles County, Calif)* February 2001;**17**(2):126–51.
69. Hong S, Gronert K, Devchand PR, Moussignac RL, Serhan CN. Novel docosatrienes and 17S-resolvins generated from docosahexaenoic acid in murine brain, human blood, and glial cells. Autacoids in anti-inflammation. *J Biol Chem* April 25, 2003; **278**(17):14677–87.

70. Serhan CN, Chiang N, Van Dyke TE. Resolving inflammation: dual anti-inflammatory and pro-resolution lipid mediators. *Nat Rev Immunol* May 2008;**8**(5):349–61.
71. Calder PC. Omega-3 polyunsaturated fatty acids and inflammatory processes: nutrition or pharmacology? *Br J Clin Pharmacol* 2012;**75**(3):645–62.
72. Jung TW, Hwang HJ, Hong HC, Choi HY, Yoo HJ, Baik SH, et al. Resolvin D1 reduces ER stress-induced apoptosis and triglyceride accumulation through JNK pathway in HepG2 cells. *Mol Cell Endocrinol* June 25, 2014;**391**(1–2):30–40.
73. Lakhani SA, Masud A, Kuida K, Porter Jr GA, Booth CJ, Mehal WZ, et al. Caspases 3 and 7: key mediators of mitochondrial events of apoptosis. *Science* February 10, 2006;**311**(5762): 847–51.
74. Cox Jr R, Phillips O, Fukumoto J, Fukumoto I, Tamarapu Parthasarathy P, Arias S, et al. Aspirin-Triggered Resolvin D1 Treatment Enhances Resolution of Hyperoxic Acute Lung Injury. *Am J Respir Cell Mol Biol* February 3, 2015;**53**(3):422–35.
75. Harrison JL, Rowe RK, Ellis TW, Yee NS, O'Hara BF, Adelson PD, et al. Resolvins AT-D1 and E1 differentially impact functional outcome, post-traumatic sleep, and microglial activation following diffuse brain injury in the mouse. *Brain Behav Immun* January 10, 2015;**47**:131–40.
76. Thomas J, Garg ML, Smith DW. Altered expression of histone and synaptic plasticity associated genes in the hippocampus of streptozotocin-induced diabetic mice. *Metab Brain Dis* July 6, 2013;**28**(4):613–8.
77. Yan Y, Jiang W, Spinetti T, Tardivel A, Castillo R, Bourquin C, et al. Omega-3 fatty acids prevent inflammation and metabolic disorder through inhibition of NLRP3 inflammasome activation. *Immunity* June 27, 2013;**38**(6):1154–63.
78. Labonte ME, Couture P, Tremblay AJ, Hogue JC, Lemelin V, Lamarche B. Eicosapentaenoic and docosahexaenoic acid supplementation and inflammatory gene expression in the duodenum of obese patients with type 2 diabetes. *Nutr J* 2013;**12**:98.
79. Ghasemi RDL, Haeri A, Moosavi M, Mohamed Z, Ahmadiani A. Brain insulin dysregulation: implication for neurological and neuropsychiatric disorders. *Mol Neurobiol* January 20, 2013;**47**(3): 1046–65.
80. Myers RR, Campana WM, Shubayev VI. The role of neuroinflammation in neuropathic pain: mechanisms and therapeutic targets. *Drug Discovery Today* January 2006;**11**(1–2):8–20.
81. Moreira JD, Knorr L, Ganzella M, Thomazi AP, de Souza CG, de Souza DG, et al. Omega-3 fatty acids deprivation affects ontogeny of glutamatergic synapses in rats: relevance for behavior alterations. *Neurochem Int* February 19, 2010;**56**(6–7):753–9.
82. Koletzko B, Larque E, Demmelmair H. Placental transfer of long-chain polyunsaturated fatty acids (LC-PUFA). *J Perinat Med* 2007; **35**(Suppl. 1):S5–11.
83. Larque E, Demmelmair H, Klingler M, De Jonge S, Bondy B, Koletzko B. Expression pattern of fatty acid transport protein-1 (FATP-1), FATP-4 and heart-fatty acid binding protein (H-FABP) genes in human term placenta. *Early Hum Dev* October 2006; **82**(10):697–701.
84. Bazan NG, Gordon WC, Rodriguez de Turco EB. Docosahexaenoic acid uptake and metabolism in photoreceptors: retinal conservation by an efficient retinal pigment epithelial cell-mediated recycling process. *Adv Exp Med Biol* 1992;**318**:295–306.
85. Martin RE, Bazan NG. Changing fatty acid content of growth cone lipids prior to synaptogenesis. *J Neurochem* July 1992;**59**(1): 318–25.
86. Lukiw WJ, Bazan NG. Inflammatory, apoptotic, and survival gene signaling in Alzheimer's disease. A review on the bioactivity of neuroprotectin D1 and apoptosis. *Mol Neurobiol* August 2010; **42**(1):10–6.
87. Bourre JM. Effects of nutrients (in food) on the structure and function of the nervous system: update on dietary requirements for brain. Part 2: macronutrients. *J Nutr Health Aging* 2006 September–October;**10**(5):386–99.
88. Su H-M. Mechanisms of n-3 fatty acid-mediated development and maintenance of learning memory performance. *J Nutr Biochem* 2010;**21**(5):364–73.
89. Chen RH, Jiang XZ, Zhao XH, Qin YL, Gu Z, Gu PL, et al. Risk factors of mild cognitive impairment in middle aged patients with type 2 diabetes: a cross-section study. *Ann Endocrinol* June 2012; **73**(3):208–12.
90. Trudeau F, Gagnon S, Massicotte G. Hippocampal synaptic plasticity and glutamate receptor regulation: influences of diabetes mellitus. *Eur J Pharmacol* April 19, 2004;**490**(1–3):177–86.
91. Sandireddy R, Yerra VG, Areti A, Komirishetty P, Kumar A. Neuroinflammation and oxidative stress in diabetic neuropathy: futuristic strategies based on these targets. *Int J Endocrinol* 2014; **2014**:674987.
92. Bergink V, van Megen HJ, Westenberg HG. Glutamate and anxiety. *Eur Neuropsychopharmacol* May 2004;**14**(3):175–83.
93. Wu A, Ying Z, Gomez-Pinilla F. Docosahexaenoic acid dietary supplementation enhances the effects of exercise on synaptic plasticity and cognition. *Neuroscience* August 26, 2008;**155**(3):751–9.
94. Akbar M, Calderon F, Wen Z, Kim HY. Docosahexaenoic acid: a positive modulator of Akt signaling in neuronal survival. *Proc Natl Acad Sci USA* August 2, 2005;**102**(31):10858–63.
95. Calon F, Lim GP, Morihara T, Yang F, Ubeda O, Salem Jr N, et al. Dietary n-3 polyunsaturated fatty acid depletion activates caspases and decreases NMDA receptors in the brain of a transgenic mouse model of Alzheimer's disease. *Eur J Neurosci* August 2005; **22**(3):617–26.
96. Elgersma Y, Sweatt JD, Giese KP. Mouse genetic approaches to investigating calcium/calmodulin-dependent protein kinase II function in plasticity and cognition. *J Neurosci* September 29, 2004;**24**(39):8410–5.
97. Crawford MA, Golfetto I, Ghebremeskel K, Min Y, Moodley T, Poston L, et al. The potential role for arachidonic and docosahexaenoic acids in protection against some central nervous system injuries in preterm infants. *Lipids* April 2003;**38**(4):303–15.
98. Dyall SC, Michael GJ, Michael-Titus AT. Omega-3 fatty acids reverse age-related decreases in nuclear receptors and increase neurogenesis in old rats. *J Neurosci Res* August 1, 2010;**88**(10):2091–102.
99. Greiner RS, Moriguchi T, Hutton A, Slotnick BM, Salem Jr N. Rats with low levels of brain docosahexaenoic acid show impaired performance in olfactory-based and spatial learning tasks. *Lipids* 1999;**34**(Suppl.):S239–43.
100. Bazan NG. Neuroprotectin D1-mediated anti-inflammatory and survival signaling in stroke, retinal degenerations, and Alzheimer's disease. *J Lipid Res* April 2009;**50**(Suppl.):S400–5.
101. Cole GM, Frautschy SA. DHA may prevent age-related dementia. *J Nutr* April 2010;**140**(4):869–74.
102. Lukiw WJ, Cui JG, Marcheselli VL, Bodker M, Botkjaer A, Gotlinger K, et al. A role for docosahexaenoic acid-derived neuroprotectin D1 in neural cell survival and Alzheimer disease. *J Clin Invest* October 2005;**115**(10):2774–83.
103. Colangelo V, Schurr J, Ball MJ, Pelaez RP, Bazan NG, Lukiw WJ. Gene expression profiling of 12633 genes in Alzheimer hippocampal CA1: transcription and neurotrophic factor down-regulation and up-regulation of apoptotic and pro-inflammatory signaling. *J Neurosci Res* November 1, 2002;**70**(3):462–73.
104. De Luis DA, Conde R, Aller R, Izaola O, Gonzalez Sagrado M, Perez Castrillon JL, et al. Effect of omega-3 fatty acids on cardiovascular risk factors in patients with type 2 diabetes mellitus and hypertriglyceridemia: an open study. *Eur Rev Med Pharmacol Sci* January–February, 2009;**13**(1):51–5.

105. Tierney AC, McMonagle J, Shaw DI, Gulseth HL, Helal O, Saris WH, et al. Effects of dietary fat modification on insulin sensitivity and on other risk factors of the metabolic syndrome—LIPGENE: a European randomized dietary intervention study. *Int J Obes (Lond)* June 2011;**35**(6):800—9.
106. Barre DE, Mizier-Barre KA, Griscti O, Hafez K. High dose flaxseed oil supplementation may affect fasting blood serum glucose management in human type 2 diabetics. *J Oleo Sci* 2008;**57**(5): 269—73.
107. Taylor CG, Noto AD, Stringer DM, Froese S, Malcolmson L. Dietary milled flaxseed and flaxseed oil improve N-3 fatty acid status and do not affect glycemic control in individuals with well-controlled type 2 diabetes. *J Am Coll Nutr* February 2010;**29**(1): 72—80.
108. Mostad IL, Bjerve KS, Bjorgaas MR, Lydersen S, Grill V. Effects of n-3 fatty acids in subjects with type 2 diabetes: reduction of insulin sensitivity and time-dependent alteration from carbohydrate to fat oxidation. *Am J Clin Nutr* September 2006;**84**(3): 540—50.
109. Mostad IL, Bjerve KS, Lydersen S, Grill V. Effects of marine n-3 fatty acid supplementation on lipoprotein subclasses measured by nuclear magnetic resonance in subjects with type II diabetes. *Eur J Clin Nutr* March 2008;**62**(3):419—29.
110. Mori TA, Woodman RJ, Burke V, Puddey IB, Croft KD, Beilin LJ. Effect of eicosapentaenoic acid and docosahexaenoic acid on oxidative stress and inflammatory markers in treated-hypertensive type 2 diabetic subjects. *Free Radic Biol Med* October 1, 2003;**35**(7):772—81.
111. Nenseter MS, Drevon CA. Dietary polyunsaturates and peroxidation of low density lipoprotein. *Curr Opin Lipidol* February 1996; **7**(1):8—13.

CHAPTER

24

Polymorphism, Carbohydrates, Fat, and Type 2 Diabetes

Jose Lopez-Miranda, Carmen Marin

Lipids and Atherosclerosis Unit, IMIBIC/Reina Sofia University Hospital, University of Cordoba and CIBER Fisiopatologia Obesidad y Nutricion (CIBERobn), Instituto de Salud Carlos III, Cordoba, Spain

1. INTRODUCTION

Type 2 diabetes (T2D) is growing exponentially and is now one of the most widespread endocrine diseases in the general population. The latest estimate from the International Diabetes Federation gives a prevalence rate of as much as 8.4% of the adult population, whereas worldwide diabetes cases hit a new record in 2013 at 382 million. However, the Centers for Disease Control and Prevention in the United States estimates that as many as 27% of individuals with T2D are undiagnosed.[1] The prevalence of diabetes for all age groups worldwide was estimated to be 2.8% in 2000 and is estimated to be 4.4% in 2030.[2] The prevalence of T2D varies with race and ethnicity.

Diabetes risk factors have been extensively studied and include clinical characteristics (e.g., age, sex, ethnicity, family history, body mass index), glucose, and biochemical parameters associated with insulin resistance (IR) and inflammation. Furthermore, T2D is associated with cardiovascular diseases (CVDs), which are among the most frequent causes of morbidity and mortality in the affected subjects, and can also cause diabetic microvascular complications (diabetic nephropathy, retinopathy, and diabetic neuropathy).[3] For this reason, the prevention and treatment of diabetes take up an enormous amount of effort and funds. In addition, T2D is associated with obesity, and it is known that adipose tissue serves as an important active endocrine organ that produces several hormone-like compounds that can increase IR.[4] Therefore, patients with T2D display both β-cell dysfunction and IR.

Genetic susceptibility plays an important role in the etiology and manifestation of T2D. Recent genome-wide association studies (GWAS) have identified several genes associated with T2D with previously unknown functions. Most of these gene loci affect insulin secretion.

On the other hand, lifestyle changes for diabetic patients, including diet, physical activity, and behavioral therapy, can be used to facilitate weight loss in conjunction with several different dietary approaches.

In light of all these findings, this chapter reviews the role of carbohydrates and dietary fat on the regulation and prevention of T2D as well as discussing the most important polymorphisms involved in the development of T2D.

2. EFFECT OF DIETARY CARBOHYDRATES AND FAT ON T2D

Each diabetic patient has a specific calorie requirement according to age, weight, sex, and exercise; the distribution of macronutrients depends on the lipid profile and renal function, schedule of intake, lifestyle, and the hypoglycemic drugs administered. Diets rich in whole grains, fruit, vegetables, legumes, and nuts; moderate in alcohol consumption; and lower in refined grains, red or processed meats, and sugar-sweetened beverages have been shown to reduce the risk of diabetes and improve glycemic control and blood lipids in patients with T2D.[5]

Previously, low-fat diets were recommended without much attention to the quality of fat; in current guidelines, however, there is general emphasis on the quality of fat.[6] For example, saturated fat worsens insulin sensitivity, whereas monounsaturated and polyunsaturated fats improve it.[7,8] Few prospective studies have used

Molecular Nutrition and Diabetes
http://dx.doi.org/10.1016/B978-0-12-801585-8.00024-5

fatty acid biomarkers to assess associations with T2D. In a recent large prospective cohort, circulating palmitic acid and stearic acid were associated with higher diabetes risk, and vaccenic acid (de novo lipogenesis fatty acid biomarkers) was associated with lower diabetes risk.[9] Experimental evidence suggests that hepatic de novo lipogenesis affects insulin homeostasis via synthesis of saturated fatty acids (SFAs) and monounsaturated fatty acids (MUFAs).

However, both the quality and quantity of fat intake affects glycemia and cardiovascular risk. A prospective cohort study showed that the high risk of developing T2D is ameliorated through adherence to a reduced fat intake from 34% to 26% of energy in subjects with impaired glucose tolerance (IGT).[10] Although dietary omega-3 (n-3) fatty acids (eicosapentaenoic acid [EPA] and docosahexaenoic acid [DHA]) may confer some cardiovascular benefits, these nutrients may also increase the risk of T2D. A recent study suggested that a high-complex carbohydrate diet supplemented with long-chain n-3 polyunsaturated fatty acid (PUFA) reduces systemic IR and improves insulin signaling in subcutaneous white adipose tissue compared with the consumption of a fat-rich diet and baseline Spanish habitual diets.[11] Djousse et al. found that increased intakes of long-chain omega-3 fatty acids of marine origin were associated with an increased risk of T2D.[12] Nevertheless, the findings of a systematic review do not support either the harmful or beneficial effects of fish/seafood or EPA + DHA on development of T2D, and suggest that alpha linoleic acid (ALA) may be associated with a slightly lower risk.[13] A recent study suggested an association between PUFA intakes, especially ALA, and the prevention of a cardiac event in patients with T2D.[14] Several studies suggest that the Mediterranean-style diet may improve glucose metabolism in patients with T2D, but the results are inconsistent. The Mediterranean diet improved glycemic control,[15] lipid profile and lower blood glucose,[16] body weight,[17] and cardiovascular risk factors in T2D patients.[18] Thus, the Mediterranean diet pattern, low in saturated fat and rich in vegetables, fruit, legumes, fish, nuts, and olive oil, can be used as therapeutic strategy in patients with T2D.[19–22] Previous studies have shown that low-carbohydrate, low glycemic index, Mediterranean, and high-protein diets may all be effective in improving various markers of cardiovascular risk in people with diabetes and could have a wider role in the management of diabetes.[23] The Mediterranean diet could, if appropriately adjusted to reflect local food availability and the individual's needs, constitute a beneficial nutritional choice for the primary prevention of diabetes.[24]

Carbohydrates are the energetic substrates that have the greatest impact on glycemia levels. The benefits of carbohydrate restriction in diabetes are immediate. Dietary carbohydrate restriction reliably reduces high blood glucose, does not require weight loss, and leads to the reduction or elimination of medication.[25–27] However, carbohydrates are the major source of energy, typically accounting for 45–70% of the total energy intake and expenditure. The diet provides carbohydrates in the form of fruit, cereals, pasta, legumes, vegetables, and tubers.

Furthermore, a high intake of whole grain cereals and their products, such as whole wheat bread, is associated with a 20–30% reduction in the risk of T2D.[28] It is possible that the benefits of whole grain consumption in reducing the risk of T2D and CVD could also be mediated by mechanisms related with postprandial metabolism. The total amount of carbohydrates is the main factor responsible for the postprandial response. Nevertheless, other variables have a role to play, such as the type of carbohydrate, richness in fiber, the freshness of the food, and how food is cooked. Similarly, ingestion of slowly absorbed isomaltose attenuates postprandial hyperglycemia by reducing the appearance of oral glucose, inhibiting endogenous glucose production, and increasing splanchnic glucose uptake compared with the ingestion of rapidly absorbed sucrose in patients with T2D.[29] Moreover, there are other factors that can also influence postprandial glycemia, such as preprandial glycemia, macronutrient distribution of the whole meal (fats and proteins), and the hypoglycemic treatment administered: oral tablets or insulin. In fact, the quality of the dietary fats and carbohydrates consumed is more crucial than the quantity of the macronutrients.[5] However, there is evidence that modifying the amount of macronutrients can improve glycemic control, weight, and lipids in people with diabetes. Recently, studies show that the consumption of tree nuts improves glycemic control in individuals with T2D.[30] It also seems that starch from legumes has a positive effect on glycemia because of their persistent effect on postprandial glycemia, with no sudden increases, and they may prevent both postprandial hyperglycemia and late hypoglycemia.[31,32] A modest reduction in dietary carbohydrate has beneficial effects on body composition, fat distribution, and glucose metabolism. After weight loss, participants who consumed the lower carbohydrate diet had 4.4% less total fat mass.[33] Weight loss also plays an important role in treating patients with T2D, because most individuals with T2D are overweight or obese. Several studies have demonstrated that modest weight loss (5–10% of body weight) is associated with significant improvements in patients' levels of glycemic control, lipids, blood pressure, and other cardiovascular risk factors. Furthermore, a modest weight loss of as little as 4.5 kg can result in reducing the glycated hemoglobin level by approximately 0.5%. Pharmacological agents, when combined with these approaches, may further augment weight loss.[34] Interventions for weight loss as obesity develops may offer the best opportunity to reduce the progression to prediabetes and T2D. Traditionally, a low-fat diet has been recommended to achieve weight loss and improve glycemic control in T2D.

Lifestyle interventions of T2D may also lead to reduced inflammation.[35] Inflammation is considered to play important role in the development of T2D as well as in further complications of the disease. Efforts to reduce systemic inflammation in patients with T2D may be crucial in preventing complications such as CVD and nephropathy. An intake of carbohydrates, in particular refined carbohydrates, has been associated with a proinflammatory response.[36] Nevertheless, a recent study has shown that advice to follow a low-carbohydrate diet, aiming at a 20% carbohydrate intake, reduces the subclinical proinflammatory state in T2D. Similarly, the consumption of a low-carbohydrate diet may be an effective strategy to improve this state in patients with T2D.[37,38]

Therefore, the review of the existing literature on low-carbohydrate, low-glycemic index, Mediterranean, and high-protein diets suggests that these diets may be effective in improving various markers of cardiovascular risk in people with diabetes and could have a wider role in the management of diabetes.[23] The most suitable composition of these diets must be <30% total fat, <10% saturated fats, >15 g/1000 kcal fiber (half soluble), 45–60% of carbohydrates, and protein intake of 15–20% of the total calories a day. Patients need to limit the intake of saturated fats to <7% of the daily calorie intake. MUFAs such as olive oil and other vegetable oils are recommended. In this way, the nutritional management of the diabetic patient is the first step in the general treatment of diabetes. A balanced diet does not only involve glycemic control (e.g., amount, type and time of food intake, insulin therapy), but it also affects the entire metabolism, preventing the progression of diabetes and its concomitant complications.

3. POLYMORPHISMS AND T2D

T2D is a common, heterogeneous, and complex disease in which genetic and environmental factors[39] interact to cause hyperglycemia, which constitutes the main hallmark of T2D.[40] Both genetic and environmental factors interact in determining impaired β-cell insulin secretion and peripheral IR in patients with T2D. Genetic association studies have identified several diabetes risk variants, and more than 1240 gene loci are associated with diabetes in humans.[41–43]

Although the heritable nature of T2D has been recognized for many years,[44] only in the past two decades have linkage analyses in families and GWAS studies in large populations begun to reveal the genetic landscape of the disease in detail. Recently, a T2D-associated locus has been identified through GWAS studies and followed through genotyping for genetic markers in family members with and without T2D.

The physiological impact of these variants is known in the majority of the risk variants which impact on β-cell function[45] (ARAP1 rs11603334,[46] insulin receptor substrate (IRS),[47–49] ABCC8,[50] MTNR1B rs10830963, GCK rs1799884, G6PC2 rs560887, GCKR rs780094),[51] ADRA2A,[52] ANKYRIN1,[53–56] CDKN2A/B,[57,58] CAPN10 rs2975760, and single nucleotide polymorphism (SNP)-43G>A,[59,60] TCF7L2 (rs290487 and rs7903146),[61–64] rather than IR[47,65] (Table 1). A previous study has even found more variants associated with insulin secretion than with IR.[47] Peripheral IR predominantly suffers from the environmental component.[52,61,66] Several gene variants involved in β-cell insulin secretion have been identified using the analysis of functional or positional candidate genes. This methodology has helped to identify new associations between T2D and genes whose function is either known or unknown,[67] and different studies have identified at least 65 genomic regions associated with T2D. A large number of variants have been identified in many of these genes, most of which may influence both hepatic and peripheral IR, adipogenesis and β-cell mass and function. A recent study showed SNPs affecting glucose metabolism variables: rs26125 (PPARGC1B) is associated to fasting glucose; rs4759277 (LRP1) to fasting insulin, C-peptide and homeostasis assessment of IR; rs184003 (AGER) to quantitative insulin sensitivity check index; rs7301876 (ABCC9) to the sensitivity index; rs290481 (TCF7L2) to acute insulin response to glucose; and rs12691 (CEBPA) is linked to disposition index[68]; *HMGA* is capable of reducing the intracellular levels of the insulin receptor gene (a gene involved in adipogenesis and β-cell insulin secretion) in insulin target tissues;[67] and rs266729 (ADIPOQ) is associated with IR. The G versus C allele of −11377C>G (rs266729) might also be a risk factor for T2D.[69]

In addition, previous studies have shown that fasting plasma glucose (FPG) levels are associated with an increased risk of T2D. Recently, several studies have identified sequence variants in the promoter region of GCK, GCKR, G6PC2, and MTNR1B gene that influence FPG.[70–73] There are genetic variants that altered IR,[74,75] in subjects with impaired fasting glucose (IFG) or newly diagnosed T2D. For example, the combined effect of AdipoQ SNPs 45T>C and 276G>T on the conversion from IGT to T2D were observed to be stronger than that of each SNP alone.[76] In addition, patients with diabetes or prediabetes are at increased risk of dyslipidemia and CVD. Thus, dyslipidemia is another risk factor for

TABLE 1 Interaction between Dietary Fat and Carbohydrates and Genes Involved in Development of T2D

Dietary intervention	Gene	Effect on glucose metabolism
Dietary carbohydrates	PSMD3 (rs709592)	T allele of rs709592 had higher HOMA-IR than C homozygotes with low carbohydrate intake.[91]
	IRS-1(rs2943641)	In women, the T allele was associated with decreased risk of T2D in the lower tertiles of carbohydrate intake.[94]
	IRS1(G972C)	Insulin sensitivity increased in GR subjects for the G972R polymorphism at the IRS1 gene locus after the intake of a carbohydrate-rich diet.[95]
	APOA5 (-1131T/C)	The replacement of refined rice with whole grains and legumes in a high-carbohydrate diet prevented diabetic hypertriglyceridemia in IFG or diabetic individuals from carrying the APOA5-1131C variant.[101]
	TCF7L2 (rs7903146)	Intake of whole grain had a protective effect on diabetes risk in CC genotype carriers of rs7903146 versus subjects carrying the T allele.[102]
	ADIPQ (276G/T)	The G allele was associated with higher fasting blood glucose in patients consuming a low-carbohydrate diet. The carriers of the T allele had greater fasting blood glucose and glycated hemoglobin concentrations, when carbohydrate intake was intermediate.[100]
Dietary protein and n-3 PUFA	rs1440581 near PPM1K	The C allele carriers, who consumed a high-fat diet, showed a smaller decrease in IR than the noncarriers. Individuals carrying the C allele of the branched-chain amino acid/aromatic amino acid ratio–associated variant rs1440581 may benefit less in weight loss and improvement of insulin sensitivity than those without this allele when undertaking an energy-restricted, high-fat diet.[105]
	GCKR (rs126 326-P446L)	Interaction between GCKR rs126326-P446L polymorphism and plasma n-3 PUFA levels modulated IR in subjects with metabolic syndrome.[108]
	PCK1 rs2179706	The carriers of the C/C genotype exhibited lower plasma concentrations of fasting insulin and HOMA-IR as compared with C/C carriers with n-3 PUFA below the median.[106]
	PIK3CA- KCNMB3 (rs7645550, rs1183319)	The n-3:n-6 PUFA ratio modulated the effects of PIK3CA- KCNMB3 variants (rs7645550 and rs1183319) on the glucose-related traits in the GOLDN study.[109]
	IRS2 (rs2289046)	The individuals carrying the rs2289046 G allele and with the highest plasma level of n-3 PUFAs showed lower fasting insulin and HOMA-IR.[107]
	Leptin receptor (rs3790433)	The indices of insulin sensitivity and IR improved in the GG homozygotes following a low-fat dietary intervention supplemented with long-chain (n-3) PUFAs.[111]
Dietary SFA	FABP2 (Ala54Thr)	FABP2 (Ala54Thr polymorphism) and PLIN1 (11482G>A polymorphism) determined insulin sensitivity and IR. The carriers of the Thr54 allele of FABP2 had higher levels of HOMA-IR in presence of a high-saturated fat diet.[112,113]
	PLIN1 (11482G>A)	When the ratio of saturated fat to carbohydrate was high, insulin and HOMA-IR were higher in minor allele carriers of PLIN1 11482G>A.[114]
	CAPN10 (rs2953171)	The G/G genotype was associated with lower fasting insulin concentrations, lower HOMA-IR, and higher glucose effectiveness in subjects with low SFA concentrations.[115]
	TCF7L2 (rs7903146, rs11196224, rs176855538, rs290481)	The carriers of the rare T allele of *TCF7L2* rs7903146 with higher plasma SFA levels, and of three other SNPs *TCF7L2* (rs11196224, rs17685538, rs290481) had lower insulin secretion.[117]
Dietary MUFA	IRS2 (rs2289046)	The rs2289046 A/A genotype was associated with lower glucose effectiveness, higher fasting insulin concentrations and higher HOMA-IR in subjects with the lowest level of plasma MUFAs.[107]

TABLE 1 Interaction between Dietary Fat and Carbohydrates and Genes Involved in Development of T2D—cont'd

Dietary intervention	Gene	Effect on glucose metabolism
	PSMD3 (rs4065321, rs709592, and rs8065443)	The major allele homozygotes of rs4065321, rs709592, and rs8065443 have lower HOMA-IR and glucose with high dietary MUFA intake.[91]
	FTO (rs9939609), MC4R (rs17782313)	The carriers of the variant alleles, FTO rs9939609 and MC4R rs17782313, had higher T2D risk when adherence to the Mediterranean diet was low, whereas a good adherence to the Mediterranean diet blunted this association.[118]
	ADIPOQ (11377C>G)	C/C homozygous men had less IR after consumption of the MUFA diet.[122]
	SCARB1	The carriers of A allele had an increase in insulin sensitivity after the consumption of a MUFA-rich diet.[123]

GOLDN, Genetics of Lipid-lowering Drugs and Diet Network; GR, genotype GR for the G972 polymorphism; HOMA-IR, homeostatic model assessment of insulin resistance; IFG, impaired fasting glucose; MUFA, monounsaturated fatty acid; PUFA, polyunsaturated fatty acid; SFA, saturated fatty acid; SNP, single nucleotide polymorphism; T2D, type 2 diabetes.

T2D and so genes regulating atherogenic dyslipidemia are promising candidate genes for T2D. The G-250A promoter polymorphism of the LIPC gene is associated with an increased risk of development of T2D in high-risk subjects with IGT.[77] Other genetic variants in vascular endothelial growth factor and KDR (rs2071559) have been related to atherosclerosis and diabetic retinopathy.[78] In addition, the CC genotype of the KDR −604T>C polymorphism is a possible risk factor for myocardial infarct in Caucasians with T2D.[79]

On the other hand, there are genetic factors involved in the development of obesity and T2D, namely ADIPOR1 and ADIPOR2,[80] GCK,[51] PCSK1,[81,82] and KCNJ11 (E23K)[83,84] as well as one of the less well-known genes, FAMI2. However, Corella et al. confirmed the association between the FAM2-rs7138803 polymorphism and obesity risk. These associations are higher in subjects with T2D, and homozygous subjects for the variant allele tended to have a higher risk of myocardial infarction than the other genotypes in T2D.[85]

The European Prospective Investigation into Cancer and Nutrition InterAct study showed that a genetic risk score for T2D is highest in those who are younger and leaner at baseline and the absolute risk of T2D is dominated by modifiable factors, particularly obesity.[86] These genetics factors include SNPs in genes that encode proteins by influencing both body composition and fat and glucose metabolism. These genes are associated with an insulin-mediated glucose profile and an abnormal expression of adipose tissue endocrine and inflammatory factors. T2D is associated with chronic inflammation that can be attributed to dysregulation of the innate immune system and this is a potential link between metabolic syndrome (MetS), diabetes, and atherosclerosis.[87] Furthermore, it has been proposed that mediators of inflammation, such as interleukins -1β, -1Ra, -18, -4, -6, and -10; tumor necrosis factor-α; and adiponectin, are involved in causing T2D. Genetic polymorphism of those cytokine genes has been implicated in the pathogenesis of diabetes and shows considerable levels of variation among ethnic groups around the world.[88]

4. INTERACTION BETWEEN CARBOHYDRATES, FAT, AND GENE POLYMORPHISMS

The major diseases (e.g., CVD, diabetes, obesity, cancers) are modulated by the interaction between genetic susceptibility and environmental factors, especially diet (Table 1). Variants in genes, including ACSL1, ADIPOQ, ADIPOR1, CLOCK, PLIN1, TCF/L2, FABP2, KCNJ11, PPARG, scavenger receptor class B type I (SCARB1), CEBPA, and LERN, have all been shown to interact with dietary fat or carbohydrate, influencing disposition index, insulin sensitivity, IR, and glucose metabolism across different populations.[89,90]

4.1 Interaction between Dietary Carbohydrates and T2D

Recent studies have shown genetic variants associated with IR that interact with dietary carbohydrates.[91] A novel candidate gene is PSMD3, a member of the 26S proteasome family. Subjects carrying the T allele of rs709592 had higher homeostatic model assessment of insulin resistance (HOMA-IR) than C homozygotes with a low carbohydrate intake. Another candidate gene was IRS1. The Genetics of Lipid-lowering Drugs and Diet Network (GOLDN) study found that genetic variations at IRS1 were associated with IR, fasting insulin, T2D, IFG/T2D, and MetS.[92] The C allele of rs2943641 has been associated with IR, hyperinsulinemia and a higher risk of T2D in French, Danish, and Finnish populations.[93,94] The IRS-1 rs2943641 showed interaction with dietary carbohydrates. Among women, the T allele was associated with decreased risk of T2D in the lower tertiles of carbohydrate intake.[94] In addition, a previous study observed that the insulin sensitivity increased in GR (Gly972Arg) subjects for the G972R polymorphism at the IRS1 gene locus after the intake of a carbohydrate-rich diet.[95]

Thus, dietary carbohydrates, which influence glucose concentration and insulin demand, may modify associations between GWAS variants and T2D.[96–98] Fiber, associated with a lower risk of T2D and which may influence the amount or rate of carbohydrates absorbed, may also modify these associations.[98,99] Hwang et al. observed that SNP276G/T polymorphism in the AdipoQ gene influenced blood glucose levels in relation to dietary carbohydrate intake.[100] The G allele was associated with higher fasting blood glucose in patients consuming a low-carbohydrate diet (<55% energy). However, when carbohydrate intake was intermediate (55–65%), carriers of the T allele had greater fasting blood glucose and glycated hemoglobin concentrations. When carbohydrate intake was high (>65%), carriers of the T allele had greater high-density lipoprotein (HDL) cholesterol concentrations. Furthermore, high-carbohydrate diet intervention with a different carbohydrate source (whole grain and legumes versus refined rice) could attenuate or mask the adverse effect of a high-carbohydrate intake in individuals with IFG or newly diagnosed T2D. Charriere et al. observed an association between diabetic dyslipidemia and APOA5-1131T/C polymorphism. A recent study found that the replacement of refined rice with whole grains and legumes in a high-carbohydrate diet may need to be considered to prevent diabetic hypertriglyceridemia in IFG or diabetic individuals found to carry the APOA5-1131C variant.[101] In addition, the influence of whole grains on postprandial glucose response and insulin demand is inversely associated with diabetes risk. Fisher et al. provided evidence that the beneficial effect of whole grain intake on diabetes risk is modified by TCF7L2 (promoted by an early insulin secretor defect) rs7903146 genotypes.[102] This study found a protective effect of whole grain intake on diabetes risk in CC genotype carriers of rs7903146 versus subjects carrying the T allele.

4.2 Interaction between the Dietary Proteins n-3 PUFA and T2D

Emerging evidence has shown that circulating amino acids may play an important role in the pathogenesis of metabolic disorders such as IR and T2D. Wang et al. identified that high levels of circulating branched-chain amino acids (BCAAs) and aromatic amino acids (AAAs) predicted T2D in two prospective cohorts.[103] Blood levels of amino acids are partially determined by genetic factors. A recent GWAS found an SNP rs1440581 near the PPM1K gene (PP2C domain-containing protein phosphatase 1K) to be associated with the ratio of BCAA to AAA (Fischer's ratio). Interestingly, the PPM1K gene was recently identified as a susceptibility gene for T2D.[104] A prospective dietary intervention trial found significant interactions between dietary fat and a genetic variant rs1440581 near the PPM1K gene associated with amino acid metabolites on weight loss and changes in IR. The C allele carriers, who consumed a high-fat diet, showed less weight loss and a smaller decrease of IR than the noncarriers compared with those subjects that consumed a low-fat diet. In addition, individuals carrying the C allele of BCAA/AAA ratio associated variant rs1440581 may benefit less in weight loss and improvement of insulin sensitivity than those without this allele when they undertake an energy-restricted high-fat diet.[105]

Observational and intervention studies suggest that the quality of dietary fats may influence IR. For example, saturated fat worsens, and MUFAs and PUFAs improve insulin sensitivity.[7,8] The phosphoenolpyruvate carboxykinase gene (*PCK1*) plays a significant role in regulating glucose metabolism, and fatty acids are key metabolic regulators, which interact with transcription factors and influence glucose metabolism. The PCK1 rs2179706 polymorphism interacts with plasma concentration of n-3 PUFA levels modulating IR. The carriers of the C/C genotype exhibited lower plasma concentrations of fasting insulin and HOMA-IR compared with C/C carriers with n-3 PUFA below the median.[106] On the other hand, several IRS-2 polymorphisms have been studied in relation to IR and T2D. The individuals carrying the rs2289046 G allele and with the highest plasma level of n-3 PUFAs showed lower fasting insulin and HOMA-IR compared with A/A subjects.[107] Another candidate gene associated with IR was GCKR, and here, a previous study demonstrated a significant interaction between GCKR rs126326-P446L polymorphism and plasma n-3 PUFA levels modulating IR in subjects with MetS.[108] One of the novel candidate genes, PIK3CA, a gene that encodes phosphatidylinositol 3-kinase class-1A p110α catalytic subunit and KCNMB3, is 5 kbp downstream of PIK3CA and is in a large linkage disequilibrium block with this gene. Zheng et al. observed that dietary PUFA, especially the n-3:n-6 PUFA ratio, modulated the effects of PIK3CA-KCNMB3 variants (rs7645550 and rs1183319) on the glucose-related traits in the GOLDN study.[109]

On the other hand, leptin, an adipocytokine that has been shown to stimulate glucose uptake and fatty acid oxidation, modulates insulin secretion and action via leptin receptors (LEPRs) that are present in β cells. Several studies have shown that LEPR polymorphisms are associated with insulin and glucose metabolism, IR, and T2D.[110] The LIPGENE-SU.VI.MAX demonstrated that a common genetic variant at the LEPR locus, rs3790433, was associated with increased risk of IR. They showed improvements to insulin sensitivity and IR index when the GG homozygotes followed a 12-week, low-fat dietary intervention supplemented with long-chain (n-3) PUFAs.[111]

4.3 Interaction between Dietary SFAs and T2D

Furthermore, there are SNPs that interact with plasma SFAs to determine insulin sensitivity[112] and IR (FABP2 Ala54Thr polymorphism and PLIN1 11482G>A).[113,114] For example, the calpain-10 protein (intracellular Ca(2+)-dependent cysteine protease) may play a role in glucose metabolism, pancreatic β-cell function, and regulation of thermogenesis. Several CAPN10 polymorphic sites have been studied for their potential use as risk markers for T2D. The rs2953171 SNP interacted with plasma total SFA concentrations, which were significantly associated with insulin sensitivity, fasting insulin, HOMA-IR, and glucose effectiveness. The G/G genotype was associated with lower fasting insulin concentrations, lower HOMA-IR, and higher glucose effectiveness in subjects with low SFA concentrations than in subjects with the minor A allele. In contrast, subjects with the G/G allele with the highest SFA concentrations had higher fasting insulin and HOMA-IR values and lower glucose effectiveness than did subjects with the A allele.[115] Recently, SNPs of the transcription factor 7-like 2 (TCF7L2) gene have been reported to affect T2D susceptibility by indirectly altering expression of GLP-1, which, in addition to insulin, plays a critical role in blood glucose homeostasis.[64,116] The carriers of the rare T allele of TCF7L2 rs7903146 with higher plasma SFA levels, and of three other SNPs TCF7L2 (rs11196224, rs17685538, rs290481) in linkage disequilibrium with rs7903146, had lower blood pressure and insulin secretion.[117]

4.4 Interaction between Dietary Monounsaturated Fatty Acids and T2D

The high-MUFA diets are a current option for diabetic patients because of their positive effects on glycemic control, IR, and HDL cholesterol levels. These effects are influenced by genetic variants and therefore the effect of IRS-2 polymorphisms on T2D could be modified by consumption of a MUFA-rich diet. Among subjects with the lowest level of plasma MUFA, the rs2289046 A/A genotype was associated with lower glucose effectiveness, higher fasting insulin concentrations, and higher HOMA-IR as compared with subjects carrying the minor G allele. In contrast, among subjects with the highest level of plasma MUFA, the A/A genotype was associated with lower fasting insulin concentrations and HOMA-IR.[107] Another candidate gene is PSMD3. Several SNPs in the PSMD3 gene interacted with dietary MUFAs, influencing glucose concentration and in HOMA-IR. The major allele homozygotes of rs4065321, rs709592, and rs8065443 appeared to have lower HOMA-IR and glucose compared with minor allele carriers with high dietary MUFA intake.[91]

The consumption of a high-MUFA diet in persons with T2D can induce favorable effects in insulin sensitivity and in other associated risk factors, such as obesity. Obesity, and in particular visceral and ectopic adiposity, increases the risk of T2D. A recent *meta*-analysis in East and South Asians concluded that the fat mass- and obesity-associated gene (FTO) rs9939609 minor allele, the risk allele for obesity, increased the risk of T2D.[118] Likewise, melanocortin-4 receptor (MC4R) rs17782313 polymorphism gene is associated with T2D,[119] depending on the diet consumed. Ortega-Azorin et al. described a significant gene–diet interaction of the FTO rs9939609 and MC4R rs17782313 with adherence to the Mediterranean diet on T2D. The FTO rs9939609 was associated with higher T2D risk in subjects with a low adherence to the Mediterranean diet, whereas a good adherence to the Mediterranean diet blunted this association.[120]

Furthermore, adiponectin (encoded by ADIPOQ) is an important adipocytokine secreted by adipocytes that plays a key role in the inflammatory response associated with obesity, insulin resistant states, and T2D. Chu et al. found that several SNPs (−11426A>G, −11391G>A, and −1177C>G) could increase T2D risk in European populations.[121] Furthermore, C/C homozygous men for the −11377C>G SNP at adipoQ gene had significantly less IR after consumption of the MUFA diet.[122]

One of the most common lipid abnormalities observed in T2D is low serum concentrations of HDL cholesterol. SCARB1 was described as the first high-density lipoprotein receptor. Increasing evidence indicates that SCARB1 plays additional roles, particularly in T2D. Steady-state plasma glucose after the consumption of a MUFA-rich diet was lower in G/A compared with G/G subjects. This study shows that carriers of the A allele have an increase in insulin sensitivity after the consumption of a MUFA-rich diet, compared with the effect observed in G/G individuals.[123]

5. FUTURE PERSPECTIVES

Large-scale studies on gene–diet interactions are necessary to thoroughly evaluate the relevance and prevention of T2D. The new recommendations for subjects with IFG, IGT, or recently diagnosed T2D can be based on the specific responses to particular dietary patterns or nutrients that, at least in part, may be modulated by various gene polymorphisms and their combinations. This knowledge will provide us with targeted dietary advice for specific groups of patients with T2D. Furthermore, it is very important to identify new therapeutic targets through nutrigenomic, nutrigenetic, and epigenetic studies to reduce the high prevalence of T2D prevalence that is predicted in the world population.

Acknowledgments

We would like to thank the CIBER Fisiopatologia de la Obesidad y Nutricion, CIBEROBN, an initiative of ISCIII, Government of Spain. Fondo Europeo de Desarrollo Regional (FEDER). It is also partly supported by research grants from the Ministerio de Ciencia e Innovacion (AGL2009-122270 to J L-M); Ministerio de Economia y Competitividad (AGL2012/39615 to J L-M); Consejeria de Salud, Junta de Andalucia (PI0193/09 to J L-M); Consejería de Innovación Ciencia y Empresa (CVI-7450 to J L-M); Plan Nacional de I + D + I 2008–2011 and ISCIII-Subdirección General de Evaluación y Fomento de la Investigación as sources of funding. C. Marin is supported by an ISCIII research contract (Programa Miguel-Servet-CP 11/00185). All decisions regarding the design, conduct, collection, analysis, or interpretation of the data and the decision to submit the manuscript for publication were made independently by the authors. None of the authors has any conflict of interest that could affect the performance of the work or the interpretation of the data.

References

1. King H, Aubert RE, Herman WH. Global burden of diabetes, 1995–2025: prevalence, numerical estimates, and projections. *Diabetes Care* September 1998;**21**(9):1414–31.
2. Wild S, Roglic G, Green A, Sicree R, King H. Global prevalence of diabetes: estimates for the year 2000 and projections for 2030. *Diabetes Care* May 2004;**27**(5):1047–53.
3. Krolewski AS, Warram JH, Freire MB. Epidemiology of late diabetic complications. A basis for the development and evaluation of preventive programs. *Endocrin Metab Clin N Am* June 1996; **25**(2):217–42.
4. Jackson MB, Ahima RS. Neuroendocrine and metabolic effects of adipocyte-derived hormones. *Clin Sci* February 2006;**110**(2):143–52.
5. Ley SH, Hamdy O, Mohan V, Hu FB. Prevention and management of type 2 diabetes: dietary components and nutritional strategies. *Lancet* June 7, 2014;**383**(9933):1999–2007.
6. Schwab U, Lauritzen L, Tholstrup T, Haldorssoni T, Riserus U, Uusitupa M, et al. Effect of the amount and type of dietary fat on cardiometabolic risk factors and risk of developing type 2 diabetes, cardiovascular diseases, and cancer: a systematic review. *Food Nutr Res* 2014;**58**.
7. Perez-Jimenez F, Lopez-Miranda J, Pinillos MD, Gomez P, Paz-Rojas E, Montilla P, et al. A Mediterranean and a high-carbohydrate diet improve glucose metabolism in healthy young persons. *Diabetologia* November 2001;**44**(11):2038–43.
8. Rivellese AA, De Natale C, Lilli S. Type of dietary fat and insulin resistance. *Ann NY Acad Sci* June 2002;**967**:329–35.
9. Ma W, Wu JH, Wang Q, Lemaitre RN, Mukamal KJ, Djousse L, et al. Prospective association of fatty acids in the de novo lipogenesis pathway with risk of type 2 diabetes: the cardiovascular health Study. *Am J Clin Nutr* January 2015;**101**(1):153–63.
10. Swinburn BA, Metcalf PA, Ley SJ. Long-term (5-year) effects of a reduced-fat diet intervention in individuals with glucose intolerance. *Diabetes Care* April 2001;**24**(4):619–24.
11. Jimenez-Gomez Y, Cruz-Teno C, Rangel-Zuniga OA, Peinado JR, Perez-Martinez P, Delgado-Lista J, et al. Effect of dietary fat modification on subcutaneous white adipose tissue insulin sensitivity in patients with metabolic syndrome. *Mol Nutr Food Res* November 2014;**58**(11):2177–88.
12. Djousse L, Gaziano JM, Buring JE, Lee IM. Dietary omega-3 fatty acids and fish consumption and risk of type 2 diabetes. *Am J Clin Nutr* January 2011;**93**(1):143–50.
13. Wu JH, Micha R, Imamura F, Pan A, Biggs ML, Ajaz O, et al. Omega-3 fatty acids and incident type 2 diabetes: a systematic review and meta-analysis. *Br J Nutr* June 2012;**107**(Suppl. 2):S214–27.
14. dos Santos AL, Weiss T, Duarte CK, Gross JL, de Azevedo MJ, Zelmanovitz T. Dietary fat composition and cardiac events in patients with type 2 diabetes. *Atherosclerosis* September 2014;**236**(1): 31–8.
15. Esposito K, Maiorino MI, Ceriello A, Giugliano D. Prevention and control of type 2 diabetes by Mediterranean diet: a systematic review. *Diabetes Res Clin Pract* August 2010;**89**(2):97–102.
16. Kastorini CM, Milionis HJ, Esposito K, Giugliano D, Goudevenos JA, Panagiotakos DB. The effect of Mediterranean diet on metabolic syndrome and its components: a meta-analysis of 50 studies and 534,906 individuals. *J Am Coll Cardiol* March 15, 2011;**57**(11):1299–313.
17. Lasa A, Miranda J, Bullo M, Casas R, Salas-Salvado J, Larretxi I, et al. Comparative effect of two Mediterranean diets versus a low-fat diet on glycaemic control in individuals with type 2 diabetes. *Eur J Clin Nutr* July 2014;**68**(7):767–72.
18. Huo R, Du T, Xu Y, Xu W, Chen X, Sun K, et al. Effects of Mediterranean-style diet on glycemic control, weight loss and cardiovascular risk factors among type 2 diabetes individuals: a meta-analysis. *Eur J Clin Nutr* November 5, 2014.
19. Salas-Salvado J, Bullo M, Babio N, Martinez-Gonzalez MA, Ibarrola-Jurado N, Basora J, et al. Reduction in the incidence of type 2 diabetes with the Mediterranean diet: results of the PREDIMED-Reus nutrition intervention randomized trial. *Diabetes Care* January 2011;**34**(1):14–9.
20. Salas-Salvado J, Martinez-Gonzalez MA, Bullo M, Ros E. The role of diet in the prevention of type 2 diabetes. *Nutr Metab Cardiovasc Dis* September 2011;**21**(Suppl. 2):B32–48.
21. InterAct C, Romaguera D, Guevara M, Norat T, Langenberg C, Forouhi NG, et al. Mediterranean diet and type 2 diabetes risk in the European Prospective Investigation into Cancer and Nutrition (EPIC) study: the InterAct project. *Diabetes care* September 2011;**34**(9):1913–8.
22. Perez-Martinez P, Garcia-Rios A, Delgado-Lista J, Perez-Jimenez F, Lopez-Miranda J. Mediterranean diet rich in olive oil and obesity, metabolic syndrome and diabetes mellitus. *Curr Pharm Des* 2011;**17**(8):769–77.
23. Ajala O, English P, Pinkney J. Systematic review and meta-analysis of different dietary approaches to the management of type 2 diabetes. *Am J Clin Nutr* March 2013;**97**(3):505–16.
24. Koloverou E, Esposito K, Giugliano D, Panagiotakos D. The effect of Mediterranean diet on the development of type 2 diabetes mellitus: a meta-analysis of 10 prospective studies and 136,846 participants. *Metab Clin Exp* July 2014;**63**(7):903–11.
25. Feinman RD, Pogozelski WK, Astrup A, Bernstein RK, Fine EJ, Westman EC, et al. Dietary carbohydrate restriction as the first approach in diabetes management: critical review and evidence base. *Nutrition* 2015;**31**(1):1–13.
26. Alssema M, Boers HM, Ceriello A, Kilpatrick ES, Mela DJ, Priebe MG, et al. Diet and glycaemia: the markers and their meaning. A report of the Unilever Nutrition Workshop. *Br J Nutr* December 2014;**11**:1–10.

27. Maki KC, Phillips AK. Dietary substitutions for refined carbohydrate that show promise for reducing risk of type 2 diabetes in men and women. *J Nutr* January 2015;**145**(1):159S–63S.
28. Gil A, Ortega RM, Maldonado J. Wholegrain cereals and bread: a duet of the Mediterranean diet for the prevention of chronic diseases. *Public Health Nutr* December 2011;**14**(12A):2316–22.
29. Ang M, Linn T. Comparison of the effects of slowly and rapidly absorbed carbohydrates on postprandial glucose metabolism in type 2 diabetes mellitus patients: a randomized trial. *Am J Clin Nutr* October 2014;**100**(4):1059–68.
30. Viguiliouk E, Kendall CW, Blanco Mejia S, Cozma AI, Ha V, Mirrahimi A, et al. Effect of tree nuts on glycemic control in diabetes: a systematic review and meta-analysis of randomized controlled dietary trials. *PloS One* 2014;**9**(7):e103376.
31. American Diabetes A, Bantle JP, Wylie-Rosett J, Albright AL, Apovian CM, Clark NG, et al. Nutrition recommendations and interventions for diabetes: a position statement of the American Diabetes Association. *Diabetes Care* January 2008;**31**(Suppl. 1): S61–78.
32. Mann JI, De Leeuw I, Hermansen K, Karamanos B, Karlstrom B, Katsilambros N, et al. Evidence-based nutritional approaches to the treatment and prevention of diabetes mellitus. *Nutr Metab Cardiovasc Dis* December 2004;**14**(6):373–94.
33. Gower BA, Goss AM. A lower-carbohydrate, higher-fat diet reduces abdominal and intermuscular fat and increases insulin sensitivity in adults at risk of type 2 diabetes. *J Nutr* January 2015;**145**(1):177S–83S.
34. Vetter ML, Amaro A, Volger S. Nutritional management of type 2 diabetes mellitus and obesity and pharmacologic therapies to facilitate weight loss. *Postgrad Med* January 2014;**126**(1):139–52.
35. Kolb H, Mandrup-Poulsen T. The global diabetes epidemic as a consequence of lifestyle-induced low-grade inflammation. *Diabetologia* January 2010;**53**(1):10–20.
36. Oliveira MC, Menezes-Garcia Z, Henriques MC, Soriani FM, Pinho V, Faria AM, et al. Acute and sustained inflammation and metabolic dysfunction induced by high refined carbohydrate-containing diet in mice. *Obesity* September 2013;**21**(9):E396–406.
37. Jonasson L, Guldbrand H, Lundberg AK, Nystrom FH. Advice to follow a low-carbohydrate diet has a favourable impact on low-grade inflammation in type 2 diabetes compared with advice to follow a low-fat diet. *Ann Med* May 2014;**46**(3):182–7.
38. Mayer SB, Jeffreys AS, Olsen MK, McDuffie JR, Feinglos MN, Yancy Jr WS. Two diets with different haemoglobin A1c and antiglycaemic medication effects despite similar weight loss in type 2 diabetes. *Diabetes Obes Metab* January 2014;**16**(1):90–3.
39. Tuomilehto J, Lindstrom J, Eriksson JG, Valle TT, Hamalainen H, Ilanne-Parikka P, et al. Prevention of type 2 diabetes mellitus by changes in lifestyle among subjects with impaired glucose tolerance. *N Engl J Med* May 3, 2001;**344**(18):1343–50.
40. Stumvoll M, Goldstein BJ, van Haeften TW. Type 2 diabetes: pathogenesis and treatment. *Lancet* June 28, 2008;**371**(9631): 2153–6.
41. Shu XO, Long J, Cai Q, Qi L, Xiang YB, Cho YS, et al. Identification of new genetic risk variants for type 2 diabetes. *PLoS Genet* September 2010;**6**(9):e1001127.
42. Lyssenko V, Laakso M. Genetic screening for the risk of type 2 diabetes: worthless or valuable? *Diabetes Care* August 2013;**36**(Suppl. 2):S120–6.
43. Dimas AS, Lagou V, Barker A, Knowles JW, Magi R, Hivert MF, et al. Impact of type 2 diabetes susceptibility variants on quantitative glycemic traits reveals mechanistic heterogeneity. *Diabetes* June 2014;**63**(6):2158–71.
44. Groop L, Forsblom C, Lehtovirta M, Tuomi T, Karanko S, Nissen M, et al. Metabolic consequences of a family history of NIDDM (the Botnia study): evidence for sex-specific parental effects. *Diabetes* November 1996;**45**(11):1585–93.
45. Andersson EA, Allin KH, Sandholt CH, Borglykke A, Lau CJ, Ribel-Madsen R, et al. Genetic risk score of 46 type 2 diabetes risk variants associates with changes in plasma glucose and estimates of pancreatic beta-cell function over 5 years of follow-up. *Diabetes* October 2013;**62**(10):3610–7.
46. Kulzer JR, Stitzel ML, Morken MA, Huyghe JR, Fuchsberger C, Kuusisto J, et al. A common functional regulatory variant at a type 2 diabetes locus upregulates ARAP1 expression in the pancreatic beta cell. *Am J Hum Genet* February 6, 2014;**94**(2): 186–97.
47. Voight BF, Scott LJ, Steinthorsdottir V, Morris AP, Dina C, Welch RP, et al. Twelve type 2 diabetes susceptibility loci identified through large-scale association analysis. *Nat Genet* July 2010;**42**(7):579–89.
48. Saxena R, Elbers CC, Guo Y, Peter I, Gaunt TR, Mega JL, et al. Large-scale gene-centric meta-analysis across 39 studies identifies type 2 diabetes loci. *Am J Hum Genet* March 9, 2012;**90**(3):410–25.
49. Perry JR, Voight BF, Yengo L, Amin N, Dupuis J, Ganser M, et al. Stratifying type 2 diabetes cases by BMI identifies genetic risk variants in LAMA1 and enrichment for risk variants in lean compared to obese cases. *PLoS Genet* May 2012;**8**(5):e1002741.
50. Haghverdizadeh P, Sadat Haerian M, Haghverdizadeh P, Sadat Haerian B. ABCC8 genetic variants and risk of diabetes mellitus. *Gene* July 25, 2014;**545**(2):198–204.
51. Wang R, Gao H, Xu W, Li H, Mao Y, Wang Y, et al. Differential expression of genes and changes in glucose metabolism in the liver of liver-specific glucokinase gene knockout mice. *Gene* March 10, 2013;**516**(2):248–54.
52. Ingelsson E, Langenberg C, Hivert MF, Prokopenko I, Lyssenko V, Dupuis J, et al. Detailed physiologic characterization reveals diverse mechanisms for novel genetic loci regulating glucose and insulin metabolism in humans. *Diabetes* May 2010;**59**(5): 1266–75.
53. Morris AP, Voight BF, Teslovich TM, Ferreira T, Segre AV, Steinthorsdottir V, et al. Large-scale association analysis provides insights into the genetic architecture and pathophysiology of type 2 diabetes. *Nat Genet* September 2012;**44**(9):981–90.
54. Harder MN, Ribel-Madsen R, Justesen JM, Sparso T, Andersson EA, Grarup N, et al. Type 2 diabetes risk alleles near BCAR1 and in ANK1 associate with decreased beta-cell function whereas risk alleles near ANKRD55 and GRB14 associate with decreased insulin sensitivity in the Danish Inter99 cohort. *J Clin Endocrinol Metab* April 2013;**98**(4):E801–6.
55. Imamura M, Iwata M, Maegawa H, Watada H, Hirose H, Tanaka Y, et al. Replication study for the association of rs391300 in SRR and rs17584499 in PTPRD with susceptibility to type 2 diabetes in a Japanese population. *J Diabetes Invest* March 18, 2013; **4**(2):168–73.
56. Imamura M, Maeda S, Yamauchi T, Hara K, Yasuda K, Morizono T, et al. A single-nucleotide polymorphism in ANK1 is associated with susceptibility to type 2 diabetes in Japanese populations. *Hum Mol Genet* July 1, 2012;**21**(13):3042–9.
57. McCarthy MI. Genomics, type 2 diabetes, and obesity. *N Engl J Med* December 9, 2010;**363**(24):2339–50.
58. Klimentidis YC, Zhou J, Wineinger NE. Identification of allelic heterogeneity at type-2 diabetes loci and impact on prediction. *PloS One* 2014;**9**(11):e113072.
59. Yan ST, Li CL, Tian H, Li J, Pei Y, Liu Y, et al. Association of calpain-10 rs2975760 polymorphism with type 2 diabetes mellitus: a meta-analysis. *Int J Clin Exp Med* 2014;**7**(10):3800–7.
60. Li YY, Gong G, Geng HY, Yang ZJ, Zhou CW, Xu J, et al. CAPN10 SNP43 G>A gene polymorphism and type 2 diabetes mellitus in the Asian population: a meta-analysis of 9353 participants. *Endocr J* November 8, 2014.
61. Ahlqvist E, Ahluwalia TS, Groop L. Genetics of type 2 diabetes. *Clin Chem* February 2011;**57**(2):241–54.

62. Zhang BC, Li WM, Zhu MY, Xu YW. Association of TCF7L2 gene polymorphisms with type 2 diabetes mellitus in Han Chinese population: a meta-analysis. *Gene* January 1, 2013;**512**(1):76–81.
63. Wang J, Zhang J, Li L, Wang Y, Wang Q, Zhai Y, et al. Association of rs12255372 in the TCF7L2 gene with type 2 diabetes mellitus: a meta-analysis. *Braz J Med Biol Res—Revista brasileira de pesquisas medicas e Biol/Sociedade Brasileira de Biofisica [et al]* April 2013;**46**(4):382–93.
64. Wang J, Li L, Zhang J, Xie J, Luo X, Yu D, et al. Association of rs7903146 (IVS3C/T) and rs290487 (IVS3C/T) polymorphisms in TCF7L2 with type 2 diabetes in 9,619 Han Chinese population. *PloS One* 2013;**8**(3):e59053.
65. Florez JC. Newly identified loci highlight beta cell dysfunction as a key cause of type 2 diabetes: where are the insulin resistance genes? *Diabetologia* July 2008;**51**(7):1100–10.
66. Doria A, Patti ME, Kahn CR. The emerging genetic architecture of type 2 diabetes. *Cell Metab* September 2008;**8**(3):186–200.
67. Brunetti A, Chiefari E, Foti D. Recent advances in the molecular genetics of type 2 diabetes mellitus. *World J Diabetes* April 15, 2014;**5**(2):128–40.
68. Delgado-Lista J, Perez-Martinez P, Solivera J, Garcia-Rios A, Perez-Caballero AI, Lovegrove JA, et al. Top single nucleotide polymorphisms affecting carbohydrate metabolism in metabolic syndrome: from the LIPGENE study. *J Clin Endocrinol Metab* February 2014;**99**(2):E384–9.
69. Han LY, Wu QH, Jiao ML, Hao YH, Liang LB, Gao LJ, et al. Associations between single-nucleotide polymorphisms (+45T>G, +276G>T, −11377C>G, −11391G>A) of adiponectin gene and type 2 diabetes mellitus: a systematic review and meta-analysis. *Diabetologia* September 2011;**54**(9):2303–14.
70. Dupuis J, Langenberg C, Prokopenko I, Saxena R, Soranzo N, Jackson AU, et al. New genetic loci implicated in fasting glucose homeostasis and their impact on type 2 diabetes risk. *Nat Genet* February 2010;**42**(2):105–16.
71. Watanabe RM. The genetics of insulin resistance: where's Waldo? *Curr Diab Rep* December 2010;**10**(6):476–84.
72. Hu C, Zhang R, Wang C, Yu W, Lu J, Ma X, et al. Effects of GCK, GCKR, G6PC2 and MTNR1B variants on glucose metabolism and insulin secretion. *PloS One* 2010;**5**(7):e11761.
73. Wang H, Liu L, Zhao J, Cui G, Chen C, Ding H, et al. Large scale meta-analyses of fasting plasma glucose raising variants in GCK, GCKR, MTNR1B and G6PC2 and their impacts on type 2 diabetes mellitus risk. *PloS One* 2013;**8**(6):e67665.
74. Matharoo K, Arora P, Bhanwer AJ. Association of adiponectin (AdipoQ) and sulphonylurea receptor (ABCC8) gene polymorphisms with type 2 diabetes in North Indian population of Punjab. *Gene* September 15, 2013;**527**(1):228–34.
75. Ouyang S, Cao D, Liu Z, Ma F, Wu J. Meta-analysis of the association of ADIPOQ G276T polymorphism with insulin resistance and blood glucose. *Endocrine* December 2014;**47**(3):749–57.
76. Chung HK, Chae JS, Hyun YJ, Paik JK, Kim JY, Jang Y, et al. Influence of adiponectin gene polymorphisms on adiponectin level and insulin resistance index in response to dietary intervention in overweight-obese patients with impaired fasting glucose or newly diagnosed type 2 diabetes. *Diabetes Care* April 2009;**32**(4):552–8.
77. Zacharova J, Todorova BR, Chiasson JL, Laakso M, Group S-NS. The G-250A substitution in the promoter region of the hepatic lipase gene is associated with the conversion from impaired glucose tolerance to type 2 diabetes: the STOP-NIDDM trial. *J Intern Med* February 2005;**257**(2):185–93.
78. Yang X, Deng Y, Gu H, Ren X, Li N, Lim A, et al. Candidate gene association study for diabetic retinopathy in Chinese patients with type 2 diabetes. *Mol Vis* 2014;**20**:200–14.
79. Kariz S, Petrovic D. Minor association of kinase insert domain-containing receptor gene polymorphism (rs2071559) with myocardial infarction in Caucasians with type 2 diabetes mellitus: case-control cross-sectional study. *Clin Biochem* November 2014; **47**(16–17):192–6.
80. Bermudez VJ, Rojas E, Toledo A, Rodriguez-Molina D, Vega K, Suarez L, et al. Single-nucleotide polymorphisms in adiponectin, AdipoR1, and AdipoR2 genes: insulin resistance and type 2 diabetes mellitus candidate genes. *Am J Therap* Jul–Aug 2013;**20**(4):414–21.
81. Gjesing AP, Vestmar MA, Jorgensen T, Heni M, Holst JJ, Witte DR, et al. The effect of PCSK1 variants on waist, waist-hip ratio and glucose metabolism is modified by sex and glucose tolerance status. *PloS One* 2011;**6**(9):e23907.
82. Choquet H, Kasberger J, Hamidovic A, Jorgenson E. Contribution of common PCSK1 genetic variants to obesity in 8359 subjects from multi-ethnic American population. *PloS One* 2013;**8**(2): e57857.
83. Jiang YD, Chuang LM, Pei D, Lee YJ, Wei JN, Sung FC, et al. Genetic variations in the Kir6.2 subunit (KCNJ11) of pancreatic ATP-sensitive potassium channel gene are associated with insulin response to glucose loading and early onset of type 2 diabetes in childhood and adolescence in Taiwan. *Int J Endocrinol* 2014;**2014**. 983016.
84. Keshavarz P, Habibipour R, Ghasemi M, Kazemnezhad E, Alizadeh M, Omami MH. Lack of genetic susceptibility of KCNJ11 E23K polymorphism with risk of type 2 diabetes in an Iranian population. *Endocr Res* 2014;**39**(3):120–5.
85. Corella D, Sorli JV, Gonzalez JI, Ortega C, Fito M, Bullo M, et al. Novel association of the obesity risk-allele near Fas Apoptotic Inhibitory Molecule 2 (FAIM2) gene with heart rate and study of its effects on myocardial infarction in diabetic participants of the PREDIMED trial. *Cardiovasc Diabetol* 2014;**13**:5.
86. Langenberg C, Sharp SJ, Franks PW, Scott RA, Deloukas P, Forouhi NG, et al. Gene-lifestyle interaction and type 2 diabetes: the EPIC interact case-cohort study. *PLoS Med* May 2014;**11**(5): e1001647.
87. Pradhan AD, Manson JE, Rifai N, Buring JE, Ridker PM. C-reactive protein, interleukin 6, and risk of developing type 2 diabetes mellitus. *JAMA* July 18, 2001;**286**(3):327–34.
88. Banerjee M, Saxena M. Genetic polymorphisms of cytokine genes in type 2 diabetes mellitus. *World J Diabetes* August 15, 2014;**5**(4): 493–504.
89. Lee YC, Lai CQ, Ordovas JM, Parnell LD. A database of gene-environment interactions pertaining to blood lipid traits, cardiovascular disease and type 2 diabetes. *J Data Mining Genomics Proteomics* January 1, 2011;**2**(1).
90. Delgado-Lista J, Perez-Martinez P, Garcia-Rios A, Phillips CM, Hall W, Gjelstad IM, et al. A gene variation (rs12691) in the CCAT/enhancer binding protein alpha modulates glucose metabolism in metabolic syndrome. *Nutr Metab Cardiovasc Dis* May 2013;**23**(5):417–23.
91. Zheng JS, Arnett DK, Parnell LD, Lee YC, Ma Y, Smith CE, et al. Genetic variants at PSMD3 interact with dietary fat and carbohydrate to modulate insulin resistance. *J Nutr* March 2013;**143**(3): 354–61.
92. Zheng JS, Arnett DK, Parnell LD, Smith CE, Li D, Borecki IB, et al. Modulation by dietary fat and carbohydrate of IRS1 association with type 2 diabetes traits in two populations of different ancestries. *Diabetes Care* September 2013;**36**(9):2621–7.
93. Rung J, Cauchi S, Albrechtsen A, Shen L, Rocheleau G, Cavalcanti-Proenca C, et al. Genetic variant near IRS1 is associated with type 2 diabetes, insulin resistance and hyperinsulinemia. *Nat Genet* October 2009;**41**(10):1110–5.
94. Ericson U, Rukh G, Stojkovic I, Sonestedt E, Gullberg B, Wirfalt E, et al. Sex-specific interactions between the IRS1 polymorphism and intakes of carbohydrates and fat on incident type 2 diabetes. *Am J Clin Nutr* January 2013;**97**(1):208–16.
95. Marin C, Perez-Martinez P, Delgado-Lista J, Gomez P, Rodriguez F, Yubero-Serrano EM, et al. The insulin sensitivity response is determined by the interaction between the G972R polymorphism of the insulin receptor substrate 1 gene and dietary fat. *Mol Nutr Food Res* February 2011;**55**(2):328–35.

96. Cornelis MC, Qi L, Zhang C, Kraft P, Manson J, Cai T, et al. Joint effects of common genetic variants on the risk for type 2 diabetes in U.S. men and women of European ancestry. *Annals Intern Med* April 21, 2009;**150**(8):541–50.
97. Cornelis MC, Tchetgen EJ, Liang L, Qi L, Chatterjee N, Hu FB, et al. Gene-environment interactions in genome-wide association studies: a comparative study of tests applied to empirical studies of type 2 diabetes. *Am J Epidemiol* February 1, 2012;**175**(3): 191–202.
98. Villegas R, Goodloe RJ, McClellan Jr BE, Boston J, Crawford DC. Gene-carbohydrate and gene-fiber interactions and type 2 diabetes in diverse populations from the National Health and Nutrition Examination Surveys (NHANES) as part of the Epidemiologic Architecture for Genes Linked to Environment (EAGLE) study. *BMC Genet* 2014;**15**:69.
99. Wannamethee SG, Whincup PH, Thomas MC, Sattar N. Associations between dietary fiber and inflammation, hepatic function, and risk of type 2 diabetes in older men: potential mechanisms for the benefits of fiber on diabetes risk. *Diabetes Care* October 2009;**32**(10):1823–5.
100. Hwang JY, Park JE, Choi YJ, Huh KB, Chang N, Kim WY. Carbohydrate intake interacts with SNP276G>T polymorphism in the adiponectin gene to affect fasting blood glucose, HbA1C, and HDL cholesterol in Korean patients with type 2 diabetes. *J Am Coll Nutr* 2013;**32**(3):143–50.
101. Kang R, Kim M, Chae JS, Lee SH, Lee JH. Consumption of whole grains and legumes modulates the genetic effect of the APOA5-1131C variant on changes in triglyceride and apolipoprotein A-V concentrations in patients with impaired fasting glucose or newly diagnosed type 2 diabetes. *Trials* 2014;**15**:100.
102. Fisher E, Boeing H, Fritsche A, Doering F, Joost HG, Schulze MB. Whole-grain consumption and transcription factor-7-like 2 (TCF7L2) rs7903146: gene-diet interaction in modulating type 2 diabetes risk. *Brit J Nutr* February 2009;**101**(4):478–81.
103. Wang TJ, Larson MG, Vasan RS, Cheng S, Rhee EP, McCabe E, et al. Metabolite profiles and the risk of developing diabetes. *Nat Med* April 2011;**17**(4):448–53.
104. Taneera J, Lang S, Sharma A, Fadista J, Zhou Y, Ahlqvist E, et al. A systems genetics approach identifies genes and pathways for type 2 diabetes in human islets. *Cell Metab* July 3, 2012;**16**(1): 122–34.
105. Xu M, Qi Q, Liang J, Bray GA, Hu FB, Sacks FM, et al. Genetic determinant for amino acid metabolites and changes in body weight and insulin resistance in response to weight-loss diets: the Preventing Overweight Using Novel Dietary Strategies (POUNDS LOST) trial. *Circulation* March 26, 2013;**127**(12):1283–9.
106. Perez-Martinez P, Garcia-Rios A, Delgado-Lista J, Gjelstad IM, Gibney J, Kiec-Wilk B, et al. Gene-nutrient interactions on the phosphoenolpyruvate carboxykinase influence insulin sensitivity in metabolic syndrome subjects. *Clin Nutr* August 2013;**32**(4): 630–5.
107. Perez-Martinez P, Delgado-Lista J, Garcia-Rios A, Tierney AC, Gulseth HL, Williams CM, et al. Insulin receptor substrate-2 gene variants in subjects with metabolic syndrome: association with plasma monounsaturated and n-3 polyunsaturated fatty acid levels and insulin resistance. *Mol Nutr Food Res* February 2012;**56**(2):309–15.
108. Perez-Martinez P, Delgado-Lista J, Garcia-Rios A, Mc Monagle J, Gulseth HL, Ordovas JM, et al. Glucokinase regulatory protein genetic variant interacts with omega-3 PUFA to influence insulin resistance and inflammation in metabolic syndrome. *PloS One* 2011;**6**(6):e20555.
109. Zheng JS, Arnett DK, Parnell LD, Lee YC, Ma Y, Smith CE, et al. Polyunsaturated fatty acids modulate the association between genetic variants and insulin resistance. *PloS One* 2013;**8**(6):e67394.
110. Chiu KC, Chu A, Chuang LM, Saad MF. Association of leptin receptor polymorphism with insulin resistance. *Eur J Endocrinol Eur Fed Endocr Soc* May 2004;**150**(5):725–9.
111. Phillips CM, Goumidi L, Bertrais S, Field MR, Ordovas JM, Cupples LA, et al. Leptin receptor polymorphisms interact with polyunsaturated fatty acids to augment risk of insulin resistance and metabolic syndrome in adults. *J Nutr* February 2010;**140**(2): 238–44.
112. Marin C, Perez-Jimenez F, Gomez P, Delgado J, Paniagua JA, Lozano A, et al. The Ala54Thr polymorphism of the fatty acid-binding protein 2 gene is associated with a change in insulin sensitivity after a change in the type of dietary fat. *Am J Clin Nutr* July 2005;**82**(1):196–200.
113. Chamberlain AM, Schreiner PJ, Fornage M, Loria CM, Siscovick D, Boerwinkle E. Ala54Thr polymorphism of the fatty acid binding protein 2 gene and saturated fat intake in relation to lipid levels and insulin resistance: the Coronary Artery Risk Development in Young Adults (CARDIA) study. *Metab Clin Exp* September 2009;**58**(9):1222–8.
114. Smith CE, Arnett DK, Corella D, Tsai MY, Lai CQ, Parnell LD, et al. Perilipin polymorphism interacts with saturated fat and carbohydrates to modulate insulin resistance. *Nutr Metab Cardiovasc Dis* May 2012;**22**(5):449–55.
115. Perez-Martinez P, Delgado-Lista J, Garcia-Rios A, Ferguson JF, Gulseth HL, Williams CM, et al. Calpain-10 interacts with plasma saturated fatty acid concentrations to influence insulin resistance in individuals with the metabolic syndrome. *Am J Clin Nutr* May 2011;**93**(5):1136–41.
116. Wang J, Hu F, Feng T, Zhao J, Yin L, Li L, et al. Meta-analysis of associations between TCF7L2 polymorphisms and risk of type 2 diabetes mellitus in the Chinese population. *BMC Med Genet* 2013;**14**:8.
117. Delgado-Lista J, Perez-Martinez P, Garcia-Rios A, Phillips CM, Williams CM, Gulseth HL, et al. Pleiotropic effects of TCF7L2 gene variants and its modulation in the metabolic syndrome: from the LIPGENE study. *Atherosclerosis* January 2011;**214**(1): 110–6.
118. Li H, Kilpelainen TO, Liu C, Zhu J, Liu Y, Hu C, et al. Association of genetic variation in FTO with risk of obesity and type 2 diabetes with data from 96,551 East and South Asians. *Diabetologia* April 2012;**55**(4):981–95.
119. Butler AA, Cone RD. Knockout studies defining different roles for melanocortin receptors in energy homeostasis. *Ann NY Acad Sci* June 2003;**994**:240–5.
120. Ortega-Azorin C, Sorli JV, Asensio EM, Coltell O, Martinez-Gonzalez MA, Salas-Salvado J, et al. Associations of the FTO rs9939609 and the MC4R rs17782313 polymorphisms with type 2 diabetes are modulated by diet, being higher when adherence to the Mediterranean diet pattern is low. *Cardiovasc Diabetol* 2012;**11**:137.
121. Chu H, Wang M, Zhong D, Shi D, Ma L, Tong N, et al. AdipoQ polymorphisms are associated with type 2 diabetes mellitus: a meta-analysis study. *Diabetes Metab Res Rev* October 2013;**29**(7): 532–45.
122. Perez-Martinez P, Lopez-Miranda J, Cruz-Teno C, Delgado-Lista J, Jimenez-Gomez Y, Fernandez JM, et al. Adiponectin gene variants are associated with insulin sensitivity in response to dietary fat consumption in Caucasian men. *J Nutr* September 2008;**138**(9):1609–14.
123. Perez-Martinez P, Perez-Jimenez F, Bellido C, Ordovas JM, Moreno JA, Marin C, et al. A polymorphism exon 1 variant at the locus of the scavenger receptor class B type I (SCARB1) gene is associated with differences in insulin sensitivity in healthy people during the consumption of an olive oil-rich diet. *J Clin Endocrinol Metab* April 2005;**90**(4):2297–300.

CHAPTER

25

Genetic Basis Linking Variants for Diabetes and Obesity with Breast Cancer

Shaik Mohammad Naushad[1], *M. Janaki Ramaiah*[1], *Vijay Kumar Kutala*[2]

[1]School of Chemical and Biotechnology, SASTRA University, Thanjavur, India; [2]Department of Clinical Pharmacology and Therapeutics, Nizam's Institute of Medical Sciences, Hyderabad, India

1. OBESITY AND BREAST CANCER

The association of obesity with breast cancer risk was explored by several researchers. A positive association of body mass index (BMI) with postmenopausal breast cancer risk was demonstrated in a *meta*-analysis of 141 articles.[1] However, in premenopausal breast cancer cases, the results were inconsistent with Asia–Pacific women showing positive association (risk ratio [RR] = 1.16; 95% confidence interval [CI], 1.01–1.32) and North American (RR = 0.91; 95% CI, 0.85–0.98) and European and Australian women (RR = 0.89; 95% CI, 0.84–0.94) showing an inverse association. To address these inconsistencies in association studies, waist-to-hip ratio (WHR) was measured instead of BMI to account for central or abdominal adiposity. Elevated WHR was found to be associated with premenopausal breast cancer risk in two *meta*-analyses.[2,3] Elevated WHR was reported to be associated with a 79% increased risk of breast cancer for premenopausal women (summary risk = 1.79; 95% CI, 1.22–2.62) and a 50% increased risk for postmenopausal women (summary risk = 1.50; 95% CI, 1.10–2.04).[2] The protective role of low WHR against breast cancer risk was demonstrated by Harvie et al.[3] These studies highlighted the association of central obesity and general obesity with pre- and postmenopausal breast cancer risk, respectively.

2. INSULIN RESISTANCE AND BREAST CANCER

Increased adiposity was reported to be associated with insulin resistance, which in turn activates insulin-like growth factor-1 (IGF-1). IGF-1 was shown to upregulate mitogen-activated protein kinase (MAPK) and phosphatidylinositol-3-kinase (PI3K) signaling.[4] IGF-1 was also reported to decrease the level of sex hormone–binding globulin and increase estrogen receptor (ER)-α expression.[4] A decrease in sex hormone–binding globulin is accompanied by increased estradiol levels that might induce oxidative DNA damage through phenylalanine hydroxylase-DNA adduct formation (Figure 1). Activation of MAPK and PI3K signaling and increased estradiol induce mitogenic effects in breast epithelium by deregulating G1-S phase of cell cycle (Figure 2).

Kang et al. have demonstrated the protective role of five intron located single nucleotide polymorphisms (SNPs; rs8032477, rs7175052, rs12439557, rs11635251, and rs12916884) of IGF1R against breast cancer risk.[5] A *meta*-analysis showed an inverse association of 19/19 genotype of IGF-1 CA repeat polymorphism with breast cancer in Caucasians, but not in Asians.[6] This observation corroborated with a study on African American and Hispanic women that showed 75% increase in breast cancer risk in women with non-19/non-19 IGF-1 CA repeat polymorphism.[7] Women harboring any rare variant in IGF-1 (rs855211, rs35765, rs2162679) and non-19/non-19 CA repeat were reported to be at risk of early onset breast cancer.[8] IGF-1 rs7965399 polymorphism was shown to be associated with 80% increased risk for breast cancer.[9] Two SNPs in IGF-1 (rs1520220 and rs10735380) were found to be significantly associated with higher IGF-1 levels[10] that might contribute to increased breast cancer risk by the mechanisms discussed above. This hypothesis was supported by the study of Tamimi et al., which reported association of rs1520220 with mammographic density.[11]

Molecular Nutrition and Diabetes
http://dx.doi.org/10.1016/B978-0-12-801585-8.00025-7

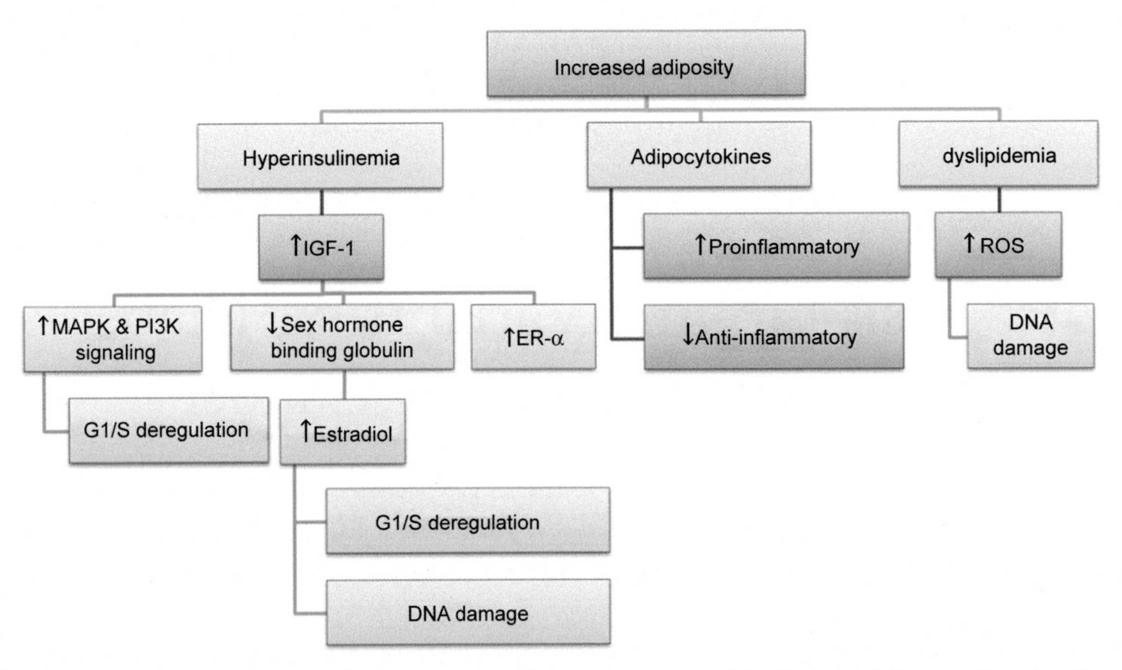

FIGURE 1 **Molecular mechanisms associated with obesity-mediated breast cancer risk.** Obesity contributes to hyperinsulinemia, altered adipokine levels, and dyslipidemia. Hyperinsulinemia upregulates insulin-like growth factor-1 (IGF-1), which in turn activates mitogen-activated protein kinase (MAPK) and phosphatidylinositol-3-kinase (PI3K) signaling, decreases sex hormone–binding globulin, and increases estrogen receptor (ER)-α expression. These events result in G1/S cell-cycle deregulation, increased estradiol, and increased estradiol-mediated oxidative DNA damage. Altered adipokine levels and dyslipidemia contribute to inflammation and free radical-mediated DNA damage, respectively. ROS, reactive oxygen species.

3. ADIPONECTIN AND ADIPONECTIN RECEPTOR 1 GENES

Adipose tissue regulates various physiological and pathological processes such as appetite, insulin sensitivity and resistance, inflammation, immunity, hematopoiesis, and angiogenesis through its function as an active endocrine organ that secretes several bioactive adipocytokines or adipokines as well as inflammatory cytokines.[12–14] Adiponectin is one such adipokine that acts as endogenous insulin sensitizer and regulates the secretion of estrogens, tumor necrosis factor, and IGF-1.[15] Human recombinant adiponectin was shown to increase IGF-1 effects on the production of estradiol by increasing IGF-IR-β subunit tyrosine phosphorylation and MAPK extracellular signal-regulated kinase 1/2 phosphorylation.[15] Lower circulating adiponectin levels were observed in obese and diabetic subjects and breast cancer patients.[16] The higher adiponectin levels were reported to be protective against breast cancer.

Kaklamani et al. have demonstrated that two ADIPOQ SNPs (rs2241766 and rs1501299) have been associated with lower circulating levels of adiponectin and with breast cancer risk (rs1501299*GG: odd ratios [OR], 1.80; 95% CI, 1.14–2.85; rs2241766*TG: OR, 0.61; 95% CI, 0.46–0.80).[17] One ADIPOR1 SNP (rs7539542), which modulates the expression of adiponectin receptor 1 messenger RNA (mRNA), was also associated with breast cancer risk (OR, 0.51; 95% CI, 0.28–0.92). Al Khaldi et al. confirmed this association by demonstrating the association of GG genotype of rs1501299 with higher levels of adiponectin (OR = 1.2, 95% CI [1.03–1.3], $P = 0.02$), and increased breast cancer risk (OR = 8.6, 95% CI(1.03–71.0), $P = 0.04$).[18] This study also confirmed the association of rs224176 with the breast cancer. Kaklamani et al. have studied the adiponectin pathways in African Americans and Hispanics and found a positive association of ADIPOQ rs1501299 with the breast cancer risk (hazard ratio [HR] for the GG/TG genotype: 1.23; 95% CI, 1.059–1.43) in African American women as well.[19]

4. LEPTIN AND LEPTIN RECEPTOR GENES

Leptin is a pleiotropic adipokine produced by adipose tissue that regulates food intake, energy expenditure, immunity, inflammation, hematopoiesis, cell differentiation, and proliferation.[20,21] Leptin overexpression was observed in breast cancer cases.[22] There are two molecular mechanisms that have been proposed to explain the association of leptin with breast cancer risk. Leptin was shown to amplify ER-α signaling that plays a pivotal role in hormone-dependent breast cancer growth and progression and also increases estrogen synthesis by upregulating the aromatase transcription.[23] Leptin and its receptor, LEPR, activate various growth and survival signaling pathways, including canonical (Janus kinase 2/signal transducer and activator of transcription 3, PI3K/Akt/mammalian target of rapamycin, extracellular signal-regulated kinase 1/2) and noncanonical (protein kinase C, JNK, p38 MAPK) pathways. Insulin also can induce leptin and LepR overexpression via PI3K and MAPK signaling pathways.[23,24] A study on healthy overweight and obese subjects showed infiltration of CD4 and CD8 T cells in visceral and subcutaneous adipose tissue with relatively higher frequency of proinflammatory T-helper (Th)-1 and elevated levels of interleukin-6 in visceral adipose tissue.[25]

The common genetic variations in the leptin (LEP), LEP receptor (LEPR), and paraoxonase 1 (PON1) genes have been considered to be implicated in the development of breast cancer.[26–31] However, the results were inconsistent. Terrasi et al. (2009) have demonstrated that the occurrence of LEP-2548G/A can enhance leptin expression in breast cancer cells via Sp1- and nucleolin-dependent mechanisms and possibly contribute to intratumoral leptin overexpression.[26] Cleveland et al. have demonstrated a modest increase in risk of breast cancer with the LEP-2548AA genotype when compared with the LEP-2548GG genotype (age-adjusted OR, 1.30; 95% CI, 1.01–1.66).[27] This association was stronger among the postmenopausal women who were obese (OR, 1.86; 95% CI, 0.95–3.64), although the interaction was of borderline statistical significance ($P = 0.07$).[27] Liu et al. have observed higher serum leptin levels in women with high-grade cancers ($P = 0.020$).[28] The LEPR-109RR genotype was more frequent in premenopausal patients with tumors larger than 2 cm ($P = 0.039$) and in premenopausal women who were overweight ($P = 0.029$).[28] Among the patients with the LEPR-109RR genotype, higher mean serum leptin concentrations were present in those with triple-negative cancers (i.e., negative for the expression of ER, progesterone receptor, and human epidermal growth factor receptor 2 [$P = 0.048$]).[28] A *meta*-analysis by Liu et al. showed null association among the LEP G2548A, LEPR Q223R, LEPR Lys109Arg, or PON1 Q192R polymorphism and breast cancer risk; however, PON1 L55M was significantly associated with breast cancer risk overall (MM versus LL: OR, 2.16; 95% CI, 1.76–2.66).[29] For LEPR Q223R polymorphism, further subgroup analysis suggested that the association was only statistically significant in East Asians (OR, 0.50; 95% CI, 0.36–0.70), but not in Caucasians (OR, 1.06; 95% CI, 0.77–1.45) or Africans (OR, 1.30; 95% CI, 0.83–2.03).[29] Galliocchio et al. (2007) have demonstrated an inverse association of PON1 (Gln192Arg) and LEPR (IVS2 + 6920) with the risk for invasive breast cancer.[30] Four LEPR polymorphisms at codons 109, 223, 656, and 1019 showed null association with breast risk in Korean population.[31]

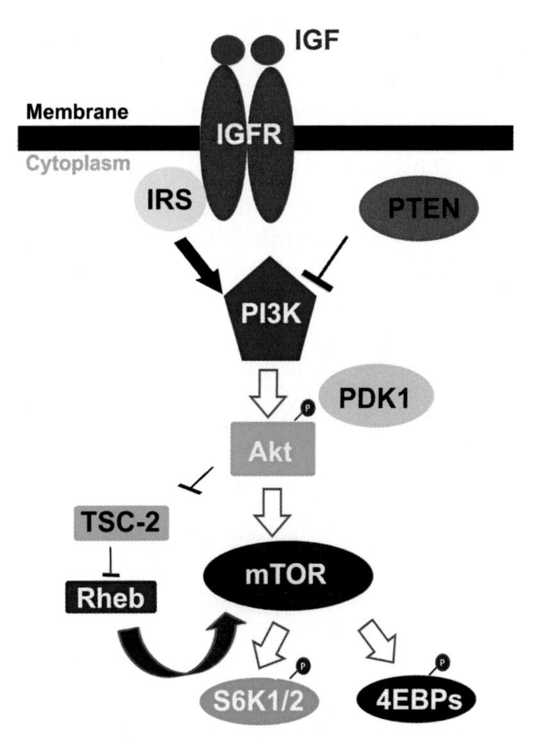

FIGURE 2 **Schema depicting the role of insulin-like growth factor-1 (IGF-1) in activating phosphatidylinositol-3-kinase (PI3K)/Akt/mammalian target of rapamycin (mTOR) pathways.** Binding of IGF-1 to its receptor activates PI3K, which localizes Akt to the membrane and phosphorylates. Phosphatase and tensin homolog antagonizes PI3K action. Activated Akt phosphorylates tuberous sclerosis complex 2 (TSC2) and inhibits its tumor suppressor function. Rheb, a small guanosine triphosphatase that is inactivated by TSC2's GAP activity, positively modulates mTOR function. mTOR phosphorylates both S6K1 and 4E-BP1 via independent pathways, resulting in the activation of S6K1 and inactivation of 4E-BP1 as a repressor of the translation initiation factor eIF4E. Increased S6K1 and eIF4E action independently promote cell growth and cell-cycle progression. S6K1 and 4E-BP1 are also phosphorylated by PI3K-dependent but mTOR-independent mechanisms.

5. FAT MASS AND OBESITY ASSOCIATED GENE

This gene is a nuclear protein of the AlkB-related non-heme iron and 2-oxoglutarate–dependent oxygenase superfamily but the exact physiological function of this gene is not known. The other non-heme iron enzymes function to reverse alkylated DNA and RNA damage by oxidative demethylation.[32]

Genome-wide association studies identified fat mass and obesity (FTO) as a novel risk factor associated with the obesity. Kaklamani et al. identified four SNPs of FTO (i.e. rs7206790, rs8047395, rs9939609, and rs1477196) to be associated with breast cancer risk.[33]

da Cunha et al. observed synergistic interactions between the FTO and MC4R (FTO rs1121980, FTO rs9939609, MC4R rs17782313) genetic variants in inflating the breast cancer risk to 4.6-fold.[34] Zhang et al. have reported five susceptibility loci (8p12, 4q34.1/ADAM29, 2q14.2, 3p24.1/TGFBR2, and 16q12.2/FTO) that correlate with breast cancer (overall and specific subtypes) in Chinese women.[35]

6. OBESITY, BREAST CANCER, AND METHYLATION

Tao et al. showed that WHR was associated with DNA promoter methylation of at least one of three genes (i.e., E-cadherin, p16, and RAR-β(2)) in postmenopausal breast tumors.[36] Stueve et al. have demonstrated that methylation of CYP19A1, a critical gene in estrogen biosynthesis, may influence timing of breast development in overweight girls.[37] Naushad et al. have observed positive association of BMI with the methylation of extracellular superoxide dismutase, Ras-association (RalGDS/AF-6) domain family member 1 (RASSF1A), and breast cancer type 1 susceptibility protein (BRCA1); and an inverse association with methylation of BCL2/adenovirus E1B 19 kDa protein-interacting protein 3 (BNIP3).[38] Further, multifactor dimensionality reduction analysis revealed a significant interaction between the BMI and catechol-O-methyl transferase H108L variant alone or in combination with cytochrome P450 1A1m1 variant, emphasizing the importance of estrogen metabolism in the molecular pathophysiology of breast cancer.[38] The hypermethylation of extracellular superoxide dismutase and hypomethylation of BNIP3 are associated with increased oxidative stress.[38] RASSF1A is a tumor suppressor that controls apoptosis, growth, and microtubule dynamics. RASSF1A methylation was reported to be associated with invasive carcinoma, advanced stage of breast cancer, lymph node involvement, and triple negative phenotype.[39] Hair et al. reported the association of obesity with differential methylation in breast cancer with hypermethylation of 21 loci among ER-positive tumors in particular, suggesting that obesity may influence the methylation of genes with known relevance to cancer.[40] Na et al. observed a U-shaped association between BMI and Alu methylation, with the lowest methylation levels occurring at BMIs of between 23 and 30 kg/m^2 in healthy Korean women, indicating that BMI-related changes in Alu methylation might play a complex role in the etiology and pathogenesis of obesity.[41] Bosviel et al. observed a positive association between the BMI and BRCA1 methylation corroborating with findings of Naushad et al.[42] BRCA1 forms BRCA1-associated genome surveillance complex in combination with tumor suppressors, DNA damage sensors, and signal transducers thus playing crucial roles in maintaining genome integrity. Reduced expression of BRCA1 induced by its methylation results in genome instability and was shown to be associated with risk for higher grade and triple negative breast cancer.[43] The hypermethylation of ER-α was shown to exhibit a positive association with BMI.[44] This hypermethylation induces epigenetic silencing of ER-α and was shown to be associated with poor prognosis.[45]

7. NUTRIGENOMICS PERSPECTIVE TO REDUCE OBESITY-MEDIATED BREAST CANCER RISK

Apple polyphenols were found to be beneficial against diet-induced obesity in animal models as they reduce Lep, Plin, and sterol regulatory element binding transcription factor 1 (Srebf1) mRNA levels and increase aquaporin 7 (Aqp7), adipocyte enhancer binding protein 1 (Aebp1), and peroxisome proliferator-activated receptor gamma coactivator 1 alpha (Ppargc1a) mRNA levels in epididymal adipocytes.[46] Barth et al. demonstrated the beneficial effects of apple juice consumption in obese men especially in subjects with CC genotype of interleukin-6 −174 G>C with significant reduction in body fat.[47] Supplementation of methyl donors (choline, betaine, folic acid, and vitamin B12) during lactation to high-fat-sucrose-fed dams was shown to confer protection in offspring against liver fat accumulation when consuming an obesogenic diet.[48]

8. CONCLUSIONS

Certain genetic variants such as IGF-1, adiponectin, adiponectin receptor, leptin, leptin receptor, and FTO link diabetes and obesity with breast cancer by deregulating cell cycle, by inducing estrogen-mediated mutagenic effects, by activating inflammatory adipocytokines, and also by inducing oxidative stress. There is growing evidence supporting obesity-mediated changes in methylome that increase susceptibility to breast cancer that also have the potential implications in dictating breast cancer phenotype. Large-scale prospective studies are warranted to study the potential effects of dietary polyphenols and methyl donors in reducing obesity-mediated breast cancer risk.

References

1. Renehan G, Tyson M, Egger M, Heller RF, Zwahlen M. Body-mass index and incidence of cancer: a systematic review and meta-analysis of prospective observational studies. *Lancet* 2008; **371**(9612):569–78.

2. Connolly BS, Barnett C, Vogt KN, Li T, Stone J, Boyd NF. A meta-analysis of published literature on waist-to-hip ratio and risk of breast cancer. *Nutr Cancer* 2002;**44**(2):127–38.
3. Harvie M, Hooper L, Howell AH. Central obesity and breast cancer risk: a systematic review. *Obes Rev* August 2003;**4**(3):157–73.
4. Vona-Davis L, Rose DP. Type 2 diabetes and obesity metabolic interactions: common factors for breast cancer risk and novel approaches to prevention and therapy. *Curr Diabetes Rev* 2012;**8**: 116–30.
5. Kang HS, Ahn SH, Mishra SK, Hong KM, Lee ES, Shin KH, et al. Association of polymorphisms and haplotypes in the insulin-like growth factor 1 receptor (IGF1R) gene with the risk of breast cancer in Korean women. *PLoS One* January 2, 2014;**9**(1):e84532.
6. He B, Xu Y, Pan Y, Li R, Gao T, Song G, et al. Differential effects of insulin-like growth factor-1 CA repeat polymorphism on breast cancer risk along with race: a meta-analysis. *Gene* August 1, 2013; **525**(1):92–8.
7. Sarkissyan M, Mishra DK, Wu Y, Shang X, Sarkissyan S, Vadgama JV. IGF gene polymorphisms and breast cancer in African-American and Hispanic women. *Int J Oncol* June 2011; **38**(6):1663–73.
8. Henningson M, Hietala M, Törngren T, Olsson H, Jernström H. IGF1 htSNPs in relation to IGF-1 levels in young women from high-risk breast cancer families: implications for early-onset breast cancer. *Fam Cancer* June 2011;**10**(2):173–85.
9. Qian B, Zheng H, Yu H, Chen K. Genotypes and phenotypes of IGF-I and IGFBP-3 in breast tumors among Chinese women. *Breast Cancer Res Treat* November 2011;**130**(1):217–26.
10. Gu F, Schumacher FR, Canzian F, Allen NE, Albanes D, Berg CD, et al. Eighteen insulin-like growth factor pathway genes, circulating levels of IGF-I and its binding protein, and risk of prostate and breast cancer. *Cancer Epidemiol Biomarkers Prev* November 2010;**19**(11):2877–87.
11. Tamimi RM, Cox DG, Kraft P, Pollak MN, Haiman CA, Cheng I, et al. Common genetic variation in IGF1, IGFBP-1, and IGFBP-3 in relation to mammographic density: a cross-sectional study. *Breast Cancer Res* 2007;**9**(1):R18.
12. Dalamaga M, Diakopoulos KN, Mantzoros CS. The role of adiponectin in cancer: a review of current evidence. *Endocr Rev* 2012; **33**:547–94.
13. Ouchi N, Kihara S, Arita Y, et al. Novel modulator for endothelial adhesion molecules: adipocyte-derived plasma protein adiponectin. *Circulation* 1999;**100**:2473–6.
14. Rose DP, Komninou D, Stephenson GD. Obesity, adipocytokines, and insulin resistance in breast cancer. *Obes Rev* 2004;**5**:153–65.
15. Chabrolle C, Tosca L, Dupont J. Regulation of adiponectin and its receptors in rat ovary by human chorionic gonadotrophin treatment and potential involvement of adiponectin in granulosa cell steroidogenesis. *Reproduction* 2007;**133**:719–31.
16. Miyoshi Y, Funahashi T, Kihara S, et al. Association of serum adiponectin levels with breast cancer risk. *Clin Cancer Res* 2003;**9**: 5699–704.
17. Kaklamani VG, Sadim M, Hsi A, Offit K, Oddoux C, Ostrer H, et al. Variants of the adiponectin and adiponectin receptor 1 genes and breast cancer risk. *Cancer Res* May 1, 2008;**68**(9):3178–84.
18. Al Khaldi RM, Al Mulla F, Al Awadhi S, Kapila K, Mojiminiyi OA. Associations of single nucleotide polymorphisms in the adiponectin gene with adiponectin levels and cardio-metabolic risk factors in patients with cancer. *Dis Markers* 2011;**30**(4): 197–212.
19. Kaklamani VG, Hoffmann TJ, Thornton TA, Hayes G, Chlebowski R, Van Horn L, et al. Adiponectin pathway polymorphisms and risk of breast cancer in African Americans and Hispanics in the women's health initiative. *Breast Cancer Res Treat* June 2013;**139**(2):461–8.
20. Moon HS, Dalamaga M, Kim SY, Polyzos SA, Hamnvik OP, Magkos F, et al. Leptin's role in lipodystrophic and nonlipodystrophic insulin-resistant and diabetic individuals. *Endocr Rev* 2013;**34**:377–412.
21. Dalamaga M, Chou SH, Shields K, Papageorgiou P, Polyzos SA, Mantzoros CS. Leptin at the intersection of neuroendocrinology and metabolism: current evidence and therapeutic perspectives. *Cell Metab* 2013;**18**:29–42.
22. Lorincz AM, Sukumar S. Molecular links between obesity and breast cancer. *Endocr Relat Cancer* 2006;**13**:279–92.
23. Ray A. Adipokine leptin in obesity-related pathology of breast cancer. *J Biosci* 2012;**37**:289–94.
24. Grossmann ME, Cleary MP. The balance between leptin and adiponectin in the control of carcinogenesis—focus on mammary tumorigenesis. *Biochimie* 2012;**94**:2164–71.
25. McLaughlin T, Liu LF, Lamendola C, Shen L, Morton J, Rivas H, et al. T-cell profile in adipose tissue is associated with insulin resistance and systemic inflammation in humans. *Arterioscler Thromb Vasc Biol* December 2014;**34**(12):2637–43.
26. Terrasi M, Fiorio E, Mercanti A, Koda M, Moncada CA, Sulkowski S, et al. Functional analysis of the −2548G/A leptin gene polymorphism in breast cancer cells. *Int J Cancer* September 1, 2009;**125**(5):1038–44.
27. Cleveland RJ, Gammon MD, Long CM, Gaudet MM, Eng SM, Teitelbaum SL, et al. Common genetic variations in the LEP and LEPR genes, obesity and breast cancer incidence and survival. *Breast Cancer Res Treat* April 2010;**120**(3):745–52.
28. Liu CL, Chang YC, Cheng SP, Chern SR, Yang TL, Lee JJ, et al. The roles of serum leptin concentration and polymorphism in leptin receptor gene at codon 109 in breast cancer. *Oncology* 2007;**72**(1–2): 75–81. Epub November 14, 2007.
29. Liu C, Liu L. Polymorphisms in three obesity-related genes (LEP, LEPR, and PON1) and breast cancer risk: a meta-analysis. *Tumour Biol* December 2011;**32**(6):1233–40.
30. Gallicchio L, McSorley MA, Newschaffer CJ, Huang HY, Thuita LW, Hoffman SC, et al. Body mass, polymorphisms in obesity-related genes, and the risk of developing breast cancer among women with benign breast disease. *Cancer Detect Prev* 2007;**31**(2):95–101.
31. Woo HY, Park H, Ki CS, Park YL, Bae WG. Relationships among serum leptin, leptin receptor gene polymorphisms, and breast cancer in Korea. *Cancer Lett* June 8, 2006;**237**(1):137–42.
32. Jia G, Fu Y, Zhao X, Dai Q, Zheng G, Yang Y, et al. N6-methyladenosine in nuclear RNA is a major substrate of the obesity-associated FTO. *Nat Chem Biol* October 16, 2011;**7**(12): 885–7.
33. Kaklamani V, Yi N, Sadim M, Siziopikou K, Zhang K, Xu Y, et al. The role of the fat mass and obesity associated gene (FTO) in breast cancer risk. *BMC Med Genet* April 13, 2011;**12**:52.
34. da Cunha PA, de Carlos Back LK, Sereia AF, Kubelka C, Ribeiro MC, Fernandes BL, et al. Interaction between obesity-related genes, FTO and MC4R, associated to an increase of breast cancer risk. *Mol Biol Rep* December 2013;**40**(12):6657–64.
35. Zhang B, Li Y, Li L, Chen M, Zhang C, Zuo XB, et al. Association study of susceptibility loci with specific breast cancer subtypes in Chinese women. *Breast Cancer Res Treat* August 2014;**146**(3): 503–14.
36. Tao MH, Marian C, Nie J, Ambrosone C, Krishnan SS, Edge SB, et al. Body mass and DNA promoter methylation in breast tumors in the Western New York exposures and breast cancer study. *Am J Clin Nutr* September 2011;**94**(3):831–8.
37. Stueve TR, Wolff MS, Pajak A, Teitelbaum SL, Chen J. CYP19A1 promoter methylation in saliva associated with milestones of pubertal timing in urban girls. *BMC Pediatr* March 20, 2014; **14**:78.

38. Naushad SM, Hussain T, Al-Attas OS, Prayaga A, Digumarti RR, Gottumukkala SR, et al. Molecular insights into the association of obesity with breast cancer risk: relevance to xenobiotic metabolism and CpG island methylation of tumor suppressor genes. *Mol Cell Biochem* July 2014;**392**(1–2):273–80.
39. Hagrass HA, Pasha HF, Shaheen MA, Abdel Bary EH, Kassem R. Methylation status and protein expression of RASSF1A in breast cancer patients. *Mol Biol Rep* January 2014;**41**(1):57–65.
40. Hair B, Troester MA, Edmiston SN, Parrish EA, Robinson WR, Wu MC, et al. Body mass index is associated with gene methylation in estrogen receptor-positive breast tumors. *Cancer Epidemiol Biomarkers Prev* January 12, 2015. pii: cebp.1017.2014.
41. Na YK, Hong HS, Lee DH, Lee WK, Kim DS. Effect of body mass index on global DNA methylation in healthy Korean women. *Mol Cells* June 2014;**37**(6):467–72.
42. Bosviel R, Garcia S, Lavediaux G, Michard E, Dravers M, Kwiatkowski F, et al. BRCA1 promoter methylation in peripheral blood DNA was identified in sporadic breast cancer and controls. *Cancer Epidemiol* June 2012;**36**(3):e177–82.
43. Timms KM, Abkevich V, Hughes E, Neff C, Reid J, Morris B, et al. Association of BRCA1/2 defects with genomic scores predictive of DNA damage repair deficiency among breast cancer subtypes. *Breast Cancer Res* December 5, 2014;**16**(6):475.
44. Pirouzpanah S, Taleban FA, Atri M, Abadi AR, Mehdipour P. The effect of modifiable potentials on hypermethylation status of retinoic acid receptor-beta2 and estrogen receptor-alpha genes in primary breast cancer. *Cancer Causes Control* December 2010;**21**(12):2101–11.
45. Ramezani F, Salami S, Omrani MD, Maleki D. CpG island methylation profile of estrogen receptor alpha in Iranian females with triple negative or non-triple negative breast cancer: new marker of poor prognosis. *Asian Pac J Cancer Prev* 2012;**13**(2):451–7.
46. Boqué N, de la Iglesia R, de la Garza AL, Milagro FI, Olivares M, Bañuelos O, et al. Prevention of diet-induced obesity by apple polyphenols in Wistar rats through regulation of adipocyte gene expression and DNA methylation patterns. *Mol Nutr Food Res* August 2013;**57**(8):1473–8.
47. Barth SW, Koch TC, Watzl B, Dietrich H, Will F, Bub A. Moderate effects of apple juice consumption on obesity-related markers in obese men: impact of diet-gene interaction on body fat content. *Eur J Nutr* October 2012;**51**(7):841–50.
48. Cordero P, Milagro FI, Campion J, Martinez JA. Supplementation with methyl donors during lactation to high-fat-sucrose-fed dams protects offspring against liver fat accumulation when consuming an obesogenic diet. *J Dev Orig Health Dis* August 1, 2014:1–11.

CHAPTER

26

Vitamin D Status, Genetics, and Diabetes Risk

Dharambir K. Sanghera[1,3], *Piers R. Blackett*[2]

[1]Department of Pediatrics, Section of Genetics, College of Medicine, University of Oklahoma Health Sciences Center, Oklahoma City, OK, USA; [2]Department of Pediatrics, Section of Endocrinology, College of Medicine, University of Oklahoma Health Sciences Center, Oklahoma City, OK, USA; [3]Department of Pharmaceutical Sciences, College of Pharmacy, University of Oklahoma Health Sciences Center, Oklahoma City, OK, USA

Vitamin D deficiency has been implicated in a variety of adverse health outcomes including skeletal and nonskeletal age-related illnesses and cardiovascular risk factors such as type 2 diabetes (T2D), obesity, and blood pressure. The purpose of this review is to summarize evidence for genetic influences on vitamin D metabolism and their association with type 1 diabetes (T1D) and T2D onset and pathophysiology. It is a systematic discussion on the epidemiology of diabetes (T1D and T2D) including clinical, environmental, and genetic risk factors that have contributed to cardiometabolic disease risk. Also included is accumulating evidence that the correction of vitamin D deficiency could contribute to the prevention of cardiometabolic diseases.

1. VITAMIN D METABOLISM AND EPIDEMIOLOGY

Vitamin D status is an excellent marker of good overall health. Beyond its well-established role in musculoskeletal health,[1] accumulating evidence suggests that vitamin D plays a crucial role in diverse physiological functions, which include several common cancers,[2] cardiovascular disease,[3] and autoimmune diseases (e.g., T1D, multiple sclerosis, Crohn's disease, rheumatoid arthritis),[4–6] Alzheimer's disease,[7] and Parkinson's disease.[8] Overwhelming evidence suggests that decreased vitamin D status increases age-related disorders including the development, progression, and severity of T2D.

Vitamin D status is known to be poor among people worldwide. Vitamin D deficiency is the most common medical condition in the world and is estimated to affect more than 1 billion people worldwide.[9] Clinical vitamin D deficiency causes rickets in children and osteomalacia in children and adults.[10] Risk factors of vitamin D deficiency include living at higher latitudes (>35°N), dark skin, old age, sunscreen use, obesity, medication, chronic disease, and genetics. Asian Indians require twice as much ultraviolet (UV) B exposure to produce vitamin D levels equal to Caucasians.[11] In addition to its detrimental effects on bones and muscle function, insufficient serum vitamin D (<50 nmol/L) has been implicated in a variety of adverse health outcomes including hyperglycemia in diagnosed T2D,[12] blood pressure,[13] and obesity.[14] Currently, the US Institute of Medicine recommends intake of vitamin D by infants (400 IU/day), children (600 IU/day), adults (600 IU/day), and the elderly (800 IU/day). Because only a small portion of vitamin D is obtained in the diet, supplementation may be necessary to achieve adequate vitamin D levels, especially in infants who are solely breastfed or in adults who avoid outdoor activities, live at higher latitudes, or have increased skin pigmentation.[3] The benefits of vitamin D supplementation have been confirmed in numerous reports. One notable investigation observed a significantly reduced risk of developing T1D among supplemented infants when compared to nonsupplemented infants.[4]

The main source of vitamin D is conversion at subcutaneous sites by the photolytic cleavage of 7-dehydrocholesterol (DHC) in response to UVB irradiance (280–315 nm) following sunlight exposure. Vitamin D status is determined by the most abundant form (25-hydroxy vitamin D_3) or 25(OH)D_3. Vitamin D activation requires two steps beginning with 25-hydroxylation in the liver followed by 1-α hydroxylation in the kidney to the active form 1,25(OH)D, the ligand for the vitamin D receptor (VDR) (Figure 2).

Molecular Nutrition and Diabetes
http://dx.doi.org/10.1016/B978-0-12-801585-8.00026-9

Expression of VDR occurs in a variety of tissues with its main action on promoting intestinal calcium absorption. However, immune cells express VDR[15] and also have 1-α-hydroxylase activity supporting a functional role.[16,17] 25(OH)D3, considered the most stable and reliable measurement of vitamin D sufficiency, tends to be low at T1D onset and was lower in men than women consistent with higher T1D rates in young men.[18] Also, children and adolescents with established T1D had lower 25(OH)D3 levels than a British control population.[19]

High prevalence of vitamin D deficiency (<25 nmol/L; 10 ng/mL) among pregnant women from ethnic minority groups is reported in Northern Europe and the United States.[20,21] There are reports on the high prevalence of vitamin D deficiency among pregnant women in the Middle East, India, Pakistan, and Ethiopia.[22,23] According to Lo et al.,[11] people from the Indian subcontinent require twice as much UVB exposure to produce 25(OH)D levels equal to Caucasians because of increased skin pigmentation. In addition, a cultural tendency to avoid direct sunlight through sunscreen usage, and customary dress that covers most of the body, may contribute to suboptimal vitamin D status, although the climate in India and Middle Eastern countries is sunny throughout the year. Ethnic and genetic differences in vitamin D metabolism may affect the vitamin D status of different populations including Chinese,[24] Asians Indians,[25] and African Americans.[26]

2. VITAMIN D DEFICIENCY AND DIABETES RISK

2.1 Type 1 Diabetes

T1D or insulin-dependent diabetes is a chronic disease characterized by the body's inability to produce insulin because of the autoimmune destruction of the β cells in the pancreas.[27] Although T1D usually appears during childhood or adolescence, it also can begin in young and middle adulthood. The incidence ranges from 4 to 40 per 100,000 in European populations and 15–25 per 100,000 worldwide.[28] There has been an alarming increase in incidence rates with doubling predicted in European children; also, the age of onset has become younger.[28] Variable environmental factors have been implicated in the pathogenesis such as viral infections and dietary components, which may act as triggers for the autoimmune process,[29] but the causative evidence is incomplete.[30]

Antibodies to islet cell antigens serve as markers of early disease progression and predict T1D and prediction is proportionate to the number of positive antibodies. Combining the titers with the β-cell's insulin response to an intravenous glucose load has enhanced prediction[31,32] (Figure 1). Individuals with a positive prediction are eligible for treatment trials designed to test strategies to offset or delay development of overt T1D. Hyperglycemia and ketone production require

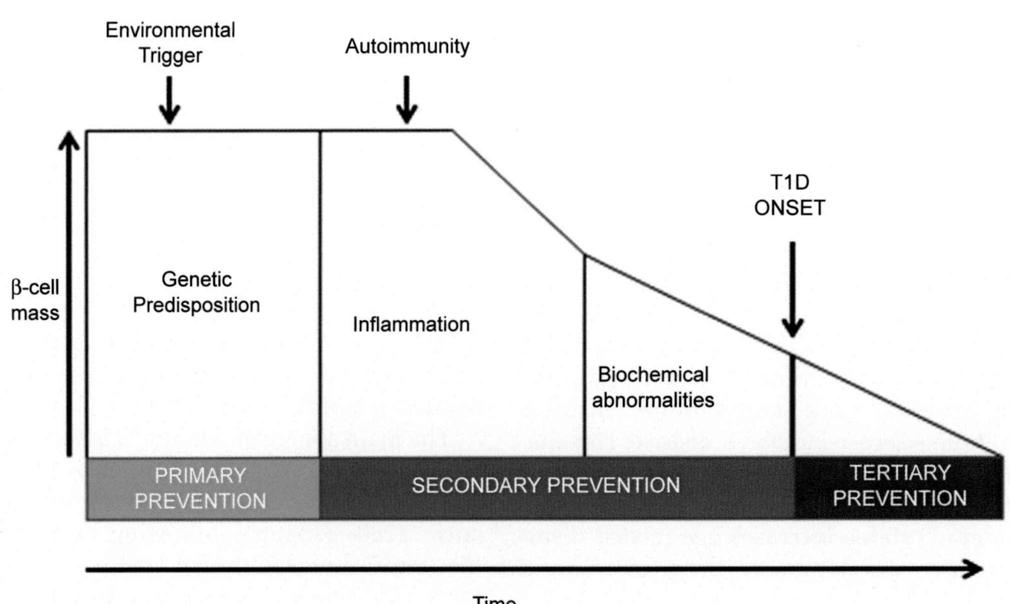

FIGURE 1 Hypothetical stages and loss of β cells during progression to type 1 diabetes (T1D) onset. Opportunities for prevention before disease onset (primary), during pathogenesis (secondary), and after onset (tertiary) are indicated in the lower panel and are associated with diminishing effects of immune modifications, possibly including vitamin D replacement alone or combined with other treatments. A fluctuating and remitting course has been confirmed by more recent observations. *Modified from* N Engl J Med *1986;**314**:1360–68.*

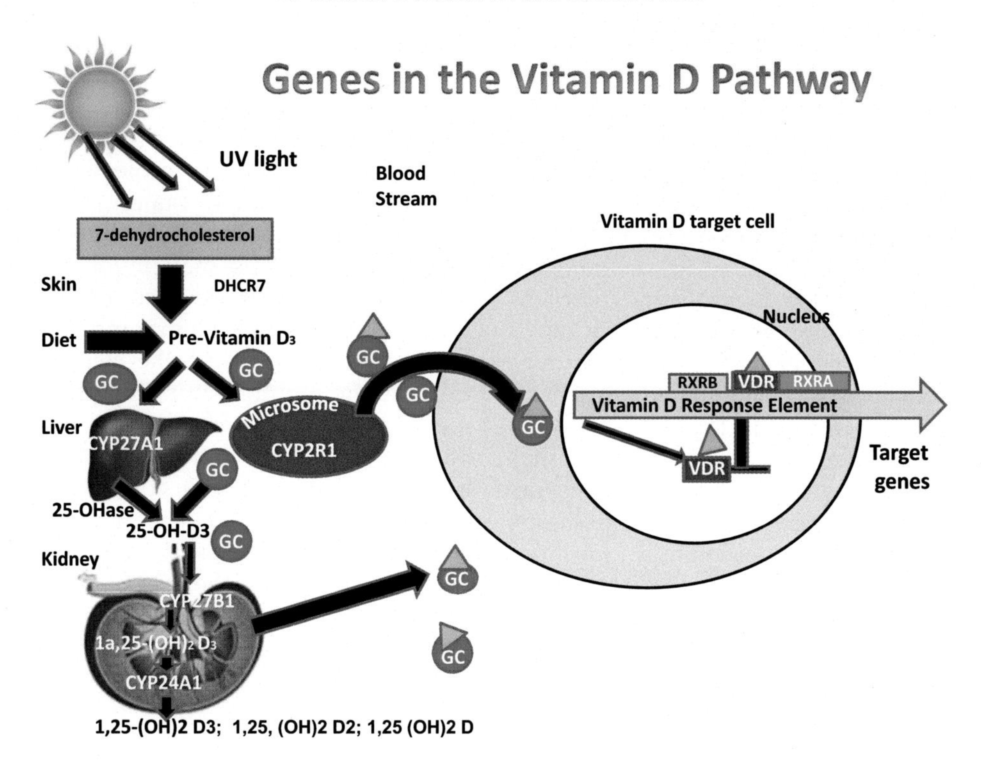

FIGURE 2 This figure illustrates the absorption, distribution, metabolism, and excretion of vitamin D. Genetic factors believed to influence circulating vitamin D levels are shown. *DHCR7* expresses a reductase that catalyzes the production of cholesterol from 7-DHC using nicotinamide adenine dinucleotide phosphate-oxidase. The *GC* gene catalyzes the formation of DBP. *CYP2R1* encodes the enzyme (25-hydroxylase) responsible for the first hydroxylation step. The enzyme responsible for the second hydroxylation in the kidneys is catalyzed by the *CYP27B1* gene product. Yellow triangles represent active 25(OH)D and green circles represent vitamin D transporter DBP. $1,25(OH)_2D_3$ is the most potent form and widely measured marker for vitamin D deficiency. DBP, vitamin D–binding protein; *GC*, group component; 25-OHase, 25-hydroxylase; 25(OH)D, 25-hydroxyvitamin D; 1-OHase, 1-hydroxylase; $1,25(OH)_2D$, 1,25-dihydroxyvitamin D; UV, ultraviolet; VDR, vitamin D receptor. *Modified from Bossé et al.* Respir Res *2009;**10**:98.*

acute management at disease onset followed by vigilant long-term control to prevent microvascular complications and cardiovascular disease.[27] The severity of presenting symptoms is variable and consists of frequent urination, thirst, and hunger. Intensified insulin therapy can result in weight gain[33] and lead to insulin resistance and β-cell loss[34] and increased cardiovascular risk according to follow-up assessment of the Epidemiology of Diabetes Interventions and Complications cohort.[35]

Currently available insulin delivery methods are often insufficient to restore normal glucose homeostasis and indices of glycemia, such as glycosylated hemoglobin, remain abnormally elevated and predict long-term complications. Consequently, research to find a cure is focused on early prevention based on modulating the autoimmune pathogenesis[29] and restoration of glucose homeostasis in diagnosed patients by β-cell replacement approaches including the use of mechanical glucose sensing and feedback to a motorized insulin pump,[36] islet transplantation,[37] and cell-based approaches.[38] Both genetic and environmental factors are responsible for increased incidence and prevalence of T1D.[30]

2.2 Type 2 Diabetes

T2D has become a significant public health problem of twenty-first century. According to International Diabetes Federation estimates, the number of people affected by T2D globally will rise from 382 million in 2011 to 592 million by the year 2035.[39] T2D is a complex disease influenced by both genetic and environmental factors. Central obesity, sedentary lifestyle, a westernized diet, and genetic predisposition are among several factors contributing to the alarming worldwide increase of T2D. The high morbidity and mortality associated with T2D presents an overwhelming health care burden necessitating improved treatment and preventative therapy.

Hyperglycemia in T2D is a consequence of complex interplay between insulin resistance (sensitivity) and abnormal insulin secretion. Unlike T1D, T2D may remain undiagnosed for many years because hyperglycemia is not severe enough to provoke noticeable symptoms. But the presence of sufficient degree of hyperglycemia can cause significant pathological and functional changes resulting in organ damage before the diagnosis is made. As for T1D, the long-term effects of T2D include microvascular and macrovascular complications, but TID has less than 10% of the overall burden consistent with the lower prevalence. Microvascular complications are progressive development of disease of the capillaries supplying blood to the kidneys contributing to hypertension, the peripheral nerves, and the retina, resulting in loss of vision. Neuropathic complications include foot ulcers, amputations, sexual dysfunction, infections, and nonhealing skin wounds. Macrovascular complications include coronary heart disease, peripheral vascular disease leading to amputations, and cerebral vascular disease leading to stroke. Certain infections such as staphylococcal sepsis are more common in diabetic patients secondary to effects on white blood cell function. Consequently, infections of the ear, nose, and throat and reactivation of tuberculosis associated with high rate of mortality and morbidity occur. Diabetes is a disproportionably expensive disease and greatly impacts quality of life. It is estimated that, between 2009 and 2034, the economic impact of T2D in the United States is expected to increase from $113 billion to $336 billion.[40]

The prevalence of T2D in the United States is higher among minorities and ethnic populations such as African Americans, Native Americans, Hispanic Americans, Asian Americans, and Pacific Islanders than in the general population.[41] High prevalence is even seen in children belonging to ethnic minorities.[42,43] Social, cultural, biological, genetic, and physical environments are implicated in health disparities but the mechanisms remain uncertain as current clinical and genetic research cannot fully explain these disparities.[44,45] Despite considerable success of genome-wide association studies (GWAS), the genetic mechanisms that predispose people to diabetes and cardiometabolic risk factors remain poorly understood.

GWAs have identified dozens of loci associated with T1D, and 90 robustly associated loci with T2D and cardiometabolic traits.[46,47] However, these genes collectively account for only 8–10% of total heritability linked with diabetes, which sharply underestimate the heritability based on family and twin studies that ranges from 26% to 69%.[48,49] It is still unclear how these genes/loci identified by GWAS contribute to clinical manifestation of diabetes and how they can be used for risk prediction of T1D and T2D.[50] It should be noted that the effects of common variants identified in most GWAS loci in Europeans are largely shared between populations.[51,52] However, these similarities in genetic association studies do not explain ethnic differences in T2D epidemiology, physiology, and disease disparity among ethnic minorities.

Vitamin D deficiency is increasing in conjunction with T1D, T2D, obesity, and cardiovascular disease. Potential impact of vitamin D deficiency in T1D is based on epidemiological studies that showed a relationship of sunlight exposure to protection from T1D. Spring births in the United States were associated with increased likelihood of T1D, but not in all regions.[53] Causal mechanisms may involve factors dependent on geographic latitude such as solar irradiance.[53] In Greece, it was observed that more children were diagnosed with T1D during the cold months as opposed to the warm months, supporting the relationship to climatic factors,[54] as supported by a large global study.[55] Studies on latitude and UVB irradiance (280–315 nm) have revealed an inverse correlation of hours of available sunshine and T1D incidence, and rates approached zero in regions worldwide with high UVB irradiance adding support for the role of vitamin D in reducing risk of the disease.[56]

The inverse relationship between circulating 25(OH) D levels and T2D has been confirmed in many observational studies. Data from the Third National Health and Nutrition Examination Survey showed a dose-dependent inverse association of vitamin D status and diabetes in non-Hispanic whites and Mexican Americans ($n = 6228$).[57] In a prospective study involving adult Australian men and women, 199 patients were newly diagnosed with T2D during the 5-year follow-up period. Higher serum 25(OH)D significantly reduced the risk of metabolic syndrome; each 25 nmol/L incremental increase in 25(OH)D levels conferred a 24% reduced risk of the metabolic syndrome.[58] A prospective study ($n = 2378$) performed by Deleskog et al.[59] concluded that vitamin D deficiency may accelerate the progression from prediabetes to T2D after an 8- to 10-year follow-up period. Another 14-year follow-up study of women ($n = 1167$) observed a 48% reduced risk of diabetes among women in the top versus bottom quartile of vitamin D.[60]

A meta-analysis of eight observational cohort studies ($n = 238{,}423$) by Mitri et al.[12] observed a 13% reduced risk of T2D among individuals with a vitamin D intake >500 IU/day compared with <200 IU/day. Individuals with the highest vitamin D status were 43% less likely to develop T2D than individuals in the lowest group. However, post hoc analysis of 11 studies (eight with normal baseline glucose tolerance and three small studies with established T2D) randomized control trials revealed no effect between vitamin D supplementation and glucose control. Vitamin D supplementation did improve insulin resistance in the other two trials whose participants were glucose intolerant at baseline.[12] Another review conducted by Khan et al.[61] comprised 14 prospective studies (seven of which were included in the review conducted by Mitri et al.[12]). The pooled results showed a 19% lower risk of developing T2D among individuals in the top quartile of vitamin D status compared to individuals in the bottom quartile. Pooled results from two studies demonstrated an inverse but nonsignificant association between insulin resistance and vitamin D status. Despite these compelling results, a causal link cannot be established between vitamin D and progressive T2D risk resulting mostly study design. Our cross-sectional study on a T2D case-control cohort from Asian Indian Sikhs showed a higher prevalence of vitamin D deficiency among patients with T2D and cardiovascular risk factors.[62] These results were in agreement with two separate case-control studies in Asian Indians showing a consistently significant inverse association between serum 25(OH)D levels and T2D.[62–64] Our Sikh study supports the decreased bioavailability of serum vitamin D observed in obese individuals in previous studies.[65,66] Wortsman et al.[66] found that obesity did not affect the skin's ability to produce vitamin D, but the increased subcutaneous fat may have altered the release of vitamin D into the circulation from adipose stores. As a fat-soluble molecule, decreased 25(OH)D levels among obese individuals may be due to enhanced uptake in adipose tissue and metabolic clearance.[65] TID and T2D have different associations with obesity—but both have known connections with vitamin D deficiency.

There is mounting evidence that proposes vitamin D plays a critical role in insulin secretion.[62,67,68] More recent randomized clinical trials have supported the hypothesis that vitamin D supplementation will improve metabolic outcomes. Two of those trials reported improved β-cell function[69] or insulin sensitivity[70] in vitamin D deficient subjects at high risk of T2D after vitamin D supplementation. A third study performed by von Hurst et al.[71] observed significantly improved insulin sensitivity but not insulin secretion in South Asian women living in New Zealand who were insulin resistant and vitamin D deficient. Several intervention studies on heterogeneous populations with varying amounts of vitamin D supplementation and duration of treatment provide little conclusive information. Many of the older trials that reported improved insulin secretion had a sample size of fewer than 25 participants.[72–77]

The biologically active form of vitamin D [$1,25(OH)_2D$] may directly stimulate insulin release by binding to the vitamin D receptors of β cells, thus influencing pathways such as calcium-mediated insulin secretion or by affecting other genes associated with β-cell metabolism and development (Figure 2). Studies suggest that vitamin D may directly enhance insulin response of peripheral tissues by stimulating expression of insulin receptors. In addition to these direct actions, vitamin D may indirectly regulate calcium homeostasis in β cells and insulin-responsive tissues.[78] Because calcium is critical for skeletal muscle function,[79] increased vitamin D status may improve muscle function,[80] thus increasing insulin sensitivity. Our findings in Asian Indian Sikhs showed a significant negative association of fasting glucose as well as a significant positive association of homeostasis model assessment for β-cell function (HOMA-B) with serum 25(OH)D. Although we did not observe a significant association of insulin resistance, assessed by HOMA for insulin resistance with serum 25(OH)D, studies suggest that vitamin D may directly enhance insulin response of peripheral tissues by stimulating expression of insulin receptors.

Cardiovascular disease remains the leading cause of death among individuals with T2D. A growing body of evidence is emerging that may implicate vitamin D deficiency as an important risk factor for hypertension and cardiovascular disease.[81–84] One proposed mechanism has linked vitamin D deficiency to cardiovascular disease in T2D patients. Our cross-sectional study conducted on Punjabi Sikh diabetic cohort in India revealed a significant increase in systolic blood pressure and pulse pressure in vitamin D–deficient males.[62] Additionally, vitamin D deficiency is proposed to increase cardiovascular disease through a secondary increase in parathyroid hormone that activates the renin-angiotensin system and stimulates systemic and vascular inflammation.[85] Oh et al.[86] found that the active form of vitamin D inhibits foam cell formation and suppresses macrophage cholesterol uptake in patients with T2D. In addition, a direct protective effect of vitamin D on the endothelium is possible because vitamin D deficiency has been associated with impaired endothelial function.[87]

3. GENETIC BASIS OF VITAMIN D DEFICIENCY

The global distribution of vitamin D receptors in the body is controlled by nearly 3000 genes,[88] which may have widespread health implications. Results of family and twin studies suggest that the variation in serum vitamin D levels, including deficiency, is under strong genetic control. Serum 25-hydroxyvitamin D [$25(OH)D_3$] is the best indicator of vitamin D status and has a high heritability ranging from 28% to 80%.[89–91] Recent in vitro evidence also suggests that vitamin D modulates the expression of hundreds of genes, many of which belong to disease-associated pathways.[88] However, the molecular basis by which vitamin D exerts effects in diverse diseases remains incompletely understood. Two rare causes of vitamin D-dependent rickets (DDRs) have been identified: type 1 and type 2. DDR-1 is an autosomal recessive point mutation in the gene (*CYP27B1*) that encodes the enzyme (1α-hydroxylase) responsible for converting the major circulating but inactive form of vitamin D, 25-hydroxyvitamin D [25(OH)D], to the biologically active form, 1,25-dihydroxyvitamin D [1,25$(OH)_2$D].[92] Children afflicted with DDR-1 have normal 25(OH)D levels but develop severe hypotonia and secondary hyperparathyroidism caused by suboptimal serum levels of calcium and 1,25$(OH)_2$D. Patients treated with 1,25$(OH)_2$D (calcitriol) and calcium carbonate have shown immediate reversal of all clinical manifestations of rickets, including a marked increase in muscle strength.[92,93] However, children with DDR-2, caused by a loss-of-function homozygous mutation in the *VDR* gene, did not respond to calcitriol treatment and were either administered intravenous or oral calcium depending on severity.[94] Nonfunctional VDRs, resulting in decreased vitamin D–dependent calcium absorption, necessitate continued calcium supplementation in these patients, as only 10–15% of calcium can be absorbed in the absence of vitamin D.[95] These clinical observations highlight the well-established role of vitamin D to maintain calcium homeostasis for optimal bone health.

3.1 Vitamin D Genetics and T1D

Variants near key 25(OH)D metabolism genes (group component (*GC*), *CYP24A1*, *CYP27B1*, *DHCR7*, and *CYP2R1*) show consistent evidence of association with T1D risk, indicating a genetic etiological role for vitamin D deficiency in T1D.[19] These genes, involved in cholesterol synthesis, hydroxylation, and vitamin D transport, affect vitamin D status. Genetic variation at these loci identifies individuals who have increased risk of vitamin D insufficiency[96] (Figure 2). Variants at three loci reached genome-wide significance for association with 25(OH)D concentrations, and were confirmed in replication cohorts: 4p12 (rs2282679 near *GC*), 11q12 (rs12785878 near *DHCR7*), and 11p15 (rs10741657, near *CYP2R1*). Participants with a genotype score (combining the three confirmed variants) in the highest quartile were at increased risk of having low 25(OH)D concentrations compared with those in the lowest quartile. A variant represented by rs6013897 near *CYP24A1* (20q13) also reached genome-wide significance in a joint meta-analysis of GWAS and replication.[96] Because the enzyme 1-alpha hydroxylase encoded by *CYP27B1* plays a key role in the second hydroxylation step forming the active vitamin D metabolite 1,25$(OH)_2$D, *CYP27B1* was studied as a candidate gene for association with T1D and compared with *CYP24A1*. The association for *CYP27B1* was significant providing evidence that genetic variation in a key vitamin D metabolic step affects susceptibility to T1D.[97]

Genetic variation in the *GC* gene encoding vitamin D–binding protein (DBP) has been studied in a case-control study to determine association with T1D characterized by positive anti-islet antibodies. DBP is the main systemic transporter of bioactive 1,25$(OH)_2$D and is essential for its subsequent endocytosis. Two known polymorphisms in exon 11 of the *GC* gene result in amino acid variants: GAT → GAG substitution replaces aspartic acid by glutamic acid in codon 416 and ACG → AAG substitution in codon 420 leads to an exchange of threonine for lysine. These variants in the *GC* gene lead to differences in affinity. Association of *GC* variant alleles with T1D have been described in different populations and frequencies of the Asp/Glu and Glu/Glu were significantly increased in diabetic subjects with detectable IA-2 antibodies.[98] This finding could lead to further studies on genetic association with the autoimmune phenotype represented by islet antibody titers first appearing at a phase in the pathogenesis when intervention might be more effective.

Association with *VDR* gene variants has been less conclusive with conflicting results. Based on evidence that four known variants (*BSML*, *FOKL*, *APAL*, and *Taq1*) in the *VDR* gene have been associated with T1D in a meta-analysis study.[99] The *BSML* variant was associated with increased risk but not *FOKL*, *APAL*, and *Taq1*. In subgroup analysis by ethnicity, the increased risk of T1D remained in an Asian subgroup for *BSML*, whereas no significant association was found in other populations.[99] A meta-analysis represented by 11 studies for *FokI*, 13 for *BsmI*, nine for *ApaI*, and seven for *TaqI*; a few family-transmission studies with analysis of these four polymorphisms showed no evidence for an association between *VDR* variants and T1D in either the case-control or family-transmission studies.[100] The T1D Genetics Consortium genotyped 38 single nucleotide polymorphisms (SNPs) in 1654 T1D nuclear families (of 6707 individuals, 3399 were affected) and also found no association with *VDR* variants, which included subgroup analyses by parent-of-origin, sex of offspring, and HLA risk.[101]

The Diabetes Autoimmunity Study of Young (DAISY), a longitudinal study showed progression to T1D in autoantibody positive children, defined as positivity for glutamic acid decarboxylase, insulin, or IA-2 autoantibodies on two or more consecutive visits. The progression was associated with the *VDR* (rs2228570) GG genotype and there was an interaction between *VDR* rs1544410 and a protein tyrosine phosphatase type 2 gene variant (*PTPN2* rs1893217), whereas in children with the *PTPN2* rs1893217 AA genotype, the *VDR* rs1544410 AA/AG genotype was associated with a protective effect.[102] This novel finding supports association with early autoimmune events and requires replication.

3.2 Vitamin D Genetics and T2D

Despite these recent advances in GWAS, genetic determinants of circulating vitamin D levels through GWAS have mainly been reported in populations of European descent.[96,103] GWAS and meta-analysis studies published in 2010 using 6722[103] and 33,996[96] participants of European descent revealed a significant association of serum 25(OH)D levels with several SNPs located on genes in the vitamin D metabolizing pathway genes (see Figure 1). Six most closely related genes in the metabolism of 25(OH)D3 pathway include *GC* associated with vitamin D transport, *DHCR7* which converts 7-dehydrocholesterol to vitamin D_3, *CYP2R1*, *CYP27A1*, and *CYP3A4*, responsible for hydroxylation of vitamin D in liver and the *VDR* gene encoding vitamin D receptor.[96] Common variants in three of these six genes (*GC* (4p12; $p = 1.9 \times 10^{-109}$), *DHCR7* (11q12; $p = 2.1 \times 10^{-27}$), and *CYP2R1* [11p15; 3.3×10^{-20})) have appeared among the top hits in GWAS and meta-analysis studies, and revealed a significantly raised risk of vitamin D deficiency.[96,103] Table 1 summarizes association of SNPs in vitamin D pathway genes with serum 25(OH)D concentrations, metabolites, and other related cardiometabolic traits among different ethnic groups. Interestingly, the association of the most robustly associated variant (rs2282679) at the *GC* gene in Europeans was consistently reproducible in Han Chinese[104] and Punjabi Sikhs,[105] South Asians and Arabs,[106] in addition to other independent European studies.[90] Targeted genotyping of vitamin D pathway SNPs, revealed through GWAS have confirmed the association of *GC* and *CYP2R1* with 25(OH)D in Han Chinese children[104] and healthy Caucasian subjects,[107] (Table 1). *DHCR7* expresses a reductase that catalyzes the production of cholesterol from 7-DHC using nicotinamide adenine dinucleotide phosphate-oxidase.[108] Mutations in *DHCR7* are known to lead to the rare Smith–Lemli–Opitz syndrome, which leads to accumulation of 7-DHC and a deficiency of cholesterol along with congenital abnormalities and intellectual disabilities.[109] *CYP2R1*, being a member of cytochrome P450 superfamily, catalyzes many reactions involved in synthesis of cholesterol, steroids, and other lipids.[110] The *GC* gene that encodes DBP belongs to the albumin gene family. It has a strong affinity to bind with 25(OH)D and its plasma metabolites and transports them to target tissues.[111] Studies suggests that *GC*-bound metabolites have longer half-life and are less susceptible to hydroxylation and degradation. A recent study conducted in 1380 European male participants identified two independent signals located in the *GC* that were highly associated with fasting serum DBP concentrations (rs7041 ($p = 1.42 \times 10^{-246}$) and rs705117 ($p = 4.7 \times 10^{-91}$) at the genome-wide level of significance.[112] Alternatively, spliced transcript variants encoding eight different isoforms have been found for this gene. Common variation in three of eight isoforms (*GC1*, *GC1F*, and *GC1S*) for influencing 25(OH)D concentrations have been reported.[113] Evidence of ethnic and geographic differences in these gene variants and their affinity with vitamin D metabolites among different ethnic groups with has been reviewed in detail previously.[108]

Genetic variation at these GWAS loci identified individuals who have increased risk of vitamin D insufficiency.[96] However, the effect size of GWAS-derived polymorphisms is small (<10%), and it is difficult to predict potential implication of these findings in clinical settings. So there is a need for further investigation to identify more putative causal variants, ideally with larger effects.

Active vitamin D (calcitriol) is thought to promote monocyte/macrophage differentiation and regulate monocyte vascular infiltration and macrophage cholesterol retention in the vessel wall.[5,6] Apolipoprotein E or *APOE* is considered to be one of the principal mediators of lipid metabolism.[7] Of the three common allelic isoforms, *APOE4* represents the highest risk of developing atherosclerosis and cardiovascular disease.[8] Initially, this was thought to be due to elevated lipid levels, but recently *APOE4* has shown to affect the biomarkers of oxidative stress and inflammation.[9,10] There has been some experimental and epidemiological evidence that *APOE4* allele protects against vitamin D deficiency mainly by better intestinal absorption and more efficient synthesis in the skin.[11] Although this association is established by a single animal and human study, its effects on the cardiovascular risk profile are largely unknown.

4. CONCLUSIONS AND FUTURE DIRECTIONS

Good vitamin D status is generally a marker for good health. The evidence continues to accumulate indicating the role of inadequate levels of vitamin D in pathogenesis of several endocrine diseases including diabetes. However, the available evidence from several observational studies is inconsistent. Main limitations in most available studies include cross-sectional design, small sample size, lack of data on prospective follow-up, and lack of nutritional status of participants represented

TABLE 1 Association of SNPs in GC, *CYP3A4*, *CYP2R1*, *DHCR7*, and *CYP24A1* with 25(OH)D Levels, Metabolites, and Cardiometabolic Traits among Different Ethnic Groups

Nearest gene(s)	Chr	SNP	Trait	GWAS *p* value	Replication *p* value	Overall *p* value	Non-GWAS *p* value	Population	References
GC	4	rs2282679	25(OH)D	4.57×10^{-63}	2.88×10^{-48}	1.9×10^{-109}		European ($n = 33{,}996$)	Wang et al.[96]
GC	4	rs3755967	25(OH)D	7.41×10^{-53}	3.0×10^{-24}	2.42×10^{-75}		European ($n = 33{,}996$)	Wang et al.[96]
GC	4	rs7041	Vitamin DBP levels	1×10^{-246}	–	1×10^{-246}		European male ($n = 1380$)	Moy et al.[112]
GC	4	rs705117	Vitamin DBP levels	5×10^{-91}	–	5×10^{-91}		European male ($n = 1380$)	Moy et al.[112]
GC	4	rs1851024	Metabolite levels	1×10^{-14}	–	1×10^{-14}		European ($n = 6608$)	Inouye et al.[114]
GC	4	rs2282679	25(OH)D	–	–	–	0.004	Han Chinese children ($n = 506$)	Zhang et al.[99]
GC	4	rs7041	25(OH)D	–	–	–	0.0003	European premenopausal women ($n = 733$)	Sinotte et al.[109]
GC	4	rs4588	25(OH)D	–	–	–	0.0001	European premenopausal women ($n = 733$)	Sinotte et al.[115]
GC	4	rs7041	25(OH)D	–	–	–	0.003	Hispanic Americans ($n = 1017$)	Engelmanet al.[116]
GC	4	rs4588	25(OH)D	–	–	–	0.004	Hispanic Americans ($n = 1017$)	Engelman et al.[116]
GC	4	rs7041	25(OH)D	–	–	–	0.025	African Americans ($n = 513$)	Engelmanet al.[116]
GC	4	rs4588	25(OH)D	–	–	–	0.007	African Americans ($n = 513$)	Engelman et al.[116]
GC	4	rs2282679	25(OH)D		3×10^{-4}	–	–	Asian Indian Sikhs ($n = 1616$)	Sapkota et al.[105]
CYP3A4	7	rs17277546	Metabolic traits	9×10^{-40}		9×10^{-40}		European ($n = 2820$)	Suhre et al.[117]
CYP2R1	11	rs10741657	25(OH)D	3.91×10^{-8}	2.09×10^{-14}	3.27×10^{-20}		European ($n = 33{,}996$)	Wang et al.[96]
CYP2R1	11	rs2060793	25(OH)D	2.69×10^{-6}	2.36×10^{-7}	1.73×10^{-11}		European ($n = 33{,}996$)	Wang et al.[96]
CYP2R1	11	rs2060793	25(OH)D	1.4×10^{-5}	1.6×10^{-17}	2.9×10^{-17}		European ($n = 6722$)	Ahn et al.[103]
CYP2R1	11	rs1993116	25(OH)D	2.94×10^{-6}	1.28×10^{-6}	6.25×10^{-11}		European ($n = 33{,}996$)	Wang et al.[96]
CYP2R1	11	rs12794714	25(OH)D	6.24×10^{-5}	8.71×10^{-7}	1.84×10^{-9}		European ($n = 33{,}996$)	Wang et al.[96]
CYP2R1	11	rs10500804	25(OH)D	7.43×10^{-5}	1.12×10^{-6}	2.67×10^{-9}		European ($n = 33{,}996$)	Wang et al.[96]
CYP2R1	11	rs7116978	25(OH)D	1.17×10^{-5}	7.59×10^{-5}	4.99×10^{-9}		European ($n = 33{,}996$)	Wang et al.[96]
CYP2R1	11	rs10832313	BMI	–	–	–	0.02	Chinese women ($n = 6922$)	Dorjgochoo et al.[118]
DHCR7	11	rs12785878	25(OH)D	1.27×10^{-12}	2.39×10^{-16}	2.12×10^{-27}		European ($n = 33{,}996$)	Wang et al.[96]
DHCR7	11	rs3829251	25(OH)D	8.8×10^{-7}	1.0×10^{-3}	3×10^{-9}		European ($n = 6722$)	Ahn et al.[103]
CYP24A1	20	rs6013897	25(OH)D	7.2×10^{-4}	8.4×10^{-8}	6×10^{-10}		European ($n = 33{,}996$)	Wang et al.[96]

BMI, body mass index; Chr, chromosome; GWAS, genome-wide association study; SNP, single nucleotide polymorphism; vitamin DBP, vitamin D–binding protein.

by vitamin D levels that are influenced by dietary factors. Also, most studies did not adjust for seasonal changes, which may influence vitamin D status, especially the populations living near the equator or above 30° N latitude. More high-quality observational studies with large sample size, studies that reflect long-term vitamin D status, and randomized controlled trials (few ongoing) in different ethnically and culturally diverse populations will be required to putatively assess vitamin D as a risk factor for diabetes and other endocrine diseases. Prospective data would define whether the vitamin D deficiency would be a consequence of diabetes rather than a cause. More genome-wide studies on vitamin D status/deficiency in non-European populations would clarify the role of genetic variation in 25(OH)D deficiency, and environmental (climate, geographic, and dietary) interactions for the associated cardiometabolic risk. We still need to identify how genetic variants impact gene products involved in vitamin D transport or vitamin D hydrolysis, and other vitamin D target genes. It is still unclear whether gene variants modify our response to sun exposure and consequently vitamin D deficiency and more work is needed to establish gene–environment interactions and the potential for variations in the requirement for vitamin D.

In summary, preliminary evidence suggests that the improved vitamin D status can be a modifying factor for endocrine and cardiometabolic health. Nutritional status with foods rich in vitamin D (fish and fortified dairy products) may have independent or synergic effects on reducing the risk of diabetes and cardiometabolic diseases. Currently ongoing trials, future genomics, and metabolomics studies will likely offer insights into both causality and treatment of diabetes and other vitamin D–related diseases.

Acknowledgments

This work was supported by National Institutes of Health grants -R01DK082766 funded by the National Institute of Health (NIDDK) and NOT-HG-11-009 funded by the National Human Genome Research Institute, and a VPR Bridge Grant from University of Oklahoma Health Sciences Center. Technical support provided by Timothy Braun, Latonya Been, Bishwa Sapkota, and Gursumeet Sanghera is duly acknowledged.

References

1. Heaney RP. Long-latency deficiency disease: insights from calcium and vitamin D. *Am J Clin Nutr* 2003;**78**:912–9.
2. Kulie T, Groff A, Redmer J, Hounshell J, Schrager S. Vitamin D: an evidence-based review. *J Am Board Fam Med* 2009;**22**:698–706.
3. Holick MF. Vitamin D: importance in the prevention of cancers, type 1 diabetes, heart disease, and osteoporosis. *Am J Clin Nutr* 2004;**79**:362–71.
4. Zipitis CS, Akobeng AK. Vitamin D supplementation in early childhood and risk of type 1 diabetes: a systematic review and meta-analysis. *Arch Dis Child* 2008;**93**:512–7.
5. Solomon AJ. Patient page. Multiple sclerosis and vitamin D. *Neurology* 2011;**77**:e99–100.
6. Cutolo M, Otsa K, Uprus M, Paolino S, Seriolo B. Vitamin D in rheumatoid arthritis. *Autoimmun Rev* 2007;**7**:59–64.
7. Sato Y, Asoh T, Oizumi K. High prevalence of vitamin D deficiency and reduced bone mass in elderly women with Alzheimer's disease. *Bone* 1998;**23**:555–7.
8. Knekt P, Kilkkinen A, Rissanen H, Marniemi J, Saaksjarvi K, Heliovaara M. Serum vitamin D and the risk of Parkinson disease. *Arch Neurol* 2010;**67**:808–11.
9. Hollick MF, Chen TC. Vitamin D deficiency a worldwide problem with health consequences. *Am J Clin Nutr* 2008;**87**:10805–68.
10. Fitzpatrick S, Sheard NF, Clark NG, Ritter ML. Vitamin D-deficient rickets: a multifactorial disease. *Nutr Rev* 2000;**58**: 218–22.
11. Lo CW, Paris PW, Holick MF. Indian and Pakistani immigrants have the same capacity as Caucasians to produce vitamin D in response to ultraviolet irradiation. *Am J Clin Nutr* 1986;**44**: 683–5.
12. Mitri J, Muraru MD, Pittas AG. Vitamin D and type 2 diabetes: a systematic review. *Eur J Clin Nutr* 2011;**65**:1005–15.
13. Vacek JL, Vanga SR, Good M, Lai SM, Lakkireddy D, Howard PA. Vitamin D deficiency and supplementation and relation to cardiovascular health. *Am J Cardiol* 2012;**109**:359–63.
14. Renzaho AM, Halliday JA, Nowson C. Vitamin D, obesity, and obesity-related chronic disease among ethnic minorities: a systematic review. *Nutrition* 2011;**27**:868–79.
15. Veldman CM, Cantorna MT, DeLuca HF. Expression of 1,25-dihydroxyvitamin D(3) receptor in the immune system. *Arch Biochem Biophys* 2000;**374**:334–8.
16. Bland R, Markovic D, Hills CE, Hughes SV, Chan SL, Squires PE, et al. Expression of 25-hydroxyvitamin D_3-1alpha-hydroxylase in pancreatic islets. *J Steroid Biochem Mol Biol* 2004;**89–90**:121–5.
17. Baeke F, Takiishi T, Korf H, Gysemans C, Mathieu C. Vitamin D: modulator of the immune system. *Curr Opin Pharmacol* 2010;**10**: 482–96.
18. Littorin B, Blom P, Scholin A, Arnqvist HJ, Blohme G, Bolinder J, et al. Lower levels of plasma 25-hydroxyvitamin D among young adults at diagnosis of autoimmune type 1 diabetes compared with control subjects: results from the nationwide Diabetes Incidence Study in Sweden (DISS). *Diabetologia* 2006; **49**:2847–52.
19. Cooper JD, Smyth DJ, Walker NM, Stevens H, Burren OS, Wallace C, et al. Inherited variation in vitamin D genes is associated with predisposition to autoimmune disease type 1 diabetes. *Diabetes* 2011;**60**:1624–31.
20. Bodnar LM, Simhan HN, Powers RW, Frank MP, Cooperstein E, Roberts JM. High prevalence of vitamin D insufficiency in black and white pregnant women residing in the northern United States and their neonates. *J Nutr* 2007;**137**:447–52.
21. van der Meer IM, Karamali NS, Boeke AJ, Lips P, Middelkoop BJ, Verhoeven I, et al. High prevalence of vitamin D deficiency in pregnant non-Western women in the Hague, Netherlands. quiz 468–9 *Am J Clin Nutr* 2006;**84**:350–3.
22. Prentice A. Vitamin D deficiency: a global perspective. *Nutr Rev* 2008;**66**:S153–64.
23. Schroth RJ, Lavelle CL, Moffatt ME. Review of vitamin D deficiency during pregnancy: who is affected? *Int J Circumpolar Health* 2005;**64**:112–20.
24. Yan L, Zhou B, Wang X, D'Ath S, Laidlaw A, Laskey MA, et al. Older people in China and the United Kingdom differ in the relationships among parathyroid hormone, vitamin D, and bone mineral status. *Bone* 2003;**33**:620–7.

25. Awumey EM, Mitra DA, Hollis BW, Kumar R, Bell NH. Vitamin D metabolism is altered in Asian Indians in the southern United States: a clinical research center study. *J Clin Endocrinol Metab* 1998;**83**:169–73.
26. Harris SS. Vitamin D and African Americans. *J Nutr* 2006;**136**: 1126–9.
27. Craig ME, Jones TW, Silink M, Ping YJ. Diabetes care, glycemic control, and complications in children with type 1 diabetes from Asia and the Western Pacific Region. *J Diabetes Complications* 2007;**21**:280–7.
28. Patterson CC, Dahlquist GG, Gyurus E, Green A, Soltesz G, Group ES. Incidence trends for childhood type 1 diabetes in Europe during 1989–2003 and predicted new cases 2005–20: a multicentre prospective registration study. *Lancet* 2009;**373**: 2027–33.
29. Bluestone JA, Herold K, Eisenbarth G. Genetics, pathogenesis and clinical interventions in type 1 diabetes. *Nature* 2010;**464**: 1293–300.
30. Tuomilehto J. The emerging global epidemic of type 1 diabetes. *Curr Diab Rep* 2013;**13**:795–804.
31. Till A, Kenk H, Rjasanowski I, Wassmuth R, Walschus U, Kerner W, et al. Autoantibody-defined risk for Type 1 diabetes mellitus in a general population of schoolchildren: results of the Karlsburg Type 1 Diabetes Risk Study after 18 years. *Diabet Med* 2015;**32**:1008–16.
32. Dotta F, Eisenbarth GS. Type I diabetes mellitus: a predictable autoimmune disease with interindividual variation in the rate of beta cell destruction. *Clin Immunol Immunopathol* 1989;**50**: S85–95.
33. Nansel TR, Lipsky LM, Iannotti RJ. Cross-sectional and longitudinal relationships of body mass index with glycemic control in children and adolescents with type 1 diabetes mellitus. *Diabetes Res Clin Pract* 2013;**100**:126–32.
34. Lauria A, Barker A, Schloot N, Hosszufalusi N, Ludvigsson J, Mathieu C, et al. BMI is an important driver of beta cell loss in type 1 diabetes upon diagnosis in 10–18 year old children. *Eur J Endocrinol* 2015;**172**:107–13.
35. Purnell JQ, Zinman B, Brunzell JD, Group DER. The effect of excess weight gain with intensive diabetes mellitus treatment on cardiovascular disease risk factors and atherosclerosis in type 1 diabetes mellitus: results from the Diabetes Control and Complications Trial/Epidemiology of Diabetes Interventions and Complications Study (DCCT/EDIC) study. *Circulation* 2013; **127**:180–7.
36. Aathira R, Jain V. Advances in management of type 1 diabetes mellitus. *World J Diabetes* 2014;**5**:689–96.
37. Balamurugan AN, Naziruddin B, Lockridge A, Tiwari M, Loganathan G, Takita M, et al. Islet product characteristics and factors related to successful human islet transplantation from the Collaborative Islet Transplant Registry (CITR) 1999–2010. *Am J Transplant* 2014;**14**:2595–606.
38. Li M, Ikehara S. Stem cell treatment for type 1 diabetes. *Front Cell Dev Biol* 2014;**2**:9.
39. Guariguata L, Whiting DR, Hambleton I, Beagley J, Linnenkamp U, Shaw JE. Global estimates of diabetes prevalence for 2013 and projections for 2035. *Diabetes Res Clin Pract* 2014;**103**:137–49.
40. Huang ES, Basu A, O'Grady M, Capretta JC. Projecting the future diabetes population size and related costs for the U.S. *Diabetes Care* 2009;**32**:2225–9.
41. Boyle JP, Thompson TJ, Gregg EW, Barker LE, Williamson DF. Projection of the year 2050 burden of diabetes in the US adult population: dynamic modeling of incidence, mortality, and prediabetes prevalence. *Popul Health Metr* 2011;**8**:29.
42. Mokdad AH, Ford ES, Bowman BA, Nelson DE, Engelgau MM, Vinicor F, et al. The continuing increase of diabetes in the US. *Diabetes Care* 2001;**24**:412.
43. Fagot-Campagna A, Burrows NR, Williamson DF. The public health epidemiology of type 2 diabetes in children and adolescents: a case study of American Indian adolescents in the Southwestern United States. *Clin Chim Acta* 1999;**286**:81–95.
44. Gholap N, Davies M, Patel K, Sattar N, Khunti K. Type 2 diabetes and cardiovascular disease in South Asians. *Prim Care Diabetes* 2011;**5**:45–56.
45. Bhopal RS. A four-stage model explaining the higher risk of Type 2 diabetes mellitus in South Asians compared with European populations. *Diabet Med* 2013;**30**:35–42.
46. Grarup N, Sandholt CH, Hansen T, Pedersen O. Genetic susceptibility to type 2 diabetes and obesity: from genome-wide association studies to rare variants and beyond. *Diabetologia* 2014;**57**:1528–41.
47. Bradfield JP, Qu HQ, Wang K, Zhang H, Sleiman PM, Kim CE, et al. A genome-wide meta-analysis of six type 1 diabetes cohorts identifies multiple associated loci. *PLoS Genet* 2011;**7**:e1002293.
48. Almgren P, Lehtovirta M, Isomaa B, Sarelin L, Taskinen MR, Lyssenko V, et al. Heritability and familiality of type 2 diabetes and related quantitative traits in the Botnia Study. *Diabetologia* 2011;**54**:2811–9.
49. Permutt MA, Wasson J, Cox N. Genetic epidemiology of diabetes. *J Clin Invest* 2005;**115**:1431–9.
50. Sanghera DK, Blackett PR. Type 2 diabetes genetics: beyond GWAS. *J Diabetes Metab* 2012;**3**.
51. Coram MA, Duan Q, Hoffmann TJ, Thornton T, Knowles JW, Johnson NA, et al. Genome-wide characterization of shared and distinct genetic components that influence blood lipid levels in ethnically diverse human populations. *Am J Hum Genet* 2013;**92**:904–16.
52. Stranger BE, Stahl EA, Raj T. Progress and promise of genome-wide association studies for human complex trait genetics. *Genetics* 2011; **187**:367–83.
53. Kahn HS, Morgan TM, Case LD, Dabelea D, Mayer-Davis EJ, Lawrence JM, et al. Association of type 1 diabetes with month of birth among U.S. youth: the SEARCH for Diabetes in Youth Study. *Diabetes Care* 2009;**32**:2010–5.
54. Kalliora MI, Vazeou A, Delis D, Bozas E, Thymelli I, Bartsocas CS. Seasonal variation of type 1 diabetes mellitus diagnosis in Greek children. *Hormones (Athens)* 2011;**10**:67–71.
55. Moltchanova EV, Schreier N, Lammi N, Karvonen M. Seasonal variation of diagnosis of Type 1 diabetes mellitus in children worldwide. *Diabet Med* 2009;**26**:673–8.
56. Mohr SB, Garland CF, Gorham ED, Garland FC. The association between ultraviolet B irradiance, vitamin D status and incidence rates of type 1 diabetes in 51 regions worldwide. *Diabetologia* 2008;**51**:1391–8.
57. Scragg R, Sowers M, Bell C. Serum 25-Hydroxyvitamin D, diabetes, and ethnicity in the third national health and nutrition examination survey. *Diabetes Care* 2004;**27**:2813–8.
58. Gagnon C, Lu ZX, Magliano DJ, et al. Serum 25-Hydroxyvitamin D, calcium intake, and risk of type 2 diabetes after 5 years: results from a national, population-based prospective study (the Australian diabetes, obesity and lifestyle study). *Diabetes Care* 2011;**34**:1133–8.
59. Deleskog A, Hilding A, Brismar K, et al. Low serum 25-hydroxyvitamin D level predicts progression to type 2 diabetes in individuals with prediabetes but not with normal glucose tolerance. *Diabetologia* 2012;**55**:1668–78.
60. Pittas AG, Sun Q, Manson JE, et al. Plasma 25-hydroxyvitamin D concentration and risk of incident type 2 diabetes in women. *Diabetes Care* 2010;**33**:2021–3.
61. Khan H, Kunutsor S, Franco OH, et al. Vitamin D, type 2 diabetes and other metabolic outcomes: a systematic review and meta-analysis of prospective studies. *Proc Nutr Soc* 2013;**72**:89–97.
62. Braun TR, Been LF, Blackett PR, Sanghera DK. Vitamin D deficiency and cardio-metabolic risk in a North Indian community with highly prevalent type 2 diabetes. *J Diabetes Metab* 2012;**3**. http://dx.doi.org/10.4172/2155-6156.1000213.

63. Kotwal SK, Laway BA, Shah ZA, et al. Pattern of 25 hydroxy vitamin D status in North Indian people with newly detected type 2 diabetes- a prospective case control study. *Indian J Endocrinol Metab* 2014;**18**:726–30.
64. Subramanian A, Priyanka Nigam P, Misra A, et al. Severe vitamin D deficiency in patients with Type 2 diabetes in north India. *Diabet Manag* 2011;**1**:477–83.
65. Liel Y, Ulmer E, Shary J, Hollis BW, Bell NH. Low circulating vitamin D in obesity. *Calcif Tissue Int* 1988;**43**:199–201.
66. Wortsman J, Matsuoka LY, Chen TC, Lu Z, Holick MF. Decreased bioavailability of vitamin D in obesity. *Am J Clin Nutr* 2000;**72**: 690–3.
67. Riachy R, Vandewalle B, Moerman E, Belaich S, Lukowiak B, Gmyr V, et al. 1,25-Dihydroxyvitamin D_3 protects human pancreatic islets against cytokine-induced apoptosis via down-regulation of the Fas receptor. *Apoptosis* 2006;**11**:151–9.
68. Chiu KC, Chu A, Go VL, Saad MF. Hypovitaminosis D is associated with insulin resistance and beta cell dysfunction. *Am J Clin Nutr* 2004;**79**:820–5.
69. Mitri J, Dawson-Hughes B, Hu FB, Pittas AG. Effects of vitamin D and calcium supplementation on pancreatic β cell function, insulin sensitivity, and glycemia in adults at high risk of diabetes: the Calcium and Vitamin D for Diabetes Mellitus (CaDDM) randomized controlled trial. *Am J Clin Nutr* 2011;**94**:486–94.
70. Nikooyeh B, Neyestani TR, Farvid M, et al. Daily consumption of vitamin D- or vitamin D + calcium-fortified yogurt drink improved glycemic control in patients with type 2 diabetes: a randomized clinical trial. *Am J Clin Nutr* 2011;**93**:764–71.
71. von Hurst PR, Stonehouse W, Coad J. Vitamin D supplementation reduces insulin resistance in South Asian women living in New Zealand who are insulin resistant and vitamin D deficient - a randomised, placebo-controlled trial. *Br J Nutr* 2010;**103**:549–55.
72. Inomata S, Kadowaki S, Yamatani T, et al. Effect of 1 alpha (OH)-vitamin D_3 on insulin secretion in diabetes mellitus. *Bone Miner* 1986;**1**(3):187–92.
73. Gedik O, Akalin S. Effects of vitamin D deficiency and repletion on insulin and glucagon secretion in man. *Diabetologia* 1985;**29**:142–5.
74. Kumar S, Davies M, Zakaria Y, et al. Improvement in glucose tolerance and beta-cell function in a patient with vitamin D deficiency during treatment with vitamin D. *Postgrad Med J* June 1994;**70**:440–3.
75. Boucher BJ, Mannan N, Noonan K, et al. Glucose intolerance and impairment of insulin secretion in relation to vitamin D deficiency in east London Asians. *Diabetologia* 1995;**38**:1239–45.
76. Allegra V, Luisetto G, Mengozzi G, et al. Glucose-induced insulin secretion in uremia: role of 1 alpha,25$(HO)_2$-vitamin D_3. *Nephron* 1994;**68**:41–7.
77. Orwoll E, Riddle M, Prince M. Effects of vitamin D on insulin and glucagon secretion in non-insulin-dependent diabetes mellitus. *Am J Clin Nutr* 1994;**59**:1083–7.
78. Pittas AG, Lau J, Hu FB, Dawson-Hughes B. The role of vitamin D and calcium in type 2 diabetes. A systematic review and meta-analysis. *J Clin Endocrinol Metab* 2007;**92**:2017–29.
79. Berchtold MW, Brinkmeirer H, Muntener M. Calcium ion in skeletal muscle: its crucial role for muscle function, plasticity, and disease. *Physiol Rev* 2000;**80**:1215–65.
80. Hamilton B. Vitamin D and human skeletal muscle. *Scand J Med Sci Sports* 2010;**20**:182–90.
81. Xiang W, Kong J, Chen S, Cao LP, Qiao G, Zheng W, et al. Cardiac hypertrophy in vitamin D receptor knockout mice: role of the systemic and cardiac renin-angiotensin systems. *Am J Physiol Endocrinol Metab* 2005;**288**:E125–32.
82. Kristal-Boneh E, Froom P, Harari G, Ribak J. Association of calcitriol and blood pressure in normotensive men. *Hypertension* 1997; **30**:1289–94.
83. Lind L, Hanni A, Lithell H, Hvarfner A, Sorensen OH, Ljunghall S. Vitamin D is related to blood pressure and other cardiovascular risk factors in middle-aged men. *Am J Hypertens* 1995; **8**:894–901.
84. Wang TJ, Pencina MJ, Booth SL, Jacques PF, Ingelsson E, Lanier K, et al. Vitamin D deficiency and risk of cardiovascular disease. *Circulation* 2008;**117**:503–11.
85. Lee JH, O'Keefe JH, Bell D, Hensrud DD, Holick MF. Vitamin D deficiency an important, common, and easily treatable cardiovascular risk factor? *J Am Coll Cardiol* 2008;**52**:1949–56.
86. Oh J, Weng S, Felton SK, Bhandare S, Riek A, Butler B, et al. 1,25$(OH)_2$ vitamin D inhibits foam cell formation and suppresses macrophage cholesterol uptake in patients with type 2 diabetes mellitus. *Circulation* 2009;**120**:687–98.
87. Ertek S, Akgul E, Cicero AF, Kutuk U, Demirtas S, Cehreli S, et al. 25-Hydroxy vitamin D levels and endothelial vasodilator function in normotensive women. *Arch Med Sci* 2012;**8**:47–52.
88. Ramagopalan SV, Heger A, Berlanga AJ, Maugeri NJ, Lincoln MR, Burrell A, et al. A ChIP-seq defined genome-wide map of vitamin D receptor binding: associations with disease and evolution. *Genome Res* 2010;**20**:1352–60.
89. Wjst M, Altmuller J, Braig C, Bahnweg M, Andre E. A genome-wide linkage scan for 25-OH-D(3) and 1,25-$(OH)_2$-D_3 serum levels in asthma families. *J Steroid Biochem Mol Biol* 2007;**103**:799–802.
90. Bu FX, Armas L, Lappe J, Zhou Y, Gao G, Wang HW, et al. Comprehensive association analysis of nine candidate genes with serum 25-hydroxy vitamin D levels among healthy Caucasian subjects. *Hum Genet* 2010;**128**:549–56.
91. Hunter D, De Lange M, Snieder H, MacGregor AJ, Swaminathan R, Thakker RV, et al. Genetic contribution to bone metabolism, calcium excretion, and vitamin D and parathyroid hormone regulation. *J Bone Miner Res* 2001;**16**:371–8.
92. Yan Y, Calikoglu AS, Jain N. Vitamin D-dependent rickets type 1: a rare, but treatable, cause of severe hypotonia in infancy. *J Child Neurol* 2011;**26**:1571.
93. Fraser D, et al. Pathogenesis of hereditary vitamin-D-dependent rickets. An inborn error of vitamin D metabolism involving defective conversion of 25-hydroxyvitamin D to 1 alpha,25-dihydroxyvitamin D. *N Engl J Med* 1973;**289**:817–22.
94. Tiosano D, Gepstein V. Vitamin D action: lessons learned from hereditary 1,25-dihydroxyvitamin-D-resistant rickets patients. *Curr Opin Endo Diab* 2012;**19**:452–9.
95. Holick MF. Vitamin D deficiency. *N Engl J Med* 2007;**357**:266–81.
96. Wang TJ, Zhang F, Richards JB, Kestenbaum B, van Meurs JB, Berry D, et al. Common genetic determinants of vitamin D insufficiency: a genome-wide association study. *Lancet* 2010;**376**: 180–8.
97. Bailey R, Cooper JD, Zeitels L, Smyth DJ, Yang JH, Walker NM, et al. Association of the vitamin D metabolism gene CYP27B1 with type 1 diabetes. *Diabetes* 2007;**56**:2616–21.
98. Ongagna JC, Pinget M, Belcourt A. Vitamin D-binding protein gene polymorphism association with IA-2 autoantibodies in type 1 diabetes. *Clin Biochem* 2005;**38**:415–9.
99. Zhang J, Li W, Liu J, Wu W, Ouyang H, Zhang Q, et al. Polymorphisms in the vitamin D receptor gene and type 1 diabetes mellitus risk: an update by meta-analysis. *Mol Cell Endocrinol* 2012;**355**: 135–42.
100. Guo SW, Magnuson VL, Schiller JJ, Wang X, Wu Y, Ghosh S. Meta-analysis of vitamin D receptor polymorphisms and type 1 diabetes: a HuGE review of genetic association studies. *Am J Epidemiol* 2006;**164**:711–24.
101. Kahles H, Morahan G, Todd JA, Badenhoop K, Type I.D.G.C.. Association analyses of the vitamin D receptor gene in 1654 families with type I diabetes. *Genes Immun* 2009;**10**(Suppl. 1): S60–3.

102. Frederiksen B, Liu E, Romanos J, Steck AK, Yin X, Kroehl M, et al. Investigation of the vitamin D receptor gene (VDR) and its interaction with protein tyrosine phosphatase, non-receptor type 2 gene (PTPN2) on risk of islet autoimmunity and type 1 diabetes: the Diabetes Autoimmunity Study in the Young (DAISY). *J Steroid Biochem Mol Biol* 2013;**133**:51–7.
103. Ahn J, Yu K, Stolzenberg-Solomon R, Simon KC, McCullough ML, Gallicchio L, et al. Genome-wide association study of circulating vitamin D levels. *Hum Mol Genet* 2010;**19**:2739–45.
104. Zhang Y, Wang X, Liu Y, Qu H, Qu S, Wang W, et al. The *GC*, *CYP2R1* and *DHCR7* genes are associated with vitamin D levels in northeastern Han Chinese children. *Swiss Med Wkly* 2012;**142**: w13636.
105. Sapkota BR, Priamvada G, Bjonnes A, Blackett PR, Saxena R, Sanghera DK. GWAS of serum 25(OH) vitamin D levels in a punjabi sikh diabetic cohort. In: *64th Annual Meeting of the American Society of human genetics, San Diego, CA*; 2014.
106. Elkum N, Alkayal F, Noronha F, Ali MM, Melhem M, Al-Arouj M, et al. Vitamin D insufficiency in Arabs and South Asians positively associates with polymorphisms in *GC* and *CYP2R1* genes. *PLoS One* 2014;**9**:e113102.
107. Engelman CD, Meyers KJ, Iyengar SK, Liu Z, Karki CK, Igo Jr RP, et al. Vitamin D intake and season modify the effects of the GC and CYP2R1 genes on 25-hydroxyvitamin D concentrations. *J Nutr* 2013;**143**:17–26.
108. Berry D, Hypponen E. Determinants of vitamin D status: focus on genetic variations. *Curr Opin Nephrol Hypertens* 2011;**20**:331–6.
109. Tint GS, Irons M, Elias ER, Batta AK, Frieden R, Chen TS, et al. Defective cholesterol biosynthesis associated with the Smith-Lemli-Opitz syndrome. *N Engl J Med* 1994;**330**:107–13.
110. Schuster I. Cytochromes P450 are essential players in the vitamin D signaling system. *Biochim Biophys Acta* 2011;**1814**:186–99.
111. Verboven C, Rabijns A, De Maeyer M, Van Baelen H, Bouillon R, De Ranter C. A structural basis for the unique binding features of the human vitamin D-binding protein. *Nat Struct Biol* 2002;**9**:131–6.
112. Moy KA, Mondul AM, Zhang H, Weinstein SJ, Wheeler W, Chung CC, et al. Genome-wide association study of circulating vitamin D-binding protein. *Am J Clin Nutr* 2014;**99**:1424–31.
113. Braun A, Bichlmaier R, Cleve H. Molecular analysis of the gene for the human vitamin-D-binding protein (group-specific component): allelic differences of the common genetic GC types. *Hum Genet* 1992;**89**:401–6.
114. Inouye M, Ripatti S, Kettunen J, Lyytikäinen LP, Oksala N, Laurila PP, et al. Novel loci for metabolic networks and multi-tissue expression studies reveal genes for atherosclerosis. *PLoS Genet* 2012;**8**(8):e1002907. http://dx.doi.org/10.1371/journal.pgen.1002907. Epub 2012 August 16.
115. Sinotte M, Diorio C, Bérubé S, Pollak M, Brisson J. Genetic polymorphisms of the vitamin D binding protein and plasma concentrations of 25-hydroxyvitamin D in premenopausal women. *Am J Clin Nutr* 2009;**89**(2):634–40.
116. Engelman CD, Fingerlin TE, Langefeld CD, Hicks PJ, Rich SS, Wagenknecht LE. Genetic and environmental determinants of 25-hydroxyvitamin D and 1,25-dihydroxyvitamin D levels in Hispanic and African Americans. *J Clin Endocrinol Metab* 2008: 3381–8.
117. Suhre K, Shin SY, Petersen AK, Mohney RP, Meredith D, Wägele B, et al. Human metabolic individuality in biomedical and pharmaceutical research. *Nature* 2011;**477**(7362):54–60.
118. Dorjgochoo T, Shi J, Gao YT, Long J, Delahanty R, Xiang YB, et al. Genetic variants in vitamin D metabolism-related genes and body mass index: analysis of genome-wide scan data of approximately 7000 Chinese women. *Int J Obes (Lond)* September 2012;**36**(9): 1252–5.

CHAPTER

27

NRF2-Mediated Gene Regulation and Glucose Homeostasis

Yoko Yagishita, Akira Uruno, Masayuki Yamamoto

Department of Medical Biochemistry, Tohoku University Graduate School of Medicine, Sendai, Japan

1. INTRODUCTION

When our bodies ingest dietary substances or environmental chemicals, xenobiotic compounds are processed by detoxification systems within the cells. Importantly, intermediary metabolites generated during the detoxification process sometimes are electrophilic, and they frequently form adducts with nucleic acids or proteins. Therefore, these intermediary metabolites potentially disturb physiological functions. In addition, when dietary nutrients are aerobically metabolized for energy production in the cells, reactive oxygen species (ROS) are generated, which are toxic to our body. Thus, our body is routinely exposed to these "environmental stressors." Because environmental stress subsequently leads to various cellular dysfunctions and diseases, detoxification and antioxidative stress response systems are essential for maintenance of our body in healthy conditions.

Nuclear factor-erythroid-related factor 2 (NRF2) plays critical roles in the regulation of genes encoding detoxification and antioxidative stress response enzymes. NRF2 contributes to the protection of our body against environmental stresses. Recently, it has also been found that NRF2 acts to maintain metabolic homeostasis. Indeed, NRF2 is intimately involved in the pathogenesis of metabolic disorders; this section will summarize recent findings in this area.

2. DETOXIFICATION PROCESSES IN CELLS

In modern society, our body is routinely exposed to various xenobiotics. To excrete these foreign substances from our body, detoxification pathways are used, including oxidation/reduction and conjugation processes. It has been well characterized that xenobiotics are metabolized by three phases of detoxification processes; these processes are called phase I, II, and III detoxification reactions (Figure 1),[1] which act coordinately to produce efflux of xenobiotics from the cells.

2.1 Three Phases of Detoxification Reactions

In the first step of xenobiotic metabolisms, foreign compounds are processed by phase I detoxification enzymes. In the phase I reaction, multiple cytochrome P450 enzymes are employed,[2] and xenobiotics are modified mainly by oxidation reactions to increase electrophilicity. These electrophilic intermediates acquire potential to form adducts with various intracellular high-molecular-weight components, such as nucleic acids and proteins, and these intermediates act as cell-toxic and carcinogenic compounds.[1,3]

However, the phase I reaction metabolites are rapidly conjugated to hydrophilic moieties, including glutathione, glucuronate, and sulfate by phase II enzymes, such as glutathione S-transferases (GSTs), glutamate cysteine ligase (GCL or γ-glutamylcysteine synthetase), uridine 5′-diphospho-glucuronosyltransferase (UDP), UDP-glucuronosyltransferases (UGTs), sulfotransferases, and nicotinamide adenine dinucleotide phosphate-oxidase:quinone oxidoreductase (NQO),[4–6] resulting in hydrophilic compounds that can be easily excreted (Figure 1).[1] Thus, while producing harmful electrophiles, the phase I reaction is a mandatory process for further detoxification.

The phase II metabolites are then excreted from the cells by using the phase III system, which are adenosine

Molecular Nutrition and Diabetes
http://dx.doi.org/10.1016/B978-0-12-801585-8.00027-0

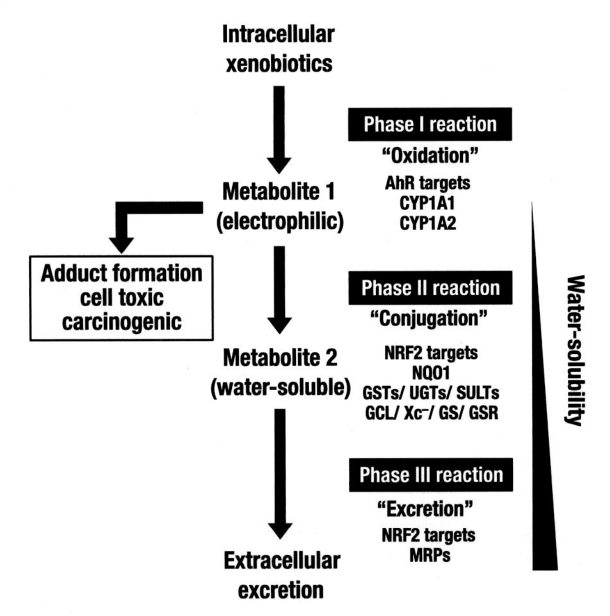

FIGURE 1 Roles of NRF2 on detoxification reactions. Xenobiotics are processed by phase I, II, and III reactions. AhR contributes to phase I reaction by inducing cytochrome P450 enzymes, and its metabolites (described as metabolite 1) are often electrophilic, and induces cell toxicity and carcinogenicity. NRF2 induces phase II enzyme expressions and helps to conjugate hydrophilic moiety to metabolite 1 and generate water-soluble metabolites (described as metabolite 2). NRF2 also contributes to the expressions of a number of phase III reaction transporters and helps excrete metabolite 2 from cells. AhR, aryl hydrocarbon receptor; GCL, glutamate cysteine ligase; GS, glutathione synthetase; GSR, glutathione disulfide reductase; GST, glutathione S-transferase; NRF2, nuclear factor-erythroid-related factor 2; SULTs, sulfotransferases; UGT, UDP-glucuronosyltransferase; Xc^-, cystine/glutamate exchange transporter.

triphosphate–binding cassette transporters,[1,7–9] including multidrug resistance-associated proteins (MRPs).

2.2 Regulations of Detoxification Enzyme Expressions

The detoxification enzymes are usually expressed at low levels in unstressed conditions, but their expressions are induced upon exposure to xenobiotics or oxidative stresses. For instance, expressions of the phase I enzyme genes (e.g., *CYP1A1*, *CYP1A2*) are regulated by aryl hydrocarbon receptor, a transcription factor binding to xenobiotic response element.[10] NRF2, in collaboration with small Maf proteins (sMaf, including MafG, MafK, and MafF), regulates the expression of a battery of phase II enzymes and phase III transporters through antioxidant/electrophile responsive element (ARE/EpRE).[11,12]

During the phase II reaction, GSTs, UGTs, and sulfotransferases act to conjugate glutathione, glucuronic acid, and sulfate to electrophiles, respectively,[1,3] which represent hydrophilic moieties exploited in the phase II reaction (Figure 2). Of the other phase II enzymes, NQO reduces quinones to hydroquinones.[13] In addition to the phase II reaction enzymes, NRF2 also induces cystine/glutamate exchange transporter (Xc^- or SLC7A11) and glutathione synthetase to support glutathione synthesis. GCL catalyzes the condensation of glutamate and cysteine to γ-glutamylcysteine, which contributes to the synthesis of glutathione. When the reduced glutathione (GSH) is used in redox reactions, it is converted to oxidized glutathione, which then is recycled by glutathione disulfide reductase (GSR) to regenerate GSH (Figure 3). Because the synthesis and recycling of GSH is critical for the phase II reaction, GCL and GSR are often categorized as phase II enzymes.

MRPs contribute to the phase III elimination of xenobiotics from cells to extracellular fluid, which are conjugated with hydrophilic moieties (Figure 3).[1,7–9] Because genes encoding both phase II[4–6,14,15] and phase III[16,17] enzymes are targets of NRF2, NRF2 makes an essential contribution to the regulation of detoxification reaction systems in our body.

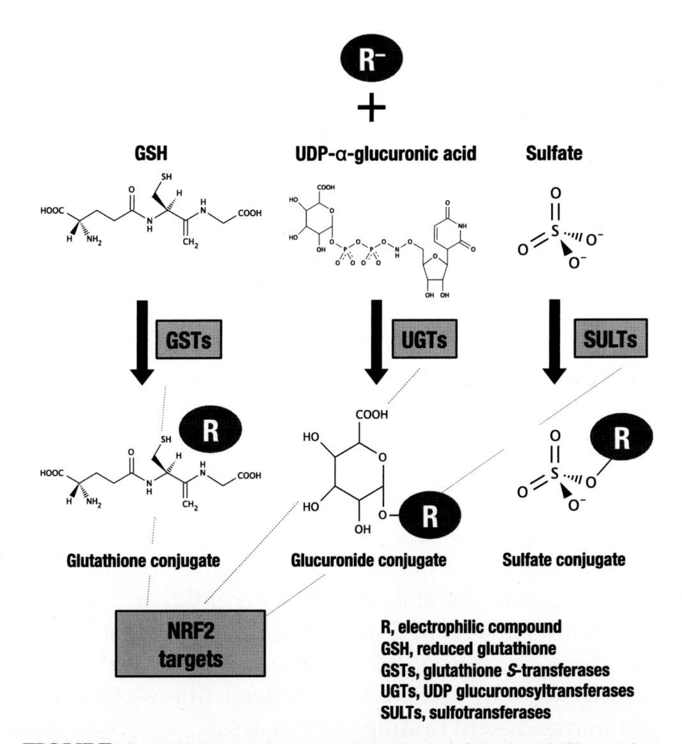

FIGURE 2 NRF2 regulates conjugations of hydrophilic moiety in detoxification processes. Electrophilic compounds are conjugated with GSH, UDP-α-glucuronic acids and sulfate, and increased water-solubility. GSTs, UGTs, and SULTs catalyze the binding of the thiol group in GST, the glucuronic acid component of UDP-α-glucuronic acids, and sulfate to electrophilic compounds, respectively. NRF2 regulates expression levels of GSTs, UGTs, and SULTs. NRF2, nuclear factor-erythroid-related factor 2; UDP, uridine 5′-diphosphoglucuronosyltransferase.

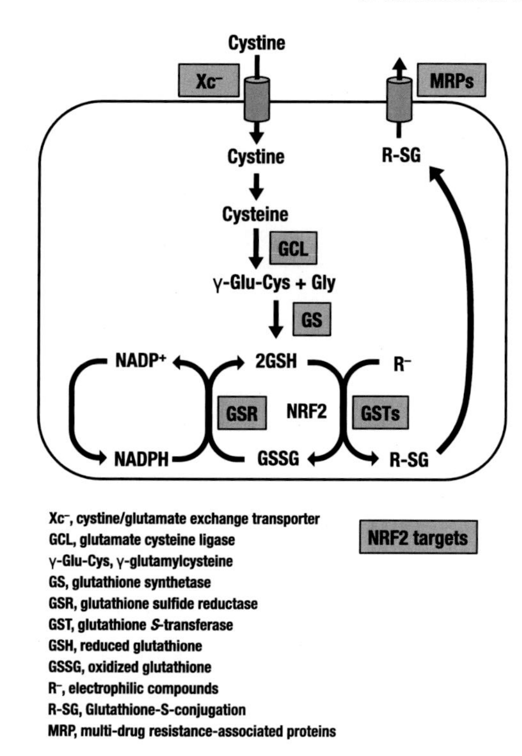

FIGURE 3 **NRF2 regulates glutathione metabolism.** NRF2 regulates the glutathione metabolism in multiple steps, including Xc⁻, γ-glutamylcysteine synthesis (GCL), GSH synthesis (GS), glutathione conjugation (GSTs), glutathione recycling (GSR), and excretions (MRPs). NRF2, nuclear factor-erythroid-related factor 2.

3. ANTIOXIDATIVE STRESS RESPONSE SYSTEMS IN CELLS

When nutrients are processed in the cells, mitochondrial redox reactions are routinely used to efficiently produce adenosine triphosphate (ATP) or cellular energy. However, this mitochondrial process is the major source of ROS, including superoxide, hydroxyl radical, and hydrogen peroxide,[18] which potentially induce various cellular dysfunctions. In addition to ROS, when nitric oxide is used in the various physiological functions and antibacterial processes, reactive nitrogen species (RNS, e.g., peroxynitrite) are generated in bodies. Therefore, antioxidative stress response systems play important roles in the protection of cells against ROS and RNS. NRF2 potently regulates various enzyme genes that use antioxidant systems in the cells.

3.1 Glutathione-Mediated Antioxidant Regulation

For the protection of the cells against ROS and RNS, glutathione plays critical roles.[19] NRF2 regulates glutathione synthesis and utilization in multiple steps, including cystine transport (Xc⁻), γ-glutamylcysteine synthesis from glutamate and cysteine (GCL), glutathione synthesis from γ-glutamylcysteine and glycine (glutathione synthetase), glutathione disulfide reduction (GSR), glutathione conjugate (GST), and excretion of conjugated metabolites (MRPs). NRF2 also induces glutathione peroxidase 2 gene expression,[20] which is the gastrointestinal type GPx catalyzing reduction process of hydrogen peroxide to water. Therefore, NRF2 intensively regulates glutathione-mediated antioxidant functions.

3.2 Thioredoxin-Mediated Antioxidant Systems

NRF2 also regulates thioredoxin (TRX)-related antioxidant systems. TRX contains two reactive cysteine residues, and these thiol groups are involved in the redox reactions. Reduced TRX provides reducing equivalents to peroxiredoxin (PRDX), which catalyzes the reaction of hydrogen peroxide breakdown to water (Figure 4).[21] In addition, the reduced TRX also reacts with various oxidized protein substrates other than PRDX. In the reducing reactions, TRX is oxidized per se; thereafter, the oxidized TRX is reduced by TRX reductase (TXNRD) to regenerate the reduced TRX.[21] TXNRD also reduces oxidized ascorbate, helping the ascorbate recycling. NRF2 induces both *TRX*[22] and *TXNRD*[23] gene expression levels to enhance the TRX-mediated antioxidative stress response system (Figure 4). Sulfiredoxin (SRX) contributes to reduction of PRDX (Figure 4).[24] Of note, NRF2 positively

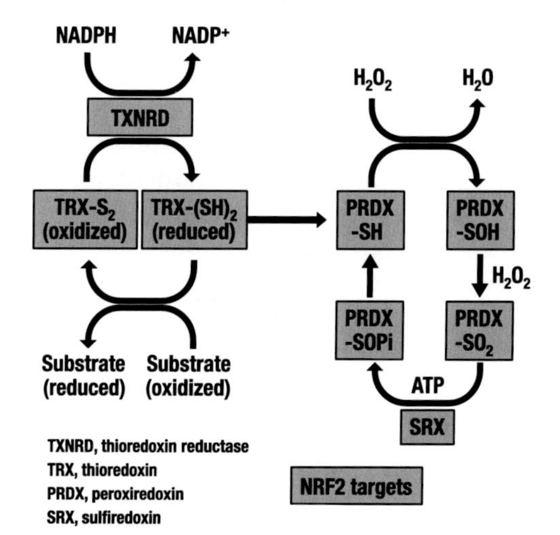

FIGURE 4 **Roles of NRF2 in the thioredoxin-mediated antioxidant reactions.** NRF2 regulates the thioredoxin (TRX)-mediated antioxidant system thorough regulation of TXNRD, TRX, PRDX, and SRX expression levels. ATP, adenosine triphosphate; H_2O, water; H_2O_2, hydrogen peroxide; $NADP^+$, pyruvate dehydrogenase; NADPH, nicotinamide adenine dinucleotide phosphate oxidase; NRF2, nuclear factor-erythroid-related factor 2; PRDX-SH, reduced peroxiredoxin; PRDX-SO_2, peroxiredoxin sulfur dioxide; PRDX-SOH, peroxiredoxin sulfuric acid; PRDX-SOPi, phosphorylated peroxiredoxin.

regulates the *PRDX*[25] and *SRX*[26] gene expressions; NRF2 acts as a regulator of TRX-mediated antioxidant systems by inducing *TRX*, *TXNRD*, *PRDX*, and *SRX* gene expression levels.

3.3 Heme Oxygenase-1, Superoxide Dismutase, and Catalase

In addition to the GSH and TRX cellular redox pathways, NRF2 induces heme oxygenase-1 (*HO-1*) gene expression.[25] HO-1 catalyzes degradation of heme to biliverdin, carbon monoxide, and iron, and biliverdin is changed to bilirubin by biliverdin reductase. All three products of HO-1 are known to contribute to the cellular protection against oxidative stresses; for instance, bilirubin contributes to the suppression of lipid peroxidation.

Superoxide dismutase (SOD) and catalase (CAT) catalyzes reducing reactions from superoxide to hydrogen peroxide and hydrogen peroxide to water, respectively. Although some reports show that NRF2 induces *SOD*[27,28] and *CAT*[28] gene expressions, this point needs further verifications. In summary, NRF2 emerges as the main player that regulates expression of genes encoding antioxidative stress response enzymes and protects our body against oxidative and nitrosative stresses.

4. ANTI-INFLAMMATORY FUNCTION OF NRF2

NRF2 also modulates inflammatory responses. By using inflammation model mice, it has been well characterized that NRF2 induction strongly suppresses inflammation (Figure 5). NRF2 induction by administration of 15-deoxy-$\Delta^{12,14}$-prostaglandin J_2 to the carrageenan-induced pleurisy model mice delays the tissue invasion and recruitment of the inflammatory cells.[29] Administration of the other NRF2 inducers have also shown to inhibit the progression of inflammation, and the underlying mechanism has been attributed to the antioxidative stress response (Figure 5). For instance, NRF2 inducer oleanolic triterpenoid 1-[2-cyano-3,12-dioxooleane-1,9(11)-dien-28-oyl] imidazole (CDDO-Im) is shown to inhibit lipopolysaccharide (LPS)-induced immune responses by suppressing cytokine and chemokine gene inductions, including tumor necrosis factor (*Tnf*), interleukin 6 (*Il6*), monocyte chemotactic protein 1 (*Mcp1* or *Ccl2*), and macrophage inflammatory protein 2 (*Mip2* or *Cxcl2*) in neutrophils.[30] Sulforaphane (SFN) suppresses pro-inflammatory cytokine and mediator expression levels, including *Tnf*, interleukin-1β (*Il1b*), inducible nitric oxide synthase

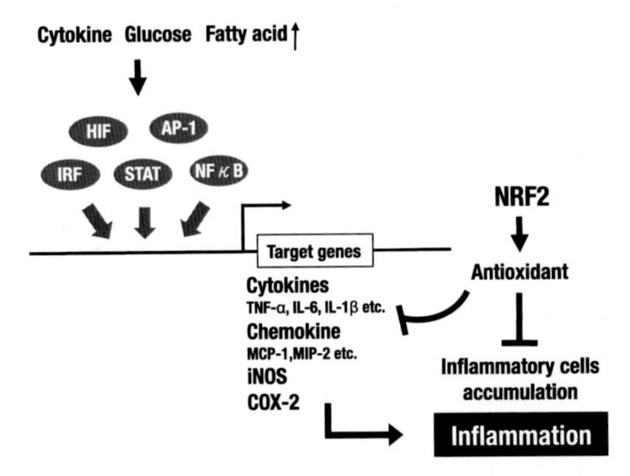

FIGURE 5 Anti-inflammatory functions of NRF2. Cytokines, hyperglycemia, and hyperlipidemia cause inflammation via activation of transcriptional factors, including HIF (hypoxia-inducible factor), AP-1 (activator protein-1), IRF (interferon regulatory factors), STAT (signal transducers and activator of transcription), and nuclear factor-κB (NF-κB). NRF2 inhibits inflammatory cell accumulation and pro-inflammatory mediators through regulation of antioxidant genes expression. COX, cyclooxygenase; IL, interleukin; iNOS, inducible nitric oxide synthase; MCP, monocyte chemotactic protein; TNF, tumor necrosis factor.

(*iNOS* or *Nos2*), and cyclooxygenase 2 (*Cox2* or *Ptgs2*) in mouse peritoneal macrophages or mouse macrophage cell line RAW 264.7 cells.[31,32]

In agreement with the inducer analyses, it is reported that *Nrf2*-deficiency aggravates inflammation in mice. Loss of NRF2 expression enhances the LPS-induced mortality and inflammatory genes expression levels, including *Tnf*, *Il6*, *Mcp1*, and *Mip2*, in *Nrf2* knockout mouse neutrophils.[30] In the dextran sulfate sodium-induced colitis model, inflammatory responses are severely exacerbated by the *Nrf2* depletion in mice.[33]

Importantly, the inflammation in adipose tissues and other metabolic organs is closely related to the pathogenesis of obesity, type 2 diabetes (T2D), and metabolic syndrome. The chronic hyperglycemia or hyperlipidemia enhances oxidative stress and promotes recruitment and/or activation of inflammatory cells through formation of the pro-inflammatory oxidized lipids and the advanced protein oxidation.[34,35] Several NRF2-inducing chemicals are reported to improve metabolic disorders in a number of rodent models. Administration of oltipraz, an NRF2 inducer, inhibits the inflammation in adipose tissues of high-fat diet–induced obesity model mice.[36] Dietary curcumin admixture also inhibits the inflammation in adipose tissues of obesity model mice.[37] In addition to the adipose tissues, supplementation of dietary ellagic acid, which is contained in fruits and nuts, induces NRF2 expression levels and attenuates inflammatory response in liver of rat.[38] NRF2 induction by SFN protects streptozotocin (STZ)-mediated pancreatic β-cell damage through inhibition of inflammatory response of the nuclear factor-κB signaling pathway.[39]

These wide-ranging data unequivocally demonstrate that NRF2 contributes to the modulation of inflammatory responses, and the NRF2 activity contributes to the resolution or suppression of the inflammation. Therefore, NRF2 is a promising target of drug development that prevents the inflammation underlying various pathological conditions that relate to or beyond the anti-oxidative stress responses.

5. MOLECULAR BASIS OF THE KEAP1-NRF2 SYSTEM FUNCTION

Because NRF2 acts as the key regulator for various genes during stress conditions, NRF2 activity as a transcription factor is sophisticatedly regulated by Kelch-like ECH-associated protein 1(KEAP1) and related mechanisms. Therefore, the regulatory mechanism is collectively referred to as the KEAP1-NRF2 system. KEAP1 is an adaptor protein of Cullin3-based ubiquitin E3 ligase complex, which rapidly ubiquitinylates NRF2 and leads the molecule to degradation via the proteasome pathway. In unstressed conditions, KEAP1 binds NRF2, so that the NRF2 activity is constitutively suppressed. In contrast, upon exposure to electrophilic stresses, KEAP1 cysteine residues are modified by the electrophiles and this modification hampers the KEAP1 activity. In this stress condition, KEAP1-mediated ubiquitination of NRF2 is largely abrogated and the induction of NRF2 activity is provoked. We refer to this induction mechanism of NRF2 activity as the derepression from the repression by KEAP1.

5.1 Identification of NRF2 and KEAP1

By the homology screening of chicken homolog of mouse erythroid transcription factor nuclear factor-erythroid 2 (NF-E2), ECH (erythroid-derived cap'n'collar homolog) was isolated; this factor was later named NRF2.[40] Concomitantly, human NRF2 complementary DNA was cloned independently by a distinct approach.[41] NRF2 belongs to the cap'n'collar family of transcription factors. Through homology analysis of chicken ECH and human NRF2, six functional domains have been detected in the NRF2 protein; these domains are called NRF2-ECH homology (Neh) domains (Figure 6(a)).[42]

NRF2 is found to turn over rapidly; degron, a regulatory region for protein degradation, for the turnover appears to be Neh2 domain—localized in the N-terminus. In the course of exploring interacting proteins with the Neh2 domain, KEAP1 was identified through yeast two-hybrid screening using the Neh2 domain as bait.[42]

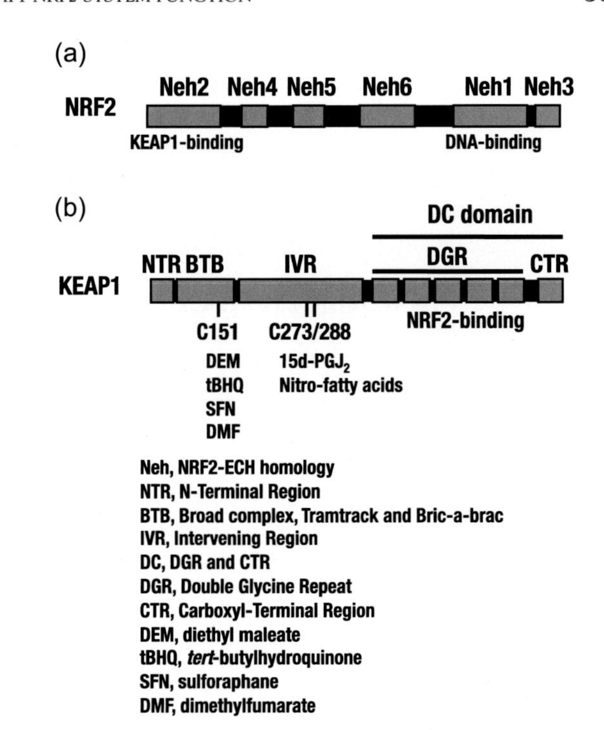

FIGURE 6 **Structure of NRF2 and KEAP1.** (a) Nuclear factor-erythroid-related factor 2 (NRF2) contains six well-conserved Neh domains, from Neh1 to Neh6. The Neh2 domain contributes to the binding with kelch-like ECH-associated protein 1 (KEAP1). The Neh1 domain is involved in demonization with small Maf proteins (sMafs) and binds DNA. (b) KEAP1 contains a BTB domain; a DC domain; and an intervening region domain. The BTB domain modulates KEAP1 homodimerization, and the DC domain is involved in the interaction with NRF2. C151, C273, and C288 are highly reactive cysteine residues that contributes to react with NRF2-inducing electrophiles, including DEM, tBHQ, SFN, 15-deoxy-$\Delta^{12,14}$-prostaglandin J_2 ($15d\text{-}PGJ_2$), and nitro-fatty acids.

KEAP1 acts as a substrate-recognizing adaptor component of Cul3-based ubiquitin ligase and as a sensor for environmental stresses.

5.2 KEAP1 Retains Multiple Cysteine Residues that Act as Sensors

KEAP1 protein harbors five domains: N-terminal region; broad complex, tramtrack, and bric-a-brac (BTB) domain; intervening region (IVR); and C-terminal DC domain that contains double glycine repeat and carboxyl-terminal region (Figure 6(b)).[43,44] KEAP1 forms homo-dimer through the BTB domain, and the homo-dimer binds to one molecule of NRF2 with two DC domains by a two-site binding mechanism.[43,44]

KEAP1 provides a unique sensor system by exploiting multiple reactive cysteine residues, which make KEAP1 possible to handle a variety of electrophilic stresses. For instance, KEAP1 senses electrophiles through cysteine residue Cys151 in the BTB domain and Cys273/288 in IVR. Intriguingly, Cys151 binds diethyl maleate, *tert*-butylhydroquinone, SFN, and dimethyl fumarate,[45] whereas Cys273/288 binds 15-deoxy-$\Delta^{12,14}$-prostaglandin J_2 and nitro-fatty acids. Thus, KEAP1 equips multiple sensor functions based on multiple cysteine residues.

6. PANCREATIC β CELLS AND OXIDATIVE AND NITROSATIVE STRESSES

Pancreatic β cells exist in clusters known as the islets of Langerhans and produce and secrete the glucose-lowering hormone insulin; therefore, pancreatic β cells exert critical functions in glucose homeostasis. Dysfunction of pancreatic β cells is a major factor for onset and progression of T2D.[46,47] It has been reported that, in both human T2D patients and rodent T2D models, oxidative stress is increased in pancreatic β cells. Indeed, the oxidative DNA damage marker 8-hydroxy-2′-deoxyguanosine is markedly increased in pancreatic β cells of T2D patients.[48] In diabetic rodent models, such as *db/db* mice and Goto-Kakizaki (GK) rats, enhancement of oxidative stresses are observed in pancreatic β cells.[49–52]

In this regard, it has been observed that expression levels of antioxidant messenger RNAs (i.e., *SOD*, *CAT*, and *GPx*) are low in pancreatic β cells compared with the other major organs or tissues,[53] and these antioxidant enzyme activities are actually low in pancreatic β cells.[54] On the other hand, it has also been observed that several antioxidant messenger RNAs, such as *TRX*, glutaredoxin, glutathione peroxidase 2, and *GPx4*, are expressed at comparable levels in β cells compared with liver.[55,56] Therefore, precise mechanisms of how various antioxidant systems play and contribute to the protection of pancreatic β cells remain to be clarified.

6.1 Source of ROS and RNS in Pancreatic β Cells

In T2D cases, hyperglycemia and hyperlipidemia induce oxidative stress in pancreatic β cells (Figure 7).[57,58] When glucose stimulates insulin release in pancreatic β cells, glucose is metabolized in mitochondria to produce ATP molecules through electron transporter chain, and ROS are inevitably generated in mitochondria of pancreatic β cells.[59] The synergetic elevation of glucose and saturated fatty acids accelerates mitochondrial ROS production in the cells.[60] In addition, ROS are also generated in plasma membrane by nicotinamide adenine dinucleotide phosphate oxidase (NADPH) oxidase-mediated redox reactions in the cells during glucose and saturated fatty acid metabolisms.[57,61]

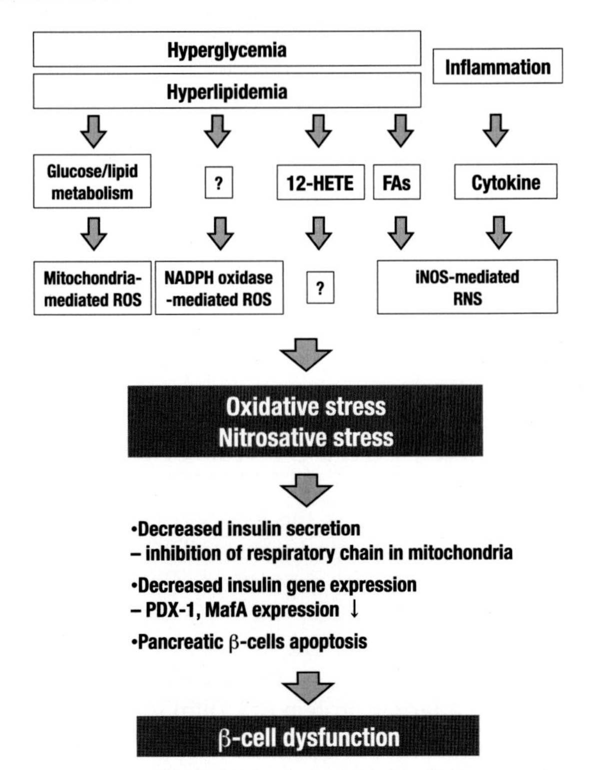

FIGURE 7 **Generation of oxidative and nitrosative stress in pancreatic β-cell dysfunction.** Hyperglycemia and hyperlipidemia stimulate oxidative and nitrosative stress in pancreatic β cells through multiple pathways, including mitochondrial glucose and lipid metabolism, nicotinamide adenine dinucleotide phosphate-oxidase, and arachidonic acid metabolites 12-hydroxyeicosatetraenoic acid (12-HETE) and saturated fatty acids (FAs). Oxidative and nitrosative stress are also generated by cytokines during immense response-mediated pancreatic β-cell damage. iNOS, inducible nitric oxide synthase; RNS, reactive nitrogen species; ROS, reactive oxygen species.

These mitochondrial metabolism and NADPH oxidase-mediated redox reaction are two representative sources of ROS in pancreatic β cells.

On the other hand, immune responses increase cytokine production in pancreatic islets, and strongly induce iNOS expression; therefore, nitrosative stresses are increased in pancreatic β cells. Thus, immune responses are the cause of RNS accumulation in pancreatic β cells of type 1 diabetes (T1D) cases.[62,63] Saturated fatty acids also increase RNS levels through induction of iNOS expression in pancreatic β cells and suppress insulin secretion from the islets.[64]

In diabetes and obesity model mice, expression level of 12-lipoxygenase and its metabolites 12-hydroxyeicosatetraenoic acid are increased in pancreatic β cells.[65,66] The 12-lipoxygenase–mediated increase of 12-hydroxyeicosatetraenoic acid generates ROS in pancreatic β cells.[67] These diverse findings indicate that lipid metabolites are alternative source of ROS and RNS, which are independent of mitochondrial metabolisms (Figure 7).

6.2 Oxidative and Nitrosative Stress Induces Pancreatic β-Cell Dysfunction

Enhancement of oxidative and nitrosative stress damages pancreatic β-cell functions (Figure 7). Oxidative and nitrosative stress inhibits respiratory chain functions in mitochondria and ATP production in pancreatic β cells, and impair insulin release.[68–70] In diabetic model mice, oxidative stress impairs functions of transcription factors, such as MafA and PDX-1, resulting in the decline of insulin gene expression.[71,72] An excess of oxidative and nitrosative stress induce apoptosis via caspase activation in the β cells (Figure 7).[73–76] Indeed, iNOS knockout mouse shows resistance to STZ-mediated β-cell damage.[77] In contrast, it is reported that oxidative stress paradoxically stimulates insulin release from β cells of diabetic model mice,[78,79] but this phenomenon needs to be rigorously tested in other contexts. Thus, in spite of the paradoxical phenomenon, it has been generally accepted that the oxidative damages are actually enhanced in T2D patients,[48] and protection of pancreatic β cells against the oxidative stresses is important for the suppression of onset and progression of T2D.

6.3 Increase of Defense against Oxidative/Nitrosative Stresses Protects β-Cell Function

Many lines of evidence support that suppression of oxidative/nitrosative stresses provides favorable effects for the pancreatic β-cell function. For instance, β cell–specific expression of antioxidant genes, such as *CAT, GPx1,* and *SOD1*, ameliorates damages in the β cells of diabetes model mice,[80–82] treatment of isolated islets with antioxidative reagents improves insulin secretion,[83–85] and supplementation of antioxidative substances ameliorates β-cell damage of *db/db* mice.[72] These observations support the contention that both the antioxidant system inherently equipped in the pancreatic β cells and genetic or pharmacological reinforcement of the antioxidant system contribute to protection of the β cells upon the progression of T2D.

Indeed, it has been reported that supplementation of natural compounds that retain antioxidative activity is one of the challenging approaches for the prevention of β-cell damage. For instance, vitamins C and E are shown to suppress β-cell damage in *db/db* mice.[72] Pancreatic β cell–derived culture cells are shown to be protected against oxidative stress in vitro by the supplementation of flavonoids contained in green tea, grapes, and cocoa; Padina arborescens extract from *Padina arborescens*; and hispidin, a derivative of mushroom component.[84,86,87]

Although efficacy of these natural antioxidant compounds for the protection of pancreatic β cells are shown in culture cells or diabetic mouse models, more rigorous verifications are mandatory upon the use of these compounds as drugs or supplements. To this end, it is necessary to identify precise cellular targets of these compounds and to clarify molecular mechanisms how these natural compounds activate or repress the targets and related signal cascades.

7. ROLES OF NRF2 ON ANTIOXIDATIVE RESPONSE IN PANCREATIC β CELLS

It has been shown that NRF2 is involved in cellular protection against oxidative and nitrosative stresses, in addition to the regulation of drug metabolism and detoxification.[25,88] In fact, a salient example of the NRF2 function to protect our body against oxidative stresses has been shown in the pancreatic β-cell protection in stress-mediated diabetes model mice.[74]

Expression level of NRF2 in pancreatic β cells was found to be enhanced in diabetic model rodents,[89,90] suggesting that NRF2 plays important roles in the protection of the cells against oxidative stresses. However, elucidation of the mechanisms how NRF2 contributes to pancreatic β-cell protection is difficult and awaited for further studies. Several elaborated genetic and pharmacological model systems have been employed to clarify the contribution of the KEAP1-NRF2 system to the protection of pancreatic β cells.

To examine whether NRF2 induces antioxidant enzyme gene expression in β cells, CDDO-Im has been administered to mice. Results show that CDDO-Im significantly induces expression levels of NRF2[74] and its target genes including antioxidant enzyme genes in β cells.[56] Showing very good agreement, genetic NRF2 induction by conditionally depleting KEAP1 has also been used,[91–93] and this β cell–specific NRF2 induction has displayed significant increase of antioxidant enzyme genes, such as *Nqo1, Hmox1, Gsta2, Gstp1,* and *Txnrd1* (encoding NQO1, HO-1, GSTα1, GSTπ1, and TXNRD1, respectively).[74] Importantly, this genetic NRF2 induction remarkably protects the β cells against damages caused by oxidative and nitrosative stress in iNOS transgenic (iNOS-Tg) mice.[74]

Consistent with the genetic and pharmacological analyses, dimerumic acid contained in Monascus-fermented rice is showed to induce NRF2 accumulation and enhances GCL expression in RINm5F cells, a rat insulinoma cell line. Dimerumic acid suppresses methyl-glyoxal-mediated RINm5F cell damage.[94] These results thus support the contention that the KEAP1-NRF2 system acts as a key regulatory system for the antioxidative stress response in pancreatic β cells (Figure 8).

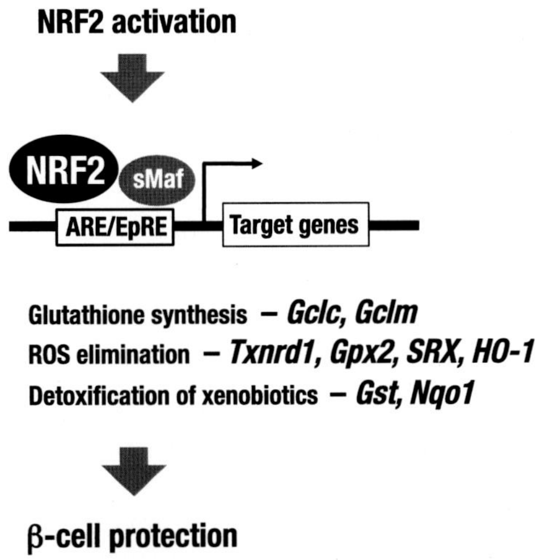

FIGURE 8 **NRF2 regulates antioxidant response and detoxifications in pancreatic β cells.** NRF2 protects pancreatic β cells from diabetic damages by inducing expression of NRF2 target antioxidant genes, including glutathione synthesis, antioxidant response, and detoxification enzyme genes. ARE/EpRE, antioxidant/electrophile responsive element; NRF2, nuclear factor-erythroid-related factor 2.

In turn, it has also been shown that the decline of NRF2 expression disrupts antioxidant functions in β cells and leads to β-cell damage.[56] For instance, heterozygous knockout of *Nrf2* gene markedly damages β cells in iNOS-Tg mice.[74] Exposure to di(2-ethylhexyl) phthalate (DEHP) suppresses NRF2 expression in INS-1 cells, a mouse β cell–derived cell line.[95] DEHP decreases antioxidant enzyme gene expressions and induces ROS levels and apoptosis of INS-1 cells. DEHP also impairs glucose uptake in adipose tissues and elevates blood glucose levels in rats.[96] DEHP is a ubiquitous environmental endocrine disruptor used as plasticizers and industrial plastic materials. Importantly, human beings are frequently exposed to DEHP, and recent studies have suggested that DEHP is correlated with T2D development.[97–99] For example, a clinical study was performed for 2350 women in the United States, and the study provides strong correlation between the urinary level of DEHP and occurrence of T2D.[99] These findings support the notion that the KEAP1-NRF2 system plays critical roles in the onset and progression of diabetes (Figure 8).

8. NRF2 REGULATION OF INFLAMMATION AND OTHER CELLULAR RESPONSES IN PANCREATIC β CELLS

Hyperglycemia and hyperlipidemia induce an inflammatory response in pancreatic β cells.[100,101] Upon damage of pancreatic β cells, pro-inflammatory cytokines are secreted and recruit inflammatory cells near pancreatic islets. Because the KEAP1-NRF2 system acts as a strong modulator of inflammation, NRF2 is expected to suppress inflammatory response in the region surrounding pancreatic islets. In fact, it has been reported that an NRF2 inducer SFN suppresses *iNOS* and *Cox2* gene expressions, and ameliorates STZ- and cytokine-mediated pancreatic β-cell damage,[39] indicating that the NRF2 induction suppresses the cytokine-mediated inflammatory responses near pancreatic islets.

Currently available evidence supports the notion that the KEAP1-NRF2 system modulates the protein degradation pathways, such as autophagy and ubiquitin-proteasome pathways. It is reported that insufficiency of the autophagy provokes damage on pancreatic β-cell function.[102–106] In pancreatic β cells, ubiquitinylated protein aggregate formation is induced by hydrogen peroxide, which leads to cell damage.[107] The aggregate formation is enhanced by chloroquine, an inhibitor of autophagy.[90] NRF2 activation is known to strongly suppress the aggregate formation and autophagy insufficiency.[90] These results thus indicate that NRF2 protects pancreatic β cells against oxidative stress; this protection seems to be attained, at least partially, through the maintenance of autophagy.[90] NRF2 also regulates expression of proteasome catalytic subunit Psmb5[108] in pancreatic β cells,[109] and contributes to the stress response of endoplasmic reticulum.[109] Thus, these results further support the notion that the KEAP1-NRF2 system regulates both autophagy and ubiquitin-proteasome pathways, the two important protein degradation pathways within the cells, and contributes to the protection and maintenance of pancreatic β-cell function.

Pancreatic islets consist of several endocrine cells, including α cells, β cells, δ cells, and pancreatic polypeptide cells. In the islets of T2D patients, the population of pancreatic β cells is decreased, but that of α cells is increased, indicating that a selective β-cell loss may contribute to onset of diabetes mellitus.[110] Although β cells are the focus of the KEAP1-NRF2 study exploiting diabetic model mice,[56] contributions of the system to the other endocrine cells in pancreatic islets remains unclear. Intriguingly, in the diabetic model pancreatic β cell–specific *FoxO1*-deficient mice, pancreatic β cells occasionally dedifferentiate into other endocrine cell linages.[111] For instance, the insulin-positive cell population is decreased compared with that of insulin-negative cells in pancreatic islets of iNOS-Tg mice.[74] Using rat insulin promoter-Cre and LoxP-based recombination-reporter system, a cell fate tracing of insulin-producing pancreatic β cells is conducted in iNOS-Tg mouse islets.[15,74] As a result, β cells are found to be frequently converted into glucagon-producing

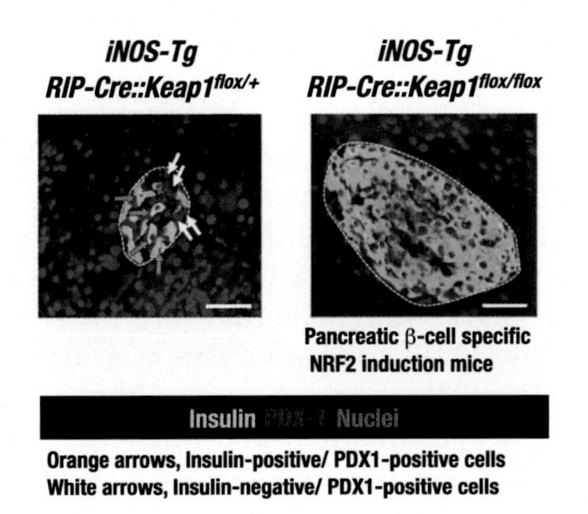

FIGURE 9 NRF2 prevents stress-mediated pancreatic β-cell conversion. In iNOS transgenic (*iNOS-Tg::RIPCre::Keap1$^{flox/+}$*) mouse islets, both normal insulin-positive/PDX-1–positive β cells (orange arrows) and abnormal insulin-negative/PDX-1–positive cells (white arrows) were observed. In contrast, in NRF2 activating iNOS transgenic (*iNOS-Tg::RIPCre::Keap1$^{flox/flox}$*) mouse islets, these abnormal insulin-negative/PDX-1–positive cells were not observed. iNOS, inducible nitric oxide synthase; NRF2, nuclear factor-erythroid-related factor 2.

α-like cells, which are insulin-negative and PDX-1–positive. These abnormal cells are absent in iNOS-Tg mouse pancreatic islets upon NRF2 induction (Figure 9). These results indicate that the NRF2 induction in pancreatic β cells prevents the dedifferentiation of β cells to α cells under conditions of the oxidative stress.

As summarized in Figure 10, these accumulating lines of evidence support the idea that the KEAP1--NRF2 system plays multiple beneficial roles for the stress response of pancreatic β cells, which include the antioxidative stress response, detoxification, anti-inflammatory response, and modulation of two important protein degradation pathways. These responses lead pancreatic β cells to the maintenance of normal functions.

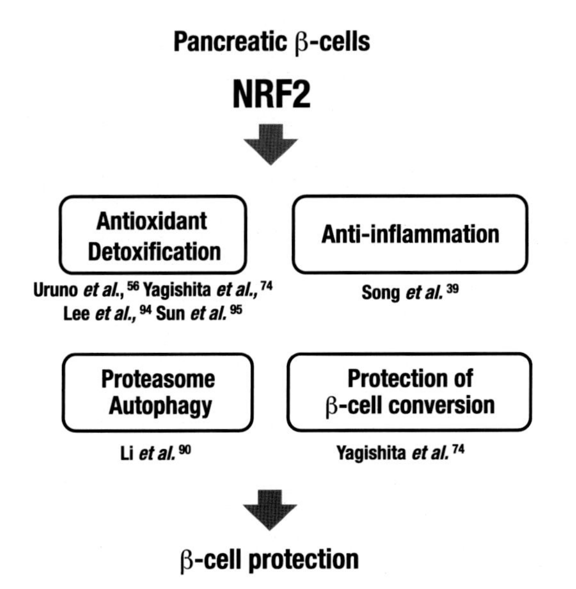

FIGURE 10 NRF2 protects pancreatic β cells though multiple pathways. Antioxidant response and detoxification pathways majorly contribute to the NRF2-mediated pancreatic β-cell protection. In addition, NRF2 elicits anti-inflammatory functions, maintains protein degradation pathways, and prevents stress-mediated cell conversion in pancreatic β cells. AMPK, adenosine monophosphate–activated protein kinase; NRF2, nuclear factor-erythroid-related factor 2.

9. GLUCOSE HOMEOSTASIS IN INSULIN-SENSITIVE TISSUES

NRF2 is also involved in homeostasis of glucose metabolism in insulin-sensitive tissues. In hepatocytes, a methionine- and choline-deficient diet induces steatosis and metabolic dysfunctions by the enhancement of oxidative stress, but the induction of NRF2 protects normal cellular metabolic regulations in the methionine- and choline-deficient diet-mediated steatosis.[112,113] Traditionally, molecular basis of this phenomenon has been attributed to the influences of oxidative stresses provoked by the methionine- and choline-deficient diet. In contrast, increasing lines of recent evidence support the notion that NRF2 directory commit the metabolic regulations beyond the antioxidant functions. Indeed, chromatin immunoprecipitation sequencing analyses have revealed that NRF2 and MafG frequently bind to metabolism-relate enzyme gene loci.[114]

9.1 NRF2 Regulates FGF21 Expression and Its Signal Transduction

It has been found that one of the factors that links NRF2 directly to metabolic regulations is fibroblast growth factor (FGF) 21. FGF21 is secreted from liver and plays important roles in metabolic regulations; the increment of plasma FGF21 concentration ameliorates obesity, diabetes mellitus, and hyperlipidemia.[115–117] Administration of FGF21 increases gene expression levels of glucose transporter 1 (*Glut1* or *Slc2a1*)[115] and hexokinase 2 (*Hk2*)[118] in mouse adipose tissues. Pharmacological and genetic inductions of NRF2 increase *Fgf21* gene expression in liver, and dramatically enhance the plasma concentration of FGF21.[118] The NRF2 inductions enhances *Slc2a1* and *Hk2* gene expression levels in mouse adipose tissues[118] and lowers blood glucose levels in diabetic mouse models.[56] Therefore, the FGF21 induction may be one of the mechanisms through which NRF2 regulates expression levels of glycolysis-related genes. Showing very good agreement with the gain-of-function evidence that NRF2 induces the *Fgf21* gene expression, loss of NRF2 suppresses FGF21 signaling in liver.[118,119] As shown in Figure 11, these results indicate that NRF2 contributes to the

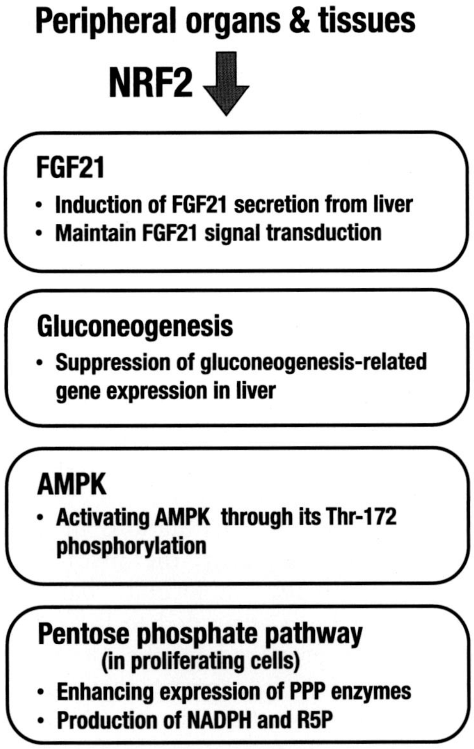

Peripheral organs & tissues

NRF2

FGF21
- Induction of FGF21 secretion from liver
- Maintain FGF21 signal transduction

Gluconeogenesis
- Suppression of gluconeogenesis-related gene expression in liver

AMPK
- Activating AMPK through its Thr-172 phosphorylation

Pentose phosphate pathway
(in proliferating cells)
- Enhancing expression of PPP enzymes
- Production of NADPH and R5P

FIGURE 11 **Roles of NRF2 induction in peripheral organs or tissues.** NRF2 induction enhances FGF21 secretion and suppresses gluconeogenesis in liver. NRF2 increases phosphorylation of AMPK Thr-172, and activates AMPK signaling pathway. In proliferating cells, NRF2 regulates expression levels of pentose phosphate pathway (PPP) enzymes, and increases production of NADPH and ribose 5-phosphate (R5P). FGF, fibroblast growth factor; NADPH, nicotinamide adenine dinucleotide phosphate oxidase; NRF2, nuclear factor-erythroid-related factor 2.

FGF21-mediated metabolic regulations through the controlling expression level and signaling transduction.

9.2 NRF2 Suppresses Gluconeogenesis

In the *db/db* T2D mouse model, NRF2 induction suppresses expression of gluconeogenesis-related enzyme genes in liver, including glucose 6-phosphatase (*G6pc*), fructose-1,6-bisphosphatase 1 (*Fbp1*), peroxisome proliferator-activated receptor γ coactivator 1-α (*Pgc1α* or *Ppargc1a*), and nuclear receptor subfamily 4, group A, member 2 (*Nr4a2*).[56] As a result, NRF2 significantly suppresses gluconeogenesis in this diabetes mouse model (Figure 11). NRF2 also downregulates the expression of gluconeogenesis-related enzyme genes in mouse hepatocyte-derived AML12 cells in the presence of cyclic adenosine monophosphate (cAMP) analogue and cAMP response element binding protein stimulation (CREB), indicating that NRF2 represses expressions of these genes via the suppression of the cAMP-CREB signaling pathway.

9.3 NRF2 Activates AMPK

Adenosine monophosphate–activated protein kinase (AMPK) is a key metabolism-regulating molecule and plays critical roles in the maintenance of energy homeostasis, and activation of AMPK results in enhancement of glucose utilization. The activity of AMPK is regulated by AMP and ATP ratio. Alternatively, the regulation is mediated by AMPK phosphorylation at threonine 172 residue (Thr-172) thorough serine/threonine kinase LKB1 or Ca^{2+}/calmodulin-dependent protein kinase kinases.[120,121]

Interestingly, both pharmacological[36,122] and genetic[123] induction of NRF2 enhance AMPK phosphorylation (Thr-172). In contrast, suppression of the NRF2 signaling attenuates the AMPK phosphorylation (Thr-172).[124] Therefore, NRF2 is suggested to regulate the AMPK activity though phosphorylation of Thr-172; in turn, AMPK contributes to the lowering effects of blood glucose level in the NRF2-induced animals (Figure 11).

In contrast to the fact that NRF2 suppresses the *Pgc1α* expression in liver, NRF2 enhances the expression level of *Pgc1α* in skeletal muscle.[56] NRF2 also increases carnitine palmitoyltransferase 1b (*Cpt1b*) gene expression and increases oxygen consumption in skeletal muscle.[56] Because AMPK regulates *Pgc1α* and *Cpt1b* gene expression levels in skeletal muscle,[125] it is speculated that AMPK is the mediator of the NRF2-mediated *Pgc1α* and *Cpt1b* gene inductions and energy consumption in skeletal.

9.4 NRF2 Regulates Pentose Phosphate Pathway Enzyme Genes in Proliferating Cells

NRF2 or *KEAP1* gene mutations that disrupt interaction between KEAP1 and NRF2 are frequently observed in cancer cells, especially in non–small-cell lung cancer cells. NRF2 signaling is actually activated in these cells, which obtain drug and stress resistance.[126,127] NRF2 binds to regulatory regions of genes encoding pentose phosphate pathway (PPP) enzymes, including glucose-6-phosphate dehydrogenase, 6-phosphogluconate dehydrogenase, transketolase, transaldolase 1, malic enzyme 1, and isocitrate dehydrogenase 1. NRF2 positively regulates expression of these enzyme genes in cancer cell lines.[128] The induction of PPP enzyme genes results in the activation of the pathway and provokes metabolic reprogramming of cancer cells in which cancer cells mainly use glucose and glutamine to produce NADPH and ribose 5-phosphate. Because NADPH plays important roles in detoxification and antioxidant responses as reducing equivalents and ribose 5-phosphate contributes to cancer cell proliferation through nucleotide synthesis, NRF2 leads cancer cells to acquire malignancy (Figure 11).[128]

Although these PPP enzymes are also strongly induced by NRF2 in forestomach and intestine, NRF2 does not enhance PPP enzyme gene expression in liver. In contrast, NRF2 potently increases the expression of PPP enzyme genes in the liver of phosphoinositide 3-kinase-Akt signaling activation model mouse. These data indicate that this metabolic reprogramming is uniquely observed in proliferating cells, including cancer cells, epithelial cells, and pathological hepatocytes, but not in quiescent cells.[128]

10. NUTRITION AND NRF2 INDUCING PHYTOCHEMICALS

As described in previous sections, upon exposure to various environmental or endogenous stresses, expressions of NRF2 and its target genes are induced within 1 h or so. In contrast, the majority of NRF2 target genes are expressed only at low levels in unstressed conditions. Therefore, it seems reasonable to expect that we can avoid the onset of various disorders, if NRF2 is pharmacologically induced in the cells that need more NRF2 and its target gene expression. In this regard, it should be noted that plants often contain chemical compounds that give rise to favorable effects on our health. These compounds are collectively referred to as phytochemicals. Because it has been reported that dietary intakes of vegetables suppress risks to develop cancer,[3,129–131] one may expect that phytochemicals activate the NRF2 signaling pathway. Indeed, recently there are accumulating reports that many phytochemicals induce the NRF2 signaling without causing significant cell toxicity.[132] Importantly, some of the NRF2 inducing phytochemicals are known to lower blood glucose levels, as will be summarized in the following section.

10.1 Sulforaphane

One of the representative NRF2-inducing phytochemicals is SFN (Figure 12). Talalay and colleagues identified SFN as a potent inducer of phase II enzymes from broccoli sprouts.[133] SFN binds to Cys151 residue of KEAP1 and induces NRF2.[45] Broccoli sprouts contain a precursor of SFN, glucoraphanin,[134] and broccoli sprout beverage[135] enhances detoxification of chemical pollutant in human.[136,137] Effectiveness of SFN administration to human diseases has been reported in autism spectrum disorder.[138]

It has also been reported that broccoli sprout extracts lowers insulin resistance in human T2D patients.[139] In addition, several studies on mice revealed the effectiveness of SFN in suppression of diabetes mellitus. For instance, SFN improves STZ-mediated pancreatic β-cell damage[39] and prevents diabetic nephropathy.[140,141]

10.2 Oleanolic Acid

Oleanolic acid is a constituent of the leaves of *Olea europaea* and *Viscum album* L (Figure 12).[142–145] Oleanolic acid is used in Chinese medicine for the treatment of liver disorders, such as viral hepatitis, and has been shown to protect mice from various hepatotoxicants that cause oxidative and electrophilic stresses, including carbon tetrachloride, acetaminophen, bromobenzene, and thioacetamide. Oleanolic acid induces expression levels of NRF2 and its target genes in mouse liver,[146] and administration of oleanolic acid exerts hypoglycemic effects in diabetes rodent models.[147]

Importantly, synthetic derivatives of oleanolic acid are shown to exert potent NRF2-inducing activity and have been developed as candidates for NRF2-targeting drugs, e.g., bardoxolone[148–150] (oleanolic triterpenoid 1-[2-cyano-3,12-dioxooleane-1,9(11)-dien-28-oyl] methyl ester), CDDO-Im,[148,151] and acetylenic tricyclic bis(cyanoeneone).[152,153] Bardoxolone has been tested in the treatment of diabetic nephropathy and has been shown to successfully increase estimated glomerular filtration rate in patients with the nephropathy.[149,150] Importantly, we found that administration of CDDO-Im protects pancreatic β-cell damage and lower blood glucose levels in *db/db* diabetes model mice by inducing NRF2.[56] Bardoxolone has also been shown to lower blood glucose levels in the *db/db* mice.[154] These findings unequivocally demonstrate that NRF2 retains activity to lower blood glucose levels.

10.3 Carnosic Acid

Carnosic acid is contained in herbal plants, including rosemary and sage.[155–157] Carnosic acid and its derivative carnosol are catechol-type electrophilic compounds (Figure 12). Carnosic acid and carnosol induce expression of *HO-1* and nerve growth factor genes in neuronal cells by inducing NRF2 and exert protective effects of neuron.[158–161] Carnosic acid also prevents proliferation of various cancer cell lines, including lung small-cell carcinoma NCI-H82 cells, prostate cancer DU-145 and PC-3 cells, hepatocellular carcinoma Hep3B cells, chronic myeloid leukemia K-562 cells, breast cancer MCF-7, and MDA-MB-231 cells.[162] Carnosic acid prevents obesity and glucose intolerance in *ob/ob* mice.[163,164] Carnosic acid activates Akt and AMPKα signaling and then enhances glucose uptake in skeletal muscle cell line L6 myotubes.[165]

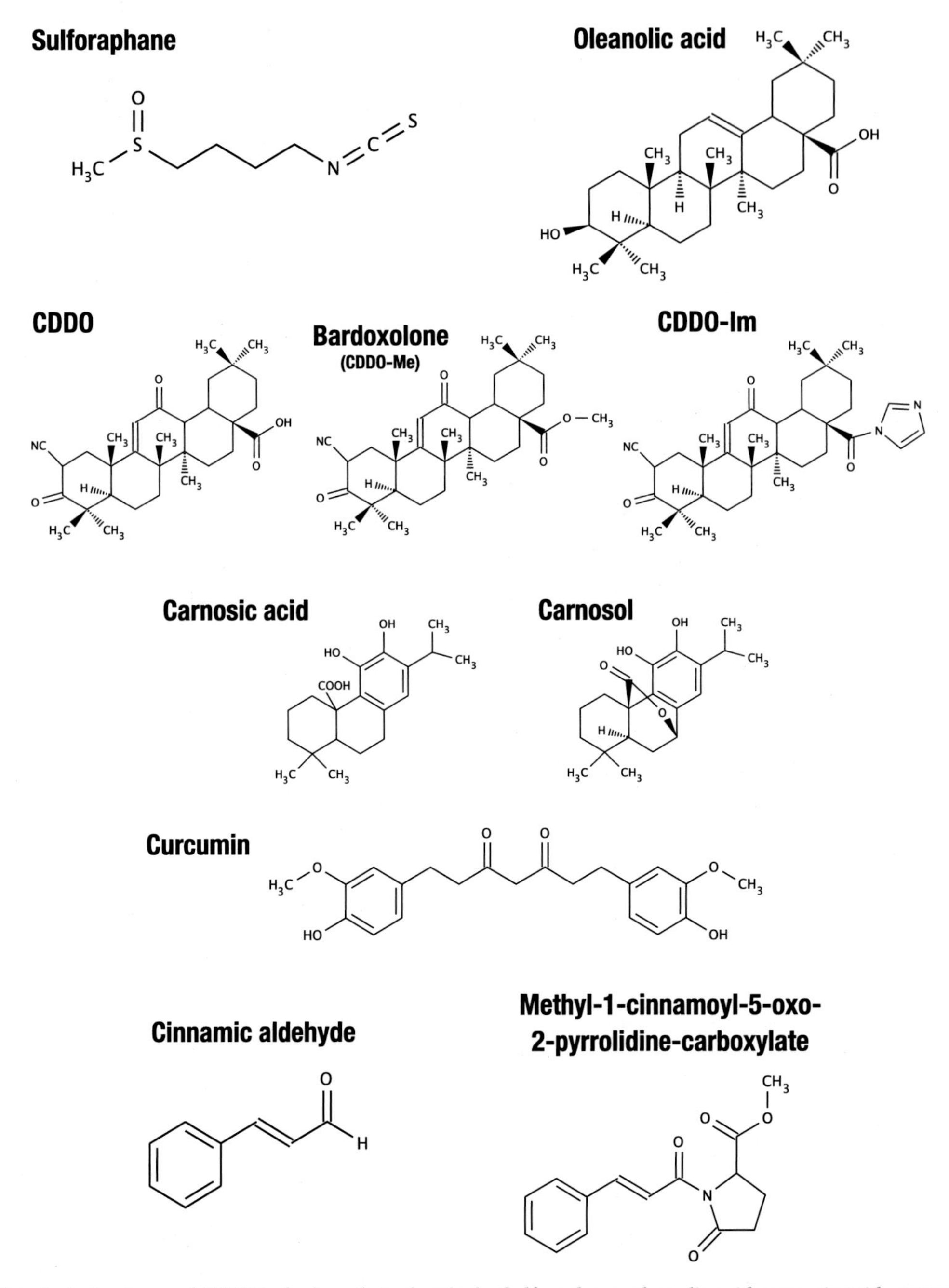

FIGURE 12 **Chemical structures of NRF2 inducing phytochemicals.** Sulforaphane, oleanolic acid, carnosic acid, carnosol, curcumin, cinnamic aldehyde, and methyl-1-cinnamoyl-5-oxo-2-pyrrolidine-carboxylate are phytochemicals that activate NRF2 signaling. In contrast, oleanolic triterpenoid 1-[2-cyano-3,12-dioxooleane-1,9(11)-dien-28-oyl (CDDO)], bardoxolone (CDDO-Me), and CDDO-Im are synthetic NRF2 inducers, which are derivatives of oleanolic acid. NRF2, nuclear factor-erythroid-related factor 2.

10.4 Curcumin

Curcumin is contained in turmeric as a yellow pigment, and it has been used in oriental medicine for treatment of cancer, lung diseases, renal diseases, neurological diseases, liver diseases, metabolic diseases, cardiovascular diseases, and various other inflammatory diseases (Figure 12). Administration of curcumin to culture cells induces expression levels of *HO-1*[166] and other NRF2 target genes.[167] Curcumin has also shown to exert anti-inflammatory response in mice through NRF2.[168]

Administration of turmeric extracts containing curcumin and its derivatives exert hypoglycemic effects in diabetes model KK-A^y mice.[169] Curcumin and its derivatives induce NRF2 and enhance *HO-1* expression in mouse pancreatic islets.[170] Dietary intake of curcumin improves inflammation in adipose tissue and obesity-mediated metabolic derangement in obesity models, including high-fat diet-induced obese mice and *ob/ob* mice.[37] In humans, curcumin treatment improves glycemic control of T2D patients.[171]

10.5 Cinnamic Aldehyde

Cinnamic aldehyde is contained in the bark of cinnamon, camphor, and cassia trees. Cinnamic aldehyde gives rise to cinnamon flavor, and it is the major content of cinnamon bark oil (Figure 12). Cinnamic aldehyde and its derivative methyl-1-cinnamoyl-5-oxo-2-pyrrolidine-carboxylate (Figure 12) induce NRF2 in cultured human skin cells, human skin Hs27 fibroblasts, and human HaCaT keratinocytes, and they enhance expression levels of NRF2 target genes, including *HO-1*, *NQO1*, *CAT*, and metallothionein 2A (*MT2A*).[172] Importantly, the administration of cinnamic aldehyde suppresses the progression of diabetic nephropathy in STZ-mediated diabetes model mice.[141]

11. CONCLUSION

In this chapter, we have reviewed that NRF2 regulates expressions of glucose metabolism-related genes as well as antioxidative stress response genes. NRF2 contributes to the maintenance of metabolic homeostasis by protecting the pancreatic β cells against oxidative stresses and by improving the insulin resistance (Figure 13). It has been reported that several phytochemicals induce NRF2 signaling, and these phytochemicals are expected to improve glucose homeostasis. In this article, we have summarized the roles that representative NRF2-inducing phytochemicals play for the glucose metabolisms. Intake of natural plant products containing NRF2-inducing phytochemicals often provide beneficial effects on glucose metabolism in human beings; they appear to improve the early metabolic abnormalities of glucose (e.g., impaired glucose tolerance and impaired fasting glucose). In addition, phytochemical-derivative synthetic NRF2 inducers have been developed and actually exert very strong NRF2 activation. These NRF2 inducers are expected to act as antidiabetic agents. In conclusion, NRF2 emerges as an important therapeutic target for diabetes mellitus and related metabolic disorders.

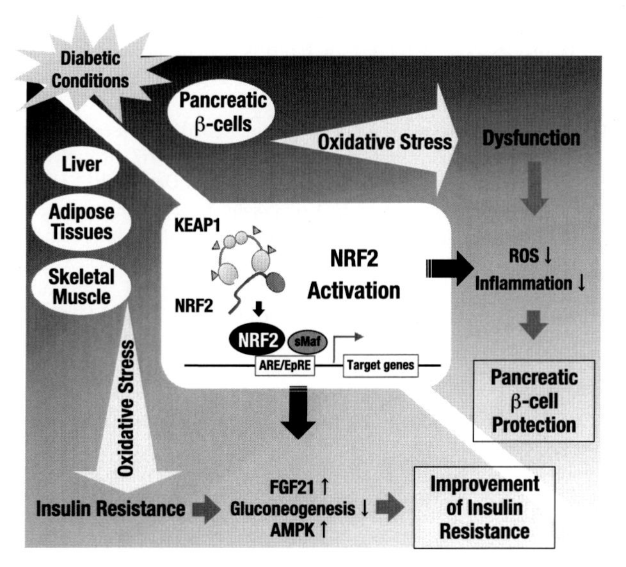

FIGURE 13 **NRF2 prevents onset and development of diabetes mellitus.** In T2D, oxidative stress is involved in progression of pancreatic β-cell dysfunction and insulin resistance in peripheral organs or tissues, such as liver, adipose tissues, and skeletal muscle. NRF2 activation decreases ROS levels in pancreatic β cells and suppresses inflammation, and protects pancreatic β cells. In addition, NRF2 improves insulin resistance by regulating FGF21 pathway, gluconeogenesis, and AMPK signals. ARE/EpRE, antioxidant/electrophile responsive element; FGF, fibroblast growth factor; KEAP1, kelch-like ECH-associated protein 1; NRF2, nuclear factor-erythroid-related factor 2; ROS, reactive oxygen species.

References

1. Dinkova-Kostova AT, Talalay P. Direct and indirect antioxidant properties of inducers of cytoprotective proteins. *Mol Nutr Food Res* 2008;**52**(Suppl. 1):S128–38.
2. Conney AH. Induction of drug-metabolizing enzymes: a path to the discovery of multiple cytochromes P450. *Annu Rev Pharmacol Toxicol* 2003;**43**:1–30.
3. Talalay P, Fahey JW. Phytochemicals from cruciferous plants protect against cancer by modulating carcinogen metabolism. *J Nutr* 2001;**131**(Suppl. 11):3027S–33S.
4. Itoh K, Chiba T, Takahashi S, et al. An Nrf2/small Maf heterodimer mediates the induction of phase II detoxifying enzyme genes through antioxidant response elements. *Biochem Biophys Res Commun* 1997;**236**(2):313–22.
5. Enomoto A, Itoh K, Nagayoshi E, et al. High sensitivity of Nrf2 knockout mice to acetaminophen hepatotoxicity associated with decreased expression of ARE-regulated drug metabolizing enzymes and antioxidant genes. *Toxicol Sci* 2001;**59**(1):169–77.
6. Kwak MK, Egner PA, Dolan PM, et al. Role of phase 2 enzyme induction in chemoprotection by dithiolethiones. *Mutat Res* 2001;**480–481**:305–15.
7. Dallas S, Miller DS, Bendayan R. Multidrug resistance-associated proteins: expression and function in the central nervous system. *Pharmacol Rev* 2006;**58**(2):140–61.
8. Cole SP. Targeting multidrug resistance protein 1 (MRP1, ABCC1): past, present, and future. *Annu Rev Pharmacol Toxicol* 2014;**54**: 95–117.
9. Leslie EM, Deeley RG, Cole SP. Toxicological relevance of the multidrug resistance protein 1, MRP1 (ABCC1) and related transporters. *Toxicology* 2001;**167**(1):3–23.
10. Nebert DW, Dalton TP, Okey AB, Gonzalez FJ. Role of aryl hydrocarbon receptor-mediated induction of the CYP1 enzymes in environmental toxicity and cancer. *J Biol Chem* 2004;**279**(23): 23847–50.
11. Rushmore TH, Morton MR, Pickett CB. The antioxidant responsive element. Activation by oxidative stress and identification of the DNA consensus sequence required for functional activity. *J Biol Chem* 1991;**266**(18):11632–9.

12. Friling RS, Bensimon A, Tichauer Y, Daniel V. Xenobiotic-inducible expression of murine glutathione S-transferase Ya subunit gene is controlled by an electrophile-responsive element. *Proc Natl Acad Sci USA* 1990;**87**(16):6258–62.
13. Dinkova-Kostova AT, Talalay P. NAD(P)H:quinone acceptor oxidoreductase 1 (NQO1), a multifunctional antioxidant enzyme and exceptionally versatile cytoprotector. *Arch Biochem Biophys* 2010;**501**(1):116–23.
14. Wu KC, Cui JY, Klaassen CD. Effect of graded Nrf2 activation on phase-I and -II drug metabolizing enzymes and transporters in mouse liver. *PLoS One* 2012;**7**(7):e39006.
15. Uruno A, Yagishita Y, Yamamoto M. The Keap1-Nrf2 system and diabetes mellitus. *Arch Biochem Biophys* 2015;**566C**:76–84.
16. Hayashi A, Suzuki H, Itoh K, Yamamoto M, Sugiyama Y. Transcription factor Nrf2 is required for the constitutive and inducible expression of multidrug resistance-associated protein 1 in mouse embryo fibroblasts. *Biochem Biophys Res Commun* 2003;**310**(3): 824–9.
17. Maher J, Yamamoto M. The rise of antioxidant signaling—the evolution and hormetic actions of Nrf2. *Toxicol Appl Pharmacol* 2010;**244**(1):4–15.
18. Murphy MP. How mitochondria produce reactive oxygen species. *Biochem J* 2009;**417**(1):1–13.
19. Cairns RA, Harris IS, Mak TW. Regulation of cancer cell metabolism. *Nat Rev Cancer* 2011;**11**(2):85–95.
20. Thimmulappa RK, Mai KH, Srisuma S, Kensler TW, Yamamoto M, Biswal S. Identification of Nrf2-regulated genes induced by the chemopreventive agent sulforaphane by oligonucleotide microarray. *Cancer Res* 2002;**62**(18):5196–203.
21. Mustacich D, Powis G. Thioredoxin reductase. *Biochem J* 2000; **346**(Pt 1):1–8.
22. Kim YC, Masutani H, Yamaguchi Y, Itoh K, Yamamoto M, Yodoi J. Hemin-induced activation of the thioredoxin gene by Nrf2. A differential regulation of the antioxidant responsive element by a switch of its binding factors. *J Biol Chem* 2001;**276**(21):18399–406.
23. Cho HY, Reddy SP, Debiase A, Yamamoto M, Kleeberger SR. Gene expression profiling of NRF2-mediated protection against oxidative injury. *Free Radic Biol Med* 2005;**38**(3):325–43.
24. Rhee SG, Jeong W, Chang TS, Woo HA. Sulfiredoxin, the cysteine sulfinic acid reductase specific to 2-Cys peroxiredoxin: its discovery, mechanism of action, and biological significance. *Kidney Int Suppl* 2007;**106**:S3–8.
25. Ishii T, Itoh K, Takahashi S, et al. Transcription factor Nrf2 coordinately regulates a group of oxidative stress-inducible genes in macrophages. *J Biol Chem* 2000;**275**(21):16023–9.
26. Soriano FX, Léveillé F, Papadia S, et al. Induction of sulfiredoxin expression and reduction of peroxiredoxin hyperoxidation by the neuroprotective Nrf2 activator 3H-1,2-dithiole-3-thione. *J Neurochem* 2008;**107**(2):533–43.
27. Chan K, Kan YW. Nrf2 is essential for protection against acute pulmonary injury in mice. *Proc Natl Acad Sci USA* 1999;**96**(22):12731–6.
28. Kwak MK, Itoh K, Yamamoto M, Sutter TR, Kensler TW. Role of transcription factor Nrf2 in the induction of hepatic phase 2 and antioxidative enzymes in vivo by the cancer chemoprotective agent, 3H-1, 2-dimethiole-3-thione. *Mol Med* 2001;**7**(2):135–45.
29. Itoh K, Mochizuki M, Ishii Y, et al. Transcription factor Nrf2 regulates inflammation by mediating the effect of 15-deoxy-$\Delta^{12,14}$-prostaglandin J_2. *Mol Cell Biol* 2004;**24**(1):36–45.
30. Thimmulappa RK, Scollick C, Traore K, et al. Nrf2-dependent protection from LPS induced inflammatory response and mortality by CDDO-Imidazolide. *Biochem Biophys Res Commun* 2006; **351**(4):883–9.
31. Lin W, Wu RT, Wu T, Khor TO, Wang H, Kong AN. Sulforaphane suppressed LPS-induced inflammation in mouse peritoneal macrophages through Nrf2 dependent pathway. *Biochem Pharmacol* 2008;**76**(8):967–73.
32. Woo KJ, Kwon TK. Sulforaphane suppresses lipopolysaccharide-induced cyclooxygenase-2 (COX-2) expression through the modulation of multiple targets in COX-2 gene promoter. *Int Immunopharmacol* 2007;**7**(13):1776–83.
33. Khor TO, Huang MT, Kwon KH, Chan JY, Reddy BS, Kong AN. Nrf2-deficient mice have an increased susceptibility to dextran sulfate sodium-induced colitis. *Cancer Res* 2006;**66**(24):11580–4.
34. de Carvalho Vidigal F, Guedes Cocate P, Gonçalves Pereira L, de Cássia Gonçalves Alfenas R. The role of hyperglycemia in the induction of oxidative stress and inflammatory process. *Nutr Hosp* 2012;**27**(5):1391–8.
35. Toma L, Stancu CS, Botez GM, Sima AV, Simionescu M. Irreversibly glycated LDL induce oxidative and inflammatory state in human endothelial cells; added effect of high glucose. *Biochem Biophys Res Commun* 2009;**390**(3):877–82.
36. Yu Z, Shao W, Chiang Y, et al. Oltipraz upregulates the nuclear factor (erythroid-derived 2)-like 2 [corrected](NRF2) antioxidant system and prevents insulin resistance and obesity induced by a high-fat diet in C57BL/6J mice. *Diabetologia* 2011;**54**(4):922–34.
37. Weisberg SP, Leibel R, Tortoriello DV. Dietary curcumin significantly improves obesity-associated inflammation and diabetes in mouse models of diabesity. *Endocrinology* 2008;**149**(7):3549–58.
38. Panchal SK, Ward L, Brown L. Ellagic acid attenuates high-carbohydrate, high-fat diet-induced metabolic syndrome in rats. *Eur J Nutr* 2013;**52**(2):559–68.
39. Song MY, Kim EK, Moon WS, et al. Sulforaphane protects against cytokine- and streptozotocin-induced β-cell damage by suppressing the NF-kappaB pathway. *Toxicol Appl Pharmacol* 2009;**235**(1): 57–67.
40. Itoh K, Igarashi K, Hayashi N, Nishizawa M, Yamamoto M. Cloning and characterization of a novel erythroid cell-derived CNC family transcription factor heterodimerizing with the small Maf family proteins. *Mol Cell Biol* 1995;**15**(8):4184–93.
41. Moi P, Chan K, Asunis I, Cao A, Kan YW. Isolation of NF-E2-related factor 2 (Nrf2), a NF-E2-like basic leucine zipper transcriptional activator that binds to the tandem NF-E2/AP1 repeat of the β-globin locus control region. *Proc Natl Acad Sci USA* 1994;**91**(21):9926–30.
42. Itoh K, Wakabayashi N, Katoh Y, et al. Keap1 represses nuclear activation of antioxidant responsive elements by Nrf2 through binding to the amino-terminal Neh2 domain. *Genes Dev* 1999;**13**(1):76–86.
43. Tong KI, Padmanabhan B, Kobayashi A, et al. Different electrostatic potentials define ETGE and DLG motifs as hinge and latch in oxidative stress response. *Mol Cell Biol* 2007;**27**(21):7511–21.
44. Tong KI, Katoh Y, Kusunoki H, Itoh K, Tanaka T, Yamamoto M. Keap1 recruits Neh2 through binding to ETGE and DLG motifs: characterization of the two-site molecular recognition model. *Mol Cell Biol* 2006;**26**(8):2887–900.
45. Takaya K, Suzuki T, Motohashi H, et al. Validation of the multiple sensor mechanism of the Keap1-Nrf2 system. *Free Radic Biol Med* 2012;**53**(4):817–27.
46. Vetere A, Choudhary A, Burns SM, Wagner BK. Targeting the pancreatic β-cell to treat diabetes. *Nat Rev Drug Discov* 2014; **13**(4):278–89.
47. Bonora E. Protection of pancreatic β-cells: is it feasible? *Nutr Metab Cardiovasc Dis* 2008;**18**(1):74–83.
48. Sakuraba H, Mizukami H, Yagihashi N, Wada R, Hanyu C, Yagihashi S. Reduced β-cell mass and expression of oxidative stress-related DNA damage in the islet of Japanese type II diabetic patients. *Diabetologia* 2002;**45**(1):85–96.
49. Lee YE, Kim JW, Lee EM, et al. Chronic resveratrol treatment protects pancreatic islets against oxidative stress in db/db mice. *PLoS One* 2012;**7**(11):e50412.
50. Fujimoto S, Mukai E, Inagaki N. Role of endogenous ROS production in impaired metabolism-secretion coupling of diabetic pancreatic β cells. *Prog Biophys Mol Biol* 2011;**107**(2):304–10.

51. Ihara Y, Toyokuni S, Uchida K, et al. Hyperglycemia causes oxidative stress in pancreatic β-cells of GK rats, a model of type 2 diabetes. *Diabetes* 1999;**48**(4):927–32.
52. Portha B, Giroix MH, Tourrel-Cuzin C, Le-Stunff H, Movassat J. The GK rat: a prototype for the study of non-overweight type 2 diabetes. *Methods Mol Biol* 2012;**933**:125–59.
53. Lenzen S, Drinkgern J, Tiedge M. Low antioxidant enzyme gene expression in pancreatic islets compared with various other mouse tissues. *Free Radic Biol Med* 1996;**20**(3):463–6.
54. Grankvist K, Marklund SL, Täljedal IB. CuZn-superoxide dismutase, Mn-superoxide dismutase, catalase and glutathione peroxidase in pancreatic islets and other tissues in the mouse. *Biochem J* 1981;**199**(2):393–8.
55. Ivarsson R, Quintens R, Dejonghe S, et al. Redox control of exocytosis: regulatory role of NADPH, thioredoxin, and glutaredoxin. *Diabetes* 2005;**54**(7):2132–42.
56. Uruno A, Furusawa Y, Yagishita Y, et al. The Keap1-Nrf2 system prevents onset of diabetes mellitus. *Mol Cell Biol* 2013;**33**(15): 2996–3010.
57. Newsholme P, Haber EP, Hirabara SM, et al. Diabetes associated cell stress and dysfunction: role of mitochondrial and non-mitochondrial ROS production and activity. *J Physiol* 2007; **583**(Pt 1):9–24.
58. Gehrmann W, Elsner M, Lenzen S. Role of metabolically generated reactive oxygen species for lipotoxicity in pancreatic β-cells. *Diabetes Obes Metab* 2010;**12**(Suppl. 2):149–58.
59. Ma ZA, Zhao Z, Turk J. Mitochondrial dysfunction and β-cell failure in type 2 diabetes mellitus. *Exp Diabetes Res* 2012;**2012**: 703538.
60. El-Assaad W, Joly E, Barbeau A, et al. Glucolipotoxicity alters lipid partitioning and causes mitochondrial dysfunction, cholesterol, and ceramide deposition and reactive oxygen species production in INS832/13 ss-cells. *Endocrinology* 2010;**151**(7): 3061–73.
61. Koulajian K, Desai T, Liu GC, et al. NADPH oxidase inhibition prevents β cell dysfunction induced by prolonged elevation of oleate in rodents. *Diabetologia* 2013;**56**(5):1078–87.
62. Jörns A, Günther A, Hedrich HJ, Wedekind D, Tiedge M, Lenzen S. Immune cell infiltration, cytokine expression, and β-cell apoptosis during the development of type 1 diabetes in the spontaneously diabetic LEW.1AR1/Ztm-iddm rat. *Diabetes* 2005;**54**(7):2041–52.
63. Pirot P, Cardozo AK, Eizirik DL. Mediators and mechanisms of pancreatic β-cell death in type 1 diabetes. *Arq Bras Endocrinol Metabol* 2008;**52**(2):156–65.
64. Michalska M, Wolf G, Walther R, Newsholme P. Effects of pharmacological inhibition of NADPH oxidase or iNOS on pro-inflammatory cytokine, palmitic acid or H_2O_2-induced mouse islet or clonal pancreatic β-cell dysfunction. *Biosci Rep* 2010;**30**(6): 445–53.
65. Laybutt DR, Sharma A, Sgroi DC, Gaudet J, Bonner-Weir S, Weir GC. Genetic regulation of metabolic pathways in β-cells disrupted by hyperglycemia. *J Biol Chem* 2002;**277**(13): 10912–21.
66. Ma Z, Ramanadham S, Corbett JA, et al. Interleukin-1 enhances pancreatic islet arachidonic acid 12-lipoxygenase product generation by increasing substrate availability through a nitric oxide-dependent mechanism. *J Biol Chem* 1996;**271**(2):1029–42.
67. Tersey SA, Maier B, Nishiki Y, Maganti AV, Nadler JL, Mirmira RG. 12-lipoxygenase promotes obesity-induced oxidative stress in pancreatic islets. *Mol Cell Biol* 2014;**34**(19): 3735–45.
68. Rebelato E, Abdulkader F, Curi R, Carpinelli AR. Low doses of hydrogen peroxide impair glucose-stimulated insulin secretion via inhibition of glucose metabolism and intracellular calcium oscillations. *Metabolism* 2010;**59**(3):409–13.
69. Rebelato E, Abdulkader F, Curi R, Carpinelli AR. Control of the intracellular redox state by glucose participates in the insulin secretion mechanism. *PLoS One* 2011;**6**(8):e24507.
70. Sakai K, Matsumoto K, Nishikawa T, et al. Mitochondrial reactive oxygen species reduce insulin secretion by pancreatic β-cells. *Biochem Biophys Res Commun* 2003;**300**(1):216–22.
71. Harmon JS, Stein R, Robertson RP. Oxidative stress-mediated, post-translational loss of MafA protein as a contributing mechanism to loss of insulin gene expression in glucotoxic β cells. *J Biol Chem* 2005;**280**(12):11107–13.
72. Kaneto H, Kajimoto Y, Miyagawa J, et al. Beneficial effects of antioxidants in diabetes: possible protection of pancreatic β-cells against glucose toxicity. *Diabetes* 1999;**48**(12):2398–406.
73. Tejedo J, Bernabé JC, Ramírez R, Sobrino F, Bedoya FJ. NO induces a cGMP-independent release of cytochrome c from mitochondria which precedes caspase 3 activation in insulin producing RINm5F cells. *FEBS Lett* 1999;**459**(2):238–43.
74. Yagishita Y, Fukutomi T, Sugawara A, et al. Nrf2 protects pancreatic β-cells from oxidative and nitrosative stress in diabetic model mice. *Diabetes* 2014;**63**(2):605–18.
75. Takamura T, Kato I, Kimura N, et al. Transgenic mice overexpressing type 2 nitric-oxide synthase in pancreatic β cells develop insulin-dependent diabetes without insulitis. *J Biol Chem* 1998; **273**(5):2493–6.
76. Oyadomari S, Takeda K, Takiguchi M, et al. Nitric oxide-induced apoptosis in pancreatic β cells is mediated by the endoplasmic reticulum stress pathway. *Proc Natl Acad Sci USA* 2001;**98**(19): 10845–50.
77. Flodström M, Tyrberg B, Eizirik DL, Sandler S. Reduced sensitivity of inducible nitric oxide synthase-deficient mice to multiple low-dose streptozotocin-induced diabetes. *Diabetes* 1999;**48**(4): 706–13.
78. Pi J, Bai Y, Zhang Q, et al. Reactive oxygen species as a signal in glucose-stimulated insulin secretion. *Diabetes* 2007;**56**(7): 1783–91.
79. Leloup C, Tourrel-Cuzin C, Magnan C, et al. Mitochondrial reactive oxygen species are obligatory signals for glucose-induced insulin secretion. *Diabetes* 2009;**58**(3):673–81.
80. Xu B, Moritz JT, Epstein PN. Overexpression of catalase provides partial protection to transgenic mouse β cells. *Free Radic Biol Med* 1999;**27**(7–8):830–7.
81. Harmon JS, Bogdani M, Parazzoli SD, et al. β-Cell-specific overexpression of glutathione peroxidase preserves intranuclear MafA and reverses diabetes in db/db mice. *Endocrinology* 2009;**150**(11): 4855–62.
82. Kubisch HM, Wang J, Bray TM, Phillips JP. Targeted overexpression of Cu/Zn superoxide dismutase protects pancreatic β-cells against oxidative stress. *Diabetes* 1997;**46**(10):1563–6.
83. Sasaki M, Fujimoto S, Sato Y, et al. Reduction of reactive oxygen species ameliorates metabolism-secretion coupling in islets of diabetic GK rats by suppressing lactate overproduction. *Diabetes* 2013;**62**(6):1996–2003.
84. Lee JH, Lee JS, Kim YR, et al. Hispidin isolated from *Phellinus linteus* protects against hydrogen peroxide-induced oxidative stress in pancreatic MIN6N β-cells. *J Med Food* 2011;**14**(11):1431–8.
85. Tang C, Han P, Oprescu AI, et al. Evidence for a role of superoxide generation in glucose-induced β-cell dysfunction in vivo. *Diabetes* 2007;**56**(11):2722–31.
86. Martín MA, Ramos S, Cordero-Herrero I, Bravo L, Goya L. Cocoa phenolic extract protects pancreatic β cells against oxidative stress. *Nutrients* 2013;**5**(8):2955–68.
87. Park MH, Han JS. Padina arborescens extract protects high glucose-induced apoptosis in pancreatic β cells by reducing oxidative stress. *Nutr Res Pract* 2014;**8**(5):494–500.
88. Uruno A, Motohashi H. The Keap1-Nrf2 system as an in vivo sensor for electrophiles. *Nitric Oxide* 2011;**25**(2):153–60.

89. Lacraz G, Figeac F, Movassat J, Kassis N, Portha B. Diabetic GK/Par rat β-cells are spontaneously protected against H_2O_2-triggered apoptosis. A cAMP-dependent adaptive response. *Am J Physiol Endocrinol Metab* 2010;**298**(1):E17–27.
90. Li W, Wu W, Song H, et al. Targeting Nrf2 by dihydro-CDDO-trifluoroethyl amide enhances autophagic clearance and viability of β-cells in a setting of oxidative stress. *FEBS Lett* 2014;**588**(12):2115–24.
91. Wakabayashi N, Itoh K, Wakabayashi J, et al. Keap1-null mutation leads to postnatal lethality due to constitutive Nrf2 activation. *Nat Genet* 2003;**35**(3):238–45.
92. Okawa H, Motohashi H, Kobayashi A, Aburatani H, Kensler TW, Yamamoto M. Hepatocyte-specific deletion of the keap1 gene activates Nrf2 and confers potent resistance against acute drug toxicity. *Biochem Biophys Res Commun* 2006;**339**(1):79–88.
93. Taguchi K, Maher JM, Suzuki T, Kawatani Y, Motohashi H, Yamamoto M. Genetic analysis of cytoprotective functions supported by graded expression of Keap1. *Mol Cell Biol* 2010;**30**(12):3016–26.
94. Lee BH, Hsu WH, Hsu YW, Pan TM. Dimerumic acid protects pancreas damage and elevates insulin production in methylglyoxal-treated pancreatic RINm5F cells. *J Funct Foods* 2013;**5**:642–50.
95. Sun X, Lin Y, Huang Q, et al. Di(2-ethylhexyl) phthalate-induced apoptosis in rat INS-1 cells is dependent on activation of endoplasmic reticulum stress and suppression of antioxidant protection. *J Cell Mol Med* 2014.
96. Rajesh P, Sathish S, Srinivasan C, Selvaraj J, Balasubramanian K. Phthalate is associated with insulin resistance in adipose tissue of male rat: role of antioxidant vitamins. *J Cell Biochem* 2013;**114**(3):558–69.
97. Stahlhut RW, van Wijngaarden E, Dye TD, Cook S, Swan SH. Concentrations of urinary phthalate metabolites are associated with increased waist circumference and insulin resistance in adult U.S. males. *Environ Health Perspect* 2007;**115**(6):876–82.
98. Svensson K, Hernández-Ramírez RU, Burguete-García A, et al. Phthalate exposure associated with self-reported diabetes among Mexican women. *Environ Res* 2011;**111**(6):792–6.
99. James-Todd T, Stahlhut R, Meeker JD, et al. Urinary phthalate metabolite concentrations and diabetes among women in the National Health and Nutrition Examination Survey (NHANES) 2001–2008. *Environ Health Perspect* 2012;**120**(9):1307–13.
100. Maedler K, Sergeev P, Ris F, et al. Glucose-induced β cell production of IL-1β contributes to glucotoxicity in human pancreatic islets. *J Clin Invest* 2002;**110**(6):851–60.
101. Böni-Schnetzler M, Boller S, Debray S, et al. Free fatty acids induce a proinflammatory response in islets via the abundantly expressed interleukin-1 receptor I. *Endocrinology* 2009;**150**(12):5218–29.
102. Jung HS, Chung KW, Won Kim J, et al. Loss of autophagy diminishes pancreatic β cell mass and function with resultant hyperglycemia. *Cell Metab* 2008;**8**(4):318–24.
103. Ebato C, Uchida T, Arakawa M, et al. Autophagy is important in islet homeostasis and compensatory increase of β cell mass in response to high-fat diet. *Cell Metab* 2008;**8**(4):325–32.
104. Rivera JF, Costes S, Gurlo T, Glabe CG, Butler PC. Autophagy defends pancreatic β cells from human islet amyloid polypeptide-induced toxicity. *J Clin Invest* 2014;**124**(8):3489–500.
105. Shigihara N, Fukunaka A, Hara A, et al. Human IAPP-induced pancreatic β cell toxicity and its regulation by autophagy. *J Clin Invest* 2014;**124**(8):3634–44.
106. Kim J, Cheon H, Jeong YT, et al. Amyloidogenic peptide oligomer accumulation in autophagy-deficient β cells induces diabetes. *J Clin Invest* 2014;**124**(8):3311–24.
107. Kaniuk NA, Kiraly M, Bates H, Vranic M, Volchuk A, Brumell JH. Ubiquitinated-protein aggregates form in pancreatic β-cells during diabetes-induced oxidative stress and are regulated by autophagy. *Diabetes* 2007;**56**(4):930–9.
108. Kwak MK, Wakabayashi N, Greenlaw JL, Yamamoto M, Kensler TW. Antioxidants enhance mammalian proteasome expression through the Keap1-Nrf2 signaling pathway. *Mol Cell Biol* 2003;**23**(23):8786–94.
109. Lee S, Hur EG, Ryoo IG, Jung KA, Kwak J, Kwak MK. Involvement of the Nrf2-proteasome pathway in the endoplasmic reticulum stress response in pancreatic β-cells. *Toxicol Appl Pharmacol* 2012;**264**(3):431–8.
110. Yoon KH, Ko SH, Cho JH, et al. Selective β-cell loss and α-cell expansion in patients with type 2 diabetes mellitus in Korea. *J Clin Endocrinol Metab* 2003;**88**(5):2300–8.
111. Talchai C, Xuan S, Lin HV, Sussel L, Accili D. Pancreatic β cell dedifferentiation as a mechanism of diabetic β cell failure. *Cell* 2012;**150**(6):1223–34.
112. Sugimoto H, Okada K, Shoda J, et al. Deletion of nuclear factor-E2-related factor-2 leads to rapid onset and progression of nutritional steatohepatitis in mice. *Am J Physiol Gastrointest Liver Physiol* 2010;**298**(2):G283–94.
113. Zhang YK, Yeager RL, Tanaka Y, Klaassen CD. Enhanced expression of Nrf2 in mice attenuates the fatty liver produced by a methionine- and choline-deficient diet. *Toxicol Appl Pharmacol* 2010;**245**(3):326–34.
114. Hirotsu Y, Katsuoka F, Funayama R, et al. Nrf2-MafG heterodimers contribute globally to antioxidant and metabolic networks. *Nucleic Acids Res* 2012;**40**(20):10228–39.
115. Kharitonenkov A, Shiyanova TL, Koester A, et al. FGF-21 as a novel metabolic regulator. *J Clin Invest* 2005;**115**(6):1627–35.
116. Wente W, Efanov AM, Brenner M, et al. Fibroblast growth factor-21 improves pancreatic β-cell function and survival by activation of extracellular signal-regulated kinase 1/2 and Akt signaling pathways. *Diabetes* 2006;**55**(9):2470–8.
117. Xu J, Lloyd DJ, Hale C, et al. Fibroblast growth factor 21 reverses hepatic steatosis, increases energy expenditure, and improves insulin sensitivity in diet-induced obese mice. *Diabetes* 2009;**58**(1):250–9.
118. Furusawa Y, Uruno A, Yagishita Y, Higashi C, Yamamoto M. Nrf2 induces fibroblast growth factor 21 in diabetic mice. *Genes Cells* 2014;**19**(12):864–78.
119. Ye D, Wang Y, Li H, et al. Fibroblast growth factor 21 protects against acetaminophen-induced hepatotoxicity by potentiating peroxisome proliferator-activated receptor coactivator protein-1α-mediated antioxidant capacity in mice. *Hepatology* 2014;**60**(3):977–89.
120. Shaw RJ, Kosmatka M, Bardeesy N, et al. The tumor suppressor LKB1 kinase directly activates AMP-activated kinase and regulates apoptosis in response to energy stress. *Proc Natl Acad Sci USA* 2004;**101**(10):3329–35.
121. Lizcano JM, Göransson O, Toth R, et al. LKB1 is a master kinase that activates 13 kinases of the AMPK subfamily, including MARK/PAR-1. *EMBO J* 2004;**23**(4):833–43.
122. Bae EJ, Yang YM, Kim JW, Kim SG. Identification of a novel class of dithiolethiones that prevent hepatic insulin resistance via the adenosine monophosphate-activated protein kinase-p70 ribosomal S6 kinase-1 pathway. *Hepatology* 2007;**46**(3):730–9.
123. Xu J, Donepudi AC, Moscovitz JE, Slitt AL. Keap1-knockdown decreases fasting-induced fatty liver via altered lipid metabolism and decreased fatty acid mobilization from adipose tissue. *PLoS One* 2013;**8**(11):e79841.
124. Meakin PJ, Chowdhry S, Sharma RS, et al. Susceptibility of Nrf2-null mice to steatohepatitis and cirrhosis upon consumption of a high-fat diet is associated with oxidative stress, perturbation of the unfolded protein response, and disturbance in the expression of metabolic enzymes but not with insulin resistance. *Mol Cell Biol* 2014;**34**(17):3305–20.

125. Cantó C, Gerhart-Hines Z, Feige JN, et al. AMPK regulates energy expenditure by modulating NAD$^+$ metabolism and SIRT1 activity. *Nature* 2009;**458**(7241):1056–60.
126. Shibata T, Ohta T, Tong KI, et al. Cancer related mutations in NRF2 impair its recognition by Keap1-Cul3 E3 ligase and promote malignancy. *Proc Natl Acad Sci USA* 2008;**105**(36):13568–73.
127. Taguchi K, Motohashi H, Yamamoto M. Molecular mechanisms of the Keap1–Nrf2 pathway in stress response and cancer evolution. *Genes Cells* 2011;**16**(2):123–40.
128. Mitsuishi Y, Taguchi K, Kawatani Y, et al. Nrf2 redirects glucose and glutamine into anabolic pathways in metabolic reprogramming. *Cancer Cell* 2012;**22**(1):66–79.
129. Colditz GA, Branch LG, Lipnick RJ, et al. Increased green and yellow vegetable intake and lowered cancer deaths in an elderly population. *Am J Clin Nutr* 1985;**41**(1):32–6.
130. Graham S. Results of case-control studies of diet and cancer in Buffalo, New York. *Cancer Res* 1983;**43**(Suppl. 5):2409s–13s.
131. Michaud DS, Spiegelman D, Clinton SK, Rimm EB, Willett WC, Giovannucci EL. Fruit and vegetable intake and incidence of bladder cancer in a male prospective cohort. *J Natl Cancer Inst* 1999;**91**(7):605–13.
132. Kumar H, Kim IS, More SV, Kim BW, Choi DK. Natural product-derived pharmacological modulators of Nrf2/ARE pathway for chronic diseases. *Nat Prod Rep* 2014;**31**(1):109–39.
133. Zhang Y, Talalay P, Cho CG, Posner GH. A major inducer of anticarcinogenic protective enzymes from broccoli: isolation and elucidation of structure. *Proc Natl Acad Sci USA* 1992;**89**(6): 2399–403.
134. Kensler TW, Chen JG, Egner PA, et al. Effects of glucosinolate-rich broccoli sprouts on urinary levels of aflatoxin-DNA adducts and phenanthrene tetraols in a randomized clinical trial in He Zuo township, Qidong, People's Republic of China. *Cancer Epidemiol Biomarkers Prev* 2005;**14**(11 Pt 1):2605–13.
135. Shapiro TA, Fahey JW, Dinkova-Kostova AT, et al. Safety, tolerance, and metabolism of broccoli sprout glucosinolates and isothiocyanates: a clinical phase I study. *Nutr Cancer* 2006;**55**(1):53–62.
136. Kensler TW, Ng D, Carmella SG, et al. Modulation of the metabolism of airborne pollutants by glucoraphanin-rich and sulforaphane-rich broccoli sprout beverages in Qidong, China. *Carcinogenesis* 2012;**33**(1):101–7.
137. Egner PA, Chen JG, Zarth AT, et al. Rapid and sustainable detoxication of airborne pollutants by broccoli sprout beverage: results of a randomized clinical trial in China. *Cancer Prev Res (Phila)* 2014;**7**(8):813–23.
138. Singh K, Connors SL, Macklin EA, et al. Sulforaphane treatment of autism spectrum disorder (ASD). *Proc Natl Acad Sci USA* 2014;**111**(43):15550–5.
139. Bahadoran Z, Tohidi M, Nazeri P, Mehran M, Azizi F, Mirmiran P. Effect of broccoli sprouts on insulin resistance in type 2 diabetic patients: a randomized double-blind clinical trial. *Int J Food Sci Nutr* 2012;**63**(7):767–71.
140. Cui W, Bai Y, Miao X, et al. Prevention of diabetic nephropathy by sulforaphane: possible role of Nrf2 upregulation and activation. *Oxid Med Cell Longev* 2012;**2012**:821936.
141. Zheng H, Whitman SA, Wu W, et al. Therapeutic potential of Nrf2 activators in streptozotocin-induced diabetic nephropathy. *Diabetes* 2011;**60**(11):3055–66.
142. Liu J, Liu Y, Klaassen CD. The effect of Chinese hepatoprotective medicines on experimental liver injury in mice. *J Ethnopharmacol* 1994;**42**(3):183–91.
143. Liu J, Liu Y, Parkinson A, Klaassen CD. Effect of oleanolic acid on hepatic toxicant-activating and detoxifying systems in mice. *J Pharmacol Exp Ther* 1995;**275**(2):768–74.
144. Liu J, Liu Y, Mao Q, Klaassen CD. The effects of 10 triterpenoid compounds on experimental liver injury in mice. *Fundam Appl Toxicol* 1994;**22**(1):34–40.
145. Liu J, Liu Y, Madhu C, Klaassen CD. Protective effects of oleanolic acid on acetaminophen-induced hepatotoxicity in mice. *J Pharmacol Exp Ther* 1993;**266**(3):1607–13.
146. Reisman SA, Aleksunes LM, Klaassen CD. Oleanolic acid activates Nrf2 and protects from acetaminophen hepatotoxicity via Nrf2-dependent and Nrf2-independent processes. *Biochem Pharmacol* 2009;**77**(7):1273–82.
147. Liu J. Pharmacology of oleanolic acid and ursolic acid. *J Ethnopharmacol* 1995;**49**(2):57–68.
148. Honda T, Honda Y, Favaloro FG, et al. A novel dicyanotriterpenoid, 2-cyano-3,12-dioxooleana-1,9(11)-dien-28-onitrile, active at picomolar concentrations for inhibition of nitric oxide production. *Bioorg Med Chem Lett* 2002;**12**(7):1027–30.
149. Pergola PE, Raskin P, Toto RD, et al. Bardoxolone methyl and kidney function in CKD with type 2 diabetes. *N Engl J Med* 2011;**365**(4):327–36.
150. de Zeeuw D, Akizawa T, Audhya P, et al. Bardoxolone methyl in type 2 diabetes and stage 4 chronic kidney disease. *N Engl J Med* 2013;**369**(26):2492–503.
151. Place AE, Suh N, Williams CR, et al. The novel synthetic triterpenoid, CDDO-imidazolide, inhibits inflammatory response and tumor growth in vivo. *Clin Cancer Res* 2003;**9**(7):2798–806.
152. Kalra S, Knatko EV, Zhang Y, Honda T, Yamamoto M, Dinkova-Kostova AT. Highly potent activation of Nrf2 by topical tricyclic bis(cyano enone): implications for protection against UV radiation during thiopurine therapy. *Cancer Prev Res (Phila)* 2012; **5**(7):973–81.
153. Liby K, Yore MM, Roebuck BD, et al. A novel acetylenic tricyclic bis-(cyano enone) potently induces phase 2 cytoprotective pathways and blocks liver carcinogenesis induced by aflatoxin. *Cancer Res* 2008;**68**(16):6727–33.
154. Saha PK, Reddy VT, Konopleva M, Andreeff M, Chan L. The triterpenoid 2-cyano-3,12-dioxooleana-1,9-dien-28-oic-acid methyl ester has potent anti-diabetic effects in diet-induced diabetic mice and Lepr(db/db) mice. *J Biol Chem* 2010;**285**(52): 40581–92.
155. Schwarz K, Ternes W. Antioxidative constituents of *Rosmarinus officinalis* and *Salvia officinalis*. I. Determination of phenolic diterpenes with antioxidative activity amongst tocochromanols using HPLC. *Z Lebensm Unters Forsch* 1992;**195**(2):95–8.
156. Schwarz K, Ternes W. Antioxidative constituents of *Rosmarinus officinalis* and *Salvia officinalis*. II. Isolation of carnosic acid and formation of other phenolic diterpenes. *Z Lebensm Unters Forsch* 1992;**195**(2):99–103.
157. Schwarz K, Ternes W, Schmauderer E. Antioxidative constituents of *Rosmarinus officinalis* and *Salvia officinalis*. III. Stability of phenolic diterpenes of rosemary extracts under thermal stress as required for technological processes. *Z Lebensm Unters Forsch* 1992;**195**(2):104–7.
158. Satoh T, Izumi M, Inukai Y, et al. Carnosic acid protects neuronal HT22 cells through activation of the antioxidant-responsive element in free carboxylic acid- and catechol hydroxyl moieties-dependent manners. *Neurosci Lett* 2008;**434**(3):260–5.
159. Satoh T, Kosaka K, Itoh K, et al. Carnosic acid, a catechol-type electrophilic compound, protects neurons both in vitro and in vivo through activation of the Keap1/Nrf2 pathway via S-alkylation of targeted cysteines on Keap1. *J Neurochem* 2008; **104**(4):1116–31.
160. Mimura J, Kosaka K, Maruyama A, et al. Nrf2 regulates NGF mRNA induction by carnosic acid in T98G glioblastoma cells and normal human astrocytes. *J Biochem* 2011;**150**(2):209–17.
161. Martin D, Rojo AI, Salinas M, et al. Regulation of heme oxygenase-1 expression through the phosphatidylinositol 3-kinase/Akt pathway and the Nrf2 transcription factor in response to the antioxidant phytochemical carnosol. *J Biol Chem* 2004;**279**(10):8919–29.

162. Yesil-Celiktas O, Sevimli C, Bedir E, Vardar-Sukan F. Inhibitory effects of rosemary extracts, carnosic acid and rosmarinic acid on the growth of various human cancer cell lines. *Plant Foods Hum Nutr* 2010;**65**(2):158–63.
163. Wang T, Takikawa Y, Satoh T, et al. Carnosic acid prevents obesity and hepatic steatosis in ob/ob mice. *Hepatol Res* 2011; **41**(1):87–92.
164. Park MY, Sung MK. Carnosic acid attenuates obesity-induced glucose intolerance and hepatic fat accumulation by modulating genes of lipid metabolism in C57BL/6J-ob/ob mice. *J Sci Food Agric* 2014.
165. Lipina C, Hundal HS. Carnosic acid stimulates glucose uptake in skeletal muscle cells via a PME-1/PP2A/PKB signalling axis. *Cell Signal* 2014;**26**(11):2343–9.
166. Balogun E, Hoque M, Gong P, et al. Curcumin activates the haem oxygenase-1 gene via regulation of Nrf2 and the antioxidant-responsive element. *Biochem J* 2003;**371**(Pt 3):887–95.
167. Shen G, Xu C, Hu R, et al. Modulation of nuclear factor E2-related factor 2-mediated gene expression in mice liver and small intestine by cancer chemopreventive agent curcumin. *Mol Cancer Ther* 2006;**5**(1):39–51.
168. Boyanapalli SS, Paredes-Gonzalez X, Fuentes F, et al. Nrf2 knockout attenuates the anti-inflammatory effects of phenethyl isothiocyanate and curcumin. *Chem Res Toxicol* 2014.
169. Kuroda M, Mimaki Y, Nishiyama T, et al. Hypoglycemic effects of turmeric (*Curcuma longa* L. rhizomes) on genetically diabetic KK-Ay mice. *Biol Pharm Bull* 2005;**28**(5):937–9.
170. Pugazhenthi S, Akhov L, Selvaraj G, Wang M, Alam J. Regulation of heme oxygenase-1 expression by demethoxy curcuminoids through Nrf2 by a PI3-kinase/Akt-mediated pathway in mouse β-cells. *Am J Physiol Endocrinol Metab* 2007;**293**(3): E645–55.
171. Chuengsamarn S, Rattanamongkolgul S, Luechapudiporn R, Phisalaphong C, Jirawatnotai S. Curcumin extract for prevention of type 2 diabetes. *Diabetes Care* 2012;**35**(11):2121–7.
172. Wondrak GT, Cabello CM, Villeneuve NF, et al. Cinnamoyl-based Nrf2-activators targeting human skin cell photo-oxidative stress. *Free Radic Biol Med* 2008;**45**(4):385–95.

CHAPTER

28

Hepatic Mitochondrial Fatty Acid Oxidation and Type 2 Diabetes

Abdelhak Mansouri[1], *Wolfgang Langhans*[1], *Jean Girard*[2,3,4], *Carina Prip-Buus*[2,3,4]

[1]Physiology and Behavior Laboratory, Institute of Food, Nutrition and Health, ETH Zurich, Schwerzenbach, Switzerland; [2]INSERM, Institut Cochin, Paris, France; [3]CNRS, Paris, France; [4]Université Paris Descartes, Paris, France

1. INTRODUCTION

The liver is a key player in glucose and lipid homeostasis, given its flexibility to metabolize carbohydrates and fatty acids in both the fed and fasted states.[1,2] In fact, after feeding a carbohydrate-rich meal, the presence of both high plasma glucose and insulin concentrations stimulates liver glucose oxidation, glycogen storage, and lipogenesis, allowing conversion of excess glucose into long-chain fatty acids (LCFAs).[3] This results in an increase in malonyl-coenzyme A (CoA), the first intermediate in lipogenesis synthetized by acetyl-CoA carboxylase. Malonyl-CoA inhibits carnitine palmitoyltransferase 1A (CPT1A), which catalyzes the rate-limiting step in the entry of cytosolic long-chain acyl-CoA (LC-CoA) into the mitochondria where LCFA β-oxidation takes place. Exogenous LCFAs taken up by the liver and endogenous LCFAs generated by lipogenesis, both are then esterified into triacylglycerols (TAGs) and partly secreted as very low-density lipoproteins to provide extrahepatic tissues with LCFAs.[4] Conversely, during fasting, the liver is crucial to provide glucose to maintain normoglycemia and, hence, to supply glucose to the whole organism. Glucagon stimulates hepatic gluconeogenesis from amino acids and glycerol,[5] and the energy and cofactors such as acetyl-CoA and nicotinamide adenine dinucleotide, required for optimal gluconeogenesis, come from the β-oxidation of LCFAs released from adipose tissue.[5] Therefore, the liver plays a crucial role in glucose homeostasis, and any defect in the metabolism of nutrients, in particular LCFA and glucose, might lead to the development of insulin resistance and type 2 diabetes (T2D).[6]

Nonalcoholic fatty liver disease (NAFLD) is considered to be the hepatic manifestation of the metabolic syndrome. NAFLD includes a broad spectrum of liver injuries, ranging from pure steatosis to nonalcoholic steatohepatitis (NASH) with possible progression to fibrosis, cirrhosis, and then hepatocellular carcinoma.[7–10] The prevalence of NAFLD in developed countries has reached alarming levels. Indeed, NAFLD is estimated to affect 15–24% of the general population,[11–14] and the prevalence rate increases in obese individuals (body mass index (BMI) $\geq$ 30 kg/m^2) and is particularly high in morbidly obese patients (BMI $>$ 40 kg/m^2). Moreover, NASH is estimated to affect more than 50% of obese people with T2D.[9,15,16]

Hepatic steatosis, characterized by an abnormal accumulation of TAGs in hepatocytes, is reversible and innocuous. However, according to the two-hit hypothesis[17] or, more recently, to the multihit hypothesis,[18] steatosic liver can progress to a more severe and pathological state (NASH) by the involvement of secondary hits such as oxidative stress and recruitment of inflammatory immune cells (Figure 1).[19]

NASH is considered to be closely associated with insulin resistance.[8,20] Many factors can contribute to the development of hepatic steatosis, namely: (1) an excessive delivery of nonesterified LCFA from adipose tissue from enhanced lipolysis induced by insulin resistance[10,21]; (2) a dysfunction in the synthesis and/or release of very low-density lipoproteins[22]; (3) an

Molecular Nutrition and Diabetes
http://dx.doi.org/10.1016/B978-0-12-801585-8.00028-2

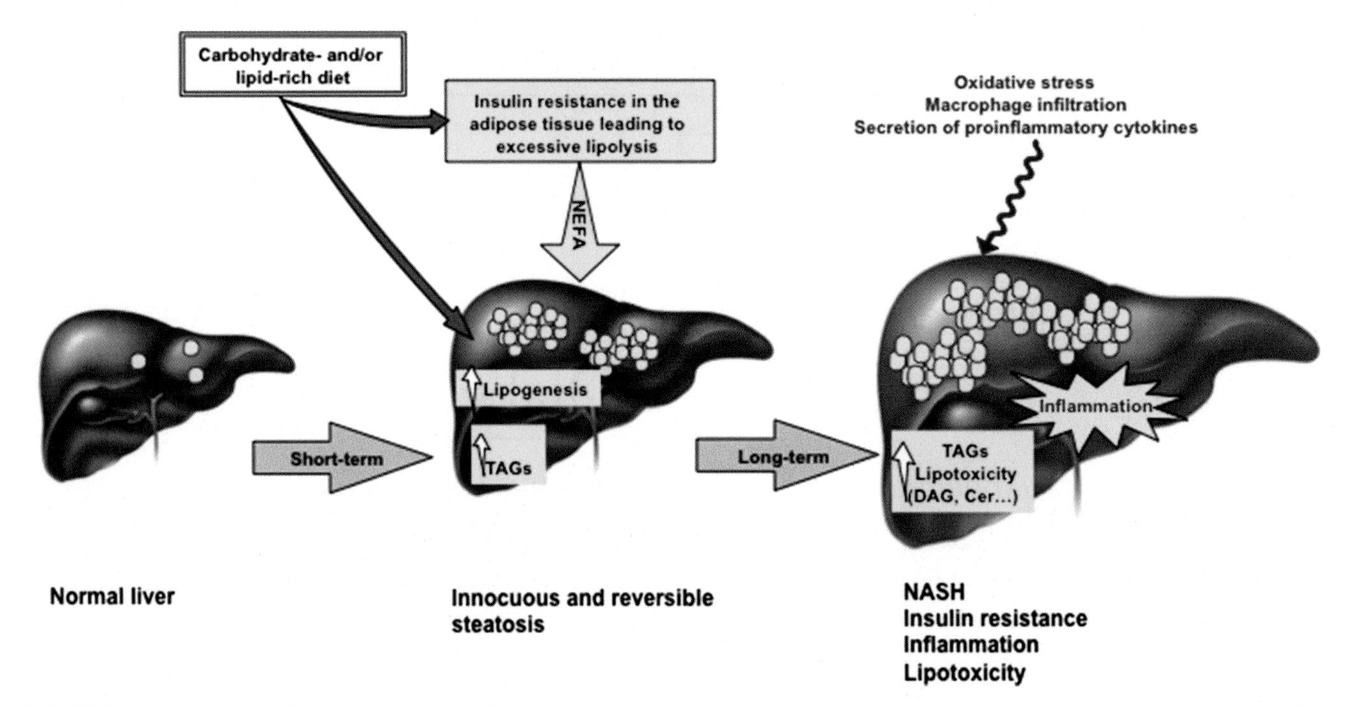

FIGURE 1 Diet-induced development of nonalcoholic steatohepatitis (NASH). Cer, ceramides; DAG, diacylglycerol; NEFA, nonesterified fatty acid; TAGs, triacylglycerols.

overactivation of hepatic lipogenesis[23,24]; and (4) an altered hepatic fatty acid oxidation capacity.[25–27]

2. LIPOGENESIS AS A TARGET TO REDUCE LIVER TRIACYLGLYCEROL CONTENT

Several epidemiological studies have shown an inverse relationship between physical activity, NAFLD, insulin resistance, and T2D in obese subjects,[28,29] suggesting that decreasing hepatic TAGs would be a viable approach to counteract hepatic steatosis and thus improve insulin sensitivity.

In this framework, several reports indicated that the liver-specific reduction of lipogenesis through inactivation of lipogenic key enzymes or transcription factors caused a substantial reduction in hepatic TAG content and improved glucose tolerance and insulin sensitivity in animal models of obesity and T2D.[23] In fact, acetyl-CoA carboxylase 1 or 2 knocking down by antisense oligonucleotide-based methodology reduced liver TAG content and hepatic glucose production and improved hepatic insulin sensitivity in high-fat (HF) diet-fed mice.[30] Inhibition of stearoyl-CoA desaturase 1, a key lipogenic enzyme catalyzing the synthesis of monounsaturated LCFA, in the liver using antisense oligonucleotide also prevented HF diet-induced hepatic steatosis, obesity, and T2D.[31] Moreover, targeting the lipogenic pathway at the transcription level by a liver-specific downregulation of the transcription factor carbohydrate responsive element binding protein in mice with genetic (leptin-deficient *ob/ob* mice)- or diet-induced obesity resulted in a substantial reduction in liver TAG content. This was accompanied with improved glucose tolerance and insulin sensitivity.[32] Interestingly, in almost all the studies targeting hepatic lipogenesis, the inhibition of de novo LCFA synthesis was associated with a decrease in the levels of malonyl-CoA, the substrate of fatty acid synthase and the physiological inhibitor of CPT1A.[33,34] In fact, disruption of hepatic lipogenesis was usually accompanied by an increase in hepatic LCFA oxidation as reflected by an increase in the circulating levels of ketone bodies represented by β-hydroxybutyrate.[32,35] Therefore, indirect stimulation of hepatic LCFA oxidation might have contributed to the reduction of hepatic steatosis and thus induced an improvement in glucose tolerance and insulin sensitivity. The decreased capacity of steatotic livers in genetic-[36] or diet-induced obese mice[37–39] to oxidize LCFA supports this interpretation.

The data from human subjects with NASH are, however, controversial. It has been reported that the activity of CPT1A in liver biopsies of patients with NASH did not differ from those of the controls.[40] However, NASH patients usually exhibit high oxidative stress, an impaired antioxidant system,[41–43] and a strong reduction in the oxidative phosphorylation capacity.[40,44,45]

In other cohorts of patients with NAFLD that included NASH and simple steatosis, the authors reported a downregulation of CPT1A at the gene level and an upregulation of the long-chain-CoA dehydrogenase.[46,47] In contrast, Nagaya et al.[48] did not observe any change in the expression of genes related to the uptake, oxidation, or export of LCFA in the livers of patients with either hepatic steatosis or NASH. Rather, they reported an impaired expression of lipogenesis enzymes (such as fatty acid synthase) and of the transcription factor sterol regulatory element-binding protein-1c.[48] In another recent study,[49] the assessment of fatty acid oxidation capacity after intralipid infusion in healthy subjects and NASH patients revealed a higher β-hydroxybutyrate/nonesterified fatty acid ratio, indicating a higher oxidative capacity in NASH patients.[49] Therefore, these human studies are inconclusive, and a study assessing the LCFA oxidation capacity of liver biopsy samples from patients diagnosed with different stages of NAFLD (simple steatosis, NASH), associated or not with T2D, is still required. Such a study would contribute to a better understanding of the relationship between hepatic fatty acid oxidation, NASH, and insulin resistance in humans.

The crucial role of LCFA oxidation for gluconeogenesis during fasting made this pathway a potential target for reducing hepatic glucose production in T2D patients characterized by uncontrolled gluconeogenesis. Strategies aiming at inhibiting LCFA oxidation have been developed by using specific inhibitors of CPT1A.[50] Long-term treatment with Teglicar, a reversible CPT1A inhibitor, improved glucose tolerance and insulin sensitivity in *ob/ob* mice, but did not reduce hepatic steatosis, and there was no evaluation of the nature of lipids in the liver after the treatment.[51]

3. STIMULATION OF THE PEROXISOME PROLIFERATOR-ACTIVATED RECEPTOR-α

Peroxisome proliferator-activated receptor (PPAR)-α is a major regulator of fatty acid oxidation at the transcriptional level. It is believed to act as a sensor of LCFAs, which are its preferential ligands. This capacity to sense LCFA availability allows to adjust LCFA oxidation depending on the feeding state.[52] The critical role played by PPAR-α in hepatic lipid metabolism has been underlined by the knockdown of PPAR-α in mice. Indeed, during fasting, these mice develop severe hepatic steatosis associated with hypoketonemia and defective fatty acid oxidation.[53] Therefore, PPAR-α appears to be a potential target to increase hepatic fatty acid oxidation and counteract TAG accumulation in the liver. PPAR-α can be stimulated by some endogenous ligands such as unsaturated LCFA[54] and oleoylethanolamide[55] and by synthetic agonists such as fibrates. Fibrates have been used as lipid-lowering drugs, showing a strong efficacy to lower circulating TAG levels and to moderately affect circulating levels of high-density lipoproteins levels.[56] In another study, the chronic treatment of mice fed a fructose-rich diet with fenofibrate prevented the development of glucose intolerance and insulin resistance.[57] This treatment increased liver fatty acid oxidation, which decreased the accumulation of TAGs and toxic lipids such as diacylglycerols and ceramides in the liver in spite of sustained lipogenesis.[57] These data support the concept that enhancing hepatic fatty acid oxidation can help to prevent hepatic insulin resistance. Nevertheless, despite the clear stimulatory effect of fenofibrate treatment on hepatic mitochondrial fatty acid oxidation, PPAR-α also controls other metabolic oxidative pathways such as peroxisomal fatty acid oxidation and microsomal fatty acid ω-oxidation catalyzed by the cytochrome P450 CYP4A subfamily.[25,58] These two metabolic pathways only account for a small part of the overall fatty acid oxidation, and mitochondrial fatty acid oxidation remains the major site for the fatty acid−derived energy. Nevertheless, sustained stimulation of PPAR-α by agonists can lead in mice to excessive generation of reactive oxygen species and dicarboxylic acids known to inhibit mitochondrial fatty acid oxidation.[25] This process can induce DNA damage, endoplasmic reticulum stress, and inflammation and then progress into hepatocyte damage and induce hepatocarcinogenesis.[59] On the other hand, it is not clear where the peripherally administered PPAR-α agonists act. Recent data obtained from rodents suggest that the beneficial effects of fibrate administration in lowering the circulating TAG levels is to a large part from enhanced fatty acid oxidation in intestinal epithelial cells.[60] Furthermore, intraperitoneal administration of either the endogenous PPAR-α agonist, oleoylethanolamide, or the synthetic one (Wy-14643) appeared to stimulate fatty acid oxidation and ketogenesis specifically in the jejunum, but not in the liver.[61,62] These findings extended our understanding about the site of action of the PPAR-α agonists and the possible contribution of intestinal fatty acid oxidation to lipid homeostasis.

4. PEROXISOME PROLIFERATOR-ACTIVATED RECEPTOR-γ COACTIVATOR-1 AS TARGET TO STIMULATE HEPATIC LONG-CHAIN FATTY ACID OXIDATION

As mentioned previously, several studies showed that NASH development is characterized by mitochondrial oxidative damage leading to impaired oxidative phosphorylation capacity.[40,44,45,63] Therefore, other strategies have been developed to counteract mitochondrial deficiency. The transcriptional factor peroxisome proliferator-activated receptor-γ coactivator (PGC)-1α appeared to be the ideal candidate given its effect on mitochondrial biogenesis.[64] Moreover, PGC-1α confers to the liver the flexibility of adaptation to food deprivation by controlling key regulatory metabolic pathways such as gluconeogenesis, mitochondrial fatty acid oxidation, and ketogenesis. This occurred via other transcriptional factors such as PPAR-α for fatty acid oxidation.[65] Morris and colleagues have shown that adenovirus-mediated expression of PGC-1α in cultured rat hepatocytes efficiently induced an increase in mitochondria content as well as in the expression of oxidative phosphorylation complexes. This effect was accompanied by a 30% increased mitochondrial fatty acid oxidation capacity. When expressed in vivo in rat liver, this resulted in a substantial decrease in hepatic TAGs content and circulating TAG levels as well as in a decreased expression of ApoB48, the main apolipoprotein of the very low-density lipoproteins.[66] Another study targeted the transcriptional factor PGC-1β, another member of the PGC-1 family. PGC-1β is also expressed in the liver, shares the effects of PGC-1α on mitochondrial biogenesis, and is also involved in lipogenesis.[67,68] It has been reported that liver-specific expression of PGC-1β in mice fed a methionine and choline–deficient diet, a diet model used to induce NASH,[69] or an HF diet, efficiently induced mitochondria biogenesis. Furthermore, an increase in hepatic fatty acid oxidation together with lipogenesis at the transcriptional level was noticed. These effects occurred together with a higher capacity to export TAGs into the circulation and with a reduced liver TAGs content. Also, the effects resulted in an attenuation of liver steatosis, oxidative stress, and inflammation.[70] This study did not, however, report any data concerning hepatic insulin resistance. Nevertheless, overall, the data from both studies suggest that targeting PGC-1s would be a potential strategy to counteract diet-induced NASH and therefore correct insulin resistance.

5. TARGETING LIVER MITOCHONDRIAL FATTY ACID OXIDATION TO IMPROVE HEPATIC INSULIN SENSITIVITY

The studies mentioned clearly demonstrated that enhancing hepatic fatty acid oxidation would be an efficient strategy to improve insulin sensitivity in hepatic steatosis. As mentioned previously, however, stimulation of either PPAR-α or PGC-1s results in an upregulation of several metabolic pathways at the same time, and sustained PPAR-α stimulation for instance might result in deleterious effects. Consequently, a different strategy was adopted by directly targeting CPT1A, the key regulatory enzyme of mitochondrial fatty acid oxidation. In fact, mitochondrial fatty acid oxidation is tightly controlled via CPT1A that catalyzes the first step in the transport of LC-CoAs across the mitochondrial membranes to be oxidized in the mitochondrial matrix. Indeed, CPT1A is sensitive to inhibition by malonyl-CoA, an intermediate of de novo fatty acid synthesis, and this prevents concurrent oxidation of glucose and fatty acids.[71,72] A first attempt to enhance hepatic fatty acid oxidation by targeting CPT1A was reported by Stefanovic-Racic et al. They overexpressed the liver isoform of CPT1A specifically in the liver using an adenovirus approach and compared its effects with those of β-galactosidase expression in rats fed either a control or HF diet. CPT1A overexpression caused an increase in hepatic palmitate oxidation flux and ketone body production in both diet groups.[73] Moreover, a reduction in hepatic TAG content was observed in the group fed an HF diet. There was, however, no improvement in insulin sensitivity in HF diet-fed rats overexpressing CPT1A.[73] The possible reason for the moderate effect on insulin sensitivity might be because the CPT1A overexpressed is sensitive to malonyl-CoA inhibition (as the endogenous CPT1A). Therefore, as long as CPT1A is posttranslationally inhibited by malonyl-CoA, an increase in the protein level of CPT1A would have only a moderate effect. The need for losing the sensitivity to malonyl-CoA became as a crucial step to be able to stimulate FAO by targeting CPT1A.

Studies aimed at deciphering the structure–function relationships of rat CPT1A identified amino acid residues critical for malonyl-CoA inhibition.[74] Mutation of methionine 593 to serine generated a CPT1A that is active, but insensitive to malonyl-CoA inhibition.[74] Such a gain-of-function mutation was used as a tool to enhance in vitro and in vivo mitochondrial fatty acid oxidation independent of the intracellular concentrations of malonyl-CoA. Indeed, cultured rat hepatocytes that expressed this mutated form of CPT1A (CPT1Am) oxidized LCFA more efficiently than hepatocytes overexpressing the wild-type form.[75] Moreover, this high capacity to oxidize LCFA was maintained even in the presence of high levels of malonyl-CoA favored by high insulin and glucose concentrations.[75] The ability of this CPT1Am model to efficiently enhance fatty acid oxidation was also proven in the L6E9[76] and C2C12[76] muscle cell lines. There, the adenovirus-mediated expression of CPT1Am in myotubes increased mitochondrial fatty acid oxidation flux[76,77] and prevented palmitate-induced apoptosis[76] and insulin resistance.[76,77]

These in vitro models demonstrated the feasibility of interfering in the CPT1A/malonyl-CoA partnership to keep a sustained fatty acid oxidation flux independent of the presence or absence of malonyl-CoA. This concept has been applied in vivo, and Orellana-Gavalda et al.[78] expressed CPT1Am in mouse liver using adeno-associated virus, allowing a long-term expression of the mutant form in mice fed a control or HF diet. Mice expressing CPT1Am exhibited an increased hepatic fatty acid oxidation flux and ketogenesis. This higher hepatic fatty acid oxidation capacity prevented TAG accumulation in the liver and improved insulin sensitivity in HF diet–fed mice. Interestingly, the stimulation of hepatic fatty acid oxidation capacity was not associated with any change in the inflammatory state or oxidative stress.[78] This study clearly demonstrates that increasing fatty acid oxidation capacity of the liver can prevent hepatic TAG accumulation and insulin resistance when the animal is challenged with an obesity-inducing diet. We further demonstrated the relevance of this strategy by showing that adenovirus-mediated hepatic expression of CPT1Am in mice fed either an HF or high-sucrose diet for 20 weeks or *ob/ob* mice reversed the insulin resistance and glucose intolerance in these models.[37] This notable improvement in insulin sensitivity was also associated with reduced hepatic oxidative stress as well as lower hepatic contents in free fatty acids, diacylglycerols, and ceramides known to impair insulin signaling. Unlike in the previous report in which CPT1Am was expressed in the liver before the HF diet challenge,[55] there was, however, no change in hepatic TAG content. These findings clearly support the view that the accumulation of neutral lipids represented by TAG is not responsible for insulin resistance[79,80]; rather, the accumulation of toxic ceramides and diacylglycerols may be the culprit.[81–84]

6. GENERAL CONCLUSION

Overall, the previously mentioned studies concerning the effect of enhancing hepatic LCFA oxidation on preventing or correcting insulin resistance in various animal models of obesity and T2D confirm two major points: (1) Hepatic lipid metabolism, and in particular mitochondrial LCFA oxidation, emerges as a valuable target for the treatment of insulin resistance; and (2) the toxic lipids (e.g., ceramides, DAG), rather than the neutral lipid (TAGs), constitute the main trigger of insulin resistance in the liver (Figure 2).

These findings reiterate that hepatic fatty acid oxidation could be a promising target in the treatment and management of T2D. More specifically, based on these findings, drugs that can interfere with the binding of malonyl-CoA to CPT1A in the liver could be a novel and potentially useful approach in this context. Previous attempts to modify fatty acid oxidation used other targets, such as the drug C81b, which was developed as a specific stimulator of CPT1 activity.[85] Furthermore, some plant extracts such as resveratrol have been reported to efficiently stimulate hepatic fatty acid oxidation and to prevent NAFLD development and T2D in animal models.[86,87]

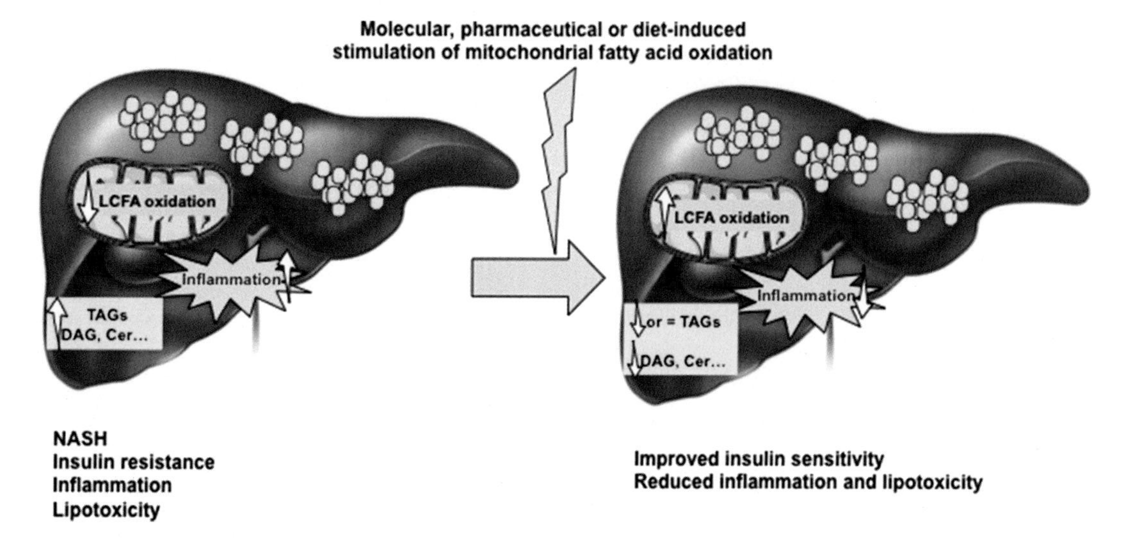

FIGURE 2 **Targeting hepatic long-chain fatty acid oxidation for the treatment of insulin resistance in the context of NAFLD.** Cer, ceramides; DAG, diacylglycerol; LCFA, long-chain fatty acid; NAFLD, nonalcoholic fatty liver disease; NASH, nonalcoholic steatohepatitis; TAGs, triacylglycerols.

References

1. McGarry JD, Foster DW. Regulation of hepatic fatty acid oxidation and ketone body production. *Annu Rev Biochem* 1980;**49**:395–420.
2. Nguyen P, Leray V, Diez M, Serisier S, Le Bloc'h J, Siliart B, et al. Liver lipid metabolism. *J Anim Physiol Anim Nutr (Berl)* 2008;**92**(3):272–83.
3. Postic C, Girard J. The role of the lipogenic pathway in the development of hepatic steatosis. *Diabetes Metab* 2008;**34**(6 Pt 2):643–8.
4. Xiao C, Hsieh J, Adeli K, Lewis GF. Gut-liver interaction in triglyceride-rich lipoprotein metabolism. *Am J Physiol Endocrinol Metab* 2011;**301**(3):E429–46.
5. Oh KJ, Han HS, Kim MJ, Koo SH. CREB and FoxO1: two transcription factors for the regulation of hepatic gluconeogenesis. *BMB Rep* 2013;**46**(12):567–74.
6. Malaguarnera M, Di Rosa M, Nicoletti F, Malaguarnera L. Molecular mechanisms involved in NAFLD progression. *J Mol Med* 2009;**87**(7):679–95.
7. Birkenfeld AL, Shulman GI. Nonalcoholic fatty liver disease, hepatic insulin resistance, and type 2 diabetes. *Hepatology* 2014;**59**(2):713–23.
8. Gruben N, Shiri-Sverdlov R, Koonen DPY, Hofker MH. Nonalcoholic fatty liver disease: A main driver of insulin resistance or a dangerous liaison? *BBA Mol Basis Dis* 2014;**1842**(11):2329–43.
9. Preiss D, Sattar N. Non-alcoholic fatty liver disease: an overview of prevalence, diagnosis, pathogenesis and treatment considerations. *Clin Sci (Lond)* 2008;**115**(5):141–50.
10. Vanni E, Bugianesi E, Kotronen A, De Minicis S, Yki-Jarvinen H, Svegliati-Baroni G. From the metabolic syndrome to NAFLD or vice versa? *Dig Liver Dis* 2010;**42**(5):320–30.
11. Farrell GC, Larter CZ. Nonalcoholic fatty liver disease: from steatosis to cirrhosis. *Hepatology* 2006;**43**(2 Suppl. 1):S99–112.
12. Ruhl CE, Everhart JE. Epidemiology of nonalcoholic fatty liver. *Clin Liver Dis* 2004;**8**(3):501–19. vii.
13. Lazo M, Clark JM. The epidemiology of nonalcoholic fatty liver disease: a global perspective. *Semin Liver Dis* 2008;**28**(4):339–50.
14. Bedogni G, Miglioli L, Masutti F, Tiribelli C, Marchesini G, Bellentani S. Prevalence of and risk factors for nonalcoholic fatty liver disease: the Dionysos nutrition and liver study. *Hepatology* 2005;**42**(1):44–52.
15. Williams CD, Stengel J, Asike MI, Torres DM, Shaw J, Contreras M, et al. Prevalence of nonalcoholic fatty liver disease and nonalcoholic steatohepatitis among a largely middle-aged population utilizing ultrasound and liver biopsy: a prospective study. *Gastroenterology* 2011;**140**(1):124–31.
16. Doycheva I, Patel N, Peterson M, Loomba R. Prognostic implication of liver histology in patients with nonalcoholic fatty liver disease in diabetes. *J Diabetes Complications* 2013;**27**(3):293–300.
17. Day CP, James OF. Steatohepatitis: a tale of two "hits"? *Gastroenterology* 1998;**114**(4):842–5.
18. Tilg H, Moschen AR. Evolution of inflammation in nonalcoholic fatty liver disease: the multiple parallel hits hypothesis. *Hepatology* 2010;**52**(5):1836–46.
19. Choudhury J, Sanyal AJ. Insulin resistance in NASH. *Front Biosci* 2005;**10**:1520–33.
20. Byrne CD, Olufadi R, Bruce KD, Cagampang FR, Ahmed MH. Metabolic disturbances in non-alcoholic fatty liver disease. *Clin Sci (Lond)* 2009;**116**(7):539–64.
21. Donnelly KL, Smith CI, Schwarzenberg SJ, Jessurun J, Boldt MD, Parks EJ. Sources of fatty acids stored in liver and secreted via lipoproteins in patients with nonalcoholic fatty liver disease. *J Clin Invest* 2005;**115**(5):1343–51.
22. Fujita K, Nozaki Y, Wada K, Yoneda M, Fujimoto Y, Fujitake M, et al. Dysfunctional very-low-density lipoprotein synthesis and release is a key factor in nonalcoholic steatohepatitis pathogenesis. *Hepatology* 2009;**50**(3):772–80.
23. Postic C, Girard J. Contribution of de novo fatty acid synthesis to hepatic steatosis and insulin resistance: lessons from genetically engineered mice. *J Clin Invest* 2008;**118**(3):829–38.
24. Diraison F, Moulin P, Beylot M. Contribution of hepatic de novo lipogenesis and reesterification of plasma non esterified fatty acids to plasma triglyceride synthesis during non-alcoholic fatty liver disease. *Diabetes Metab* 2003;**29**(5):478–85.
25. Reddy JK, Rao MS. Lipid metabolism and liver inflammation. II. Fatty liver disease and fatty acid oxidation. *Am J Physiol Gastrointest Liver Physiol* 2006;**290**(5):G852–8.
26. Begriche K, Massart J, Robin MA, Bonnet F, Fromenty B. Mitochondrial adaptations and dysfunctions in nonalcoholic fatty liver disease. *Hepatology* 2013;**58**(4):1497–507.
27. Nassir F, Ibdah JA. Role of mitochondria in nonalcoholic fatty liver disease. *Int J Mol Sci* 2014;**15**(5):8713–42.
28. Goncalves IO, Oliveira PJ, Ascensao A, Magalhaes J. Exercise as a therapeutic tool to prevent mitochondrial degeneration in nonalcoholic steatohepatitis. *Eur J Clin Invest* 2013;**43**(11):1184–94.
29. Zelber-Sagi S, Ratziu V, Oren R. Nutrition and physical activity in NAFLD: an overview of the epidemiological evidence. *World J Gastroenterol* 2011;**17**(29):3377–89.
30. Savage DB, Choi CS, Samuel VT, Liu ZX, Zhang D, Wang A, et al. Reversal of diet-induced hepatic steatosis and hepatic insulin resistance by antisense oligonucleotide inhibitors of acetyl-CoA carboxylases 1 and 2. *J Clin Invest* 2006;**116**(3):817–24.
31. Jiang G, Li Z, Liu F, Ellsworth K, Dallas-Yang Q, Wu M, et al. Prevention of obesity in mice by antisense oligonucleotide inhibitors of stearoyl-CoA desaturase-1. *J Clin Invest* 2005;**115**(4):1030–8.
32. Dentin R, Benhamed F, Hainault I, Fauveau V, Foufelle F, Dyck JR, et al. Liver-specific inhibition of ChREBP improves hepatic steatosis and insulin resistance in ob/ob mice. *Diabetes* 2006;**55**(8):2159–70.
33. McGarry JD, Leatherman GF, Foster DW. Carnitine palmitoyltransferase I. The site of inhibition of hepatic fatty acid oxidation by malonyl-CoA. *J Biol Chem* 1978;**253**(12):4128–36.
34. Bonnefont JP, Djouadi F, Prip-Buus C, Gobin S, Munnich A, Bastin J. Carnitine palmitoyltransferases 1 and 2: biochemical, molecular and medical aspects. *Mol Aspects Med* 2004;**25**(5–6):495–520.
35. Choi CS, Savage DB, Abu-Elheiga L, Liu ZX, Kim S, Kulkarni A, et al. Continuous fat oxidation in acetyl-CoA carboxylase 2 knockout mice increases total energy expenditure, reduces fat mass, and improves insulin sensitivity. *Proc Natl Acad Sci USA* 2007;**104**(42):16480–5.
36. Perfield 2nd JW, Ortinau LC, Pickering RT, Ruebel ML, Meers GM, Rector RS. Altered hepatic lipid metabolism contributes to nonalcoholic fatty liver disease in leptin-deficient Ob/Ob mice. *J Obes* 2013;**2013**:296537.
37. Monsenego J, Mansouri A, Akkaoui M, Lenoir V, Esnous C, Fauveau V, et al. Enhancing liver mitochondrial fatty acid oxidation capacity in obese mice improves insulin sensitivity independently of hepatic steatosis. *J Hepatol* 2012;**56**(3):632–9.
38. Vial G, Dubouchaud H, Couturier K, Cottet-Rousselle C, Taleux N, Athias A, et al. Effects of a high-fat diet on energy metabolism and ROS production in rat liver. *J Hepatol* 2011;**54**(2):348–56.
39. Garcia-Ruiz I, Solis-Munoz P, Fernandez-Moreira D, Grau M, Colina F, Munoz-Yague T, et al. High-fat diet decreases activity of the oxidative phosphorylation complexes and causes nonalcoholic steatohepatitis in mice. *Dis Models Mech* 2014;**7**(11):1287–96.

40. Perez-Carreras M, Del Hoyo P, Martin MA, Rubio JC, Martin A, Castellano G, et al. Defective hepatic mitochondrial respiratory chain in patients with nonalcoholic steatohepatitis. *Hepatology* 2003;**38**(4):999–1007.
41. Koruk M, Taysi S, Savas MC, Yilmaz O, Akcay F, Karakok M. Oxidative stress and enzymatic antioxidant status in patients with nonalcoholic steatohepatitis. *Ann Clin Lab Sci* 2004;**34**(1): 57–62.
42. Videla LA, Rodrigo R, Orellana M, Fernandez V, Tapia G, Quinones L, et al. Oxidative stress-related parameters in the liver of non-alcoholic fatty liver disease patients. *Clin Sci (Lond)* 2004; **106**(3):261–8.
43. Perlemuter G, Davit-Spraul A, Cosson C, Conti M, Bigorgne A, Paradis V, et al. Increase in liver antioxidant enzyme activities in non-alcoholic fatty liver disease. *Liver Int* 2005;**25**(5):946–53.
44. Cortez-Pinto H, Chatham J, Chacko VP, Arnold C, Rashid A, Diehl AM. Alterations in liver ATP homeostasis in human nonalcoholic steatohepatitis: a pilot study. *JAMA* 1999;**282**(17):1659–64.
45. Chiappini F, Barrier A, Saffroy R, Domart MC, Dagues N, Azoulay D, et al. Exploration of global gene expression in human liver steatosis by high-density oligonucleotide microarray. *Lab Invest* 2006;**86**(2):154–65.
46. Nakamuta M, Kohjima M, Morizono S, Kotoh K, Yoshimoto T, Miyagi I, et al. Evaluation of fatty acid metabolism-related gene expression in nonalcoholic fatty liver disease. *Int J Mol Med* 2005; **16**(4):631–5.
47. Kohjima M, Enjoji M, Higuchi N, Kato M, Kotoh K, Yoshimoto T, et al. Re-evaluation of fatty acid metabolism-related gene expression in nonalcoholic fatty liver disease. *Int J Mol Med* 2007;**20**(3): 351–8.
48. Nagaya T, Tanaka N, Suzuki T, Sano K, Horiuchi A, Komatsu M, et al. Down-regulation of SREBP-1c is associated with the development of burned-out NASH. *J Hepatol* 2010;**53**(4):724–31.
49. Dasarathy S, Yang Y, McCullough AJ, Marczewski S, Bennett C, Kalhan SC. Elevated hepatic fatty acid oxidation, high plasma fibroblast growth factor 21, and fasting bile acids in nonalcoholic steatohepatitis. *Eur J Gastroenterol Hepatol* 2011;**23**(5):382–8.
50. Ceccarelli SM, Chomienne O, Gubler M, Arduini A. Carnitine palmitoyltransferase (CPT) modulators: a medicinal chemistry perspective on 35 years of research. *J Med Chem* 2011;**54**(9): 3109–52.
51. Conti R, Mannucci E, Pessotto P, Tassoni E, Carminati P, Giannessi F, et al. Selective reversible inhibition of liver carnitine palmitoyl-transferase 1 by teglicar reduces gluconeogenesis and improves glucose homeostasis. *Diabetes* 2011;**60**(2):644–51.
52. Desvergne B, Wahli W. Peroxisome proliferator-activated receptors: nuclear control of metabolism. *Endocr Rev* 1999;**20**(5):649–88.
53. Hashimoto T, Cook WS, Qi C, Yeldandi AV, Reddy JK, Rao MS. Defect in peroxisome proliferator-activated receptor alpha-inducible fatty acid oxidation determines the severity of hepatic steatosis in response to fasting. *J Biol Chem* 2000;**275**(37):28918–28.
54. Grygiel-Gorniak B. Peroxisome proliferator-activated receptors and their ligands: nutritional and clinical implications—a review. *Nutr J* 2014;**13**:17.
55. Fu J, Gaetani S, Oveisi F, Lo Verme J, Serrano A, Rodriguez De Fonseca F, et al. Oleylethanolamide regulates feeding and body weight through activation of the nuclear receptor PPAR-alpha. *Nature* 2003;**425**(6953):90–3.
56. Watts GF, Dimmitt SB. Fibrates, dyslipoproteinaemia and cardiovascular disease. *Curr Opin Lipidol* 1999;**10**(6):561–74.
57. Chan SM, Sun RQ, Zeng XY, Choong ZH, Wang H, Watt MJ, et al. Activation of PPARalpha ameliorates hepatic insulin resistance and steatosis in high fructose-fed mice despite increased endoplasmic reticulum stress. *Diabetes* 2013;**62**(6):2095–105.
58. Reddy JK, Hashimoto T. Peroxisomal beta-oxidation and peroxisome proliferator-activated receptor alpha: an adaptive metabolic system. *Annu Rev Nutr* 2001;**21**:193–230.
59. Misra P, Reddy JK. Peroxisome proliferator-activated receptor-alpha activation and excess energy burning in hepatocarcinogenesis. *Biochimie* 2014;**98**:63–74.
60. Uchida A, Slipchenko MN, Cheng JX, Buhman KK. Fenofibrate, a peroxisome proliferator-activated receptor alpha agonist, alters triglyceride metabolism in enterocytes of mice. *Biochim Biophys Acta* 2011;**1811**(3):170–6.
61. Karimian Azari E, Ramachandran D, Weibel S, Arnold M, Romano A, Gaetani S, et al. Vagal afferents are not necessary for the satiety effect of the gut lipid messenger oleoylethanolamide (OEA). *Am J Physiol Regul Integr Comp Physiol* 2014.
62. Karimian Azari E, Leitner C, Jaggi T, Langhans W, Mansouri A. Possible role of intestinal fatty acid oxidation in the eating-inhibitory effect of the PPAR-alpha agonist Wy-14643 in high-fat diet fed rats. *PLoS One* 2013;**8**(9):e74869.
63. Morris EM, Rector RS, Thyfault JP, Ibdah JA. Mitochondria and redox signaling in steatohepatitis. *Antioxid Redox Signal* 2011; **15**(2):485–504.
64. Liang HY, Ward WF. PGC-1alpha: a key regulator of energy metabolism. *Adv Physiol Educ* 2006;**30**(4):145–51.
65. Lin JD, Handschin C, Spiegelman BM. Metabolic control through the PGC-1 family of transcription coactivators. *Cell Metab* 2005; **1**(6):361–70.
66. Morris EM, Meers GM, Booth FW, Fritsche KL, Hardin CD, Thyfault JP, et al. PGC-1alpha overexpression results in increased hepatic fatty acid oxidation with reduced triacylglycerol accumulation and secretion. *Am J Physiol Gastrointest Liver Physiol* 2012; **303**(8):G979–92.
67. Lin J, Yang R, Tarr PT, Wu PH, Handschin C, Li S, et al. Hyperlipidemic effects of dietary saturated fats mediated through PGC-1 beta coactivation of SREBP. *Chem Phys Lipids* 2005;**136**(2):95.
68. Liu C, Lin JD. PGC-1 coactivators in the control of energy metabolism. *Acta Biochim Biophys Sin* 2011;**43**(4):248–57.
69. Rinella ME, Elias MS, Smolak RR, Fu T, Borensztajn J, Green RM. Mechanisms of hepatic steatosis in mice fed a lipogenic methionine choline-deficient diet. *J Lipid Res* 2008;**49**(5):1068–76.
70. Bellafante E, Murzilli S, Salvatore L, Latorre D, Villani G, Moschetta A. Hepatic-specific activation of peroxisome proliferator-activated receptor gamma coactivator-1beta protects against steatohepatitis. *Hepatology* 2013;**57**(4):1343–56.
71. Zammit VA. Carnitine palmitoyltransferase 1: central to cell function. *IUBMB Life* 2008;**60**(5):347–54.
72. McGarry JD, Woeltje KF, Kuwajima M, Foster DW. Regulation of ketogenesis and the renaissance of carnitine palmitoyltransferase. *Diabetes Metab Rev* 1989;**5**(3):271–84.
73. Stefanovic-Racic M, Perdomo G, Mantell BS, Sipula IJ, Brown NF, O'Doherty RM. A moderate increase in carnitine palmitoyltransferase 1a activity is sufficient to substantially reduce hepatic triglyceride levels. *Am J Physiol Endocrinol Metab* 2008; **294**(5):E969–77.
74. Morillas M, Gomez-Puertas P, Bentebibel A, Selles E, Casals N, Valencia A, et al. Identification of conserved amino acid residues in rat liver carnitine palmitoyltransferase I critical for malonyl-CoA inhibition. Mutation of methionine 593 abolishes malonyl-CoA inhibition. *J Biol Chem* 2003;**278**(11):9058–63.
75. Akkaoui M, Cohen I, Esnous C, Lenoir V, Sournac M, Girard J, et al. Modulation of the hepatic malonyl-CoA-carnitine palmitoyltransferase 1A partnership creates a metabolic switch allowing oxidation of de novo fatty acids. *Biochem J* 2009;**420**(3): 429–38.

76. Sebastian D, Herrero L, Serra D, Asins G, Hegardt FG. CPT I overexpression protects L6E9 muscle cells from fatty acid-induced insulin resistance. *Am J Physiol Endocrinol Metab* 2007;**292**(3):E677–86.
77. Henique C, Mansouri A, Fumey G, Lenoir V, Girard J, Bouillaud F, et al. Increased mitochondrial fatty acid oxidation is sufficient to protect skeletal muscle cells from palmitate-induced apoptosis. *J Biol Chem* 2010;**285**(47):36818–27.
78. Orellana-Gavalda JM, Herrero L, Malandrino MI, Paneda A, Sol Rodriguez-Pena M, Petry H, et al. Molecular therapy for obesity and diabetes based on a long-term increase in hepatic fatty-acid oxidation. *Hepatology* 2011;**53**(3):821–32.
79. Sun Z, Lazar MA. Dissociating fatty liver and diabetes. *Trends Endocrinol Metab* 2013;**24**(1):4–12.
80. Benhamed F, Denechaud PD, Lemoine M, Robichon C, Moldes M, Bertrand-Michel J, et al. The lipogenic transcription factor ChREBP dissociates hepatic steatosis from insulin resistance in mice and humans. *J Clin Invest* 2012;**122**(6):2176–94.
81. Carobbio S, Rodriguez-Cuenca S, Vidal-Puig A. Origins of metabolic complications in obesity: ectopic fat accumulation. The importance of the qualitative aspect of lipotoxicity. *Curr Opin Clin Nutr Metab Care* 2011;**14**(6):520–6.
82. Monetti M, Levin MC, Watt MJ, Sajan MP, Marmor S, Hubbard BK, et al. Dissociation of hepatic steatosis and insulin resistance in mice overexpressing DGAT in the liver. *Cell Metab* 2007;**6**(1):69–78.
83. Alkhouri N, Dixon LJ, Feldstein AE. Lipotoxicity in nonalcoholic fatty liver disease: not all lipids are created equal. *Expert Rev Gastroenterol Hepatol* 2009;**3**(4):445–51.
84. Samuel VT, Shulman GI. Mechanisms for insulin resistance: common threads and missing links. *Cell* 2012;**148**(5):852–71.
85. Aja S, Landree LE, Kleman AM, Medghalchi SM, Vadlamudi A, McFadden JM, et al. Pharmacological stimulation of brain carnitine palmitoyl-transferase-1 decreases food intake and body weight. *Am J Physiol Regul Integr Comp Physiol* 2008;**294**(2):R352–61.
86. Carpene C, Gomez-Zorita S, Deleruyelle S, Carpene MA. Novel strategies for preventing diabetes and obesity complications with natural polyphenols. *Curr Med Chem* 2015;**22**(1):150–64.
87. Xiao J, Fai So K, Liong EC, Tipoe GL. Recent advances in the herbal treatment of non-alcoholic Fatty liver disease. *J Tradit Complement Med* 2013;**3**(2):88–94.

CHAPTER

29

Current Knowledge on the Role of Wnt Signaling Pathway in Glucose Homeostasis

Tianru Jin

Division of Advanced Diagnostics, Toronto General Research Institute, University Health Network, Toronto, ON, Canada

1. INTRODUCTION OF THE WNT SIGNALING PATHWAY

In 1982, Nusse and Varmus presented their study on the first Wnt ligand-encoding gene *INT1* in their breast cancer research.[1] This proto-oncogene was later renamed as *WNT1* because it shares a strong amino acid sequence homology with the *Drosophila Wingless*, which is important for segment polarity of this insect.[2] Scientists in embryological development studies in frogs (*Xenopus*) and other organisms have also made important contributions in illustrating the function and biological importance of this evolutionarily conserved signaling pathway.[3–9] The physiological role of the Wnt signaling cascade in metabolic homeostasis and its implication in the development and progression of metabolic disorders including type 2 diabetes (T2D), however, have received broad appreciation only in the past decade.[10–12]

In mammals, 19 genes encode different Wnt ligands. The first receptor family for Wnt ligands is called the Frizzled proteins, which are seven-transmembrane receptors. In addition to 10 known genes that encode the Frizzled receptors, smoothened, a G protein–coupled receptor, also belongs to the Frizzled receptor family. Smoothened is an essential component of the hedgehog signaling cascade.[13] In *Drosophila*, the long single-pass transmembrane protein Arrow was shown to interact with Wnt ligands and is genetically required for the Wingless signaling.[14] Low-density lipoprotein receptor-related protein 5/6 (LRP5/6) is the homolog of the Arrow protein in mammals.[15]

When a Wnt ligand binds to a Frizzled receptor on the cell membrane, it leads to the activation of either the canonical (β catenin–dependent) or noncanonical (β catenin–independent) Wnt signaling pathway. The latter can be further subdivided into the planar cell polarity and Ca^{2+} pathways, which are important for polarization of cells along the embryonic axis and cell motility and behavior, respectively.[16] For this review, only the physiological importance of the canonical or β catenin–dependent Wnt signaling pathway (defined as Wnt pathway hereafter, unless further clarification is necessary) in glucose homeostasis will be discussed.

The key effector of the Wnt signaling pathway is the bipartite transcription factor β-catenin/TCF (or cat/TCF), formed by free β-catenin (β-cat) and a member of the TCF transcription factor family (TCF7L2 [= TCF-4],[10,17,18] TCF7, TCF7L1, and LEF-1). TCFs possess the high-mobility group (HMG) DNA-binding domain, whereas β-cat provides the transcriptional activation domains. As shown in Figure 1(a), in resting cells, the cytoplasmic free β-cat level is tightly controlled by the proteasome-mediated degradation process. This involves the actions of the tumor suppressor adenomatous polyposis coli (APC) and Axin/Conductin as well as the serine/threonine kinases glycogen synthase kinase-3 (GSK-3) and casein kinase Iα (CK-Iα).[19,20] Whereas Axin and APC serve as the scaffold, GSK-3 and CK-Iα phosphorylate several serine residues at an N-terminal portion of β-cat, including the S33 residue. β-cat phosphorylated at these residues will then undergo proteasome degradation. Because the S33 residue phosphorylation is important for its proteasome degradation, the S33Y β-cat mutant serves as a constitutively active molecule and has been broadly used in Wnt signaling pathway study.[10,18,21] In response to Wnt ligand stimulation, an association is established between the Wnt receptor and *Dishevelled*,

Molecular Nutrition and Diabetes
http://dx.doi.org/10.1016/B978-0-12-801585-8.00029-4

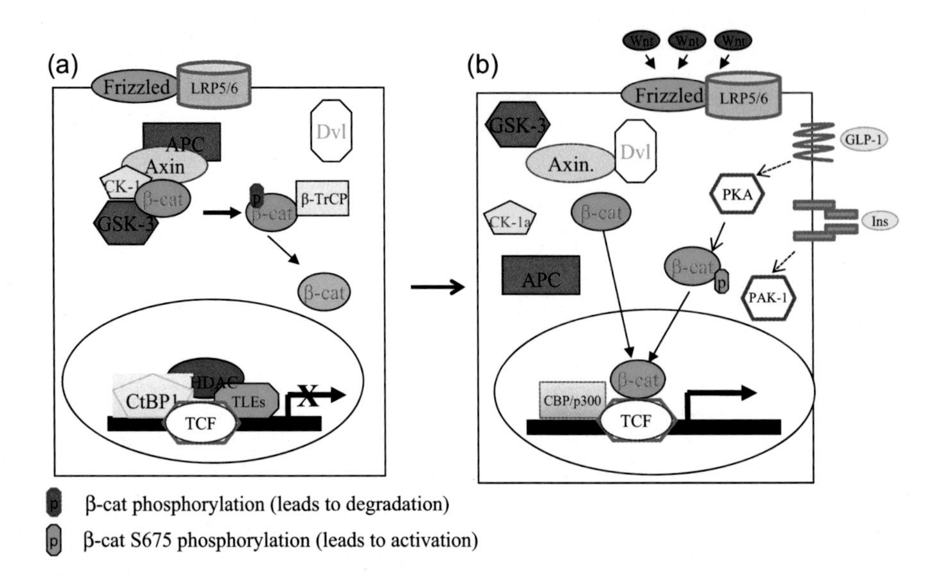

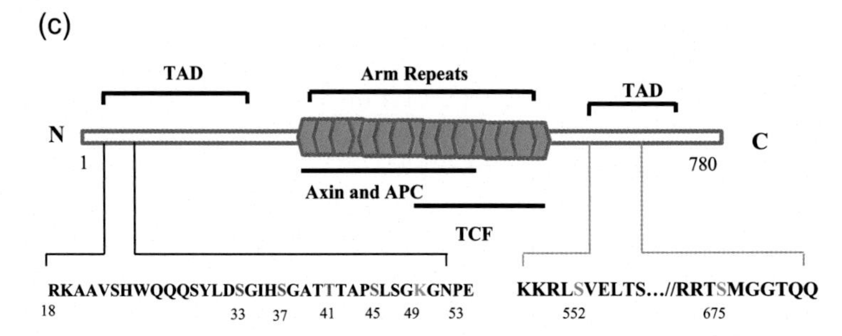

FIGURE 1 **Illustration of Wnt signaling pathway and its effector β-cat.** (a) In the absence of Wnt ligand, β-cat is phosphorylated at its N-terminal portion, leading to its proteasome degradation. This process involves the scaffold proteins Axin and APC as well as GSK-3 and CK-Iα. TCFs will recruit nuclear corepressors CtBP1, TLEs, and HDAC, repressing Wnt target gene expression. (b) After Wnt ligand stimulation, the degradation complex is dissociated. β-cat will enter into the nucleus, forming the bipartite transcription factor cat/TCF, stimulating Wnt target gene expression. (c) Illustration of functional domains of β-cat and its phosphorylation residues at the N-terminal portion and C-terminal portion. APC, adenomatous polyposis coli; β-cat, β-catenin; CtBP1, C-terminal-binding protein 1; GSK, glycogen synthase kinase; HDAC, histone deacetylases; TAD, transactivation domains; TLEs, transducin-like enhancers of split (the *Drosophila* ortholog is Groucho); PAK-1, P21-activated kinase 1.

an event that triggers the dissociation of the degradation complex of APC, Axin, and GSK-3. Free β-cat will then accumulate and enter the nucleus, resulting in the formation of the cat/TCF bipartite transcription factor and the activation of cat/TCF (or Wnt signaling) downstream target gene expression (Figure 1(b)). In the absence of β-cat, however, TCFs may repress Wnt target gene expression via recruiting certain nuclear corepressors, including histone deacetylases, C-terminal-binding protein 1, and transducin-like enhancer of split (the *Drosophila* ortholog is Groucho) (Figure 1(a)).

Studies in cancer research and related disciplines have shown the existence of cross talk between the Wnt pathway and the insulin/insulin growth factor-1 (IGF-1) signaling pathways. Although these two signaling pathways share the common component GSK-3, mechanistic explorations have revealed that the cross talk may not be mainly mediated by this protein kinase.[18,22,23] Investigations have demonstrated that insulin or IGF-1 may stimulate β-cat phosphorylation within its C-terminal portion, including the serine 675 and serine 552 residues (S675 and S552).[24–26] This event is also known to be facilitated by protein kinase A (PKA) and positively associated with the cat/TCF activity.[27–29] Recent investigations by our team and others have revealed that insulin and IGF-1 may stimulate β-cat S675 and S552 phosphorylation via activating P21-activated protein kinase-1 (Figure 1(b)).[28,30–33] Furthermore, Liu and Habener have demonstrated the stimulatory effect of glucagon-like peptide-1 (GLP-1) on β-cat S675 phosphorylation[34] (Figure 1(b)). Figure 1(c) shows the overall structure of β-cat, including the serine and threonine residues at the N-terminal portion that can be phosphorylated by GSK-3 and Ck-Iα as well as the two serine residues at the C-terminal portion, of which the phosphorylation stimulates the cat/TCF activity.

β-cat serves as the cofactor for Forkhead box (FoxO) as well, serving as effectors of the stress and aging signaling pathway.[35] Thus, the developmental Wnt signaling and the stress/aging signaling is connected by the competition between TCFs and FoxOs for a limited pool of β-cat as their common cofactor. Insulin and several growth factors, however, can activate PKB or serum- and glucocorticoid-regulated protein kinase, resulting in FoxO phosphorylation and nuclear exclusion. We and others have presented our view in several review articles on how these three signaling cascades orchestrate the homeostasis in health and diseases.[22,31,36–41]

2. RECOGNITION OF WNT SIGNALING PATHWAY COMPONENTS AS DIABETES RISK GENES

Numerous in vitro and in vivo investigations have shown that many components of the Wnt signaling pathway play important roles in pancreatic β-cell proliferation,[42–45] normal cholesterol metabolism, glucose-stimulated insulin secretion (GSIS),[46] and the production and function of the incretin hormones.[10] Genome-wide association studies (GWAS) have been generating tremendous impact on identifying multiple disease genes, including that of T2D.[47–50] During the past decade, GWAS have indicated at least 38 single nucleotide polymorphisms (SNPs) that are associated with T2D and an additional two dozen other SNPs associated with glycemic traits. A milestone GWAS discovery on the importance of the Wnt signaling pathway in T2D, however, was made in 2006 by Grant et al.[51] This study reveals the strong genetic association between TCF7L2 SNPs and the risk of T2D.[51] This finding has drawn intensive attention globally during the past 9 years. In the following section, the Wnt ligand WNT5B, the Wnt coreceptor LRP5, and a few potential Wnt target genes are discussed. TCF7L2 will be discussed in a separate section.

2.1 Wnt Ligand WNT5B

Kanazawa et al. conducted GWAS investigations on the role of WNT genes in conferring susceptibility to T2D in a Japanese population.[52] They first screened 40 SNPs within 11 Wnt genes among 188 T2D patients and 564 control subjects. The study allowed them to identify six positive SNPs. They then focused on these six SNPs in a second cohort study, with 733 T2D patients and 375 control subjects. This allowed them to further define the association between an SNP locus within *WNT5B* and T2D susceptibility.[52] Evidently, WNT5B is expressed in metabolic tissues, including adipose tissue, pancreas, and liver. They have also reported that overexpression of WNT5B in preadipocytes resulted in the promotion of adipogenesis and the enhancement of adipocytokine-gene expression.[52] The association between *WNT5B* SNPs and T2D was then demonstrated by another study in the United Kingdom in Caucasian subjects.[53] The stimulatory effect of mouse Wnt5b on adipogenesis was also reported by another group with the mouse 3T3-L1 cells, indicating that the stimulation is likely mediated via increasing the expression of peroxisome proliferator-activated receptor-γ and adipocyte protein 2.[54] It is worth pointing out that WNT5B may not activate canonical Wnt signaling pathway, whereas the repression of canonical Wnt signaling pathway is considered as a prerequisite for adipogenesis.[52,54,55] In addition, Wnt5b and Wnt5a are involved in coordinating chondrocyte differentiation and proliferation in mice.[56]

2.2 Wnt Coreceptor LRP5

The Wnt signaling pathway coreceptor LRP-5/6 also plays important role in lipid and glucose metabolism.[14,57] LRP5 interacts with Axin, an inhibitory factor of the Wnt signaling pathway (Figure 1(a)). An early investigation demonstrated that loss-of-function mutations of *LRP5* are associated with the development of the autosomal recessive disorder osteoporosis-pseudoglioma syndrome.[58] The *LRP5* gene is located within a type 1 diabetes risk region on chromosome 11q13.[59] Recent GWAS have also demonstrated that LRP5 polymorphisms are associated with the obesity phenotypes.[60]

When the mouse *Lrp5* complementary DNA was expressed in fibroblasts, the Lrp5 protein did not exert an appreciable stimulatory effect on Wnt signaling, unless Wnt ligand was also provided.[61] However, when the extracellular domain of Lrp5 was deleted, it served as a constitutively active molecule on Wnt signaling. Furthermore, Wnt ligands trigger the translocation of Axin to the cell membrane and enhanced the interaction between Lrp5 and Axin. In addition, the Axin interaction domain of Lrp5 is also required for cat/TCF-mediated transcriptional regulation. Thus, the binding of Axin by Lrp5 and its translocation to the cell membrane is an important part of Wnt signal pathway activation.[41]

Lrp5$^{-/-}$ mice exhibited markedly impaired glucose tolerance on chow diet feeding and showed increased plasma cholesterol levels on high-fat diet feeding. Furthermore, Wnt3a- and Wnt5a-stimulated insulin secretion in mouse islets was attenuated in *Lrp5*$^{-/-}$ mice. Very recently, Go et al. have demonstrated that

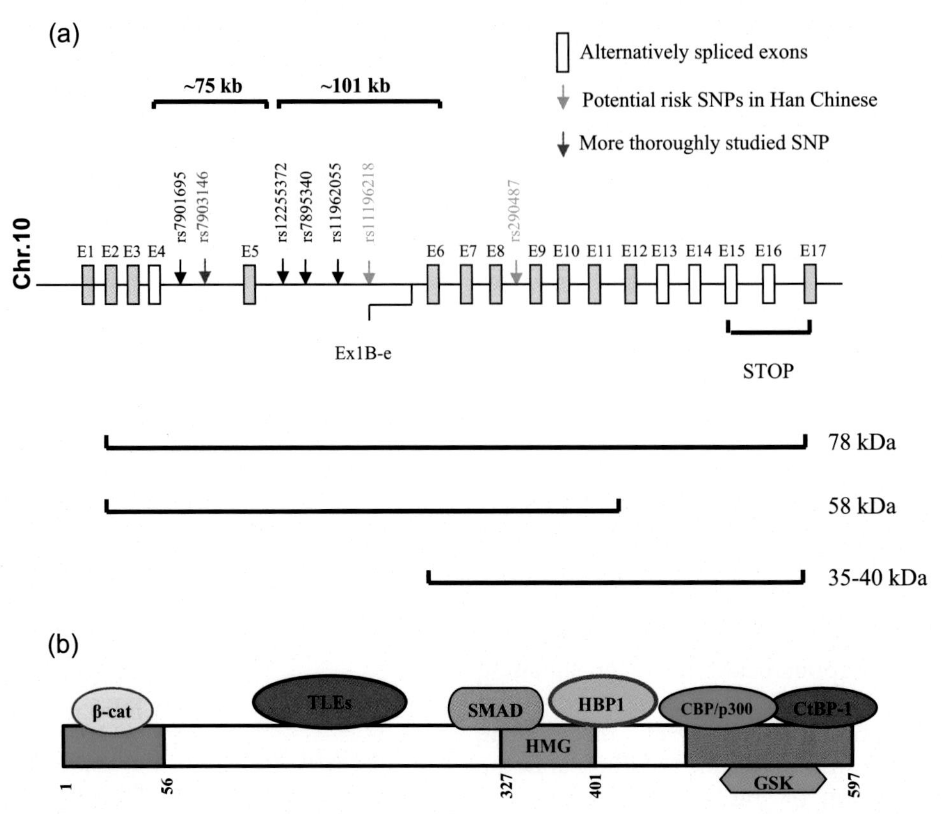

FIGURE 2 **Schematic of human TCF7L2 gene and functional domains.** (a) Illustration of type 2 diabetes risk SNPs and the sizes of its protein products. Ex1Be is an alternatively used promoter in the brain, leading to the generation of a native dominant negative TCF7L2 (35–40 kDa). (b) Illustration of functional domains of TCF7L2. GSK, glycogen synthase kinase; HBP1, HMG-box transcription factor 1; HMG, high-mobility group; SMAD, Sma- and Mad-related protein; SNP, single nucleotide polymorphism; TLEs, transducin-like enhancers of split.

the defect in *Lrp5*$^{-/-}$ mice on lipid metabolism can be attenuated by in vivo Wnt3a administration.[62]

2.3 Other Wnt-Associated Genes or Wnt Target Genes

In addition to *TCF7L2*, *WNT5B*, and *LRP5*, GWAS have suggested the implication of a number of other genes with T2D, of which the expression could be directly or indirectly regulated by the Wnt pathway effector cat/TCF.

Fat mass and obesity-associated protein (FTO) is an enzyme that is encoded by the *FTO* gene, located on chromosome 16. Certain variants of *FTO* are correlated with the development of obesity.[63,64] In mice, a 1.6-Mb deletion mutation results in the deletion of the locus of *FTO*, *FTS*, and *FTM* and three homeobox genes known as *Irx3*, *Irx5*, and *Irx6*. Among them, FTS interacts with Akt, which upregulates the Wnt signaling pathway activity via the inactivation of GSK-3.[65] Furthermore, Wnt activation was shown to stimulate the expression of *Irx3*.[66] Very recently, a study by Osborn et al. show that Wnt pathway is compromised in the absence of FTO in *zebrafish* in vivo and in vitro in *Fto*$^{-/-}$ mouse embryonic fibroblasts and the human HEK293T cell line. In these systems, canonical Wnt signaling was shown to be downregulated by abrogated β-cat translocation to the nucleus, whereas noncanonical Wnt/Ca^{2+} pathway was activated via its key signal mediators CaMKII and PKCδ.[67] Thus, FTO functions as a protein regulator of the balanced activation between canonical and noncanonical Wnt signaling cascades.

Neurogenic locus notch homolog protein 2 (Notch2) is also among the known diabetes risk genes identified by GWAS.[68] The cross talk between Notch and Wnt signaling pathway has been recognized for many years. Notch2 is likely involved in the early branching morphogenesis of the pancreas.[69] As discussed previously, GSK-3 is a common component of Wnt as well as insulin signaling cascades. It has been shown that GSK-3 phosphorylates Norch2 and hence represses Notch target gene expression.[70,71]

GWAS have also identified IGF-2 binding protein 2 as a T2D risk gene.[68] It regulates the translation of IGF-2, a member of the insulin family of polypeptide growth factors, involved in pancreatic development of insulin action. Wnt1 was shown to protect 3T3-L1 preadipocytes from apoptosis via inducing IGF-2 expression.[72]

Hematopoietically expressed Homeobox (HHEX) encodes the homeodomain protein that regulates hematopoietic cell proliferation and specification. Mouse HHEX has been shown to function as a corepressor for β-cat in attenuating Wnt signaling cascade.[73,74] Zhao et al. have examined 20 polymorphisms at 18 T2D loci. They revealed that HHEX is a locus that influences pediatric body mass index.[68]

Additional information on the Wnt pathway and diabetes risk can be found in excellent review articles elsewhere.[75–81]

3. TCF7L2 AS A DIABETIC RISK GENE AND ITS ROLE IN GLUCOSE HOMEOSTASIS

3.1 Introduction of TCF7L2 Genome-Wide Association Studies

It has been well-documented that a region on chromosomal 10q is strongly linked with T2D susceptibility.[82,83] In 2006, Grant et al. presented their important discovery on further mapping T2D risk within this region. They revealed that a few SNPs within the intronic regions of *TCF7L2* show robust association with the susceptibility of T2D.[51] This discovery has been replicated among different ethnic populations by numerous groups.[48,51,84–91] For further reading, please see review articles elsewhere.[92–96] Next the structure of TCF7L2 and then a summary of recent mechanistic explorations with various mouse models will be presented.

Figure 2(a) illustrates the structure of the human *TCF7L2* gene. The positions of five SNPs within intron 4 and intron 5 are presented, which were initially investigated by Grant et al.[51] Subsequent studies determined that rs7903146 has the greatest effect in Caucasian population. In the Chinese Han population, two other risk SNPs have been reported.[89,97] Several studies have shown that these T2D risk SNPs are also associated with the vascular complications of T2D, atherosclerosis, or coronary heart diseases as well as the severity and mortality of coronary artery disease.[98–102]

Another feature of *TCF7L2* is the existence of intensive alternative splicing.[103–111] However, the major protein products are 78 kD and 58 kD in size (Figure 2). These two different-sized proteins differ at their C-terminus. Thus, both the HMG DNA binding and β-cat interaction domains remain the same in majority of, if not all, protein products of TCF7L2 (Figure 2). Although an alternative promoter was located upstream of exon 6, namely Ex1b-e, which drives the transcription of messenger RNA (mRNA) that encodes a native dominant negative TCF7L2 (TCF7L2DN) in embryonic brain neurons,[112] its expression in peripheral metabolic tissues in adulthood and its function in metabolic homeostasis remain largely unknown. Artificially generated TCF7L2DN (Figure 2), however, has been created and used in numerous in vitro as well as in vivo investigations for more than 15 years,[21,34,55,113–116] which will be discussed later.

After the initial discovery of the association between *TCF7L2* SNPs and T2D risk in 2006,[51] enormous effort has been made to explore molecular mechanisms underlying the function of this Wnt signaling pathway effector in pancreatic β cells and elsewhere.[36,85,86,117,118] A considerable amount of effort has been made on pancreatic β cells since Schafer et al. demonstrated that impaired GLP-1–induced insulin secretion occurred in the carriers of T2D risk *TCF7L2* SNPs.[86] The function of TCF72 in the brain, liver, and adipose tissues has also been investigated intensively. Next, the main findings on the function of TCF7L2 in these four major organs are summarized. The dispute on its function in the liver as well as in pancreatic β cells will also be briefly discussed.

3.2 Role of TCF7L2 in Gut GLP-1 Production

The incretin hormone GLP-1 is encoded by the proglucagon gene and is produced in the distal ileum and the colon endocrine L-type cells as well as in certain neuronal cells in the brain. Although proglucagon gene transcription and GLP-1 synthesis can be stimulated by PKA activation, deleting or mutating the cyclic adenosine monophosphate (cAMP)-response element (CRE) on the proglucagon gene promoter does not block the stimulatory effect of PKA on its transcription. To explore mechanisms underlying this CRE-independent activation, we have examined the role of Wnt signaling pathway on proglucagon gene expression and GLP-1 synthesis using the large intestinal GLP-1–producing endocrine L cell line GLUTag[119] and small intestinal STC-1 cell line.[10] We first demonstrated the stimulatory effect of lithium treatment and constitutively active S33Y β-cat cotransfection on proglucagon gene promoter activity. The stimulation on endogenous proglucagon mRNA expression and GLP-1 synthesis was then confirmed in GLUTag cells and in the fetal rat intestinal cell cultures.[10,18] We then localized a TCF binding motif within the E2 enhancer element of the proglucagon gene promoter that mediates the Wnt stimulatory effect and confirmed the participation of TCF-4 (= TCF7L2) in this activation.[18] Interestingly, García-Martínez et al. found that the expression of another incretin hormone, gastric inhibitory polypeptide (GIP), was also positively regulated by the Wnt signaling pathway, involving the binding of β-cat/LEF-1 to a TCF binding site within the proximal promoter region of the GIP gene.[120]

Importantly, Liu and Habener reported that the function of GLP-1 can also be mediated by the Wnt signaling pathway effector cat/TCF7L2. They found that in isolated mouse pancreatic islets and the clonal INS-1 cell line, both GLP-1 and its agonist exendin-4 induced Wnt signaling activation. The inhibition of Wnt signaling by small interfering RNA (siRNA)-mediated knockdown of β-cat or the expression of TCF7L2DN decreased both basal and exendin-4–stimulated β-cell proliferation.[34] Thus, Wnt signaling pathway regulates both the production and function of the incretin hormone GLP-1.[121] We have also tested the repressive effect of TCF7L2DN in a transgenic mouse model in the proglucagon producing cells, showing the essential role of Wnt signaling in driving proglucagon gene expression in the gut and brain, but not in pancreatic α cells. Finally, we confirmed the role of cat/TCF7L2 in mediating the function of brain GLP-1.[116]

3.3 Role of TCF7L2 in Pancreatic β Cells

Using both overexpression and genetic knockdown strategies in isolated human islets, Shu et al. demonstrated that TCF7L2 plays a protecting role for β cells against cytokine-induced apoptosis, stimulates β-cell proliferation, and mediates GSIS.[122] The same group then conducted further exploration in human islets, showing that siRNA-medicated *TCF7L2* knockdown reduced the expression of genes that encode the receptors for GLP-1 and GIP (GLP-1R and GIPR), resulted in attenuated response to incretin stimulation.[123] This beneficial effect of TCF7L2 in human islets observed by Shu et al., however, was in disagreement with an early study conducted by Lyssenko et al., who showed that TCF7L2 overexpression reduced GSIS.[85] This, however, was in agreement with the study by da Silva Xavier et al., showing that, in both rodent pancreatic islets and β-cell lines, siRNA-mediated knockdown of TCF7L2 expression decreased glucose-induced but not KCl-induced insulin secretion. They also found that in TCF7L2-silenced β cells, syntaxin 1A mRNA expression was increased but Munc18-1 and ZnT8 mRNA expression was reduced.[118]

Tcf7l2$^{-/-}$ mice die at the age of 8–10 days after their birth.[17] Several transgenic mouse models have been generated, targeting the function of TCF7L2 in whole body or specifically in pancreatic β cells. Savic et al. reported that the 92-kb genomic interval associated with T2D risk contains enhancer elements that regulate the temporal/spatial expression patterns of TCF7L2 in various metabolic tissues. They developed a *Tcf7l2* copy-number allelic series in mice. The *TCF7L2* null mice displayed enhanced glucose tolerance coupled to reduced plasma insulin levels, whereas transgenic mice harboring multiple *Tcf7l2* copies and overexpressing this gene displayed reciprocal phenotypes, including glucose intolerance.[117] These observations would suggest that TCF7L2 plays a deleterious effect on metabolic homeostasis.

The use of various strategies to delete *Tcf7l2* specifically in pancreatic β cells has been performed recently. Boj et al. presented their study, showing that β-cell *Tcf7l2* deletion generated no effects on metabolic homeostasis,[124] whereas da Silva Xavier et al. and Mitchell et al. found that *Tcfl2* deletion in β cells resulted in an impairment of glucose homeostasis and β-cell function.[125,126]

Our team and another group have used the functional knockdown approach to explore the role of Tcf7l2 and Wnt signaling pathway in pancreatic β cells.[114,115] Takamoto et al. generated a transgenic mouse model in which the short isoform of *Tcf7l2DN* (58 kDa) was driven by the *Ins2* promoter. These transgenic mice showed impaired glucose homeostasis, reduced β-cell mass, and insulin secretion.[115] We reported very recently the study with the longer isoform of human TCF7L2DN both in vitro and in vivo.[114] First of all, we found that in the pancreatic Ins-1 cell line, TCF7L2DN expression repressed β-cell proliferation, β cell–specific gene expression, and GSIS. We then created the fusion gene in which *TCF7L2DN* expression is driven by P_{TRE3G} and generated the transgenic mouse line $TCF7L2DN_{Tet}$. The double transgenic mouse line was then created by mating *Ins2-rtTA with* $TCF7L2DN_{Tet}$, designated as *βTCFDN*. Inducing *TCF7L2DN* expression with doxycycline feeding in *βTCFDN* during adulthood or immediately after weaning generated no appreciable metabolic defect. We then feed the pregnant mothers with doxycycline. The *βTCFDN* offspring showed glucose intolerance and reduced response to GLP-1, associated with reduced β-cell mass and altered β-cell gene expression. It is worth pointing out that these *βTCFDN* offspring also showed reduced expression levels of the homeobox gene *Isl-1*, in agreement with observations made by Zhou et al. that the TCF7L2-Isl-1 regulatory transcriptional network is responsible for TCF7L2 in exerting its pancreatic function.[127]

3.4 Role of TCF7L2 in Hepatocytes

In 2007, Lyssenko et al. demonstrated that the CT/TT genotypes of SNP rs7903146 were also associated with enhanced rates of hepatic glucose production (HGP).[85] Pilgarrd et al. demonstrated that in young healthy man, the T allele of rs7903146 was associated with impaired insulinotropic effect of incretins, reduced 24-h profiles of plasma insulin and glucagon, and increased hepatic glucose production.[91] The association between this risk allele and insulin resistance was also documented by Wagner et al.[128] Thus, the liver is also a target organ for assessing the role of TCF7L2 and Wnt signaling in glucose and lipid homeostasis.

A few investigations, including one made by our group, have indicated the repressive effect of Wnt signaling pathway or TCF7L2 in hepatic gluconeogenesis.[36,129–131] Norton et al. observed that silencing TCF7L2 in hepatocytes induced a marked increase in basal HGP, associated with increased gluconeogenic gene expression, whereas TCF7L2 overexpression reversed this phenotype and reduced HGP.[129] An in vivo study in transgenic mice presented by Oh et al. also suggested a crucial role of TCF7L2 in reducing HGP.[130] We demonstrated the expression of three TCF members (TCF7, TCF7L1, and TCF7L2) in hepatocytes and revealed that hepatic TCF7L2 expression can be stimulated by feeding in C57BL/6 mice or by insulin treatment in mouse primary hepatocytes in vitro, whereas Wnt3a treatment repressed gluconeogenesis in mouse primary hepatocytes.[36] The repressive effect of TCF7L2 on gluconeogenic gene expression was also reported very recently by Neve et al.[131]

Interestingly, Boj et al. found that liver-specific knockout of TCF7L2 reduced HGP, whereas hepatic overexpression of TCF7L2 increased HGP.[124] We hence have conducted a transgenic mouse work to address the role of Wnt signaling and TCF7L2 in hepatic gluconeogenesis from a different angle, with TCF7L2DN. We generated the transgenic mice in which TCF7L2DN expression is driven by the liver-specific albumin promoter. These transgenic mice, designated as *LTCFDN*, exhibited a progressive impairment in pyruvate and glucose tolerance, whereas *LTCFDN* hepatocytes showed elevated glucose production and attenuated response to Wnt3a treatment. The stimulatory effect of TCF7L2DN on gluconeogenesis was also verified in vitro in mouse primary hepatocytes infected with TCF7L2DN adenovirus. Furthermore, the stimulatory effect of TCF7L2DN on gluconeogenic gene expression was observed both in vitro in primary hepatocytes and in vivo in the *LTCFDN* mouse liver.[113]

3.5 The Role of TCF7L2 in Adipocytes

In 2000, Ross et al. reported that Wnt pathway activation, likely through the canonical Wnt ligand Wnt10b, maintained preadipocytes in an undifferentiated state via inhibiting adipogenic transcription factors including peroxisome proliferator-activated receptor-γ.[55] They have also shown that the expression of TCF7L2DN or the Wnt pathway inhibitor Axin stimulated preadipocyte differentiation. TCF7L2DN and Axin were also shown to trigger the transdifferentiation of myoblasts into adipocytes.[55] On the other hand, CHIR99021, a GSK-3 inhibitor, was able to block adipogenesis.[132] Gustafson and Smith demonstrated that another Wnt pathway inhibitor Dickkopf 1 promoted adipogenesis in cells with a low degree of differentiation, and Dickkopf 1 and bone morphogenetic protein 4 exert additive effects on adipocyte differentiation.[133] These investigations have defined the fundamental role of Wnt signaling pathway in repressing adipogenesis, a chronic differentiation process.[134]

In mature adipocytes, however, Wnt signaling pathway components, including several Wnt ligands, receptors, coreceptors, and effectors are abundantly expressed.[135]The expression of these components in visceral adipose tissue was also shown to be modulated by chronic hypoadiponectinemia.[136] Of interest, Schinner et al. demonstrated that human adipocyte-derived Wnt ligands can stimulate β-cell insulin secretion, the expression of glucokinase gene, and β-cell proliferation.[77,136] These observations suggest that Wnt ligands secreted by mature adipocytes may serve as "endocrine" factors.

We investigated the expression and function of Tcf7l2 in both rodent epididymal fat tissues and rat mature adipocytes. *Tcf7l2* mRNA expression in C57BL/6 mouse fat tissue was upregulated by feeding but downregulated by intraperitoneal insulin injection.[137] In high-fat diet fed mice, db/db mice and Zucker (fa/fa) rats, epididymal fat *Tcf7l2* mRNA levels were lower than the corresponding controls. Mature adipocytes received a 100-nM insulin treatment showed reduced *Tcf7l2* mRNA and protein levels associated with reduced expression of the peptide hormone leptin. When 1 nM insulin was applied, however, both *Tcf7l2* and *leptin* mRNA levels were increased. Further investigation on such stimulation allowed us to conclude that in mature adipocytes, acute low levels of insulin treatment or Wnt pathway activation stimulates leptin gene transcription, and the stimulation is at least partially dependent on an evolutionarily conserved CRE-binding motif within the proximal region of the leptin gene promoter. Thus, we suggest the existence of a dual function of Wnt signaling pathway in adipose tissue, including the well-known repressive effect on adipogenesis and the stimulation of leptin synthesis in mature adipocytes in an acute manner in response to nutritional challenge.

4. SUMMARY AND PERSPECTIVES

The identification of TCF7L2 and other Wnt signaling pathway components as T2D risk genes by GWAS has greatly facilitated the exploration of the role of Wnt signaling pathway in metabolic homeostasis. The observations made in different metabolic organs allowed us to suggest that in adulthood, Wnt pathway mediates the function of metabolic hormones and other hormonal factors in response to nutritional and environmental changes in physiological and pathophysiological conditions. Next is our view on the recognition of Wnt signaling pathway effector cat/TCF as the mediator of metabolic hormones in regulating glucose homeostasis.

4.1 TCF Expression Can Be Regulated by Metabolic Hormones

As the key component of the Wnt signaling pathway effector, the expression levels of TCFs significantly influence the transcriptional activity of the bipartite transcription factor cat/TCF. It has been mentioned previously that TCF7L2 expression in three different metabolic organs/cell lineages (i.e., pancreatic β cells, hepatocytes, and mature adipocytes) can be regulated by insulin. It appears that insulin exerts its effect on TCF7L2 expression bidirectionally in mature adipocytes.[137] At a lower dosage, probably close to the physiological levels, insulin activates TCF7L2 expression in adipocytes. At higher or pathological levels, insulin inhibits TCF7L2 expression. Whether this is also true for pancreatic β cells needs further investigation because we have only observed the repressive effect of 100 nM of insulin on β-cell TCF7L2 expression, without testing the effect of insulin at lower dosages.[138] In pancreatic β cells, the expression of other two TCF members, TCF7 and TCF7L1, was also shown to be repressed by 100 nM insulin.[138] We propose that this regulation is at least partially because of the existence of TCF7L2 autoregulation.[139]

4.2 Metabolic Hormones Control Cat/TCF Activity via Regulating Posttranscriptional Modifications of β-Catenin

In 1993, Rubinfeld et al. identified the physical interaction between APC and β-cat.[140] Subsequent studies have defined the role of cat/TCF in Wnt pathway activity as well as the inactivation of cat/TCF by stimulating β-cat N-terminal serine residue phosphorylation, with the participation of APC, Axin, GSK-3, and CK-Iα. The phosphorylation on the C-terminal serine residues (S675) was not recognized until 2005. Hino et al. found that several PKA activators, including prostaglandin E1 and dibutyryl cAMP, increases cytoplasmic and nuclear β-cat levels, and stimulated TCF-dependent transcription via β-cat.[27] They have also demonstrated that PKA does not directly interact with any of the components of the destructive complex of β-cat, including APC, Axin, GSK-3, and CK-Iα, but directly phosphorylated β-cat at S765. This phosphorylation stabilizes β-cat, preventing its ubiquitin-mediated degradation. Subsequent studies have revealed that this phosphorylation can also be activated by metabolic hormones that use PKA as the effector, such as GLP-1,[34] and that insulin and IGF-1 treatment can stimulate this phosphorylation as well.[26] One target kinase that mediated the effect of insulin for β-cat S675 phosphorylation is P21-activated protein kinase-1.[28,32,37] Furthermore, insulin may also stimulate β-cat phosphorylation at S552 residue.[24] Thus, metabolic hormones can facilitate Wnt signaling activity via increase β-cat posttranslational modification.

4.3 Overview and Perspectives

Based on TCF7L2 studies summarized above in the gut, pancreatic β cells, liver, and adipocytes, our suggestions on the role of Wnt signaling pathway in glucose homeostasis are summarized. The Wnt signaling pathway effect on cat/TCF is essential for GLP-1 production and function as well as pancreatic β-cell genesis and function. It is also a negative regulator of hepatic gluconeogenesis and a stimulator of adipocyte leptin production. Importantly, cat/TCF mediates the function of metabolic hormones in response to nutritional and environmental changes. Figure 3 illustrates this function with the response to food intake as an

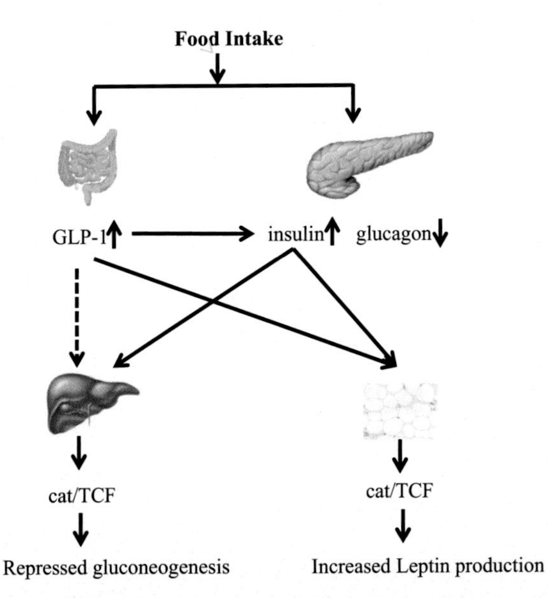

FIGURE 3 Proposed mechanisms underlying the effect of cat/TCF in regulating glucose homeostasis. Plasma insulin and GLP-1 levels increase after feeding. This involves the role of cat/TCF in the gut on GLP-1 production and in pancreatic β cells on insulin secretion. Insulin will stimulate hepatic TCF7L2 expression and β-cat S675 phosphorylation, leading to repressed hepatic gluconeogenesis. Insulin will also stimulate cat/TCF7L2 activity in adipocytes, leading to increased leptin production and release, which exerts its effects on metabolic homeostasis in the brain and peripheral tissues. Whether GLP-1 has a direct effect in the liver remains to be further explored (indicated by dotted line). Cat, catenin; GLP, glucagon-like peptide.

example. After food intake, plasma insulin and GLP-1 levels increase, whereas glucagon level drops. In pancreatic β cells, TCF7L2 works with β-cat in mediating insulin secretion. In the gut, cat/TCF7L2 is important for GLP-1 production, which is indirectly implicated in insulin secretion. Elevated plasma insulin levels will result in the stimulation of cat/TCF activity in hepatocytes. This is achieved by activating TCF7L2 expression and stimulating β-cat S675 and S552 phosphorylation. Increased hepatic cat/TCF activity contributes to the repression of hepatic gluconeogenesis by insulin. In adipocytes, the stimulation of cat/TCF activation by insulin leads to increased leptin gene expression and secretion, which exerts its metabolic effects in both the brain and peripheral tissues, including inducing satiety and sensitizing insulin signaling.

There are still many questions that remain to be addressed or further explored in the near future. Below are few examples. First, because GLP-1 stimulates cat/TCF7L2 activity in pancreatic β cells, can it directly exert the stimulatory effect on cat/TCF activation in the liver and adipose tissue and hence represses gluconeogenesis, without the participation of insulin? Second, both GLP-1 and glucagon use cAMP as their second messenger and PKA as their downstream effector. These two hormones, however, exert opposite effects on glucose production in the liver. Will they activate cat/TCF activity the same way or differently in the liver? Third, what is the negative feedback mechanism to attenuate the stimulatory effect of Wnt activation in pancreatic β cell, liver, and adipocytes? Fourth, how does insulin resistance in T2D and other metabolic diseases affect this novel signaling cascade? Finally, how does this signaling network contribute to lipid metabolic homeostasis? We speculate that most of these questions, if not all, will be addressed, or partially addressed, within the next few years.

Acknowledgments

The studies in Jin laboratory have been supported by operating grants from the Canadian Institutes of Health Research (CIHR, MOP-89987, and MOP-97790) and Canadian Diabetes Association (CDA, OG-3-10-3040). The author regrets not being able to cite all of the excellent contributions in the field due to space limitations.

References

1. Nusse R, Varmus HE. Many tumors induced by the mouse mammary tumor virus contain a provirus integrated in the same region of the host genome. *Cell* 1982;**31**:99–109.
2. Rijsewijk F, Schuermann M, Wagenaar E, Parren P, Weigel D, Nusse R. The Drosophila homolog of the mouse mammary oncogene int-1 is identical to the segment polarity gene wingless. *Cell* 1987;**50**:649–57.
3. Kinzler KW, Nilbert MC, Su LK, Vogelstein B, Bryan TM, Levy DB, et al. Identification of FAP locus genes from chromosome 5q21. *Science* 1991;**253**:661–5.
4. Kinzler KW, Vogelstein B. Lessons from hereditary colorectal cancer. *Cell* 1996;**87**:159–70.
5. van Es JH, Barker N, Clevers H. You Wnt some, you lose some: oncogenes in the Wnt signaling pathway. *Curr Opin Genet Dev* 2003;**13**:28–33.
6. He TC, Sparks AB, Rago C, Hermeking H, Zawel L, da Costa LT, et al. Identification of c-MYC as a target of the APC pathway. *Science* 1998;**281**:1509–12.
7. Morin PJ, Sparks AB, Korinek V, Barker N, Clevers H, Vogelstein B, et al. Activation of beta-catenin-Tcf signaling in colon cancer by mutations in beta-catenin or APC. *Science* 1997; **275**:1787–90.
8. McMahon AP, Moon RT. Ectopic expression of the proto-oncogene int-1 in Xenopus embryos leads to duplication of the embryonic axis. *Cell* 1989;**58**:1075–84.
9. Wolda SL, Moody CJ, Moon RT. Overlapping expression of Xwnt-3A and Xwnt-1 in neural tissue of *Xenopus laevis* embryos. *Dev Biol* 1993;**155**:46–57.
10. Ni Z, Anini Y, Fang X, Mills G, Brubaker PL, Jin T. Transcriptional activation of the proglucagon gene by lithium and beta-catenin in intestinal endocrine L cells. *J Biol Chem* 2003;**278**:1380–7.
11. Doble BW, Woodgett JR. GSK-3: tricks of the trade for a multi-tasking kinase. *J Cell Sci* 2003;**116**:1175–86.
12. Clevers H, Nusse R. Wnt/beta-catenin signaling and disease. *Cell* 2012;**149**:1192–205.
13. Zhang XM, Ramalho-Santos M, McMahon AP. Smoothened mutants reveal redundant roles for Shh and Ihh signaling including regulation of L/R asymmetry by the mouse node. *Cell* 2001;**105**: 781–92.
14. Wehrli M, Dougan ST, Caldwell K, O'Keefe L, Schwartz S, Vaizel-Ohayon D, et al. Arrow encodes an LDL-receptor-related protein essential for wingless signalling. *Nature* 2000;**407**: 527–30.
15. Pinson KI, Brennan J, Monkley S, Avery BJ, Skarnes WC. An LDL-receptor-related protein mediates Wnt signalling in mice. *Nature* 2000;**407**:535–8.
16. Gordon MD, Nusse R. Wnt signaling: multiple pathways, multiple receptors, and multiple transcription factors. *J Biol Chem* 2006;**281**:22429–33.
17. Korinek V, Barker N, Moerer P, van Donselaar E, Huls G, Peters PJ, et al. Depletion of epithelial stem-cell compartments in the small intestine of mice lacking Tcf-4. *Nat Genet* 1998;**19**:379–83.
18. Yi F, Brubaker PL, Jin T. TCF-4 mediates cell type-specific regulation of proglucagon gene expression by beta-catenin and glycogen synthase kinase-3beta. *J Biol Chem* 2005;**280**:1457–64.
19. Zeng X, Tamai K, Doble B, Li S, Huang H, Habas R, et al. A dual-kinase mechanism for Wnt co-receptor phosphorylation and activation. *Nature* 2005;**438**:873–7.
20. Dierick H, Bejsovec A. Cellular mechanisms of wingless/Wnt signal transduction. *Curr Top Dev Biol* 1999;**43**:153–90.
21. Kolligs FT, Hu G, Dang CV, Fearon ER. Neoplastic transformation of RK3E by mutant beta-catenin requires deregulation of Tcf/Lef transcription but not activation of c-myc expression. *Mol Cell Biol* 1999;**19**:5696–706.
22. Jin T, George Fantus I, Sun J. Wnt and beyond Wnt: multiple mechanisms control the transcriptional property of beta-catenin. *Cell Signal* 2008;**20**:1697–704.
23. Ding VW, Chen RH, McCormick F. Differential regulation of glycogen synthase kinase 3beta by insulin and Wnt signaling. *J Biol Chem* 2000;**275**:32475–81.
24. Fang D, Hawke D, Zheng Y, Xia Y, Meisenhelder J, Nika H, et al. Phosphorylation of beta-catenin by AKT promotes beta-catenin transcriptional activity. *J Biol Chem* 2007;**282**:11221–9.

25. Yi F, Sun J, Lim GE, Fantus IG, Brubaker PL, Jin T. Cross talk between the insulin and Wnt signaling pathways: evidence from intestinal endocrine L cells. *Endocrinology* 2008;**149**:2341–51.
26. Sun J, Khalid S, Rozakis-Adcock M, Fantus IG, Jin T. P-21-activated protein kinase-1 functions as a linker between insulin and Wnt signaling pathways in the intestine. *Oncogene* 2009;**28**:3132–44.
27. Hino S, Tanji C, Nakayama KI, Kikuchi A. Phosphorylation of beta-catenin by cyclic AMP-dependent protein kinase stabilizes beta-catenin through inhibition of its ubiquitination. *Mol Cell Biol* 2005;**25**:9063–72.
28. Zhu G, Wang Y, Huang B, Liang J, Ding Y, Xu A, et al. A Rac1/PAK1 cascade controls beta-catenin activation in colon cancer cells. *Oncogene* 2012;**31**:1001–12.
29. Taurin S, Sandbo N, Qin Y, Browning D, Dulin NO. Phosphorylation of beta-catenin by cyclic AMP-dependent protein kinase. *J Biol Chem* 2006;**281**:9971–6.
30. Chiang YA, Shao W, Xu XX, Chernoff J, Jin T. P21-activated protein kinase 1 (Pak1) mediates the cross talk between insulin and beta-catenin on proglucagon gene expression and its ablation affects glucose homeostasis in male C57BL/6 mice. *Endocrinology* 2013;**154**:77–88.
31. Chiang YT, Ip W, Jin T. The role of the Wnt signaling pathway in incretin hormone production and function. *Front Physiol* 2012;**3**:273.
32. Chiang YT, Ip W, Shao W, Song ZE, Chernoff J, Jin T. Activation of cAMP signaling attenuates impaired hepatic glucose disposal in aged male p21-activated protein kinase-1 knockout mice. *Endocrinology* 2014;**155**:2122–32.
33. Lim GE, Xu M, Sun J, Jin T, Brubaker PL. The rho guanosine 5′-triphosphatase, cell division cycle 42, is required for insulin-induced actin remodeling and glucagon-like peptide-1 secretion in the intestinal endocrine L cell. *Endocrinology* 2009;**150**:5249–61.
34. Liu Z, Habener JF. Glucagon-like peptide-1 activation of TCF7L2-dependent Wnt signaling enhances pancreatic beta cell proliferation. *J Biol Chem* 2008;**283**:8723–35.
35. Essers MA, de Vries-Smits LM, Barker N, Polderman PE, Burgering BM, Korswagen HC. Functional interaction between beta-catenin and FOXO in oxidative stress signaling. *Science* 2005;**308**:1181–4.
36. Ip W, Shao W, Chiang YT, Jin T. The Wnt signaling pathway effector TCF7L2 is upregulated by insulin and represses hepatic gluconeogenesis. *Am J Physiol Endocrinol Metab* 2012;**303**:E1166–76.
37. Chiang YT, Jin T. p21-Activated protein kinases and their emerging roles in glucose homeostasis. *Am J Physiol Endocrinol Metab* 2014;**306**:E707–22.
38. Manolagas SC, Almeida M. Gone with the Wnts: beta-catenin, T-cell factor, forkhead box O, and oxidative stress in age-dependent diseases of bone, lipid, and glucose metabolism. *Mol Endocrinol* 2007;**21**:2605–14.
39. Manolagas SC. From estrogen-centric to aging and oxidative stress: a revised perspective of the pathogenesis of osteoporosis. *Endocr Rev* 2010;**31**:266–300.
40. Manolagas SC. Wnt signaling and osteoporosis. *Maturitas* 2014;**78**:233–7.
41. Hoogeboom D, Burgering BM. Should I stay or should I go: beta-catenin decides under stress. *Biochim Biophys Acta* 2009;**1796**:63–74.
42. Murtaugh LC, Law AC, Dor Y, Melton DA. Beta-catenin is essential for pancreatic acinar but not islet development. *Development* 2005;**132**:4663–74.
43. Papadopoulou S, Edlund H. Attenuated Wnt signaling perturbs pancreatic growth but not pancreatic function. *Diabetes* 2005;**54**:2844–51.
44. Heiser PW, Lau J, Taketo MM, Herrera PL, Hebrok M. Stabilization of beta-catenin impacts pancreas growth. *Development* 2006;**133**:2023–32.
45. Rulifson IC, Karnik SK, Heiser PW, ten Berge D, Chen H, Gu X, et al. Wnt signaling regulates pancreatic beta cell proliferation. *Proc Natl Acad Sci USA* 2007;**104**:6247–52.
46. Fujino T, Asaba H, Kang MJ, Ikeda Y, Sone H, Takada S, et al. Low-density lipoprotein receptor-related protein 5 (LRP5) is essential for normal cholesterol metabolism and glucose-induced insulin secretion. *Proc Natl Acad Sci USA* 2003;**100**:229–34.
47. Frayling TM. Genome-wide association studies provide new insights into type 2 diabetes aetiology. *Nat Rev Genet* 2007;**8**:657–62.
48. Florez JC. The new type 2 diabetes gene TCF7L2. *Curr Opin Clin Nutr Metab Care* 2007;**10**:391–6.
49. Billings LK, Florez JC. The genetics of type 2 diabetes: what have we learned from GWAS? *Ann N Y Acad Sci* 2010;**1212**:59–77.
50. Groop L. Open chromatin and diabetes risk. *Nat Genet* 2010;**42**:190–2.
51. Grant SF, Thorleifsson G, Reynisdottir I, Benediktsson R, Manolescu A, Sainz J, et al. Variant of transcription factor 7-like 2 (TCF7L2) gene confers risk of type 2 diabetes. *Nat Genet* 2006;**38**:320–3.
52. Kanazawa A, Tsukada S, Sekine A, Tsunoda T, Takahashi A, Kashiwagi A, et al. Association of the gene encoding wingless-type mammary tumor virus integration-site family member 5B (WNT5B) with type 2 diabetes. *Am J Hum Genet* 2004;**75**:832–43.
53. Salpea KD, Gable DR, Cooper JA, Stephens JW, Hurel SJ, Ireland HA, et al. The effect of WNT5B IVS3C>G on the susceptibility to type 2 diabetes in UK Caucasian subjects. *Nutr Metab Cardiovasc Dis* 2009;**19**:140–5.
54. van Tienen FH, Laeremans H, van der Kallen CJ, Smeets HJ. Wnt5b stimulates adipogenesis by activating PPARgamma, and inhibiting the beta-catenin dependent Wnt signaling pathway together with Wnt5a. *Biochem Biophys Res Commun* 2009;**387**:207–11.
55. Ross SE, Hemati N, Longo KA, Bennett CN, Lucas PC, Erickson RL, et al. Inhibition of adipogenesis by Wnt signaling. *Science* 2000;**289**:950–3.
56. Yang Y, Topol L, Lee H, Wu J. Wnt5a and Wnt5b exhibit distinct activities in coordinating chondrocyte proliferation and differentiation. *Development* 2003;**130**:1003–15.
57. Tamai K, Semenov M, Kato Y, Spokony R, Liu C, Katsuyama Y, et al. LDL-receptor-related proteins in Wnt signal transduction. *Nature* 2000;**407**:530–5.
58. Gong Y, Slee RB, Fukai N, Rawadi G, Roman-Roman S, Reginato AM, Osteoporosis-Pseudoglioma Syndrome Collaborative Group, et al. LDL receptor-related protein 5 (LRP5) affects bone accrual and eye development. *Cell* 2001;**107**:513–23.
59. Twells RC, Mein CA, Phillips MS, Hess JF, Veijola R, Gilbey M, et al. Haplotype structure, LD blocks, and uneven recombination within the LRP5 gene. *Genome Res* 2003;**13**:845–55.
60. Guo YF, Xiong DH, Shen H, Zhao LJ, Xiao P, Guo Y, et al. Polymorphisms of the low-density lipoprotein receptor-related protein 5 (LRP5) gene are associated with obesity phenotypes in a large family-based association study. *J Med Genet* 2006;**43**:798–803.
61. Mao J, Wang J, Liu B, Pan W, Farr 3rd GH, Flynn C, et al. Low-density lipoprotein receptor-related protein-5 binds to Axin and regulates the canonical Wnt signaling pathway. *Mol Cell* 2001;**7**:801–9.
62. Go GW, Srivastava R, Hernandez-Ono A, Gang G, Smith SB, Booth CJ, et al. The combined hyperlipidemia caused by impaired Wnt-LRP6 signaling is reversed by Wnt3a rescue. *Cell Metab* 2014;**19**:209–20.

63. Jia G, Fu Y, Zhao X, Dai Q, Zheng G, Yang Y, et al. N6-methyladenosine in nuclear RNA is a major substrate of the obesity-associated FTO. *Nat Chem Biol* 2011;**7**:885–7.
64. Loos RJ, Yeo GS. The bigger picture of FTO: the first GWAS-identified obesity gene. *Nat Rev Endocrinol* 2014;**10**:51–61.
65. Braun MM, Etheridge A, Bernard A, Robertson CP, Roelink H. Wnt signaling is required at distinct stages of development for the induction of the posterior forebrain. *Development* 2003;**130**: 5579–87.
66. Petri A, Ahnfelt-Ronne J, Frederiksen KS, Edwards DG, Madsen D, Serup P, et al. The effect of neurogenin3 deficiency on pancreatic gene expression in embryonic mice. *J Mol Endocrinol* 2006;**37**:301–16.
67. Osborn DP, Roccasecca RM, McMurray F, Hernandez-Hernandez V, Mukherjee S, Barroso I, et al. Loss of FTO antagonises Wnt signaling and leads to developmental defects associated with ciliopathies. *PLoS One* 2014;**9**:e87662.
68. Zhao J, Bradfield JP, Zhang H, Annaiah K, Wang K, Kim CE, et al. Examination of all type 2 diabetes GWAS loci reveals HHEX-IDE as a locus influencing pediatric BMI. *Diabetes* 2010;**59**:751–5.
69. Lee KM, Yasuda H, Hollingsworth MA, Ouellette MM. Notch 2-positive progenitors with the intrinsic ability to give rise to pancreatic ductal cells. *Lab Invest* 2005;**85**:1003–12.
70. Nakhai H, Siveke JT, Klein B, Mendoza-Torres L, Mazur PK, Algul H, et al. Conditional ablation of Notch signaling in pancreatic development. *Development* 2008;**135**:2757–65.
71. Espinosa L, Ingles-Esteve J, Aguilera C, Bigas A. Phosphorylation by glycogen synthase kinase-3 beta down-regulates Notch activity, a link for Notch and Wnt pathways. *J Biol Chem* 2003;**278**: 32227–35.
72. Longo KA, Kennell JA, Ochocinska MJ, Ross SE, Wright WS, MacDougald OA. Wnt signaling protects 3T3-L1 preadipocytes from apoptosis through induction of insulin-like growth factors. *J Biol Chem* 2002;**277**:38239–44.
73. Foley AC, Mercola M. Heart induction by Wnt antagonists depends on the homeodomain transcription factor Hex. *Genes Dev* 2005;**19**:387–96.
74. Zamparini AL, Watts T, Gardner CE, Tomlinson SR, Johnston GI, Brickman JM. Hex acts with beta-catenin to regulate anteroposterior patterning via a Groucho-related co-repressor and Nodal. *Development* 2006;**133**:3709–22.
75. Liu Z, Habener JF. Wnt signaling in pancreatic islets. *Adv Exp Med Biol* 2010;**654**:391–419.
76. Welters HJ, Kulkarni RN. Wnt signaling: relevance to beta-cell biology and diabetes. *Trends Endocrinol Metab* 2008;**19**:349–55.
77. Schinner S. Wnt-signalling and the metabolic syndrome. *Horm Metab Res* 2009;**41**:159–63.
78. Wagner R, Staiger H, Ullrich S, Stefan N, Fritsche A, Haring HU. Untangling the interplay of genetic and metabolic influences on beta-cell function: examples of potential therapeutic implications involving TCF7L2 and FFAR1. *Mol Metab* 2014;**3**:261–7.
79. Hakonarson H, Grant SF. Planning a genome-wide association study: points to consider. *Ann Med* 2011;**43**:451–60.
80. Grant SF, Hakonarson H, Schwartz S. Can the genetics of type 1 and type 2 diabetes shed light on the genetics of latent autoimmune diabetes in adults? *Endocr Rev* 2010;**31**:183–93.
81. Xia Q, Grant SF. The genetics of human obesity. *Ann N Y Acad Sci* 2013;**1281**:178–90.
82. Duggirala R, Blangero J, Almasy L, Dyer TD, Williams KL, Leach RJ, et al. Linkage of type 2 diabetes mellitus and of age at onset to a genetic location on chromosome 10q in Mexican Americans. *Am J Hum Genet* 1999;**64**:1127–40.
83. Reynisdottir I, Thorleifsson G, Benediktsson R, Sigurdsson G, Emilsson V, Einarsdottir AS, et al. Localization of a susceptibility gene for type 2 diabetes to chromosome 5q34-q35.2. *Am J Hum Genet* 2003;**73**:323–35.
84. Florez JC. Newly identified loci highlight beta cell dysfunction as a key cause of type 2 diabetes: where are the insulin resistance genes? *Diabetologia* 2008;**51**:1100–10.
85. Lyssenko V, Lupi R, Marchetti P, Del Guerra S, Orho-Melander M, Almgren P, et al. Mechanisms by which common variants in the TCF7L2 gene increase risk of type 2 diabetes. *J Clin Invest* 2007; **117**:2155–63.
86. Schafer SA, Tschritter O, Machicao F, Thamer C, Stefan N, Gallwitz B, et al. Impaired glucagon-like peptide-1-induced insulin secretion in carriers of transcription factor 7-like 2 (TCF7L2) gene polymorphisms. *Diabetologia* 2007;**50**:2443–50.
87. Gonzalez-Sanchez JL, Martinez-Larrad MT, Zabena C, Perez-Barba M, Serrano-Rios M. Association of variants of the TCF7L2 gene with increases in the risk of type 2 diabetes and the proinsulin:insulin ratio in the Spanish population. *Diabetologia* 2008;**51**:1993–7.
88. Lyssenko V, Jonsson A, Almgren P, Pulizzi N, Isomaa B, Tuomi T, et al. Clinical risk factors, DNA variants, and the development of type 2 diabetes. *N Engl J Med* 2008;**359**: 2220–32.
89. Ng MC, Park KS, Oh B, Tam CH, Cho YM, Shin HD, et al. Implication of genetic variants near TCF7L2, SLC30A8, HHEX, CDKAL1, CDKN2A/B, IGF2BP2, and FTO in type 2 diabetes and obesity in 6,719 Asians. *Diabetes* 2008;**57**: 2226–33.
90. Cornelis MC, Qi L, Kraft P, Hu FB. TCF7L2, dietary carbohydrate, and risk of type 2 diabetes in US women. *Am J Clin Nutr* 2009;**89**: 1256–62.
91. Pilgaard K, Jensen CB, Schou JH, Lyssenko V, Wegner L, Brons C, et al. The T allele of rs7903146 TCF7L2 is associated with impaired insulinotropic action of incretin hormones, reduced 24 h profiles of plasma insulin and glucagon, and increased hepatic glucose production in young healthy men. *Diabetologia* 2009;**52**:1298–307.
92. Perry JR, Frayling TM. New gene variants alter type 2 diabetes risk predominantly through reduced beta-cell function. *Curr Opin Clin Nutr Metab Care* 2008;**11**:371–7.
93. Palomaki GE, Melillo S, Marrone M, Douglas MP. Use of genomic panels to determine risk of developing type 2 diabetes in the general population: a targeted evidence-based review. *Genet Med* 2013;**15**:600–11.
94. Basile KJ, Johnson ME, Xia Q, Grant SF. Genetic susceptibility to type 2 diabetes and obesity: follow-up of findings from genome-wide association studies. *Int J Endocrinol* 2014;**2014**: 769671.
95. Jin T, Liu L. The Wnt signaling pathway effector TCF7L2 and type 2 diabetes mellitus. *Mol Endocrinol* 2008;**22**:2383–92.
96. Jin T. The WNT signalling pathway and diabetes mellitus. *Diabetologia* 2008;**51**:1771–80.
97. Chang YC, Chang TJ, Jiang YD, Kuo SS, Lee KC, Chiu KC, et al. Association study of the genetic polymorphisms of the transcription factor 7-like 2 (TCF7L2) gene and type 2 diabetes in the Chinese population. *Diabetes* 2007;**56**:2631–7.
98. Bielinski SJ, Pankow JS, Folsom AR, North KE, Boerwinkle E. TCF7L2 single nucleotide polymorphisms, cardiovascular disease and all-cause mortality: the Atherosclerosis Risk in Communities (ARIC) study. *Diabetologia* 2008;**51**:968–70.
99. Sousa AG, Marquezine GF, Lemos PA, Martinez E, Lopes N, Hueb WA, et al. TCF7L2 polymorphism rs7903146 is associated with coronary artery disease severity and mortality. *PLoS One* 2009;**4**:e7697.
100. Kucharska-Newton AM, Monda KL, Bielinski SJ, Boerwinkle E, Rea TD, Rosamond WD, et al. Role of BMI in the association of the TCF7L2 rs7903146 variant with coronary heart disease: the Atherosclerosis Risk in Communities (ARIC) study. *J Obes* 2010; **2010**.

101. Pechlivanis S, Scherag A, Muhleisen TW, Mohlenkamp S, Horsthemke B, Boes T, et al. Coronary artery calcification and its relationship to validated genetic variants for diabetes mellitus assessed in the Heinz Nixdorf recall cohort. *Arterioscler Thromb Vasc Biol* 2010;**30**:1867–72.
102. Lombardi R, Dong J, Rodriguez G, Bell A, Leung TK, Schwartz RJ, et al. Genetic fate mapping identifies second heart field progenitor cells as a source of adipocytes in arrhythmogenic right ventricular cardiomyopathy. *Circ Res* 2009;**104**:1076–84.
103. Duval A, Busson-Leconiat M, Berger R, Hamelin R. Assignment of the TCF-4 gene (TCF7L2) to human chromosome band 10q25.3. *Cytogenet Cell Genet* 2000;**88**:264–5.
104. Duval A, Rolland S, Tubacher E, Bui H, Thomas G, Hamelin R. The human T-cell transcription factor-4 gene: structure, extensive characterization of alternative splicings, and mutational analysis in colorectal cancer cell lines. *Cancer Res* 2000;**60**:3872–9.
105. Osmark P, Hansson O, Jonsson A, Ronn T, Groop L, Renstrom E. Unique splicing pattern of the TCF7L2 gene in human pancreatic islets. *Diabetologia* 2009;**52**:850–4.
106. Prokunina-Olsson L, Welch C, Hansson O, Adhikari N, Scott LJ, Usher N, et al. Tissue-specific alternative splicing of TCF7L2. *Hum Mol Genet* 2009;**18**:3795–804.
107. Prokunina-Olsson L, Kaplan LM, Schadt EE, Collins FS. Alternative splicing of TCF7L2 gene in omental and subcutaneous adipose tissue and risk of type 2 diabetes. *PLoS One* 2009;**4**:e7231.
108. Prokunina-Olsson L, Hall JL. Evidence for neuroendocrine function of a unique splicing form of TCF7L2 in human brain, islets and gut. *Diabetologia* 2010;**53**:712–6.
109. Weise A, Bruser K, Elfert S, Wallmen B, Wittel Y, Wohrle S, et al. Alternative splicing of Tcf7l2 transcripts generates protein variants with differential promoter-binding and transcriptional activation properties at Wnt/beta-catenin targets. *Nucleic Acids Res* 2010;**38**:1964–81.
110. Mondal AK, Das SK, Baldini G, Chu WS, Sharma NK, Hackney OG, et al. Genotype and tissue-specific effects on alternative splicing of the transcription factor 7-like 2 gene in humans. *J Clin Endocrinol Metab* 2010;**95**:1450–7.
111. Le Bacquer O, Shu L, Marchand M, Neve B, Paroni F, Kerr Conte J, et al. TCF7L2 splice variants have distinct effects on beta-cell turnover and function. *Hum Mol Genet* 2011;**20**:1906–15.
112. Vacik T, Stubbs JL, Lemke G. A novel mechanism for the transcriptional regulation of Wnt signaling in development. *Genes Dev* 2011;**25**:1783–95.
113. Ip W, Shao W, Song Z, Chen Z, Wheeler MB, Jin T. Liver-specific expression of dominant negative transcription factor 7-like 2 causes progressive impairment in glucose homeostasis. *Diabetes* 2015;**64**:1923–32.
114. Shao W, Xiong X, Ip W, Xu F, Song Z, Zeng K, et al. The expression of dominant negative TCF7L2 in pancreatic beta cells during the embryonic stage causes impaired glucose homeostasis. *Mol Metab* 2015;**4**:344–52.
115. Takamoto I, Kubota N, Nakaya K, Kumagai K, Hashimoto S, Kubota T, et al. TCF7L2 in mouse pancreatic beta cells plays a crucial role in glucose homeostasis by regulating beta cell mass. *Diabetologia* 2014;**57**:542–53.
116. Shao W, Wang D, Chiang YT, Ip W, Zhu L, Xu F, et al. The Wnt signaling pathway effector TCF7L2 controls gut and brain proglucagon gene expression and glucose homeostasis. *Diabetes* 2013;**62**:789–800.
117. Savic D, Ye H, Aneas I, Park SY, Bell GI, Nobrega MA. Alterations in TCF7L2 expression define its role as a key regulator of glucose metabolism. *Genome Res* 2011;**21**:1417–25.
118. da Silva Xavier G, Loder MK, McDonald A, Tarasov AI, Carzaniga R, Kronenberger K, et al. TCF7L2 regulates late events in insulin secretion from pancreatic islet beta-cells. *Diabetes* 2009;**58**:894–905.
119. Drucker DJ, Jin T, Asa SL, Young TA, Brubaker PL. Activation of proglucagon gene transcription by protein kinase-A in a novel mouse enteroendocrine cell line. *Mol Endocrinol* 1994;**8**:1646–55.
120. Garcia-Martinez JM, Chocarro-Calvo A, Moya CM, Garcia-Jimenez C. WNT/beta-catenin increases the production of incretins by entero-endocrine cells. *Diabetologia* 2009;**52**:1913–24.
121. Gustafson B, Smith U. WNT signalling is both an inducer and effector of glucagon-like peptide-1. *Diabetologia* 2008;**51**:1768–70.
122. Shu L, Sauter NS, Schulthess FT, Matveyenko AV, Oberholzer J, Maedler K. Transcription factor 7-like 2 regulates beta-cell survival and function in human pancreatic islets. *Diabetes* 2008;**57**:645–53.
123. Shu L, Matveyenko AV, Kerr-Conte J, Cho JH, McIntosh CH, Maedler K. Decreased TCF7L2 protein levels in type 2 diabetes mellitus correlate with downregulation of GIP- and GLP-1 receptors and impaired beta-cell function. *Hum Mol Genet* 2009;**18**:2388–99.
124. Boj SF, van Es JH, Huch M, Li VS, Jose A, Hatzis P, et al. Diabetes risk gene and Wnt effector Tcf7l2/TCF4 controls hepatic response to perinatal and adult metabolic demand. *Cell* 2012;**151**:1595–607.
125. da Silva Xavier G, Mondragon A, Sun G, Chen L, McGinty JA, French PM, et al. Abnormal glucose tolerance and insulin secretion in pancreas-specific Tcf7l2-null mice. *Diabetologia* 2012;**55**:2667–76.
126. Mitchell RK, Mondragon A, Chen L, McGinty JA, French PM, Ferrer J, et al. Selective disruption of Tcf7l2 in the pancreatic beta cell impairs secretory function and lowers beta cell mass. *Hum Mol Genet* 2014.
127. Zhou Y, Park SY, Su J, Bailey K, Ottosson-Laakso E, Shcherbina L, et al. TCF7L2 is a master regulator of insulin production and processing. *Hum Mol Genet* 2014;**23**:6419–31.
128. Wegner L, Hussain MS, Pilgaard K, Hansen T, Pedersen O, Vaag A, et al. Impact of TCF7L2 rs7903146 on insulin secretion and action in young and elderly Danish twins. *J Clin Endocrinol Metab* 2008;**93**:4013–9.
129. Norton L, Fourcaudot M, Abdul-Ghani MA, Winnier D, Mehta FF, Jenkinson CP, et al. Chromatin occupancy of transcription factor 7-like 2 (TCF7L2) and its role in hepatic glucose metabolism. *Diabetologia* 2011;**54**:3132–42.
130. Oh KJ, Park J, Kim SS, Oh H, Choi CS, Koo SH. TCF7L2 modulates glucose homeostasis by regulating CREB- and FoxO1-dependent transcriptional pathway in the liver. *PLoS Genet* 2012;**8**:e1002986.
131. Neve B, Le Bacquer O, Caron S, Huyvaert M, Leloire A, Poulain-Godefroy O, et al. Alternative human liver transcripts of TCF7L2 bind to the gluconeogenesis regulator HNF4alpha at the protein level. *Diabetologia* 2014.
132. Bennett CN, Ross SE, Longo KA, Bajnok L, Hemati N, Johnson KW, et al. Regulation of Wnt signaling during adipogenesis. *J Biol Chem* 2002;**277**:30998–1004.
133. Gustafson B, Smith U. The WNT inhibitor Dickkopf 1 and bone morphogenetic protein 4 rescue adipogenesis in hypertrophic obesity in humans. *Diabetes* 2012;**61**:1217–24.
134. Li L, Tam L, Liu L, Jin T, Ng DS. Wnt-signaling mediates the anti-adipogenic action of lysophosphatidic acid through cross talking with the Rho/Rho associated kinase (ROCK) pathway. *Biochem Cell Biol* 2011;**89**:515–21.
135. Wada N, Hashinaga T, Otabe S, Yuan X, Kurita Y, Kakino S, et al. Selective modulation of Wnt ligands and their receptors in adipose tissue by chronic hyperadiponectinemia. *PloS One* 2013;**8**:e67712.
136. Schinner S, Ulgen F, Papewalis C, Schott M, Woelk A, Vidal-Puig A, et al. Regulation of insulin secretion, glucokinase gene transcription and beta cell proliferation by adipocyte-derived Wnt signalling molecules. *Diabetologia* 2008;**51**:147–54.

137. Chen ZL, Shao WJ, Xu F, Liu L, Lin BS, Wei XH, et al. Acute Wnt pathway activation positively regulates leptin gene expression in mature adipocytes. *Cell Signal* 2015;**27**:587–97.
138. Columbus J, Chiang Y, Shao W, Zhang N, Wang D, Gaisano HY, et al. Insulin treatment and high-fat diet feeding reduces the expression of three Tcf genes in rodent pancreas. *J Endocrinol* 2010;**207**:77–86.
139. Harano Y, Ohgaku S, Kosugi K, Yasuda H, Nakano T, Kobayashi M, et al. Clinical significance of altered insulin sensitivity in diabetes mellitus assessed by glucose, insulin, and somatostatin infusion. *J Clin Endocrinol Metab* 1981;**52**:982–7.
140. Rubinfeld B, Souza B, Albert I, Muller O, Chamberlain SH, Masiarz FR, et al. Association of the APC gene product with beta-catenin. *Science* 1993;**262**:1731–4.

Index

Note: Page numbers followed by "f" indicate figures and "t" indicate tables.

Printed in the United States
By Bookmasters